CONTINENTAL DEFORMATION

RELATED PERGAMON TITLES OF INTEREST

Books

CONDIE
Plate Tectonics and Crustal Evolution, 3rd edition

DRESEN AND RÜTER
In-Seam Seismics

GHOSH
Structural Geology: Fundamentals and Modern Developments

LISLE
Geological Strain Analysis: A Manual for the Rf/Ø Technique

LISLE
Geological Structures and Maps: A Practical Guide

Journals

Continental Shelf Research
Journal of African Earth Sciences (and the Middle East)
Journal of Geodynamics
Journal of South American Earth Sciences (including Central America, The Caribbean and the Antarctic Peninsula)
Journal of Southeast Asian Earth Sciences
Journal of Structural Geology

Full details of all Pergamon publications/free specimen copy of any Pergamon journal available on request from your nearest Pergamon office.

CONTINENTAL DEFORMATION

Edited by

Paul L. Hancock

*Reader in Structural Geology,
University of Bristol, U.K.*

PERGAMON PRESS
Oxford · New York · Seoul · Tokyo

U.K.	Pergamon Press Ltd, Headington Hill Hall, Oxford OX3 0BW, England
U.S.A.	Pergamon Press, Inc, 660 White Plains Road, Tarrytown, New York 10591-5153, U.S.A.
KOREA	Pergamon Press Korea, KPO Box 315, Seoul 110-603, Korea
JAPAN	Pergamon Press Japan, Tsunashima Building Annex, 3-2-12 Yushima, Bunkyo-ku, Tokyo 113, Japan

Copyright © 1994 Pergamon Press Ltd

All Rights Reserved. No part of this publication may be reproduced, stored in a retrieval system or transmitted in any form or by any means: electronic, electrostatic, magnetic tape, mechanical, photocopying, recording or otherwise, without permission in writing from the publishers.

First edition 1994

Library of Congress Cataloging in Publication Data
A catalogue record for this book is available from the Library of Congress.

British Library Cataloguing in Publication Data
A catalogue record for this book is available from the British Library.

ISBN 0 08 037931 1 Hardcover
ISBN 0 08 037930 3 Flexicover

Printed in Great Britain by BPCC Wheatons Ltd., Exeter

Contents

List of Contributors		vii
Preface		ix

Part 1 Deformation processes and deformation analysis

Chapter 1	Ductile deformation processes P. F. Williams, L. B. Goodwin and S. Ralser	1
Chapter 2	Palaeostrain analysis R. J. Lisle	28
Chapter 3	Brittle crack propagation T. Engelder	43
Chapter 4	Fault slip analysis and palaeostress reconstruction J. Angelier	53
Chapter 5	Palaeostress analysis of small-scale brittle structures W. M. Dunne and P. L. Hancock	101
Chapter 6	Linked fault systems; extensional, strike-slip and contractional I. Davison	121
Chapter 7	Prelithification deformation A. Maltman	143
Chapter 8	Advances in salt tectonics M. P. A. Jackson and C. J. Talbot	159

Part 2 Deformation styles and tectonic settings

Chapter 9	Arc-trench tectonics J. Charvet and Y. Ogawa	180
Chapter 10	Craton tectonics, stress and seismicity R. G. Park and W. Jaroszewski	200
Chapter 11	Continental extensional tectonics A. Roberts and G. Yielding	223
Chapter 12	Continental strike-slip tectonics N. H. Woodcock and C. Schubert	251

Chapter 13	Continental collision M. Coward	264
Chapter 14	Inversion tectonics M. Coward	289
Chapter 15	Suspect terranes W. Gibbons	305
Chapter 16	Tectonosedimentation: with examples from the Tertiary-Recent of southeast Japan K. T. Pickering and A. Taira	320
Chapter 17	Archaean tectonics J. S. Myers and A. Kröner	355
Chapter 18	Neotectonics I. S. Stewart and P. L. Hancock	370

Thematic Index   411

List of Contributors

J. ANGELIER
Département de Géotectonique,
Tectonique Quantitative,
Université Pierre et Marie Curie,
T.26, E1.4 Place Jussieu,
75252 Paris Cedex 05, France

J. CHARVET
Laboratoire de Géologie Structurale,
Départment des Sciences de la Terre,
Université d'Orleans, B.P. 6759,
45067 Orleans Cedex 2, France

M. COWARD
Department of Geology,
Royal School of Mines,
Imperial College,
Prince Consort Road,
London SW7 2BP, U.K.

I. DAVISON
Department of Geology,
Royal Holloway and Bedford New College,
University of London, Egham,
Surrey TW20 0EX, U.K.

W. M. DUNNE
Department of Geological Sciences,
306 G & G Building,
University of Tennessee,
Knoxville, Tennessee 37996, U.S.A.

T. ENGELDER
Department of Geosciences,
336 Deike Building,
The Pennsylvania State University,
University Park,
Pennsylvania 16802, U.S.A.

W. GIBBONS
Department of Geology,
University of Wales,
PO Box 914,
Cardiff CF1 3YE, U.K.

L. B. GOODWIN
Department of Geoscience,
New Mexicco Institute of Mining and Technology,
Socoro,
New Mexico 87801, U.S.A.

P. L. HANCOCK
Department of Geology,
University of Bristol,
Wills Memorial Building,
Queen's Road,
Bristol BS8 1RJ, U.K.

M. P. A. JACKSON
Bureau of Economic Geology,
University of Texas at Austin,
University Station,
Box X, Austin,
Texas 78713, U.S.A.

W. JAROSZEWSKI
Institute of Geology,
Warsaw University,
Al. Aworki i Wigury 93,
02-089 Warszawa, Poland

A. KRÖNER
Institut für Geowissenschaften,
Johannes Gutenberg-Universität,
Saarstrasse 21, Postfach 3980,
D-6500 Mainz, Germany

R. J. LISLE
Department of Geology,
University of Wales,
P.O. Box 914,
Cardiff CF1 3YE, U.K.

A. MALTMAN
Institute of Earth Studies,
University of Wales,
Llandinam Building,
Aberystwyth,
Dyfed SY23 3DB, U.K.

J. S. MYERS
Geological Survey of Western Australia,
100 Plain Street,
East Perth, Western Australia 6004,
Australia

Y. OGOWA
Institute of Geoscience,
The University of Tsukuba,
Ibaraki 305, Japan

R. G. PARK
Department of Geology,
University of Keele,
Keele, Staffordshire ST5 5BG, U.K.

K. T. PICKERING
Department of Geology,
University of Leicester,
University Road,
Leicester LE1 7RH, U.K.

S. RALSER
Department of Geoscience,
New Mexico Institute of Mining and Technology,
Socoro,
New Mexico 87801, U.S.A.

A. ROBERTS
Badley Earth Sciences Ltd,
Winceby House,
Winceby,
Horncastle, Lincolnshire LN9 6BP, U.K.

C. SCHUBERT
Centro de Ecologia, I.V.I.C.,
Apartado 21827,
Caracas 1020-A, Venezuela

I. S. STEWART
Division of Geography and Geology,
West London Institute (College of Brunel University),
Lancaster House,
Borough Road,
Isleworth, Middlesex TW7 5DU, U.K.

A. TAIRA
Ocean Research Institute,
University of Tokyo,
1-15-1 Minamidai,
Nakano-ku, Tokyo 164, Japan

C. J. TALBOT
Institute of Geology,
Uppsala University,
Box 555,
S-751 22 Uppsala, Sweden

P. F. WILLIAMS
Department of Geology,
University of New Brunswick,
Fredericton,
New Brunswick,
Canada E3B 5AE

N. H. WOODCOCK
Department of Earth Sciences,
University of Cambridge,
Downing Street,
Cambridge CB2 3EQ, U.K.

G. YIELDING
Badley Earth Sciences Ltd,
Winceby House,
Winceby,
Horncastle,
Lincolnshire LN9 6BP, U.K.

Preface

THIS book is the outcome of Jean Dercourt's idea that in the Developing World there is a need for a text that explains the new concepts and terms that have arisen in structural geology and tectonics during the last fifteen to twenty years. When Jean first expressed this idea to me (December 1987) he was Chairman of the Commission on Tectonics of the International Union of Geological Sciences, and had just asked me to head its Subcommission on Tectonic Nomenclature. Jean's vision was for a book with the working title *New Concepts in Tectonics* and a subtitle *New Concepts: New Words*. I was enthusiastic about Jean's proposal, thinking that the phrase "You cannot know what you cannot name" is an especially appropriate maxim for structural geologists. Thus, all contributors to this book were asked to define, or introduce, key words and phrases (italics in the text). I decided to drop *"New"* from the title because of the realisation that in a few years time some of the ideas expressed here would be well-established, if not old. The eventual choice of the title *Continental Deformation* reflects the final contents of the book as a result of editorial policy and, regrettably, the non-submission of some planned chapters or parts of chapters. Although some of the rocks discussed in Chapters 9 and 16 started life on the oceanic crust they have either been accreted to continents or, are in the process of being accreted, and hence, I believe, fit comfortably with those of continents.

The compilation of a multi-author book, such as this one, is a task that those with wisdom advise one not to undertake: I failed to take this good advice. At least 38 contributors, or potential contributors, were involved in the project at some stage. The period of editing of the book coincided with the introduction into most institutions of the facsimile machine — a device that permitted more than one author to fax their copy within days of the final publisher's deadline — a nerve racking editorial experience. I am, however, most grateful to all 27 authors for putting up with my editorial prejudices, and constant cajoling.

The book has been divided into two parts; the first emphasizes deformational processes, the second the operation of those processes in tectonic settings. Because each chapter was written by a different author, or authors, and because I wished each chapter to be more or less self contained, there is some overlap of material between chapters. I hope that readers will forgive this relatively small amount of repetition. Where authors of different chapters were unable to agree on ideas or nomenclature they were left to express their own opinions.

In addition to thanking contributors, especially those who delivered on time, I also owe a debt of gratitude to Peter Henn, the Earth Sciences' Commissioning Editor at Pergamon Press. He has been more patient than duty demands. Likewise, his staff, especially Diana Gallannaugh, the book's production editor, are thanked for their efforts. Jon Venn did much of the work involved in compiling the index. I edited most of this book, and wrote some of it, while on sabbatical leave at the 'Center for Neotectonic Studies' (CNS) at the University of Nevada, Reno. Steve Wesnousky and his colleagues at CNS provided me with friendship, stimulus and excellent facilities. The stunning views from my office window of the Sierra Nevada and the Virginia Mountains, on the western edge of the Basin and Range province, also worked their magic, as did the realization that I was living in a region of active continental deformation. Anne Becher has given me constant support and encouragement throughout the years it has taken to edit the book.

Paul Hancock, Bristol, England

Part 1
Deformation Processes and Deformation Analysis

CHAPTER 1

Ductile Deformation Processes

PAUL F. WILLIAMS, LAUREL B. GOODWIN and STEVEN RALSER

STRAIN

Introduction

STRAIN (also see Chapter 2) is concerned with a change in the shape of a body and simply describes the final shape in terms of the initial shape. The description is independent of the way in which the change occurred. There are *homogeneous* and *heterogeneous strains* (Fig. 1.1) where a homogeneous strain is one in which all initially straight marker lines are straight after *straining*; all other strains are heterogeneous. A geometrical consequence of the definition of homogeneous strain is that any circular or spherical marker in the initial state will be elliptical or ellipsoidal in the final state. The only exception to this statement is where the shape change is simply an isotropic area or volume change. In that case the initial circle or sphere will still be a circle or sphere in the final state but will be larger or smaller. By convention, volume change is considered to be isotropic but it may be only one component of a general strain. When considering homogeneous strain a *general two-dimensional strain* is one in which a circle is strained to an ellipse, which in the most general case will be different in area to the initial circle. Similarly a *general three-dimensional strain* results in a triaxial ellipsoid which may be different in volume to the initial sphere.

The ellipse that results from the homogeneous strain of an initial circle is called the *strain ellipse* and similarly there is a *strain ellipsoid*. Both are unique for a given two- or three-dimensional strain and are used to represent the strain. Their axes are generally different in length to the diameter of the initial circle or sphere and the difference is a measure of the strain magnitude. Thus, if we consider a three-dimensional strain, for example, in which there has been no volume change, at least one axis of the strain ellipsoid will be longer than it was originally, and at least one will be shorter than originally. The third axis may be different or the same in length, and the product of the three axes must be equal to the cube of the diameter of the original sphere. The three axes of the ellipsoid are orthogonal to three symmetry planes and are referred to as the *principal axes of strain*. The material lines parallel to them are perpendicular to one another in both the final strained and initial unstrained states.

The magnitude of a given strain can be expressed completely in terms of changes in the lengths of the principal axes of the strain ellipse or ellipsoid. Three conventions are commonly used for this purpose and they may also be used for representing the strain of any material line. They are known as the elongation (ε), the quadratic elongation (λ), and the stretch (S) and are defined as follows:

$$\varepsilon = (\ell - \ell_0)/\ell_0$$

$$\lambda = (\ell - \ell_0)^2$$

$$S = \ell/\ell_0$$

where ℓ is the length of the axis or any line in the strained state and ℓ_0 is its length in the unstrained state.

In any general strain all lines that were perpendicular in the initial state, other than those parallel to the principal axes, are non-parallel in the final state (Fig. 1.2). This change in angular relationship between a given line (ℓ) and the line initially normal to it (ℓ') is expressed as the shear strain parallel to the given line, where shear strain (γ) is defined as follows:

$$\gamma = \tan(90-\theta)$$

where θ is the angle between the line and its original normal after straining. The shear strain may also be expressed as an *angular shear strain* (sometimes denoted by ψ) which is simply defined as follows:

$$\psi = 90 - \theta.$$

It may be mathematically convenient to express strain in terms of external co-ordinates but the *strain axes* (i.e. the principal axes of strain) are an adequate reference frame to describe homogeneous strain.

Three factors influence the final shape of a strained

body. They are the initial shape, the strain as expressed by the strain ellipsoid and the orientation of the strain ellipsoid with respect to the initial shape. This point is demonstrated for a two-dimensional example where two initial squares share a common marker circle (Fig. 1.3a) which has been strained to an ellipse (Fig. 1.3b). Because of their different orientations with respect to the strain ellipse the strained squares are different in shape; one is a rectangle and the other is a parallelogram. In some situations it is useful to describe strain in terms of the change in shape of squares or cubes (the *strain rectangles* of Means *et al.* 1984) but the orientation factor limits its use to special situations. In the general case a circle or sphere is much more useful because of its isotropic shape and resultant independence of the orientation factor.

Heterogeneous strain (sometimes referred to as *inhomogeneous strain*) can be represented by an array of approximately homogeneous domains (Fig. 1.4) where the accuracy of the approximation is determined by the size of the domains relative to the scale of the heterogeneity.

Deformation used as a noun is similar to strain but provides a little more information. Like strain it simply relates an initial to a final state and provides no information about the intermediate steps. However, whereas strain can be described completely in terms of a co-ordinate system (viz. the strain axes) attached to the strained body, deformation must be described in terms of an arbitrarily defined external co-ordinate system.

Deformation has three components: *strain*, *translation* and *rotation*, where the translation and rotation are described as appropriate to a rigid body. This means that these two components are independent of any translation or rotation resulting from the strain component. They are most conveniently described in terms of the translation and rotation of the strain axes (Fig. 1.5) with respect to the external co-ordinate system.

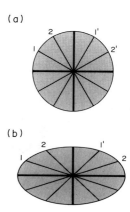

Fig. 1.2. Homogeneous straining of a circle to an ellipse. Initially perpendicular diametrical lines (e.g. 1/1' and 2/2' in (a)) are no longer perpendicular after straining, with the exception of the major and minor axes (heavy lines) of the ellipse which remain orthogonal (b). These axes are referred to as the principal axes of strain and they are defined as the material lines that are perpendicular before and after straining.

Strain and deformation paths

Whereas strain and deformation simply relate initial and final states, *strain paths* and *deformation paths* are descriptions of the intermediate stages during the process of straining or deformation. Strain paths are divided into coaxial and non-coaxial. A *coaxial strain path* is one in which the strain axes are parallel to the same material lines throughout the straining. An example is given in Fig. 1.6, where particle paths for some material points are also shown. This figure represents the only possible path for a constant-volume, two-dimensional coaxial straining of the magnitude shown, except for the situation where there is swapping of the axes (i.e. where the extensional axis has been the shortening axis for part of the strain history and vice versa). In three dimensions, a greater number of paths is possible because a given shortening parallel to one axis can be accommodated by a variety of responses parallel to the other two axes. For example, both axes may extend equally or, alternatively, one may remain constant in length while the other extends. If there is a volume change the number of possibilities is further

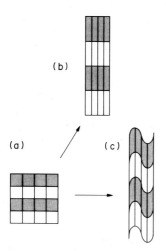

Fig. 1.1. Distortion (straining) of a square (a) to other shapes. In (b) all initially straight lines are still straight and the strain is therefore homogeneous. In (c) some initially straight lines are now curved and the strain is therefore heterogeneous.

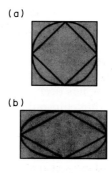

Fig. 1.3. Homogeneous straining of two squares showing how the final shape (in this case rectangle vs parallelogram) is dependent on the orientation of the marker squares with respect to the strain axes.

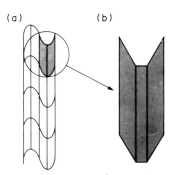

Fig. 1.4. Approximation of a heterogeneous strain (a) by homogeneous domains (b).

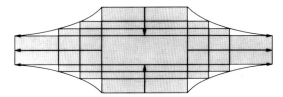

Fig. 1.6. Three incremental steps in the *pure shear* straining of a square to a rectangle. The arrows represent the movement paths for selected particles in the initial square. The particle paths are independent of the initial shape of the straining body; they depend only on the initial particle position and the strain.

increased. A *pure shear* is commonly defined as a constant volume, coaxial strain path in which one axis of the strain ellipsoid is constant in length. However, some writers use it synonymously with a constant volume, coaxial strain path which allows all three axes to vary in length.

A *non-coaxial strain path* is one in which the strain axes are parallel to different material lines during each infinitesimal increment of straining. There is an infinite variety of non-coaxial strain paths, but one special case is given prominence in geological literature. It is referred to as *simple shear* and is analogous to the behaviour of a deck of playing cards that is deformed by sliding the cards over one another in a single direction (Fig. 1.7). Simple shear results in a constant volume, *plane strain*, which means that one strain axis is constant in length throughout straining. It is characterized by the fact that one plane remains constant in area throughout the straining. In the card deck model the plane of constant area is parallel to the cards. In Fig. 1.8 intermediate stages along a simple shear strain path are represented and the axes for the strain ellipse for each increment are shown. The material lines parallel to the axes in each incremental diagram have been traced on the initial square. It can be seen that for each increment the material lines are perpendicular before and after straining, but the lines are differently oriented for each increment, as required by the definition of a non-coaxial strain path.

Because many geological materials have the planar mechanical anisotropy of a deck of cards, that is, they deform readily by slip on planes of weakness (e.g. basal planes of mica or foliation in schist), simple shear may be a particularly appropriate strain path in geological situations. It is also a simple strain path to describe and can be combined with a pure shear component oriented orthogonal to the shear plane to describe any non-coaxial strain path (Fig. 1.9). Strain is totally independent of path and all plane strains can be achieved by pure shear or simple shear (Fig. 1.10).

Non-coaxial strain paths are said to have *vorticity*. That is, material lines have an average angular velocity. All material lines rotate during non-coaxial straining with respect to the strain axes. In pure shear, for every line rotating clockwise there is a line rotating by precisely the same amount counterclockwise (Fig. 1.2); thus the average rotation is zero. If straining is non-coaxial, the fact that the strain axes rotate with respect to material lines means that the high symmetry of the pure shear situation is lost and material lines rotate through the strain axes. There is thus an average angular velocity of material lines with respect to strain axes during straining, and the strain path therefore has vorticity. This component of vorticity, which is due to the straining, is referred to as *shear induced vorticity* (SIV).

Deformation path is analogous to strain path and like deformation can only be expressed relative to an external co-ordinate system. It has four components: *strain, translation, SIV* and *spin*. *Spin*, a component of vorticity, is an average angular velocity appropriate to a

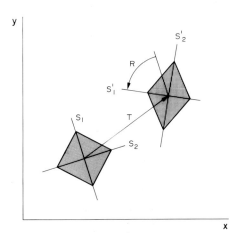

Fig. 1.5. General two-dimensional deformation. An initial square has undergone a change of shape (shortening and extension parallel to strain axes S_1 and S_2 respectively) and a translation (T) and rotation (R) of the strain axes with respect to the external co-ordinate system (xy).

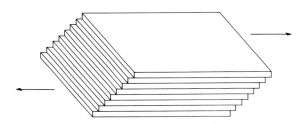

Fig. 1.7. Diagrammatic representation of simple shear by straining of a 'deck of cards'.

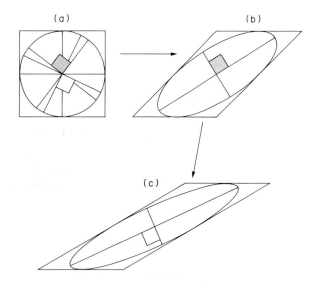

Fig. 1.8. Two incremental steps in the straining of a square to a parallelogram by homogeneous simple shear. The appropriate *strain* ellipse has been inscribed for both increments and the material lines coincident with the major and minor axes (strain axes) have been inscribed in the initial square. It should be noted that the lines are orthogonal before and after straining for both increments, as for all increments, but the axes for the first increment are not coincident with the axes for the second increment (a condition of non-coaxial straining). The differently shaded boxes indicate right angles and distinguish the material lines parallel to the axes for the strain ellipses for each increment.

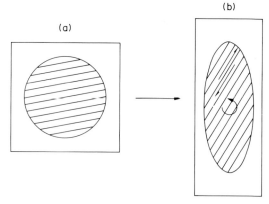

Fig. 1.10. Model demonstrating that any homogeneous plane strain can be achieved by either a pure shear or a simple shear. The square represents a block of material capable of deforming continuously in pure shear; a cylindrical hole in its centre is filled by a stack of rigid plates. The surfaces between the plates and between the plates and the surrounding block are effectively frictionless. The only deformation mechanism available to the rigid plates is slip plus rotation and they can mimic the shape of any strain ellipse by this mechanism.

rigid body. Thus in a deformation path that involves no strain there can be no SIV, but if the body is rotating (Fig. 1.11), material lines will still have vorticity and that component is called the spin.

If a deformation path involves straining then the spin is the vorticity of the strain axes and the total vorticity is the sum of the spin and the SIV.

Vorticity = spin + SIV.

Thus if spin and SIV are equal and opposite the vorticity is zero. All permutations of the two are possible as shown in Fig. 1.12. A deformation path that has vorticity is said to be *rotational* and one that does not is said to be *irrotational*. Some writers confusingly use these terms synonymously with non-coaxial and coaxial respectively, but we consider the terminology presented here more useful since it covers all possibilities and has no redundancy.

Several terms are used to qualitatively describe the magnitude of strain or portions of a strain path. For example, in considering a cleaved rock we might divide

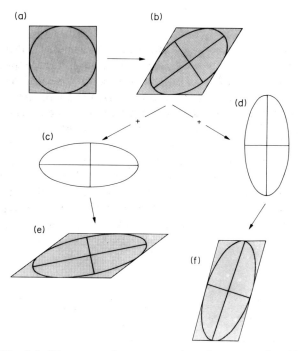

Fig. 1.9. Diagrammatic representation of transpression (e) and transtension (f) by adding a simple shear (a-b) and two differently oriented pure shears (c & d). See Chapter 12 for definitions of *transpression* and *transtension*.

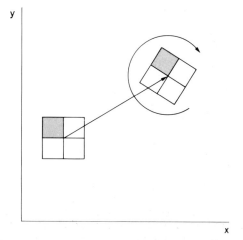

Fig. 1.11. Rotation of a rigid body or the spin component of vorticity (see text for discussion).

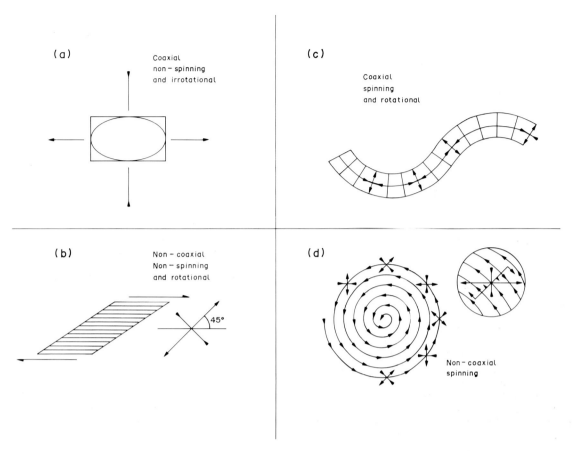

Fig. 1.12. Diagrammatic representation of the various combinations of spin and SIV (after Lister & Williams 1983). Coaxial and non-coaxial are strain path terms related to SIV and irrotational and rotational are deformation terms related to spin. The crossed lines divide the fields and represent external coordinates. (a) Pure shear of a rectangle in which the strain axes remain constant in orientation with respect to the external coordinates. (b) Simple shear of a rectangle in which the strain axes for every infinitesimal increment are inclined to the horizontal shear plane at 45°. (c) Idealized buckling of a beam in which each rectangle undergoes a pure shear and the strain axes for all rectangles except those in the fold hinges rotate progressively. The rectangles in the hinges undergo an irrotational deformation. (d) An idealized vortex in which an initial small rectangle undergoes simple shear so that its infinitesimal strain axes are constantly inclined at 45° to the flow lines, and therefore progressively change orientation with respect to external coordinates.

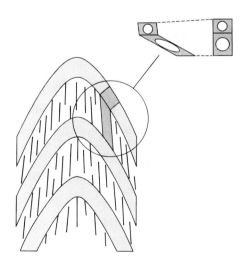

Fig. 1.13. Diagrammatic representation of strain and strain path partitioning. An alternating sequence of shale (bold lines) and sandstone (stippled) is folded and the sandstone has deformed largely by rotation. The shale has undergone layer-parallel shear and has consequently maximized SIV and undergone a larger strain than the sandstone. See text for additional discussion.

the strain into an increment (*strain increment* resulting from an *incremental strain*) associated with compaction of the initial sediment and an increment associated with the development of the cleavage (also see Chapter 7). Both increments would be *finite strains* which loosely means that they are perceptible to an observer and together they would comprise the *total strain* or the *finite strain*. Each increment could be subdivided into finer and finer increments and, in the limit, such an increment constitutes an *infinitesimal strain* which may also be described as an *instantaneous strain*. The term *progressive strain* is used to describe a portion of the strain path that is believed to have been continuous. In

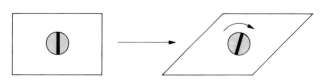

Fig. 1.14. Diagrammatic representation of a rigid body in a matrix undergoing simple shear showing how the rigid body will maximize the spin (i.e. it will rotate). See text for additional discussion.

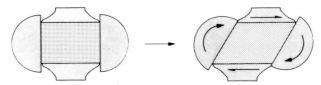

Fig. 1.15. Schematic diagram of a simple shear deformation rig (see Price & Torok 1989). Rotating semi-circular pistons and top and bottom platens, free to translate horizontally, constrain a rectangular specimen to deform in simple shear.

this context continuous does not necessarily mean continuous in time. For example, the deformation in a shear zone over a period of time might have followed a dextral simple shear path throughout the history of the zone. There may have been times of active shear and times of quiescence, but the strain path would be considered continuous and the strain progressive. On the other hand, if at some time in its history the shear zone reversed and became, say, a sinistral *transpressive zone** for the rest of its history, the strain path would be discontinuous and would comprise two progressive strain increments. In nature, strain is generally heterogeneous and where some domains are strongly deformed and others weakly deformed, the strain may be said to be *partitioned* between the different domains. For example, when a sequence of alternating sandstones and shales is folded the magnitude of strain is generally greater in the shales than in the sandstones (Fig. 1.13) and the strain is said to be partitioned between the different layers. More significantly, there may be partitioning of strain or deformation path. For example, in Fig. 1.13 the strain path in the shales is non-coaxial and transpressive whereas in the sandstones it is a coaxial extension parallel to the layering. Thus, there is strain path partitioning and the shales can be said to have maximized SIV whereas the sandstones have minimized SIV. Similarly, because an essentially rigid porphyroblast in a matrix undergoing simple shear will rotate rather than strain, the porphyroblast is said to maximize the spin; it converts the SIV of the matrix into a rigid body rotation (Fig. 1.14).

Stress can be represented by an ellipsoid and is a symmetrical quantity. If it acts on an isotropic material it can be expected to result in a series of smoothly varying strain ellipsoids with their axes parallel to the same material lines for all increments, that is, it can be expected to result in a coaxial strain path or pure shear (Fig. 1.6). A non-coaxial strain path is asymmetrical and for it to be the product of a symmetrical stress there has to be some constraint acting on the straining body. There are two possible constraints; one is external to the body and can be referred to as the steel box constraint. It applies, for example, when a specimen is placed in the chamber of an experimental rig that during the experiment changes shape, progressively following, for example, a simple shear path and thereby forcing the specimen to follow the same path (Fig. 1.15). This is what happens in shear zones where two relatively rigid bodies move past one another and the weaker material between, being attached to the two rigid bodies, is constrained to undergo a simple shear (Fig. 1.15). The infinitesimal strain axes are inclined to the boundary of the shear zone at 45° for every infinitesimal increment while the finite strain axes rotate (compare Figs 1.8 and 1.12).

The other constraint that causes materials to undergo non-coaxial straining is a material property, anisotropy. For example, in Fig. 1.10 the strong frictionless plates forming the cylinder can mimic the change in the shape of the hole that they occupy most easily by sliding over one another. From a geometrical point of view the same shape change could be achieved by the plates becoming longer and thinner but such a mechanism would require a much larger stress. Thus the planar anisotropy of the cylinder-fill constrains it to undergo a non-coaxial straining. This simple-minded model has considerable relevance for geological materials as they commonly have planar anisotropy.

DEFORMATION MECHANISMS AND PROCESSES

Introduction

Our knowledge of deformation mechanisms in rocks is based primarily on experimental, light-microscope and electron-microscope observations. Deformation experiments provide information about the macroscopic behaviour of the sample tested, which may be either a polycrystalline aggregate or a single crystal. Traditionally, *brittle behaviour* was defined at the scale of the experimental sample by the formation of fractures or faults (generally at less than 5% strain), while *ductile flow* was indicated by the accumulation of strain (typically >5%) without, or before, brittle failure (Heard 1960). The use of the terms 'brittle' and 'ductile' has since been extended to other scales, requiring specification of the level at which such behaviour is observed (Paterson 1969, Rutter 1986, Means 1990). For example, microscopically brittle behaviour may produce macroscopically ductile flow. The inability to simply define a boundary reflects the broad transition between macroscopically brittle and ductile deformational behaviour which is accommodated by a wide range of deformation mechanisms. Macroscopically ductile flow, which may be affected by microscopically brittle and/or ductile deformation mechanisms, accommodates shape change in rocks, and will be the main focus of this section. We use Paterson's (1969) definition of (macroscopic) *ductility*: "the capacity for undergoing permanent change of shape without fracturing."

The microstructure of an experimentally or naturally deformed sample can be characterized by light or electron microscopy. The microstructure includes the

---

*As defined by Woodcock and Schubert in Chapter 12 of this book *transpression* is "deformation in a zone of oblique shortening." By contrast, transtension is defined by these authors as "deformation in a zone of oblique extension."

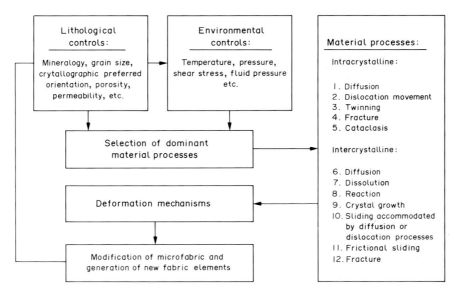

Fig. 1.16. Flow diagram modified from Knipe (1988) illustrating the variables affecting the material behaviour of a deforming rock body, and the interaction between these variables.

ways in which grains are organized with respect to one another; the grain shapes and character of grain boundaries; and the types and arrangements of internal structures such as twins and dislocations (cf. Hobbs *et al.* 1976). Determination of which microstructures result from the operation of a given deformation mechanism or mechanisms has received increasing attention in the past two decades. The correlation of microstructures with deformation mechanisms has been aided considerably by 'see-through' experiments (see review by Means 1989, and references therein). Such experiments allow observation of *in situ* deformation of analogue materials with a light microscope, and therefore permit accurate identification of deformation mechanisms, and association of these mechanisms with the resulting microstructures. Along the same lines, a recent experiment in deforming common rock-forming minerals — muscovite and biotite — in a special stage in a transmission electron microscope (Meike 1989) has allowed the direct observation of the deformation process within a mineral.

A complex picture emerges from studies of deformation processes. A given mineral will deform by specific mechanisms under specific conditions of grain size, temperature, pressure, strain rate, fluid activity, etc. The way in which a rock accommodates strain will be dictated both by the deformation mechanisms operating in its individual mineral constituents, and the ease of activation of the deformation mechanisms of different phases relative to one another (e.g. Knipe & Wintsch 1985). The ease with which a given rock deforms will be enhanced by the presence of a phase that deforms easily (i.e. is *incompetent*) under specific geological conditions, and inhibited by the presence of a relatively 'hard', or *competent*, phase. The least competent phase in a rock may change over time as geological conditions change; thus, strain may be partitioned over space and/or over time (Knipe & Wintsch 1985) at all scales. A deforming rock is therefore a dynamic system, sensitive to a number of variables (Fig. 1.16), which more often than not behaves in a spatially and temporally heterogeneous manner.

The microstructure of an individual mineral grain records only its most recent geological history. Multiple grains collectively record the most recent experiences of their host rock. This record consists of the rock's *microstructure*, the *crystallographic preferred orientation* (CPO) of individual phases, the *dimensional preferred orientation* of the mineral constituents and the distribution of mineral phases throughout the rock. Of these features, those that are pervasive compose the *microfabric* of the rock, and will define the macroscopically visible record of deformation — foliations and lineations. It is important to note that the microfabric itself will influence the additional response of a rock to stress (e.g. Ralser *et al.* 1991), rendering it more competent (*strain hardening*) or less competent (*strain softening*) over time.

Strain softening processes (see reviews by White *et al.* 1980, Kirby & Kronenberg 1987, Hobbs *et al.* 1990) are required to localize deformation within shear zones. Throughout this section, mention will be made of the various means by which localization of deformation may be accomplished and perpetuated. It should be noted that the record of processes by which strain is localized within a shear zone is often erased by subsequent deformation. However, we are aided in the study of deformation processes by the typically heterogeneous character of natural deformation of rocks, which allows various stages of deformation events of interest to be preserved.

Mechanisms and processes

Heard (1960) first systematically studied the transition from macroscopically brittle to ductile behaviour as a function of temperature and confining pressure. He experimentally deformed Solenhofen limestone which, for a given confining pressure, became increasingly ductile with increasing temperature. That is, at higher temperatures the samples studied exhibited larger strains before failure. Similarly, at a set temperature, ductility could be increased by increasing confining pressure. In passing from the brittle to the ductile regime, transitional behaviour was observed in which brittle failure along discrete surfaces was replaced by deformation within broader zones in which failure followed small amounts of ductile strain. As either temperature or confining pressure was increased within the ductile field, deformation became increasingly more distributed (as opposed to localized) until the deformation was nearly homogeneous. Since both temperature and pressure increase with depth in the Earth, we can correlate experimental with natural conditions and state that rocks generally become increasingly ductile with depth. We also expect that deformation becomes more homogeneous with depth.

Classification of deformation behaviour with depth requires generalization; the transition behaviour is, of course, material dependent. For example, salt will deform by crystal-plastic flow at shallow depths where most other rock-forming minerals deform by fracture. Feldspar may fracture under conditions in which quartz changes shape through crystal-plastic flow. Despite their limitations, generalizations provide a practical context within which deformation mechanisms may be considered in greater detail.

It should be noted that decreasing the strain rate has an effect similar to increasing the temperature, and vice versa, a fact exploited for years by experimentalists. This relationship can be understood by considering that although diffusion rates (diffusivities) are temperature dependent, diffusion distances depend on both temperature, through the diffusivity, and time. That is, deformation regimes where diffusive mass transfer is the rate-controlling factor will, at a given temperature, typically exhibit an upper strain rate limit, above which alternative deformation processes will be preferentially activated.

The many deformation mechanisms by which minerals can accommodate macroscopically ductile flow can be broadly grouped into three categories (cf. Rutter 1976, Knipe 1988): (1) *fracture, frictional grain-boundary sliding* and *cataclastic flow*; (2) *crystal-plastic flow*, where grains deform by dislocation glide or twinning; and (3) *diffusive mass transfer*, which results in the redistribution of material by diffusion processes during deformation. From (a) to (c), the three categories of deformation mechanisms are favoured by increasing temperature and/or decreasing strain rate. Additionally, the first category is favoured by low confining pressures, while crystal-plastic flow is favoured at higher confining pressures. Crystal-plastic flow and diffusive mass transfer operate together in the process of dynamic recrystallization, with the dominant recrystallization processes reflecting the predominant deformation mechanisms. These three categories of deformation mechanisms and the process of dynamic recrystallization will be discussed briefly below under appropriate headings. It is important to remember that the division of mechanisms into groups is done for convenience. In general, no single process operates alone and the operation of one mechanism will influence the operation of others.

A detailed discussion of deformation mechanisms is beyond the scope of this chapter. The reader is referred to Poirer (1985) for a more comprehensive discussion of material processes and deformation mechanisms, to Knipe (1989) for a review of the determination of deformation mechanisms from microstructures in natural tectonites, and to Drury & Urai (1990) for a review of deformation-related recrystallization processes. Papers in Barber & Meredith (1990) include state-of-the-art knowledge of various deformation processes.

Frictional grain-boundary sliding, fracturing and cataclasis. Deformation by any or all of these mechanisms will leave crystal structures largely undistorted, although some slip through movement of dislocations (see later section) may precede (Rovetta *et al.* 1987) or post-date (e.g. Fitz Gerald *et al.* 1991) fracture. A comprehensive review of these mechanisms is provided by Knipe (1989); a simplified summary is presented below.

Fracture occurs through a number of mechanisms (see reviews of Atkinson 1982, 1989 and Chapters 3–7 for more general aspects). All involve the nucleation and propagation of cracks. The cracks behave as free surfaces along which displacement may subsequently occur as deformation continues. *Tensile fracturing*, for which the maximum amount of data is available, may be controlled by: (1) pre-existing cracks; (2) cracks that nucleate when dislocations or glide twins pile up at grain boundaries; (3) intergranular cracks generated to accommodate general plasticity or grain boundary sliding, which subsequently promote intragranular cracks; and (4) nucleation of cracks on grain or phase boundaries, resulting in intergranular creep fracture. Crack propagation may be facilitated by the presence of fluids (e.g. liquid water, water vapour or some other reactive species) in the crack tip, which can promote weakening. This *stress corrosion* is believed to be the most important chemical effect on fracture.

Frictional grain-boundary sliding is the sliding of grains past one another. Sliding occurs when the frictional strength between grains is overcome and cohesion between grains is lost. Not surprisingly, the deformation is dependent on the amount and strength of intergranular cement; sliding may be preceded by the nucleation and propagation of cracks on grain boundaries. Frictional grain-boundary sliding, with the associated loss of cohesion between grains, is probably gradational with, but can be distinguished from, *grain-boundary sliding* at higher temperatures. In the latter process, cohesion is maintained and sliding is effected by defect motion (e.g. movement of dislocations and vacancies; see later sections). Frictional grain-boundary sliding is facilitated at low effective stresses (low confining pressure and/or high pore pressure) where cohesion between grains is easier to overcome. It is also aided by fluid-assisted diffusion, which expedites the removal of asperities (e.g. Williams 1990). The presence of fluids therefore greatly increases the range of conditions under which frictional grain-boundary sliding can effectively operate.

Cataclasis involves fine-scale fracturing, movement along fractures, frictional grain-boundary sliding, and fragment rotation. The opening of fractures necessarily results in dilatancy, as does the subsequent fragment rotation; however, cataclasis may cause a decrease in porosity (and the formation of a *deformation band*) and therefore a decrease in the volume of porous rocks (Borg *et al.* 1960). As a process involving dilatancy, cataclasis is pressure sensitive; that is, it is favoured by lower confining pressures. Cataclasis may be localized or may accommodate macroscopically ductile deformation over a broader area (*cataclastic flow*; cf. Paterson 1978). Whereas cataclasis confined to a narrow fault zone (*localized cataclasis*) is macroscopically brittle, cataclastic flow over a broad area is macroscopically ductile and typically will be favoured over localized cataclasis at greater depth (higher pressures and temperatures).

Paterson & Weiss (1968) described folding and boudinage of quartz-rich layers in phyllite experimentally deformed at 500 MPa confining pressure and room temperature. This macroscopically ductile behaviour was accommodated by cataclasis within the quartz-rich layers. Cataclasis may also be localized within a single grain, as observed by Goodwin & Wenk (1990) through petrographic and transmission electron microscope studies of biotite. As shown by these authors, *intracrystalline cataclasis* visible with the electron microscope (in zones varying in size from c. 0.05 to 6 nm wide) results in ductile deformation at the scale of the petrographic microscope. Cataclasis is therefore a process that occurs at all scales, from that made visible by the electron microscope to that seen with the naked eye. Intracrystalline cataclasis is also preceded by intracrystalline folding and there is evidence of associated intracrystalline slip, so deformation takes place through a complex 'brittle–ductile' process (Goodwin & Wenk 1990). These examples illustrate the importance of specifying scale when discussing whether behaviour is brittle or ductile.

Transient frictional behaviour (including frictional grain-boundary sliding, fracture and cataclasis) may occur episodically within a zone which is dominated by microscopically ductile flow. Such phenomena have been observed in natural rocks (e.g. Sibson 1980, White & White 1983) as well as during *in situ* deformation of analogue materials (Means 1989). Within a shear zone, evidence of transient frictional behaviour may be partially or completely erased by later deformation.

Crystal-plastic flow. Crystal-plastic flow takes place by two mechanisms: twinning and intracrystalline slip. Both mechanisms accommodate shape change by shear of one section of a crystal with respect to another. Twinning induced by deformation is termed *mechanical, deformation* or *glide twinning*, and is favoured over intracrystalline slip at lower temperatures and faster strain rates (Vernon 1983). A number of common rock-forming minerals exhibit deformation twins. In plagioclase, experimental work indicates that deformation twins follow both the albite and pericline twin laws (Borg *et al.* 1959, Borg & Heard 1969, 1971). Naturally deformed microcline also exhibits glide twins which follow the albite and pericline twin laws (White & Barnett 1990). In natural rocks, these can be distinguished from growth twins on the basis of shape (e.g. Vance 1961, Vernon 1965, Brown 1989a). Mechanical twins tend to be lenticular and do not commonly extend across the deformed crystal, but taper to a point within it. Growth twins typically have planar interfaces; where they do not extend throughout the grain in which they occur, they terminate abruptly. Simple shear may produce pseudo-twins in plagioclase, from which true twins can subsequently be developed through rearrangement of the crystal lattice by diffusion (Starkey 1964, Brown 1989b and references therein).

Carbonate minerals commonly exhibit deformation twins, which have been produced experimentally in calcite (Turner *et al.* 1954a) and dolomite (Turner *et al.* 1954b). Deformation of quartz may produce either Brazil-law (McLaren *et al.* 1967) or Dauphiné-law twins (Tullis 1968, 1970, Tullis & Tullis 1972); twins in quartz may not be visible with a petrographic microscope. Deformation twins in biotite have been observed with a transmission electron microscope (Goodwin & Wenk 1990).

The subsequent discussion and definitions follow Hobbs *et al.* (1986) and Poirier (1985). *Intracrystalline slip* is accomplished by the movement of *dislocations*. A *dislocation loop* is defined as a line which separates a closed area which has experienced slip from an external, unslipped region (Fig. 1.17). The loop (and the region of slip) expands with increasing shear strain. The direction and magnitude of slip is quantified by the *Burgers vector*, **b**, which is constrained to be a vector of the Bravais lattice; a dislocation loop normally expands in all directions.

A section of the dislocation loop which is normal to **b** is said to have *edge character*, and the dislocation line will lie at the edge of an extra half plane in the crystal lattice (Fig. 1.17). A segment of a dislocation parallel to **b** has *screw character*, and distorts the adjacent lattice planes in the shape of a helicoid. A section of a dislocation oblique to **b** has *mixed character*. Within a region a few Burgers-vectors'-wide, adjacent to the dislocation line, the *core* of the dislocation, the crystal lattice is distorted and highly defective. Within this core, bonds in the crystal lattice are broken during slip. Slip along a dislocation line occurs between planes with low Miller indices; that is, between planes of closely packed atoms. Movement along these *slip planes* requires a minimum amount of energy. A *slip system* is

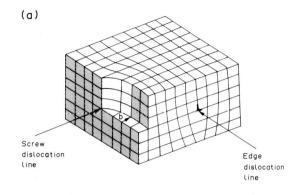

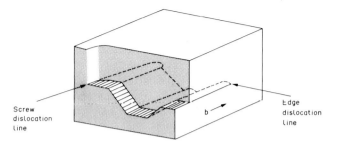

Fig. 1.18. Schematic illustration of cross-slip, modified from Vernon (1983). The orientation and magnitude of the Burgers vector is shown by the arrow labelled **b**. Two sets of slip planes are shown: the first is parallel to the base of the block-shaped crystal, while the second is oriented at 45° to the first. As illustrated, cross-slip of the screw segment of the dislocation has moved the dislocation out of the first set of slip planes and into the second set. Continued propagation of the edge portion of the dislocation is indicated by bulges in the edge dislocation line in the first set of slip planes. Note that the edge dislocation cannot cross-slip, thus continued propagation of the edge component in the second set of slip planes is impossible.

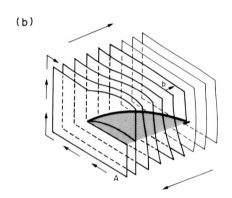

Fig. 1.17. (a) Diagram schematically illustrating the orientation and magnitude of the Burgers vector, **b**, with respect to the segments of the dislocation loop with edge character (edge dislocation line) and screw character (screw dislocation line). Material lines show distortion of the crystal lattice due to the dislocation. The upward facing step introduced in the lower left hand portion of the crystal by slip will propagate through the crystal, leaving a downward-facing step of length b on the opposite side (after Hobbs et al. 1976). (b) Shading indicates slipped section of crystal; dislocation loop is highlighted by heavy bold line; **b** is the Burgers vector. Subvertical lattice planes are schematically illustrated. Shear experienced by the crystal lattice is indicated by long bold arrows. If the boundaries of the lattice planes which are marked by thicker solid and dashed lines are traced from A to the extra half plane (begin by following string of short bold arrows), they will be shown to define a helicoid around the screw portion of the dislocation. Adapted from Poirier (1985) and Hobbs et al. (1976).

described by a slip plane and a slip direction (the direction of **b**). A dislocation with edge character is constrained to move within the slip (or *glide*) plane defined by the dislocation line and **b**; in contrast, a dislocation with screw character may theoretically move along any plane parallel to **b**. Movement of a dislocation with screw character out of one glide plane into another is termed *cross-slip* (Fig. 1.18).

Crystal-plastic flow may be accompanied by a component of grain-boundary sliding. Gifkins (1976) identified a regime in which grain-boundary sliding is accommodated by dislocation flow in the grain boundary and grain mantle (the rim of the grain in which recrystallization is generally initiated), as well as a regime in which sliding is accommodated by dislocation flow in the body of the grain.

For a given mineral, crystal-plastic flow is generally associated with greater depths than cataclastic flow. Like cataclasis, crystal-plastic deformation may or may not be localized (shear zone *vs* homogeneous crystal-plastic flow) (Fig. 1.19). Homogeneous crystal-plastic flow is to be anticipated at greater depths than is flow localized within a shear zone. Since decreasing the temperature has the same effect as increasing the strain rate, both decreasing temperature and increasing strain rate tend to favour localization of deformation. Localization may also result with increasing strain, or

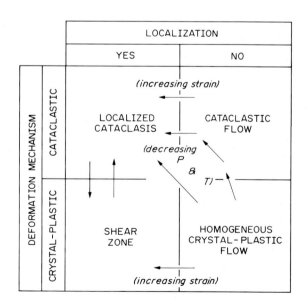

Fig. 1.19. Arrows indicate potential mode of failure transitions; controlling variables are, in most cases, indicated. The transition between cataclastic faulting and crystal-plastic flow in shear zones (illustrated by vertical arrows) can be brought on by changes in strain rate, temperature and/or pore fluid pressure. In all cases, changes induced by increasing temperature can be caused equally by decreasing strain rate, and vice versa. Modified from Rutter (1986).

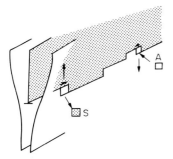

Fig. 1.20. Schematic illustration of dislocation climb. Transfer of matter toward (A) or away (S) from the dislocation core varies the length of the half plane, causing dislocation climb. Transfer is accomplished by adding (A) or subtracting (S) atoms (shaded squares) in exchange for vacancies (white-filled squares). Direction of climb is indicated by bold, vertical arrows. Modified from Poirier (1985).

by other processes which will be discussed later. These relationships are summarized in Fig. 1.19, which illustrates the ways in which deformation mechanisms may shift within a deforming region with changes in environmental and lithologic conditions. The positions of the boundaries between deformation regimes in temperature–pressure space will differ from mineral to mineral. For example, a shear zone may develop in salt at shallow crustal levels, while shear zones may also develop in dunite in the mantle.

A transition from localized cataclasis to homogeneous crystal-plastic flow, though theoretically possible, has not been demonstrated in natural rocks. It would, in fact, be difficult to recognize overprinting of brittle structures by homogeneous ductile flow, as the latter would effectively erase evidence of the former. However, cases of ductile shear localized in zones of initially brittle failure have been recognized (Goodwin 1986, Segall & Simpson 1986).

Dislocation climb, subgrains and rotation recrystallization. Stepping of a dislocation of edge character from one to another of a set of (parallel) glide planes, *dislocation climb*, is accomplished when material is transferred to or away from the dislocation core, allowing the length of the extra half plane to increase or decrease (Fig. 1.20). Material transfer is accomplished by *solid-state diffusion*, which occurs by exchange of atoms for vacancies in the crystal lattice. Dislocation climb is only possible when slow strain rates or high temperatures allow solid-state diffusion to keep pace with dislocation movement. The combination of dislocation glide (crystal-plastic flow) and dislocation climb (a diffusion process) results in *dislocation creep*. Both dislocation climb and cross-slip aid movement of dislocations through the deforming crystal, particularly because they provide methods by which obstacles (other defects) in the crystal lattice can be circumvented.

If dislocations are generated more rapidly than they can be propagated through a crystal, the density of dislocations in the crystal will increase. Several things can happen when dislocations build up in a crystal. They may, particularly at relatively low temperatures

where climb is inhibited or impossible, form disorganized tangles. *Dislocation tangles* inhibit further slip, causing strain hardening. It is also possible for *dislocations of opposite sign* travelling in a single plane to meet and annihilate (Fig. 1.21a). Dislocations are considered to have *opposite sign* when they have the same slip plane, but have half planes which fall above and below the plane of slip. Dislocations of opposite sign may climb into a single slip plane; climb therefore facilitates annihilation. Dislocations of opposite sign travelling in adjacent slip planes may also meet to produce a line of vacancies (Fig. 1.21b). With sufficient mobility, dislocations may aggregate within stable, low-energy arrays; *dislocation walls* (Fig. 1.21c) and

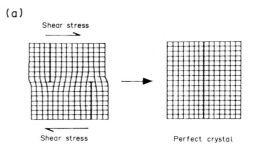

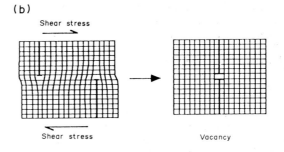

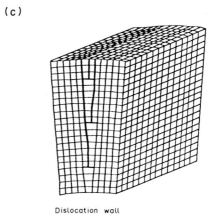

Fig. 1.21. Diagram showing possible interactions of dislocations in a crystal lattice. Extra half planes are shown as bold, upright lines. (a) Intersection of dislocations of opposite sign moving on a single slip plane results in a perfect crystal. (b) Dislocations of opposite sign on adjacent slip planes produce a line of vacancies (shown as a single vacancy in two dimensions). (c) Alignment of a series of edge dislocations, resulting in a simple tilt boundary. After Hobbs *et al.* (1976).

dislocation networks are examples of such arrays. Dislocation networks are similar in appearance to chicken wire. The rearrangement of defects (e.g. dislocations, vacancies, etc.) to lower the internal strain energy of a crystal lattice is termed *recovery*. If recovery rates balance strain hardening rates, the crystal can continue to deform. If recovery rates outstrip strain hardening rates, microstructural evidence of dislocation movement may be erased.

Subgrain formation occurs when a segment of a crystal becomes separated from the main crystal lattice by dislocation arrays. The subgrain boundaries, defined by the dislocation arrays, accommodate up to 3–5° misorientation with respect to the host crystal lattice in albite (Fitz Gerald *et al.* 1983) and 4–5° in perthite (White & Mawer 1988). With progressive misorientation, the boundaries may evolve from low- to high-angle boundaries (>*c*. 3–5°), in which case the subgrain becomes an independent grain. The transition from subgrain to grain in these examples is generally abrupt, involving significant increases in the misorientation of boundaries and in grain size (Fitz Gerald *et al.* 1983, White & Mawer 1988). The transition may also be accompanied by a noticeable decrease in dislocation density and straightening of the boundaries (Fitz Gerald *et al.* 1983). This method of forming new grains from old ones is called *rotation recrystallization*, one of the two main processes by which *dynamic recrystallization* occurs (Drury & Urai 1990). The other process, *grain boundary migration*, will be discussed in the next section. The exact mechanism(s) by which progressive misorientation of the subgrain boundaries takes place during rotation recrystallization is not completely understood, but two general processes are possible. In one, build-up and subsequent reorganization of the dislocations along a stationary boundary create the misorientation. In the other, a mobile subgrain boundary migrates through a region over which the lattice changes orientation (Urai *et al.* 1986). Boundary mobility necessarily involves diffusive mass transfer (see below).

Diffusive mass transfer. Diffusive mass transfer involves the transport of material from one site to another within a rock body. The material transfer may be intracrystalline, intercrystalline, or both. The mechanism(s) driving diffusion reflect the presence of a chemical potential gradient, which can occur as a result of either existing compositional variations and/or differential stress. In this section, we deal only with diffusion as a deformation mechanism, which is fundamentally a result of the latter.

Intracrystalline diffusion involves the transport of matter in a direction opposite to the movement of vacancies as atoms replace vacancies in the crystal lattice. There is evidence that diffusion in silicates occurs more rapidly along dislocation cores than within undistorted crystal (*pipe diffusion*; Yund *et al.* 1981). As noted previously, diffusive mass transfer is necessary to both dislocation climb and subgrain boundary migration; the relative contributions of crystal-plastic

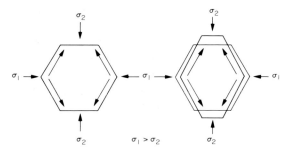

Fig. 1.22. Schematic, two-dimensional illustration of bulk transport directions in a hexagonal grain deforming through intercrystalline diffusive mass transport. Material moves away (through either solid-state diffusion or transport in solution) from areas subjected to high intergranular normal stress (σ_1) toward regions of relatively low intergranular stress (σ_2) along pathways illustrated by arrows. Subsequent change in shape of grain is shown from left to right (original grain shape is indicated in right section of diagram for comparison). After White & White (1981).

flow and diffusion to deformation will vary with temperature. For a given strain rate, crystal-plastic flow will dominate at lower temperatures and diffusion will be more effective at higher temperatures.

Processes of intercrystalline diffusion are influenced by temperature, the presence or absence of fluid and the character of the grain boundaries. We use the term fluid to include liquid, vapour and supercritical fluid. Fluid inclusions in minerals commonly contain combinations of water, water vapour and CO_2; chlorides, particularly NaCl, are common solutes in such inclusions. The *effective* grain boundary includes not only the grain margin proper, but also the zone of lattice distortion and/or electronic impurities resulting from grain–grain contacts, which extends into the grains. In metals, the effective grain boundary is approximately two Burgers vectors in width (**2b**), but studies of natural rocks suggest different values. In natural tectonites the effective grain boundary appears to be 10–30 nm in width — at least an order of magnitude larger than in metals (White & White 1981). In addition, White & White (1981) show that typical grain-boundary features include isolated voids and grain junction tubules. Such features are available as fluid pathways, shortening effective diffusion distances and accelerating intercrystalline transport. Urai *et al.* (1986) described similar grain-boundary structures in natural salt. Following experimental deformation, these structures were replaced by a thin (<100 nm) intergranular fluid film. Grain boundary tubules were subsequently re-established, inferred to develop by the breakdown of the original subcontinuous film, presumably by a solution-precipitation process (Urai *et al.* 1986). In contrast, Ricoult & Kohlstedt (1983) report structural widths of 5.9–8.8 nm for subgrain boundaries in natural olivine. The boundaries are shown by transmission electron microscopy to comprise dislocation arrays. Their findings suggest that: (a) there is a wide variation in thickness of the effective grain boundaries in natural rocks, and/or (b) the zone of lattice distortion adjacent to subgrain boundaries is significantly less than that adjacent to true grain boundaries. Both are probably true; there is no reason to suspect that subgrain boundaries should be exactly analogous to grain boundaries, nor should boundary behaviour be the same in every rock. In particular, one would expect boundaries between different minerals to be different in character than boundaries between grains of the same mineral.

Intercrystalline diffusion involves transfer of material from a source, which is a zone of high intergranular normal stress, to a sink, which is a site of low normal stress (cf. Rutter 1983) (Fig. 1.22). Along the potentially distorted and disordered solid – solid grain boundaries, processes may be similar to, but more rapid than, *intracrystalline diffusion* (Poirier 1985). Diffusion is also more rapid along a migrating grain boundary than along a stationary one (Smidoda *et al.* 1978). Diffusive mass transfer is necessary for grain boundary migration in the same way that it affects migration of subgrain boundaries. As noted previously, grain boundary migration and rotation recrystallization are the two processes by which dynamic recrystallization proceeds.

Where fluid is present along grain boundaries, it will enhance transport of material, with the potential transport distances being related to the composition of the fluid, the volume of the fluid and the degree of interconnectedness of fluid pathways throughout the rock. The most commonly cited mechanism of fluid-assisted mass transfer in rocks, *pressure solution* (see review by Rutter 1983), involves the diffusion of matter along stress-induced chemical potential gradients. Electron microscope study of microstructures associated with 'pressure solution' seams in limestone (also see Chapter 5) indicates that their development may also involve other processes (Meike & Wenk 1988). Meike & Wenk (1988) suggest that selective dissolution of grains with high dislocation densities, with dislocations arranged in tangles, is a primary mechanism by which solution cleavage may form. There is also evidence that the seams propagate along crack tips. *Dislocation-enhanced selective dissolution* is favoured at low temperatures, where dislocation mobility is inhibited (Meike 1990), and slow strain rates, where dissolution can keep pace with deformation (Meike & Wenk 1988).

Transfer of material in solution is also fundamental to the formation of differentiated layering in low prograde metamorphic rocks such as slates and metaturbidites (Williams 1990). In turbidites deformed primarily through solution-assisted grain-boundary sliding, phyllosilicates play an important role in focusing both deformation (as the least competent phase in the rocks) and fluid flow (because of their strong anisotropy). Focused fluid flow may have occurred through transient fracture formation associated with shear along phyllosilicate domains (Williams 1990: Model 1) or by channelling of fluids along the well aligned, platy phyllosilicate grains in the steep limbs of crenulation folds (Wiliams 1990: Model 2). In both cases, fluids would either have transported material from one part of the system to another, or removed material from the system altogether. The net result is a well developed, mesoscopically visible compositional layering. The latter is also an example of diffusion-assisted grain-boundary sliding, which is possible with or without the presence of fluid. Diffusion rates increase with increasing temperature, though diffusive mass transfer is clearly not limited to high-temperature regimes because of the important role of fluids. In fact, the effects of diffusion are generally most significant where fluid is present.

Fluids and deformation

In general, fluids have a strain-softening effect on deforming rocks (see e.g. Carter *et al.* 1990). This is accomplished in several ways.

(1) Reactions may not actively involve fluids as participants (either reactants or products), but are generally enhanced by the presence of a fluid phase to aid diffusion and transport of mineral constituents. If the products are more competent than the reactants, the reaction will have a strain-hardening effect. *Reaction softening* occurs in one of two ways. (a) When deformation takes place under conditions that favour the reaction of a mineral or mineral assemblage to less competent mineral(s), the rock itself will become less competent. Many reactions involving fluids are essentially hydration reactions, producing relatively incompetent sheet silicates. (b) A new phase developed as the product of a reaction can inhibit the growth to equilibrium grain size of the other mineral constituents of a rock, facilitating the operation of grain-size-sensitive deformation mechanisms (Brodie & Rutter 1985; see later).

(2) Fluids greatly accelerate diffusive mass transfer, as indicated previously.

(3) High pore-fluid pressures decrease the effective stress, enhancing deformation mechanisms such as frictional grain-boundary sliding and fracturing.

(4) The presence of water significantly weakens some minerals, notably quartz (Griggs & Blacic 1965, Hobbs *et al.* 1972, review in Carter *et al.* 1990). Until recently, this 'water-weakening' effect was believed to result from water-related defects diffusing through the quartz crystal lattice (e.g. Hobbs 1981). More recent studies (Kronenberg *et al.* 1986, Rovetta & Holloway 1986) established the importance of the development and subsequent healing of cracks to incorporating water into the quartz crystal lattice, while continuing to emphasize the importance of subsequent diffusion of defects to 'water-weakening'. Since that time, solution precipitation and microcracking have been shown to cause 'water-weakening' in an experimental study of quartz deformation (Den Brok & Spiers 1991). An electron microscope study by Fitz Gerald *et al.* (1991) on samples deformed by Ord & Hobbs (1986) confirmed the findings of Kronenberg *et al.* (1986) and Rovetta & Holloway (1986) that water is incorporated along

cracks. However, Fitz Gerald *et al.* (1991) also presented evidence that dislocations necessary for crystal-plastic flow nucleate on fractures, so microscopically ductile deformation mechanisms are also involved. Whether or not any of the deformation behaviour described above occurs in naturally deformed rocks is an important question. If it does, are the associated microstructures erased during the subsequent crystal-plastic flow? And do water-related defects diffusing outward from microcracks play a role?

The influence of grain size

Creep. Creep describes macroscopically ductile, continuous flow in which the active deformation mechanisms can accommodate a constant stress. Experimental studies of creep behaviour have produced strain/time curves with three characteristic segments (cf. Turner & Weiss 1963) exhibiting: (1) decreasing strain rate (*primary creep; transient* or *plastic flow*), (2) constant minimum strain rate, resulting in permanent strain (*secondary,* or *steady-state creep*) and (3) accelerating strain rate, ultimately leading to rupture (*tertiary creep*). Steady-state creep can be accommodated by various of the mechanisms and processes described in previous sections, or by combinations thereof. Each of these *creep regimes* can be modelled mathematically; the *constitutive equations* or *flow laws* describing creep typically relate strain rate to stress and are fundamentally Arrhenius equations. That is, the equations express an exponential increase in strain rate with increasing temperature. A number of other dependencies are also accounted for: the activation energy, the stress dependence of strain rate, and a frequency factor for the active deformation mechanism(s) are all empirically determined. Pressure effects may be incorporated by relating the pressure sensitivity of creep to the influence of pressure on the melting temperature (see Handy 1989 and references therein).

Flow laws that describe the behaviour of a material of interest under varying conditions of temperature, pressure, etc. allow the construction of deformation regime maps such as that shown in Fig. 1.23(a). The fields within the map and the positions and morphologies of boundaries between these fields will vary with the rock studied. The positions of the boundaries are determined by the constitutive equations for the different flow regimes; the boundaries represent areas of equal contribution to the flow from the dominant deformation mechanisms to either side. There are broad zones of overlap between the various fields; the fields simply indicate volumes within which a deformation mechanism or mechanisms are dominant. Several important points that were mentioned earlier are illustrated by Fig. 1.23(a). The field in which brittle flow dominates (which includes processes such as fracture, frictional grain-boundary sliding and cataclasis) shrinks with increasing pressure, but expands with increasing strain rate. Ductile flow becomes dominant with an increase in homologous temperature ($T/T_{melting}$). For constant grain size and pore pressure (the conditions represented in Fig. 1.23a), grain-size-

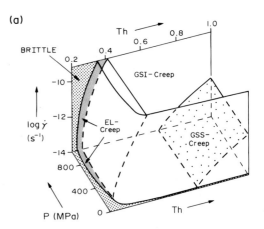

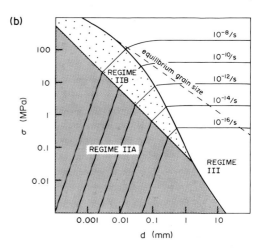

Fig. 1.23. (a) Schematic diagram based on experimental data for hydrous polycrystalline olivine (dunite), showing the extent of deformation regimes in pressure — homologous temperature — shear strain rate space with grain size and pore pressure held constant (Handy 1989). The brittle regime includes fracture, cataclasis and frictional grain-boundary sliding. EL-Creep is exponential law creep (referring to the flow law describing the behaviour), which is accomplished by dislocation glide and incipient dislocation creep. GSI-Creep (grain-size-insensitive creep) occurs through dislocation creep. GSS-Creep (grain-size-sensitive creep) takes place through diffusion-accommodated grain-boundary sliding. (b) Deformation mechanism map for quartz at 500°C, showing relative positions of deformation fields with varying stress (σ) and grain size (d) (Etheridge & Wilkie 1979). Flow laws (*not* typical exponential relationships) used to calculate the boundaries for different deformation regimes from Gifkins (1976). Regime IIA involves grain-boundary sliding accommodated by grain-boundary diffusion. In regime IIB, grain-boundary sliding is accommodated by dislocation glide and climb along grain boundaries and within grain mantles. Regime III represents grain-boundary sliding accommodated by dislocation glide and climb throughout the grains. The dashed line illustrates the relationship between equilibrium grain size and stress based on Twiss' (1977) equation. Thin solid lines show strain rates.

sensitive creep is favoured by relatively low strain rates and high homologous temperature.

Grain size sensitivity. Some deformation mechanisms, such as dislocation glide, are not affected by variations in grain size. However, deformation mechanisms influenced by the width of the effective grain boundary relative to the grain diameter are sensitive to changes in grain size. In particular, deformation regimes dependent on diffusion are strongly favoured by a decrease in grain size. Smaller grains have larger surface area/volume ratios, and therefore shorter intracrystalline diffusion paths; finer-grained rocks have a greater volume of grain boundaries, facilitating intergranular diffusion. Diffusion-accommodated grain-boundary sliding is arguably the most important grain-size-sensitive deformation process; if the grain size decreases, or temperature increases, sufficiently for a rock to deform in this regime, it results in a significant increase in strain rate (Gifkins 1976) and can affect superplastic flow. *Superplastic flow* was initially described in metals, and refers to materials which can accommodate an unusually large strain (c.1000%) without necking when subjected to extension.

Figure 1.23(b), which shows the relative positions of creep regimes IIA, IIB and III of Gifkins (1976) with changing stress and grain size, illustrates the effect of variation in grain size on deformation. Regime III (no shading) is a region in which deformation takes place mainly by grain-boundary sliding accommodated by dislocation glide and climb throughout the deforming grains. Strain rate curves within this field are virtually horizontal, indicating a lack of dependence on grain size. In contrast, regime IIA (fine dark shading), grain-boundary sliding accommodated by grain-boundary diffusion, exhibits a significant dependence on grain size: an order of magnitude decrease in grain size results in greater than an order of magnitude increase in strain rate. Grain-boundary sliding accommodated by dislocation glide and climb along grain boundaries and within the grain mantle (regime IIB; light stipple) shows a dependence on grain size intermediate between the response of regime III and that of regime IIA.

It is obvious from Fig. 1.23(b) that an increase in strain rate will result if the grain size is reduced sufficiently to move deformation into regime IIA. It should be noted that Etheridge & Wilkie (1979) constructed this diagram using a grain boundary width of 2b; if the order of magnitude larger values determined by White & White (1981) are more relevant, the curves shown would be shifted to the right. In the latter case, diffusion-assisted grain-boundary sliding (regime IIA) would become important at larger grain sizes than those shown in Fig. 1.23(b).

If the diagram in Fig. 1.23(a) were reconstructed for a smaller grain size, the region of grain-size-sensitive creep would be shifted to the left (Handy 1989). In other words, grain-size-sensitive creep becomes important at increasingly lower homologous temperatures with decreasing grain size. Clearly, grain size reduction can alter the material behaviour of a rock profoundly. There are two main processes by which the grain size in a rock is reduced: the first is dynamic recrystallization, the second cataclasis. These will be discussed in order below.

In a monomineralic rock or sector of a rock, the grain size of a mineral experiencing dynamic recrystallization will approach an equilibrium value inversely proportional to the stress. The relationship between stress and grain size was expressed mathematically by Twiss (1977), who developed the equation used to construct the equilibrium grain size line in Fig. 1.23(b); the reader will note that this falls almost entirely within regime III. That is, once an equilibrium grain size is attained, deformation will be dominated by grain-boundary sliding accommodated by dislocation movement. Production of an equilibrium grain size through dynamic recrystallization of a monomineralic domain ensures that diffusion-accommodated grain-boundary sliding will not be favoured (Etheridge & Wilkie 1979). It should be noted, again, that if Fig. 1.23(b) were reconstructed with an effective grain boundary width greater than 2b, the curves would be shifted to the right; the equilibrium grain size line, though, would remain in the same position. In other words, the larger the effective grain boundary width, the more likely it is that deformation at equilibrium grain size will occur by grain-size-sensitive deformation mechanisms. Etheridge & Wilkie (1979) also pointed out that very small, non-equilibrium grain sizes may be maintained during dynamic recrystallization in a multiphase rock, where the growth to equilibrium grain size of the mineral of interest is inhibited by another phase. As noted earlier, the additional phase or phases may be product(s) of metamorphic reaction(s). Natural instances of diffusion-accommodated grain-boundary sliding in rocks where a fine grain size has been stabilized due to the presence of two phases have been documented (e.g. Boullier & Gueguen 1975, Behrmann & Mainprice 1987). The presence of multiple phases will also accelerate strain because of the enhanced diffusivity between unlike phases (Etheridge & Wilkie 1979).

Reduction in grain size through cataclasis can be significant. Reference to Fig. 1.23(a) indicates that an increase in strain rate can shift deformation from grain-size-insensitive creep to the brittle field, enabling cataclasis to occur. As indicated by Fig. 1.23(b), the subsequent reduction in grain size can shift deformation into a grain-size-sensitive regime (either IIA or IIB). Such a deformation history has been suggested in natural rocks (Goodwin 1986). It is also significant that reduction in the grain size of either a monomineralic or a polymineralic rock, through any process, can promote fluid flow (Sibson 1977) and potentially cause strain-softening (see previous section).

In general, grain-size-sensitive deformation mechanisms result in little to no crystallographic preferred orientation (e.g. Boullier & Gueguen 1975, Behrmann & Mainprice 1987). In contrast, grain-size-insensitive

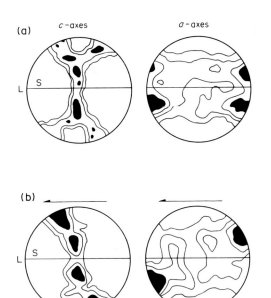

Fig. 1.24. Two examples of typical crystallographic preferred orientation developed in quartzites (Law *et al.* 1986). (a) *c*-axis and *a*-axis preferred orientations developed in quartzites having a coaxial deformation history. (b) *c*-axis and *a*-axis preferred orientations developed in quartzites having a non-coaxial deformation history (in this example, sinistral shear).

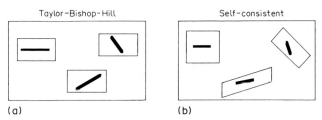

Fig. 1.25. (a) Schematic representation of homogeneous deformation as required by the Taylor–Bishop–Hill model. Note how the strain of individual grains is identical, and is the same as the overall strain. (b) Schematic representation of heterogeneous deformation if the influence of the slip system orientations is taken into account in the viscoplastic self-consistent model (Wenk *et al.* 1989). Note that the strain of individual grains is different, with the amount of strain dependent on the orientation of the dominant slip system. The bold lines indicate the orientation of the dominant slip system in each grain.

deformation, generally dominated by dislocation movement, results in strong crystallographic preferred orientations (see below).

Crystallographic preferred orientations

As a rock deforms, the crystallographic orientation of each constituent grain changes. The change can result from a number of different mechanisms, including rotation of the whole grain by intergranular movement, intragranular movement (e.g. dislocation glide, twinning, and climb) and recrystallization. As a result of this deformation, the pattern of crystallographic orientations of all grains in a rock also changes (Fig. 1.24). A rock having an initially random orientation of grains can develop a strong preferred orientation during deformation. An initial preferred orientation of grains will become modified during deformation. In quartzites, which we will use as an illustrative example, crystallographic preferred orientations can provide information about the deformation history (whether the deformation was coaxial or non-coaxial) and the deformation mechanisms responsible for the crystallographic preferred orientation.

In the following we will consider a situation where microscopically ductile deformation occurs at relatively low temperatures (Fig. 1.23a), with the deformation accommodated by dislocation glide. Deformation of a monomineralic rock must be accommodated by at least five independent slip systems to be homogeneous (von Mises criterion: see Paterson 1969). Five independent slip systems are rarely observed in natural rocks. The resulting crystallographic preferred orientation reflects the reorientation along all slip systems.

Many different models have been developed in an attempt to replicate crystallographic preferred orientations recorded in natural rocks, and thereby elucidate the underlying processes. Crystallographic preferred orientation development can be modelled using a single slip system, assuming homogeneous stress throughout the rock (Sachs 1928, Etchecopar 1977). Such a model is unrealistic as incompatibilities develop at grain boundaries due to the assumption of homogeneous stress. If the strain is assumed to be homogeneous throughout the rock, the *Taylor–Bishop–Hill model* (Taylor 1938, Bishop & Hill 1951) can be used to model the development of crystallographic preferred orientations (see Lister *et al.* 1978, Lister & Hobbs 1980 for the application of this model to quartzites). The assumption of homogeneous strain requires that all grains deform by the same amount (Fig. 1.25a). Figure 1.26 illustrates how a crystallographic preferred orientation could develop according to the Taylor–Bishop–Hill model in two dimensions (where only two slip systems are needed for homogeneous deformation). Crystallographic preferred orientations developed using the Taylor–Bishop–Hill model are reasonably similar to crystallographic preferred orientations observed in naturally deformed rocks (Price 1985). The strict requirement for five slip systems can be relaxed if other deformation mechanisms are present (e.g. diffusion, grain boundary recrystallization); this also allows the requirement for homogeneous strain through the specimen to be relaxed. In certain circumstances (e.g. flat grains) a modified version of the Taylor–Bishop–Hill model can be used to model crystallographic preferred orientation development using less than five slip systems (Kocks & Canova 1981, Kocks & Chandra 1982, Hobbs 1985). Other deformation mechanisms (e.g. grain boundary recrystallization) are assumed to operate to allow overall homogeneous strain. A more recent

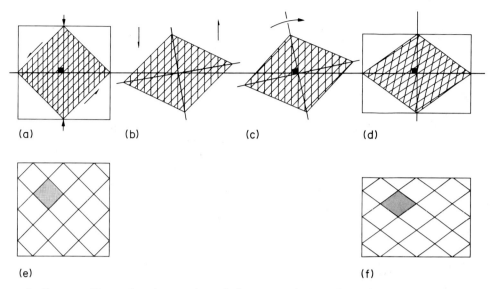

Fig. 1.26. Schematic diagrams illustrating the rotation of slip systems in two dimensions according to the Taylor–Bishop–Hill model. Three stages can be recognized in the development of a CPO, although it must be stressed that these operate simultaneously. (a) Slip on one system produces an asymmetrical parallelogram. Because lattice orientation varies from grain to grain and because the fit of grain boundaries must be preserved, only a symmetrical strain is compatible with the bulk strain (e & f). Thus, the second slip system is activated (b) to distort the parallelogram to a symmetrical form (c) which undergoes a rigid body rotation to bring its strain axes into coincidence with the bulk axes (d). The crystallographic preferred orientation which develops results from the rigid body rotation, rotating the slip systems towards parallelism with the extension axis. Note that the angle between the two slip systems remains constant throughout because the two lines represent crystal lattice planes.

development in the modelling of crystallographic preferred orientations is the *large-strain viscoplastic 'self-consistent' theory* for plastic deformation (Molinari *et al.* 1987, Wenk *et al.* 1989). In this model the deformation of each grain is a function of the orientation of all grains, with the number of grains influencing the deformation behaviour; as opposed to the Taylor–Bishop–Hill model in which the behaviour of each grain is determined separately. Incompatibilities are allowed to occur on a grain scale, with grains that have slip systems in orientations suitable for slip deforming more than grains with slip systems in unsuitable orientations (Fig. 1.25b). Such a model predicts a *c*-axis maximum parallel to the intermediate strain axes (Wenk *et al.* 1989); such a maximum is often observed in natural rocks.

Simplistically, the slip systems which operated during deformation in natural rocks can be determined by observing which crystallographic directions are subparallel to the inferred flow direction (e.g. mylonitic lineation), and which crystallographic planes are subparallel to the inferred flow plane (Schmid & Casey 1986, see Wenk & Christie 1991, for discussion). Measurement of the preferred orientation of quartz, for example, shows that <*a*> is nearly parallel to the flow direction and that the basal, prism, and positive and negative rhomb planes are nearly parallel to the flow plane (Fig. 1.24). Such observations are considered to indicate that the <*a*> direction is the only slip direction, with slip occurring on the basal, prism, and positive and negative rhomb planes (Schmid & Casey 1986). It must be stressed that this simplification does not take into account the simultaneous operation of several slip systems (Wenk & Christie 1991). Both the Taylor–Bishop–Hill model and the 'self-consistent' model show that, for simple shear, slip systems do not reach stable end orientations (i.e. slip plane parallel to the shear plane, slip direction parallel to the shear direction). The slip systems rotate throughout the deformation; the crystallographic preferred orientation observed reflects regions where this rotation is slowest (Wenk & Christie 1991). These models, as well as experimental studies (Dell'Angelo & Tullis 1989), show that the prominent slip plane is not aligned parallel to the inferred flow plane (and can be up to 20° off this direction). These observations indicate that extreme care must be taken in interpreting crystallographic preferred orientations using the simplistic approach described above.

The deformation history recorded by the crystallographic preferred orientations of a mineral in a rock can, in part, be determined by examining the relationship between the crystallographic preferred orientation and the microfabric (foliation and lineation). A symmetrical distribution of *c*-axes with respect to the microfabric indicates a coaxial deformation path (Fig. 1.24a). An asymmetric distribution of *c*-axes with respect to the foliation and lineation indicates a non-coaxial deformation history (Fig. 1.24b). In quartzites, for example, the sense of asymmetry can be used to determine the sense of shear. Figure 1.24(b) shows the crystallographic preferred orientation of a quartzite showing a sinistral sense of shear.

In rocks with a coaxial deformation history, grains with slip systems in orientations unsuitable for slip will remain as comparatively undeformed augen throughout the deformation (e.g. Ralser 1990, figs 7b & c). However, during a non-coaxial deformation history these comparatively strong grains will be progressively rotated so that slip systems become oriented in easy directions for slip, allowing all grains to deform (e.g. Ralser 1990, fig. 7a). The rotation of grains into easy orientations for slip will cause a strain softening effect. This is one of the more important processes by which strain remains localized within an active shear zone.

SHEAR ZONES

Introduction

Here, we are concerned with the use of kinematic indicators to determine the sense of movement in *shear zones*. A more extensive review of the topic is given in Hanmer & Passchier (1991). *Kinematic indicators* are a product of non-coaxial straining and comprise a number of structures, microstructures and fabrics that have an asymmetry reflecting the sense of vorticity. They vary in reliability, and wherever possible, different types of kinematic indicators should be used in concert. When using porphyroblasts we suggest that numerous observations should be combined, and if they lack consistency in the sense of vorticity, they should not be used as an indicator for an assumed simple deformational history.

The use of kinematic indicators is complicated by the knowledge that once a shear zone is established it is likely to be reactivated. Such reactivation does not generally follow the same movement pattern. Thus, for example, thrusts are commonly reactivated as low-angle normal faults, and reactivation of transcurrent shear zones commonly results in a reversal of the sense of movement. The situation is probably further complicated by the fact that shear zones separate bodies of rock that are themselves undergoing deformation, albeit at a much lower strain rate than that within the shear zone. When considering faulting, it is customary to assume that a given fault separated rigid blocks (see Chapter 4). While this may be a reasonable approximation it cannot be generally true because many faults die within the rock body, and thus have to be accommodated by deformation within the surrounding rocks (Fig. 1.27). This becomes a greater problem when we consider shear zones and it is likely that most shear zones are analogous to what Means (1989, 1990) has called *stretching faults*, i.e. they are zones of high strain rate within a more slowly deforming rock body. This leads to various problems. For example, *stretching lineations*, which are defined by alignment of minerals and clasts approximately parallel to the extension direction, are generally assumed to represent the movement direction of the shear zone in which they occur. However, a lineation in the margin of such a zone may in fact be sharply inclined to the movement vector in the centre of the zone. Further, it is possible for the sense of shear within such a zone to be the reverse of that indicated by displaced markers (Williams & Schoneveld 1981). Such problems are beyond the scope of this discussion and interested readers are referred to the literature. Here, we restrict ourselves to description and discussion of kinematic indicators and assume that the shear zone is an uncomplicated one that has only moved in one direction.

Kinematic indicators

The use of kinematic indicators is usually associated with the study of shear zones. However, because the indicators are simply a product of rotational deformation they can be used anywhere where the deformation can be shown to have had a vortical component. Care must be exercised in their interpretation, however, as very similar structures may

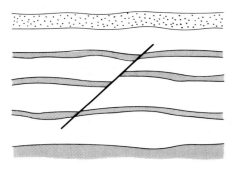

Fig. 1.27. Fault within a layered sequence showing how the fault dies out at either end. Note that if only the upper end of the fault is exposed it is indistinguishable from a normal growth fault.

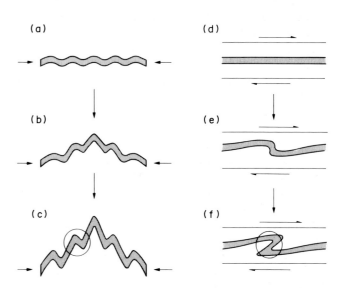

Fig. 1.28. Two models for the development of asymmetrical folds. (a–c) Folds are formed by layer-parallel shortening. Those forming early are symmetrical; when larger folds form later the early folds that lie on the limbs of the former are constrained to become asymmetrical. (d–f) Folds developed by rotation of layering by layer-parallel shear, or 'drag', in a shear zone situation. Note the similarity between the folds (circled) resulting from the two processes.

be a product of quite different strain histories. For example, *parasitic folds* used to be referred to as *drag folds* because they were believed to develop by the sliding of one layer over another on the limb of a larger fold. This explanation has been discarded in favour of a mechanism (Figs 1.28a – c) demonstrated experimentally by Ramberg (1963). The problem with the old explanation was not the principle but the fact that the magnitude of shear on a fold limb is generally too small to generate the folds. In shear zones, however, where the magnitude may be orders greater, the mechanism is considered viable (Lister & Williams 1983), and it produces folds with the same appearance but a very different history (Figs 1.28d – f). The indicators described here, in addition to being a product of non-coaxial strain histories, mostly represent large magnitudes of strain: magnitudes that are generally only recognized in shear zones.

Porphyroblasts. Porphyroblasts are large grains, relative to their matrix, that grow as a product of metamorphism. They are commonly rich in inclusions and if they grow during deformation in a shear zone, trails of inclusions may preserve evidence of the vorticity. The best examples are *snowball garnets*; so called because they have an overall spirally layered geometry similar to a large snowball made by rolling a small starter in snow so that it adds a continuous layer of snow as it rolls. Garnets behave as relatively rigid bodies in shear zones even when growing during deformation, and therefore they respond to the vorticity by maximizing spin (Fig. 1.14).

Snowball garnets have been described in considerable detail by Schoneveld (1979). He designed a simple apparatus that demonstrated very clearly the development of a growing porphyroblast. It models the garnet two dimensionally and depicts a diametrical plane normal to the rotation axis (Fig. 1.29). Viewed in that plane, a typical snowball garnet contains a spiral of quartz inclusions that represents overgrown pressure-shadow material, and a double spiral that crosses the quartz spiral and is commonly defined by graphite (Fig. 1.30).

The double spiral is not cylindrical in three dimensions but is a more complex shape resembling a rolled up sheath. Schoneveld (1979) demonstrated this very well by serial sampling through individual garnets. By building up a detailed picture of the complete three-dimensional geometry he was able to show how the orientation of different sections with respect to the rotation axis can be estimated.

The classical view of snowball garnets as expounded by Schoneveld (1979) has been challenged by Bell & Johnson (1989). They claim that the garnets do not rotate, but overgrow an alternating sequence of horizontal and vertical foliations to give a structure that is not spiral but can resemble a spiral. They base their hypothesis on imperfect spirals that are cited as evidence that spirals do not exist. Many spirals are imperfect, but the imperfections are not incompatible with the rotation model. They can be explained very simply by the rate of

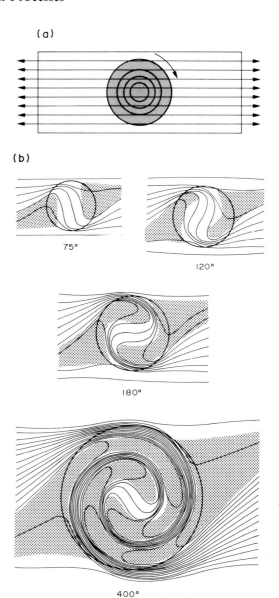

Fig. 1.29. (a) Diagram of a simple apparatus designed by Schoneveld (1979). The rectangular board has a circular disc cut from its centre. The disc (stippled) is coated with plasticine and replaced in the hole. Strings are stretched across the board and held tight by weights (not shown). A set of nesting rings (heavy-line circles) completes the apparatus. The smallest ring is pressed into the plasticine in the centre of the disc locking some of the strings to the disc. The disc is then rotated a little and the next larger ring is pressed into the plasticine and so forth. In this way a rotated garnet is simulated with the outer ring representing the garnet boundary and the strings representing the internal and external foliation. (b) Four stages in the simulation of a snowball garnet traced from the model. The traces of the strings which are smoothed to give a more realistic appearance represent the trace of the internal foliation; their gross distribution is exactly as in the model. The spacing of the strings simulates the relative spacing of mica as seen in nature and the 'mica-poor' areas (stippled) represent the pressure shadows outside of the garnet and the double spirals of quartz commonly seen inside of snowball garnets. After Schoneveld (1979).

rotation at times being high relative to the rate of growth. The Bell and Johnson model, however, is incapable of explaining the perfect spirals that are

Fig. 1.30. Tracing of a photograph from Schoneveld (1979) of a snowball garnet from the Northern Adula region, Switzerland. Rotation is about 380°. Garnet is solid black, quartz is white, mica is dark grey and other minerals are pale grey.

sometimes observed. It is also a two-dimensional model that, unlike the rotational model, cannot explain the observed three-dimensional structure of snowball garnets. Other lines of evidence against Bell and Johnson's model are as follows. (1) The sequence of vertical and horizontal foliations required by their model has not been demonstrated in areas where snowball garnets occur. (2) Snowball garnets occur exclusively in shear zones where very larger strains and vorticity can be demonstrated. This is unexplained by the Bell and Johnson model and in fact is difficult to reconcile with their 'series-of-foliations' requirement. On the other hand it is precisely what is to be expected from the rotation model. On the basis of this evidence we reject the Bell and Johnson model.

Anyone wishing to use snowball garnets as a kinematic indicator should refer to Schoneveld (1979) to ensure that they have a section normal to the rotation axis as oblique sections can look similar to Fig. 1.30 and can give rise to spurious inferences. Also, when a section is clearly not oriented normal to the rotation axis Schoneveld's diagrams can be used to determine the orientation of the appropriate section with reasonable accuracy, thus minimizing the amount of trial and error necessary.

Porphyroclast systems. Mylonites commonly consist of relatively large relict grains (*porphyroclasts*) in a fine-grained matrix (where the grain size has been reduced during deformation). Feldspars, quartz and micas all commonly form porphyroclasts. The suffix 'clast' implies that a porphyroclast forms by fragmentation of an original grain. In fact, reduction in the size of the porphyroclast continues throughout the deformation history either through cataclasis or, more commonly, through dynamic recrystallization, or through a combination of both processes (cf. Goodwin & Wenk 1990). The smaller grains produced from a porphyroclast form tails that lead away from the porphyroclast (Fig. 1.31); the tails and porphyroclast together forming a *porphyroclast system* (Passchier & Simpson 1986).

A porphyroclast system, in which the tails are derived from the deforming porphyroclast, may superficially resemble a porphyroblast or porphyroclast with pressure fringes. There is, however, an important distinction between the two types of structure. Pressure fringes generally comprise material that is mineralogically distinct from, and not genetically related to, the grain they surround (e.g. quartz pressure fringes around a garnet porphyroblast or a feldspar porphyroclast). The material forming the tails of a porphyroclast system is derived from the porphyroclast and may be: (1) mineralogically similar to the relict grain, though potentially chemically distinct (e.g. a calcic plagioclase porphyroclast with more sodic plagioclase tails), or (2) derived from the relict grain by reaction (e.g. quartz and muscovite tails to a feldspar porphyroclast).

In general, porphyroclasts rotate during deformation, just as the previously described synkinematic porphyroblasts do. Again, there is an important distinction. Structures such as snowball garnets develop when porphyroblasts grow during deformation; porphyroclast systems develop as crystals experience grain-size reduction during deformation. The rotation of the porphyroclast defines the geometry and symmetry of the tails with respect to the porphyroclast (Passchier & Simpson 1986). A discussion of types of porphyroclast system, experimental study of their formation, and the use of such systems as kinematic indicators in naturally deformed rocks is presented in detail in Passchier & Simpson (1986). A brief summary

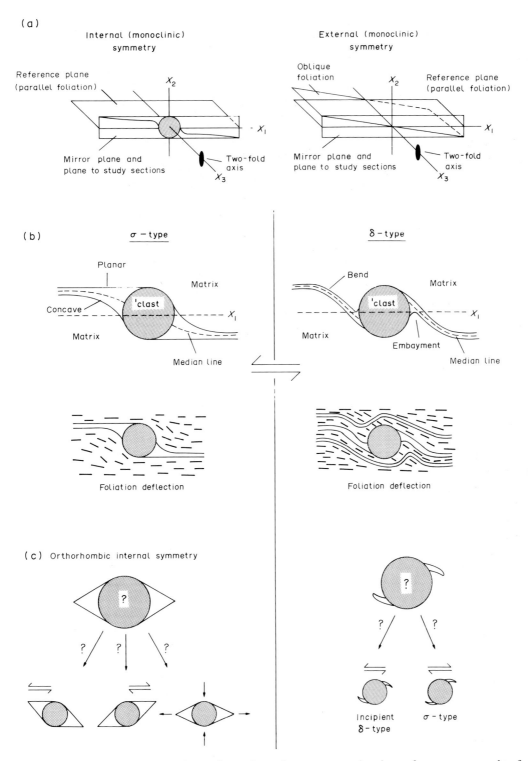

Fig. 1.31. (a) Internal and external symmetry of porphyroclast systems, showing reference axes and reference plane, symmetry elements and plane containing maximum information for petrographic study. (b) Features of σ- (left) and δ- (right) type porphyroclast systems with monoclinic internal symmetry in sinistral shear. Note that tails of both types step up to the left with respect to the trace of the reference plane (x_1), and that both deflect the foliation in a similar way. Further discussion in text. (c) Porphyroclast systems which lack distinctive morphology in their tails. To the left, σ-type porphyroclast systems with orthorhombic internal symmetry must be interpreted with respect to external symmetry; the three possible interpretations are shown. To the right, an apparent δ-type system with tails too short to distinguish characteristic bends may be an incipient δ-type, or may be a σ-type porphyroclast system. The sense of shear indicated by each is different. Adapted from Passchier & Simpson (1986).

of their description and use of porphyroclast systems follows; factors governing the formation of the systems are not discussed.

The internal symmetry of a porphyroclast system may be described with respect to a reference plane defined by the axis of rotation of the porphyroclast (x_3) and the normal to x_3 which is parallel to the straight sections of the tails (x_1; Fig. 1.31). The reference plane is parallel to a foliation (typically C-surfaces to which compositional layering is parallel) which contains a stretching lineation, generally parallel to x_1. The external symmetry can also be described with respect to these reference elements (Fig. 1.31a); for example, a second foliation oblique to the first will result in monoclinic external symmetry. The description of external symmetry does not include the way in which the foliation is deflected around the porphyroclast system (Fig. 1.31b). A thin section made at right angles to x_3 and the reference plane will be parallel to a mirror plane, and will provide the maximum information about the kinematics of the system. The mirror plane is present no matter what the combination of internal and external symmetry (e.g. monoclinic internal and external symmetry, orthorhombic internal and monoclinic external symmetry, etc.).

Two types of porphyroclast system can be described. In both types, the symmetry of 'stair-stepping' (cf. Lister & Snoke 1984) of tails with respect to the reference plane will be identical for a given shear sense (cf. Fig. 1.31b), but the morphology of the tails will differ in detail. The stair-stepping symmetry of either type of porphyroclast system can best be described by the position of a median line drawn in the tails relative to the reference plane (Fig. 1.31b). In both types, the median line will step up in the direction of movement with respect to the reference plane. In δ-type porphyroclast systems, the median line will also cross the reference plane next to the porphyroclast at the beginning of the step.

σ-type porphyroclast systems (Figs 1.31b & c) comprise a central porphyroclast from which wedge-shaped tails of fine-grained material derived from the porphyroclast extend. Two classes of σ-type porphyroclast systems have been distinguished: $σ_a$ porphyroclasts are surrounded by matrix material and are widely separated in the host rock; $σ_b$ porphyroclast systems tend to occur in groups, and are bounded by shear bands (see later) along which their tails extend. The outermost (furthest from the reference plane) margins of σ-type tails are planar, while the innermost boundaries vary in character: if the internal symmetry is orthorhombic, they are also planar (Fig. 1.31c), but if the internal symmetry is monoclinic, they are concave toward the reference plane (Fig. 1.31b). The tails thin away from the porphyroclast parallel to the reference plane and the foliation. Rotation of the porphyroclast also results in a characteristic deflection of the foliation around the porphyroclast system (Fig. 1.31b).

The deflection of foliation around a δ-type system is similar to that around a σ-type system (Fig. 1.31b), but the morphology of the tails is different. δ-type porphyroclast systems are distinguished by tails which are not wedge-shaped, but which maintain a relatively constant thickness throughout their extent. Whereas the external margins of the σ-type tails are planar, those of the δ-type tails are embayed (Fig. 1.31b). As the δ-type tails are extended parallel to the reference plane, they develop a distinctive bend (Fig. 1.31b).

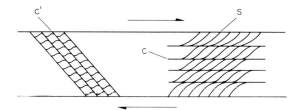

Fig. 1.32. Relationship between foliations developed in shear zones and the sense of shear. S is a shape fabric and is generally penetrative. C is a less penetrative foliation representing narrow zones in which shear strain is high relative to the rest of the shear zone. S generally curves into C indicating the sense of shear. C' is a crenulation cleavage which may or may not be penetrative. It can form in S or in C, and tends to rotate them counter to the sense of shear of the zone. The sense of shear on C' is the same as for the zone.

Where the tails are not sufficiently long, a reference plane cannot be defined. In such a circumstance, interpretation of the system may still be possible if distinctive morphological features (σ-type: planar and concave margins to the tail; δ-type: embayments and bends in the tail) are present. If the system has orthorhombic symmetry, interpretation still may be possible with respect to the external symmetry — the symmetry of the porphyroclast system relative to surrounding foliation(s). Other complications involve folded or overturned porphyroclast systems, which usually can be interpreted with caution (Passchier & Simpson 1986), and systems with flattened tails, which generally cannot be interpreted. Polydeformed regions, of course, present problems of their own. It is therefore important to remember that there will be circumstances where unambiguous interpretation is not possible (Fig. 1.31c). Interpretation should be based on multiple observations, petrographic study is imperative, and, whenever feasible, different types of kinematic indicators should be checked against one another.

Foliations. Three types of foliation are associated with the development of shear zones. In addition, there may be inherited foliations as well as younger surfaces overprinting the zone. Here we are only concerned with the three that actually develop as a result of shearing.

During shear all markers such as grains and grain aggregates deform and become elongated in the direction of extension and thereby define a foliation that is inclined to the boundary of the shear zone (e.g. Ramsay & Graham 1970). This *foliation* is generally referred to as the *shape fabric* or *S-surface* (Berthé et al. 1979). If it is defined by markers such as xenoliths in a deforming granitoid (Ramsay & Graham 1970) the foliation will rotate progressively towards the shear plane and at large strains will be essentially parallel to the shear zone boundary. If defined by grain shape the foliation can be modified by recrystallization which will tend to randomize the fabric. Thus during deformation there will be two competing mechanisms, one (strain) tending to produce a grain-shape fabric and another (recrystallization) tending to annihilate that fabric. Because statistically the old grains are the most rotated, and therefore most nearly parallel to the finite strain axes, and as they are also the most deformed, and therefore the most likely to recrystallize, the shape fabric does not keep pace with the rotating strain axes. Instead, for a given set of deformation conditions the fabric will tend to become steady state and therefore constant in orientation. We therefore do not support the use of this fabric for determining the magnitude of shear as has been suggested by some writers. However, its asymmetry with respect to the shear zone boundary is certainly useful for determining the sense of shear (Fig. 1.32).

The second type of foliation developed in shear zones forms approximately parallel to the shear zone boundary and is interpreted as an array of local shear surfaces. It may be that some such surfaces represent transient brittle features — certainly such structures are observed during the ductile field, experimental deformation of organic salts (unpublished work by P.F.W. using a see-through deformation rig designed by J. Urai and W.D. Means). Alternatively the surfaces may be narrow zones in which the shear strain is higher than in the intervening rock (Fig. 1.32). This foliation may be less penetrative than the S-fabric; it is commonly penetrative at hand specimen scale but not at microscopic scale. It is referred to as the *C-surface* after the French word cisaillement, meaning shearing (Berthé *et al.* 1979).

Layering is commonly parallel to the C-surface because earlier layering, veins and any other type of compositional features are progressively rotated towards the shear plane. Thus at high strains these compositional domains are smeared out into a layering that is essentially parallel to the shear zone boundary. Weaker layers are then likely to become the locus of high-shear strain rates and thus become C-surfaces. Consequently, in most high-strain shear zones the C-surfaces and layering are coincident and both are parallel to the shear zone boundary. The C- and S-surfaces commonly occur together, forming what is known as a *C/S-fabric* (Fig. 1.32).

The third shear zone foliation is referred to by several names including *strain-band cleavage, shear-band cleavage, Riedel shear, C'-surface* and *extensional crenulation cleavage*. It has the appearance of a crenulation cleavage and like the latter only forms in rocks which already have a good planar fabric. It is generally developed where there is a good S- or C-fabric. It may be a brittle or brittle/ductile transition feature and is inclined to the earlier foliation at an angle approaching 45°. It has been shown experimentally to

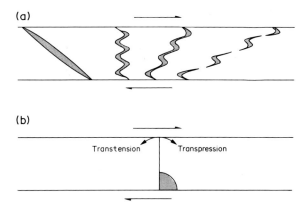

Fig. 1.34. (a) Progressive deformation of a tension gash vein in a shear zone. Four successive stages are shown going from left to right. (b) The same shear zone showing the quadrant in which all lines are extended during simple shear (shaded area). In transtension and transpression the extensional axes lie in the same quadrant as in simple shear, but the angle in which all lines undergo extension is different. One line limiting this angle is still defined by the shear surface; the other line is anticlockwise of the normal to the shear surface for transtension and clockwise of the normal for transpression.

form only where there is shortening orthogonal with the earlier foliation, so that if the latter is parallel to the shear zone boundary it implies that the zone is transpressional (Williams & Price 1990). Its relationship to the sense of shear is shown in Fig. 1.32. If shear parallel to the earlier foliation is large compared to the shortening perpendicular to it, then only the one set of surfaces develops (Fig. 1.32). However, if the shear component is small compared to the shortening component a conjugate pair of bands will form (Williams & Price 1990).

C'-surfaces may be penetrative at the microscopic scale but commonly are hand-specimen or outcrop-scale fabric elements. They also tend to be localized in occurrence, and it is not uncommon for them to occur singularly or in small groups.

In areas of poor outcrop it may be difficult to distinguish C/S-fabrics from C'/S-fabrics. However, from the point of view of determining the sense of shear this is not important as the interpretation will be the same in both cases.

Folds. So called 'drag folds' have long been used for determining sense of shear, and asymmetrical folds in shear zones are probably one of the most reliable kinematic indicators if used with caution. This statement is supported by the observation that in many shear zones the folds are consistently of the same

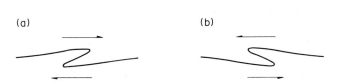

Fig. 1.33. Diagrammatic representation of the relationship between 'drag-fold' asymmetry and sense of shear in shear zones.

Fig. 1.35. Diagram showing how folds with sinistral asymmetry can develop in dextral shear zones.

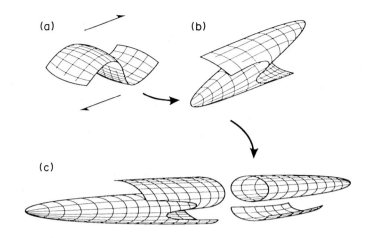

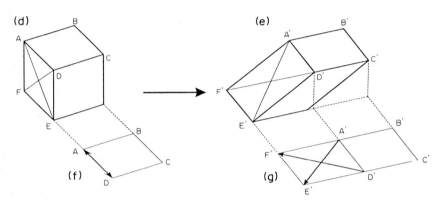

Fig. 1.36. (a–c) Diagrams showing the development of a sheath fold (c) from a doubly plunging fold (a) by simple shear. The way in which the axes rotate is further shown in (d) and (e). In (d) the axes of the doubly plunging fold are represented by the lines AE and DF. Simple shear causes these lines to rotate towards parallelism as is seen clearly in the plan views of the two blocks.

asymmetry. The relationship between asymmetry and shear sense is shown in Fig. 1.33.

There are two potential pitfalls that can be avoided if care is taken. (1) Not all folds in a shear zone are

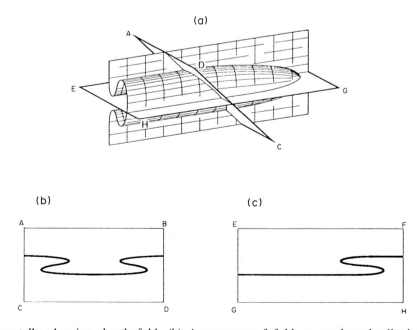

Fig. 1.37. (a) Horizontally plunging sheath fold. (b) Appearance of fold on moderately dipping surface ABCD. (c) Appearance of fold on horizontal surface EFGH. The asymmetry viewed on the horizontal surface indicates that the fold formed in response to sinistral shear.

necessarily 'drag folds' (i.e. folds that are a direct product of the vorticity). For example, the folds in Fig. 1.34 are simply a product of shortening. As the vein rotates it is constrained to shorten and the resulting folds may have the wrong asymmetry. Similarly, in shear zones where the magnitude of strain is sufficiently small, earlier layering may be steeply inclined to the zone and may be shortened and folded in the same manner as the vein (Fig. 1.35). This problem is avoided by only using folds that are developed in layering or veins that are approximately parallel to the shear zone boundary. (2) Once formed, folds may rotate in shear zones such that their hinge lines converge on the extension direction. Because fold hinges are not generally straight, different segments of a single hinge line commonly rotate in different directions (Fig. 1.36). At large strains the product of this process is a *sheath fold* (Figs 1.36c and 1.37a). In areas of gently plunging sheath folds it is possible to have two-dimensional, sloping exposures of the same fold pair that appear to indicate opposite senses of shear (Fig. 1.37b). This problem can be avoided by always considering the geometry of the fold pair as it appears on a section parallel to the stretching direction (Fig. 1.37c). As well as being useful for determining the sense of shear, sheath folds uniquely define the direction of shear. The latter must lie within the sheath and thus, the tighter the sheath, the more accurately the direction of shear is defined.

Veins. Veins in shear zones are very useful kinematic indicators because they commonly enable us to determine the direction of extension (also see Chapter 5). Both the instantaneous and finite extensional axes lie in the same quadrant (Fig. 1.34), defined by the shear plane and its normal, throughout the strain history, for all types of shear zones (transtensional, simple shear or transpressional). Thus by identifying the extension direction we can determine the sense of shear.

It is commonly possible to demonstrate that certain veins were emplaced during shearing, but it is generally not a prerequisite for being able to use them to determine the sense of shear. All veins, irrespective of origin, will rotate during shearing towards the extension direction which in any shear zone points in the direction of shear (Fig. 1.34). Figure 1.34 shows what can happen to a tension vein formed in response to the stress field driving the shear zone. The vein forms at an angle approaching 45° to the shear zone boundary and during shearing rotates through the normal to the shear zone and converges on the finite extension axis. The latter makes a very small angle with the shear plane at large strains. Such a vein is likely to fold initially and then experience boudinage. In many shear zones various stages in this sequence are preserved in veins of different age, and the overall pattern can be used to determine the sense of shear (Fig. 1.34). However, the complete sequence is not necessary; just the presence of boudinaged veins is generally sufficient to define the extensional quadrant. The exception occurs when a zone is strongly transpressional. It is then possible for a vein to be back-rotated (to rotate in the direction opposite to the sense of shear) if its initial orientation is close to that of the shear plane. Because of problems of strain compatibility between the shear zone and its margins the strain magnitude in such zones is not likely to be very large. Thus in zones where veins are boudinaged they can be used to identify the extension direction and thereby the shear direction (Fig. 1.34).

Acknowledgements — We thank J. C. White for critical comments on the second section of this chapter; his suggestions improved the manuscript.

REFERENCES

Atkinson, B. K. 1982. Subcritical crack propagation in rocks: theory, experimental results and applications. *J. Struct. Geol.* **4**, 41–56.

Atkinson, B. K. (ed.) 1989. *Fracture Mechanics of Rock*. Academic Press Geology Series.

Barber, D. J. & Meredith, P. G. 1990. *Deformation Processes in Minerals, Ceramics and Rocks*. The Mineralogical Society Series 1, Unwin Hyman, London.

Behrmann, J. H. & Mainprice, D. 1987. Deformation mechanisms in a high-temperature quartz-feldspar mylonite: evidence for superplastic flow in the lower continental crust. *Tectonophysics* **140**, 297–305.

Bell, T. H. & Johnson, S. E. 1989. Porphyroclast inclusion trails: the key to orogenesis. *J. Met. Geol.* **7**, 279–310.

Berthé, D., Choukroune, P. & Jegouzo, P. 1979. Orthogneiss, mylonite and non coaxial deformation of granites: the example of the South Armorican Shear Zone. *J. Struct. Geol.* **1**, 31–42.

Borg, I. Y. & Heard, H. C. 1969. Mechanical twinning and slip in experimentally deformed plagioclases. *Contrib. Mineral. Petrol.* **23**, 128–135.

Borg, I. Y. & Heard, H. C. 1971. Experimental deformation of plagioclases. In: *Experimental and Natural Rock Deformation* (edited by Paulitsch, P.). Springer, Berlin, 375–402.

Borg, I. Y., Handin, J. & Higgs, D. V. 1959. Experimental deformation of plagioclase single crystals. *J. geophys. Res.* **64**, 1094.

Borg, I., Friedman, M., Handin, J. & Higgs, D. V. 1960. Experimental study of St. Peter Sand: a study of cataclastic flow. In: *Rock Deformation* (edited by Griggs, D. & Handin, J.). *Mem. geol. Soc. Am.* **79**, 133–192.

Boullier, A. M. & Gueguen, Y. 1975. SP-mylonites: origin of some mylonites by superplastic flow. *Contrib. Mineral. Petrol.* **50**, 93–104.

Brodie, K. & Rutter, E. H. 1985. On the relationship between deformation and metamorphism with special reference to the behaviour of basic rocks. In: *Kinetics, Textures and Deformation* (edited by Thompson, A. B. & Rubie, D.). *Adv. Phys. Geochem.* **4**, 138–179.

Brown, W. L. 1989a. Glide twinning and pseudotwinning in peristerite: Si, Al diffusional stabilization and implications for the peristerite solvus. *Contrib. Mineral. Petrol.* **102**, 313–320.

Brown, W. L. 1989b. Glide twinning and pseudotwinning in peristerite: twin morphology and propagation. *Contrib. Mineral. Petrol.* **102**, 306–312.

Carter, N. L., Kronenberg, A. K., Ross, J. V. & Wiltschko, D. V. 1990. Control of fluids on deformation of rocks. In: *Deformation Mechanisms, Rheology and Tectonics* (edited by Knipe, R. J. & Rutter, E. H.). *Spec. Publs geol. Soc. Lond.* **54**, 1–14.

Dell'Angelo, L. N. & Tullis, J. 1989. Fabric development in

experimentally sheared quartzites. *Tectonophysics* **169**, 1–21.

Den Brok, S. W. J. & Spiers, C. J. 1991. Experimental evidence for water weakening of quartzite by microcracking plus solution-precipitation creep. *J. geol. Soc. Lond.* **148**, 541–548.

Drury, M. R. & Urai, J. L. 1990. Deformation-related recrystallization processes. *Tectonophysics* **172**, 235–253.

Etchecopar, A. 1977. A plane kinematic model of progressive deformation in a polycrystalline aggregate. *Tectonophysics* **39**, 121–139.

Etheridge, M. A. & Wilkie, J. C. 1979. Grainsize reduction, grain boundary sliding and the flow strength of mylonites. *Tectonophysics* **58**, 159–178.

Fitz Gerald, J. D., Etheridge, M. A. & Vernon, R. H. 1983. Dynamic recrystallization in a naturally deformed albite. *Textures Microstructure* **5**, 219–237.

Fitz Gerald, J. D., Boland, J. N., McLaren, A. C., Ord, A. & Hobbs, B. E. 1991. Microstructures in water-weakened single crystals of quartz. *J. geophys. Res.* **96**, 2139–2155.

Gifkins, R. C. 1976. The effect of grain size and stress upon grain-boundary sliding. *Metall. Trans. A* **8A**, 1507–1516.

Goodwin, L. B. 1986. Structural facies within the Santa Rosa mylonite zone, southern California. *Geol. Soc. Am. Abstr. Prog.* **18**, 618.

Goodwin, L. B. & Wenk H.-R. 1990. Intracrystalline folding and cataclasis in biotite of the Santa Rosa mylonite zone: HVEM and TEM observations. *Tectonophysics* **172**, 201–214.

Griggs, D. T. & Blacic, J. D. 1965. Quartz: anomalous weakness of single crystals. *Science* **147**, 292–295.

Handy, M. R. 1989. Deformation regimes and the rheological evolution of fault zones in the lithosphere: the effects of pressure, temperature, grainsize and time. *Tectonophysics* **163**, 119–152.

Hanmer, S. & Passchier, C. 1991. Shear-sense indicators: a review. *Geol. Surv. Pap. Can.* **90–17**, 1–72.

Heard, H. C. 1960. Transition from brittle fracture to ductile flow in Solenhofen limestone as a function of temperature, confining pressure, and interstitial fluid pressure. In: *Rock Deformation* (edited by Griggs, D. & Handin, J.). *Mem. geol. Soc. Am.* **79**, 193–226.

Hirsch, P. B. 1981. Plastic deformation and electronic mechanisms in semiconductors and insulators. *J. Phys. Colloq.* **42**(C3), 149–159.

Hobbs, B. E. 1981. The influence of metamorphic environment upon the deformation of minerals. *Tectonophysics* **78**, 335–383.

Hobbs, B. E., McLaren, A. C. & Paterson, M. S. 1972. Plasticity of single crystals of synthetic quartz. In: *Flow and Fracture of Rocks* (edited by Heard, H. C., Borg, I. Y., Carter, N. I. & Raleigh, C. B.). *Am. Geophys. Union Monogr.* **16**, 29–53.

Hobbs, B. E., Means, W. D. & Williams, P. F. 1976. *An Outline of Structural Geology*. John Wiley, New York.

Hobbs, B. E., Mülhaus, H.-B. & Ord, A. 1990. Instability, softening and localization of deformation. In: *Deformation Mechanisms, Rheology and Tectonics* (edited by Knipe, R. J. & Rutter, E. H.). *Spec. Publs. geol. Soc. Lond.* **54**, 143–165.

Kirby, S. H. & Kronenberg, A. K. 1987. Rheology of the lithosphere: selected topics. *Rev. Geophys.* **25**, 1219–1244.

Knipe, R. J. 1989. Deformation mechanisms — recognition from natural tectonites. *J. Struct. Geol.* **11**, 127–146.

Knipe, T. J. & Wintsch, R. P. 1985. Heterogeneous deformation, foliation development, and metamorphic processes in a polyphase mylonite. In: *Metamorphic Reactions: Kinetics, Textures, and Deformation* (edited by Thompson, A. B. & Rubie, D. C.). *Adv. phys. Geochem.* **4**, 180–201.

Kronenberg, A. K., Kirby, S. H., Aines, R. D. & Rossman, G. R. 1986. Solubility and diffusional uptake of hydrogen in quartz at high water pressures: implications for hydrolytic weakening. *J. geophys. Res.* **91**, 12723–12744.

Law, R. D., Casey, M. & Knipe, R. J. 1986. Kinematic and tectonic significance of microstructures and crystallographic fabrics within quartz mylonites from the Assynt and Eriboll regions of the Moine thrust zone, NW Scotland. *Trans. R. Soc. Edinb.: Earth Sci.* **77**, 99–125.

Lister, G. S. & Hobbs, B. E. 1980. The simulation of fabric development during plastic deformation and its application to quartzite: the influence of deformation history. *J. Struct. Geol.* **1**, 355–370.

Lister, G. S. & Snoke, A. W. 1982. S–C mylonites. *J. Struct. Geol.* **6**, 617–638.

Lister, G. S. & Williams, P. F. 1983. The partitioning of deformation in flowing rock masses. *Tectonophysics* **92**, 1–33.

Lister, G. S., Paterson, M. S. & Hobbs, B. E. 1978. The simulation of fabric development in plastic deformation and its application to quartzite: the model. *Tectonophysics* **45**, 107–158.

McLaren, A. C., Retchford, J. A., Griggs, D. T. & Christie, J. M. 1967. Transmission electron microscope study of Brazil twins and dislocations experimentally produced in natural quartz. *Phys. Stat. Sol.* **19**, 631–644.

Means, W. D. 1989a. Stretching faults. *Geology* **17**, 893–896.

Means, W. D. 1989b. Synkinematic microscopy of transparent polycrystals. *J. Struct. Geol.* **11**, 163–174.

Means, W. D. 1990. Kinematics, stress, deformation and material behaviour. *J. Struct. Geol.* **12**, 953–971.

Means, W. D., Williams, P. F. & Hobbs, B. E. 1984. Incremental deformation and fabric development in a KCl/mica mixture. *J. Struct. Geol.* **6**, 391–398.

Meike, A. 1989. *In situ* deformation of micas: a high-voltage electron-microscope study. *Am. Miner.* **74**, 780–796.

Meike, A. 1990. Dislocation enhanced selective dissolution: an examination of mechanical aspects using deformation-mechanism maps. *J. Struct. Geol.* **12**, 785–794.

Meike, A. & Wenk, H.-R. 1988. A TEM study of microstructures associated with solution cleavage in limestone. *Tectonophysics* **154**, 137–148.

Passchier, C. W. & Simpson, C. 1986. Porphyroclast systems as kinematic indicators. *J. Struct. Geol.* **8**, 831–843.

Paterson, M. S. 1969. The ductility of rocks. In: *Physics of Strength and Plasticity* (edited by Argon, Ali S.). The M.I.T. Press, Cambridge, MA, 377–392.

Paterson, M. S. 1978. *Experimental Rock Deformation: The Brittle Field*. Springer, Berlin.

Paterson, M. S. & Weiss 1968. Folding and boudinage of quartz-rich layers in experimentally deformed phyllite. *Bull. geol. Soc. Am.* **79**, 795–812.

Poirier, J.-P. 1985. *Creep of Crystals: High-temperature Deformation Processes in Metals, Ceramics, and Minerals*. Cambridge University Press.

Price, G. P. 1985. Preferred orientations in quartzites. In: *Preferred Orientation in Deformed Metals and Rocks: An Introduction to Modern Texture Analysis* (edited by Wenk, H.-R.). Academic Press, New York, 385–406.

Price, G. P. & Torok, P. A. 1989. A new simple shear deformation apparatus for rocks and soils. *Tectonophysics* **158**, 291–309.

Ralser, S. 1990. Shear zones in an experimentally deformed quartz mylonite. *J. Struct. Geol.* **12**, 1033–1045.

Ralser, S., Hobbs, B. E. & Ord, A. 1991. Experimental deformation of a quartz mylonite. *J. Struct. Geol.* **13**, 837–850.

Ramberg, H. 1963. Evolution of dragfolds. *Geol. Mag.* **100**, 97–106.

Ramsay, J. G. & Graham, R. H. 1970. Strain variation in shear belts. *Can. J. Earth Sci.* **7**, 786–813.

Ricoult, D. L. & Kohlstedt 1983. Structural width of low-angle grain boundaries in olivine. *Phys. Chem. Miner.* **9**, 133–138.

Rovetta, M. R. & Holloway, J. R. 1986. Solubility of hydroxyl in natural quartz annealed in water at 900°C and 1.5 GPa. *Geophys. Res. Lett.* **13**, 145–148.

Rovetta, M. R., Blacic, J. D. & Delaney, J. R. 1987. Microfracture and crack healing in experimentally deformed peridotite. *J. geophys. Res.* **92**, 12902–12910.

Rutter, E. H. 1976. The kinetics of rock deformation by pressure solution. *Phil. Trans. R. Soc. Lond.* **A283**, 203–219.

Rutter, E. H. 1983. Pressure solution in nature, theory and experiment. *J. geol. Soc. Lond.* **140**, 725–740.

Rutter, E. H. 1986. On the nomenclature of mode of failure transitions in rocks. *Tectonophysics* **122**, 381–387.

Segall, P. & Simpson, C. 1986. Nucleation of ductile shear zones on dilatant fractures. *Geology* **14**, 56–59.

Schmid, S. M. & Casey, M. 1986. Complete texture analysis of commonly observed quartz c-axis patterns. In: *Mineral and Rock Deformation: Laboratory Studies* (edited by Hobbs, B. E. & Heard, H. C.). *Am. Geophys. Union Monogr.* **36**, 263–286.

Schoneveld, Chr. 1979. The geometry and the significance of inclusion patterns in syntectonic porphyroblasts. Ph.D. Thesis, University of Leiden, The Netherlands.

Sibson, R. H. 1977. Fault rocks and fault mechanisms. *J. geol. Soc. Lond.* **133**, 191–213.

Sibson, R. H. 1980. Transient discontinuities in ductile shear zones. *J. Struct. Geol.* **2**, 165–171.

Smidoda, K., Gottschalk, W. & Gleiter, H. 1978. Diffusion in migrating interfaces. *Acta Metall.* **26**, 1833–1836.

Starkey, J. 1964. Glide twinning in the plagioclase feldspars. In: *Deformation Twinning* (edited by Reed-Hill, R. E. *et al.*). Gordon & Breach, New York, 177–191.

Tullis, J. A. 1968. Preferred orientation in experimental quartz mylonites. *Trans. Am. Geophys. Union* **39**, 755.

Tullis, J. A. 1970. Quartz: preferred orientation in rocks produced by Dauphiné twinning. *Science* **168**, 1342–1344.

Tullis, J. & Tullis, T. 1972. Preferred orientation of quartz produced by mechanical Dauphiné twinning: thermodynamics and axial experiments. In: *Flow and Fracture of Rocks* (edited by Heard, H. C., Borg, I. Y., Carter, N. L. & Raleigh, C. B.) *Am. Geophys. Union Monogr.* **16**, 67–82.

Turner, F. J. & Weiss, L. E. 1963. *Structural Analysis of Metamorphic Tectonites*. McGraw-Hill, San Francisco.

Turner, F. J., Griggs, D. T. & Heard, H. 1954a. Experimental deformation of calcite crystals. *Bull. geol. Soc. Am.* **65**, 883–934.

Turner, F. J., Griggs, D. T., Heard, H. & Weiss, L. E. 1954b. Plastic deformation of dolomite rock at 380°C. *Am. J. Sci.* **252**, 477–488.

Twiss, R. J. 1977. Theory and applicability of a recrystallized grain size paleopiezometer. *Pure appl. Geophys.* **115**, 227–244.

Urai, J. L., Means, W. D. & Lister, G. S. 1986a. Dynamic recrystallization of minerals. *Am. Geophys. Union Monogr.* **36**, 161–199.

Urai, J. L., Spiers, C. J., Zwart, H. J. & Lister, G. S. 1986b. Weakening of rock salt by water during long-term creep. *Nature* **324**, 554–557.

Vance, J. A. 1961. Polysynthetic twinning in plagioclase. *Am. Miner.* **46**, 1097–1119.

Vernon, R. H. 1983. *Metamorphic Processes*. George Allen & Unwin, Sydney.

Vernon, T. H. 1965. Plagioclase twins in some mafic gneisses from Broken Hill, Australia. *Miner. Mag.* **35**, 488–507.

Wenk, H.-R. & Christie, J. M. 1991. Comments on the interpretation of deformation textures in rocks. *J. Struct. Geol.* **13** 1091–1110.

Wenk, H.-R., Canova, G., Molinari, A. & Kocks, U. F. 1989. Viscoplastic modelling of texture development in quartzite. *J. geophys. Res.* **94**, 17895–17906.

White, J. C. & Barnett, R. L. 1990. Microstructural signatures and glide twins in microcline, Hemlo, Ontario. *Can. Miner.* **28**, 757–769.

White, J. C. & Mawer, C. K. 1988. Dynamic recrystallization and associated exsolution in perthites: evidence of deep crystal thrusting. *J. geophys. Res.* **93**, 325–337.

White, J. C. & White, S. H. 1981. On the structure of grain boundaries in tectonites. *Tectonophysics* **78**, 613–628.

White, J. C. & White, S. H. 1983. Semi-brittle deformation within the Alpine fault zone, New Zealand. *J. Struct. Geol.* **5**, 579–589.

White, S. H., Burrows, S. E., Carreras, J., Shaw, N. D. & Humphreys, F. J. 1980. On mylonites in ductile shear zones. *J. Struct. Geol.* **2**, 175–187.

Williams, P. F. 1990. Differentiated layering in metamorphic rocks. *Earth Sci. Rev.* **29**, 267–281.

Williams, P. F. & Schoneveld, Chr. 1981. Garnet rotation and the development of axial plane crenulation cleavage. *Tectonophysics* **78**, 307–334.

Yund, R. A., Smith, B. M. & Tullis, J. 1981. Dislocation-assisted diffusion of oxygen in albite. *Phys. Chem. Miner.* **7**, 185–189.

CHAPTER 2

Palaeostrain Analysis

RICHARD J. LISLE

INTRODUCTION*

PALAEOSTRAIN analysis is a sub-discipline of structural geology concerned with quantifying and interpreting the shape changes or distortions of a rock body which have been brought about by tectonic activity. Although concepts were pioneered last century, the subject has seen an explosion of development prompted by the work of Cloos (1947), Breddin (1956) and, particularly, Ramsay (1967). Since 1967 a large number of techniques have been devised for the analysis of strain from the variety of strained objects found in rocks. Besides these advances in methodology, modern strain analysis has found a wider range of applications including, for example, the calculation of the restored, pre-deformation geometry of rock bodies. In addition, attempts to quantify strains have produced indirect benefits such as a source of data on rock rheology at the time of deformation.

In this review space does not allow the description of all methods in detail. Instead the emphasis will be placed on the more widely-used and recent methods as well as on providing examples of the range of deductions that can be drawn from their application to deformed rocks. For more specific details of particular methods the reader is referred to the following excellent texts: Ramsay & Huber (1983), Ragan (1985) and Simpson (1988).

THE SCOPE OF STRAIN ANALYSIS

Viewed in a narrow sense, strain analysis is concerned only with the estimation, at various points in a rock body, of the form and orientation of the *finite strain ellipsoid*, a geometrical representation of the total tectonic strain which has built up throughout the entire deformation history. A complete analysis requires the estimation of six parameters; three describing the dimensions of the principal axes of the ellipsoid (the *principal stretches*, $S_1 \geq S_2 \geq S_3$) and three describing the orientation of those axes.

In practice a complete analysis is rarely possible.

Usually the restricted data available allow only the ratios of the principal stretches (the *strain ratios*, $S_1 : S_2 : S_3$) to be calculated. Frequently, the data permit only a two-dimensional analysis which yields the shape of a single section through the ellipsoid, the *strain ellipse*.

Strain analysis, defined more broadly, also includes the interpretation of data relating to the strain history, that is, the evolution of the strain ellipsoid and its progressive rotation with respect to external or internal (material) coordinates.

LEVELS OF ANALYSIS

Strain analysis requires the presence of *strain markers*. A complete analysis yielding the strain ellipsoid depends on the availability of markers which provide quantitative data on the three-dimensional state of strain. If, as is often the case, the data fall short of these requirements only a partial or lower level of analysis is possible.

One-dimensional analyses

This restricted form of analysis is all that is possible if the available markers allow the calculation of length change (*stretch*, S) and/or the angular distortion of a perpendicular (ψ, angular shear strain) of only a single direction in the rock (Fig. 2.1).

An example is the calculation of the *percentage shortening* ($100(1 - S)\%$) in a direction parallel to the enveloping surface of bedding by restoring folded and faulted beds to their pre-deformational geometry (Fig. 2.2a). This calculation, which involves the assumption of the conservation of bed-length during deformation, is clearly appropriate only to structures of a specific tectonic style.

Compaction related to sediment loading, a strain which produces an oblate strain ellipsoid parallel to bedding, is usually quantified as a one-dimensional measure of shortening. This can be estimated from the differential thinning of beds and enclosed assumed rigid objects, for example, concretions or folded sandstone dykes (Fig. 2.2b). The calculation of stretch requires the

---

*Editor's note: This chapter should be read after Chapter 1 in which some basic definitions relating to strain are given.

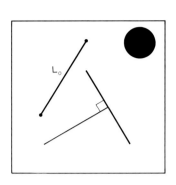
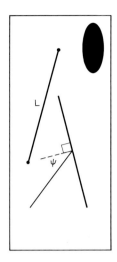

Fig. 2.1. Longitudinal strain and shear strain. Longitudinal strain of a line is expressed by the stretch S, the ratio of deformed to undeformed lengths of a line ($S = L/L_o$). The shear strain of a line is quantified by the angular shear ψ, the angular distortion of an original right angle.

measurement of deformed and undeformed lengths of the *same material line*. Because the perpendicular to bedding is usually not the same material line throughout the deformation, percentage stratigraphic thinning (or thickening) is generally not a measure of the stretch of the line orthogonal to bedding in the deformed state. Schwerdtner (1978) discusses this problem.

Equivalent methods exist for extended lines and these allow calculation of *percentage extension* ($100(S - 1)\%$). In concept, the methods used for normal faults which assume constant bed-length (Fig. 2.2c) (Wernicke & Burchfiel 1982) are identical to those which calculate the percentage extension by piecing together the fragments of brittle strain markers (Fig. 2.2d) such as belemnites, conodont denticles, stretched tourmalines, arsenopyrite crystals and rutile needles enclosed in quartz grains. Ferguson (1981) proposed a method of calculating extension from such objects which allows for the fact that individual fractures separating the boudins are likely to have formed at different stages in the extensional history. The merits of this method in relation to earlier ones used by Ramsay (1967, p. 248) and Hossain (1979) are discussed by Ford & Ferguson (1985).

Attempts are sometimes made to derive a one-dimensional measure of strain from one estimated in another direction. Chamberlin's method (Fig. 2.2e) (see Hossack 1979) allows the overall shortening in a structural cross-section to be found from the amount of vertical thickening or thinning. The calculation involves an additional assumption; namely, the conservation of cross-sectional area. This is a substantial simplification since theoretically the area of a plane of general orientation must change, even during plane strain and other *isochoric* (constant volume) *deformations*. The extension represented by *pinch-and-swell structure* can be calculated from such a constant area method (Fig. 2.2f).

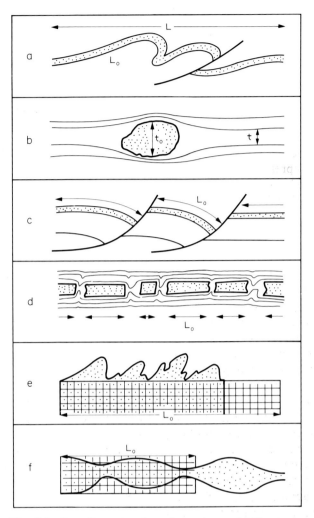

Fig. 2.2. One-dimensional strain analyses. (a) Shortening estimated by 'unrolling' folds. (b) Vertical shortening estimated from differential compaction around pebbles, concretions etc. (c) Extension deduced from bed-lengths in normal fault structures. (d) Extension found by piecing together boudinaged layers. (e) Chamberlin's method — shortening calculated from the sectional area of a deformed zone of known original thickness. (f) The equivalent of Chamberlin's method applied to extensional structures (pinch-and-swell).

Ironically, nearly all of the above mentioned methods of calculating strain in one dimension rely on the presence of competence contrasts on the scale of the strain markers. Markers of two- and three-dimensional strains on the other hand, ideally possess the same competence as the enclosing rock.

Two-dimensional analyses

Methods of determining the strain ellipse on a plane fall into two broad categories; those that involve the integration of several one-dimensional measures of strain and those that directly determine the strain utilizing two-dimensional markers such as deformed circular or elliptical markers.

Methods which construct the strain ellipse from strain parameters (S, ψ) of several lines in a plane. A large variety of constructions exist for the strain ellipse depending on the type and quantity of data available. These are fully described in Ramsay (1967, pp. 69–81 and 228–251), Ragan (1985, pp. 175–200) and Ramsay & Huber (1983). Many of them involve the graphical derivation of the strain ellipse with the aid of the *Mohr circle construction*; a graphical device which involves a special transformation which makes ellipses plot as circles. The concept of the *pole of the Mohr circle* (Mandl & Shippam 1981) clarifies the relationship between the geometry of the deformed plane and its representation in Mohr space and leads to simplification of many of the standard constructions (Treagus 1987, Lisle 1991).

Fossils, trace fossils and sedimentary structures provide an important source of strain data but their diverse morphology has meant that a wide range of specific constructions have had to be devised for their analysis. A review of relevant methods appears elsewhere (Lisle 1991).

Strain rosettes consist of a collection of differently oriented lines for which the stretches (S) or relative stretches (kS) are known, for example, several stretched belemnites on a single bedding plane (Fig. 2.3). A rosette made up of only three such lines is theoretically sufficient for the determination of the strain ellipse. Several different graphical and algebraic solutions for the so-called three-stretch problem have been described (Ramsay & Huber 1983, p. 96, Ding Zhong-Yi 1984, Ragan 1987, De Paor 1988a, Lisle & Ragan 1988).

Rosettes with more than three lines overdefine the strain ellipse and in such circumstances some best-fit ellipse is usually sought. The approach of Hossain (1979) and Mukhopadhyay (1980) is to transform the variables in the ellipse equation so as to give the latter a linear form, permitting linear regression-type methods to be used to find the ellipse which best fits the stretches. Both methods require prior (independent) knowledge of the orientation of the principal stretches. Siddans (1976), Sanderson (1977a), Kanagawa (1990) and Erslev & Ge (1990) suggest other ways of deriving the best-fit ellipse from such data.

If the additional assumption of original isotropy of line orientations is made, several other methods can be used. Panozzo (1984) has adopted an approach to the analysis of deformed line data which involves summing the projections of each line onto reference lines of varying orientations. The sum of the projected lengths is a direct function of the stretch in the direction of the reference line. If the lines measured come from a population which defined an isotropic fabric before straining, the directions of greatest and least summed projection length will be orthogonal and the ratio of these extreme values of projection length will be equal to the strain ratio.

Sanderson & Phillips (1987) used a resultant vector approach for the analysis of samples of deformed lines. Because each line direction lacks polarity it is necessary to double the angles describing the azimuth to give each line an unambiguous direction. The vector mean direction yields an unbiased estimate of the S_1 principal stretch direction. The magnitude (R) of the resultant vector when normalized by dividing by the total combined length (ΣL) of the deformed lines can be used to estimate the strain ratio. Sanderson & Phillips (1971) present tables to convert the calculated values of $R/\Sigma L$ into strain ratio (R_s) values. The theoretical relationship between strain and $R/\Sigma L$ has yet to be discovered. Wheeler (1989) finds that by first weighting the measured vectors according to the square of their length, Sanderson & Phillips' (1971) method yields a resultant vector which is a simple function of the strain ratio.

The above methods cannot be applied if the available deformed line data consist only of their orientations, that is, no lengths are known. Applicable methods to data of this type are based on the fact that the orientation distributions of populations of lines are modified by the imposition of strain. Sanderson (1973) investigated strain-modified Gaussian distributions of line orientations and applied the findings for determining strain from the mean and standard deviations of samples of fold axis orientations. Lloyd (1983) used two parameters (skewness and kurtosis) in order to characterize strained distributions whereas Soto (1991) relied on the maximum frequency of orientations.

These methods clearly rely on the knowledge of the nature of the orientation distribution in the pre-strain state. The uniform distribution has received most attention; due more to its simplicity than any other reason. Borradaile (1976), for example, estimated the strain by carrying out a range of computer-simulated de-strainings of samples of deformed line directions until the distribution was a statistically satisfactory approximation to a uniform orientation distribution. Sanderson (1977b), treating the deformed lines as unit vectors, demonstrated theoretically that the vector mean

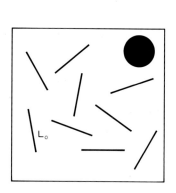
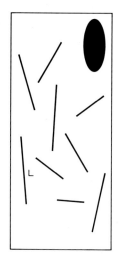

Fig. 2.3. The straining of collections of line elements. If original lengths of the lines are known, stretches (S) can be calculated for directions parallel to each of the deformed lines. If the lines all had the same, but unknown, length before straining only the relative stretches (kS) for the lines can be found. Data consisting of S or kS measured in a variety of directions define a strain rosette.

orientation and magnitude (again calculated using doubled angles) can be used to estimate the orientation and axial ratio of the strain ratio. For large samples the strain ratio will approach the value $(n + R)/(n - R)$ where R is the length of the resultant vector calculated from n lines (Sanderson & Phillips 1987). Because of the similarity of the strain-modified uniform distribution and the circular normal distribution, Sanderson (1977b) was able to suggest how approximate confidence limits for the estimated strain ratio can be made. Trayner (1986) applied the Sanderson (1977b) and Panozzo (1984) methods to calcite grain boundaries and compared the results.

Harvey & Laxton (1980) also treated each line as a unit vector and used the orientations in the sample to construct the following matrix:

$$\begin{vmatrix} a & b \\ c & d \end{vmatrix}$$

where $a = \Sigma\cos^2\theta_i$, $b = \Sigma\cos\theta_i\sin\theta_i$, $c = b$, $d = \Sigma\sin^2\theta_i$ and θ_i ($i = 1$ to n) are the azimuths of the lines recorded with respect to a fixed reference line. They show that the ratio of the eigenvalues of this matrix is a direct estimate of the strain ratio, that is,

$$R_s = \frac{a + d + ((a-d)^2 + 4b^2)^{0.5}}{a + d - ((a-d)^2 + 4b^2)^{0.5}}.$$

This simplifies to

$$R_s = (n + R)/(n - R)$$

where $R = ((\Sigma\cos 2\theta)^2 + (\Sigma\sin 2\theta)^2)^{0.5}$. Thus, apparently, the methods of Sanderson (1977b) and Harvey & Laxton (1980) are identical. Harvey & Laxton (1980) used Monte Carlo trials to produce confidence limits of the strain ratio, R_s.

As the strain ratio increases, originally uniformly distributed line orientations will become clustered around the S_1 direction of the strain ellipse. For instance, 50% of the lines will be expected to lie in a 20° wide sector on either side of the S_1 direction when the strain ratio is 3.0. De Paor (1981) suggested using the angles corresponding to the upper and lower quartiles of the distribution of azimuths to estimate the strain ratio.

Summarizing, we note that numerous methods exist for treating deformed lines and some of these have the benefits of simplicity and/or mathematical elegance. The reality though is that their results are only as good as the assumptions they incorporate; particularly those concerning the passive behaviour of lines and their original distribution of orientations. The most satisfactory methods are those which include built-in checks on these assumptions.

Deformed circular markers. The vast majority of strain determinations to date have used the deformed shape of circular or elliptical markers, usually the cross-sections of grains, aggregates and other rock particles. If these markers are circular or nearly so in their unstrained condition and are mechanically similar to the rock containing them, they provide simple strain gauges, their shape matching that of the strain ellipse. As the analysis is based on the shapes of several markers some sort of average shape is used as an estimate of the strain.

The slope method (Cloos 1947, Ramsay 1967, p. 193) uses the gradient of the best-fit line on a graph plotting long-axis length versus short-axis length as an estimate of the strain ratio, R_s. Mukhopadyay (1973) plotted quartz grain data in the above manner and used regression analysis to find the best-fit straight line passing through the origin:

$$R_s = \text{slope of regression line} = \Sigma(x_iy_i)/\Sigma(x_i^2)$$

where x_i, y_i ($i = 1$ to n) are the short and long dimensions of markers, respectively. This calculation (see, e.g. Williams 1986) assumes that the long-axis length is the independent variable. Because in this case, the assumption of dependent and independent variables is difficult to make, a reduced major axis line may be preferable (Van der Pluijm 1987). The slope of this line when forced through the origin (by adding for each particle in the sample another 'ghost' grain with negative axial dimensions) becomes

$$R_s = \sqrt{(\Sigma(y_i^2)/\Sigma(x_i^2))}.$$

Some of these different variants of the slope method have been employed by Tobisch *et al.* (1977), Graham (1978) and Srivastava (1985).

For shapes which deviate from ellipses, or where long axes are difficult to visually select, the grain radii in three non-principal directions can be measured, summed for all the grains and combined by one of the three-stretch constructions to calculate an average ellipse (Ramsay 1967, p. 195). These constructions are more precise when the radii are chosen 120° apart. Because of the summation involved, this calculation becomes weighted in favour of grains with the greater cross-sectional area.

Deformed elliptical markers. When homogeneously deformed markers show a wide variation in aspect ratio, together with fluctuation in the orientation of their long axes, this can be interpreted to mean that the markers had non-circular shapes in their pre-strained state. The relative abundance of elliptical markers in unstrained particulate rocks (including the outlines of clastic grains) has meant that several methods have been devised for their analysis.

R_f/ϕ *methods.* The deformed aspect ratio (R_f) and orientation (ϕ) of an elliptical marker depend not only on the strain ratio (R_s) but also on the starting orientation of the marker (θ) and its original axial ratio (R_i). Because the variables R_s, R_i and θ are unknowns in the problem it is clearly not possible to solve for the strain from the R_f/ϕ measurements of a single deformed

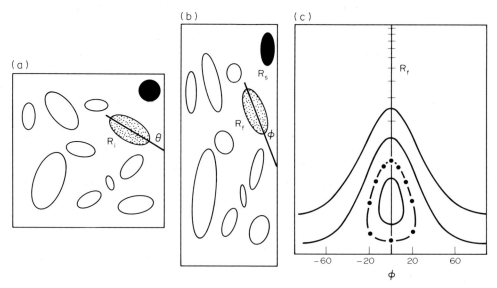

Fig. 2.4. Terminology of the R_f/ϕ method. An elliptical marker of initial shape ratio R_i and orientation θ transforms after the imposition of a strain with ratio R_s into a marker with a final axial ratio R_f and final orientation ϕ. (a) A suite of particles of constant R_i. (b) The deformed markers. (c) R_f/ϕ plot of the deformed markers. The shapes of the onion curves are diagnostic of R_s and R_i.

marker. Even by combining data from several markers in the same section no solution is available unless some additional restrictions exist on the range of R_i or θ values present in the sample.

The original description of the R_f/ϕ method (Ramsay 1967, pp. 204–211) lays down the theoretical framework of the method which is based on the fact that a group of elliptical markers sharing the same initial eccentricity will deform to give a characteristic curve when plotted on a graph of R_f vs ϕ (Fig. 2.4). The shape of this curve is diagnostic of both the strain ratio and the R_i values of the markers. The comparison of R_f/ϕ plots of deformed markers with these theoretical 'onion' curves forms the basis of the method as developed by Dunnet (1969). In practice the elliptical markers (e.g. sedimentary clasts) often have variable initial shapes and these result in R_f/ϕ data which do not plot on a single onion curve but on several, leading to a diffuse cloud of data. The onion curves cannot be compared to such data because the former do not theoretically have the significance of density contours for the data.

The method is improved by the use of another set of reference curves to characterize the deformed marker data. These are known as *theta curves* (Fig. 2.5) and they depict on the R_f/ϕ diagram all strained markers which possessed the same initial orientation θ (cf. 50% data curve employed by Dunnet & Siddans 1971). A set of reference theta curves constructed for each strain consists of curves for constant intervals of theta. If the

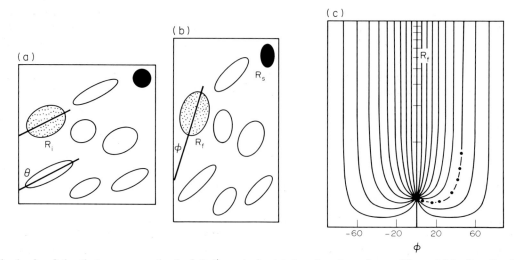

Fig. 2.5. The basis of the theta curve method of R_f/ϕ analysis. (a) A suite of markers with variable R_i ratios but identical orientations. (b) The markers shown in (a) in their deformed condition. (c) The R_f/ϕ plot of the markers shown in (b). The markers plot on a single theta curve. If it is now assumed that all orientations (θ) are equally represented in the undeformed state, the R_f/ϕ plot will consist of points equally distributed throughout the cells defined by the theta curves in (c). This is the criterion used to select the appropriate set of reference theta curves and hence to determine the strain ratio.

assumption that the sampled markers come from a uniform orientation distribution is correct, a set of curves should exist which divides the data equally. Lisle (1977a) and Robin (1977) described a goodness-of-fit test for selecting the most appropriate set of curves and hence objectively determining the strain ratio. Full details of the method and sets of reference curves are published in Lisle (1985). Peach & Lisle (1979) offer a computer program to perform the theta-curve method.

Attempts have been made to devise an R_f/ϕ plot on which the whole cloud of points representing the strained markers translates bodily across the graph as a strain is imposed upon them. The potential advantage of such a plot is that it would allow certain pre-tectonic fabrics to be readily identified in the deformed state. No one has succeeded in devising such a plot. In Elliott's (1970) so-called *shape factor grid* the points change their relative position with respect to neighbours during translation, so that patterns of point density are different on the undeformed and deformed plots. This drawback exists with the plot invented by Wheeler (1984), but to a lesser extent.

Nevertheless the traditional cartesian R_f/ϕ plot is not the most natural manner of displaying the azimuths of marker long axes. De Paor's (1988b) hyperbolic net, in common with the above graphical representations, benefits from a polar-coordinate layout. Yu & Zheng's (1984) method, which linearizes the theoretical relationship between the R_f/ϕ variables by employing hyperbolic trigonometric functions, is applicable to sets of markers with the same initial eccentricity, such as certain deformed fossils (Rajlich 1989).

All R_f/ϕ analyses effectively determine the strain by applying a series of retrodeformations until a fabric is produced which approximates to that assumed to exist in the pre-tectonic state. In most applications the reciprocal strain is identified as that which returns the markers' long axes to a uniform orientation distribution. Harvey & Ferguson (1981) suggested that other criteria could be used, such as the strain which minimizes the average eccentricity (R_i) of the restored markers. Schultz-Ela (1990), on the other hand, attempted to utilize the frequency distribution of R_i values. Essentially, the assumed isotropy of the initial marker fabric can be defined geometrically in a number of ways and each will have associated with it a statistical check which is most appropriate for the purpose of recognizing the destrained condition.

As Siddans (1980a) pointed out, these methods are valid for all planes through a rock (not just principal planes). It has been demonstrated that R_f/ϕ techniques are valid when the deformation history was non-coaxial (Le Theoff 1979), or when the total deformation accumulated as a result of two deformation events; one of these deformation events could be compaction.

In addition to the above graphical methods, Shimamoto & Ikeda (1976) presented a simple algebraic method for the analysis of R_f/ϕ data. They represented each elliptical marker by a matrix, termed a shape matrix. By averaging the respective components of the shape matrices of all particles in the sample, a matrix is arrived at which defines the 'average' ellipse. Shimamoto & Ikeda (1976) showed that this average marker behaves under straining as a single ellipse. On the assumption that isotropy existed in the undeformed state, that is, the average marker was a circle, the strain ellipse shape is indicated directly by that of the average deformed marker. Isotropy in this instance is defined in terms of both shape and orientation.

Clearly, there is a need for more data to be collected on the nature of fabrics in undeformed sediments. At the present time it is difficult to decide which definition of isotropy is most appropriate for particular sedimentary fabrics and hence which R_f/ϕ method is likely to be the most accurate estimator of strain.

Some degree of fabric anisotropy exists in all particulate rocks in the undeformed state and this necessarily creates a source of error to simple R_f/ϕ methods. Elliott (1970) and Dunnet & Siddans (1971) were pioneers in the field of strain analysis from rocks with pre-tectonic fabrics. However, discussion still continues on how best to tackle this thorny topic (Matthews *et al.* 1974, Seymour & Boulter 1979, Holst 1982, Wheeler 1986, Borradaile 1987). It is frequently assumed that the primary fabric on a two-dimensional section will be parallel to the bedding trace. Unfortunately, contrary to widely-held belief the presence of primary fabrics cannot always be detected from an asymmetry in the R_f/ϕ point distribution (Lisle 1985, p. 90).

Furthermore, Siddans (1980b) showed that these methods, unlike the R_f/ϕ methods which assume initial isotropy, are valid only for coaxial strain histories. Bell (1981) has attempted to factorize the total fabric into tectonic and pre-tectonic components.

In a number of situations it may not be practical or time-efficient to undertake one of the above R_f/ϕ methods. Rapid methods estimate the strain from the deformed markers exhibiting the extreme axial ratios (Ramsay 1967, p. 210), the axial ratio of the deformed marker with the extreme fluctuation in orientation (Lisle 1985, pp. 5–6), and the average axial ratio of all markers (Hossack 1968). In the case of the latter, the harmonic mean of R_f has been proven to produce reliable estimates, especially in situations where the average fluctuation is low and where application of the R_f/ϕ type methods is therefore hampered (Lisle 1985).

An attractively simple form of analysis which requires prior knowledge of the directions of the strain axes is that of Robin (1977). For each grain, the ratio of two diameters, one measured parallel to each principal axis, is calculated. The average ratio for all grains equals the strain ratio. The method assumes initial fabric isotropy in the sense that the equivalent average ratio in the undeformed state is unity. Robin (1977) suggested the geometric mean as the appropriate average to calculate, though isotropy could equally be defined in terms of other types of mean ratio (e.g. arithmetic, harmonic, etc). Prospective users of the techniques reviewed above will doubtless profit from referring to comparative

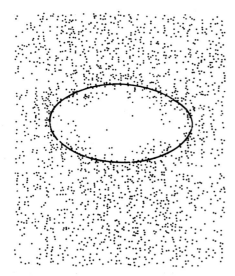

Fig. 2.6. An all object separation plot produced by the Fry method applied to an oolitic limestone. From Dunne *et al.* (1990).

studies of the different methods carried out by Hanna & Fry (1979a), Paterson (1983), Babaie (1986), Kanagawa (1990) and Schultz-Ela (1990).

Methods using the spatial distribution of markers. Strain alters the distance between pairs of neighbouring grains. During a deviatoric strain the distances to a grain's neighbours will change depending on the orientation of the line linking the grains. By assuming that a condition of isotropy existed with respect to inter-particle distances in the pre-tectonic state, Ramsay (1967, p. 195) designed a method for finding the strain ratio from the ratio of the maximum and minimum neighbour distances (which theoretically ought to correspond to measurements made in orthogonal directions). The method employs a cartesian graph of inter-neighbour distance versus θ, the orientation of the join line of the centres. The advantage of the method is that it offers the possibility of determining the whole-rock strain which may not be the same as the strain recorded by the markers themselves. For example, the method has been applied to rocks affected by pressure solution which exhibit heterogeneous deformation on a grain scale, and therefore contravene the assumptions of R_f/ϕ type methods. Hanna & Fry (1977), Schmid & Paterson (1977) and Pfiffner (1980) have applied this method to deformed oolites. The technique is, however, not without its problems. The presentation of azimuths of the grain-to-grain join lines on a cartesian based graph is unnatural. In addition there are technical difficulties when deciding which grains are neighbours and which are not.

The *'all-object-separations' plot of Fry* (1979a) overcomes some of these difficulties. This plot shows centre-to-centre distances for all permutations of pairs of markers, whether neighbours or not, as a function of direction in polar co-ordinates. The method is particularly simple in execution and involves placing the graph over the image of the deformed markers and, with the origin centred on each grain in turn, marking on an overlay the position of all other centres. The resulting plot (Fig. 2.6) commonly possesses a region around the origin devoid of plotted centres.

This central vacancy, the shape and orientation of which reflects the strain ellipse, arises when the pre-tectonic grain centres show some tendency for a lower end cut-off in grain neighbour distances. As Fry (1979a) pointed out, where this tendency is absent, such as in the situation where grain centres are positioned randomly and independently of the position of neighbouring markers, the technique fails. The crispness with which the elliptical hole is defined therefore depends on the initial degree of 'anticlustering' exhibited by the markers. This dependence has been demonstrated by means of geometrical simulations by Crespi (1986), on the basis of which optimum sizes of the marker samples are suggested.

Bhattacharyya & Longiaru (1986) suggested how poor anticlustering characteristics can arise in two dimensions from the effect of sectioning a well-ordered three-dimensional structure and Erslev (1988) suggested a correction for this effect. This involves normalization of the inter-marker distance by dividing by the radii of the markers. Although this normalization procedure improves the definition of the elliptical vacancy, the incorporation of particle dimensions into the method is likely to produce results which are dependent on the competence contrast between markers and their matrix. Practical applications of the method reveal that the markers in many rock-types show that the property of 'anticlustering' is sufficiently developed for at least a crude estimate of the strain to be made.

One of the main areas of subjectivity of the Fry method is deciding on the shape of the inscribed ellipse which encloses the central 'hole'. In some cases the hole is surrounded by a high-density rim of points which itself suggests an ellipse (Fig. 2.6); in other cases the rim is absent. Lacassin & van den Driessche (1983) suggested how this halo pattern depends on the marker size and shape.

When deciding on the shape of the vacancy, the observer probably does this by visually contouring the densities of plotted points. Theoretically, it can be shown that, although the shape of point density contours on the Fry plot may be circular in the undeformed state, these contours will not themselves strain in a passive way to yield the shape of the strain ellipse unless point densities are determined using a counting area which has the shape of the (unknown) strain ellipse.

Several workers have suggested ways of taking out this human source of error by numerical analysis of point distribution on a Fry diagram. Although simple methods can be devised to express quantitatively the shape of the halo, for example, moment of inertia calculations, these are hampered by the fact that the outer boundary of the elliptical halo needs to be defined. Ribeiro *et al.* (1983) discussed this problem of edge effects and used a variance−covariance matrix constructed from the points on the Fry plot. Robin

(1983) used a circular outer boundary and discovered that the strain can be estimated by the eigenvectors/eigenvalues of the following matrix constructed from the coordinates of the points:

$$\begin{vmatrix} \Sigma x^2 & \Sigma xy \\ \Sigma xy & \Sigma y^2 \end{vmatrix}$$

It is suggested that the ratio of the eigenvalues relates to the principal stretches S_1, S_2, the number of points on the plot, N, and the radius of the boundary circle, R, by

$$\lambda_1/\lambda_2 = (S_1 - NR^2/4) / (S_2 - NR^2/4).$$

Erslev & Ge (1990) numerically selected the best-fit ellipse using a least-squares fit to the innermost points.

The Fry technique has opened up the possibility of strain estimation in a wider range of materials; for example, quartz grains, granite plutons, sand volcanoes, chondrules, feldspars in gneisses, salt domes.

The principal advantage is the method's potential capability to calculate the bulk strain of the rock in cases where a competence contrast exists between particles and matrix, or where the strain is discontinuous, as in the case of pressure solution deformation. Before this potential can be realized there are still problems to be overcome. Onasch (1986) examines the performance of the Fry method when applied to simulated fabrics affected by pressure solution. The operation of disjunctive deformation mechanisms such as pressure solution means that centres of markers in the deformed state do not correspond to the centres before deformation. This effect is considered by Dunne *et al.* (1990) who concluded that this leads to strain estimates which are too low.

Strain from flattened buckle folds. Second in importance only to 'particulate' methods of strain analysis are methods which use the profile geometry of folded layers. A complete analysis of folding strains is complex as it involves calculating the strain increments which accumulated at every stage of fold development, from the early buckling-dominated stages through to the latter stages where kinematic components of folding become dominant. Such analyses have only rarely been undertaken (Sherwin & Chapple 1968, Holst 1987). The majority of strain determinations using folds have estimated only the magnitude of the strains corresponding to the last stage of folding which, for simplicity, is usually assumed to be a homogeneous strain leading to the shape modification of buckles (with class 1B geometry) into flattened (class 1C) folds.

The most popular way of assessing the amount of this post-buckle flattening is to use the procedure described by Ramsay (1962) and Ramsay (1967, pp. 411–413). This procedure uses measurements of layer thicknesses (t_α) at positions around the fold with different limb dips (α). Determination of the strain is carried out by comparing these data with theoretically determined t_α/α curves computed for various flattening strain ratios.

Ramsay (1967, p. 413) and Milnes (1971) published different variants of these theoretical curves.

Hudleston (1973) makes simpler the comparison of layer thickness measurements with the theoretical curves by applying transformations to the t_α and α variables so as to make the theoretical strain contours become straight lines. The above method has been widely applied (e.g. Themistocleous & Schwerdtner 1977, Hossack 1978) and assumes that the long axis of the strain ellipse in the fold profile plane is parallel to the axial surface trace, allowing the strain to be estimated from a single fold limb.

Gray & Durney (1979) described a modification to the t_α/α method which, by analyzing the data from two limbs of a fold together, determines the direction as well as the amount of flattening.

Lisle (1992) simplifies the method still further. If, as is implicit in homogeneous strain, the area change is uniform at all points around the folded layer, there must be an inverse proportionality between the layer thickness (t_α) and the stretch of the line parallel to the tangent with dip α (Fig. 2.7), that is,

$$s_\alpha = k(1/t_\alpha).$$

Therefore on a graph of $1/t_\alpha$ against α in polar coordinates, the data should define an ellipse, the axial ratio and orientation of which matches the strain ellipse

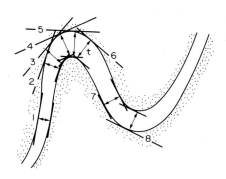

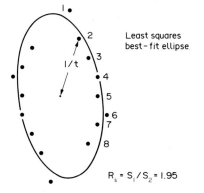

Fig. 2.7. The use of layer thickness of folded layers to determine the shape of the strain ellipse which represents the post-buckle flattening. The method is based on the fact that the stretch in the direction of each tangent to the layer is inversely proportional to the layer thickness. See text for details. From Lisle (1992).

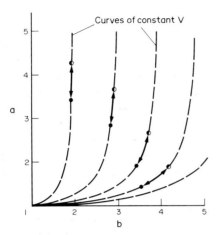

Fig. 2.8. Flinn diagram showing the nature of the errors induced by errors in the axial ratios of ellipses measured on four or more non-principal plane sections. Errors in orientation of the sectional ellipses are assumed to be absent so that the V parameter of the ellipsoid can be correctly estimated. The tie lines join the true and estimated ellipsoids.

of the flattening. The latter can be found by least-squares fitting.

Three-dimensional methods

Three-dimensional strain by combining two-dimensional data. The most convenient way to construct the strain ellipsoid is to collect two-dimensional strain data from principal planes since these planes reveal the principal axial ratios directly. Fortunately, usually a set of joint/cleavage planes exists which is sub-parallel to the principal planes. If the strains are of low magnitude it may be more difficult to select the principal planes. Cutler (1985) assesses theoretically the errors introduced by unwittingly measuring on planes which are oblique to the principal axes.

If strain ellipses have been determined on non-principal planes a number of procedures are available for combining them to construct the strain ellipsoid (Milton 1980, Gendzwill & Stauffer 1981, De Paor 1990). Although three random planes are sufficient for this purpose it is clear that the accuracy of the solution will increase if more planes are measured which have a wide variety of orientations (Owens 1984).

One approach to quantifying the inaccuracies involved is to isolate separate sources of error. If errors in axial ratios only are considered, the orientations of four or more sectional ellipses will allow the accurate calculation of the strain ellipsoid orientation and will constrain the ellipsoid shape to one of the V curves on the *Flinn diagram* (Flinn 1962) in Fig. 2.8 (Lisle 1986).

It is the position of the calculated ellipsoid on its V curve which depends on the measured ellipse axial ratios. In Fig. 2.8, it can be seen that for most possible types of strains the main result of errors in the eccentricities of the sectional strains is to produce a much more dramatic inaccuracy in the ellipsoid's $a = S_1/S_2$ ratio than in its $b = S_2/S_3$ ratio. Stauffer & Burnett (1979) suggest a rapid way to estimate the form of the strain ellipsoid which involves the oblique down-plunge viewing of the markers exposed on planar cross-sections.

Three-dimensional strain markers. In some instances, strain markers such as pebbles can be extracted from the rock which contains them, allowing their three-dimensional form to be determined directly. Serial sectioning is another way of deriving the three-dimensional form of included markers. Both situations provide data of a type which can be interpreted directly in terms of the strain ellipsoid, making the intermediate stage of determining two-dimensional strains unnecessary. For initially spherical passively-deformed objects the analysis is trivial.

For objects like sedimentary clasts, techniques exist which take into account the effects of an initial shape factor. Ramsay (1967, pp. 211–216) explains how the deformed objects which display extreme axial ratios can be used to calculate the strain ratios. Burns & Spry (1969) and Srivastava (1985) use the modal pebble shape for the same purpose. Martin Escorza (1978) utilizes these methods to treat deformed igneous enclaves. A method of determining strain from the maximum range of axial orientations of inclusions is outlined by Lisle (1979). Although potentially useful, especially in rocks in which marker/matrix competence contrasts are important, no three-dimensional implementation of centre-to-centre techniques has so far been attempted.

Deformed distributions of lines and planes. The theory of March predicts how an initially uniform orientation distribution of passive lines or planes will become modified by homogeneous strain. Owens (1973) generalized the March predictions by showing how any given starting orientation distribution is modified. Strain analysis involves inverting the problem, that is, using deformed patterns of preferred orientation to estimate strains. Oertel (1974) used the March prediction that, for a modified uniform distribution of lines, the stretch in any direction is given by:

$$S = \rho^{0.3333}$$

where ρ is the spherical density of lines in that direction.

For preferred orientations of planes the equivalent relation is:

$$S = \rho^{-0.333}$$

where ρ is the density of plane normals. The extreme densities, found by contouring, can therefore be used to furnish estimates of the principal stretches directly. Wood & Oertel (1980) have determined principal strains in slates this way from the (001) poles of muscovites measured with the X-ray goniometer.

When the measured planar or linear elements are less abundant it seems logical to attempt to use the attributes of the entire data set, rather than relying solely on the estimates of the maximum and minimum densities. One approach, adopted by Sanderson & Meneilly (1980), is to firstly characterize the total fabric by means of the orientation tensor, a moment of inertia description of directional data which treats each line or plane's normal as a unit mass on a sphere (Woodcock 1977). The tensor **T** is composed of:

$$\begin{vmatrix} \Sigma\ l^2 & \Sigma\ lm & \Sigma\ ln \\ \Sigma\ lm & \Sigma\ m^2 & \Sigma\ mn \\ \Sigma\ ln & \Sigma\ mn & \Sigma\ n^2 \end{vmatrix}$$

where l, m and n are the direction cosines of each line/plane normal. The next stage is then to use the eigenvectors and eigenvalues of **T** for estimating the strain ellipsoid. However, whilst the eigenvectors of **T** can be used directly to estimate the principal strain axes, the link between the eigenvalues of **T** and the principal stretches is more complex. Harvey & Laxton (1980) considered this relationship in detail and presented tables for converting eigenvalue ratios into strain ratios. Sanderson & Meneilly (1980) determined the strains from preferred orientations of andalusites in this manner.

Cobbold & Gapais (1979) have found a more direct way to estimate strain ratios from such preferred orientations. They proposed that, instead of **T**, a modified orientation tensor should be used. The latter is constructed by assigning a length to each of the measured directions equal to its quadratic elongation (the square of the stretch). The weighted orientation tensor becomes identical to the strain tensor and has eigenvalues equal to the principal quadratic elongations. When, as is often the case, the quadratic elongations of the measured directions are not known, the weighting factor can be based on the contoured density of directions in those directions, which according to March's theory is proportional to the stretch. Kanagawa & Yoshida (1988) used the weighted orientation tensor method to analyze strain from fabrics measured by texture goniometry.

Talbot's method. The geometry of the strain ellipsoid provides a graphic illustration of the following property of three-dimensional strain; namely that a line will suffer elongation, contraction or retain its length depending on its orientation relative to the principal strain axes. Extended lines will lie in orientations close to the S_1 axis whilst contracted lines will define an orientation field centred on the S_3 axis. The transition between these fields of positive and negative extension will be marked by lines with orientations that define the conical *surface of no finite elongation*, the angular dimensions of which depend on the axial ratio of the strain ellipsoid and the volume change (Flinn 1962).

Talbot (1970) devised an ingenious method by which the shape of the surface of no finite elongation can be estimated stereographically by plotting the orientations of extended and contracted elements. From the form of this surface the axial ratios of an ellipsoid can be deduced. Such a method is ideally suited to the analysis of variably oriented line elements. However, Talbot's (1970, 1987) application of the method to planar data (e.g. veins which have been folded or boudinaged) is complicated by the fact that such planes should often contain both extended and contracted lines.

Ramsay (1976) examined in detail the problem of deducing the strain ellipsoid from the angular dimensions of the surface of no finite elongation and found, depending on the volume change, that the form of the surface is often not diagnostic of a single ellipsoid. The existence of multiple solutions obviously means that Talbot's method should be used with care. On the other hand, if independent strain data are available to help decide which of the solutions is geologically valid, the application of Talbot's method offers the potential of calculating tectonic volume changes (Barr & Coward 1974). Hutton (1982), Talbot (1987) and Passchier (1990) have attempted to extend this type of analysis to detect non-coaxial strain histories.

Marker density method. The usual outcome of strain, even constant volume strains, is to produce area changes in two-dimensional cross-sections. It is straightforward to show from volume considerations that for any given ellipsoid, there is an inverse proportionality between the area of the strain ellipse on a specified section plane and distance between the ellipsoid's pair of tangent planes, which are parallel to the section plane (Fig. 2.9). From this it can be deduced that the area of the strain ellipse on any section is directly proportional to the radius of the *reciprocal strain ellipsoid* measured in the direction of the normal to the section.

These properties of the strain ellipsoid form the basis of the technique invented by Mimran (1976) and revised by Fry (1979b), which determines the strain ellipsoid from the relative area distortions on a number (at least six) of section planes through a tectonite; the relative dilations having been estimated on the basis of measured strain marker densities, that is, the number of marker objects per unit cross-sectional area. Mimran (1976) applied this method to deformed chalk by measuring the densities of calcispheres intersected on cross-sections.

Strains from twinned calcite grains. Mechanical twinning in calcite grains is a feature which is common in deformed carbonate rocks. Deformation twins are narrow zones of homogeneous simple shear with

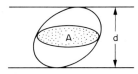

Fig. 2.9. Distortions of areas on planes through the strain ellipsoid form the basis of marker density methods of strain analysis. An inverse proportionality exists between the area, A, of the strain ellipse on the section plane and the distance, d, between the tangent planes which are parallel to the section plane.

orientations that are governed by the crystallography of the mineral. Such twins occur in parallel sets and the strain in a grain is the combined result of the strains of the twins belonging to one or more sets. Although the shear strain of an individual twin has a magnitude fixed by the crystallography, a set of twins is associated with a strain whose magnitude depends on the proportion of the grain which has experienced twinning. Groshong (1972) devised a widely-used method of assessing the bulk strain in mildly deformed rocks which involves the measuring of the directions and magnitudes of shear strains produced by twin sets within calcite grains. The inversion procedure for locating the principal strains treats each twin set as a strain gauge of angular shear. The assumed behaviour of the twinned structures as passive indicators of shear strain exposes the method to criticism. Nevertheless its application to experimentally deformed limestones has confirmed its usefulness.

Methods based on strain-induced anisotropies. The fact that traditional ways of estimating strains are usually labour intensive and rely on the availability of suitable markers has stimulated the development of several types of methods which estimate strain from the anisotropy of physical properties of rocks. Attempts have been made to relate finite strains to the anisotropy of thermal conductivity, seismic velocity and magnetic susceptibility. The latter has received much attention after encouraging early results on slate rocks. However, work to date suggests that no general rule exists governing the relationship of strain to rock anisotropy, but that the nature of the relationship will always depend on factors such as how the rock fabric is defined, on the components of the rock which impart the anisotropy and the deformation mechanisms leading to its formation. In spite of these complications, 'local' relationships have been established for particular suites of rocks and, once calibrated by comparison with independent strain estimates, can provide useful empirical indicators of strain orientations and magnitudes. The subject of rock anisotropy has been recently reviewed by Kern & Wenk (1985).

APPLICATIONS OF STRAIN ANALYSIS

This review has so far concentrated on methodology because this aspect is of greater help to the beginner embarking on strain analysis for the first time. It needs to be emphasized though that the discipline is not solely concerned with techniques. Applications of the various methods are numerous and account for a voluminous literature. The examples of these applications given below are classified according to the type of information they provide.

Restorations

A knowledge of the state of strain provides an opportunity for reconstructing the original geometry of rock bodies. For example, the pre-tectonic thicknesses of stratigraphic units can be estimated (Tobisch *et al.* 1977, Hossack 1978, Borradaile 1979) using, for example, the procedures outlined by Ramsay (1969) and Schwerdtner (1978). Windley & Davies (1978) have reconstructed the shape of volcanoes in Archaean rocks.

De-straining (or removing the distortions produced by the measured strains) of bedded rocks allows the possibility of calculating the amount of rigid-body rotation (Ramsay 1969, Cobbold & Percevault 1983) which is information helpful for deciphering the strain history at a given locality (Schwerdtner 1989). An example of pre-tectonic restoration is that performed on a folded bed by Oertel (1974) who unfolded it by reversing the measured strains and body rotations. Recently, attempts have also been made to incorporate strain data into the procedure of restoring balanced cross-sections (Woodward *et al.* 1986, Protzman & Mitra 1990).

Mapping structures

Regional strain surveys involve the systematic mapping of the state of strain throughout an area. Such surveys indicate the presence and kinematic characteristics of important structures such as thrusts and shear zones (Coward 1976). The presence of thrusts and other faults has been characterized by spatial gradients of strain magnitude (Milton & Williams 1981) or discontinuities in the strain pattern (Lisle 1984). Paterson *et al.* (1989) have delineated terrane boundaries on the basis of a regional strain survey in the Sierra Nevada, California. The distribution of finite strain can provide an essential mapping criterion in some gneissic regions (Sutton & Watson 1951).

Understanding tectonic processes

Strain analyses on tectonites have increased our understanding of the processes leading to the formation of a variety of deformational structures. For example, strain data have provided vital input for the discussion about the origin of foliation; *slaty cleavage* (Ramsay & Wood 1973, Tullis & Wood 1975), *pencil cleavage* (Reks & Gray 1982) and *fracture cleavage* (Lisle 1977a). Theories of fabric development have been evaluated by comparing natural fabrics with measured strains, for example, the studies of quartz fabrics (Marjoribanks 1976, Law 1986), magnetic fabrics and vitrinite reflectance fabrics. Strain studies have helped elucidate folding mechanisms (Hudleston & Holst 1984), plutonic intrusion mechanisms and diapirism (Miller 1983, Talbot 1987, Ramsay 1989), and mechanisms of thrusting (Hossack 1968, Chapman *et al.* 1979).

On the scale of individual rock samples, the discrepancies between strain estimates yielded by methods which calculate strain from intra-granular features, by methods which use overall grain shape, and by methods which use inter-particular distances have permitted estimates to be made of the partitioning of the total strain between several deformation mechanisms (Mitra 1976, Borradaile 1979, Pfiffner 1980, White 1982, Mosar 1989).

Providing a source of rheological data. Probably the most significant factor which affects the accuracy of the majority of strain determination techniques is the competence difference which exists between the strain marker objects and the enclosing rock. The existence of competence contrasts means that the strains obtained are systematic under- or over-estimates of the whole-rock strains. Equally significant from the strain analyst's point of view is the fact that the deformed geometry of the marker/matrix system will no longer depend solely on the state of finite strain but also on the strain history.

The physics of the deformation of 'conglomerates' is complex even when the inclusions are linear viscous, ellipsoidal in form, and isolated from their neighbours. Nevertheless attempts to tackle the problem (Gay 1969, Bilby *et al.* 1975, Freeman 1987) represent major strides in the development of suitable forms of analysis.

The additional reward for attempting to calculate strain in rocks containing inclusions of different rock-types is the acquisition of quantitative data on the relative competence (the so-called 'viscosity contrasts') of different lithologies (Gay 1969, Lisle *et al.* 1983, Norris & Bishop 1990). The potential of obtaining such rheological data from naturally-deformed rocks can be seen as an important spin-off of strain analysis work.

CONCLUSIONS

Practical strain analysis research has shown the concept of the strain ellipsoid as the unique representation of the local state of finite strain as inadequate. Each method of palaeostrain analysis based on different aspects of the geometry, using different fabric elements, on different scales produces its own distinct strain ellipsoid. Although this discovery can be seen as a set-back by those whose main aim is to quantify the strain, it also broadens the scope of the discipline because it allows strain analysis, if performed in a critical way using a variety of methods, to contribute to discussions on deformation processes on various scales. With this discovery also comes a sense of realism concerning the significance of the numerical results of analyses. Methods which produce imprecise estimates together with realistic error bars, probably viewed by many as weak methods, are to be preferred to those that deliver an unqualified strain ratio of 4.68!

Palaeostrain analysis has entered a new and rewarding stage in its development. The strains deduced from local observations are being integrated on a regional scale and are providing a powerful tool for deducing the kinematics and for reconstructing the pre-tectonic geometry of a region.

REFERENCES

Babaie, H. A. 1986. A comparison of two-dimensional strain analysis methods using elliptical grains. *J. Struct. Geol.* **8**, 585–587.

Barr, M. & Coward, M. P. 1974. A method of measuring volume change. *Geol. Mag.* **111**, 293–296.

Bell, A. M. 1981. Strain factorisations from lapilli tuff, English Lake District. *J. geol. Soc. Lond.* **138**, 463–474.

Bhattacharyya, T. & Longiaru, S. 1986. Ability of Fry method to characterize pressure solution deformation — discussion. *Tectonophysics* **131**, 199.

Bilby, B. A., Eshelby, J. D. & Kundu, A. K. 1975. The changes in shape of a viscous ellipsoidal region embedded in a slowly deforming matrix having a different viscosity. *Tectonophysics* **28**, 265–274.

Borradaile, G. J. 1976. A strain study of a granite–granite gneiss transition and accompanying schistosity formation in the Betic orogenic zone, S.E. Spain. *J. geol. Soc. Lond.* **132**, 417–428.

Borradaile, G. J. 1979. Strain study in the Islay region, SW Scotland: implications for strain histories and deformation mechanisms in greenschists. *J. geol. Soc. Lond.* **136**, 77–88.

Borradaile, G. J. 1987. Analysis of strained sedimentary fabrics: review and tests. *Can. J. Earth Sci.* **24**, 442–455.

Breddin, H. 1956. Die tektonische Deformation der Fossilien im Rheinischen Schiefergebirge. *Deut. Geol. Ges. Z.* **106**, 261–269.

Burns, K. L. & Spry, A. H. 1969. Analysis of the shape of deformed pebbles. *Tectonophysics* **7**, 177–196.

Chapman T. J., Milton N. J. & Williams G. D. 1979. Shape fabric variations in deformed conglomerates at the base of the Laksefjord Nappe, Norway. *J. geol. Soc. Lond.* **136**, 683–691.

Cloos, E. 1947. Oolite deformation in the South Mountain Fold, Maryland. *Bull. geol. Soc. Am.* **58**, 843–918.

Cobbold, P. R. & Gapais, D. 1979. Specification of fabric shapes using an eigenvalue method. *Bull. geol. Soc. Am.* **90**, 310–312.

Cobbold, P. R. & Percevault, M.-N. 1983. Spatial integration of strains using finite elements. *J. Struct. Geol.* **5**, 299–305.

Coward, M. P. 1976. Archean deformation patterns in southern Africa. *Phil. Trans. R. Soc.* **A283**, 313–331.

Crespi, J. M. 1986. Some guidelines for the practical application of Fry's method of strain analysis. *J. Struct. Geol.* **8**, 799–808.

Cutler, J. M. 1985. Error due to strain measurement in non-principal sections. *Tectonophysics* **113**, 185–190.

De Paor, D. G. 1981. Strain analysis using deformed line distributions. *Tectonophysics* **73**, T9–T14.

De Paor, D. G. 1988a. Strain determination from three known stretches — an exact solution. *J. Struct. Geol.* **10**, 639–642.

De Paor, D. G. 1988b. R_f/ϕ strain analysis using an orientation net. *J. Struct. Geol.* **10**, 323–333.

De Paor, D. G. 1990. Determination of the strain ellipsoid from sectional data. *J. Struct. Geol.* **12**, 131–137.

Ding Zhong-Yi 1984. Some formulae for calculating parameters of the strain ellipse. *Tectonophysics* **110**, 167–175.

Dunne, W. M., Onasch, C. M. & Williams, R. T. 1990. The problem of strain-marker centers and the Fry method. *J. Struct. Geol.* **12**, 933–938.

Dunnet, D. 1969. A technique of finite strain analysis using elliptical particles. *Tectonophysics* **7**, 117–136.

Dunnet, D. & Siddans, A. W. B. 1971. Non-random sedimentary fabrics and their modification by strain. *Tectonophysics* **12**, 307–325.

Elliott, D. 1970. Determination of finite strain and initial shape from deformed elliptical objects. *Bull. geol. Soc. Am.* **81**, 2221–2236.

Erslev, E. A. 1988. Normalized center-to-center strain analysis of packed aggregates. *J. Struct. Geol.* **10**, 201–209.

Erslev, E. A. & Ge, H. 1990. Least-squares center-to-center and mean object ellipse fabric analysis. *J. Struct. Geol.* **12**, 1049–1059.

Ferguson, C. C. 1981. A strain reversal method for estimating extension from fragmented rigid inclusions. *Tectonophysics* **79**, T43–T52.

Flinn, D. 1962. On folding during three dimensional progressive deformation. *Q. J. geol. Soc. Lond.* **118**, 385–428.

Ford, M. & Ferguson, C. C. 1985. Cleavage strain in the Variscan Fold Belt, County Cork, Ireland, estimated from stretched arsenopyrite rosettes. *J. Struct. Geol.* **7**, 217–223.

Freeman, B. 1987. The behaviour of deformable ellipsoidal particles in three-dimensional slow flows: implications for geological strain analysis. *Tectonophysics* **132**, 297–309.

Fry, N. 1979a. Random point distributions and strain measurements in rocks. *Tectonophysics* **60**, 89–105.

Fry, N. 1979b. Density distribution techniques and strained line length methods for determination of finite strains. *J. Struct. Geol.* **1**, 221–229.

Gay, N. C. 1969. The analysis of strain in the Barberton Mountain Land, Eastern Transvaal, using deformed pebbles. *J. Geol.* **77**, 377–396.

Gendzwill, D. J. & Stauffer, M. R. 1981. Analysis of triaxial ellipsoids: their shapes, plane sections and plane projections. *Math. Geol.* **13**, 135–152.

Graham, R. H. 1978. Quantitative deformation studies in the Permian rocks of Alpes-Maritime. *Bur. Rech. Geol. Min. Mem.* **91**, 219–238.

Gray, D. R. & Durney, D. W. 1979. Investigations on the mechanical significance of crenulation cleavage. *Tectonophysics* **58**, 35–79.

Groshong, R. H. 1972. Strain calculated from twinning in calcite. *Bull. geol. Soc. Am.* **83**, 2025–2038.

Hanna, S. S. & Fry, N. 1979. A comparison of methods of strain determination in rocks from southwestern Dyfed (Pembrokeshire) and adjacent areas. *J. Struct. Geol.* **1**, 155–162.

Harvey, P. K. & Ferguson, C. C. 1981. Directional properties of polygons and their application to finite strain estimation. *Tectonophysics* **74**, T33–T43.

Harvey, P. K. & Laxton, R. R. 1980. The estimation of finite strain from the orientation distribution of passively deforming linear markers: eigenvalue relationships. *Tectonophysics* **70**, 285–307.

Holst, T. B. 1982. The role of initial fabric on strain determination from deformed ellipsoidal objects. *Tectonophysics* **82**, 329–350.

Holst, T. B. 1987. The analysis of buckle folds from the Proterozoic of Minnesota. *Am. J. Sci.* **287**, 616–634.

Hossack, J. R. 1968. Pebble deformation and thrusting in the Bygdin area (Southern Norway). *Tectonophysics* **5**, 315–339.

Hossack, J. R. 1978. The correction of stratigraphic sections for tectonic finite strain in the Bygdin area, Norway. *J. geol. Soc. Lond.* **135**, 229–241.

Hossack, J. R. 1979. The use of balanced cross-sections in the calculation of orogenic contraction: a review. *J. geol. Soc. Lond.* **136**, 705–711.

Hossain, K. M. 1979. Determination of strain from stretched belemnites. *Tectonophysics* **60**, 279–288.

Hudleston, P. J. 1973. An analysis of single layer folds developed experimentally in viscous media. *Tectonophysics* **16**, 189–124.

Hudleston, P. J. & Holst, T. B. 1984. Strain analysis and fold shape in a limestone layer and implications for layer rheology. *Tectonophysics* **106**, 321–347.

Hutton, D. H. W. 1982. A tectonic model for the emplacement of the main Donegal granite. *J. geol. Soc. Lond.* **139**, 615–631.

Kanagawa, K. 1990. Automated two-dimensional strain analysis from deformed elliptical markers using an image analysis system. *J. Struct. Geol.* **12**, 139–143.

Kanagawa, K. & Yoshida, S. 1988. Utility of the orientation tensor method for quantitative representation of preferred orientations of phyllosilicates and amphiboles measured with X-ray texture goniometer. *J. Fac. Sci. Univ. Tokyo* **21**, 447–465.

Kern, H. & Wenk, H.-R. 1985. Anisotropy in rocks and the geological significance. In: *Preferred Orientation in Deformed Metals and Rocks: An Introduction to Modern Texture Analysis* (edited by Wenk, H. R.). 537–555.

Lacassin, R. & van den Driessche, J. 1983. Finite strain determination of gneiss: application of Fry's method to porphyroid in the southern Massif Central (France). *J. Struct. Geol.* **5**, 245–253.

Law, R. D. 1986. Relationships between strain and quartz crystallography fabrics in the Roche Maurice quartzite of Plougastel, western Brittany. *J. Struct. Geol.* **3**, 129–142.

Le Theoff, B. 1979. Non coaxial deformation of elliptical particles. *Tectonophysics* **53**, T7–T13.

Lisle, R. J. 1977a. Clastic grain shape and orientation in relation to cleavage from the Aberystwyth Grits, Wales. *Tectonophysics* **39**, 381–395.

Lisle, R. J. 1977b. Estimation of tectonic strain ratio from the mean shape of deformed elliptical markers. *Geol. Mijnb.* **56**, 140–144.

Lisle, R. J. 1979. Strain analysis using deformed pebbles: the influence of initial pebble shape. *Tectonophysics* **60**, 263–277.

Lisle, R. J. 1984. Strain discontinuities within the Seve–Koli Nappe Complex, Scandinavian Caledonides. *J. Struct. Geol.* **6**, 101–110.

Lisle, R. J. 1985. *Geological Strain Analysis: A Manual for the R_f/ϕ Method*. Pergamon Press, Oxford.

Lisle, R. J. 1986. The sectional strain ellipse during progressive coaxial deformations. *J. Struct. Geol.* **8**, 809–818.

Lisle, R. J. 1991. Strain analysis by simplified Mohr constructions. *Ann. Tect.* **5**, 102–117.

Lisle, R. J. 1992. Strain estimation from flattened buckle folds. *J. Struct. Geol.* **14**, 369–371.

Lisle, R. J. & Ragan, D. M. 1988. Strain determination from three measured stretches — a simple Mohr circle solution. *J. Struct. Geol.* **10**, 905–906.

Lisle, R. J., Rondeel, H. E., Doorn, D., Brugge, J. & Van de Gaag, P. 1983. Estimation of the viscosity contrast and finite strain from deformed elliptical inclusions. *J. Struct. Geol.* **5**, 603–609.

Lloyd, G. E. 1983. Strain analysis using the shape of expected and observed continuous frequency distributions. *J. Struct. Geol.* **5**, 225–231.

Mandl, G. & Shippam, G. K. 1981. Mechanical model of thrust sheet gliding and imbrication. In: *Thrust and Nappe Tectonics* (edited by McClay, K. R. & Price, N. J.). *Spec. Publs. geol. Soc. Lond.* **9**, 79–98.

Majoribanks, R. W. 1976. The relation between microfabric and strain in a progressively deformed quartzite sequence from central Australia. *Tectonophysics* **32**, 269–293.

Martin Escorza, C. 1978. Estructura y deformacion de los enclaves microgranulares negros (gabarros) del Alto de los Leones, Guadarrama. *Bol. R. Soc. espan. Hist. Nat. (geol.)* **76**, 57–87.

Matthews, P. E., Bond, R. A. B. & Van den Berg, J. J. 1974. An algebraic method of strain analysis using elliptical markers. *Tectonophysics* **24**, 31–67.

Miller, D. M. 1983. Strain on a gneiss dome in the Albion

Mountains metamorphic core complex, Idaho. *Am. J. Sci.* **283**, 605–632.

Milnes, A. G. 1971. A model for analyzing the strain history of folded competent layers in deeper parts of orogenic belts. *Eclog. geol. Helv.* **64**, 335–342.

Milton, N. J. 1980. Determination of the strain ellipsoid from measurements on any three sections. *Tectonophysics* **64**, T19–T27.

Milton, N. J. & Williams, G. D. 1981. The strain profile above a major thrust fault, Finnmark, N Norway. In: *Thrust and Nappe Tectonics* (edited by McClay, K. R. and Price, N. J.). *Spec. Publs. geol. Soc. Lond.* **9**, 235–239.

Mimran, Y. 1976. Strain estimation using a density-distribution technique and its application to deformed Upper Cretaceous Dorset chalks. *Tectonophysics* **31**, 175–192.

Mitra, S. 1976. A quantitative study of deformation mechanisms and finite strain in quartzites. *Contrib. Miner. Petrol.* **59**, 203–226.

Mosar, J. 1989. Deformation interne dans les Prealpes medianes (Suisse). *Eclog. geol. Helv.* **82**, 765–794.

Mukhopadhyay, D. 1973. Strain measurements from deformed quartz grains in the slaty rocks from the Ardennes and the northern Eifel. *Tectonophysics* **16**, 279–296.

Mukhopadhyay, D. 1980. Determination of finite strain from grain centre measurements. *Tectonophysics,* **67**, T9–T12.

Norris, R. J. & Bishop, D. G. 1990. Deformed conglomerates and textural zones in the Otago Schists, South Island, New Zealand. *Tectonophysics* **174**, 331–349.

Oertel, G. 1974. Unfolding of an antiform by the reversal of observed strains. *Bull geol. Soc. Am.* **85**, 445–450.

Onasch, C. M. 1986. Ability of the Fry method to characterize pressure solution deformation. *Tectonophysics* **122**, 187–193.

Owens, W. H. 1973. Strain modifications of angular density distributions. *Tectonophysics* **16**, 249–261.

Owens, W. H. 1984. The calculation of best-fit ellipsoid from elliptical sections on arbitrarily oriented planes. *J. Struct. Geol.* **6**, 571–578.

Panozzo, R. 1984. Two-dimensional strain from the orientations of lines in a plane. *J. Struct. Geol.* **6**, 215–221.

Passchier, C. W. 1990. A Mohr circle construction to plot the stretch history of material lines. *J. Struct. Geol.* **12**, 513–515.

Paterson, S. R. 1983. A comparison of the methods used in measuring finite strains from ellipsoidal objects. *J. Struct. Geol.* **5**, 611–618.

Paterson, S. R., Tobisch, O. T. & Bhattacharyya, T. 1989. Regional, structural and strain analysis of terranes in Western Metamorphic Bely, Central Sierra Nevada. *J. Struct. Geol.* **11**, 255–274.

Peach, C. J. & Lisle, R. J. 1979. A Fortran IV program for the analysis of tectonic strain using deformed elliptical markers. *Comput. Geosci.* **5**, 325–334.

Pfiffner, O. A. 1980. Strain analysis in folds (Infrahelvetic Complex, central Alps). *Tectonophysics* **61**, 337–362.

Protzman, G. M. & Mitra, G. 1990. Strain fabric associated with the Meade thrust sheet: implications for cross-section balancing. *J. Struct. Geol.* **12**, 403–417.

Ragan, D. M. 1985. *Structural Geology: An Introduction to Geometrical Techniques.* 3rd ed. Wiley, New York, 393 pp.

Ragan, D. M. 1987. Strain from three measured stretches. *J. Struct. Geol.* **9**, 897–898.

Rajlich, P. 1989. Strain analysis of Devonian fossils from the NE part (Rhenohercynian zone) of the Bohemian Massif. *Ann Tect.* **3**, 44–51.

Ramsay, J. G. 1967. *Folding and Fracturing of Rocks.* McGraw-Hill, New York, 1–568.

Ramsay, J. G. 1969. The measurement of strain and displacement in orogenic belts. In: *Time and Place in Orogeny* (edited by Kent, P. E. *et al.*). *Spec. Publs. geol. Soc. Lond.* **3**, 43–79.

Ramsay, J. G. 1976. Displacement and strain. *Phil. Trans. R. Soc.* **A283**, 3–25.

Ramsay, J. G. 1989. Emplacement kinematics of a granite diapir: the Chindamora batholith, Zimbabwe. *J. Struct. Geol.* **11**, 191–209.

Ramsay, J. G. & Huber, M. I. 1983. *The Techniques of Modern Structural Geology. Volume 1: Strain Analysis.* Academic Press, London.

Ramsay, J. G. & Wood, D. S. 1973. The geometrical effects of volume change during deformation processes. *Tectonophysics* **16**, 263–277.

Reks, I. J. & Gray, D. R. 1982. Pencil structure and strain in weakly deformed mudstone and siltstone. *J. Struct. Geol.* **4**, 161–176.

Ribeiro, A., Kullberg, M. C. & Possolo, A. 1983. Finite strain estimation using anti-clustered distributions of points. *J. Struct. Geol.* **5**, 233–244.

Robin, P.-Y. F. 1977. Determination of geologic strain using randomly oriented strain markers of any shape. *Tectonophysics* **42**, T7–T6.

Robin, P.-Y. F. 1983. Algebraic calculation of two dimensional strain from point distribution data. *J. Struct Geol.* **5**, 552.

Sanderson, D. J. 1973. The development of fold axes oblique to the regional trend. *Tectonophysics* **16**, 55–70.

Sanderson, D. J. 1977a. The algebraic evaluation of two-dimensional finite strain rosettes. *Math. Geol.* **9**, 483–496.

Sanderson, D. J. 1977b. The analysis of finite strain using lines with an initial random orientation. *Tectonophysics* **43**, 199–211.

Sanderson, D. J. & Meneilly, A. W. 1981. Analysis of three-dimensional strain modified uniform distributions: andalusite fabrics from a granite aureole. *J. Struct. Geol.* **3**, 109–116.

Sanderson, D. J. & Phillips, S. J. L. 1987. Strain analysis using weighting of deformed line elements. *J. Struct. Geol.* **9**, 511–514.

Schmid, S. & Paterson, M. S. 1977. Strain analysis in an experimentally deformed oolitic limestone. In: *Energetics of Geological Processes* (edited by Saxena, S. K. & Battacharji, S.). Springer, New York, 67–93.

Schultz-Ela, D. D. 1990. A method for estimating errors in calculated strains. *J. Struct. Geol.* **12**, 939–943.

Schwerdtner, W. M. 1978. Determination of stratigraphic thickness of strained units. *Can. J. Earth Sci.* **15**, 1379–1380.

Schwerdtner, W. M. 1989. The solid-body tilt of deformed palaeohorizontal planes: application to an Archean transpressive zone, Southern Canadian shield. *J. Struct. Geol.* **11**, 1021–1028.

Seymour, D. B. & Boulter, C. A. 1979. Tests of computerised strain analysis methods by the analysis of simulated deformation of natural unstrained sedimentary fabrics. *Tectonophysics* **58**, 221–235.

Sherwin, J. A. & Chapple, W. M. 1968. Wavelengths of single layer folds: a comparison between theory and observation. *Am. J. Sci.* **266**, 167–179.

Shimamoto, T. & Ikeda, Y. 1976. A simple algebraic method for strain estimation from deformed strain ellipsoidal objects — 1. Basic theory. *Tectonophysics* **36**, 315–337.

Siddans, A. W. B. 1976. Deformed rocks and their textures. *Phil. Trans. R. Soc.* **A283**, 43–54.

Siddans, A. W. B. 1980a. Analysis of three-dimensional homogeneous finite strain using ellipsoidal fabrics. *Tectonophysics* **64**, 1–16.

Siddans, A. W. B. 1980b. Elliptical markers and non-coaxial strain increments. *Tectonophysics* **67**, 21–25.

Simpson, C. 1988. Strain analysis. In: *Basic Methods of Structural Geology* (edited by Marshak, S. & Mitra, G.). Prentice Hall.

Soto, J. I. 1991. Strain analysis method using the maximum frequency of unimodal orientation distributions: applications to gneissic rocks. *J. Struct. Geol.* **13**, 329–335.

Srivastava, D. C. 1985. The deformation and strain analysis of the Gandemara conglomerate from the Archean Iron-Ore Group rocks, District Singhbhum, Bihar. *J. geol. Soc. Ind.* **26**, 233–244.

Stauffer, M. R. & Burnett, 1979. Down-plunge viewing: a rapid method for estimating the strain ellipsoid for large clasts in deformed rocks. *Can. J. Earth Sci.* **16**, 290–304.

Sutton, J. & Watson, J. V. 1951. The pre-Torridonian metamorphic history of the Loch Torridon and Scourie areas in the North-west Highlands. *Q. J. geol. Soc. Lond.* **106**, 241–307.

Talbot, C. J. 1970. The minimum strain ellipsoid using deformed quartz veins. *Tectonophysics* **9**, 47–76.

Talbot, C. J. 1987. Strains and vorticity beneath a tabular batholith in the Zambesi belt, northeast Zimbabwe. *Tectonophysics* **138**, 121–158.

Themistocleous, S. G. & Schwerdtner, W. M. 1977. Estimates of distortional strain in mylonites from the Grenville Front Tectonic Zone, Tomiko area, Ontario. *Can. J. Earth Sci.* **14**, 1708–1720.

Tobisch, O. T., Fiske, R. S., Sacks, S. & Taniguchi, D. 1977. Strain in measured volcaniclastic rocks and its bearing on the evolution of orogenic belts. *Bull. geol. Soc. Am.* **88**, 23–40.

Trayner, P. M. 1986. A comparison of Sanderson's and Panozzo's strain measurement methods using calcite grain boundaries from the Variscan fold and thrust belt in Ireland. *J. Struct. Geol.* **8**, 205–207.

Treagus. S. H. 1987. Mohr circles for strain, simplified. *Geol. J.* **22**, 119–132.

Tullis, T. E. & Wood, D. S. Correlation of finite strain from both reduction bodies and preferred orientation of mica in slate from Wales. *Bull. geol. Soc. Am.* **86**, 632–638.

Van der Pluijm, B. A. Note on analysis of quartz grain dimensions in foliated greywackes. *Geol. Rund.* **76**, 851–855.

Wernicke, B. & Burchfiel, B. C. 1982. Modes of extensional tectonics. *J. Struct. Geol.* **4**, 105–115.

Wheeler, J. 1984. A new plot to display the strain of elliptical markers. *J. Struct. Geol.* **6**, 417–423.

Wheeler, J. 1986. Strain analysis in rocks with pretectonic fabrics. *J. Struct. Geol.* **8**, 887–896.

Wheeler, J. 1989. A concise algebraic method for assessing strain in distributions of linear objects. *J. Struct. Geol.* **11**, 1007–1010.

White, J. C. 1982. Quartz deformation and the recognition of recrystallization regimes in the Flinton Group conglomerates, Ontario. *Can. J. Earth Sci.* **19**, 81–93.

Williams, R. B. G. 1986. *Intermediate Statistics for Geographers and Earth Scientists*. Macmillan, London.

Windley, B. F. & Davies, F. B. 1978. Volcanoes and lithospheric/crustal thickness in the Archean. *Earth Planet. Sci. Lett.* **38**, 291–297.

Wood, D. S. & Oertel, G. 1980. Deformation in the Cambrian slate belt of Wales. *J. Geol.* **88**, 309–326.

Woodcock, N. H. 1977. Specification of fabric shapes using an eigenvalue method. *Bull. geol. Soc. Am.* **88**, 1231–1236.

Woodward, N. B., Gray, D. R. & Spears, D. B. 1986. Including strain data in balanced cross-sections. *J. Struct. Geol.* **8**, 313–324.

Yu, H. & Zheng, Y. 1984. A statistical analysis applied to the R_f/ϕ method. *Tectonophysics* **110**, 151–155.

CHAPTER 3

Brittle Crack Propagation

TERRY ENGELDER

INTRODUCTION

RECENT advances in our understanding of brittle deformation processes have included the definition of geological conditions favouring either crack propagation or shear rupture. *Crack propagation* involves the parting of rock within a narrow process zone with a motion normal to the plane of parting, whereas shear rupture involves a complex network of cracking within intact rock, with the wall rock on either side of the rupture zone simultaneously displaced in shear. The mesoscopic products of crack propagation and shear rupture are *joints* and *shear fractures*, respectively (see discussion of these terms in Chapter 5). Shear fracturing along with reactivation of joints by frictional slip are examples of a general class of brittle process called *faulting*. Much of the early work on brittle deformation processes was influenced by the perception that shear fractures were a very common mode of brittle failure in the upper crust (e.g. Badgley 1965). This perception arose because joints were commonly mistaken for shear fractures (e.g. Bucher 1920, Scheidegger's (1982) discussion of Engelder 1982), particularly if the joints were reactivated in frictional slip.

Early models for earth stress were constrained by the strength of intact rocks under large compressive, differential stresses (e.g. Handin & Hager 1957). Rock strength is the differential stress a rock can sustain without developing a shear fracture. Later, geologists appreciated that an intact rock was not a good model for the upper crust which was cut by pervasive sets of joints and faults. This observation led to the inevitable conclusion that differential stress in the crust was more closely constrained by the frictional strength of pre-existing joints, shear fractures, and faults (e.g. Byerlee 1978, Brace & Kohlstedt 1980). Because the upper crust is so completely pervaded by fractures, differential stress throughout much of the upper crust is regulated by frictional slip and, thus, differential stress remains at a lower level than required to rupture intact rock by shear fracturing. However, this lower differential stress does not preclude crack development which occurs under an effective tensile stress (Etheridge 1983).

An appreciation of the importance of crack propagation in rock arose out of detailed studies of shear fracture development. Although less attention was focused on microcrack development during early work on rock strength, microscopy suggested that crack propagation was an important precursor to shear rupture (Bombolakis & Brace 1963, Scholz 1968). With the certain knowledge that microcrack propagation was an important mechanism leading to faulting of intact rock, theories for shear failure were developed based on the growth of microcracks (McClintock & Walsh 1962). At the same time, field work showed that the formation of joints by crack propagation was a very common mode of failure of intact rock in the crust (Secor 1965, Nickelsen & Hough 1967). During the past decade many of the significant strides in understanding the brittle failure of crustal rocks have come from the analysis of crack propagation using linear elastic fracture mechanics.

A review can capture only the bare essence of our understanding of the cracking processes which has three basic stages: initiation; propagation; and arrest. My review starts with a discussion of the fundamental concepts of *linear elastic fracture mechanics* (LEFM) by focusing on the contributions of Inglis, Griffith, and Irwin. Then I discuss the basics of crack initiation using a couple of examples of crack propagation under the influence of high fluid pressure. Such examples are appropriate because one of the most appealing mechanisms for crack propagation deep in the crust is the fluid drive mechanism which acts to relieve high fluid pressure. However, the rules of fracture mechanics developed for fluid-driven cracks apply to all situations involving crack propagation such as those associated with near-surfce jointing where net tensile stresses develop (e.g. Hancock & Engelder 1989). Because of space limits, I only touch on the issues associated with crack propagation and arrest which have been covered in books (e.g. Lawn & Wilshaw 1975, Broek 1987, Atkinson 1987, Rossmith 1983) and in review articles (e.g. Anderson & Grew 1977, Atkinson 1984, Pollard & Aydin 1988).

ANALYSES OF CRACKS IN ELASTIC MATERIALS

The study of crack propagation in rock has its origin in the earliest part of the twentieth century. Crack propagation stems from stress concentrations which develop around holes in elastic materials. Rocks are a composite of grains, matrix, and cement with pore space commonly found at grain boundaries. Large stress concentrations develop around pore space and elastic mismatches between grains and cement when the rock aggregate is subject to boundary tractions (e.g. Gallagher *et al.* 1974). Some of the sharp corners on pore space and inclusions are so oriented that large tensile stresses develop, and it is at these sharp corners that crack propagation initiates.

Inglis' contribution

An understanding of stress concentration in an elastic material is illustrated using a circular hole in an elastic plate. Assume that the circular hole is nowhere near the edge of the plate. If opposite edges of the elastic plate are pulled by a force large enough to cause a tensile stress, σ^r, within the plate, then this remote stress is concentrated three times (i.e. the hoop stress = $3\sigma^r$) at the point (assume that the plate is infinitely thin) on the edge of the circular hole 90° from the direction of the force on the edges of the plate (Fig. 3.1). Tensile stress is negative in sign. Pore space and other stress concentrators in rocks are more likely to be elliptical in cross section. Inglis (1913) introduced a solution for the state of stress in the vicinity of the tip of an elliptical hole with major axis, c, and minor axis, b. In its limiting case, the elliptical hole represents a crack with a very small minor axis, b. If an elastic plate with an elliptical crack is loaded in tension normal to the major axis of the crack, so that stress within the plate far from the crack is σ^r, then the local tensile stress acting at right angles to the major axis of the crack rises to several times that of σ^r. Inglis (1913) found that

$$\sigma^c = \sigma^r \left(1 + 2\frac{c}{b}\right) \quad (1)$$

where σ^c is the stress concentration at the 'crack' tip.

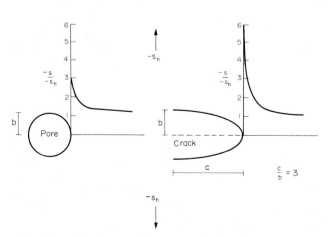

Fig. 3.1. The stress concentration around circular and elliptical holes. See text for details.

The stress concentration decreases rapidly with distance from the edge of the elliptical crack. As the elliptical crack becomes longer with a larger aspect ratio, $\frac{c}{b}$, the stress concentration at the tip of the crack may be approximated by

$$\sigma^c \approx 2\sigma^r \left(\frac{c}{\rho}\right)^{\frac{1}{2}} \quad (2)$$

where $\rho = \frac{b^2}{c}$ is the radius of curvature at the tip of the crack. Note that for cracks with a large aspect ratio, the first term in equation (1) becomes very small relative to the second term. The significance of Inglis' solution is that the high tensile stress necessary for crack propagation will develop at the tips of cracks with large aspect ratios, when the cracks are subject to a remote tensile stress or internal fluid pressure which is only a small fraction of an intact rock's tensile strength.

Griffith's contribution

After Inglis' solution, the second step in understanding crack initiation in rock was Griffith's (1921, 1924) analysis of cracking as a thermodynamic system. A crack subject to high fluid pressure in a rock might have similar properties to a steam piston-cylinder which is capable of expanding against an external force. A steam piston-cylinder consists of one internal component, gas, which is characterized by an equation of state, the ideal gas law. In contrast, a rock consists of two internal components: the crack defined by its long axis, $2c$, and the solid rock defined by its elastic properties (Fig. 3.2). Fluid pressure within the crack, as well as confining pressure on the rock, is equivalent to the load on the piston of the steam cylinder. In the steam piston-cylinder, adiabatic work by the system on the load (F), W is positive.

$$\Delta U_T = -\Delta U_{ST} + W = nRT \int_{V_b}^{V_a} \frac{dV}{V} + F\Delta\ell = 0 \quad (3)$$

where the external force on the loading device (i.e. the piston) is compressive. U_T is the total energy of the system, ΔU_{ST} is the change in internal energy of the steam and ℓ is the displacement of the load, F. In driving a crack within a dry rock, one must subject the outer boundary of the rock to a net pull (i.e. a tensile force), so that the boundary moves outward under tension. In a sense, when the exterior walls of the rock displace there is a decrease in potential energy of the loading device which is any boundary traction that might cause the generation of tensile stress, and as a consequence $dW_R < 0$. For example, such a loading situation occurs during the stretching of a rock layer within a thrust sheet in the vicinity of a lateral ramp. The subscript R designates work on surrounding rock across boundaries away from the crack. The work to propagate the crack is positive and defined as the

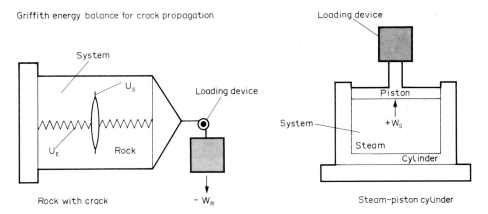

Fig. 3.2. The thermodynamic behaviour of a crack in a rock (after Pollard 1989). See text for details.

increase in surface energy, dU_s. As the crack propagates, the rock will undergo a change in strain energy, dU_E. The total change in energy for crack propagation is

$$\Delta U_T = \Delta U_s - W_R + \Delta U_E. \qquad (4)$$

Equation (4) is equivalent to equation (3), but (4) accounts for the complexity of crack propagation. Like the steam piston-cylinder, Griffith (1924) recognized that crack propagation could take place without changing the total energy of the rock-crack system. This is known as the Griffith energy-balance concept where the standard equilibrium requirement is that for an increment of crack extension dc,

$$\frac{dU_T}{dc} = 0. \qquad (5)$$

The mechanical and surface energy terms within the rock-crack system must balance over a crack extension, dc. During crack propagation the crack walls move outward to some new lower energy configuration upon removal of the restraining tractions across an increment of crack. In effect, the motion of the crack walls represents a decrease in mechanical energy while work must be expended to remove the restraints across the crack increment. The work to remove the restraints is the surface energy for incremental crack propagation.

To evaluate the three energy terms relating to crack propagation, Griffith cited a theorem of elasticity which states that, for an elastic body under a constant load, the boundaries will displace from the unstressed state to the equilibrium state so that

$$W_R = 2U_E. \qquad (6)$$

For a thin plate (i.e. a sheet of rock) containing an elliptical crack with a major axis perpendicular to a uniform tension, Griffith calculated that

$$U_E = \left[\frac{\pi(1-\nu^2)}{E}\right] c^2 (\sigma^r)^2 \qquad (7)$$

where E is the Young's modulus for the rock and ν is the Poisson ratio. For the surface energy of the crack, Griffith defined crack length as $2c$, and recognized that crack propagation produces two crack faces. Therefore,

$$U_s = 4c\gamma. \qquad (8)$$

Substituting equations (6), (7), and (8) into (4) and then applying equation (5), Griffith solved for the critical condition for crack propagation

$$\sigma^r = \left[\frac{2E\gamma}{\pi(1-\nu^2)c}\right]^{1/2}. \qquad (9)$$

A crack driven by a tensile traction on an external boundary is one of two end member cases for joint propagation in rocks. The other end member is a crack driven by an internal fluid pressure. In this latter case, the rock is usually in a state of compression from boundary tractions and would, therefore, contain a strain energy due to a compressive external load, σ^r

$$U_E^r = \frac{\pi c^2 (\sigma^r)^2}{E}. \qquad (10)$$

If the crack is vertical, the vertical, outer boundaries of the local rock-crack system may not displace during crack propagation because the local system is surrounded by adjacent rock-crack systems which are simultaneously attempting to expand. In this case, the rock is in a state of uniaxial strain. If the vertical boundaries do not displace then the rock-crack system does no work against horizontal earth stress during crack propagation, so

$$dW_R = 0. \qquad (11a)$$

Therefore, the strain energy term from remote boundary tractions (equation 10) does not enter into Griffith's thermodynamic calculation. In fracture mechanics textbooks this is the 'fixed-grips' case where the loading device does not move the outer boundary of the rock-crack system (e.g. Lawn & Wilshaw 1975).

If we assume that the crack in the rock is internally pressurized but that the pore space behind the crack remains dry by virtue of an impermeable membrane at the wall of the crack, then we can solve for the crack driving pressure, P_i. When the crack propagates, the crack walls displace, as was the case for crack propagation under tensile boundary tractions. However, in the latter case there was no internal pressure on the crack wall so that no work was done during the displacement of the crack wall. For the case with internal fluid pressure, work is done to move the crack wall by the fluid in the crack

$$dW_c = 2U_E^c. \tag{11b}$$

Note that because work was done on the rock-crack system by the crack fluid, the sign of the work term is again negative. Examination of the forces acting on the crack suggest that the crack walls will not part under the influence of P_i until there is a net outward force which happens when $P_i > |\sigma^r|$. The strain energy for movement of the crack wall depends on the net outward force or effective stress within the crack which is the difference between the total stress on the outer boundary of the crack system, σ^r, and fluid pressure inside the crack, P_i,

$$U_E^c = \frac{\pi c^2 (P_i - \sigma^r)^2}{E}. \tag{12}$$

We solve equation (6) for the internal pressure necessary to drive the crack when the outer boundary of the rock-crack system is held in a fixed position

$$P_i = \left[\frac{2E\gamma}{\pi(1-\nu^2)c}\right]^{1/2} + \sigma^r. \tag{13}$$

This is Secor's (1969) solution to crack propagation under the influence of very high fluid pressures. However, this equation applies only if there is an impermeable membrane between the crack wall and the pore space behind the crack, so that fluid from the crack is not allowed to drain into the pore space. Such a situation is unrealistic.

LINEAR ELASTIC FRACTURE MECHANICS

Conditions defining tensile strength, shear fracturing, and frictional slip on faults are illustrated using the Coulomb–Mohr failure envelope (e.g. Engelder & Marshak 1988, also see Chapters 4 and 5 this volume). Although the Coulomb–Mohr failure envelope serves as a good empirical gauge for the stresses at shear failure, it provides information on neither the rupture path nor post-failure behaviour. Likewise, Griffith's analysis using Inglis' stress concentration factor served well to predict the initiation of crack propagation but proved unsatisfactory as a propagation criterion. This is largely because both the stress field in the vicinity of the crack tip and the radius of the crack tip are poorly defined.

Irwin's contribution

To satisfy the need for a propagation criterion, Irwin (1958) noted that the stress field, σ_{ij}, in the vicinity of a sharp crack tip in an elastic body was approximately proportional to K, the stress intensity factor where

$$\sigma_{ij} \approx \frac{Kf_{ij}(\theta)}{(2\pi r)^{1/2}}, \tag{14}$$

r and θ are polar coordinates centred at the crack tip. The trigonometric functions, $f_{ij}(\theta)$, vary slightly from unity near the crack tip. These lengthy functions are given in fracture mechanics textbooks (e.g. Broek 1987, Lawn & Wilshaw 1975). Equation (14) for the stress field in the vicinity of the crack tip is remarkable in that the dependence of σ_{ij} on the applied load and geometry of the crack are all incorporated in the stress intensity factor, K. For $r \ll c$, then $\sigma_{ij} \gg \sigma^r$. In general, the equation for K at the crack tip depends on the loading stresses, the length of the crack, $2c$, and the geometry of the body containing the crack which is specified by the dimensionless modification factor, Y,

$$K = Y\sigma^r(c)^{1/2}. \tag{15}$$

For a penny-shaped crack $Y = \frac{2}{\sqrt{\pi}}$, whereas for a tunnel (i.e. blade-shaped) crack $Y = \sqrt{\pi}$. Here a crack will propagate only if the net loading stress is tensile. The engineering literature assigns a positive value to the stress intensity, K, when the net loading stress is tensile. This convention is confusing because it is not consistent with the geological convention of assigning a negative sign to tensile stresses.

Griffith defined the *tensile strength of a rock* in terms of a balance among the work by the loading system, the strain energy within the rock, and the crack surface energy. The tensile strength can also be specified in terms of the critical value of the crack-tip stress intensity at the time of crack propagation

$$K_{Ic} = K \tag{16}$$

where K_{Ic} is called the fracture toughness of the rock. If K_{Ic} is known, then the internal crack fluid pressure necessary to initiate crack propagation is derived from equation (15)

$$P_i = \frac{K_{Ic}}{Y(c)^{1/2}} + \sigma^r \tag{17}$$

where the remote stress is compressive (positive in sign). The net loading stress is actually tensile because $P_i > \sigma^r$. Again this equation assumes that the wall of the crack is impermeable. Experiments on crack propagation have shown that cracks will propagate when $K < K_{Ic}$ at the crack tip (Anderson & Grew 1977, Atkinson 1984). This phenomenon is a process called *subcritical crack growth*. Subcritical crack growth is permitted by chemical reactions at the crack tip, known as *stress corrosion*, which act to weaken the rock in the vicinity of the tip. Under the influence of stress corrosion, crack propagation may take place at velocities less than 1 mm/s. When cracks propagate under tip stresses equal to K_{Ic}, the cracks travel unstably at speeds which may approach the shear wave velocity of the host rock.

Equation (15) for brittle failure using K_{Ic} should reconcile with Griffith's energy balance (equation 9) (Irwin 1957). If no remote displacements are assumed, the fixed-grips case for crack propagation, then Irwin (1957) showed that the reduction of strain energy in the rock with respect to an increase in crack length is a measure of the energy available for crack propagation

$$G = -\frac{\partial U_E}{\partial c} \qquad (18)$$

where G is the energy release rate per unit length of crack tip. For a crack propagating in its own plane G is related to K at the crack tip (Irwin, 1957)

$$G = K^2 \left[\frac{(1-\nu^2)}{E} \right]. \qquad (19)$$

The propagation of a crack is resisted by a surface tension force, 2γ, where the cutting of such a crack requires the supply of an amount of energy equivalent to

$$dU_s = 2\gamma dc. \qquad (20)$$

For the fixed-grips case where $W_R = 0$, equations (18) and (20) may be substituted into the Griffith energy-balance equation (4)

$$dU_T = -Gdc + 2\gamma dc. \qquad (21)$$

At Griffith equilibrium $dU_T = 0$, crack propagation begins when $G = G_c$, so the critical energy release rate is related to the surface energy of the rock

$$G_c = 2\gamma. \qquad (22)$$

CRACK PROPAGATION AND PORE PRESSURE

Secor's (1965, 1969) classic model for jointing under the influence of pore pressure postulates that joints initiate from randomly orientated small cracks or flaws which are loaded internally by pore fluid within the rock mass. This form of *fluid-induced joint growth* is akin to fracture propagation during *oil well hydraulic fracturing* (OWHF) and is called *natural hydraulic fracturing* (NHF) (Engelder & Lacazette 1990). There are two important differences between an OWHF and NHF. First, to a rough approximation an OWHF boosts the internal pressure of a borehole (i.e. the initial flaw) without an accompanying increase in pore pressure in the rock behind the borehole wall. Although there is some infiltration through the mudcake on the wall of the borehole, infiltration is confined to the immediate vicinity of the borehole. The mudcake acts like an impermeable membrane. Prior to the initiation of NHF, internal pressure within the initial flaw increases at the same rate as pore pressure in the rock behind the wall of the flaw. For an OWHF a pore pressure increase would decrease breakdown pressure. In contrast, if poroelastic behaviour applies for extension of a penny-shaped flaw during NHF, an increase in pore pressure adds to the load on the crack walls and, in a sense, increases the internal pressure necessary for vertical crack initiation. Second, the crack driving stress for OWHF does not drop abruptly after crack initiation because the borehole is continually charged from surface pumps. In contrast, crack driving stresses for NHF drop either immediately or soon after crack initiation (Secor 1969). Both of these differences impact on our understanding of NHF.

The following are reasons for making the assumption that $P_i = P_p$ at the initiation of NHF. In sedimentary rocks, even with a very low permeability, fluid pressure in one pore will tend to equilibrate with pressure in surrounding pores, provided that the pores are interconnected. There is no known mechanism for suddenly increasing pressure in one pore relative to its neighbours as is the case for an OWHF where $P_i > P_p$. As abnormally high fluid pressures develop in sedimentary basins, pore pressure increases uniformly in the immediate vicinity of the incipient joint, and so the mechanism for the initiation of NHF must operate even though $P_i = P_p$.

The effect of poroelasticity

Price (in Fyfe *et al.* 1978) and more recently Gretener (1981) have correctly pointed out that Secor's model neglects the role of pore pressure (i.e. poroelasticity) in increasing the total stress on the crack wall. Although Fyfe *et al.* (1978) appear to recognize that Secor's (1965) mechanism is viable if the general law of effective stress (Nur & Byerlee 1971) is used in place of the simple law, their analysis needs clarification (Engelder & Lacazette 1990).

The poroelastic effect is visualized using a force-balance model (Fig. 3.3). Grain boundaries are characterized by the elastic grain–grain contacts and pore space between the grains. All the pores are connected so that the aggregate has a relatively high permeability. An initial flaw or small crack is introduced into the aggregate by a cut through the centre of the model. Two halves of the aggregate are compressed together in a container with rigid walls. Initially the aggregate is dry so that the rigid walls of the container press the initial flaw together with an average force per unit area, S_t. S_t represents a rock stress normal to the flaw. The aggregate is then filled with a pore fluid at a relatively low pressure P_p. If the face of the initial

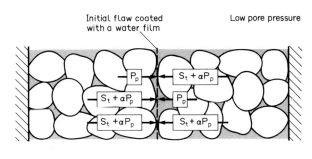

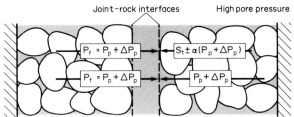

Fig. 3.3. Poroelastic model for a rock with an initial flaw and constrained by rigid boundaries on all sides. The model consists of grains, elastic grain–grain contacts, and interconnected pore space. Vectors are shown to represent the balance of forces along the initial flaw interface and the joint-rock interface. Symbols explained in text.

flaw is considered as an impermeable interface, the addition of pore fluid within the flaw would cause a uniform increase in the stress along the interface, yet the addition of pore fluid within the aggregate would not cause a uniform increase in stress from the aggregate side of the interface. Where pore fluid was in direct contact with the interface from the aggregate side, the pressure on the interface would increase by P_p. Where grains were in contact with the interface the normal stress on the interface would increase by a fraction, α, of P_p. This fractional increase in normal stress arises because the grain contacts are cemented and act like springs which take-up part of the force exerted by pore fluid inside pores. This partial transfer of pore pressure to the aggregate boundary is known as the *poroelastic effect* (Biot 1941). Without an impermeable interface, any increase in P_p within the initial flaw will act to open the initial flaw, but this increase is partially counterbalanced by a poroelastic expansion within the aggregate which acts to keep the initial flaw closed.

Too initiate crack propagation for both OWHF and NHF, internal fluid pressure must counterbalance the total least principal stress, σ_3 (i.e. $S_t + \alpha P_p$ in the model shown in Fig. 3.3). Poroelastic behaviour applies to the development of vertical joints, where the total least horizontal stress, S_h is equal to σ_3. In saturated rocks, total stress may be divided into two components: the stress carried by grain–grain contacts under dry conditions (i.e. S_t in Fig. 3.3) and stress generated by fluid pressure within the pore space of the rock (i.e. αP_p in Fig. 3.3). By the poroelastic effect an increase in P_p will cause an increase in total stress provided that the rock is constrained by rigid boundaries. One type of rigid boundary behaviour is called *uniaxial strain* which is a common model for horizontal strain in sedimentary basins (Geertsma 1957):

$$\varepsilon_H = \varepsilon_h = 0, \quad (23a)$$

$$\varepsilon_v \neq 0 \quad (23b)$$

ε_H and ε_h are principal strains in the horizontal direction. The effect of a change in pore pressure in sedimentary basins may be seen by solving Biot's (1941) elasticity equations for uniaxial strain. Biot's elasticity equations are given by Rice & Cleary (1976) as

$$2\,\Gamma \varepsilon_{ij} = \langle \sigma_{ij} \rangle - \frac{\nu}{1+\nu} \langle \sigma_{kk} \rangle \delta_{ij} \quad (24a)$$

where

$$\langle \sigma_{ij} \rangle = \sigma_{ij} + \alpha P_p . \quad (24b)$$

In Biot's equations, α is Biot's poroelastic term defined as $\{1 - C_i/C_b\}$ with the intrinsic compressibility of the uncracked solid, C_i (i.e. the compressibility of the solid grains in Fig. 3.2), and the bulk compressibility of the solid with cracks and pores, C_b (i.e. the compressibility controlled largely by the 'springs' in Fig. 3.3) (Nur & Byerlee 1971). Γ and ν are the shear modulus and Poisson ratio of the rock when it is deformed under 'drained' conditions.

For this analysis of vertical joints and veins, the crack-normal stress is the least horizontal stress, S_h. Little is known about the least horizontal stress in a sedimentary basin where vertical joints are forming except the obvious; the total horizontal stress, $S_h < S_v$. The total vertical stress is

$$S_v = \rho_t g z \quad (25)$$

where ρ_t is the integrated density of the rock to the depth, z, of interest and g is the acceleration of gravity. Although the state of stress was probably more complicated, I make the simplifying assumption that S_h was equal to that found in a tectonically relaxed basin.

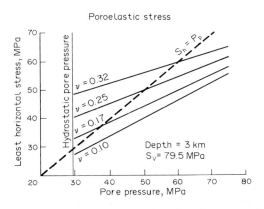

Fig. 3.4. Relationship between S_h and P_p in a tectonically relaxed basin assuming poroelastic behaviour. In a tectonically relaxed basin, horizontal stresses are due solely to the overburden load. The poroelastic effect is strongly dependent on Poisson's ratio, ν, as indicated by the four curves for various values of ν. This calculation assumes conditions at a depth of burial of 3 km.

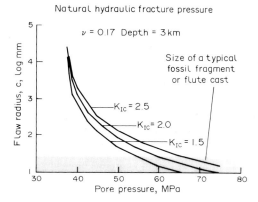

Fig. 3.5. Relationship between flaw or crack length and pore pressure required to initiate cross-fold joint propagation in a tectonically relaxed basin assuming poroelastic behaviour. The flaw length at initiation of propagation is moderately dependent on the K_{Ic} of the rocks within which crack propagation takes place. This calculation assumes a penny-shaped flaw, a burial depth of 3 km and a ν of 0.17 for a siltstone.

A tectonically relaxed basin is one in which S_h is proportional to S_v through the uniaxial elastic strain model:

$$S_h = \frac{\nu}{1-\nu} S_v . \qquad (26)$$

Solving Biot's elasticity equations for uniaxial strain

$$S_h = \frac{\nu}{1-\nu} S_v + \frac{(1-2\nu)}{(1-\nu)} \alpha P_p . \qquad (27a)$$

Terms in this equation may be rearranged to appear in the same form as equation (13) derived by Anderson *et al.* (1973) for fracture pressure at a borehole

$$S_h = \frac{\nu}{1-\nu} (S_v - \alpha P_p) + \alpha P_p . \qquad (27b)$$

The first term on the right-hand side of equation (27b) is equivalent to but not the same as S_t in Fig. 3.3.

The internal fluid pressure, P_i, necessary to initiate crack propagation and thereby cause NHF is a function of several general parameters including total rock stress normal to the crack plane, S_h, and the elastic properties of the rock. To understand the variation of P_i, it is appropriate to consider the variation of S_h due to the poroelastic effect in a tectonically relaxed basin (Fig. 3.4). The calculations for Fig. 3.4 assume 3 km of overburden having a density of 2.7 g/cc so that $S_v = 79.5$ MPa. An α of 0.7 is also assumed. S_h can vary by as much as 50% of the overburden weight depending on ν and P_p. A larger S_h is generated in rock with a higher ν. The field of interest in Fig. 3.4 is defined by $P_p > S_h$ for it is within this field that NHF occurs. In rocks with a very low ν, conditions favouring NHF occur even at hydrostatic pore pressure whereas for rocks with a high ν, conditions for NHF are suppressed until a much higher P_p.

The initiation of NHF

For several reasons it is not intuitively obvious that a net tensile stress (i.e. a crack driving stress) is generated along an initial flaw: (1) a compressive stress, S_h, increases as a function of P_p, (2) pores of the rock behind the flaw are also subject to the same pressure, and (3) fluids can readily drain from the flaw to the pore space. The poroelastic behaviour of rock, for which $\alpha < 1$, is responsible for the generation of a net tensile stress against the face of a flaw.

In addition to total rock stress, the internal fluid pressure, P_i necessary to initiate joint propagation is a function of the fracture toughness of the rock, K_{Ic}, the crack length, $2c$, and the shape of the crack, Y. K_I, a measure of the stress concentration at a crack tip, increases with an increase in net tensile stress on the crack. K_{Ic}, the *critical stress intensity factor* or *fracture toughness* is a material property that indicates the ease with which a rock will fracture. K_{Ic} is a laboratory measure of the pull normal to a crack plane at the time the crack tip propagates rapidly. *Joint initiation* occurs only when the crack (i.e. initial flaw) walls are pulled apart or subject to a net tensile stress as a consequence of the poroelastic effect. The linear elastic fracture mechanics equation for the rapid growth of a joint is

$$P_i = \left\{ \frac{K_{Ic}}{Yc^{1/2}} \right\} . \qquad (28a)$$

This is the condition if the walls of the crack are not supported by an earth stress. If the walls of the crack are forced closed by S_h, then

$$P_i = \left\{ \frac{K_{Ic}}{Yc^{1/2}} \right\} + \frac{\nu}{1-\nu} S_v + \frac{(1-2\nu)}{(1-\nu)} \alpha P_p . \qquad (28b)$$

As equation (28b) indicates, P_i can vary significantly depending on the size of the pre-existing crack. In Fig. 3.3 the difference between the length of the vector for normal stress across a grain and the vector for fluid pressure is the vector for $K_{Ic}/Yc^{1/2}$.

The equations developed above may be used in conjunction with measured rock properties to constrain the stress and pore conditions under which a given set of natural hydraulic fractures initiated. Under the assumption that joints formed by NHF started propagating rapidly from small cracks or flaws when $P_i = P_p$, equation (28b) may be rewritten to give an indication of flaw length leading to initiation of joints.

$$c = \left[\frac{K_{Ic}}{Y \left\{ P_i - \frac{\nu}{1-\nu} S_v - \frac{(1-2\nu)}{1-\nu} \alpha P_p \right\}} \right]^2 . \qquad (29)$$

To illustrate the initiation and propagation of joints under high fluid pressures, the properties of a siltstone from the Appalachian Plateau are used (Engelder & Lacazette 1990). Typically K_{Ic} of rocks varies between 2.5 MPam$^{1/2}$ and 1.5 MPam$^{1/2}$ (Atkinson 1984).

Laboratory measurements of ν for siltstones of the Appalachian Plateau suggest that $\nu = 0.17$ is reasonable (Evans et al. 1989). All flaws in the siltstone are assumed to be penny-shaped cracks which have $Y = 1.13$. Using equation (29) the flaw radius for NHF initiation within the siltstone varies as a function of pore pressure at a depth of 3 km. These calculations assume three arbitrary values of K_{Ic} for joint initiation (2.5, 2.0, and 1.5 MPam$^{1/2}$). (Fig. 3.5). Figure 3.5 shows that joints will initiate from larger flaws at lower P_p.

At an early stage in their development rocks have no large joints but contain either microcracks in the form of pore space and grain boundaries, or flaws in the form of fossil and/or rock fragments and sedimentary structures such as flute casts. Unfractured siltstone has two types of flaws: grain boundary microcracks and larger structures such as flute casts, concretions, and fossil fragments. Grain-boundary microcracks are on the scale of individual grains less than 0.1 mm in diameter. In contrast, fossil fragments and flute casts are roughly 1–3 cm in diameter. The plumose surface morphology on the surface of cross-fold joints in the siltstone from the Appalachian Plateau allows the joint propagation to be traced back to origin flaws which are commonly 1–3 cm structures. From this observation in flaw size, $2c$, is known for the initiation of NHF. Assuming conditions in a tectonically relaxed basin at a depth of 3 km, pore pressure at the initiation of NHF was on the order of 65 MPa or higher.

Figures 3.5 and 3.6 illustrate several points concerning the effect of both ν and K_{Ic} on the flaw length for the initiation of joints. First, grain boundary microcracks are too small to account for the initial propagation of NHF, even if siltstones have an extremely low ν. Second, abnormal pore pressures, significantly above hydrostatic, were necessary for the initiation of joints at depth. Third, at $\nu = 0.17$, typical fossil fragments or flute casts were large enough flaws

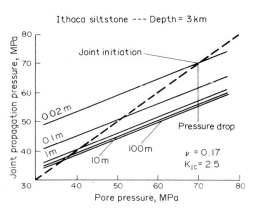

Fig. 3.7. Relationship between crack propagation pressure and pore pressure for cross-fold joints of lengths between 0.02 and 100 m. This calculation assumes a tunnel crack, a burial depth of 3 km, a K_{Ic} of 2.5 MPam$^{1/2}$, and a ν of 0.17 for a siltstone. See text for details.

to favour the initiation of vertical joints at 3 km. Fourth, in a bedded siltstone–shale sequence, the initiation of NHF is favoured in a rock with a lower ν (i.e. a siltstone) relative to a rock with higher ν (i.e. a shale).

Once joints have initiated from small flaws, less severe internal pressures are necessary for further growth. Equation (29) also gives the incremental crack propagation pressure. Suppose that a vertical joint initiates from an initial flaw with a radius of about 1 cm. At initiation $P_p = P_i = 68$ MPa (Fig. 3.5). By the time a joint has run to a length of 30 cm, an internal pressure 20 MP less than P_p is required for reinitiation of joint propagation. Reinitiation pressure decreases to approach the total stress normal to the crack wall. Figure 3.7 illustrates the difference between crack initiation pressure and crack propagation pressure. Crack initiation pressure is indicated by the dashed line cutting across the propagation pressure – pore pressure lines for joints of various lengths. Once a joint propagates to a length of more than one metre, the crack propagation pressure changes very little. However, that pressure may be more than 20 MPa less than the crack initiation pressure.

OTHER DETAILS OF THE MECHANICS OF JOINTING

Joint-propagation path

The *joint-propagation path* is that direction which the rupture takes as it leaves the present *joint front or tip* of the crack. The joint-tip stress field, as given in equation (14), controls the joint propagation path at the joint front. One method for charting the crack path through a rock is to consider the stress intensity for small increments of growth. To a first approximation, the crack will propagate in a direction normal to the least principal stress. This is the orientation which will produce the maximum propagation energy, G (Pollard

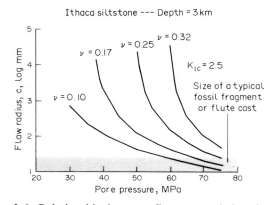

Fig. 3.6. Relationship between flaw or crack length and pore pressure required to initiate crack propagation within a siltstone. The flaw length at initiation of cross-fold joints is strongly dependent on the Poisson's ratio (ν) of the rocks within which crack propagation takes place. This calculation assumes a penny-shaped flaw, a tectonically relaxed basin with a poroelastic response to changes in pore pressure, a burial depth of 3 km, and a K_{Ic} of 2.5 MPam$^{1/2}$ for a siltstone.

& Aydin 1988). If the crack is subject to pure opening mode loading (i.e. stress normal to the crack plane on a *Mode 1* crack) each increment of crack propagation will be in the plane of the initial crack. As long as the orientation of the extensional stress field does not change at the tip of the crack a planar joint is produced. If the crack is subject to a shear couple, then each propagation increment will turn and the crack will continue in a path so that the subsequent crack plane is normal to the extension direction. Such a turn is known as out-of-plane propagation of which there are two types, a twist and a tilt. A *tilt* is a turn of the crack rupture about an axis parallel to the joint front and this occurs when the crack is subject to *Mode 2* loading. A *twist* takes place as the joint front turns about an axis perpendicular to the joint front. During a tilt the joint maintains its continuity but during a twist the joint breaks into en échelon segments.

Joint arrest

Joint arrest is best modelled by considering the energy release rate per unit crack propagation, G (equation 19). The energy release rate is proportional to the driving stress and the square root of the crack length. There is a critical value of *G* for which crack propagation will take place, G_c. Some authors call this critical rate the *crack growth resistance, R* (Broek 1987). Once $G = G_c$, the crack will continue to propagate unless the driving stress decreases. In the case of a fluid driven crack the volume increase of the crack is proportional to the crack opening displacement or COD

$$\text{COD} = \frac{4\sigma(1-\nu^2)c}{E}. \quad (30)$$

As the volume of the crack increases with crack tip displacement the pressure of the fluid within the crack must decrease as indicated by the equation of state of the fluid. Propagation will arrest once the driving pressure has dropped, so that *G* is less than *R*.

Mechanical interaction

A very common outcrop pattern consists of two joints overlapping each other slightly. Each joint is planar until the joints approach closely, at which point the joints may tend to diverge and then rotate sharply toward each other. This pattern reflects the mechanical interaction between joints as their tips propagate toward each other. As joint tips approach, each joint induces a tensile stress in the vicinity of its neighbour's tip and therefore enhances the propagation energy of the neighbour. At the same time each joint induces a shear stress in the neighbour's propagation plane. This shear stress will at first cause the neighbours to deflect away from each other. As the en échelon joints pass, each joint induces a compressive stress field with shear stress oriented so that the joints will then propagate toward each other (also see Chapter 5).

This same type of mechanical interaction is seen for joints of different lengths which are completely overlapping. A longer joint will shield the shorter joint from the overall tensile stress field which is driving the joint. As the length of the longer joint increases the tensile stress on the neighbouring shorter joint decreases by this shielding effect. In an outcrop of many joints a few long joints will prevent the extension of many neighbouring shorter joints. This mechanical interaction is largely responsible for joint spacing within rocks.

CONCLUSIONS

Recent advances in the understanding of brittle deformation processes include the application of linear elastic fracture mechanics to geological problems. This review illustrated this application with examples concerning the effect of high fluid pressure on crack propagation in rocks.

Acknowledgements — This paper was written in conjunction with support from the Gas Research Institute under contract # 5088-260-1746. Alfred Lacazette is thanked for his review of an early version of this paper.

REFERENCES

Anderson, O. L. & Grew, P. C. 1977. Stress corrosion theory of crack propagation with applications to geophysics. *Rev. Geophys. Space Phys.* **15**, 77–104.

Anderson, R. A., Ingram, D. S. & Zanier, A. M. 1973. Determining fracture pressure gradients from well logs. *J. Petrol. Technol.* **26**, 1259–1268.

Atkinson, B. K. 1984. Subcritical crack growth in geological materials. *J. geophys. Res.* **89**, 4077–4114.

Atkinson, B. K. (ed.) 1987. *Fracture Mechanics of Rock*. Academic Press, Orlando.

Badgley, P. C. 1965. *Structural and Tectonic Principals*, Harper & Row, New York.

Biot, M. A. 1941. General theory of three-dimensional consolidation. *J. appl. Phys.* **12**, 155–164.

Brace, W. F. & Bombolakis, E. G. 1963. A note on brittle crack growth in compression. *J. geophys. Res.* **68**, 3709–3713.

Brace, W. F. & Kohlstedt, D. L. 1980. Limits on lithospheric stress imposed by laboratory experiments. *J. geophys. Res.* **85**, 6248–6252.

Broek, D. 1987. *Elementary Engineering Fracture Mechanics*. Martinus Nijhoff, Dordrecht, The Netherlands.

Bucher, W. H. 1920. The mechanical interpretation of joints. *J. Geol.* **28**, 707–730.

Byerlee, J. 1978. Friction in rocks. *Pure appl. Geophys.* **116**, 615–626.

Engelder, T. 1982. Is there a genetic relationship between selected regional joints and contemporary stress within the lithosphere of North America? *Tectonics* **1**, 161–177.

Engelder, T. & Lacazette, A. 1990. Natural hydraulic fracturing. In *Rock Joints* (edited by Barton, N. & Stephansson, O.). A. A. Balkema, Rotterdam, 35–44.

Engelder, T. & Marshak, S. 1988. The analysis of data from rock-deformation experiments. In: *Basic Methods of Structural Geology* (edited by Marshak, S. & Mitra, G.). Prentice-Hall, Englewood Cliffs, 193–212.

Etheridge, M. A. 1983. Differential stress magnitudes during regional deformation and metamorphism: upper bound imposed by tensile fracturing. *Geology* **11**, 231–234.

Evans, K., Oertel, G. & Engelder, T. 1989. Appalachian Stress Study 2: Analysis of Devonian shale core: Some implications for the nature of contemporary stress variations and Alleghanian deformation in Devonian rocks. *J. geophys. Res.* **94**, 7155–7170.

Fyfe, W. S., Price, N. J. & Thompson, A. B. 1978. *Fluids in the Earth's Crust*. Elsevier, New York.

Gallagher, J. J., Friedman, M., Handin, J. & Sowers, G. 1974. Experimental studies relating to microfracture in sandstone. *Tectonophysics* **21**, 203–247.

Geertsma, J. 1957. The effect of fluid pressure decline on volumetric changes of porous rock. *Trans. AIME* **210**, 331–339.

Griffith, A. A. 1920. The phenomena of rupture and flow in solids. *Phil. Trans. R. Soc. Lond.* **A221**, 163.

Griffith, A. A. 1924. Theory of rupture. *Proc. First Int. Cong. appl. Mech. Delft*, 55–63.

Gretener, P. E. 1981. *Pore Pressure: Fundamentals, General Ramifications, and Implications for Structural Geology*. American Association of Petroleum Geologists Educational Course Note Series **4**, 1–131.

Hancock, P. L. & Engelder, T. 1989. Neotectonic Joints. *Geol. Soc. Am. Bull.* **101**, 1197–1208.

Handin, J. & Hager, R. V. 1957. Experimental deformation of sedimentary rocks under confining pressure: tests at room temperature on dry samples. *Bull. Am. Assoc. Petrol. Geol.* **41**, 1–50.

Inglis, C. E. 1913. Stresses in a plate due to the presence of cracks and sharp corners. *Trans. Inst. Naval. Archit.* **55**, 219.

Irwin, G. R. 1957. Analysis of stress and strains near the end of a crack traversing a plate. *J. appl. Mech.* **24**, 361–364.

Lawn, B. R. & Wilshaw, T. R. 1975. *Fracture of Brittle Solids*. Cambridge University Press, Cambridge.

McClintock, F. A. & Walsh, J. B. 1962. Friction of Griffith cracks under pressure. *4th U.S. nat. Cong. appl. Mech. Proc.*, 1015–1021.

Nickelsen, R. P. & Hough, V. D. 1967. Jointing in the Appalachian Plateau of Pennsylvania. *Bull. geol. Soc. Am.* **78**, 609–630.

Nur, A. & Byerlee, J. D. 1971. An exact effective stress law for elastic deformation of rocks with fluids. *J. geophys. Res.* **76**, 6414–6419.

Pollard, D. D. & Aydin, A. 1988. Progress in understanding jointing over the past century. *Bull. geol. Soc. Am.* **100**, 1181–1204.

Rice, J. R. & Cleary, M. P. 1976. Some basic diffusion solutions for fluid-saturated elastic porous media with compressible constituents. *Rev. Geophys. Space Phys.* **14**, 227–241.

Rossmanith, H. P. (ed.) 1983. *Rock Fracture Mechanics*. Springer, New York.

Scheidegger, A. E. 1982. Comment on "Is there a genetic relationship between selected regional joints and contemporary stress within the lithosphere of North America?" by T. Engelder. *Tectonics* **1**, 463–464.

Scholz, C. H. 1968. Microfacturing and the inelastic deformation of rock in compression. *J. geophys. Res.* **73**, 1417–1432.

Secor, D. T. 1965. Role of fluid pressure in jointing. *Am. J. Sci.* **263**, 633–646.

Secor, D. T. 1968. Mechanics of natural extension fracturing at depth in the earth's crust. *Geol. Surv. Pap. Can.* **68–52**, 3–48.

Voight, B. & St. Pierre, B. H. P. 1974. Stress history and rock stress. *Adv. Rock Mech. Proc 3rd Cong. ISRM* **II**, 580–582.

CHAPTER 4

Fault Slip Analysis and Palaeostress Reconstruction

J. ANGELIER

INTRODUCTION

FAULTS are fractures which separate blocks across which (1) the total displacement is visible to the naked eye, and (2) most of the displacement occurs parallel to the fault plane. This definition has two consequences which deserve attention. First, the naked-eye criterion implies that fractures of identical origin, but with microscopic displacement, can exist, and had the movement on them been larger, they would have been called faults (Fig. 4.1). Thus, so-called shear joints (for discussion of this controversial nomenclature see Chapter 5) differ from faults in that the latter generally bear slickenside lineations (Fig. 4.2) or other indications of finite displacement. As a result, the total relative motion of fault blocks can be accurately reconstructed in terms of direction and sense (even sometimes in terms of amplitude and path), whereas *joints* display none of these features. Hence, the observation and analysis of fault sets commonly provides an unambiguous, albeit in many cases complicated, key to understanding the deformation of the brittle upper crust.

Second, the dominance of the component of motion parallel to the fracture plane, inherent in the definition of a fault, does not imply that no other component of motion can occur across the fault surface. Some faults display significant components of either dilatation or convergence across their surfaces (Fig. 4.3), or even both motions in the case of irregular fault surfaces (Fig. 4.4). As a result, instead of considering faults, veins, pressure solution seams and joints (see Chapter 5) as independent features, we can view them as phenomena belonging to an array of brittle structures (Fig. 4.5): pure faults (i.e. surfaces with shear motion on them and without a significant component of motion perpendicular to the fracture plane) denote simple shear, while pure *dilatant veins* and *pressure solution seams* (i.e. surfaces on which there was no significant component of motion parallel to the structure plane), respectively, yield extension and compression directions. In addition to these end members, numerous intermediate ones exist, such as faults with noticeable opening, veins with some lateral displacement, and so on.

In terms of detailed geological observation in outcrop, faults and veins are easily identifiable due to the presence of slickenside lineations or offsets (for faults) and mineral infills (for veins). In contrast, as discussed in Chapter 5, the genetic classification of fractures without any perceivable macroscopic displacement on them is a controversial issue (see e.g. Hancock 1985).

COMPONENTS OF FAULT SEPARATION AND CLASSIFICATION OF FAULTS

A *fault* is a discontinuity which results from slip on a narrow zone. As a consequence, there is generally some mismatch in geological structure across a fault. This mismatch, which is used to recognize and locate fault zones, is commonly obvious due to differences in lithology on either side of the fault. A common consequence of these differences is a contrast in morphology between the two sides of a fault, due to (1) their vertical relative displacement, and (2) different erosion rates in different rock formations.

Fault morphology

Figure 4.6 illustrates three cases which are commonly observed in the field. In the first case (Fig. 4.6a), denudation has been negligible (or the same on both sides of the fault); this is a common situation along recent faults which displace bedrocks, because erosional phenomena cannot modify the scarp during a short time span (see Chapter 18). In the second case (Fig. 4.6b), the upthrown side of the fault is still the highest one despite denudation; however, the difference in elevation and the vertical displacement can differ widely, due to differences in erosion rates which depend on relief distribution, and on variations in lithology. In the third case (Fig. 4.6c), the upthrown side of the fault is now the lowest because erosion was very active in soft rocks after destruction of the uplifted hard layer. These examples demonstrate that there is no *a priori* relationship between the *facing direction* or height of a fault scarp, and the vertical displacement of the blocks.

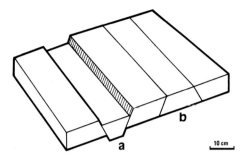

Fig. 4.1. Contrast between faults (a) and so-called shear joints (b) that are mechanically consistent with each other. Faults exhibit signs of macroscopic displacement, whereas joints do not.

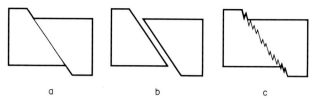

Fig. 4.3. Faults with pure slip (a), vein opening (b) and pressure solution (c). Although the dip and apparent offset of the fault are the same in each case, (b) and (c), respectively, imply more, and less, horizontal lengthening than that caused by the simple slip shown in (a).

Although relations between vertical displacement and fault relief are complex (Fig 4.6), the presence of faults is usually revealed by their morphological expression. In addition, fault zones are generally much more eroded than surrounding rocks, because fault-rocks are weaker, and because underground water circulation is generally concentrated along such discontinuities and results in weathering, which makes rocks still weaker. As a consequence, many faults, especially major ones, are easily detected in the landscape; for instance, there are commonly valleys along fault zones. However, they are in turn difficult to observe in outcrops, because they are commonly hidden by valley infills and soils. In tectonic studies, it is common to infer the fault-slip geometry of a large fault (which cannot be observed directly) from the observation of minor faults in neighbouring outcrops, assuming that the mechanisms are similar, despite differences in size.

General constraints on slip orientation

The present surface morphology of an old fault zone can reveal little about the slip direction. This is illustrated in Fig. 4.7, which depicts a fault scarp similar to that shown in Fig. 4.6(a). The same scarp may correspond either to a fault without significant lateral displacement (Fig. 4.7a) or to a fault where lateral displacement prevails (Fig. 4.7b). However, the morphology of *active fault zones* that move during earthquakes (sudden slip) or slowly (progressive slip, or creep), can provide valuable information on slip orientation (see also Chapter 18). For instance, with a similar vertical displacement, the surface expression of an active fault can differ widely depending on the absence (Fig. 4.8a) or the presence (Fig. 4.8b) of a lateral component of motion. Figure 4.9 is a simple way of illustrating the relationship between shear motion and the opening of *en échelon tension cracks*, as, for example, those shown in Fig. 4.8(b) (also see Chapter 5 for additional details).

In some cases, the orientation of the slip vector on a major fault is determined without ambiguity because there are enough geometric constraints. This is the case shown in Fig. 4.10, where the offset of horizontal layers indicates the amount of vertical displacement, while the amount of lateral displacement is revealed by an offset vertical dyke. In more complex situations (dipping layers, oblique dykes, and so on), this type of reasoning can allow complete determination of fault slip geometry provided that a unique line can be identified on both sides of the fault. In Fig. 4.10, this line is the intersection between the dyke and a given bed. In other situations, it may be related to unconformities, volcanic intrusions, fossil reefs, dunes, etc. In any case, the identification on both sides of a fault of a single plane (e.g. a layer or a dyke) does not allow the determination of the slip orientation although it constrains one of its components (e.g. the vertical offset of an horizontal layer or the lateral displacement of a vertical dyke).

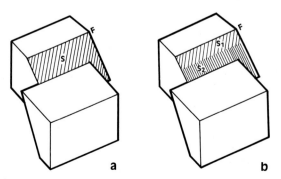

Fig. 4.2. Reconstruction of fault slip from slickenside lineations (s) on fault plane (F). (a) Unique fault slip. (b) Successive slips (s_1 then s_2) on a single fault.

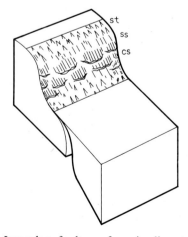

Fig. 4.4. Irregular fault surface in limestone showing stylolitic striae (st) where pressure solution occurs, calcite steps (cs) where dilatation occurs, and simple slickenside lineations (ss) in intermediate segments with simple shear.

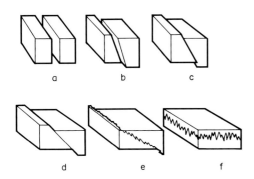

Fig. 4.5. Variety of brittle structures with macroscopic displacement. (a) Pure dilatation (vein). (b) Oblique dilatation. (c) and (d) Simple slip (with increasing friction). (e) Oblique stylolites. (f) Pure convergence (stylolite). Compression vertical, extension from right to left.

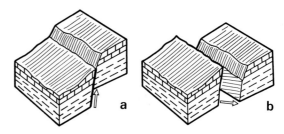

Fig. 4.7. Similar morphological expression of a fault despite major differences in slip geometry. (a) Fault without lateral movement. (b) Fault with lateral displacement greater than vertical displacement. Vertical displacements identical. Note that after erosion scarps are similar. Arrows: relative displacement of block on the right.

Slip orientation and slickenside lineations

A convenient way to determine the slip orientation on faults is provided by the observation of *slickenside lineations* on fault surfaces. Most fault surfaces are striated, and the attitude of *striae* reveals the orientation of slip. Furthermore, in most cases, various asymmetric features associated with slickenside lineations reveal the sense of relative motion (see discussion later in this chapter and in Chapter 5). Thus, the geometry of a fault can be entirely described at one point through measurement of three angles and one distance, as follows (Fig. 4.11).

(a) The first angle, d, describes the *dip direction of the fault plane* (Fig. 4.11a). This angle is an azimuth, measured from the north and positive clockwise. It consequently ranges from 0 to 360°. For instance, for a fault that strikes N15°E and dips to the west, $d = 285°$.

(b) The second angle, p, is the *dip of the fault plane* (ranging from 0 to 90°). The two values d and p entirely describe the orientation of the fault plane (Fig. 4.11a).

(c) The third angle, i, is the *pitch of the slickenside lineation* which indicates slip direction (Fig. 4.11b). Because relative motion can occur in opposite senses, this angle ranges from 0 to 360° and a sense convention is required. I propose for example that the angle i is measured positive counterclockwise from the right side of the fault plane dipping towards the observer (Fig. 4.11b).

(d) The *distance* is any component of the total fault displacement, as discussed later.

These definitions (Fig. 4.11) imply that the sense of dip of the fault, as well as the sense of slip, are entirely constrained by three numerical values (i.e. d, p and i). However, for convenience, measurements carried out in the field are commonly more complicated as they include alphabetic indications rather than pure numerical values. For instance, a fault is said to have a N72°E strike, a dip of 48°N, a pitch of 27°W and to be reverse-sinistral. The corresponding values of d, p and i, which describe this fault slip are 342°, 48° and 207°. These values are more suitable for additional calculations, whereas data collection in the field is made easier and safer with an alphanumeric system.

The relationship between the pitch of the slickenside lineation considered above (Fig. 4.11) and the sense of motion on the fault, as usually described, is summarized in Fig. 4.12: the upper rows of diagrams in the figure corresponding to normal slips, the lower rows to reverse ones. The column on the left corresponds to left-lateral (sinistral) slips, the column on the right to right-lateral

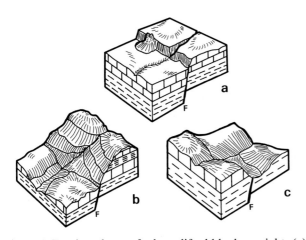

Fig. 4.6. Erosion along a fault, uplifted block on right. (a) Scarp height and facing direction consistent with the vertical relative displacement of blocks. (b) Deeper erosion of the downthrown block, resulting in a fault scarp higher than the vertical displacement. (c) Deeper erosion in the uplifted block, after denudation of a hard layer, so that the highest side of the fault corresponds to the downthrown block (i.e. relief inversion).

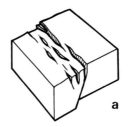 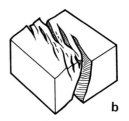

Fig. 4.8. Differences in surface rupture expression related to differences in slip along active faults. Vertical displacement is the same in both cases. (a) Fault without lateral displacement. (b) Fault with significant lateral displacement. Note that in (b) a systematic en échelon arrangement of open surface fractures and scarps is present.

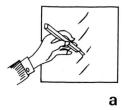 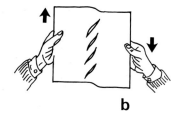

Fig. 4.9. Simple experiment illustrating the opening of en échelon cracks in relation to a shear couple. (a) The en échelon gashes are cut in a sheet of paper. (b) To open the gashes it is necessary to move the sides parallel to the arrows.

(dextral) ones. The faults at the four corners of the figure are oblique-slip, whereas the central situations are dip-slip (normal on top, reverse at base) or strike-slip (dextral on right, sinistral on left). One finally observes that the slip sense is related to increasing values of the angle i, as follows:

$i = 0°$, pure normal dip-slip;
$0° < i < 90°$, normal-dextral oblique-slip;
$i = 90°$, pure dextral strike-slip;
$90° < i < 180°$, reverse-dextral oblique-slip;
$i = 180°$, pure reverse dip-slip;
$180° < i < 270°$, reverse-sinistral oblique-slip;
$i = 270°$, pure sinistral strike-slip;
$270° < i < 360°$, normal-sinistral oblique-slip;

Components of fault displacement

The relationships between the different components of slip are illustrated in Fig. 4.13. The *total displacement (net separation)* on a fault is characterized by a vector **D**, which has three perpendicular components (along fault strike, along the dip direction, and vertical). **V** is the *vertical offset*, **T** the *transverse horizontal displacement* and **L** the *lateral horizontal displacement*. By definition, in vectorial notation, **D** = **T** + **V** + **L**. Incidentally, one may distinguish the component of displacement along the slope, **S** (**S** = **T** + **V** and **S** + **L** = **D** in vectorial notation). Let p be the fault dip (as in Fig. 4.11a) and i_0 the pitch ranging from 0 to 90°

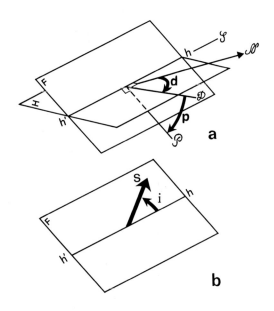

Fig. 4.11. Complete description of fault slip orientation with three angles d, p and i ($0 \leq d < 360$, $0 \leq p < 90$, $0 \leq i < 360$, in degrees). (a) Orientation of fault plane. (b) Orientation of the slip vector in the fault plane. In (a): H, horizontal plane (for reference). F, fault plane. Line $h-h'$ is the direction of the fault (corresponding to strike, S, relative to north, N: this fault strikes approximately N20°W). P indicates the slope of the fault plane, D is the dip direction. D is perpendicular to S in the horizontal plane, H: this fault dips approximately 40° towards N70°E. Angles d and p completely describe fault orientation (in this case, $d = 70°$ and $p = 40°$; see text for details). In (b) the fault is same as in (a). The slip vector describes the direction and sense of slip of the lower block relative to the upper one, and the corresponding angle, i, is counted positive counterclockwise from the 'right side', h, of the faut plane. In this case, i averages 35°, meaning that the sense of fault slip is similar to that shown in Fig. 4.10, that is, both normal and dextral (also see Fig. 4.12).

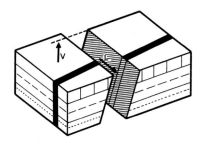

Fig. 4.10. Vertical and lateral components of displacement on an oblique-slip fault, as revealed by the vertical offset of horizontal layers and the horizontal displacement of a dyke (black). Arrows define the vertical (V) and lateral (L) components of the relative motion of the block on the right.

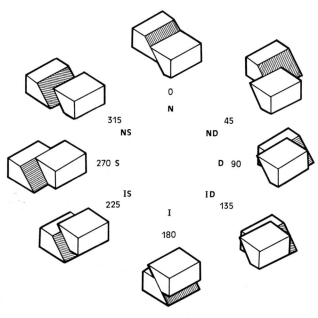

Fig. 4.12. Typical geometries of fault slip, for fault planes with the same dip. Numerical values correspond to the angle i as defined in Fig. 4.11. Capitals refer to the sense of slip (N, normal; I, reverse; D, dextral; S, sinistral).

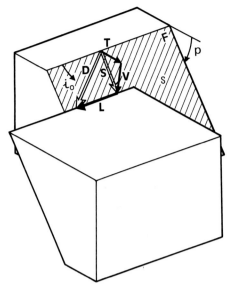

Fig. 4.13. Components of fault slip: D, total displacement (net separation). S, displacement along slope. T, transverse horizontal component of dispacement. V, vertical offset. L, lateral horizontal component of displacement. F, fault plane. s, slickenside lineation. p, fault dip. i_0, pitch of slickenside lineations, from 0 to 90° (it differs from i as defined in Figs 4.11 and 4.12, an angle which ranges from 0 to 360°). Sense of arrows (D, S, T, V and L) refer to relative movement of downthrown block.

regardless of sense (Fig. 4.13, not as in Fig. 4.11b where the angle i is measured from 0 to 360°).

One thus observes that the components of displacement depend on p and i_0, so that $V = D \sin i_0 \sin p$, $T = R \sin i_0 \cos p$, $L = R \cos i_0$ $V = S \sin p$, and so on. As a consequence, any component of the total displacement can be used to determine the other components provided that the angles p and i_0 are known. For instance, knowing either the vertical offset (as indicated by bedding in Fig. 4.10) or the lateral displacement (the dyke offset shown in Fig. 4.10) is sufficient to determine all characteristics of fault slip, taking angles i_0 and p into account.

Tectonic regimes in relation to fault types

The components **T**, **V** and **L** of the total displacement of a fault play a major role in terms of deformation style in brittle tectonics, because both vertical deformation (uplift versus subsidence) and the horizontal deformation (lengthening versus shortening) are considered. It is important to observe, at this stage, that the transverse horizontal component may be positive (lengthening and normal faulting) or negative (shortening and reverse faulting). Likewise, the lateral component may be dextral or sinistral. These senses were implicit when a pitch in the 0–360° range was used, (i, as discussed before: see Fig. 4.12) but they are not with the pitch, i_0, in the 0–90° range (as in Fig. 4.13). Pure dip-slip normal faulting (i.e. $i = 90°$) corresponds to crustal thinning, and consequently to elongation in the horizontal plane. Conversely, pure dip-slip reverse faulting ($i = 270°$) corresponds to crustal thickening, and to shortening in the horizontal plane. Pure strike-slip ($i = 0$ or 180°) results in plane horizontal deformation without variation in crustal thickness (at least in theory). Oblique-slip faulting generally involves more complex deformation. For these reasons, it is worthwhile to consider in a statistical way the different components of fault displacements in a given region.

Figure 4.14 shows a simple way to estimate the relative contribution of a given fault to vertical and horizontal deformation (Angelier 1979a). For instance, in Fig. 4.14(a), for a constant pitch of 45°, the magnitude of the strike-slip component relative to vertical offset increases dramatically when the fault dip decreases from 60° or more (V and L nearly equal) to 30° or less (L is larger than twice the vertical offset). In a similar way, Fig. 4.14(b) shows that the ratio between the horizontal transverse slip (i.e. the component of shortening perpendicular to the strike of a reverse fault or of lengthening perpendicular to that of a normal fault) and lateral slip (i.e. the dextral or sinistral component of motion) depends on the same angles. For example, for the same constant pitch of 45°, the lateral slip increases dramatically when the fault dip increases from 30° or less (T and L nearly equal) to 60° or more (L is larger than $2L$ and close to $8L$ for a fault dip of 80°).

Classification of fault slip directions

Taking into account these relationships (Fig. 4.14), as well as the fault slip senses already discussed (Fig. 4.12), it is possible to present a classificiation of faults based on dip-pitch relationships. This classification is summarized in Fig. 4.15, where six major domains have been distinguished within the same frame, as in Fig. 4.14.

(a) *Flat faults* lie along the left side of the diagram, say, fault dips gentler than 10–15°. Pitches have little significance in this case, because fault strike is highly variable.

(b) *Vertical faults* with predominant vertical motion correspond to the upper half along the right side of the diagram, say, fault dips steeper than 75–80° and pitches of slickenside lineations steeper than 45°. Note that pitches and plunges of striae are nearly equal for very steep fault dips.

(c) *Pure dip-slip faults* correspond to the uppermost portion of the remaining area, say, ratios of T/L larger than 4–8.

(d) *Pure strike-slip faults* correspond to the lowermost and lower-right portions of the diagram, say, ratios of L/T larger than 4–8.

(e) The remaining central area corresponds to various *oblique-slip faults* and can be divided into two sub-areas: lateral slips smaller than transverse ones (upper portion), or larger (lower portion). For instance, a fault will be called normal-dextral in the upper sub-area and dextral-normal in the lower one, in order to indicate which horizontal component of motion predominates.

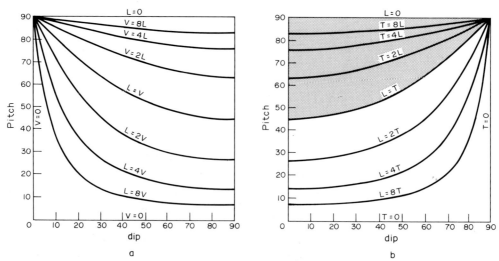

Fig. 4.14. Graphical determination of the lateral vs. vertical (a), and lateral vs. transverse (b), components of fault displacement, based on fault dip (abscissae) and pitch of fault slip (ordinates) in degrees. In (a) : V and L are the vertical and lateral horizontal components of displacement, respectively. In (b) : L and T are the lateral horizontal and transverse horizontal components, respectively.

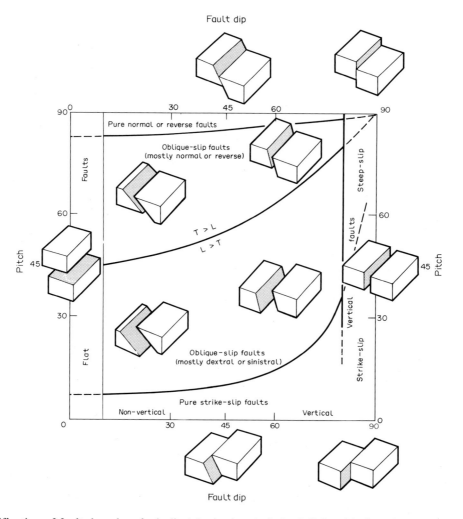

Fig. 4.15. Classification of faults based on fault dip (abscissa) and pitch of slickenside lineation (ordinates). Compare with Fig. 4.14b. See text for details.

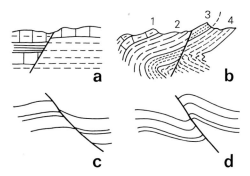

Fig. 4.16. Sense of fault displacement inferred from stratigraphic offset (a and b) or from associated folding (c and d). (a) Common situation: younger rocks crop out in the downthrown block. (b) Exception: older rocks crop out on the downthrown side of the fault because the stratigraphic pile (1–4, from older to younger) had been inverted before faulting. (c) Normal deflection. (d) Reverse deflection.

The general classification shown in Fig. 4.15 is valid regardless of the sense of motion (normal/reverse or dextral/sinistral).

Criteria for identifying sense of slip

The main criteria used to identify sense of fault slip are: (1) stratigraphic separation or, more generally, the offset of various markers predating fault motion, (2) so-called drag folds near the fault, and (3) a large variety of asymmetric features observable on the fault surface. The first two criteria are easy to identify and interpret despite a few exceptions (Fig. 4.16). The third type of criterion deserves more discussion (also see Chapters 5 and 18).

The major criteria used for determining the sense of slip on a fault surface will be described hereafter as 'positive' or 'negative' depending on the relative amounts of felt friction which occurs when the observer's hand moves in opposite senses on the fault surface, parallel to a slickenside lineation. A 'positive' criterion means that the easiest hand motion (i.e. the least friction) corresponds to the actual sense of displacement of the absent block. A 'negative' criterion means that when the hand moves in the same sense as did the lacking fault side, the friction felt by the observer is greater. Basic criteria (Angelier 1979a) are illustrated in Fig. 4.17, with the same numbers as in the following description.

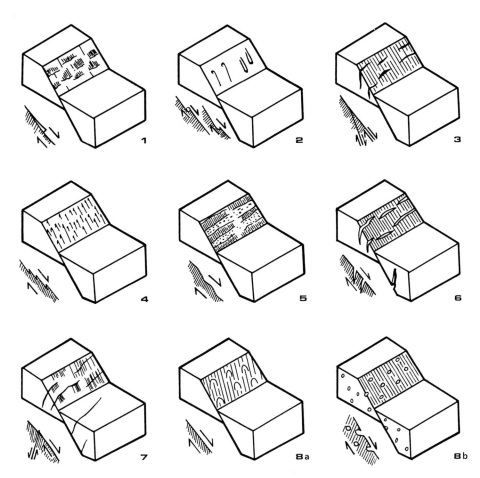

Fig. 4.17. Criteria for determining the sense of motion on a fault surface. The examples shown are dip-slip normal faults, but the criteria are valid regardless of fault slip orientation. Numbers between parentheses refer to description in the text. (1) Mineral steps. (2) Tectonic tool marks. (3) Riedel shears. (4) Stylolitic peaks. (5) Alternating polished (or crushed, and/or striated) and rough facets. (6) Tension gashes. (7) Conjugate shear fractures. (8) Miscellaneous criteria: (a) parabolic marks, and (b) deformed bubbles in lava.

(1) *Accretionary mineral steps* which develop due to crystal growth fibres or other grains being crystallized during fault slip are common. Most are made of calcite or quartz, but other minerals, such as gypsum, may crystallize in such a way. Similarly, fibrous minerals developing along slickenside lineations form steps that indicate the sense of motion in the same way (e.g. asbestos in sheared serpentinites). This criterion is 'positive' and 100% reliable.

(2) *Tectonic tool marks* (pebbles or other clasts) can occur either in relief on a fault surface or as asymmetric grooves (depending on the side observed). The tool can be a small quartz grain or a large boulder, and be present or absent at the distal end of the tool mark. In about 85% of cases, and 100% where the tool is still observable, this criterion is reliable and 'negative'. There are some exceptions where the tool was progressively destroyed during fault motion.

(3) *Riedel shears* commonly intersect fault surfaces. They make angles of 5–25° with the fault plane and their sense of motion, where observable, is the same as that on the main plane. Their intersection with the fault surface is nearly perpendicular to slickenside lineations. This criterion, despite numerous convincing examples (75%), can be ambiguous, especially where shear lenses develop along fault surfaces. It is 'negative' in terms of the crest of the acute dihedron formed by the main fault plane and the Riedel shear.

(4) *Stylolitic peaks*, or *lineations*, produced by solution phenomena on frictional facets, provide a criterion which is 100% reliable and common in limestones. It is typically 'negative'. One may commonly observe on a single fault surface, accretionary calcite steps (on facets that tended to open during fault motion) and stylolitic peaks (on facets that underwent shortening), the carbonate transfer having been facilitated by water circulation (Fig. 4.4). It is worthwhile to note that such fault movements correspond to creep because pressure-solution phenomena need time to occur, in contrast to sudden seismic fault slip.

(5) *Polished and rough facets*; the latter tend to open during fault motion, while the former experience friction. This criterion is extremely common in all types of rocks, but it is especially useful in non-calcitic ones, more generally where pressure-solution phenomena cannot occur so that criteria (1) or (4) are absent. This criterion is thus widely used in sandstones, basalts, etc. The facets have various shapes; in many cases, they are elongated perpendicular to slip direction. The facets which experience friction are either polished and striated, or crushed and in some cases made white by frictional effects (in the latter case, a hasty observer might think that they correspond to mineral steps and thus determine a wrong sense of slip). The facets which tend to open are either simply rough, or covered with rock fragments, iron oxides or crystallized minerals (calcite, quartz, etc.). This criterion is 'positive' and 80% reliable.

(6) *Tension gashes* whose intersection with a fault surface is approximately perpendicular to a slickenside lineation, make an acute angle of 30–50° with the fault surface so that the corresponding crest defines a 'negative' polarity, as with Riedel shears. This criterion is 70% reliable.

(7) *Conjugate shear fractures* or small faults at 40–70° to the main fault are a 'negative' criterion which is 70% reliable. As with Riedel shears and tension gashes, the intersection of such features on the main fault plane is approximately perpendicular to the slickenside lineation.

(8) Other criteria, although interesting, are generally of lesser use and some of them can be ambiguous (such as shear lenses or a rotated block along a fault plane). One may also mention the *parabolic marks* commonly found on polished faults, which define a 'positive' criterion despite the absence of relief. Many criteria are highly dependent on the type of rock (e.g. *deformed bubbles in lavas* which are strained close to the fault surface and thus indicate the sense of motion, as a 'positive' criterion); these criteria are too numerous to be reviewed here.

COMMON FAULT PATTERNS AND THEIR MECHANICAL INTERPRETATION

Faults are commonly associated in systems with particular symmetries. Here, a *fault family* (partial synonym *fault set*) is defined based on a common geometry (within the range of uncertainties and natural variations), in terms of dip direction, dip and pitch of slickensides (the angles d, p and i as defined in Fig. 4.11). For instance, faults with the same attitude and contrasting orientations of slickenside lineations, as in Fig. 4.12, do not belong to the same family in terms of fault movement.

A *fault system* is composed of two or more families which moved under the same tectonic regime so that their mechanisms are consistent (a concept that will be discussed later). In contrast, a *fault pattern* is a geometrical assemblage which may include several systems and result from several tectonic events (see Chapter 6 for a discussion of linked fault systems).

Comparison with rock mechanics

Early rock mechanics experiments highlighted the development of fractures under a given stress regime, because in such experiments the boundary forces and displacements applied to the sample were known and controlled by the observer so that their relation to the orientation of the resulting system of fractures was unambiguous. In turn, the results of these studies have been used in order to reconstruct the orientation of forces which were active in the past based on present-day observation of fault systems. The most famous early experiments were done by Daubrée (1879) who compared the results to actual fracture patterns. The clearest analysis of fault and dyke systems based on analogy with rock mechanics was presented later by

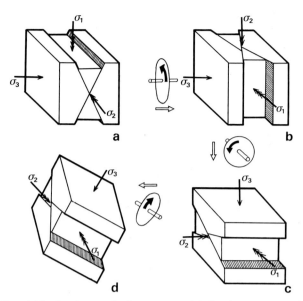

Fig. 4.18. Conjugate fault patterns. (a) Conjugate normal dip-slip faults. (b) Conjugate strike-slip faults. (c) Conjugate reverse dip-slip faults. (d) Conjugate oblique-slip faults. Principal stress axes shown with triple, double and single arrow-heads (σ_1, σ_2 and σ_3, respectively). Wheels with black arrows indicate rotations necessary to obtain (b) from (a), (c) from (b), and (d) from (c). Each of the first two rotations equals 90°.

Anderson (1942). The aim of this section is to present an interpretation of common fault systems rather than to discuss analogies with rock mechanics experiments. The reader should also refer to Chapters 3, 5 and 6 and to classics of the rock mechanics literature, such as Jaeger (1978).

Let us consider a simple case first (Fig. 4.18a): a typical system of conjugate normal faults including two families of dip-slip normal faults with parallel strikes and opposite dips. This kind of system develops in rock mechanics experiments, due to vertical compression; extension being horizontal and perpendicular to fault strike. Where a similar fault system is observed in nature, one can reconstruct the orientations of related stresses.

Stress and palaeostress

At this stage, it is necessary to introduce the concepts of *stress* and *palaeostress* which have proved to be extremely successful and thus extensively used in order to interpret fault slip patterns during the last decades. Considering a planar element in a rock (which can be represented as a weakness plane), the stress vector, σ, acting on this element is defined as $\sigma = dF/dS$, where dF is the force (a vector) and dS the surface (Fig. 4.19a). The stress has the same physical dimension as a pressure (i.e. a force divided by surface area).

Stresses that act in solids are characterized by anisotropic orientations, whereas pressure in a fluid is, of course, isotropic. As a consequence, the state of stress in a rock is characterized by three *principal stress axes* which define the *stress ellipsoid* (Fig. 4.19b): the maximum compressional stress σ_1, the intermediate stress σ_2, and the minimum stress σ_3. Note that as geologists we consider compression positive and extension negative ($\sigma_1 \geq \sigma_2 \geq \sigma_3$).

Geometry of conjugate fault systems

In Fig. 4.18(a), illustrating *normal dip-slip faults*, the σ_2 axis is parallel to the intersection direction between the fault planes, the σ_1 axis is vertical and bisects the acute angle between them, and the σ_3 axis is horizontal and bisects the obtuse angle between the faults. In Fig. 4.18(b), both faults are vertical and motion is pure strike-slip. The axis σ_2 is now vertical. The only difference between the conjugate systems of normal faults (Fig. 4.18a) and *strike slip faults* (Fig. 4.18b) is a matter of orientation: this difference is simply accounted for by a rotation of 90° around their common σ_3 axis. The system of conjugate *reverse dip-slip faults* (Fig. 4.18c) is interpreted in the same manner. The vertical axis is now σ_3. As before, a simple rotation of 90° around the common axis σ_1 accounts for the difference with the strike-slip conjugate system. Any rotation can be applied to these conjugate faults; for example, the *oblique-slip faults* of Fig. 4.18(d) are obtained through a rotation of about 35° around the oblique axis shown in Fig. 4.18(c).

Identification and interpretation of conjugate faults

Although conjugate systems of pure reverse faults (Fig. 4.18c), strike-slip faults (Fig. 4.18b) and normal faults (Fig. 4.18a) are easy to interpret, the case of oblique-slip conjugate faults (Fig. 4.18d) demonstrates that the geometrical properties of conjugate systems are intrinsic and independent of orientation. For a *conjugate fault system* to be identified, the following requirements must be satisfied.

(1) There are two families of faults.

(2) Both sets of slickenside lineations are perpendicular to the fault intersection direction.

(3) The rotation couples implied by the two fault slips should be of opposite sense, in such a way that shortening occurs in the acute angle between the faults.

(4) The angle between the faults should be consistent with the mechanical properties of the rock at the time of faulting.

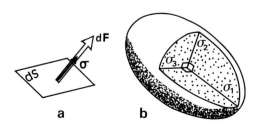

Fig. 4.19. (a) Stress vector. (b) Stress ellipsoid. Surface element dS, force dF and stress vector σ ($\sigma = dF/dS$). Principal axes of the stress ellipsoid shown as σ_1, σ_2 and σ_3 ($\sigma_1 > \sigma_2 > \sigma_3$).

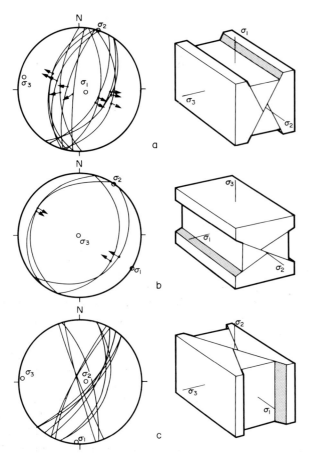

Fig. 4.20. Common examples of conjugate fault systems shown as cyclographic traces (great circles) with slip vectors shown as poles (points with arrows). (a) Normal dip-slip faults. (b) Reverse dip-slip faults. (c) Vertical strike-slip faults. On left: stereograms (Schmidt's projection, lower hemisphere). Principal stress axes: σ_1, σ_2 and σ_3 as in Fig. 4.18. Data from the Late Cenozoic of southern Greece.

Provided that these requirements are met, the palaeostress axes can be reconstructed as follows:

(a) The σ_1 axis bisects the acute angle between the faults.

(b) The σ_2 axis corresponds to the intersection direction of faults.

(c) The σ_3 axis bisects the obtuse angle between the faults.

Because of the dispersion of orientation data, reflecting both measurement uncertainties and natural irregularities, the identification of conjugate fault systems, as well as the determination of principal stress axes, should be made in terms of average orientations. Figure 4.20 illustrates some simple examples of characteristic conjugate fault systems, with the actual data shown in stereograms on left, and corresponding to simplified models on right.

Why is one stress axis generally vertical?

The situations illustrated in Figs 18(a–c), can be considered exceptional, whereas the oblique system (Fig. 4.18d) apparently represents a more general case.

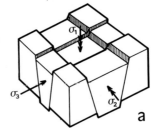 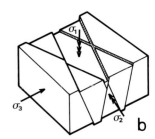

Fig. 4.21. Examples of neoformed fault systems which differ from simple conjugate fault systems. (a) Two perpendicular systems of dip-slip normal faults accommodate major extension (parallel to σ_3) and minor perpendicular extension, under a single regional stress regime. Note that each system may be interpreted as a single conjugate system (there is a single σ_1 axis and a permutation between σ_2 and σ_3). (b) A system of normal oblique-slip faults accommodate major extension (parallel to σ_3) and very little perpendicular extension, under a single stress regime. Note that among the four sets of faults no conjugate system can be identified. The orthorhombic symmetry is maintained despite obliquity of the slip directions (e.g. the σ_1 σ_3 plane is a plane of symmetry).

However, numerous field studies have demonstrated not only that this case of obliquity (relative to vertical) is in fact unusual, but also that it generally results from rotation of a pre-existing non-oblique fault system. The examples shown in Fig. 4.20, which are very common ones, effectively display one horizontal (or nearly horizontal) plane of symmetry, so that one principal axis is vertical or approximately vertical. This effect results from gravity and the Earth's free surface, which constrains one axis to be vertical (the vertical stress being directly dependent on rock weight and fluid pressure) and the tectonic stress to be a plane horizontal stress, at least in most cases.

Exceptions occur, for instance, on the edges of diapirs and magmatic intrusions or in accretionary prisms where large-scale thrusting occurs. In most cases, such as in continental platforms, palaeostress reconstructions based on analyses of fault systems (including conjugate ones) confirm that deviations of one stress axis from the vertical remain surprisingly small (usually less than 10°).

Neoformed faults

It is important to note that the analysis of conjugate fault systems can be carried out by simple geometrical means because a single tectonic stress is responsible for both the fracturing process (fault formation) and the continuing fault slip. Where weakness planes of various orientations are present in the rock mass, their activation as faults can occur and the development of genuine conjugate systems may be inhibited; in such cases, the mechanical analysis is necessarily more complex than a simple search for symmetries (this problem will be discussed later in this chapter). Conjugate faults thus belong to a particular class of faults: *neoformed faults*, which form and move in a single stress regime. Some systems of neoformed faults are more complicated than those already described: they include either two perpendicular systems (e.g. two systems of dip-slip normal faults with perpendicular trends), or two oblique systems with oblique slips that form a complex system with *orthorhombic symmetry* (Fig. 4.21). For a more detailed discussion of such systems, including experimental insights, see Reches (1983).

Extensional, compressional and strike-slip tectonic regimes

A fuller description of tectonic regimes is provided in Chapters 6, 11, 12 and 13. It is worthwhile, however, briefly mentioning the most common associations of faults in contrasting tectonic regimes. *Extensional tectonics* is characterized by *normal faulting*, such as in continental or oceanic rifts. Conjugate systems of dip-slip normal faults are common, although more complex systems can exist where extension is oblique to rift axes. Strike-slip faulting is also present; in addition, *transfer fault* movements result from irregularly distributed extension across adjacent rift segments. *Compressional tectonics* is dominated by *reverse faulting (thrusting)*, generally consistent with a single direction of compression. As in the previous case, oblique slip occurs where compression is oblique to structural trends. *Lateral ramping* accommodates differences in the distribution of shortening across adjacent segments of a fold-and-thrust belt. *Strike-slip faulting* can also dominate, with local occurrences of extension or compression, or both (depending on the geometry of the strike-slip pattern). In all these situations, numerous regional analyses have revealed that conjugate fault systems are common and allow reliable local reconstruction of palaeostress axes by simple graphical means. Thus, the search for conjugate systems is an efficient tool in the regional analysis of brittle deformation.

ACTIVATION OF WEAKNESS PLANES AS FAULTS

The simple geometrical relationships between stress axes and conjugate faults are valid because the faults are neoformed. Such situations commonly occur where brittle deformation affects rocks which were previously undeformed. Other fault patterns are usually more complicated because successive tectonic events are overprinted, resulting in an increasing complexity of brittle structures. It is thus necessary to examine the activation (or reactivation) of weakness surfaces which might have any orientation.

Neoformed and inherited faults

In many situations, faults related to a given tectonic event occur in rocks which are much older than the faults and have undergone earlier brittle deformation,

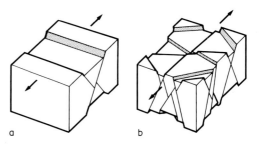

Fig. 4.22. (a) Neoformed faults. (b) Inherited faults. Black arrows indicate direction of extension.

so that numerous pre-existing planes of weakness were reactivated as faults during the event under consideration. For instance, tectonic analyses of the Rhine-graben system, a segment of the West European rift system, have revealed that most faults in the syn-rift Oligocene sediments are conjugate. As a result, the palaeostress regime related to Oligocene rifting can be easily reconstructed. Similar analyses of Mesozoic rocks along the graben margins show a more complex sequence of faulting: for example, some older strike-slip faults, probably of Eocene age, were reactivated as normal faults during the Oligocene so that their dips are now much steeper than expected for normal faults. In the Palaeozoic rocks of the rift shoulders, fracture and fault slip patterns are still more complicated because several tectonic events, including Hercynian ones, affected these rocks before Oligocene rifting. Numerous preexisting weakness planes were thus reactivated as faults with various orientations during the Oligocene.

Weakness planes may be present in rocks even when there has not been brittle tectonic deformation: this is typically the case for cooling joints in plutons, and bedding joints. Bedding joints are generally nearly horizontal, hence approximately perpendicular to one principal stress axis. As a consequence, the resulting anisotropy of mechanical rock properties has little effect on faulting as long as deformation remains minor and mechanical decoupling does not occur across the bedding joints, slip on such surfaces being inhibited by friction. Exceptions occur where strata are tilted or

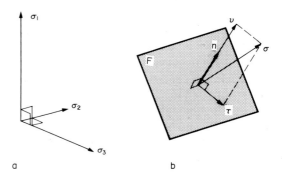

Fig. 4.23. (a) Stress state. (b) Weakness plane activated as fault. F, weakness plane; $\mathbf{n}$, unit vector perpendicular to weakness plane; σ, stress vector acting on F; $\mathbf{v}$, normal stress (perpendicular to F); τ, shear stress (parallel to F).; σ_1, σ_2 and σ_3, principal stress axes. The stress vector σ depends on both $\mathbf{n}$ and σ_1, σ_2 and σ_3.

folded, and where weakness zones (e.g. clays) and/or fluid pressure (e.g. hydroplastic faulting) induce mechanical decoupling, and hence décollement along stratigraphic interfaces. In the case of cooling joints, reactivation as faults is likely because in most cases a large variety of joint orientations is available.

Figure 4.22 contrasts *neoformed faults* and *inherited faults*. Neoformed fault systems are geometrically simple (Fig. 4.22a) so that determination of principal palaeostress axes is easy (Figs 18 and 20), whereas inherited faults possess various attitudes (Fig. 22b), so that it may be difficult to determine the orientations of the stress axes. Where there are neoformed fault systems, fault plane attitudes are related to stress orientation (Fig. 4.18) and slickenside lineations confirm that this interpretation is valid (Fig. 4.20). In contrast, where there are inherited faults, the distribution of orientations of reactivated weakness planes does not provide firm constraints on the orientations of stresses. In this case, the only key to deciphering the distribution of fault slip directions is provided by the stress-slip relationships analyzed in the next subsection.

Stress-slip relationships on weakness planes

Let us consider a pre-existing weakness plane in a rock mass and the stress which acts on it at a later stage. The stress state is characterized by the three principal stress axes σ_1, σ_2 and σ_3 (Figs 23 and 19b). There is no *a priori* special relationship between the orientation of the weakness plane and the stresses. According to solid mechanics, the stress vector, σ, (Fig. 4.19a) acting on the weakness plane F in Fig. 4.23 can be thought of as two perpendicular stress components, the normal stress, ν perpendicular to F, and the shear stress, τ, parallel to F; $\sigma = \nu + \tau$ in vectorial notation. The orientation of σ depends on both the attitude of F, conveniently described by its normal unit vector **n** (Fig. 4.23a) and the principal stresses (Fig. 4.23b).

It is important to note at this stage that the normal stress ν tends either to open the weakness plane or to close it. If this normal stress, ν_1 acts alone (e.g. when the stress vector σ, is perpendicular to the weakness plane) and is large enough, a tension gash (e.g. a vein or dyke) or a stylolitic seam (see Chapter 5) forms. In contrast, the shear stress, τ, tends to induce slip along the mechanical discontinuity, that is faulting. Because the amount of friction depends on both the sign (extension or compression) and the amplitude of the normal stress, the shear stress should not be considered alone, that is, slip depends on both the shear stress, and the normal stress, through the friction laws.

If slip occurs, and provided that the stress/fault system shown in Fig. 4.23 can be considered independently of its environment, slip should be parallel to, and in the same sense as, the shear stress. This hypothesis was proposed firstly by Wallace (1951) and secondly by Bott (1959). The *Wallace–Bott hypothesis* has provided a basis for additional studies, despite the fact that no rigorous interpretation of inherited fault systems in terms of palaeostress reconstruction was proposed until 1974 (this will be discussed later in the chapter).

Are faults independent and stresses uniform?

The main criticism of the *Wallace–Bott hypothesis* is concerned with fault interaction and stress deviations in brittle rocks. Faults commonly intersect and abut each other in linked systems so that geometrical and mechanical interaction occurs (see Chapter 6). The simple presence and shear activation of a mechanical discontinuity may induce significant deviations of stress. Consequently, one should not expect uniform distributions of stress in rock masses. This conclusion contradicts the hypothesis underlying the situation shown in Fig. 23: a single stress state acting on a weakness plane in a rock mass.

We should also consider whether the concept of a single stress acting on a weakness plane, so that slip is parallel to the shear stress (the Wallace–Bott hypothesis *stricto sensu*) is valid. Furthermore, is the concept of a unique stress inducing independent slips on weakness planes with various orientations (a more general hypothesis formulated later by Carey & Brunier 1974) acceptable?

Answers to both questions came firstly from empirical observations, and secondly from theoretical analyses. Computer-based determinations of palaeostress states have become more and more numerous since the 1970s. All the methods for such fault slip data inversions (discussed later in this chapter) were based on these assumptions, and all users of these methods pointed out repeatedly that the internal consistency of the results supported the validity of the basic assumptions. Had the assumptions been unreasonable, large average angular misfits between observed slips and computed shear stresses would have been noticed where fault slip data sets are large and fault orientations varied: this was never the case. On the contrary, the average angular misfits remained surprisingly low, taking measurement uncertainties and natural dispersion into account.

The second answer to these basic questions came from numerical modelling, performed more recently in order to calculate the stress distribution in faulted rocks (Dupin *et al.* 1992). The results of these theoretical analyses, which aimed at taking all sources of dispersion into account, showed firstly that the shear stress vectors and the slip vectors on a single isolated fault plane vary little in orientation; the average slip remaining parallel to the average shear stress, thus agreeing with the Wallace–Bott assumption (in detail, most variations occur near fault edges and remain minor). Secondly, the results demonstrated that where two faults intersect, the amplitude of variations depends on the geometry of the system, but remains statistically minor in comparison with the uncertainties inherent in data collection. Fault interaction is commonly an asymmetric process: the slip

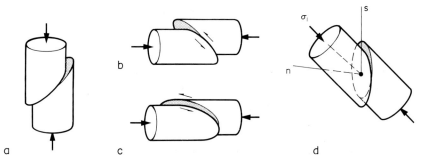

Fig. 4.24. Simple experiments illustrating stress-slip relationships: rock sample with discontinuity submitted to axial compression (solid arrows). (a), (b), (c) and (d) Different orientations of the same device (angle between compression axis and weakness plane constant). σ_1, compressional stress axis; s, slip direction; n normal to discontinuity.

on one fault fitting the extended Wallace–Bott hypothesis, whereas the slip on the other fault is less consistent with the hypothesis. However, the discrepancy usually remains minor, depending on the orientation of both faults relative to the stresses. In summary, the hypothesis of fault slips that occur independently and in agreement with a single stress state should be considered valid as a first approximation.

Stress-slip relationship: a geometrical approach

Let us consider a cylindrical rock sample containing an oblique planar discontinuity (weakness plane) and undergoing compression parallel to its axis (Fig. 4.24). When the sample axis is vertical, slip is normal dip-slip (Fig. 4.24a). When the sample axis is horizontal, the orientation of slip depends on the attitude of the discontinuity, which may vary through rotations of the sample around its axis. For instance, slip is reverse dip-slip along a discontinuity which strikes perpendicular to sample axis (Fig. 4.24b), whereas it is strike-slip along a vertical discontinuity (Fig. 4.24c). Note that the angle between the weakness plane and the axis of the cylinder is the same in both cases. All the experiments shown in Fig. 4.24 are identical: differences result merely from changes of viewpoint.

The same experiment, shown obliquely in Fig. 4.24(d), illustrates the *intrinsic relationship between stress and slip in compression*. Slip occurs parallel to the orthogonal projection of the compression axis on the weakness plane. In other words, the slip direction coincides with the intersection direction between two planes: the discontinuity and the perpendicular plane which contains the compression axis. Thus, the slip direction, s, the normal to the discontinuity, n, and the compression axis, σ_1, lie in a single plane (Fig. 24d).

This reasoning can be easily transferred to the case of axial extension (Fig. 4.25). Provided slip occurs, requiring an addition of lateral pressure in order to avoid dilation, it occurs parallel to the perpendicular projection of the extension axis, σ_3, on the weakness plane. Note that the direction of slip is the same as in compression (Fig. 4.24d), whereas the sense is opposite.

Shape of the stress ellipsoid

Slip-stress relationships are simple in cases of uniaxial compression (Fig. 4.24) or uniaxial extension (Fig. 4.25). However, these situations must be considered very special because in each case the corresponding stress ellipsoid (a concept illustrated in Fig. 4.19b) is one of revolution. In the case of Fig. 4.24(d), the *prolate stress ellipsoid* is cigar-shaped, around the σ_1 axis (*uniaxial compression*, with $\sigma_2 = \sigma_3$: see Fig. 4.26a). Considering Fig. 4.25, the revolution axis is σ_3, which implies an *oblate stress ellipsoid* (pie-shaped) (*uniaxial extension*, with $\sigma_1 = \sigma_2$: see Fig. 26b).

The shape of a stress ellipsoid is partly described by the magnitude of the intermediate principal stress, σ_2, relative to the extreme principal stresses σ_1 and σ_3. In the two uniaxial cases discussed above, this magnitude

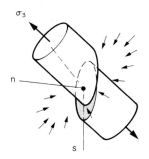

Fig. 4.25. Same experiment as shown in Fig. 4.24(d), with traction (solid arrows) instead of compression (σ_3, extensional stress axis). Lateral pressure added in order to avoid dilatation. Other symbols as in Fig. 4.24.

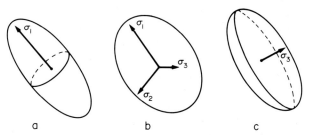

Fig. 4.26. Stress ellipsoids. (a) Prolate uniaxial compressional ellipsoid, $\sigma_2 = \sigma_3$, $\Phi = 0$. (b) Triaxial ellipsoid, $\sigma_1 > \sigma_2 > \sigma_3$, $0 < \Phi < 1$. (c) Oblate uniaxial extensional ellipsoid, $\sigma_1 = \sigma_2$, $\Phi = 1$. $\Phi = (\sigma_2 - \sigma_3)/(\sigma_1 - \sigma_3)$, ranging from 0 to 1. See text for details.

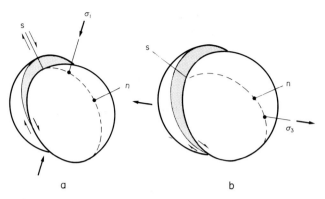

Fig. 4.27. Model of a faulted sphere showing the slip geometry reconstructed for two extreme cases. (a) $\sigma_2 = \sigma_3$. (b) $\sigma_1 = \sigma_2$. **n**, normal to fault; S, slip direction; dashed line, great circle containing n, s and σ_1 (a), or σ_3 (b). Note that σ_1 (in a) and σ_3 (in b) are perpendicular.

equals σ_1 or σ_3. Figure 4.27 shows the same sphere twice, with the same planar discontinuity and a single frame of stress axes, σ_1, σ_2 and σ_3. However, the σ_2 and σ_3 axes are not illustrated in Fig. 4.27(a) because the corresponding magnitudes are equal so that these axes are undefined (as in Fig. 4.26a). Conversely, the single axis shown in Fig. 27(b) is σ_3 because the magnitudes of σ_1 and σ_2 are equal (Fig. 4.26b).

Slip direction depends on the shape of stress ellipsoid

It is obvious from Fig. 4.27 that slip directions (parallel to shear stress according to the Wallace–Bott hypothesis) differ, depending on whether $\sigma_2 = \sigma_3$ or $\sigma_1 = \sigma_2$. In both cases, the principal stress axis, the normal to the fault plane and the slip vector intersect the sphere on a great circle because these three lines lie in a single plane as shown in Figs 4.24(d) and 25. In intermediate cases ($\sigma_1 > \sigma_2 > \sigma_3$, see Fig. 4.26b), the slip directions display intermediate orientations.

The ratio Φ between principal stress differences

The shape of the stress ellipsoid (Fig. 4.26) is partly but conveniently described by a single number which ranges between 0 and 1, the *ratio* Φ of the stress differences defined as follows (Angelier 1975):

$$\Phi = (\sigma_2 - \sigma_3)/(\sigma_1 - \sigma_3).$$

Because $\sigma_1 \geq \sigma_2 \geq \sigma_3$, both stress differences are positive and $0 \leq \Phi \leq 1$. The extreme cases shown in Figs 4.26(a & c) are thus characterized by extreme values of Φ, 0 and 1, respectively. Ratios between stress differences have been called *R ratios* by some workers, but definitions of R vary whereas that of Φ is unique.

In terms of a Mohr diagram (Fig. 4.28a), the ratio Φ has a simple geometrical significance. The *Mohr diagram* is a plot of normal stress vs shear stress relationships in a rectangular frame. Firstly, any multiplication of all the principal stress magnitudes by a positive constant k, an operation which corresponds to a scale variation of the Mohr diagram (Fig. 4.28b), does not affect the value of Φ. Secondly, any addition of isotropic stress of value l (positive for compression, negative for extension), which corresponds to a shift of the Mohr circles along the normal stress axis, does not affect the value of Φ (Fig. 4.28c). Coming back to the instance of uniaxial compressional stresses (Figs 24, 26a and 27a), the corresponding value of Φ is 0. A value $\Phi = 1$ characterizes the uniaxial extensional stresses shown in Figs 25, 26(b) and 27(b).

Extreme shear stress orientations for $\Phi = 0$ and $\Phi = 1$

Figure 4.29 shows, in stereoplots, relationships between the stresses (described in terms of principal stress directions and the ratio of stress differences) and the slip on a given weakness plane (discontinuity). For easier reading, one axis has been chosen as the vertical

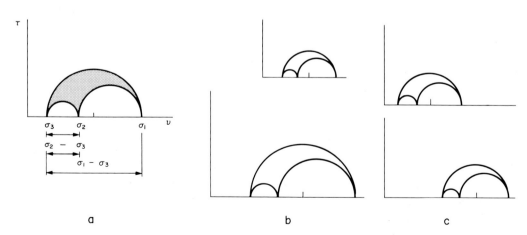

Fig. 4.28. Mohr circles and the ratio Φ (a) Definition. v, normal stress magnitude. τ, shear stress magnitude. σ_1, σ_2 and σ_3, principal stress magnitudes. All stress vectors acting on faces of various orientations correspond to points in stippled area (also see Fig. 4.23b). Points on circles correspond to faces containing the σ_2 axis (largest circle), the σ_1 axis ($\sigma_2 - \sigma_3$ circle) or the σ_3 axis ($\sigma_1 - \sigma_2$ circle). Points on the normal stress axis correspond to faces perpendicular to principal stress axes σ_1, σ_2 or σ_3. The ratio Φ increases from 0 to 1 while σ_2 increases from σ_3 to σ_1, because $\Phi = (\sigma_2 - \sigma_3)/(\sigma_1 - \sigma_3)$. (b) The ratio Φ does not depend on diagram scale. (c) The ratio Φ does not depend on shift along the normal stress axis.

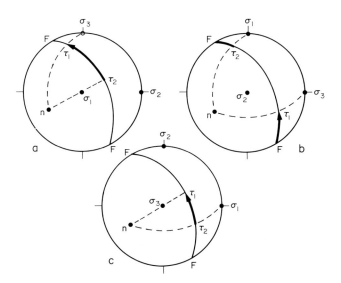

Fig. 4.29. Stress axes, the ratio Φ and slip on a fault. F, fault plane (thin line); **n**, pole to F (normal vector); σ_1, σ_2 and σ_3, stress axes; black arrow, rotation of shear stress from τ_0 to τ_1 as Φ increases from 0 to 1 (i.e. σ_2 increases from σ_3 to σ_1).

(σ_1 in Fig. 4.29a, σ_2 in Fig. 4.29b, σ_3 in Fig. 4.29c). The cyclographic trace and the pole of the weakness plane are also shown. The extreme positions of shear stress, τ_0 for Φ = 0 and τ_1 for Φ = 1, are plotted in Fig. 4.29 according to the same geometrical rules as in Figs 4.24, 4.25 and 4.27. The bold arrows describe the rotation of shear stress from τ_0 to τ_1, as Φ increases from 0 to 1 (Angelier 1979a,b). The slip (assumed to be parallel to shear stress and consistent in sense) may vary in orientation but remains normal in the situation shown in Fig. 4.29(a) with a vertical σ_1 axis, and remains reverse in Fig. 29(c) with a vertical σ_3 axis. However, in the case of Fig. 4.29(b) with a vertical σ_2 axis, the slip is reverse-dextral or reverse-sinistral depending on Φ values; this explains why the shear stress path (bold arrow) intersects the horizontal plane (the projection circle): the only way to change motion from reverse to normal on a given fault plane (see also Fig. 4.12).

Possible orientations of shear stress

Simple geometrical constructions, as in Fig. 4.29, enable the rapid constraint of the possible variations of shear stress orientation (presumed to be parallel to slip) for any attitude of the planar discontinuity (i.e. the inherited fault), provided that the orientations of the stress axes are known. Note that 'forbidden orientations' are thus defined, because shear stress cannot lie outside the sector defined by τ_0 and τ_1. For example, right-lateral components of normal slip are thus ruled out in the case of Fig. 29(a), as well as dip-slip in Fig. 29(b) or right-lateral components of reverse slip in Fig. 29(c).

These geometrical constructions also indicate that the angle δ between the extreme positions of shear stress (τ_0 and τ_1, see Fig. 4.29) is variable, depending on the attitude of the fault plane. One can demonstrate that $\cos\delta = \mathrm{tg}\alpha_1 \cdot \mathrm{tg}\alpha_3$, where α_1 and α_3 respectively, describe the angles between the fault plane and the principal stress axes σ_1 and σ_3 (Angelier 1979a). This formula as well as the construction in the stereoplots, indicates that τ_0 and τ_1 coincide (i.e. δ = 0) on a plane containing the intermediate stress axis σ_2 (in this case $\mathrm{tg}\alpha_1 = \mathrm{cotg}\alpha_3$ and the orthogonal projections of σ_1 and σ_3 on the fault planes coincide). Conjugate fault systems (Figs 4.18 and 4.20) are not affected by variations of the ratio Φ because the intermediate stress axis is parallel to the fault intersection direction. In turn, this indicates that the ratio Φ cannot be reliably determined based solely on observation of conjugate fault systems.

Oblique inherited faults allow determination of Φ

In contrast to conjugate faults, oblique-slip inherited faults allow determination not only of stress axes, but also of the ratio Φ, a ratio that plays an essential role in stress-slip relationships. However, despite their interest, geometrical constructions do not allow easy determination of shear stress in the most general case (i.e. the fault is oblique to all stress axes and the ratio Φ varies from 0 or 1). Such a construction exists, but it is too complicated to be useful routinely: thus, it is now indispensable to examine the stress-slip relationships using mathematical formulation.

Mathematical aspects of stress-shear relationships

The stress ellipsoid is entirely described by a *stress tensor*, **T**, which can be described in matricial form containing six independent variables:

$$\begin{bmatrix} a & d & f \\ d & b & e \\ f & e & c \end{bmatrix}.$$

In the coordinate system defined by the principal stress axes, the stress tensor has a much simpler expression with three variables, the magnitudes of the principal stresses:

$$\begin{bmatrix} \sigma_1 & 0 & 0 \\ 0 & \sigma_2 & 0 \\ 0 & 0 & \sigma_3 \end{bmatrix}.$$

The three variables which have been eliminated correspond to the orientations of the three principal axes. These orientations, in a general rectangular coordinate frame, correspond to three perpendicular unit vectors along the axes σ_1, σ_2 and σ_3, which are respectively given by:

$$\begin{bmatrix} x_1 \\ y_1 \\ z_1 \end{bmatrix}, \begin{bmatrix} x_2 \\ y_2 \\ z_2 \end{bmatrix} \text{ and } \begin{bmatrix} x_3 \\ y_3 \\ z_3 \end{bmatrix}.$$

These nine values correspond to three independent variables solely because they describe unit vectors (three relationships) which are mutually perpendicular (three more relationships). As a result, the general expression of the stress tensor shown first contains three independent variables related to the principal stress orientations, and three other independent variables related to principal stress magnitudes. Thus, the terms a to f are computed according to the following matrix product, which relates the general form to the *non-rotated stress tensor*:

$$\begin{pmatrix} a & d & f \\ d & b & e \\ f & e & c \end{pmatrix} = \begin{pmatrix} x_1 & x_2 & x_3 \\ y_1 & y_2 & y_3 \\ z_1 & z_2 & z_3 \end{pmatrix} \cdot \begin{pmatrix} \sigma_1 & 0 & 0 \\ 0 & \sigma_2 & 0 \\ 0 & 0 & \sigma_3 \end{pmatrix} \cdot \begin{pmatrix} x_1 & y_1 & z_1 \\ x_2 & y_2 & z_2 \\ x_3 & y_3 & z_3 \end{pmatrix}.$$

Note that this equation is simply the matricial expression of a tensor rotation, that is, the transfer from the system of principal axes into a general system of cartesian coordinates.

The stress vector σ acting on a fault plane characterized by its normal unit vector **n** (Fig. 4.23) is given, in vectorial and matricial notations successively, by the following equations (note that bold characters refer to vectors, whereas normal characters refer to scalars):

$$\sigma = \mathbf{Tn}$$

that is:
$$\begin{pmatrix} \sigma_x \\ \sigma_y \\ \sigma_z \end{pmatrix} = \begin{pmatrix} a & d & f \\ d & b & e \\ f & e & c \end{pmatrix} \cdot \begin{pmatrix} x \\ y \\ z \end{pmatrix}.$$

The modulus of the normal stress v (Fig. 4.23) is given by the scalar product of the stress vector by the unit normal vector:

$$|v| = \sigma \cdot \mathbf{n}$$

or:

$$|v| = x\sigma_x + y\sigma_y + z\sigma_z.$$

The normal stress vector itself, v, is thus easily obtained:

$$v = |v|\mathbf{n},$$

or:

$$\begin{pmatrix} v_x \\ v_y \\ v_z \end{pmatrix} = |v| \begin{pmatrix} x \\ y \\ z \end{pmatrix}.$$

Knowing the stress vector σ and the normal stress vector v, one obtains the shear stress vector, τ:

$$\sigma = v + \tau$$

that is:
$$\begin{pmatrix} \tau_x \\ \tau_y \\ \tau_z \end{pmatrix} = \begin{pmatrix} \sigma_x \\ \sigma_y \\ \sigma_z \end{pmatrix} - \begin{pmatrix} v_x \\ v_y \\ v_z \end{pmatrix}.$$

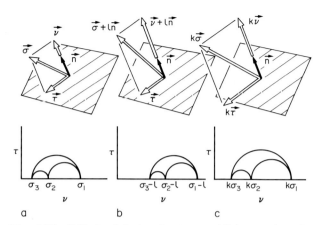

Fig. 4.30. Effect of isotropic stress addition and scale-factor multiplication on the components of stress acting on any plane. Above: fault plane (striated) and its normal unit vector **n**; σ, stress vector; v, normal stress; τ, shear stress. Below: corresponding Mohr circles (also see Fig. 4.28). σ_1, σ_2 and σ_3, principal stresses. Variables k and l are discussed in the text.

Stress orientation and stress magnitude

Let us examine in more detail the effects of the operations described in Figs 4.28(b) and (c): that is, the multiplication of stress by a positive scale factor, and the addition of an isotropic stress tensor (compression or tension of hydrostatic type). Multiplying the stress tensor by a positive factor k implies that the stress vector σ is multiplied by the same factor and becomes $k\sigma$ (Figs 4.30a & c). Adding an isotropic stress $l\mathbf{I}$ (l is any number, positive for compression or negative for tension, and **I** describes the isotropic unit compression tensor) implies that a vector $l\mathbf{n}$ is added to the stress vector σ which becomes $\sigma + l\mathbf{n}$ (Figs 4.30a & b).

In terms of a Mohr diagram, these changes have already been discussed (Figs 4.28b & c). The simple constructions shown in Fig. 4.30, and vectorial operations similar to the preceding ones, confirm that neither the orientation nor the sense of shear stress on any weakness plane is modified by such changes. Furthermore, it has already been demonstrated that these operations do not affect the ratio Φ.

The four variables related to shear orientation

It should be noted again that the magnitudes of stress, especially the magnitudes of stress differences, play a major role in rupture and friction processes. However, because most field observations deal with fault slip orientations and slip senses (rather than slip amplitude and normal-shear stress magnitudes), numerous constraints are imposed in terms of orientation and sense, whereas few are available in terms of magnitude. As a result, rupture and friction laws are generally very poorly constrained by data (exceptions will be discussed later).

Consequently, the six variables of the stress tensor are of unequal interest: four variables play an essential role because they constrain the orientation of the stress ellipsoid and reveal information about its shape. The two remaining unknowns, k and l, are as discussed above (Fig. 4.30). Among the four major variables, three describe the orientations of stress axes (they are implicit in the rotation matrices noted earlier); the fourth variable is Φ.

The reduced stress tensor

Let us consider, again, the stress tensor in the system of its principal axes themselves:

$$\begin{bmatrix} \sigma_1 & 0 & 0 \\ 0 & \sigma_2 & 0 \\ 0 & 0 & \sigma_3 \end{bmatrix}.$$

Adding an isotropic stress defined by $l = -\sigma_3$, then multiplying the tensor by the positive constant $k = 1/(\sigma_1 - \sigma_3)$, one obtains:

$$\begin{bmatrix} 1 & 0 & 0 \\ 0 & \Phi & 0 \\ 0 & 0 & 0 \end{bmatrix}.$$

These operations, consistent with the information shown in Fig. 4.30, indicate that both tensors are equivalent in terms of directions and senses of shear stresses. Of course, any rotation (which involves the three orientation variables already discussed) does not modify this conclusion. The resulting stress tensor (Angelier 1979a) is written as follows, and contains four independent variables (as discussed before, taking into account the six relationships between coordinates x_i, y_i and z_i):

$$\begin{bmatrix} x_1 & x_2 & x_3 \\ y_1 & y_2 & y_3 \\ z_1 & z_2 & z_3 \end{bmatrix} \cdot \begin{bmatrix} 1 & 0 & 0 \\ 0 & \Phi & 0 \\ 0 & 0 & 0 \end{bmatrix} \cdot \begin{bmatrix} x_1 & y_1 & z_1 \\ x_2 & y_2 & z_2 \\ x_3 & y_3 & z_3 \end{bmatrix}.$$

This tensor, which simply depends on the orientation of principal stress axes and the ratio Φ, is called a *reduced stress tensor* because it contains less independent variables than the complete stress tensor (i.e. four instead of six).

The impossibility of determining k and l from slip orientation data

A major implication of the above arguments is the impossibility of determining the two variables k and l based on analyses of fault slip orientations and senses. As a consequence, as far as fault slip data are limited to this kind of information in the absence of definite rupture and friction laws, only four among the six independent variables of the stress tensor can be determined. This limitation plays an important role in the discussion of fault slip data inversion discussed later in this chapter.

It should be pointed out that all the slip-stress relationships discussed in this section, in the case of inherited faults, are *a fortiori* valid for the neoformed faults discussed in the previous section. However, note that although considering neoformed faults in the same way as inherited ones is correct it implies that some information is lost (i.e the relationship between stress and fracture orientation).

FAULT SLIP CONSTRAINTS ON STRESS ORIENTATION

In the preceding section, the relationship between stress and slip on a weakness plane was discussed, regardless of the origin of the fault (neoformed or inherited). The discussion was based on simple geometrical considerations (Figs 4.24, 4.25, 4.27 and 4.29), and on tensor calculation (Figs 4.23 and 4.26).

Note that we addressed the direct problem which can be formulated as follows: how does the stress influence slip on a weakness plane? The inverse problem, determining the palaeostress from fault slip data, was not addressed.

How many fault slip directions and senses are needed to constrain palaeostress?

It was shown earlier that the stress can be entirely determined at one point by knowing six variables (called a to f in the most general expression of the stress tensor), and that two of the unknowns cannot be determined from directions and senses of slip solely, in the absence of rupture-friction data (unknowns called k and l, see Fig. 4.30). It was also assumed that the stress is uniform in the rock mass (the extended Wallace–Bott hypothesis) and that fault slip directions and senses are independent within this volume (as formulated by Carey & Brunier 1974).

As a consequence, one cannot determine more than four unknowns at this stage. The *reduced stress tensor* discussed above effectively contains four unknowns, which fully constrain the orientations of the three principal stress axes and the ratio Φ (Figs 26 and 28). Because each fault slip corresponds to one relationship between the stress tensor and the shear stress, at least four distinct pieces of fault slip data are required in order to determine a reduced stress tensor.

Thus, a single fault slip imposes a constraint on stress variables, but does not allow determination of stress. In this section, I first aim to discuss the constraints that can be inferred from a single fault slip, and then describe a method for taking into account such constraints for a complete data set.

How does a single fault constrain palaeostress?

This problem can be addressed in a simple geometrical way provided that one stress axis is assumed to be vertical, a situation that is assumed to hold where there has not been tilting or folding. Accepting this, it is easy to demonstrate that there is a relationship between

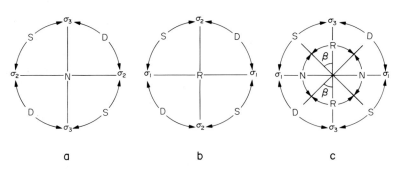

Fig. 4.31. Senses of transverse and lateral components of horizontal fault motions in relation to stress state. (a) Extensional stress regime with σ_1 vertical. (b) Compressional stress regime with σ_3 vertical. (c) Strike-slip regime with σ_2 vertical. Outer circular arrows refer to senses of lateral motion (D, dextral; S, sinistral) relative to fault strike. Inner circular arrows in (c) indicate senses of the transverse motion component (N, normal; R, reverse), also relative to fault strike; this component is uniformly normal in (a) and reverse in (b). The angle β (in c) is a function of Φ (see text).

the orientation and sense of fault slip and the trends of horizontal principal stress axes. However, because the relationship also depends on the value of the ratio Φ (Fig. 4.29), a single fault slip merely constrains a range of possible trends for the principal stress axes.

With a few exceptions, an extensional tectonic regime with σ_1 vertical induces fault slips with normal components of motion, whereas a compressional regime with σ_3 vertical induces fault slips with reverse components. However, a strike-slip tectonic regime with σ_2 vertical, can induce various lateral and transverse components of motion depending on the orientation of the fault plane (Fig. 4.31).

For a vertical σ_1 axis (Fig. 4.31a), all faults are normal and the sense of lateral motion depends on fault strike. For a vertical σ_3 axis (Fig. 4.31b), all faults are reverse and their right- or left-lateral sense is determined in the same manner. For a vertical σ_2 axis, the sense of lateral component of motion also depends on the strike of the fault, but the sense of the transverse component, reverse or normal, is variable as a function of the angle between this strike and, say, the σ_3 axis. The threshold angle, β, between normal and reverse components depends on the ratio Φ as follows (Fig. 4.31c):

$$\operatorname{tg}^2\beta = \frac{1-\Phi}{\Phi}.$$

Conversely, for a given fault slip, the stereoplots shown in Fig. 4.32 allow determination of the possible trends of horizontal axes, assuming that one is vertical. First, for a fault slip with a normal component, the vertical axis may be either σ_1 or σ_2 (Figs 4.32a & b, respectively); in each case, the range of σ_3 trends is constrained but the trend cannot be determined accurately unless the value of Φ is known (except for dip-slip). A similar reasoning is applied to fault slip with a reverse component: the vertical axis being σ_3 (Fig. 4.32c) or σ_2 (Fig. 4.32d), the possible trends of the σ_1 axis are thus constrained. In both cases, the limit between the two sub-ranges of trends corresponds to an extreme value of Φ (i.e. 1 in Figs 4.32a & b, and 0 in Figs 4.32c & d), whereas the opposite extreme value corresponds to normal or reverse dip-slip, provided that the vertical axis is not σ_2 (Figs 4.32a & c, respectively).

The right dihedra (or P and T dihedra)

Geometrical considerations such as those shown in Figs 4.31 and 4.32 are certainly helpful, but constraints are highly dependent on the choice of the vertical principal stress axis, and, furthermore, the assumption that one axis is vertical is disputable.

In the general case, however, it is possible to demonstrate that the extreme principal axes, σ_1 and σ_3, cannot display any orientation provided that fault slip is consistent with the basic Wallace–Bott hypothesis. Let us simply consider the reduced stress tensor with prinicpal stresses 1, Φ and 0 as described earlier (equivalent to σ_1, σ_2 and σ_3 in terms of shear stress orientations and senses, Fig. 4.30). Based on formulae already given, the scalar product between the shear stress and a unit vector $\mathbf{N}_i$ along each of the three principal axes σ_i is obtained as follows:

$$\begin{aligned}\boldsymbol{\tau}\cdot\mathbf{N}_1 &= \mathbf{n}\cdot\mathbf{N}_1\,(1 - n_1^2 - n_2^2\,\Phi) \\ \boldsymbol{\tau}\cdot\mathbf{N}_2 &= \mathbf{n}\cdot\mathbf{N}_2\,(\Phi - n_1^2 - n_2^2\,\Phi) \\ \boldsymbol{\tau}\cdot\mathbf{N}_3 &= \mathbf{n}\cdot\mathbf{N}_3\,(-n_1^2 - n_2^2\,\Phi)\quad.\end{aligned}$$

The unit vector $\mathbf{n}$ is perpendicular to the fault plane, while n_1 and n_2 are the direction cosines of $\mathbf{n}$ along the σ_1 and σ_2 axes, respectively (i.e. $\mathbf{n}\cdot\mathbf{N}_1 = n_1$ and $\mathbf{n}\cdot\mathbf{N}_2 = n_2$). The expression between parentheses is positive in the first line, whereas it is negative in the third one, demonstrating that $\boldsymbol{\tau}\cdot\mathbf{N}_1$ and $\mathbf{n}\cdot\mathbf{N}_1$ have the same sign whereas $\boldsymbol{\tau}\cdot\mathbf{N}_3$ and $\mathbf{n}\cdot\mathbf{N}_3$ have opposite signs. In contrast, the sign of the expression between parentheses in the second line can vary according to variations in fault orientation and the value of Φ.

The geometrical significance of these sign constraints (for σ_1 and σ_2) is fairly simple: both the angles $(\boldsymbol{\tau}, \mathbf{N}_1)$ and $(\mathbf{n}, \mathbf{N}_1)$ are acute (plus signs) or obtuse (minus signs), whereas for angles $(\boldsymbol{\tau}, \mathbf{N}_3)$ and $(\mathbf{n}, \mathbf{N}_3)$ one must be acute, while the other is obtuse. In other words, the orientation of the σ_1 axis (as shown by unit vector $\mathbf{N}_1$) must belong to the compressional dihedron (P) of Fig. 4.33(a), whereas the orientation of the σ_3 axis ($\mathbf{N}_3$)

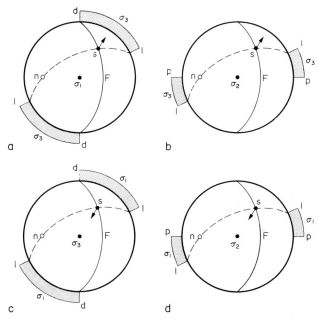

Fig. 4.32. Possible directions of extreme horizontal principal stresses, σ_1 and σ_3, for a given fault slip direction (one axis assumed to be vertical) in stereoplots (lower hemisphere). (a) Normal-sinistral faulting, σ_1 vertical. (b) Same fault slip, σ_2 vertical. (c) Reverse-dextral faulting, σ_3 vertical. (d) Same fault slip, σ_2 vertical. F, fault plane; n, pole to fault plane; s, slip (slickenside lineation); d, fault strike; p, dip direction of fault. Possible trends of σ_3 axis, for (a) and (b), and of σ_1 axis, for (c) and (d), shown stippled.

belongs to the extensional dihedron (T). There is no additional constraint on the orientations of these axes except (1) their obvious perpendicularity, and (2) an additional relationship discovered by Lisle (1987).

The focal mechanism of a fault

Figure 4.33 summarizes the major characteristics of a *fault focal mechanism* from a perspective view (Fig. 4.33a) and a stereoplot (Fig. 4.33b). The two right dihedra, P and T, are separated by the fault plane and by the *auxiliary plane* perpendicular to the shear stress (consistent with the slip vector according to the Wallace–Bott assumption). Note that the fault plane is a real shear discontinuity in the rock mass whereas the auxiliary plane does not exist except in the geometrical construction. Two right dihedra are thus unambiguously defined taking the sense of slip into account; in Fig. 4.33(b), the T dihedron is shown in black whereas the P dihedron is shown stippled. The demonstration discussed above indicates that the σ_1 axis should be searched for in the P dihedron and its perpendicular, the σ_3 axis, in the T dihedron. In contrast, there is no firm constraint about the orientation of the intermediate stress axis, σ_2.

The focal mechanism of an earthquake

There is a striking resemblance between a *fault focal mechanism* (Fig. 4.33b), and the classical representation of a double-couple *earthquake focal mechanism* as used by seismologists (Fig. 4.34) (also see Chapter 18). In both cases, two perpendicular planes, called *nodal planes* by seismologists, separate P and T dihedra. As a result, in contrast with the unambiguous observation of a fault plane with slickenside lineations by the field geologist (Fig. 4.35a), the geophysicist has to choose the actual fault plane among the two nodal planes (Fig. 4.35b).

Because stress-shear relationships have been discussed in some detail, it should be pointed out that the two possible solutions for an earthquake focal mechanism are generally far from being equivalent in terms of consistency with stress. This property is illustrated in Fig. 4.36, which shows the two solutions for a single earthquake focal mechanism. In one case, the slip vector on the fault plane is perpendicular to the other nodal plane which represents the *auxiliary plane* as defined earlier for faults (Fig. 4.33a). However, the shear stress theoretically exerted on this auxiliary plane (considered as a virtual discontinuity) has no reason to be perpendicular to the fault plane. Had the other solution been adopted, the reciprocal situation would prevail (cf. Figs 4.36a & b). Thinking that both slip directions, the real one and the 'virtual' one, are simultaneously perpendicular to nodal planes (respectively) would be a mistake. As a consequence, the choice of a focal mechanism solution of an earthquake, as in Figs 4.35(b) and 4.36, might have significant consequences in terms of relations to stress. This problem does not exist for geologists (Fig. 4.35a).

What situation corresponds to the absence of such an ambiguity for a focal mechanism of an earthquake? Considering the same reduced stress tensor as before, and the condition for both shear stresses, real and 'virtual', to be perpendicular to nodal planes, one simply obtains, with the same symbols as before:

$$n_1 n_2 n_3 \, \Phi \, (1 - \Phi) = 0 \, .$$

This equation shows that the choice among nodal planes is unnecessary in the following cases: a stress ellipsoid of revolution ($\Phi = 0$, Fig. 4.26a, or $\Phi = 1$, Fig. 4.26c), or a fault plane parallel to a stress axis. The latter condition indicates that where seismic faulting occurs on a fault belonging to a conjugate set (Figs 4.18 and 4.20), there is no need to choose a nodal plane for the 'corresponding earthquake' because $n_2 = 0$. As such geometries are common, convenient simplifications can be made. However, in the general case, or where conjugate faulting remains undetected, the choice among nodal planes remains indispensable.

The right dihedra method: principle

It has been demonstrated that the σ_1 axis belongs to the compressional dihedron (P in Fig. 4.33) while the σ_3 axis belongs to the extensional dihedron (T in Fig. 4.33). As a consequence, where two or more fault slips have occurred in the same stress regime (i.e. the same orientations of principal stress axes and the same

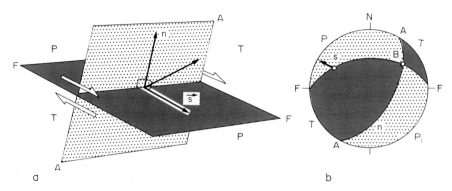

Fig. 4.33. Right dihedra and the fault slip mechanism. (a) Perspective view. (b) Stereoplot, lower hemisphere. *F*, fault plane; *A*, auxiliary plane; *n*, normal to fault plane (unit vector); *s*, unit slip vector in (a), reverse-dextral slickenside lineation in (b); *B*, intersection of planes A and F; *P*, compressional dihedron; *T*, extensional dihedron.

ratio Φ), and provided that the extended Wallace–Bott hypothesis is acceptable, the extreme stress axes must belong to the corresponding dihedra for all the fault mechanisms. In other words, the σ_1 axis should belong to the range of orientations to all *P* dihedra, and the σ_3 axis should be common to all *T* dihedra.

This is the principle of the *right dihedra method of fault analysis* (Angelier & Mechler 1977), which is illustrated in Fig. 4.37 for two fault mechanisms only. This method is easy to apply by hand using stereoplots: in projection, all areas which are not homogeneous in terms of *P* and *T* dihedra (respectively) are progressively erased. As a result, the residual *P* and *T* areas (respectively stippled and black in the upper row) indicate the range of possible orientations for the principal stress axes, σ_1 and σ_3, respectively.

Numerical application of the right dihedra method

Where faults are numerous, it often happens that no residual area remains in the final diagram, which is left entirely white according to the pattern convention as shown in Fig. 4.37. Such a disappointing result can have various causes: errors in determinations of fault slip senses (which result in an exchange of *P* and *T* dihedra in the corresponding mechanism), presence of faults related to another stress regime (e.g. undetected polyphase faulting), natural dispersion of fault slips (i.e. all data do not fit the requirements of the extended Wallace–Bott hypothesis), and errors and uncertainties of angular measurements. Due to the 'yes-or-no' logic involved in the elimination process (upper row of Fig. 4.37), a single error or misfit can result in the disappearance of common *P* and *T* orientation domains; the situation is hopeless as long as the source of the anomalies remains undetected.

Instead of cancelling non-common domains, one can conceive of a different process: all mechanisms are first plotted individually on transparent sheets with *T* dihedra in light grey and *P* dihedra left white, all these diagrams are accurately superposed and examined together in front of a light: the darker region of the resulting stereoplot corresponding to the range of common σ_3 orientations for most fault mechanisms, while the paler grey area indicates the range of common σ_1 orientations. If present, a minor proportion of contradictory data will not destroy the reconstruction.

This process is applicable in a simple numerical way: extreme values, say, 0 and 100, are respectively associated with the *P* and *T* dihedra of a fault mechanism (Fig. 4.37, lower row). Assignment of these values is made throughout a projection grid which represents all directions in space, given a certain resolution. When two fault mechanisms are combined, these percentages are added and averaged, so that one obtains values 0 (common compressional directions), 100 (common extensional directions) and 50 (for conflicting directions); combining three mechanisms results in values of 0, 33, 66 and 100; and so on. The final result, easily obtained using a small computer, allows determination of best-fit domains; the degree of compatibility between the fault slip data. For instance, 10% and 90% as extreme percentages suggest that 10% of data are inconsistent in terms of *P* and *T* directions.

The results of the numerical application of the right dihedra method may be displayed in various ways: as a plot of numbers (Fig. 4.38b), as graphical patterns

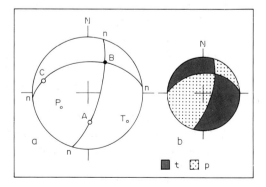

Fig. 4.34. Earthquake focal mechanism (normal double-couple). Schmidt's projection, lower hemisphere. (a) Individual geometrical axes. (b) *P* and *T* right dihedra. *n*, nodal planes; *B*, intersection of nodal planes; *A* and *C*, axes perpendicular to nodal planes; *P* and *T*, axes bisecting the right dihedra. *A*, *C*, *P*, *T* are all perpendicular to *B* and enclose angles of 45°. *t* and *p*, in (b): *T* (extension) and *P* (compression) dihedra, respectively.

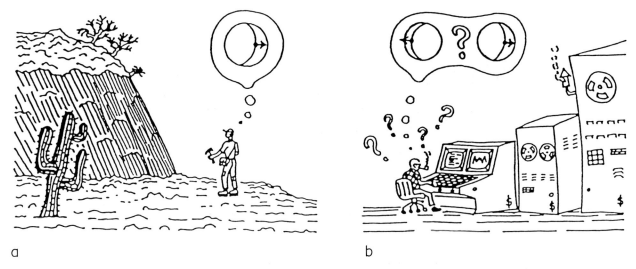

Fig. 4.35. (a) Geological observation of a fault. (b) Geophysical analysis of a focal mechanism of earthquake.

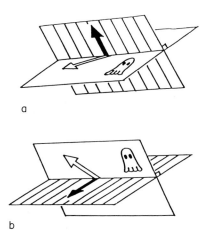

Fig. 4.36. The two possible solutions, (a) or (b), for an earthquake focal mechanism (double-couple). The actual fault plane is striated, with the shear stress as a black arrow. A ghost indicates the auxiliary plane, with the virtual shear stress as an open arrow. Discussion in text.

(Fig. 4.38c) and as average axes that correspond to barycentres of best compatibility domains on the sphere (Fig. 4.38d). It is necessary to point out that such average axes have little value where residual P and T domains are large, because the stress axes σ_1 and σ_3 have no reason to occupy a central orientation in these solid angles, and can lie near their edges as well. However, where residual domains are very small, the average axes and the actual stress axes (as determined by other means) are fairly consistent.

The right dihedra method is directly applicable to earthquakes

Where fault slip data are collected by geologists (Fig. 4.35a), it is necessary to reconstruct the auxiliary plane, perpendicular to slickenside lineations, prior to the characterization of the P and T dihedra according to slip sense (Fig. 4.33). In contrast, earthquake mechanism data, as reconstituted after seismological analysis (Fig. 4.35b), consist of P and T dihedra separated by perpendicular nodal planes, so that focal mechanisms are directly processed without preliminary transformation.

The use of the right dihedra method is thus particularly appropriate for earthquake analysis in terms of stress consistency, because, by definition, choices among nodal planes are unnecessary, and the ambiguity previously pointed out (Fig. 4.36) is harmless. In other words, the available data are fully exploited even though some realistic constraints are ignored, as will be discussed below.

The right dihedra method implies some loss of geological data

The transformation of a fault slip datum (Fig. 4.23a) into a focal mechanism (Fig. 4.34b) implies the loss of an important piece of geological information: the identity of the actual fault plane, which cannot be distinguished from the auxiliary plane any longer. The two perpendicular planes of a fault mechanism are strictly equivalent in terms of right dihedra analysis, whereas they are not in terms of shear stresses, whose distribution is generally asymmetric, as Fig. 4.36 shows.

This contrast between the symmetry of a focal mechanism and the asymmetry of shear stresses exerted on a fault plane (actual) and an auxiliary plane (virtual) explains how and why the geological fault slip data are not fully exploited, contrary to conventional double-couple earthquake focal mechanisms. This loss of geological information has a counterpart. Although the right dihedra method allows reliable determination of ranges of compatibility directions for compression and extension, which are, in many cases, tightly constrained, the best possible fit between shear data and stress is not ensured. More complete analysis will be discussed later, but contrary to the right dihedra method, simple graphical techniques will not be applied any more but pure numerical means will be used.

The accuracy of the results is obviously dependent on the variety of geometrical orientations of fault slip data:

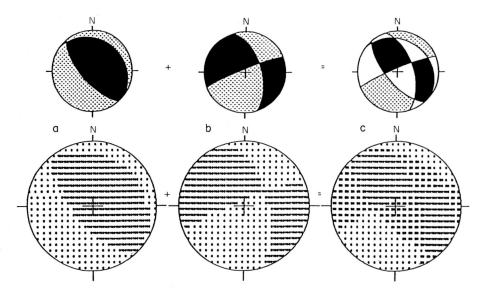

Fig. 4.37. Principle of the right dihedra method. Schmidt's projection, lower hemisphere. (a) and (b) Simple case of two faults with the result shown in (c). Upper row: graphic application, with fault or earthquake mechanisms shown as in Fig. 4.34 (P dihedron stippled, T dihedron in black); incompatibility domains left white in (c). Lower row: numerical application, with percentages (full compatibility for compression = 0, full compatibility for extension = 100, maximum incompatibility = 50).

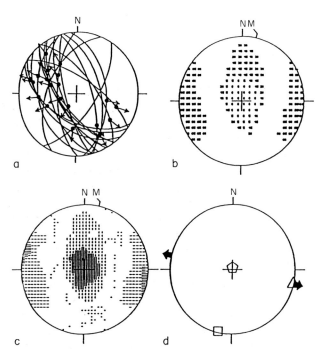

Fig. 4.38. Example of application of the right dihedra method. Plio-Quaternary normal faulting in Late Cenozoic sediments, Baja California, Mexico; site with 20 normal fault slips. Schmidt's projection, lower hemisphere. (a) Field data: faults planes as cyclographic traces, slickenside lineations as dots with arrows, all slips normal. (b) Numerical expression of results, same caption as for the lower row of Fig. 4.37; intermediate values not plotted. (c) Simplified graphic expression: horizontal hatch pattern for high values (extension), vertical hatch pattern for low values (compression), small crosses indicating maximum incompatibility (50%); hatched patterns denser in maximum compatibility domains. (d) Corresponding average axes (triangle for extension, pentagon for compression, square for intermediate axis); black arrows indicate extension.

the more diversified the fault slip orientations, the tighter the constraints on compatibility directions for compression and tension. Thus, one should not only collect relatively uniform data, such as for conjugate fault systems, but also search for oblique-slip faults. The importance of faults oblique to stress axes in palaeostress reconstructions is additionally highlighted in a later section.

The right dihedra method and the ratio Φ

The *ratio* Φ is one of the four critical variables of the reduced stress tensor. This ratio is not taken into account explicitly in the right dihedra method, although it plays a major role in shear-stress relationships (Fig. 4.29). The basic assumption which underlies the right dihedra method can be formulated as follows: fault slips occurring under a single uniform state of stress must have some directions of compression (and tension) in common, and the stress axis σ_1 corresponds to one of these common directions of compression (conversely, σ_3 and tension). It must be emphasized that nothing is stated about the ratio Φ. In other words, despite the fact that the concept of the *reduced stress tensor* includes Φ as a major variable controlling slip (Fig. 4.29), the state of stress, as implicitly regarded in this method, deals with only the orientations of the three principal stress axes. As a result, there is in the right dihedra method no assumption about the uniqueness of the Φ value, whereas there are major constraints about the uniqueness of each stress axis (provided that data orientations are varied enough). The value of Φ is thus implicitly considered independently for each fault slip datum.

In turn, this additional degree of freedom explains why the results, although reliable, are incomplete. The use of other methods, assuming that Φ is unique for a given stress state, will allow more complete determination of stress as discussed below. The effects of variations in Φ ratios, and the relationships between the value of Φ and the shapes of residual compatibility domains in the sphere, have been discussed by Angelier & Mechler (1977): in many cases, the distribution of P and T domains provides some general information about approximate values of Φ.

FAULT SLIP DATA INVERSION

A striking feature of fault analysis since about 1980 is the development of numerical palaeostress reconstructions using fault slip data sets. These methods are based on the stress-shear relationship described by Wallace (1951) and Bott (1959) and discussed earlier. The *inverse problem*, which consists of determining the stress tensor knowing the direction and the sense of slip on numerous faults of various orientations, was first solved by Carey & Brunier (1974). Various methodological developments and improvements have been proposed since then (e.g. Angelier 1975, Carey 1976, Armijo & Cisternas 1978, Etchecopar et al. 1981, Angelier et al. 1982, Michael 1984, Reches 1987). Under certain conditions, these computational methods apply to populations of focal mechanisms of earthquakes (Angelier 1984, Gephart & Forsyth 1984, Mercier & Carey-Gailhardis 1989). They also allow reconstruction of palaeostress based on analyses of calcite mechanical twins (Laurent et al. 1981, Lacombe et al. 1990, 1992).

The development of methods

The pioneering work of Arthaud (1969) deserves attention; it played an essential role despite the mechanical misconception which underlay the proposed method (i.e. the concept of principal axes with independent influences on fault slips), because it was the first attempt at solving the inverse problem of palaeostress reconstruction based on the graphical analysis of slips on weakness planes with all possible orientations. Later, Mercier (1976) demonstrated that Arthaud's method cannot be applied in the general case, but remains valid for particular stress states (i.e. $\sigma_1 = \sigma_2$ or $\sigma_2 = \sigma_3$: that is, stress ellipsoids of revolution, Fig. 4.26). Likewise, the analysis applied by Arthaud & Choukroune (1972) was quite successful but aimed at solving a particular case (strike-slip movements on weakness planes). During the late 1960s to early 1970s, numerous attempts at reconstructing palaeostresses based on analyses of inherited brittle structures in particular cases were successful but (1) the general inverse problem was not formulated properly, and (2) the search for geometrical rather than numerical techniques impeded discovery of the general solution. As pointed out earlier, the method of right dihedra is graphically applicable and mechanically correct, but does not put fault slip data to the best use. The additional information obtained from other structures such as tectonic joints (Price 1966, Hancock 1985), tension gashes and stylolites (Arthaud & Mattauer 1969, Mattauer 1973) is also important but is discussed in Chapter 5.

Carey & Brunier (1974) formulated the inverse problem in a proper way for the first time; they reversed Bott's reasoning and proposed computing a palaeostress tensor by inversion of a data set including the directions of motion shown by striae on fault planes of various attitudes. Mathematical aspects were described in more detail by Carey (1976). This analysis represented a fundamental step in the advancement of tectonic studies of fault populations, and allowed numerous subsequent improvements and additional analyses to be made. Most of these methods involve numerical computations of reduced stress tensors by solving the inverse problem for four independent unknowns. Among the major improvements that followed Carey and Brunier's proposal were the consideration of fault slip sense (Angelier 1975), the estimation of errors (Angelier et al. 1982), the search for more unknowns of the stress tensor (Reches 1987), and the direct inversion (Angelier 1990). None of these analyses, however, contradicts the initial formulation of the inverse problem.

The inverse problem

The direct problem consists of determining the orientation and sense of slip knowing the orientation of a fault plane, for a given stress tensor, **T**. The inverse problem consists of determining the mean stress tensor, **T**, knowing the orientations and senses of slip on numerous faults. In both cases, the basic assumption is that each fault slip (indicated by a slickenside lineation) has the direction and sense of shear stress that corresponds to a single common stress tensor. However, data collection involves errors, dispersion occurs in local stress patterns, and fault movements influence one another. In practice, one searches for the best fit between all fault slip data that belong to a given tectonic event and a common unknown stress tensor.

The assumption that all faults which moved during the same tectonic event were moving independently but consistently within a single stress tensor is an obvious approximation. However, as mentioned earlier, application to numerous actual cases has shown high levels of consistency as indicated by small values of average angles (s, τ). Note that (s, τ) is the angle between the observed slickenside lineation or unit slip vector, **s**, and the theoretical shear stress vector, τ, shown in Fig. 4.23(a), and derived from the stress tensor solution of the problem.

The actual stress tensor has six degrees of freedom. Neither adding an isotropic stress, nor multiplying the tensor by a positive constant, can modify the direction and the sense of slip on any fault (Fig. 4.30). As a consequence, the actual tensor being **T***, any tensor **T** equally solves the problem:

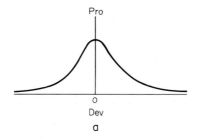 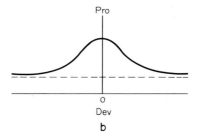

Fig. 4.39. Deviations (dev) of actual data relative to theoretical model: simple probabilistic (pro) distribution models. (a) Gaussian distribution. (b) Modified distribution: more large deviations acceptable.

$$\mathbf{T} = k\mathbf{T}^* + l\,\mathbf{l}$$

where k and l designate any constants (k positive) and $\mathbf{l}$ the unit stress matrix. The tensor $\mathbf{T}$ has four degrees of freedom so that one can adopt a particular form, which is called the *reduced stress tensor*. Because the number of unknowns is four, whereas the number of equations (i.e. the number of fault slips) is much larger, the inverse problem is clearly over constrained, and an appropriate simple statistical model must be adopted in order to ensure the best fit between actual slips and theoretical shear stresses.

The least-square criterion

Most authors adopted a least-square criterion in the minimization process. It is well known that the use of this criterion makes algebraic derivations easier. It also implies that the dispersion of data relative to the basic model of stress-slip relationships is of Gaussian type. That is, the probability of finding very large deviations should tend to zero (Fig. 4.39a). This choice is questionable because high proportions of anomalous data may be present in fault slip data sets so that the Gaussian distribution is no longer acceptable. Such anomalous data can correspond to fault slips that belong to a different, undetected, stress regime, or to the effects of local stress deviations or geometrically constrained fault movements. Where these sources of dispersion are present, they result in abnormally high proportions of large deviations that simple errors and uncertainties in data collection cannot explain. In such occurrences, a more realistic probabilistic model should be adopted (Fig. 4.39b).

When the inversion technique involves the use of a simple least-square criterion, implying a Gaussian-type distribution of individual misfits, the presence of too numerous anomalous data may vitiate the stress determination. A simple but efficient cure consists of removing the data with unacceptable deviations, before starting the calculation again. This process can be progressive in order to avoid sudden changes, taking into account the fact that a 'bad' datum may become a 'good' one in a further step, because the stress tensor solution of the problem changes as the data set is modified. The user must be aware that such a rejection of data should remain limited and geologically controlled. Rejected data should be never forgotten, but carefully examined in order to detect the source of anomalies, which in many cases reveal polyphase faulting.

The use of non-Gaussian distributions (e.g. Fig. 4.39b) generally results in heavier calculations and ridiculously long runtimes. Many experiments that I have done with actual fault slip data sets (unpublished results) reveal that no spectacular improvement of results is thus obtained. The simple removal of anomalous data, as described above, yields equally significant results. I conclude that under careful geological control the adoption of the least-square criterion constitutes a reasonable approximation.

The minimization function and misfit criterion

Adopting the least-square criterion as discussed above, the average *reduced stress tensor*, $\mathbf{T}$, which best fits the K fault slip data of the set considered, that is, the solution of the inverse problem, is obtained by minimizing a sum S, written as follows:

$$S_m = \sum_{k=1}^{k=K} w_k (F_{mk})^2 .$$

In this general expression, w_k is the weight of the datum number k, and F_k is a function which expresses the deviation of this datum ($F_k = 0$ corresponds to a perfect fit with the average stress tensor, and increasing values of F_k indicate increasing misfits). The additional index m refers to the particular form adopted for F_k and consequently S.

Different misfit criteria have been adopted and their uses yielded similar results in well-constrained cases. The simplest one is given by the angle between the actual unit slip vector, $\mathbf{s}$, and the shear stress, τ, computed as a function of the stress tensor (Figs 4.23 and 4.40). Note that in the following expressions of F_m, the index k is omitted for the sake of simplicity.

$$F_1 = (\mathbf{s}, \tau).$$

Similar results were obtained with faster computation using a different formula (Angelier 1975):

$$F_2 = \sin\frac{(\mathbf{s}, \tau)}{2} .$$

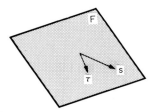

Fig. 4.40. Actual slip, **s**, and theoretical shear stress, τ, on a fault plane, *F*. See also Fig. 4.23.

Whereas F_1 varies from 0 to 180°, F_2 varies from 0 to 1. The half-angle is adopted in order to take slip sense into account; had the angle (**s**, τ) been adopted, misfit angles of 20 and 160° would have been equivalent. In addition, keeping the choice of the least-square criterion in mind, the square of the function F_2 has a simple expression:

$$F_2^2 = \frac{1}{2} - \frac{1}{2} \cos(\mathbf{s}, \tau).$$

Another function, albeit discontinuous, also allows reliable and fast determination of the stress tensor in terms of slip-shear angle:

$F_3 = \text{tg}(\mathbf{s}, \tau)$ for an angle smaller than 45°.
$F_3 = 1$ for an angle larger than 45°.

One must observe that functions such as F_1, F_2 and F_3, which solely depend on the angle slip-computed shear, lack significance for a stress vector, σ, very close to the direction of the normal to the fault, **n**. In this particular case, small variations of the direction of σ can result in large variations of the direction of τ, so that for particular orientations of the weakness plane relative to stress axes small changes of the stress tensor may have much influence on the misfit criterion.

A different criterion was proposed and extensively discussed by Angelier (1984, 1990)

$$F_4 = |\lambda \mathbf{s} \cdot \tau|^2.$$

With the scalar product denoted by a point the square of this criterion can be written:

$$F_2^4 = \lambda^2 + |\tau|^2 - 2\lambda(\mathbf{s} \cdot \tau).$$

This function, F_4, which was called function upsilon,

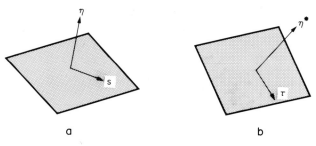

Fig. 4.41. (a) Real and (b) theoretical fault slip, **n**, normal to fault plane; **s**, unit slip vector; τ, computed shear stress; **n***, normal to best-fitting fault plane in which τ is computed. Orientations and senses of **n** and **n***, and of **s** and τ, should be as close as possible simultaneously. Compare with Fig. 4.40, where no variation in fault plane orientation is foreseen.

has the same dimension as a stress, contrary to F_2 and F_3 which are dimensionless. The parameter λ is related to the value of the largest possible shear stress, which solely depends on the form of the reduced stress tensor. The properties and advantages of function upsilon in tensor calculation have been extensively discussed elsewhere (Angelier 1990). It was pointed out that its dependence on both the angle (**s**, τ) and the modulus |τ| of shear stress corresponds to quite a reasonable requirement: that is, shear stress levels are large enough to induce fault slip, despite rock cohesion and friction.

Although mathematical aspects of the inverse problem are not presented here, it is worth pointing out that, strictly speaking, the misfits should not be considered in simple terms of slip-shear relationships in a perfectly defined fault plane (Fig. 4.40). The errors and uncertainties affecting the fault plane orientation should also be taken into account (Fig. 4.41): the fault slip datum being characterized by a couple of perpendicular unit vectors, **n** and **s**, the theoretical fault slip corresponding to a couple of vectors, **n*** and τ, whose respective orientations are as close as possible to those of **n** and **s**. This kind of method was presented by Angelier *et al.* (1982), the algebra being more complicated than for the previous functions. The advantage of this heavier method, called the *total inversion method* is that it takes account of all error estimates.

The reduced stress tensor

The form of the *reduced stress tensor* (with four unknowns) may be chosen according to the mathematical requirements of the method, which depend on numerous factors such as the form of the misfit criterion discussed above and the searching technique discussed below. Among an infinity of possible forms, some have interesting properties (depending on methodological choices), such as:

$$\begin{pmatrix} x_1 & x_2 & x_3 \\ y_1 & y_2 & y_3 \\ z_1 & z_2 & z_3 \end{pmatrix} \cdot \begin{pmatrix} 1 & 0 & 0 \\ 0 & \Phi & 0 \\ 0 & 0 & 0 \end{pmatrix} \cdot \begin{pmatrix} x_1 & y_1 & z_1 \\ x_2 & y_2 & z_2 \\ x_3 & y_3 & z_3 \end{pmatrix};$$

$$\begin{pmatrix} \cos\psi & \alpha & \gamma \\ \alpha & \cos(\psi + \frac{2\pi}{3}) & \beta \\ \gamma & \beta & \cos(\psi + \frac{4\pi}{3}) \end{pmatrix};$$

$$\begin{pmatrix} x_1 & x_2 & x_3 \\ y_1 & y_2 & y_3 \\ z_1 & z_2 & z_3 \end{pmatrix} \cdot \begin{pmatrix} \cos\psi & 0 & 0 \\ 0 & \cos(\psi + \frac{2\pi}{3}) & 0 \\ 0 & 0 & \cos(\psi + \frac{4\pi}{3}) \end{pmatrix} \cdot \begin{pmatrix} x_1 & y_1 & z_1 \\ x_2 & y_2 & z_2 \\ x_3 & y_3 & z_3 \end{pmatrix};$$

$$\begin{pmatrix} x_1 & x_2 & x_3 \\ y_1 & y_2 & y_3 \\ z_1 & z_2 & z_3 \end{pmatrix} \cdot \begin{pmatrix} 2-\Phi & 0 & 0 \\ 0 & 2\Phi-1 & 0 \\ 0 & 0 & -1-\Phi \end{pmatrix} \cdot \begin{pmatrix} x_1 & y_1 & z_1 \\ x_2 & y_2 & z_2 \\ x_3 & y_3 & z_3 \end{pmatrix}.$$

The three last reduced tensors are deviatoric, whereas the first one is not. In the first and fourth expressions, the ratio Φ is explicit, whereas in the second and third expressions it is not (although it is related to ψ in a

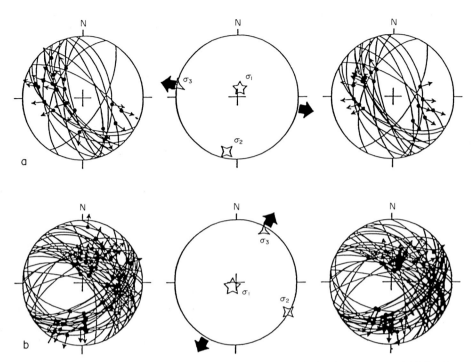

Fig. 4.42. Determination of average palaeostress tensors with singlephase fault slip data sets. (a) Plio-Quaternary extension, site MD1, Arroyo Montado, Baja California, Mexico: 20 faults in Late Cenozoic sediments. (b) Tertiary extension, site KAM, Kamogawa, Japan: 50 faults in Tertiary ophiolite. Diagrams on left (raw data) and on right (fitted slips): fault planes as thin lines, slickenside lineations as dots with small arrows (all slips normal). Diagrams in centre: computed stress axes σ_1, σ_2 and σ_3 shown as 5-, 4- and 3-branched stars; large black arrows indicate the direction of extension. Numerical parameters: see Table 4.1 (these determinations) and Table 4.2 (same data, different method). See also Figs 4.38 and 4.43: same site as in (a). All diagrams are Schmidt's projection, lower hemisphere.

simple way in the third form). In the second expression of the stress tensor, which was used by Angelier & Goguel (1979) who presented a first direct inversion method, the four variables of the reduced stress tensor, α, β, γ and ψ, are explicit. In the first, third and fourth expressions, the value related to the ratio of stress differences, Φ or ψ, is isolated and the three remaining variables are contained in the rotation matrices; in other words, the orientation unknowns and the shape unknown of the stress ellipsoid are clearly separated. Note, incidentally, that for a given stress, the values of ψ in the second and third expressions do not coincide except in particular cases, because ψ in the second expression depends on the orientations of stress axes whereas in the third expression it does not.

Numerical techniques

The first method of solving the inverse problem consists of comparing numerous attempts through a 4-D search; that is, different values of the four variables of the reduced stress tensor are used until a satisfactory solution is obtained, with smaller and smaller numerical intervals. After Carey & Brunier (1974) had introduced the method, such numerical searches were widely used in determinations of palaeostress tensors. Differences in search process have little influence on the final results, but the runtime may vary widely. Adoption of grids with decreasing mesh size during the numerical search hastens the process (e.g. Angelier 1979a).

A quite different method of solving the inverse problem consists of setting the partial derivatives of the sum S (defined above) to zero in order to compute directly the extreme limits of S, especially the minimum value, by analytical means. This kind of mathematical analysis was adopted by Angelier & Goguel (1979) and Angelier (1990), using the expression of the reduced stress tensor with α, β, γ and ψ as four independent variables.

$$\delta S/\delta \alpha = 0, \quad \delta S/\delta \beta = 0, \quad \delta S/\delta \gamma = 0, \delta S/\psi = 0.$$

This technique allows fast computation of the stress tensor: in contrast with iterative processes which need time, the result is obtained in a few seconds on a small PC. In counterpart, the system of four equations mentioned above must be solved by analytical means, which requires a particular form of the minimization function and misfit criterion, such as F_4 (described in an earlier subsection).

Weighting of data

Observations on faults are made at all possible scales; thus there is no reason to apply the term microtectonics to these methods. The size and the net separation of a fault measured in the field can vary from centimetres to kilometres. By weighting fault measurements by fault size and throw, more significance is given to major structures in the data inversion. Repeated measurement

Table 4.1. Results of stress tensor determinations with the new direct inversion method (Angelier 1990)

Site	Axis σ_1		Axis σ_2		Axis σ_3		Ratio Φ	Number of data	Average RUP	n_1	n_2
	trend	plunge	trend	plunge	trend	plunge					
MD1	17	78	191	12	281	1	0.52	20	44%	2	5
KAM	218	83	119	1	29	7	0.31	50	43%	5	11

Same examples as in Fig. 4.42. MD1, Arroyo Montado; KAM, Kamogawa. Angles in degrees (trends and plunges of stress axes σ_1, σ_2 and σ_3). Ratio Φ and estimator RUP defined in text. n_1, number of data with RUP>75%; n_2, number of data with 75%⩾RUP>50%.

from place to place along major faults not only provides an accurate knowledge of the fault geometry, but also ensures a kind of empirical and approximate weighting, the smallest faults rarely being measured twice.

Finally, experience demonstrates that for large data sets, various determinations of reduced stress tensors generally provide identical or very similar results, whether computation takes weights into account or not. As a consequence, the constancy of results with the weighting mode can be considered as an additional criterion for evaluating the consistency of fault mechanisms at various scales, and the quality of data collection. The weight of each fault slip datum in palaeostress reconstruction also depends on the accuracy of individual angle measurements (strike, dip, and pitch).

Case examples

Figure 4.42 illustrates simple examples of fault slip data inversion; two data sets are described, with different origins and numbers of faults. Raw data are shown on the left, in stereoplots. For each row, the central diagram shows the reconstructed axes of palaeostress, σ_1, σ_2 and σ_3; Table 4.1 provides more detail, especially about the reliability and accuracy of results. The axes shown in Fig. 4.42 and the results in Table 4.1 are obtained through the *new direct inversion method* (INVD), and the *quality estimator* (RUP) reported in Table 4.1 ranges from 0 (perfect fit) to 200%, proportional to the misfit function upsilon previously described as F_4 (for details, see Angelier 1990). In Fig. 4.42, diagrams on the right show the same fault planes with the orientations and senses of shear stress computed according to the *average stress tensor*; in other words, these diagrams illustrate perfect artificial data sets. A comparison between these computed slips and actual data shows that despite the acceptable average fit, individual misfits vary significantly.

With the same data, the palaeostress reconstruction referred to in Table 4.2 was carried out with the 4-D search method R4DT based on the misfit criterion previously described as F_3; the results obtained with the same technique and criterion F_2 (R4DS) are similar (not shown). The relevant quality estimator, ANG, ranges from 0 (perfect fit) to 180°; it simply represents the angle between observed slip and computed shear stress (Fig. 4.40). Comparison between Tables 4.1 and 4.2 shows that the results do not differ markedly, despite the strong contrast between the methods.

Figure 4.43 shows the principal palaeostress axes computed at site MD1 (Fig. 4.42a) with a third method based on a *non-linear least-squares technique* (Angelier *et al.* 1982). The axes are the same within a range of few degrees; after the determination of the covariance matrices, confidence ellipses were plotted around stress axes (Fig. 4.43). In this case, three average estimators are displayed: they describe the average values of strike, dip and pitch deviations of theoretical fault slip data relative to actual ones (3, 1 and 2 degrees, respectively, in site 'MD1').

Finally, these results should be compared with the constraints on the positions of stress axes obtained with the right dihedra method already described for site MD1. As in all cases where fault slip data satisfactorily constrain the result, results obtained in different ways are consistent (Figs 4.42 and 4.43, Tables 4.1 and 4.2).

Application to focal mechanisms of earthquakes

Figure 4.44 shows a group of focal mechanisms of microearthquakes determined by Chinese seismologists in the Longitudinal Valley of eastern Taiwan (Yu & Tsai

Table 4.2. Results of stress tensor determinations with the 4-D exploration method (Angelier 1975), to be compared with results shown in Table 1

Site	Axis σ_1		Axis σ_2		Axis σ_3		Ratio Φ	Number of data	Average ANG	n_1	n_2
	trend	plunge	trend	plunge	trend	plunge					
MD1	19	72	194	17	285	1	0.57	20	19°	2	6
KAM	246	84	118	4	28	5	0.32	50	19°	5	15

Same examples as in Fig. 4.42: MD1, Arroyo Montado; KAM, Kamogawa. Legends as in Table 4.1, except for estimator ANG (angle shear-striae, in degrees; see text). n_1, number of data with ANG⩾45°; n_2, number of data with 45⩾ANG>22.5°.

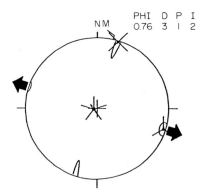

Fig. 4.43. Determination of average palaeostress tensor by the non-linear least-square inversion method. Same fault slip data set as in Fig. 4.42(a). Schmidt's projection, lower hemisphere. Computed stress axes σ_1, σ_2 and σ_3 shown as 5-, 4- and 3-branched stars respectively, with confidence ellipses added (ellipses around σ_2 and σ_3 clearly visible, ellipse around σ_1 very small); large black arrows indicate the direction of extension. See also Figs 4.38 and 4.42(a).

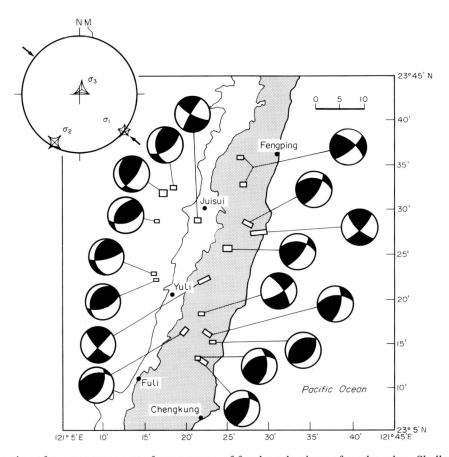

Fig. 4.44. Determination of average stress state from a group of focal mechanisms of earthquakes. Shallow microearthquakes near Yuli, Taiwan. Data from Yu & Tsai (1982). Schmidt's projection, lower hemisphere. Focal mechanisms with T (extension) dihedra in black and P (compression) dihedra left white. On top left: average σ_1, σ_2 and σ_3 axes (5-, 4- and 3-branched stars, respectively) and corresponding direction of compression (black arrows). See also Table 4.3.

1982); the corresponding stress axes were reconstructed with the 4-D search method R4DT, based on the criterion F_3 already described. Because of the problem of the choice among nodal planes for each focal mechanism (Figs 4.35b and 4.36), different combinations were used in several experiments, five of which are mentioned in Table 4.3 (for more detail see Angelier 1984). The first experiment was made with the complete set of 34 nodal planes. The next four experiments implied different choices among nodal planes; subsets are complementary two by two (i.e. each nodal plane is used in two experiments, and both the nodal planes of an earthquake mechanism are never used in the same experiment). All results are quite similar, suggesting that in this case the choice between nodal planes has little significance (Table 4.3).

This conclusion, however, has no general value: the value of Φ determined in all Yuli experiments is low (0–2), so that the particular relationship discussed in an earlier section is nearly satisfied. In the general case,

Table 4.3. Comparison of different stress tensor determinations with the data set of Yuli earthquakes (see also Fig. 4.44)

Experiment	Axis σ_1		Axis σ_3		Ratio Φ	Average ANG	Percentage ANG<45°	Average ANG<45°
	trend	plunge	trend	plunge				
	121	05	249	82	0.2	30	79°	15°
1	130	02	028	81	0.0	26	85°	12°
2	131	02	031	82	0.1	32	76°	14°
3	130	04	001	83	0.0	21	94°	11°
4	128	07	019	71	0.1	22	94°	13°
5	133	01	231	82	0.2	30	76°	10°

Angles in degrees.

choices among nodal planes must be supported by independent geophysical or geological data. Note that in Table 4.3 two subsets were selected taking into account the attitude of active faults near Yuli, as well as information on geological structure, which implies that the other two subsets fit these requirements poorly. Other examples of stress determinations from sets of focal mechanisms of earthquakes have been discussed in detail by Mercier & Carey-Gailhardis (1989) (also see Chapter 18).

Quality estimators

With the methods referred to in Tables 4.1 and 4.2, the simplest individual or average quality estimators are called RUP (from 0 to 200%) and ANG (from 0 to 180°). The latter simply describes the misfit angle as shown in Fig. 4.40, whereas the former depends on both the misfit angle and the relative shear stress magnitude.

The approximate relationship between the estimators ANG (for the method minimizing simple functions of the slip-shear stress angle) and RUP (for the new direct inversion method) is summarized in Fig. 4.45. The rough equivalences between the limits adopted for ANG and RUP in order to detect inhomogeneities in data sets with both methods (see n_1 and n_2 in Tables 4.1 and 4.2) are thus explained. Numerous fault slips generally display approximate proportionality between individual estimators AND and RUP (Fig. 4.45).

Consistency of results

The basic principle of stress tensor determinations (finding a reduced stress tensor that best accounts for the observed directions and senses of motion on a variety of fault planes in a rock mass) implies that for each fault slip datum, the direction and the sense of actual motion are, ideally, identical with those of the shear stress that is induced on a fault plane by the common stress tensor. Sources of variation in stress orientation, and the ratio Φ within the rock mass are thus neglected, including the effects of fault interaction (e.g. Price 1966). However, it is possible to estimate the significance of this assumption by determining, for each fault, the misfit between the actual slickenside lineation and the theoretical shear stress computed from the average stress tensor (compare diagrams on left and right in Fig. 4.42).

This type of check has been made with several hundred fault populations, many of them being large (50–200 measurements) and geometrically complex. In most cases, the average deviation of computed shear stress from actual striae remains smaller than the sum of instrumental and observational errors. When measuring faults in the field, one makes small instrumental errors (± 1° for each of the three angles, with a precise and large compass and clinometer) and larger observation errors that correspond, for example, to irregularities of fault surfaces and striations (± 5° – 15° in common cases) (Stewart & Hancock 1991). Seismological uncertainties in the determination of focal mechanisms of earthquakes are generally larger.

The average deviation of computed shear stress from actual slip direction generally remains acceptable in comparison to all possible error sources in measurements. As a consequence, the basic principle that relates the geometry of each independent fault movement to the common stress tensor must be considered valid as a

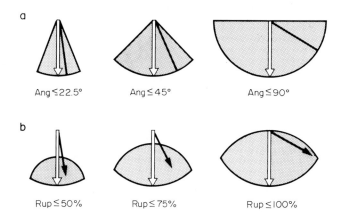

Fig. 4.45. Approximate relationship between shear-slip angle in fault plane [(a): individual estimator ANG as in Table 4.2] and ratio upsilon [(b): individual estimator RUP as in Table 4.1]. Actual slip vector: open arrow; its length (e.g. √3/2) depends on the reduced tensor form (Angelier 1990). Computed shear stress: black arrow (b) or thick line (a). Note that shear stress magnitude has importance for (b), but not for (a). Shaded area in (b): possible location of shear stress. Values larger than 90° (ANG) or 100% (RUP) not illustrated.

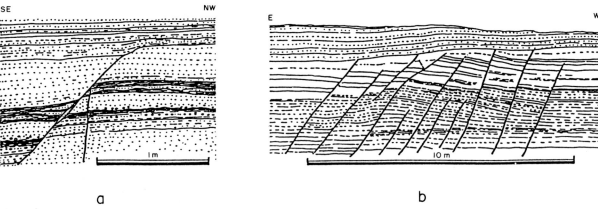

Fig. 4.46. Stratigraphic dating of faults. Uppermost layers in these sections are not affected by faulting. (a) Rainbow Canyon. (b) Panaca Basin. Late Cenozoic sediments and extension, southern Nevada, U.S.A.

first approximation, as far as large homogeneous data sets with various fault orientations are concerned. Individual anomalies occur, but do not vitiate the final result, provided that three characteristics are present in the inversion technique: (1) consideration of the sense of fault slip in the basic formula, (2) choice of functions that emphasize the minimization of acceptable angles and not that of anomalous values, and (3) adoption of robust research algorithms.

The method discussed above enable determination of an average stress applied to the periphery of the studied rock mass, regardless of local variations that occur within this mass. Such internal variations remain small and appear to cancel one another, provided that the size and variety of the whole data set are large enough. In all cases, the determination of individual and average angular deviations of computed shear stress from actual striae allows a control to be made.

CHRONOLOGY OF FAULTING

Dating of faulting events involves different techniques. Firstly, the age of fault movements can be bracketed, or in some cases more accurately determined, by taking into account the ages of rock formations affected or unaffected by faults. Secondly, geometrical relationships between fault motions allows reconstruction of relative chronology. Thirdly, the mechanical consistency between fault slips is widely used in order to correlate fault mechanisms, combined with the preceding techniques.

Stratigraphic dating

The best way to determine the *age of faulting* is illustrated in Fig. 4.46: fault movement post-dates deposition of the lower sedimentary rocks in the sections, whereas the age of the upper layers assigns an upper limit to that of faulting. In the cases shown in Fig. 4.46, the determination of the age of faulting is accurate because both the lower and the upper layers have approximately the same age. Unfortunately, in

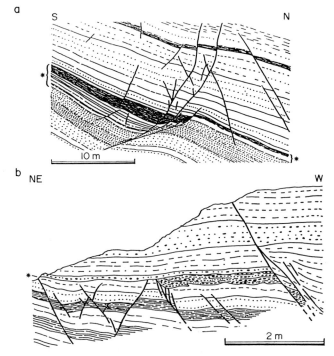

Fig. 4.47. Syndepositional faults. Some fault movements predate the upper layers in these sections, whereas other ones postdate all observable layers. Note the variation of thickness of the layer shown by an asterisk in (a), and the local uncomformity also shown by an asterisk in (b). Late Cenozoic sediments and extension, southern Nevada, U.S.A. (a) Mesquite Basin. (b) Rainbow Canyon.

many other cases, age bracketing is rather imprecise; for instance, in large regions of the European platform where extensive palaeostress analyses from faults have been made (Letouzey 1986, Bergerat 1987), at a majority of sites the sedimentary rocks are of Mesozoic age (especially Jurassic), so that the age of faulting is difficult to constrain in the absence of Tertiary sediments.

Valuable information is obtained where *syndepositional faults* can be recognized because the age of the formation is the same as that of faulting (Fig. 4.47). Age determinations of igneous rocks, if available, also help constraining the age of faulting where geometrical relationships between fault offsets and intrusions (or

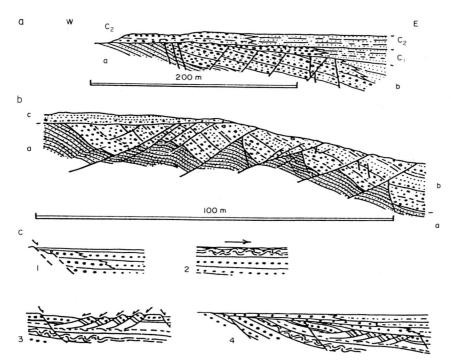

Fig. 4.48. Synsedimentary deformation across the margin of an extensional basin. Overton, Mesquite Basin, southern Nevada, U.S.A. (a) Normal faulting in layers [a] and [b], unconformably overlain by layers [C1] and [C2]. Note the transition between coarse clastics and thinner deposits, from the basin edge (on left) towards depocentres (on right). (b) Small tilted blocks and antithetic normal faults in layers [a] and [b], unconformably overlain by layer [c] as in (a). (c) Tectonic instability and syndepositional evolution of basin margin: (1) Initial stage. (2) Slumping. (3) Block tilting and faulting. (4) Deposition of upper clastic layers above an uncomformity.

lava flows) are observable. Such relationships are commonly used in magmatic provinces such as Iceland, where sediments are scarce.

The margins of faulted basins commonly allow observation of syndepositional faults which reveal continuing tectonic instability. In Fig. 4.48, the evolution of such a margin showing a complicated sequence of slumps, shallow normal faults associated with tilted blocks, and unconformities associated with rapid facies changes is illustrated. These structures may, or may not, reflect deep-seated tectonism; careful comparisons with results obtained in sites external to the basins can help to solve the problem.

On a wider scale, *syndepositional faulting* can be accurately reconstructed where fault movement and deposition have interacted during a longer time interval, so that several stages of fault movements are reconstructed based on increasing offsets of strata of increasing age. This is the case for the *growth faults* illustrated in Fig. 4.49; deep downcutting by rivers allows observation of normal faults in canyon walls (also see Figs 4.46–4.48) (Angelier *et al.* 1987), and successions of ash-flow tuffs that have recorded continued normal faulting during the Late Miocene.

The examples described above are of extensional faulting but syndepositional compressional deformation also occurs, as demonstrated, for instance, by the development of piggy-back basins and foredeeps associated with fold-and-thrust belts (Chapter 13). In such situations, reverse faulting, folding and thrusting control the development and migration of depocentres. Case studies in the southern Apennines, based on a combination of structural analysis, seismic reflection profiles and palaeostress reconstructions, are described by Hippolyte (1992).

Geometrical relationships and tectonic relative chronology

Successive fault slips along different directions on a single fault surface are common (Fig. 4.50a). Cross-cutting relationships are also frequent; in many cases, a younger fault clearly cuts and offsets an older fault (Fig. 4.50b). In other situations, however, the relationships between fault movements are ambiguous, because both the older and the younger fault have been active during the second event, resulting in complicated reactivations. Synchronous faults can also intersect, in such a way that successive reciprocal offsets result in complicated patterns (e.g. Fig. 4.50c). The relative chronology of brittle deformation is also commonly based on relationships between stratigraphically datable faults and other structures, such as veins, dykes, pressure solution seams and joints (see Chapter 5). Such observations can reveal a sequence of two fault mechanisms belonging either to a single tectonic event or to two distinct events. Because of uncertainties in the recognition of tectonic events based on mechanical reasoning, and multiple sources of errors, such an assemblage of relative chronology data must be as large

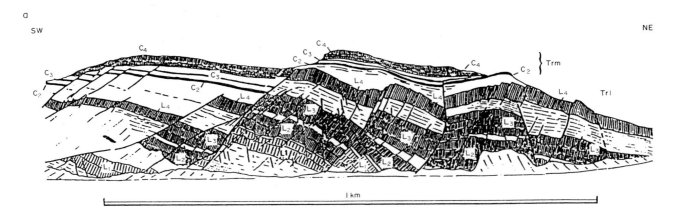

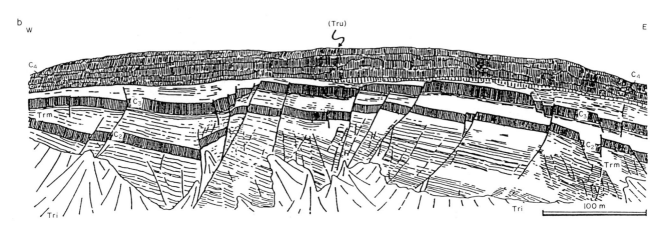

Fig. 4.49. Examples of Miocene syndepositional normal faults. (a) Growth faults in the wall of Rainbow Canyon, southern Nevada, U.S.A. Several ash-flow tuff units are identifiable (L1, L2, L3 and L4 from base to top) in the lower formation (Trl). Characteristic welded ash-flow tuff layers are also observable (C2, C3 and C4) in the upper information (Trm). (b) Growth faults in the upper formation (Trm) of Kershaw Canyon, same area as (a). The youngest formation, Tru, is not affected by these faults.

as possible in order to constrain the sequence of tectonic episodes. A powerful numerical method for deciphering such large sets of chronological data is discussed in a later subsection.

Mechanical consistency

Figure 4.51 illustrates a common situation. Initially, extensional block tilting was associated with antithetic normal faulting and some strike-slip faulting (Fig. 4.51a). At a later stage (Fig. 4.51b), normal faulting continued with limited block tilting and

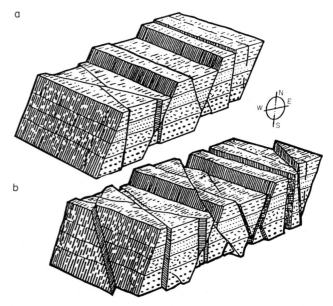

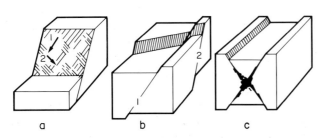

Fig. 4.50. Evaluation of the relative chronology of faulting. (a) Successive fault slips, 1 and 2. (b) Older fault, 1, offset by younger fault, 2 (c) Synchronous fault slips during conjugate faulting. Note than in (a) and (b) successive motions may belong to a single event.

Fig. 4.51. Schematic block diagrams summarizing the main aspects of tectonic evolution at Hoover Dam, U.S.A. (a) Main tilting stage (NE–SW extension). (b) Post-tilt stage (WNW–ESE extension).

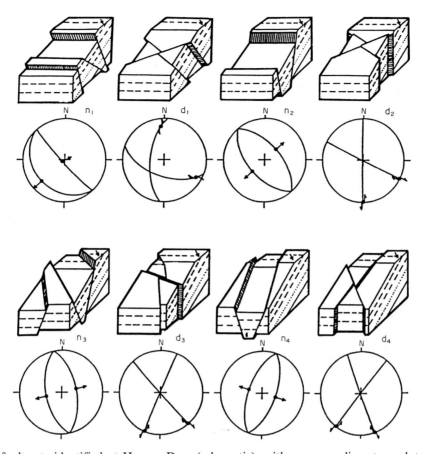

Fig. 4.52. Conjugate fault sets identified at Hoover Dam (schematic), with corresponding stereoplots (Schmidt projection, lower hemisphere). Present attitude shown by a small arrow on top of a faulted block which indicates north. The arrangement of conjugate sets relative to tilted bedding reflects the faulting-tilting chronology. n_1 and d_1, N55°E extension: pre-tilt normal and strike-slip faults (respectively). n_2 and d_2, n_3 and d_3, n_4 and d_4: post-tilt normal and strike-slip faults related to N55°E, N80°E and N105°E trends of extension, respectively.

clockwise rotation of the direction of extension so that the older dip-slip normal faults underwent oblique, right-lateral and normal, reactivation. Newly formed dip-slip normal faults with oblique trends also developed during this major stage. Some strike-slip faulting continued.

This sequence of extensional faulting events produces a complex pattern of: (1) older faults with little reactivation; (2) neoformed faults initiated during the first stage and obliquely reactivated during the second stage, and (3) neoformed faults that formed during this second event. The presence of widespread block tilting, principally during the first stage, and strike-slip faulting, during both the first and the second stages, resulted in a complex pattern of cross-cutting and reactivated faults. This tectonic sequence was reconstructed from an extensive analysis of fault slip data collected from Miocene rocks in the Colorado River canyon walls at Hoover Dam, 40 km southeast of Las Vegas, Nevada (see details in Angelier et al. 1985, and Angelier 1989). The distribution of fault slips was extremely complicated, but the identification of conjugate fault sets and analysis of their chronological relationships allowed a preliminary clarification of the local tectonic evolution (Fig. 4.52). The two major stages of late Cenozoic tectonic evolution at Hoover Dam are: (1) dip-slip and strike-slip faulting, with associated striatal tilting affecting fault blocks as a result of a N50°E direction of extension; and (2) post-tilt dip-slip, oblique-slip and strike-slip faulting, with a dominant N105°E direction of extension (Fig. 4.51).

In more detail, four sub-stages are distinguishable at the Hoover Dam site. The first two correspond to the same direction of extension, N50°E, and they were separated on a geometrical basis as being mainly pre-tilt and post-tilt, respectively (upper row, Fig. 4.52). Each sub-stage includes two palaeostress states, one with predominantly dip-slip normal faulting, the other with predominantly strike-slip faulting. These first two sub-stages can be grouped into a single tectonic event (the first stage) which induced extensive block faulting and tilting. Qualitative evaluations of fault slip chronology based on numerous observations (Fig. 4.50) indicated that during each sub-stage, dip-slip and strike-slip faulting probably alternated in time. The last two sub-stages (the second stage) involved solely post-tilt activity, with the direction of extension rotating clockwise from N80°E to N105°E (lower row, Fig. 4.52). Again, each sub-stage probably includes oscillations in time between dip-slip and strike-slip faulting (as shown by a complex, intricate sequence of dip-slip and strike-slip movements).

Finally, the palaeostress history revealed by the fault slip data analyzed at the Hoover Dam site includes four pairs of main stress states, each pair corresponding to a sub-stage illustrated in Fig. 4.52 by two conjugate fault systems, dip-slip and strike-slip. I define four pairs rather than eight individual stress states, because for each pair the minimum stress axis, σ_3, is stable while the other two axes interchange. These pairs are called $n_i - d_i$, where n and d denote predominantly normal dip-slip and strike-slip, respectively, and numbers, i, indicate succession from 1, older, to 4, younger (Fig. 4.52).

For each pair, $n_i - d_i$, determinations of stress tensors allow reliable separation of two types of stress: one type corresponds to predominant normal dip-slip faulting (n_i), the other type, with σ_2 close to the vertical corresponds to dominantly strike-slip faulting (d_i). However, the common orientation of extension (compare n and d for each pair in Fig. 4.52) and the complex and intricate fault slip relative chronologies suggest that these two types of stress alternated in time within each tectonic sub-stage.

Automatic separation of stress states

The identification of eight systems of conjugate faults at Hoover Dam was difficult as a consequence of the wide variety of fault slip orientations. It was thus necessary to split the complete data set into several groups according to mechanical reasoning based on the consistency of the slip-stress relationships discussed earlier. A method for an *automatic separation of fault slip data subsets* within an inhomogeneous data set (and corresponding determinations of average palaeostress tensors) was proposed by Angelier & Manoussis (1980), its basic algorithm being summarized in Fig. 4.53 (Angelier 1984).

The separation of the different stress tensors and related classes is made using an adaptation of a general algorithm which was proposed first by Diday (1971), the *dynamic clustering method*. Computation time and memory size being taken into consideration, this method is more efficient than more elaborate cluster analyses. This property was important in the problem of distinguishing between fault slip data classes, because the criterion that enables classification of each measurement is more complicated than the usual criteria in Cartesian coordinates. Each step requires the computation of stress tensors prior to the determination of individual deviations of theoretical fault movements from actual ones. The functions used for computing reduced tensors and individual deviations must be identical. As when studying homogeneous data sets, misfit functions $F_1 - F_4$ are suitable.

The application of the method to the complete Hoover Dam data set is illustrated in Fig. 4.54. The four data subsets (a) to (d) shown in the figure correspond to the stress states referred to as n_2, d_2, n_4 and d_4 in Fig. 4.52. The other four subsets are not shown. Plotting the fault slip data in the diagrams shown in the upper row of Fig. 4.54 only aims at illustrating the variety of fault orientations: the stereographic projection of fault planes and slickenside lineations resulting in diagrams which look like balls of string. However, conjugate fault patterns have been found and provide a simpler key to decipher complex fault slip distributions (Fig. 4.52). Stress axes are shown in the lower row of diagrams in Fig. 4.54; Figs 4.52 and 4.54 showing that for each substage, normal and strike-slip faulting modes are related through simple permutations between the principal stress axes σ_1 and σ_2.

Stress regimes and tectonic events

Separation of fault slip data subsets based on mechanical consistency with several unknown stress tensors (Figs 4.53 and 4.54) allow distinction of tectonic regimes. The distribution of all faults within a subset is complex because of natural dispersion in orientations and the presence of inherited fault slips (upper row, Fig. 4.54). A simpler illustration is obtained after identification of the conjugate fault patterns included in the subsets (Fig. 4.52).

For each major direction of extension, the internal set of relative chronologies between normal and strike-slip fault types (e.g. n_2 and d_2, Fig. 4.52) lacks consistency, which suggests that dominantly normal and strike-slip movements alternated in time. In other words, two

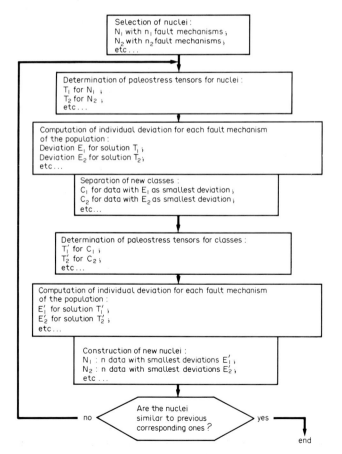

Fig. 4.53. Principle of separation of homogeneous classes and related stress tensors within an inhomogeneous fault population.

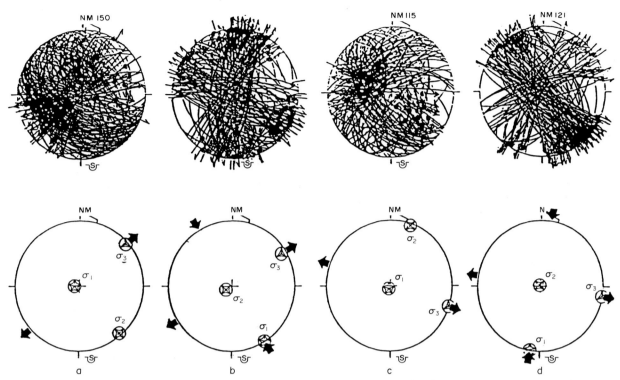

Fig. 4.54. Example of separation of fault slip data subsets within an heterogeneous data set, with determination of corresponding average palaeostress tensors. Schmidt's projection, lower hemisphere. Miocene volcanic formations at Hoover Dam, Nevada, Arizona, U.S.A. (Angelier et al. 1985). Upper row of diagrams: the four homogeneous subsets separated according to the principle summarized in Fig. 4.53. Fault planes shown as thin lines, slickenside lineations as small dots with single or double thin arrows (mostly normal or strike-slip, respectively). Lower row of diagrams: corresponding axes of stress tensors. (a) 158 faults, $\Phi = 0.2$, ANG = 15° (parameters defined in the last section of this chapter); same regime as for n_2 in Fig. 4.52. (b) 131 faults, $\Phi = 0.3$, ANG = 11°; same regime as for d_2 in Fig. 4.52. (c) 115 faults, $\Phi = 0.3$, ANG = 16°: same regime as for n_4 in Fig. 4.52. (d) 127 faults, $\Phi = 0.3$, ANG = 10°; same regime as for d_4 in Fig. 4.52.

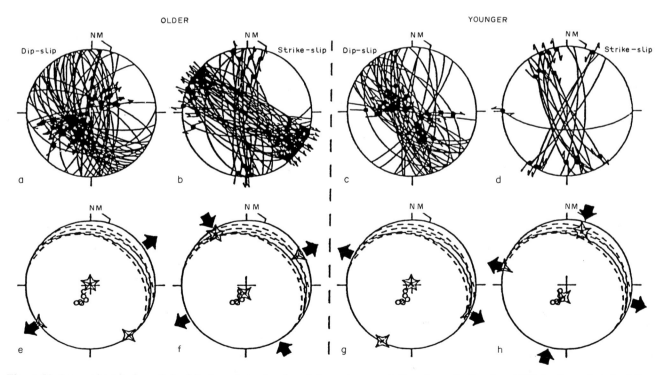

Fig. 4.55. Lower-hemisphere Schmidt steroplots showing data and computational results for about 200 faults and six bedding planes from the area between Kershaw-Ryan Park and Sawmill graben, Rainbow Canyon area, Nevada, U.S.A. Upper row shows projections of fault planes (lines) and striae (dots). Lower row shows projections of bedding planes (dashed lines), poles to bedding (open circles), and computed palaeostress axes (three-, four-, and five-branched stars respectively indicate σ_1, σ_2 and σ_3 axes). Compare with Fig. 4.54.

stress states could be reliably separated for each direction of extension, but there is no reason to interpret this duality in terms of successive events. Rather, it should be interpreted in terms of oscillatory stress regimes characterized by repeated permutations between two stress axes (σ_2 and σ_1 in this case).

Despite the identification of consistent relative chronologies of fault slips, the first two pairs of stress states do not correspond to distinct events either, but simply to block tilting, so that the oldest fault systems are tilted ($n_1 - d_1$, Fig. 4.52) whereas the youngest are not ($n_2 - d_2$, Figs 4.52 and 4.54). The apparent clockwise rotation between the last two stress states ($n_3 - d_3$ and $n_4 - d_4$, Fig. 4.52) might correspond either to a real clockwise rotation of stress (one or two events) or to a simple counterclockwise block rotation during a single event involving WNW–ESE extension (Fig. 4.53).

Do the eight stress states identified at the Hoover Dam site correspond to a single extensional event? This is probably not so for several reasons, including a consistency with similar results independently obtained in several other areas of the Basin and Range province. Figure 4.55, for instance, illustrates the analysis of an inhomogeneous data set collected in the Rainbow Canyon area (Michel-Noël et al. 1990), that reveals the same contrasting directions of extension as shown in Fig. 4.54 despite the difference in location and structural setting. As at Hoover Dam, each event includes two states of stress linked through a simple permutation of stress axes, σ_2 and σ_1, resulting in a complex mixture of normal and strike-slip faults.

Summarizing, the separation methods based on a search for mechanical consistencies within inhomogeneous sets of fault slip data allow reliable identification of distinct stress regimes. Whether or not these stress regimes correspond to distinct stress events raises an additional problem, which must be solved taking into account various sources of geological information.

Successive stress states, folding and tilting

Figures 4.51 and 4.52 illustrate the influence of block tilting, which can result in successive sub-stages of extensional deformation and different states of stress during a single tectonic event that combines *domino-like tilting* and *antithetic normal faulting*. Similarly, in fold-and-thrust belts, compressional deformation combining folding and reverse/strike-slip faulting can result in an apparent complexity of palaeostresses. Geometrical relationships between two compressional events are illustrated in contrasting situations in Fig. 4.56. The two compressions occur successively, their compressional trends making an angle of about 30°; compression associated with folding occurs second in Fig. 4.56(a), and first in Fig. 4.56(b), resulting in different relationships between faults (strike-slip and reverse) and folds. As with the succession of extensional episodes previously described, one must address carefully the problem of two compressional stresses, which may reflect either a rotation of stress or a rotation of fold-and-thrust blocks (in the opposite sense) in a single compression.

The complicated fault pattern shown in Fig. 4.57 (from the Plio-Quaternary collisional belt of Taiwan), results from a single compressional event combining stages of strike-slip faulting (Figs 57a & e) and reverse faulting (Figs 4.57b & d) with folding and bedding slip (Fig. 4.57c). Four contrasting stress regimes can be identified and related to a single compression through σ_2/σ_3 stress permutations (cf. Figs 4.57a & b, d & e) and tilting due to folding (cf. Figs 4.57a-b, d-e).

Increments of extension can be analyzed in detail in a series of faulted tilt-blocks (Fig. 4.58) because the progressive tilting resulted in the activation of vertical tension fractures as normal faults. Studies of fault slip sequences in terms of palaeostress reconstructions provided a key to understanding the structural development of such systems and distinguishing successive steps.

The chronological matrix analysis

Two or several tectonic events can generally be identified at many sites, and if numerous, their analysis allows complete reconstruction of a regional succession of events. Apart from interpretational errors, sources of confusion are numerous (undetected rotations, similar events, tectonic type changes through space or time). A quantitative tool is thus needed in order to visualize the complete set of local chronological data. A number, *n*, of events is recognized in an area; a number as large as

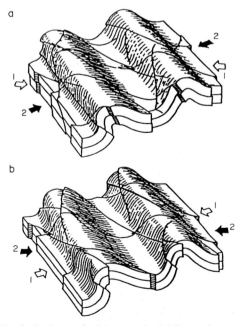

Fig. 4.56. Polyphase faulting and folding. Summary of geometrical relationship between two compressional events on a decametric-kilometric scale. (a) Strike-slip faulting prior to folding, with reverse faulting during each event; arrows indicate successive σ_1 axes (1 – old, 2 – new). (b) Strike-slip and reverse faulting after folding.

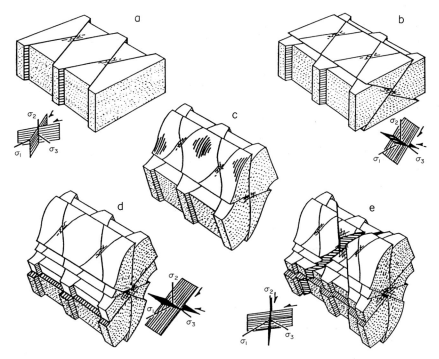

Fig. 4.57. Singlephase compressional faults and folds on a metric to decametric scale. Reconstruction of five stages of deformation. (a) Early conjugate strike-slip faulting. (b) Early conjugate reverse faulting. (c) Folding with bedding slip. (d) Late conjugate reverse faulting. (e) Late conjugate strike-slip faulting. σ_1, σ_2, σ_3: principal stress axes.

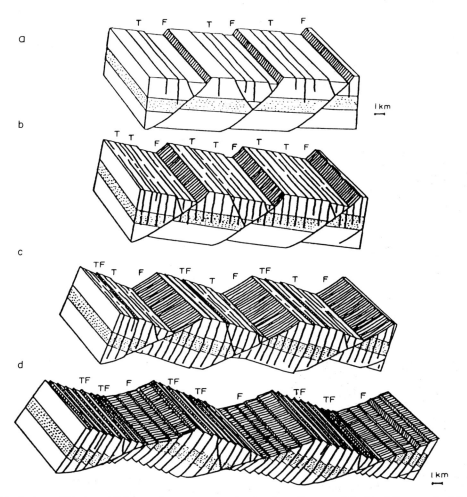

Fig. 4.58. (a – d) Evolution of faults with increasing amounts of extension. F, normal faults oblique to bedding (especially first-order faults); T, tension gashes and fractures; TF, normal faults approximately perpendicular to bedding (second- and third-order faults).

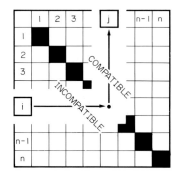

Fig. 4.59. Matrix of relative chronologies. Line index *i* (on left) for first event; column index *j* (on top) for second event. Black dot thus designates the chronology *i* before *j*.

possible is used in order to classify these events. The basic datum is the succession order of two events.

We must consider first that the number of possible successions for *n* events is *n*! Classification of five events or more through an analysis of binary chronologies might thus be complicated, taking into account the possible existence of contradictory evidence. Rigorous means are necessary in order to quantify and minimize such contradictions.

For a number, *n*, of events, the number of different possible binary chronologies is $n(n-1)$. The problem of classifying *n* events finally consists of finding, among the *n*! possible solutions, the solution that corresponds to the smallest number of contradictions among the $n(n-1)$ possible kinds of binary chronologies. A square matrix is built, in such a way that one column and one line with the same rank correspond to a single event (Fig. 4.59). Because the complete succession order is unknown, the choice of ranks is arbitrary (Fig. 4.60a). In this kind of matrix, lines correspond to the first event and columns to the second one (Fig. 4.59). Thus, the upper-right triangle of the matrix contains all chronologies consistent with the classification adopted, whereas the lower-left triangle is the domain of incompatibility. The classification shown in Fig. 4.60(a) is not satisfactory, because the lower-left triangle is not empty. This example was obtained from 115 relative chronologies at 57 sites in the Cretaceous basin of Kyongsang, Korea. Events E1–E4 are extensional, while events C1–C3 are compressional; for each mode, events have been distinguished based on their contrasting directions of compression or extension.

By definition, the *succession order solution* of the problem corresponds to a data set as large as possible in the upper-right half-matrix, and as small as possible in the lower-left half-matrix. The search for this solution is done through successive permutations between lines and simultaneous corresponding permutations between columns. In the case illustrated, the best possible succession accounts for 112 consistent binary chronologies and includes 3 inconsistent ones (Fig. 4.60b). From a mathematical point of view, the problem consists of building a triangular matrix through successive numbering permutations for which a microcomputer program is easily written. Major advantages of this kind of resolution are the estimation of amounts of contradiction between data, and comparisons between solutions. Multiple solutions are frequent, meaning that the total set of binary chronologies does not fully constrain the problem. In such cases, additional collection of chronological data in the field is necessary. The use of a *chronology matrix* allows complete quantification of the problem (Angelier 1991) in order to obtain the succession order of multiple tectonic events. This method is especially useful where stratigraphic constraints are few.

PALAEOSTRESS TRAJECTORIES MAPPED

Palaeostress determinations and dating of tectonic events result in the construction of *palaeostress maps* which allow consideration of relationships between stress distributions and regional kinematics.

Reconstructon of events at the regional scale

In Fig. 4.61(a), local determinations of compressional trends in the Western Foothills of the Taiwan fold-and-thrust belt are shown (Angelier *et al.* 1986). Each bar in Fig. 4.61 corresponds to the nearly horizontal σ_1 axis of a local stress tensor determination from fault slip data. At most sites, two events were reconstructed, with contrasting trends of σ_1 axes (NW–SE and E–W in southwestern Taiwan; NNW–SSE and NW–SE in northeastern Taiwan).

The chronology of these events was established based on collection of relative chronology data at each site, and confirmed by stratigraphic information on the age of the faulted sedimentary formations (all Late Cenozoic). The relative chronology of the most important two events is indicated at each major polyphase site shown in Fig. 4.61(a); it reveals a consistent apparent counterclockwise rotation of compressional palaeostress axes (a single exception may result from misinterpretation or a more complicated local sequence).

Both the systematic determination of palaeostresses

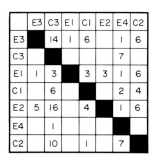 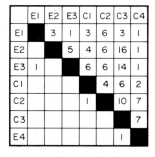

Fig. 4.60. Example of application. 57 sites, 7 major tectonic events, 115 binary chronologies. (a) Chronology matrix obtained with a succession order chosen at random. (b) Chronology matrix corresponding to the best possible solution. Note that matrix infill is irregular for (a) and concentrated within the upper right triangle for (b).

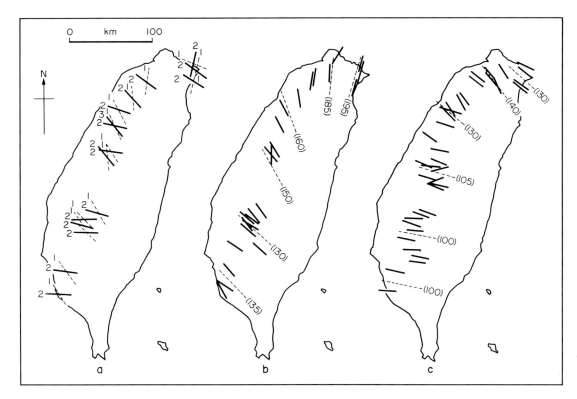

Fig. 4.61. Trends of compressional palaeostress in the Plio-Quaternary Taiwan collision belt. (a) Major polyphase sites; 1 – 2 – 3 refer to local relative chronologies of stress regimes. (b) First major compressional event, with σ_1 axes as bars and average regional trends as dotted lines; azimuths between parentheses in degrees. (c) Second major compressional event.

and the analysis of relative tectonic chronology thus allow complete classification of compressional palaeostress tensors attributable to these two Plio-Quaternary events, including compressional trends identified at singlephase sites. The result is shown in Figs 4.61(b) and (c), which respectively correspond to the first and second events separated in Fig. 4.61(a). Note that although identification of events is made on a directional basis where no dating is available, the compressional trends vary widely at the scale of the entire island. For the first major event (Fig. 4.61b), they change from N45°W in southwestern Taiwan to N15°E in northeastern Taiwan. For the second event (Fig. 4.61b), they change from N80°W to N50°W.

Although there is no need to discuss here the geodynamic significance of these results, it should be pointed out that in northern Taiwan, the larger clockwise deviation of σ_1 trends for the first event (Fig. 4.61b) is partly attributable to a clockwise rotation of the northernmost segment of the belt, independently determined from palaeomagnetic analyses. This rotation having been corrected (i.e. the northeastern segment of the belt is back-rotated counterclockwise for about 20°), the approximate trends of σ_1 for the first event vary from N45°W (south) to N5°W (north). As a consequence, the amplitude of the progressive clockwise variation of σ_1 trends along the fold-and-thrust belt (from south to north) is approximately the same for both events (i.e. 30–40°). The average counterclockwise change of σ_1 trends with time, however, is spectacular (about 30–40° in central Taiwan, see Figs 4.61b & c).

Relation to plate kinematics

Palaeostress analysis on both sides of the collision belt of Taiwan (Barrier 1985, Angelier *et al.* 1986) allowed comparison between distributions of compressional trends throughout the deformed zone in the general context of the kinematics of the Philippine Sea plate – Eurasia plate convergence (Chapters 9 and 13). The smoothed stress trajectories shown in Fig. 4.62 were derived from hundreds of stress tensor determinations, including those summarized in Fig. 4.61. For the most recent event (Fig. 4.62b), there is good correspondence between the average trend of σ_1 in north-central Taiwan (approximately N110°E) and the direction of plate convergence independently reconstructed by geophysical means. For the earlier event, the direction of plate convergence is unknown but geometrical reasoning suggests, and finite-element analysis confirms, that to be consistent with compressional deformation in Taiwan at this time, it should have been approximately N25°W.

On a wide scale, the reconstruction of palaeostress tensors for several Cenozoic events based on analyses of fault slip data sets throughout the West European platform allows correlation between average palaeostress trends and major aspects of Africa – Eurasia plate convergence (Bergerat 1985,

1987). This relationship is illustrated in Figs 63(a) & (b) for two major events; the late Eocene N–S compression and the late Miocene NW–SE compression, respectively. These compressions are related to different locations of the Africa/Eurasia pole of rotation (Fig. 4.63), and to different configurations of plate boundaries (Dercourt *et al.* 1986, Le Pichon *et al.* 1988). Note that most palaeostress determinations were obtained in the stable platform areas of the Eurasian and African forelands (Fig. 4.63), thus providing further confirmation that weakly deformed regions record major events very well, albeit in attenuated form (also see Chapter 5).

Figure 4.64 illustrates the case of large-scale extensional deformation in the continental arc/back-arc geodynamic setting of southern Aegea. Based on the same approach as for the compressional events discussed above, a sequence of two major extensional stages was reconstructed for the Plio-Quaternary in southern Greece (see discussion in Angelier *et al.* 1982). A comparison between regional palaeostress distributions shown in Figs 4.64(a) & (b) demonstrates firstly that a major clockwise rotation of average σ_3 trends with time has occurred in this region (e.g. from N–S to NE–SW in western Anatolia). Secondly, the palaeostress trajectory pattern shown in Fig. 4.64(b) generally resembles that of Fig. 4.64(a), with some major characteristics in common such as the geographic permutation between σ_2 and σ_3 axes along the southeastern branch of the Hellenic arc. From a geodynamic point of view, this σ_2/σ_3 permutation is probably related to the distribution and orientation of suction forces that differ along the southwestern and southeastern branches of the Hellenic subduction zone.

Stress permutations and deviations

The fan-shaped pattern of Plio-Quaternary compressional stress trajectories shown in Fig. 4.62 is related to the shape of the plate boundary near Taiwan (i.e. the northwestern corner of the Philippine Sea plate indenting the Eurasian margin). A similar fan-shaped pattern of σ_1 trends is observable in Fig. 4.63(b) along the arcuate fold-and-thrust belt of the Western Alps, for the Late Miocene compressional event. The Plio-Quaternary extensional stress trajectories in southern Aegea are strongly influenced by the arcuate shape of the Hellenic subduction zone (Fig. 4.64), which results in several perturbations of σ_3 trajectories.

For extensional/strike-slip faulting (Fig. 4.65a), stress tensors may remain generally homogeneous in terms of σ_3 orientation despite the variety of fault slip patterns. In Fig. 4.65(a), diagrams 1, 2 and 3, illustrate normal dip-slip faulting, normal-dextral oblique-slip faulting and pure strike-slip faulting respectively, nevertheless all indicate a consistent E–W direction of extension, according to geological observations in the extensional strike-slip tectonic setting of the Gulf of California. In contrast, palaeostress analyses of normal dip-slip faulting within the dextral pull-apart basins of the same region revealed a NW–SE direction of extension, parallel to the trend of major strike-slip faults. That such a local occurrence of extension related to the direction of strike-slip faulting, rather than to

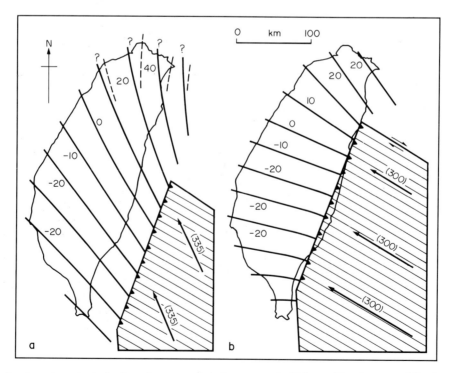

Fig. 4.62. Smoothed palaeostress trajectories of compressional stress, σ_1 in Taiwan, for the same Plio-Quaternary events as in Figs 4.61(b) & (c). Hatched pattern: northwestern corner of the Philippine Sea plate. Large open arrows, with azimuths between parentheses, indicate the motion of the Philippine Sea plate relative to Eurasia. In Taiwan, σ_1 trajectories as lines, with their deviation in degrees (+ clockwise, − counterclockwise) relative to a fan axis. In (a), dashed trajectories correspond to initial pattern before torsion of the northeastern segment of the belt. (a) First event. (b) Second event.

regional stresses, illustrates the influence of structural pattern on the distribution of tectonic palaeostresses.

Other examples of the regional deflection of palaeostress trajectories were described from the Oligocene of the West European rift system, especially in the Rhine-Saône transform zone (Lacombe et al. 1990).

Stress permutations are common at local (Figs 4.52, 4.54, 4.55 and 4.57) and regional (Fig. 4.66) scales (Angelier and Bergerat 1983). For instance, where extensional faulting occurs, such permutations result either in perpendicular systems of conjugate normal faults (σ_2/σ_3 permutation, Fig. 4.66a) or in mixed conjugate sets of normal and strike-slip faults (σ_1/σ_2 permutation, Fig. 4.66b). The σ_1/σ_2 permutation in extensional deformation was described from the Hoover Dam area (Figs 4.52 and 4.54); cross-cutting relationships may reveal intricate sequences from normal dip-slip to strike-slip (Fig. 4.66b), or from strike-slip to normal dip-slip (Fig. 4.66c).

Palaeostress and the ratio Φ

Figure 4.67 illustrates different extensional modes: perpendicular extension as in the Suez rift (Fig. 4.67a), oblique extension as in the Gulf of California (Fig. 4.67b; also see Fig. 4.65), and multidirectional extension as in Crete, Greece (Fig. 4.67c; also see Fig. 4.64). Whereas simple extension generally corresponds to relatively high values of Φ (e.g. 0.5), multidirectional extension is characterized by low values which make stress permutation (σ_2/σ_3) easier.

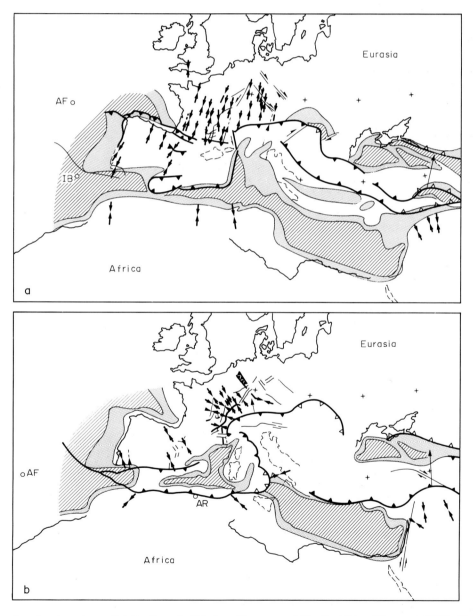

Fig. 4.63. Compressional palaeostress distributions in Western Europe and North Africa (Bergerat 1987). Trends of σ_1 axes as pairs of convergent arrows. Oceanic crust shown hatched, intermediate crust stippled, and continental crust blank. Major convergent boundaries as thick lines, with triangles in the overthrust plate. AF and IB, poles of rotation of Africa and Iberia with respect to Eurasia. (a) Late Eocene. (b) Late Miocene.

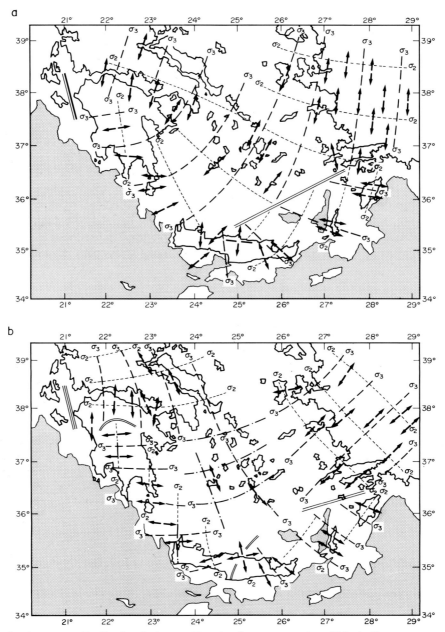

Fig. 4.64. Extensional palaeostress distribution in Aegea. Trends of σ_3 axes as pairs of black divergent arrows. Trajectories of σ_3 dashed, trajectories of σ_2 dotted. Deepest (>2000 m) portions of the Eastern Mediterranean basin are stippled. Double lines correspond to geographic permutation of σ_2/σ_3 principal stresses. Triple lines correspond to a transition to a compressional stress regime. (a) Late Pliocene – Early Quaternary. (b) Middle-Late Pleistocene-Holocene.

A similar effect accounts for reverse/strike-slip changes of faulting mode in compressional deformation (e.g. Fig. 4.57): low values of Φ generally correspond to situations characterized by σ_2/σ_3 permutations. These considerations also highlight the major role played by the ratio between principal stress differences: the importance of reverse/strike-slip mixed modes of faulting increases as Φ decreases in a compressional stress regime. Conversely, a decrease in the ratio Φ in an extensional stress regime results in more irregular trajectories of σ_3 and local permutations of σ_2/σ_3; fault slip analyses in Crete revealed low values of Φ (Fig. 4.67c) and frequent permutations between σ_2 and σ_3 axes (Fig. 4.64).

STRESS MAGNITUDES: THE COMPLETE STRESS TENSOR

In the palaeostress determinations described here there is no access to the values of k and l discussed earlier. As a result, only four variables of the stress tensor were discussed: the *orientations of the three principal axes* and *the ratio* Φ. This meant that in order to solve the *inverse problem* it was convenient to adopt a special form of the *stress tensor*, with four variables (the unknowns of the problem) through an arbitrary choice of k and l: *the reduced stress tensor*. To reconstruct the complete stress tensor, one must consequently determine two values (corresponding to k and l) after

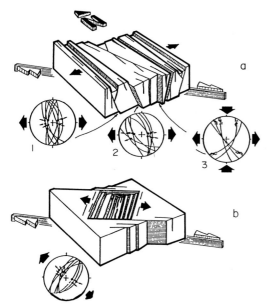

Fig. 4.65. Diagrammatic models of two distinct fault mechanisms induced by the same N 120–125°E dextral transform motion in northwestern Mexico during the Plio-Quarternary. Half-white arrows: transform movement. Black arrows: direction of extension (σ_3). (a) Mechanisms in continental blocks of northwestern Mexico with strike-slip and oblique-slip movements along older N 135–170°E structures, and E–W extension on N–S normal faults. (1), (2), (3): Schmidt's projections of N–S normal faults and NNW–SSE strike-slip faults corresponding to an E–W extension. (b) Mechanisms within the Gulf of California with transform faults and opening of NE–SW oceanic basins. 1: Schmidt's projection of NW–SE normal faults corresponding to a N 120–125°E extension in pull-apart basins.

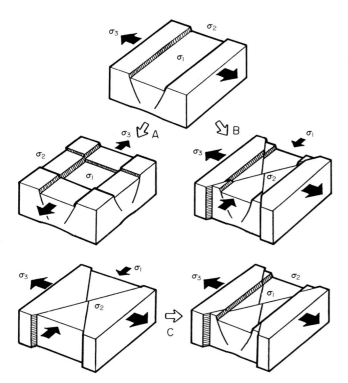

Fig. 4.66. Permutations between stress axes and rift opening. (a) σ_2/σ_3 permutation (perpendicular extension). (b) σ_1/σ_2 permutation (from normal fault extension to strike-slip regime). (c) σ_1/σ_2 permutation (from strike-slip regime to normal fault extension).

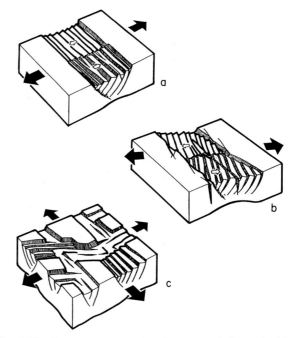

Fig. 4.67. Important extensional pattern (schematic block-diagrams). Large black arrows: direction of extension. Small open arrows: dip direction of tilted faulted block systems. (a) Transverse uni-directional extension. (b) Oblique uni-directional extension. (c) Multi-directional extension.

having determined a reduced stress tensor. The position and the size of the Mohr circles are thus fixed (Figs 4.28 and 4.30). Three different criteria can be used in order to determine the two variables that have been left unknown in the reduced stress tensor. These three criteria are illustrated schematically in Fig. 4.68 and they can be discussed in more detail considering the Hoover Dam area (Figs 4.51, 4.52 and 4.54) as a case example.

The failure criterion

A typical *failure envelope* is shown in the *Mohr diagram* of Fig. 4.68(a); the presence of neoformed conjugate faults being easily detected within the fault slip data set. By definition, the normal and shear stress magnitudes for conjugate faults correspond to a point on the largest Mohr circle, and the characteristic dihedral angle (2θ) between these faults is easily read from the diagram.

In theory, the geometrical relationship shown in Fig. 4.68(a) between a Mohr circle and a failure envelope, should be sufficient to completely determine the magnitudes of the extreme principal stresses, σ_1 and σ_3, because the failure envelope is non-linear so that knowledge of the 2θ angle constrains the gradient of the rupture curve at the point of tangency. This is not true in practice, because the slope of the failure curve varies little (much of the envelope is straight), except for low values of normal stress. As a result, in general, knowing the angle 2θ simply enables one to check the compatibility between this characteristic angle for the

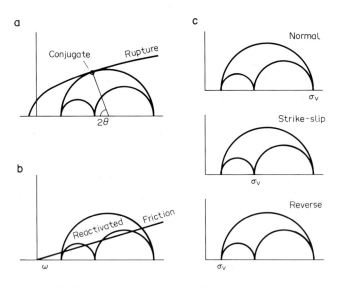

Fig. 4.68. Three criteria for determining the magnitudes of the principal stresses knowing the orientations of the stress axes and the stress ratio, Φ. Mohr diagrams (shear stress on ordinates, normal stress on abscissae). (a) Neoformed conjugate faults, with characteristic 2θ angle, define a point where the largest Mohr circle ($\sigma_1 - \sigma_3$) is tangent to the failure envelope. (b) Inherited faults resulting from the reactivation of older faults (or activation of older rock discontinuities as faults) correspond to points between the three Mohr circles; these points should be above the friction line characterized by the angle ω, thus fixing the ratio Ψ between the maximum and minimum stresses. (c) Knowing palaeodepth during a tectonic event, and hence lithostatic pressure, fixes the magnitude of the vertical principal palaeostress σ_v (σ_1, σ_2, or σ_3 depending on the tectonic regime).

conjugate system, and the average gradient of the failure envelope, within a certain range of normal stress values.

To summarize, using the geometrical relationship shown in Fig. 4.68(a), it is possible to reliably determine one relationship between the extreme principal stress magnitudes σ_1 and σ_3, provided that neoformed conjugate faults are observed and the rupture law of the rock mass is known. Because the failure envelope is not a straight line, this relationship is non-linear. Another relationship is still necessary to determine σ_1 and σ_3.

The friction law

A typical *initial friction curve* is shown in Fig. 4.68(b). From experimentation, the law is considered linear and no initial resistance to slip is assumed, so that this curve is a straight line intersecting at the origin. All inherited faults correspond to preexisting rock discontinuities (e.g. bedding planes, joints, older faults and fractures) which may slip under the stress considered. Slip on such discontinuities is controlled by friction laws (Jaeger 1969).

In the Mohr diagram, if the *initial friction law* is known, all points that represent fault slips should lie above the corresponding friction line (Fig. 4.68b); otherwise, slip will not occur. This requirement obviously imposes a constraint on the size of the Mohr circle. If the gradient of the friction line is unknown, it can be found provided that inherited faults are numerous enough to define the lower boundary of the mass of related points in the Mohr diagram. In both cases, considering friction laws assigns geometrical constraints on the Mohr circles. A second relationship between extreme principal magnitudes, σ_1 and σ_3, is thus obtained.

Combining rupture and friction laws

Finally, the two geometrical constraints shown in Figs 4.68(a) and (b) fix the magnitudes of extreme principal stresses, σ_1 and σ_3, provided that the mechanical properties of the rock mass in terms of rupture and friction laws are known (shear stress *versus* normal stress).

Because the ratio Φ has already been determined while computing the reduced stress tensor, the magnitude of the intermediate principal stress, σ_2, is immediately obtained. Inherited faults, especially faults oblique to all stress axes, thus play a major role in the determination of stress magnitudes. First, they provide the only way to determine the value of the stress ratio Φ while reconstructing the reduced stress. Second, they bring constraints in terms of initial friction.

The vertical stress

Determining the *lithostatic load* gives the value of the vertical stress, σ_v. Practical analyses show, and theoretical reasoning suggests, that one of the principal stress axes is generally vertical during a tectonic event (rotations can occur later). Depending on the rank (σ_1, σ_2 or σ_3) of this principal stress axis, dominantly normal, strike-slip or reverse faults, respectively, develop (Fig. 4.68c). Provided that the *palaeodepth* at the time of deformation can be determined, as well as the average density of overlying rocks, one thus obtains additional information (the value of one principal stress). Erosional exhumation or sedimentary burial must be taken into account.

Actual values

To conclude, at sites where the reduced stress tensor has been determined, rock mechanics considerations enable one to determine three independent relationships between the magnitudes of σ_1 and σ_3 (Fig. 4.68). Therefore, not only are the two remaining unknowns of the stress tensor firmly constrained, but also possible discrepancies between rupture, friction, and depth data are detected. In sites where rupture, friction or depth data are absent, the information can be inferred from this analysis but cannot be checked. For instance, any determination of principal stress magnitudes using the rupture and friction rock mechanics criteria shown in Figs 4.68(a) and (b) implicitly assigns a certain value to the depth of overburden.

Typical rupture and friction laws are contained in the rock mechanics literature (e.g. Jaeger & Cook 1969, Byerlee 1978). A compilation of published rock mechanics analyses, not discussed here, has resulted in selecting empirical relationships rather than theoretical laws. Hoek & Brown (1980) and Hoek & Bray (1981) have thus presented a failure criterion that satisfactorily accounts for the results of experiments. Figure 4.69(a) illustrates corresponding failure curves for volcanic rocks (the Hoover Dam case), with rock-mass quality decreasing from intact rock (curve A) to rock with closely spaced and heavily weathered joints (curve F).

Figure 4.69(b) shows the *initial friction law* and *maximum friction law* that have been proposed by Byerlee (1978), based on a compilation of experimental results. Byerlee has shown that, whereas at low normal stress (up to 5 MPa), the shear stress required for sliding depends on surface roughness, at higher normal stress this effect diminishes and friction is nearly independent of rock type. In the shear stress – normal stress diagram, most points corresponding to friction at normal stresses up to 600 MPa plot in the stippled area between the two lines shown in Fig. 4.69(b). The lower boundary (initial friction) corresponds to a minimum gradient of about 0.3, while the upper boundary (maximum friction) corresponds to gradients of about 0.85 and 0.5 (below and above 200 MPa, respectively). The most interesting property of these experimentally determined relationships between normal and shear stresses is their linear, or almost linear, character (Fig. 4.69b). In addition, observe in Fig. 4.69 that for a rock mass of poor quality (heavily jointed and weathered), the failure criterion of Hoek & Brown (1980) may require smaller shear stress magnitudes for slip to occur, than the friction criterion of Byerlee (1978) for rock samples.

Analysis of dimensionless Mohr diagrams

Figure 4.70 summarizes the results of Mohr diagram analyses at Hoover Dam. The fault slip data are the same as those shown in Fig. 4.54. However, 19% of faults, with individual angles (s, τ) larger than 20°, were eliminated before plotting the Mohr diagrams. The results obtained for the four tectonic sub-stages are given separately for normal and strike-slip faulting modes (Fig. 4.52). Despite data dispersion, a reasonable approximation for the minimum friction line is in good agreement with initial friction data, as comparison with Fig. 4.69(b) shows. This fit allows determination of the abscissa origin in Fig. 4.70, because the friction lines intersect the origin.

The only remaining unknown is the scale factor. This factor, k, is determined using the relationship between conjugate faults and the failure envelope (Fig. 4.68a). Neoformed conjugate faults were identified in fault slip data sets (Fig. 4.52); they correspond to point concentrations on the largest Mohr circle (Fig. 4.70). The corresponding 2θ angle averages 61° for normal as well as strike-slip fault systems.

Rock mechanics analyses of rocks from the Hoover Dam site indicate that the uniaxial compressional strength, σ_c, of the most common rocks (including ash-flow tuff units from which most of the structural data were collected) averages 40 MPa. With this value, the empirical criterion of Hoek & Brown (1980) for volcanic rocks has been adopted with appropriate parameters (i.e. good quality jointed rock mass: empirical constants in Hoek & Bray 1981). One thus obtains the failure envelope (as in Fig. 4.69a). Because most fractures developed very early in the history of the rock mass, no variation of the failure envelope with time was assumed.

With the Mohr diagram patterns shown in Fig. 4.70, and the position of the origin already determined using friction data, the best fit between this failure envelope and the largest Mohr circle is obtained for stress magnitudes σ_3 and σ_1 that, respectively, average 0.5 and 7 – 9 MPa for normal faulting, and 6 – 8 and 25 – 33 MPa for strike-slip faulting. The magnitude of the intermediate stress, σ_2, is thus fixed (2 – 2.5 and

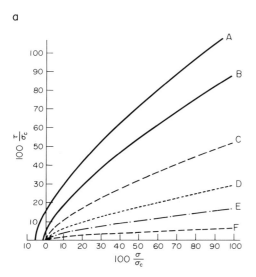

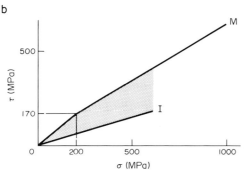

Fig. 4.69. Rupture and friction laws. (a) Curves obtained using the empirical failure criterion of Hoek & Brown (1980). Abscissa and ordinate: dimensionless normal stress, σ, and shear stress, τ, respectively (σ_c, compressive strength). Curves A – F refer to decreasing rock-mass quality (A, intact rocks; F, heavily jointed and weathered rock mass). Curves plotted according to parameters tabulated by Hoek & Bray (1981) for fine-grained igneous rocks. (b) Initial friction (I) and maximum friction (M) curves (adapted from Byerlee 1978). Normal stress, σ, shear stress τ, in MPa. Most friction data points should be in the stippled area between I and M.

10–14 MPa, for normal and strike-slip faulting, respectively).

Palaeodepth and lithostatic load

An additional constraint given by geological reasoning about the depth of overburden and the corresponding *lithostatic load* should now be discussed.

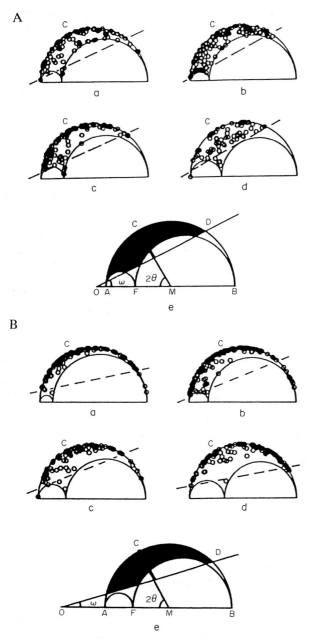

Fig. 4.70. Analysis of dimensionless Mohr diagrams derived from data collected at Hoover Dam. A (upper half of figure): dominantly normal faulting. B (lower half): dominantly strike-slip faulting. Diagrams (a), (b), (c) and (d) refer to the tectonic sub-stages defined in Table 4.1; numerical results and reference numbers are listed in Table 4.2. Individual fault slip data are plotted as small open circles. Lower boundary of the mass of representative points is shown as a pecked line. C, points corresponding to conjugate faults. Diagram (e) depicts the geometrical significance of parameters given in Table 4.2, and discussed in the text (ratios Φ and Ψ, angle ω and 2θ), with average values for each faulting mode.

A value of between 200 and 800 m is a reasonable estimate for the palaeodepth of the site during most of the tectonic events considered. With these depths and an average rock density of $2.6-2.7$ g cm^{-3}, one obtains lithostatic pressures of 5 and 20 MPa, respectively. These values are direct estimates of the maximum principal stress, σ_1, (for normal faulting) or intermediate principal stress, σ_2, (for strike-slip faulting). The determinations discussed above, which are solely based on rupture-friction analyses, have yielded values of 7–9 MPa for σ_1 during the normal faulting, and 10.5–14 MPa for σ_2 during the strike-slip faulting. These estimates are reasonably compatible with the range of 5–20 MPa obtained on the basis of independent lithostatic pressure determinations.

Stress magnitudes

Taking all these results into account, as well as the additional constraint that σ_1 during normal faulting should be equal to σ_2 during strike-slip faulting (these modes have alternated very rapidly, so that no significant change in vertical stress, σ_v, in dry conditions can have occurred), reasonable *average stress magnitudes* of σ_1, σ_2 and σ_3 at Hoover Dam are obtained: 10, 3 and 0.6 MPa, respectively, during normal faulting, and 25, 10 and 6 MPa, respectively, during strike-slip faulting. These low values suggest that the depth during most faulting was closer to the shallowest case (200 m), than to the deepest one (800 m): a vertical stress of 10 MPa corresponds to a depth of about 400 m in dry conditions.

All determinations have been made implicitly assuming that faulting occurred in a dry rock mass, that is, with zero pore pressure. Considering the hydrostatic pore pressure case, that is rock saturated with water, means that effective stress magnitudes can be computed by removing the hydrostatic head. This results in dramatically diminishing the stress levels required for failure. However, the least principal stress should remain larger than the pore pressure: otherwise, the pore pressure would hydrofracture the rock resulting in the opening of tension cracks instead of sliding on neoformed and inherited faults (see Chapter 5). Tension cracks are more numerous at Hoover Dam in the upper layers than at the level where the fault slip data were collected.

Pore pressure

A crucial factor in stress magnitude determination is the existence of *pore pressure*. Because all determinations have been discussed assuming no hydrostatic head, the values calculated should be considered as maximum values of effective stress. Actual magnitudes of effective stresses obviously lie between the values computed for the zero and hydrostatic pore pressure cases. A better approximation would require determination of the nature and amount of rock porosity in order to compute the actual pore pressure, and hence the actual effective stress.

How many unknowns in palaeostress determinations?

As discussed above, the determination of the *complete stress tensor* commonly includes two steps (Angelier 1989). Firstly, a *reduced stress tensor*, with four unknowns is computed; secondly, the remaining two unknowns are determined based on taking into account friction and failure parameters, and the depth of overburden (Fig. 4.68). It is also possible to determine more than four unknowns in the inversion process provided that a common friction-failure criterion is adopted for both the inherited and the neoformed faults (Reches 1983). In particular, the problem of palaeostress determination based on inversion of mechanical calcite twin data include five unknowns of the stress tensor (instead of four), because differential stress magnitudes depend on a nearly constant value of yield stress for twinning to occur (Laurent *et al.* 1981, Lacombe & Laurent 1992).

REFERENCES

Anderson, E. M. 1942. *The Dynamics of Faulting* (1st edn). Oliver & Boyd, Edinburgh.

Angelier, J. 1975. Sur l'analyse de mesures recueillies dans des sites faillés: l'utilité d'une confrontation entre les méthodes dynamiques et cinématiques. *C.R. Acad. Sci. Paris. D* **281**, 1805–1808. (Erratum: Ibid (*D*) 1976, **283**, 466.)

Angelier J. 1979a. Néotectonique de l'arc égéen. *Soc. géol. Nord. Pub.* **3**, 418.

Angelier, J. 1979b. Determination of the mean principal directions of stresses for a given fault population. *Tectonophysics* **56**, T17–T26.

Angelier, J. 1984. Tectonic analysis of fault slip data sets. *J. geophys. Res.* **89**, 5835–5848.

Angelier, J. 1989. From orientation to magnitudes in paleostress determinations using fault slip data. *J. Struct. Geol.* **11**, 37–50.

Angelier, J. 1990. Inversion of field data in fault tectonics to obtain the regional stress. III. A new rapid direct inversion method by analytical means. *Geophys. J. Int.* **103**, 363–376.

Angelier, J. 1991. Analyse chronologique matricielle et succession régionale des événements tectoniques. *C.R. Acad. Sci. Paris*, t.312, Sér.II, 1633–1638.

Angelier, J. & Bergerat, F. 1983. Systèmes de contrainte en extension intracontinentale. *Bull. Centr. Rech. Expl. Prod. Elf-Aquitaine* **7**, 137–147.

Angelier, J., Barrier, E. & Chu, H. T. 1986. Plate collision and palaeostress trajectories in a fold-thrust belt: the Foothills of Taiwan. *Tectonophysics* **125**, 161–178.

Angelier, J., Colletta, B. & Anderson, R. E. 1985. Neogene paleostress changes in the Basin and Range: a case study at Hoover Dam, Nevada-Arizona. *Bull. geol. Soc. Am.* **96**, 347–361.

Angelier, J., Faugere, E., Michel-Noel, G. & Anderson, R. E. 1987. Bassins en extension et tectonique synsédimentaire. Exemples dans les "Basin and Range" (U.S.A.). *Notes et Mémoires, Total-C.F.P.*, **21**, 51–72.

Angelier, J. & Goguel, J. 1979. Sur une méthode simple de détermination des axes principaux des contraintes pur une population de failles. *C.R. Acad. Sci., Paris, D* **288**, 307–310.

Angelier, J., Lyberis, N., Le Pichon, X., Barrier, E. & Huchon, P. 1982. The tectonic development of the Hellenic arc and the Sea of Crete: a synthesis. *Tectonophysics* **86**, 159–196.

Angelier, J. & Manoussis, S. 1980. Classification automatique et distinction de phases superposées en tectonique cassante. *C.R. Acad. Sci. Paris D* **290**, 651–654.

Angelier, J. & Mechler, P. 1977. Sur une méthode graphique de recherche des contraintes principales également utilisable en tectonique et en séismologie: la méthode des dièdres droits. *Bull. Soc. géol. Fr.* **7**, 1309–1318.

Angelier, J., Tarantola, A., Manoussis, S. & Valette, B. 1982. Inversion of field data in fault tectonics to obtain the regional stress. 1: single phase fault populations: a new method of computing the stress tensor. *Geophys. J. R. astr. Soc.* **69**, 607–621.

Armijo, R. & Cisternas, A. 1978. Un problème inverse en microtectonique cassante. *C.R. Acad. Sci., Paris D* **287**, 595–598.

Arthaud, F. 1969. Méthode de détermination graphique des directions de raccourcissement, d'allongement et intermédiaire d'une population de failles. *Bull. Soc. géol. Fr.* **7**, 729–737.

Arthaud, F. & Choukroune, P. 1972. Méthode d'analyse de la tectonique cassante à l'aide des microstructures dans les zones peu déformées. Exemple de la plate-forme nord-aquitaine. *Rev. Inst. fr. Pétr.* **XXVII (5)**, 715–732.

Arthaud, F. & Mattauer, M. 1969. Exemples de stylolites d'origine tectonique dans le Languedoc, leurs relations avec la tectonique cassante. *Bull. Soc. géol. Fr.* **7**, 738–744.

Barrier, E. 1985. Tectonique d'une chaîne de collision active: Taiwan. Thèse de Doctorat d'Etat ès Sci. nat. *Mém. Sc. Terre*, Univ. P. & M. Curie, Paris, 85–29, 2 vols.

Bergerat, F. 1985. Déformations cassantes et champs de contrainte tertiaires dans la plate-forme européene. Thèse Doct. Etat ès-Sci. *Mém. Sc. Terre*, Univ. P. & M. Curie, Paris, 1–315.

Bergerat, F. 1987. Stress fields in the European platform at the time of Africa-Eurasia collision. *Tectonics* **6**, 99–132.

Bott, M. H. P. 1959. The mechanisms of oblique slip faulting. *Geol. Mag.* **96**, 109–117.

Byerlee, J. D. 1978. Friction of rocks. *Pure appl. Geophys.* **116**, 615–626.

Carey, E. 1976. Analyse numérique d'un modèle mécanique élémentaire appliqué à l'étude d'une population de failles: calcul d'un tenseur moyen des contraintes à partir des stries de glissement. Thèse. 3ème cycle. Univ. Paris-Sud, Orsay, 1–138.

Carey, E. & Brunier, B. 1974. Analyse théorique et numérique d'un modèle mécanique élémentaire appliqué à l'étude d'une population de failles. *C.R. Acad. Sci. Paris, D* **279**, 891–894.

Daubree, M. 1879. Application de la méthode expérimentale à l'étude des déformations et des cassures terrestres. *Bull. Soc. géol. Fr.* **3**, 108–141.

Dercourt, J., Zonenshain, L. P., Ricou, L. E., Kazmin, V. G., Le Pichon, X., Knipper, A. M., Grnadjacquet, C., Sborshikov, I. M., Geyssant, J., Lepvrier, C. & Pechersky, D. H. 1986. The geological evolution of the Tethys belt from the Atlantic to the Pamira since the Liassic. *Tectonophysics* **123**, 241–315.

Diday, E. 1971. Une nouvelle méthode de classification automatique et reconnaissance des formes: la méthode des nuées dynamiques. *Rev. Stat. Appl.* **19**, 283–300.

Dupin, J. M., Sassi, W. & Angelier, J. 1993. Homogeneous stress hypothesis and actual fault slip: a distinct element analysis. *J. Struct. Geol.* **15**, 1033–1043.

Etchecopar, A., Vasseur, G. & Daignieres, M. 1981. An inverse problem in microtectonics for the determination of stress tensors from fault striation analysis. *J. Struct. Geol.* **3**, 51–65.

Gephart, J. W. & Forsyth, D. W. 1984. An improved method for determining the regional stress tensor using earthquake focal mechanism data: an application to the San Fernando earthquake sequence. *J. geophys. Res., B* **89**, 9305–9320.

Hancock, P. L. 1985. Brittle microtectonics: principles and practice. *J. Struct. Geol.* **7**, 437–457.

Hippolyte, J. C. 1992. Tectonique de l'Apennin méridional: structures et paléocontraintes d'un prisme d'accrétion continental. Thèse de Doctorat de l'Université Paris 6. *Mém. Sci. Terre,* Univ. P. M. Curie, Paris, 1–340.

Hoek, E. & Bray, J. W. 1981. *Rock Slope Engineering* (5th edn). The Institution of Mining and Metallurgy, London.

Hoek, E. & Brown, E. T. 1980. *Underground Excavations in Rock.* The Institution of Mining and Metallurgy, London.

Jaeger, J. C. 1969. *Elasticity, Fracture and Flow with Engineering and Geological Applications.* Chapman & Hall, London.

Jaeger, J. C. & Cook, N. G. W. 1969. *Fundamentals of Rocks Mechanics.* Methuen, London.

Lacombe, O., Angelier, J. & Laurent, P. 1992. Determining paleostress orientations from faults and calcite twins: a case study near the Sainte-Victoire Range (southern France). *Tectonophysics* **201**, 141–156.

Lacombe, O., Angelier, J., Laurent, P., Bergerat, F. & Tourneret, C. 1990. Joint analyses of calcite twins and fault slips as a key for deciphering polyphase tectonics: Burgundy as a case study. *Tectonophysics* **182**, 279–300.

Lacombe, O. & Laurent, Ph. 1992. Determination of principal stress magnitudes using calcite twins and rock mechanics data. *Tectonophysics* **202**, 83–93.

Laurent, Ph., Bernard, Ph., Vasseur, G. & Etchecopar, A. 1981. Stress tensor determination from the study of e twins in calcite: a linear programming method. *Tectonophysics* **78**, 651–660.

Le Pichon, X., Bergerat, F. & Roulet, M. J. 1988. Kinematics and tectonics leading to alpine belt formation: a new analysis. *Spec. Pap. geol. Soc. Am.* **218**, Processes in continental lithospheric deformation, pp. 111–131.

Letouzey, J. 1986. Cenozoic paleo-stress pattern in the Alpine Foreland and structural interpretation in a platform basin. *Tectonophysics* **132**, 215–231.

Lisle, R. J. 1987. Principal stress orientations from faults: an additional constraint. *Annls. Tecton* **1**, 155–158.

Mattauer, M. 1973. *Les Déformations des Matériaux de l'Écorce Terrestre.* Hermann édit., Paris.

Mercier, J. 1976. La néotectonique, ses méthodes et ses buts. Un exemple: l'arc égéen (Méditerranée orientale). *Rev. Géog. phys. Géol. dyn.* **2** (XVIII), 323–346.

Mercier, J. L. & Carey-Gailhardis, E. (1989). Regional state of stress and characteristic fault kinematics instabilities shown by aftershock sequence: the aftershock sequences of the 1978 Thessaloniki (Greece) and 1980 Campania-Lucania (Italy) earthquakes as examples. *Earth Planet. Sci. Lett.* **92**, 247–264.

Michael, A. 1984. Determination of stress form slip data: fault and folds. *J. geophys. Res., B* **89**, 11517–11526.

Michel-Noel, G., Anderson, R. E. & Angelier, J. 1990. Fault kinematics and estimates of strain partitioning of a Neogene extensional fault system in southeastern Nevada. *Mem. geol. Soc. Am.* **176**, 155–180.

Price, N. J. 1966. *Fault and Joint Development in Brittle and Semi-Brittle Rocks.* Pergamon Press, Oxford.

Reches, Z. 1983. Faulting of rocks in three-dimensional strain fields. II. Theoretical analysis. *Tectonophysics* **95**, 133–156.

Reches, Z. 1987. Determination of the tectonic stress tensor from slip along faults that obey the Coulomb yield criterion. *Tectonics* **6**, 849–861.

Stewart, I. S. & Hancock, P. L. 1991. Scales of structural heterogeneity within neotectonic normal fault zones in the Aegean region. *J. Struct Geol.* **13**, 191–204.

Wallace, R. E. 1951. Geometry of shearing stress and relation to faulting. *J. Struct. Geol.* **59**, 118–130.

Yu, S. B. & Tsai, Y. B. 1982. A study of microseismicity and crustal deformation of the Kuanfu-Fuli area in eastern Taiwan. *Bull. Inst. Earth Sci. Acad. Sinica* **2**, 1–18.

CHAPTER 5

Palaeostress Analysis of Small-Scale Brittle Structures

W. M. DUNNE and P. L. HANCOCK

RATIONALE

This chapter describes and discusses methodologies of inferring sequences of palaeostress axis orientations from naturally occurring brittle and semi-brittle structures of small-scale that cut sedimentary rocks. We exclude small faults from consideration because the procedures for analyzing palaeostresses from them are identical to those discussed in Chapter 4. Thus, we focus attention on veins, stylolites, shear zones, sediment-filled fractures and joints. These structures occur both in association with large faults and independently of them. Indeed, in some regions of weak deformation they are the only tectonic structures in the rock mass. Despite the development of stylolites, shear zones and some veins not involving brittle deformation processes *sensu stricto*, these structures are included in this account because of their intimate spatial and temporal association with truly brittle structures.

Although many brittle deformation studies focus on the major faults that achieve large translations, and cause the majority of stress release during large magnitude earthquakes, the analysis of small structures can yield a reliable palaeostress history (e.g. Eyal & Reches 1983). Futhermore, in some platform settings, small structures record more stress episodes than are expressed by map-scale structures. In addition, vertical dykes can also be used as palaeo- and neotectonic stress indicators (see pp. 397–400, Chapter 18 for details).

The order in which structures in this chapter are treated reflects their generally perceived quality from the perspective of palaeostress analysis. Thus, the chapter begins with accounts of veins and stylolites and concludes with a discussion of joints, the palaeostress value of which remains controversial. We follow the convention used in Chapter 4 and refer to the maximum principal stresses as σ_1, the intermediate principal stress as σ_2, and the minimum principal stress as σ_3. Other stress parameters are defined at an appropriate place in the text.

VEINS

Introduction

Veins are mineral bodies that form from chemical precipitation of fluids. *Replacement veins* have precipitated minerals that chemically replace pre-existing country rock (Fig. 5.1a). *Dilatant veins* have minerals that were precipitated into pre-existing or propagating fractures (Fig. 5.1b). These two vein types are easily distinguishable. Dilatant veins have matching walls produced during fracture propagation, offset features in the country rock and lack ghost fabrics. In contrast, replacement veins produce non-matching walls from country rocks with different chemical solubilities, do not offset features in the country rock, and preserve *ghost fabrics* that are relic country-look fabric.

Vein fills in fractures

The most common vein minerals are quartz, calcite, dolomite and gypsum. These minerals can have a *massive-fill* or *granular texture* (Fig. 5.2a) of irregular shaped grains with variable grain size and no crystallographic preferred orientation. This fill fabric is not a kinematic or palaeostress indicator for fracturing. In contrast, another vein-fill type, *parallel fibrous mineral grains*, can be a kinematic and palaeostress indicator because fibre-growth directions may be parallel to vein-opening directions during precipitation.

The two common fibrous fills are syntaxial (Fig. 5.2b) and antitaxial (Fig. 5.2c) (Durney & Ramsay 1973). *Syntaxial-vein fibres* (Fig. 5.2b) grow in optical continuity from grains in the vein walls, meet unevenly at a central seam, and the oldest parts of fibres are adjacent to vein walls. *Antitaxial-vein fibres* (Fig. 5.2c) grow incrementally from the vein into the vein walls, lack a central seam and are not continuous from wall grains, unlike syntaxial veins. Instead, fibres are continuous between vein walls, differ in mineralogy from grains in vein walls, and the youngest parts of fibres are at the vein walls.

Antitaxial vein fibres form by cycles of fracturing and mineral precipitation known as *crack-seal* (Ramsay 1980a). Fracturing is between fibre ends and vein wall,

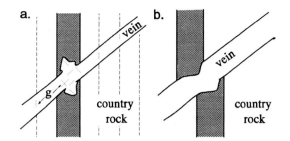

Fig. 5.1. (a) Replacement vein with irregular walls from different chemical solubilities. g, ghost fabric; cross hatch, marker bed. (b) Dilation vein.

but fragments of wall grains break off and are incorporated as inclusions in the fibres by vein precipitation. The *wall-rock inclusions* either develop as wall-parallel layers of *inclusion bands* with each layer representing a single fracture cycle, or as *inclusion trails* that are parallel to the vein-opening direction from repeated breaking of a wall grain during a series of cycles. The fibre length between wall inclusions may record the amount of vein accretion for a single crack-seal cycle.

Other less common fibrous fills are stretched and composite. *Stretched fibres* (Fig. 5.2d) connect pairs of wall-grain fragments that were whole before vein formation. The fibres form by a crack-seal mechanism like antitaxial fibres, but grow continuously from wall-rock grains like syntaxial fibres. *Composite veins* consist of alternating layers of syntaxial and antitaxial fibres.

Recent work has shown that fibres without wall inclusions may not be reliable kinematic indicators (Cox 1987, Urai *et al.* 1991). For example, veins with planar walls can precipitate wall-perpendicular fibres for all dilations, while the true opening direction is only recorded by inclusion trails (Fig. 5.3). Consequently, surface roughness of the vein wall can influence fibre geometry. Reliable kinematic interpretations may only be made with undeformed primary fibres connecting markers across the vein that have (1) inclusion trails and (2) microstructural evidence for localized cracking and mineral precipitation (Urai *et al.* 1991).

Vein fills in pressure shadows

Where surrounding rock grains decouple from larger, more rigid grains during extension (because of rheological differences) the resultant voids are sites for fibrous vein precipitation called *pressure shadows* (Fig. 5.4). Alternatively, pressure shadows can be replacement veins (Brantley *et al.* 1989) where fibres grow in the

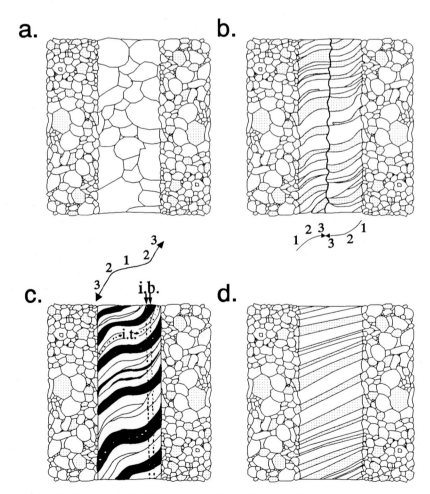

Fig. 5.2. Textures of vein fills (stipple and white ornament for vein mineralogy that is the same as wall rock; black and white ornament is used for different mineralogies). (a) Massive or granular fill. (b) Syntaxial fill. (c) Antitaxial fill. i.b., inclusion band; i.t., inclusion trail. (d) Stretched fill. All parts modified from Ramsay & Huber (1983).

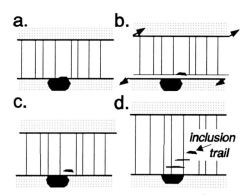

Fig. 5.3. (a) Antitaxial vein with adjacent wall grain (black). (b) Oblique dilation with cracking of wall grain. (c) Fibre precipitation where fibres only record dilation and the inclusion displacement records the true oblique dilation. (d) Oblique inclusion trail recording increments of the true oblique dilation.

extension direction but replace country rock adjacent to the rigid grain without fracturing. Fibre geometry depends on: (1) whether the fibres grew from the rigid grain or surrounding grains; (2) whether the fibres deformed during growth; and (3) the strain history during growth (Ramsay & Huber 1983, Ellis 1986, Etchecopar & Vasseur 1987).

Pressure shadow fibres that grow continuously from surrounding grains towards the rigid grain are *pyrite-type fibres* (Figs 5.4a & c) and the youngest parts of fibres are adjacent to the rigid grain. In contrast, pressure shadows that grow continuously from the rigid grain towards surrounding grains are *crinoid-type fibres* (Figs 5.4b & d) and the oldest parts are adjacent to the rigid grain. Hence, the growth sequence of crinoid-type fibres is exactly opposite to that of pyrite-type fibres (cf. syntaxial and antitaxial fibres). A geometric consequence of the opposite growth sequences is that crinoid-type and pyrite-type fibres have the opposite curvature for the same rotational strain history (Figs 5.4c & d).

Palaeostress determination

Although the attitude and form of pressure shadow fibres can reflect relative changes in differential stress and the geometric relationship between maximum principal stress and cleavage (Fisher 1990), absolute principal stress directions cannot be directly inferred from shadows. Fibres in fractures provide a better basis for determining these directions. For purely dilational fractures (i.e. Mode 1 cracks, see Chapter 3), fibres are perpendicular to fracture walls and parallel to σ_3, confirming that the veins formed perpendicular to a minimum effective tensional principal stress and in the plane of σ_1 and σ_2. Where multiple extensional fibrous vein sets intersect, so that younger vein fill cross-cuts older vein fill, the fracture sequence of geometries for causative stress-fields can be determined (Dunne & North 1990). A fracturing sequence that records the progression of alternation of directions σ_3 (Fig. 5.5), possibly indicating that σ_1 was parallel to the intersect between sets, provides a series of stress directions that may be correlated to known regional stress events to determine the timing of fracturing. A fracture grid-lock of orthogonal joints forms in a similar manner (Hancock *et al.* 1987, Caputo & Caputo 1988, and later in this chapter).

Fibres in *fibre sheets* of *shear veins* can indicate a fracture-opening direction, but the fibre sheets do not simply correlate to stress directions (Ramsay & Huber 1983, 1987). Consequently, shear veins have little previous history of success as palaeostress indicators. However, the advent of fluid inclusion studies has provided a means for determining the normal stress at the time of formation across shear veins. These inclusions in the vein minerals contain fluids that are believed to be representative of the mineralizing fluids and the ambient pressure–temperature conditions. Experimentally determined *homogenization* and *freezing temperatures* for fluids in natural inclusions yield the fluid composition and pressure–temperature conditions for phase changes (Srivistava & Engelder 1990, Foreman & Dunne 1991). These conditions, corrected for fluid pressure, inclusion radius and fracture toughness of the vein-filling minerals, become the *trapping pressure* or normal stress across the fracture during inclusion formation. When applied to originally horizontal bed-parallel veins, the trapping

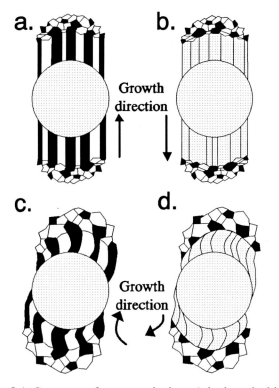

Fig. 5.4. Geometry of pressure shadows (stipple and white ornament indicates shadow mineralogy that is the same as the inclusion; black and white ornament used to show different mineralogies) from growth directions (shown by arrows). (a) Straight pyrite-type fibres. (b) Straight crinoid-type fibres. (c) Curved pyrite-type fibres. (d) Curved crinoid-type fibres. Modified from Ramsay & Huber (1983).

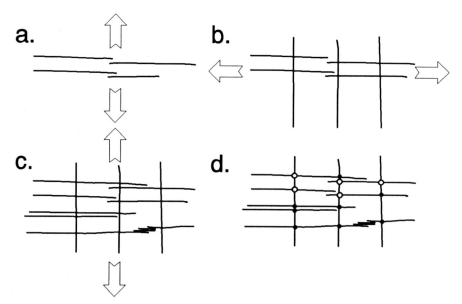

Fig. 5.5. Orthogonal fracture system produced by alternations (a–c) in the direction of the tensional minimum principal stress (open arrows). (d) Resultant distribution of relative fracture ages at vein intersections. Open circle, N–S fractures younger; solid circle, E–W fractures younger.

pressure is often equivalent to overburden stress, including any tectonic load (Srivastava *et al.* 1990).

STYLOLITES

Introduction

Stylolites are solution structures where shortening is mostly accommodated by volume loss (Park & Schot 1968). Deformation involves: (1) *pressure-driven* or *free-face solution* (Engelder & Marshak 1985); (2) *chemical diffusional* (Rutter 1976) or *physical advective transfer* (Geiser & Sansone 1981) of solute in a fluid phase; and (3) either local precipitation of solute in veins and fibrous overgrowths as a closed system, or solute export through an open system. Evidence for solution across stylolites includes: mutual penetration of fossil fragments, pit development in conglomerates, and elongation of clastic quartz grains by silica overgrowth (Groshong 1975, Ramsay 1977). However, stylolites do not pervade all shortened rocks because rock solution decreases with increasing temperature, increasing strain rate, increasing deviatoric stress, increasing grain size, decreasing clay content, and decreasing mineral solubility (Rutter 1976, Kerrich *et al.* 1977, Engelder & Marshak 1985, Houseknecht 1987).

Stylolitic morphology and geometry

Stylolite seams have a distinctive wave-form morphology and consist of two interpenetrating surfaces that are separated by an insoluble residue of clay minerals and oxides from rock dissolution. The wave forms are defined by *first-order stylolite columns* (Park & Schot 1968) with *teeth* and *sockets* of hummocky, sutured, rectangular or peaked shape (Fig. 5.6a) that contain smaller *parasitic* or *second-order stylolite columns* (Fig. 5.6b). The parallel orientations of first-order columns define the *stylolitic lineation*, whereas the second-order columns only point towards the culminations of their host first-order (Fig. 5.6b). The stylolitic lineation is parallel to the maximum shortening across stylolites. Commonly, stylolites accommodate less than 4% shortening and have greater then 5 cm spacing (Alvarez *et al.* 1978). For greater shortening, stylolites are absent, and smooth or anastomosing solution structures develop.

Stylolites do not have a unique geometry to bedding. They can be bedding-parallel structures that formed during burial in response to lithostatic load, or they can

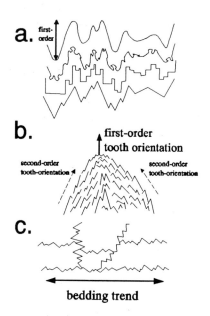

Fig. 5.6. (a) Different stylolite morphologies: from top to bottom — hummocky, sutured, rectangular and peaked. (b) Single first-order stylolite tooth or column with smaller second-order teeth. (c) Common geometries of stylolites with respect to bedding.

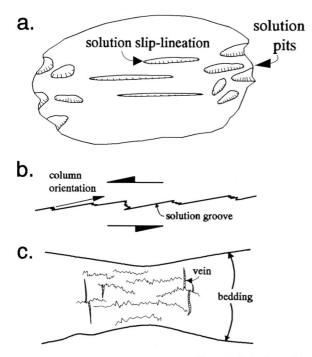

Fig. 5.7. (a) Pitted pebble. (b) Profile of slickolite with shear sense shown by coupled arrows. (c) Necking of a bed by styloboudinage.

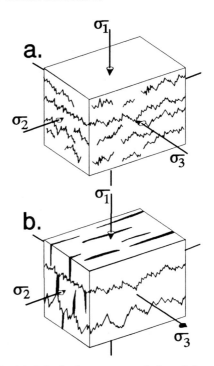

Fig. 5.8. (a) Principal stress axes inferred from stylolite seams with surface-normal columns. (b) Principal stress axes inferred from coeval orthogonal stylolite seams and dilational veins.

form obliquely or normal to bedding in response to tectonic compression (Fig. 5.6c). A *disjunctive spaced cleavage* can be one expression of such *through-rock stylolites*.

Related structures

In clast-supported conglomerates where the clasts are in contact, transmit stress, and dissolve locally to yield interpenetrated pebbles, the following structures can form: (1) *solution pits* or *sockets* that develop in more soluble pebbles or those with broader radii of curvature; (2) *slip-lineations* from oblique solution or frictional wear between passing clasts; and (3) *fractured insoluble pebbles* (McEwen 1981, Schrader 1988) (Fig. 5.7a). Intraclast fracturing also occurs when the strain rate exceeds that of solution-related shortening.

Slickolites (*oblique stylolites*) develop along faults by *pressure-solution slip* (Nitecki 1962, Elliott 1976) (also see Chapter 4). They have columns, but are dominated by solution grooves that are parallel to fault motion (Fig. 5.7b). Slickolites are shear structures that accommodate displacements at very small, aseismic strain rates.

Styloboudins form where stylolites and dilational veins grown interactively, achieving layer-parallel segmentation (Mullenax & Gray 1984, De Paor *et al.* 1991). The veins cause layer-segmentation, whereas stylolites cause 'pinching' and can be reactivated bedding-parallel stylolites (Fig. 5.7c).

Palaeostress determination

Stylolite seams have been interpreted to have an *anticrack* origin (Fletcher & Pollard 1981) and to have propagated normal to σ_1 (Tapp & Cook 1988), that is, to be exactly the opposite of Mode 1 dilational cracks. Also, first-order stylolite columns are interpreted to form parallel to σ_1. Both geometries are well displayed where first-order stylolite columns are normal to their host surfaces (Fig. 5.8a). However, where columns are oblique to a solution surface, they are not normal to σ_1. One cause for this oblique columnar geometry is solution along pre-existing fractures that act as conduits for the dissolving fluids but are oblique to the direction of σ_1 (Geiser & Sansone 1981). Stylolitic columns on a slickolite seam are roughly parallel to σ_1, but can be refracted away from σ_1 towards the solution surface.

Where coeval orthogonal stylolites and fibrous dilational veins form together (Fig. 5.8b), palaeostress geometry is completely defined (Nelson 1981). The inferred direction of σ_1 is parallel to the stylolite columns, σ_2 is parallel to the intersects between stylolites and veins, and tensional σ_3 is parallel to vein fibres. The distribution of these paired dynamic indicators can be used to infer stress trajectories around faults (Rispoli 1981) and stress conditions during burial and halokinesis (Watts 1983) in limestones.

SHEAR ZONES

Introduction

Tabular rock bands containing locally concentrated shear deformation are *shear zones*. The deformation behaviour of the zones ranges from completely brittle and discontinuous (Fig. 5.9a) to completely ductile and continuous (Fig. 5.9c). Only *semi-brittle* or

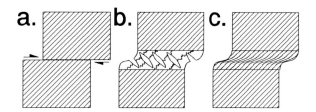

Fig. 5.9. (a) Brittle shear zone (fault). (b) Brittle–ductile shear zone with en échelon veins and stylolitic cleavage. (c) Ductile shear zone modifying foliation of country rock. Modified from Ramsay (1980b).

brittle–ductile shear zones will be discussed in this section because *brittle shear zones* (i.e. faults) as palaeostress indicators are discussed in Chapter 4 and *ductile shear zones* have only limited value as palaeostress indicators. Likewise, *deformation bands* (*shear bands*) are not discussed here because they too are products of ductile deformation. The distinctive signature of brittle–ductile shear zones is an *array of en échelon veins* (Fig. 5.9b). The zones are commonly developed in dolomites, sandstones and clean limestones, particularly where these rocks are interbedded with mudstones or slates. Their environment of formation is commonly of lower grade than greenschist facies, thus precluding the intragranular crystal-plastic deformation mechanisms that produce continuous behaviour in *ductile shear zones* (Knipe & White 1979).

Internal zone geometry

Brittle–ductile shear zones are the sum of their parts or secondary structures, because these features accommodate all displacement across zone boundaries. The secondary structures include dilational veins, secondary cleavage, grain-shape fabrics, faults, and Riedel shear fractures (Fig. 5.10). Sub-perpendicular cleavage and veins that accommodate shortening with volume loss and extension with volume increase, respectively, are a common association of secondary structures in shear zones. *En échelon dilational veins* face against the shear directions, whereas *secondary en échelon cleavage surfaces* face with shear directions (Fig. 5.10). Depending on strain history, veins may be younger and truncate cleavage (Fig. 5.10a), age equivalent and mutually cross-cut cleavage (Fig. 5.10b), or older and be offset or truncated by cleavage (Fig. 5.10c). Veins commonly contain a fibrous fill that is antitaxial, syntaxial or stretched, and fibres may be straight or curved. The cleavage is commonly sinuous or stylolitic, and the result of pressure solution.

A *secondary grain-shape fabric* develops in the *rock bridges* between veins (Fig. 5.10d), and consists of elongate grains that are sub-parallel to cleavage and sub-perpendicular to veins. The elongate grains commonly contain *inclusion planes* that are perpendicular to the long axes of grains and indicate that elongation was a brittle process achieved incrementally by a crack-seal mechanism (Knipe & White 1979). *Zone-parallel faults* commonly offset veins and cleavage in some shear zones (Fig. 5.10e), indicating a transition to more brittle behaviour with increasing deformation. The faults are usually located near zone centres and have the same shear sense as the zone. Arrays of en échelon *Riedel fractures* are secondary shear fractures that can develop alone in brittle–ductile shear zones (Fig. 5.10f) and commonly form at an angle of about 15° to shear zone boundaries (also see Chapters 4, 6 and 12).

Stress or strain?

The use of en échelon veins in brittle–ductile shear zones to determine internal palaeostress (Rickard & Rixon 1983) is fraught with problems simply because agreement is lacking about whether vein development is controlled by stress or strain. One view considers that they form perpendicular to the maximum extension direction (Ramsay 1980b). The initial angle between veins and shear zone boundaries is a function of the components of volume change, pure shear and simple shear occurring across the zone (Fig. 5.11). Veins and other secondary structures change abundance, the degree of overlap as the bulk deformation changes (Ramsay & Huber 1987, session 26) (Fig. 5.12). Originally planar veins become *sigmoidal veins* and multiple generations of en échelon veins develop

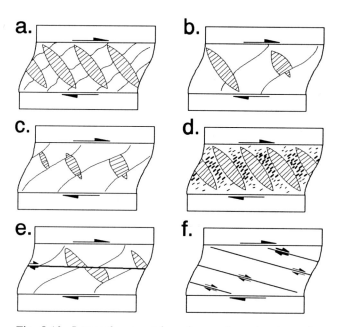

Fig. 5.10. Internal geometries of secondary structures in brittle–ductile shear zones (all right-lateral as shown by shear couples on boundaries of zones). (a) Veins (elliptical shapes with internal lines showing geometry of vein fibres) younger than cleavage (curvilinear lines within zone). (b) Veins and cleavage same age. (c) Veins older than cleavage. (d) Veins with elongate grains (tick marks between veins). (e) Late-stage fault (solid line with shear couple) offsetting cleavage and veins. (f) Riedel fractures (parallel lines with shear couples).

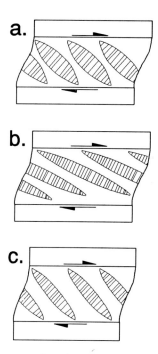

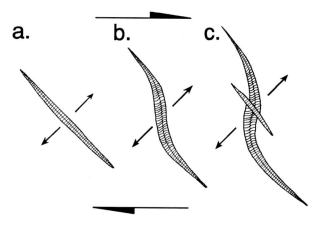

Fig. 5.13. Progressive vein development in a brittle – ductile shear zone with syntaxial fibres. (a) Planar vein with straight fibres. (b) Vein rotation producing sigmoidal veins with curved fibres. (c) Initiation of younger cross-cutting planar vein to accommodate additional strain. After Durney & Ramsay (1973).

Fig. 5.11. Geometry of en échelon veins (elliptical shapes with internal straight lines to show fibre geometry) within brittle – ductile shear zone related to different deformation regimes. (a) Simple shear. (b) Simple shear with volume increase (analogous to transpression on a large scale). (c) Simple shear with zone-parallel pure shear that elongates normal to zone boundaries (analogous to transtension on a large scale).

(Fig. 5.13) during progressive deformation (Ramsay & Graham 1970), so that vein shape can be used to determine total shear strain. Consequently, vein geometries controlled by strain should not be used to infer palaeostress geometries.

An alternative view is that en échelon veins develop as dilatant fringes to a parent *Mode 1 crack* that forms perpendicular to the direction of σ_3, which is effectively tensional because of fluid pressure (Pollard et al. 1982, Nicholson & Pollard 1985). These en échelon veins can be regarded as analogous structures to twist hackles on joints (see later section) or en échelon segments at the ends of igneous dykes (Fig. 5.14). Again, veins should not be used as indicators of internal palaeostress attitude. Instead the parent vein is the palaeostress indicator, and if demonstrably Mode 1 from its vein fill, it is normal to the minimum principal stress.

Olson & Pollard (1991) have employed a fracture mechanics approach to demonstrate the view that arrays of en échelon Mode 1 veins may arise from a pre-existing population of randomly oriented flaws. A dilated fracture generates a local stress field of *crack-tip stresses* as a function of crack length, crack width, crack orientation, and crack-tip curvature (Fig. 5.15) (also see Chapter 3). This field aids crack propagation by intensifying the tension at its tips for an appropriately oriented regional stress field and internal fluid pressure (Pollard & Segall 1987). Also, crack-tip stresses locally

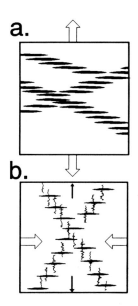

Fig. 5.12. Change in abundance, overlap and attitude of en échelon veins (black elliptical shapes) in shear zones between (a) bulk volume increase (white arrows show direction of increase), and (b) bulk volume decrease with pure shear (white arrows show direction of decrease, small black arrows show direction of elongation for pure shear, and curvilinear lines are stylolites). Modified from Ramsay & Huber (1983).

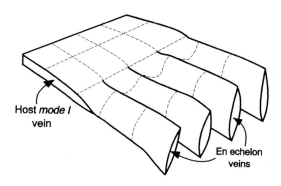

Fig. 5.14. En échelon veins at the termination of a planar Mode 1 vein.

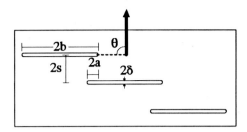

Fig. 5.15. Quantitative parameters for an array of en échelon fractures. *Symbols*: 2b, crack length; 2a, overlap between adjacent cracks; 2s, space between adjacent cracks; 2δ, wall perpendicular dilation; θ, angle to the regional minimum principal stress.

perturb the regional stress field and can aid the propagation of adjacent fractures by intensifying the tensional stress at their tips sufficiently to exceed the tensile strength of the country rock. The intensity of this interaction is enhanced where the fractures are closely spaced and are just approaching an overlapping arrangement. This effect decreases markedly after partial overlap is achieved. Consequently, arrays of en échelon fractures can be generated by a series of interactive crack-propagation events where a series of pairs of propagating veins achieve overlap and accumulate to form an array. An implication following from this view is that fracture curvature, such as that displayed by sigmoidal veins, is not the result of simple shear. Instead, cracks propagate in curves in response to the changing geometry of the crack-tip stress fields, and as a result of changing relative positions of the crack tips with increasing fracture overlap. Again, these types of veins should not be used as indicators of internal palaeostress directions because they develop in response to initial and transitory interactions between regional and crack-tip stresses, rather than from some steady-state internal palaeostress field (Nicholson & Pollard 1985).

These three views accord in that en échelon veins do not represent internal palaeostress geometries in brittle–ductile shear zones. Yet, it remains necessary to assess which, if any, view is appropriate to determine whether array development should be considered in terms of stress or strain. An effective assessment might be to determine if rock bridges between veins are deformed by displacements involving shear strain (Knipe & White 1979), or rigidly rotated with bending during vein dilation controlled by stress (Nicholson & Ejiofor 1987).

Palaeostress determination from conjugate arrays

Conjugate brittle–ductile shear zones, rather than individual veins, reliably indicate regional stress directions. Unlike conjugate faults, which have an inward-moving acute wedge with a *dihedral* (2θ) *angle* of about 60°, the inward-moving wedge between conjugate shear zones can enclose angles ranging from 40 to 130° (Fig. 5.16). Thus, shear sense for conjugate shear zones must be established to identify inward-moving blocks and hence stress directions: (1) the bisector of inward-moving blocks is parallel to σ_1; (2) the intersect between shear zones is parallel to σ_2; and (3) the bisector of outward-moving blocks is parallel to σ_3.

FISSURES AND SEDIMENT-FILLED FRACTURES

Types

A *fissure* is a barren surficial crack with gaping walls that developed either from new (i.e. neoformed in the sense of Angelier 1989) dilation related to surficial movements or widening by weathering of pre-existing fractures (Fig. 5.17a). *Sediment-filled fractures* include: (1) *clastic* and *intrusive sedimentary dykes and sills* (Fig. 5.17b) containing sediment injected at a high fluid pressure from below into pre-existing or propagating cracks (Winslow 1983); (2) *neptunian dykes* (Fig. 5.17c) containing sediment deposited from above into a pre-existing crack, possibly enhanced by dilation from pre-depositional surficial movements or weathering; and (3) *vein structures* that contain sediment derived from the adjacent walls as 'fill' and developed only in accretionary wedges (Fig. 5.17d).

Distinctive features of clastic dykes are a direct underlying connection to a source-rock bed, wall-rock fragments and diminished thickness of the source rock beneath the dykes. Bedding-parallel clastic sills are more difficult to detect because they look similar to mélange horizons or sedimentary rip-up deposits. Three distinguishing characteristics are (1) lack of sedimentary

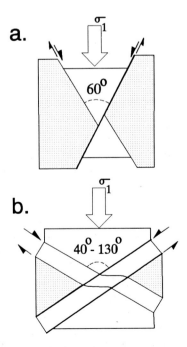

Fig. 5.16. Conjugate geometries. (a) Brittle shear zones (faults). (b) Brittle–ductile shear zones. Outward moving blocks are stippled and motion sense is shown by shear couples. Modified from Ramsay (1980b).

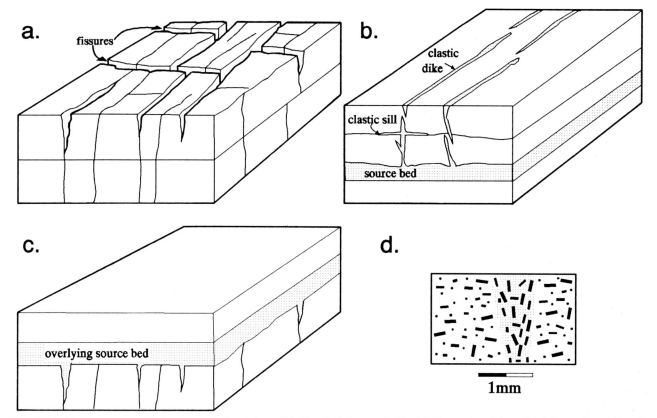

Fig. 5.17. (a) Fissures modifying pre-existing joints. (b) Clastic dykes and sills. (c) Neptunian dykes. (d) Schematic profile of rock fabric in vein structure from accretionary wedge. Modified from Knipe (1986).

structures, (2) fragments derived from the overlying bed that could not have resulted from sedimentary erosion, and (3) direct connection to clastic dykes (Winslow 1983). In contrast, neptunian dykes connect directly to overlying beds, so that sediment composition and fabric change continuously upwards into the overlying bed (Lundberg & Moore 1986). Vein structures are only found in the higher parts of accretionary wedges on the upper trench slope (see Chapters 7, 9 and 16). They contain *in situ* sediment from the wall rock with phyllosilicates parallel to vein walls, and some disaggregation with additional cement from passing fluids (Knipe 1986, Lundburg & Moore 1986). Such veins appear to be extensional or shear structures that formed when fluid pressures were high.

Palaeostress determination

These structures have several fundamental weaknesses that limit their use as stress indicators. (1) Frequently, they exploit pre-existing fractures that are geometrically unrelated to the causative stress field. (2) Dilation of the structures may only be related to local stresses generated by topographic curvature or slope, and not regional tectonic stresses. (3) Walls commonly do not match because of weathering or injection attrition, preventing the determination of a true opening direction. Possible exceptions would be clastic dykes or vein structures that form in propagating Mode 1 dilational fractures that are related to regional stresses. For confidence in the value of such fractures, they should be parallel, in order to preclude topographic effects, and they should not show signs of shear offset (Knipe 1986) to preclude principal stress geometries at an unknown obliqueness to the structures. If these conditions are fulfilled, the structures can then be assumed to have formed normal to σ_3.

JOINTS

Preamble

The word joint has been used since the eighteenth century to express the idea that the parts of a rock mass are joined together across fractures. The structure could just as well be called a crack. Some British and North American workers disagree about the meaning of the word joint. For example, Hancock (1985, p. 445) defined a *joint* as "a fracture which, at the scale of observation possible in the field, is a barren closed crack on which it is not possible to detect any evidence for there having been offset related to shear, dilation or shortening". By contrast, Pollard & Aydin (1988, p. 1186) recommended that the term joint should be "restricted to those fractures with field evidence for dominantly opening displacements", and they argue that joints are extension fractures, that is, *Mode 1 cracks* (see Chapter 3). Both definitions have their drawbacks. Classing a structure as a joint in the field means that with additional evidence (e.g. a thin section) it might be renamed a fault (shear fracture), vein or

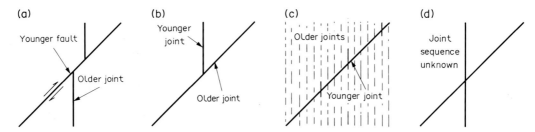

Fig. 5.18. (a) An older joint trace offset by a younger fault. (b) A younger joint trace abutting an older joint trace. (c) Short traces of older sealed joints cut by the long trace of a younger joint. (d) Crossing traces of joints, the order of formation of which is not determinable.

solution seam. Conversely, deciding the origin of a structure in order to describe it is unsound practice. Here, we use the word joint as a field term.

Some descriptive qualifiers to the name joint have been added recently. Ramsay & Huber (1987) employ the term *sealed joint* to describe a crack across which the wall rocks are now welded by a thin (say <1 mm) fill of vein material. Joints that have been reactivated in shear and on which displacement is detectable have been called *faulted joints* by Cruikshank *et al.* (1991) and Zhao & Johnson (1992). The latter authors also introduced the term *jointed fault*.

Despite the general lack of kinematic indicators on joints, their potential for palaeostress determinations stems from their ubiquitousness in indurated rocks and the knowledge that identical palaeostress axes can be inferred from some populations of faults and joints (e.g. Caputo 1991). Confidence in the value of joints for tracking palaeostress trajectories is especially great where they are regularly arranged throughout most of a large region of simple structure, such as a platform (see Chapter 10). However, even within platforms, small faults can locally perturb the stress fields responsible for jointing (Rawnsley *et al.* 1992). Many of the joints in major deformation zones, such as thrust-fold belts, are also related to local stresses.

Although traditionally joint orientations have been the main focus of joint surveys, it is also important to understand *joint style*, an expression used to embrace attributes such as joint dimensions, surface morphology, spacing, and terminations (e.g. Stubbs & Wheeler 1975). Contrasting attributes can characterize joints of different genetic class, and, as Grout & Verbeek (1983) have pointed out, the styles of joints in different sets are commonly different.

Environments of jointing

Engelder (1985) concluded on the basis of a study of joints in the Appalachian Plateau that there are at least four types of regionally extensive joints. (1) *Tectonic joints* form at depth in the Earth's crust before rocks are uplifted and denuded. Joint initiation and propagation is a response to the combined action of abnormally high pore-fluid pressures generated during the tectonic compaction of the sediments, and stresses of tectonic origin. (2) *Hydraulic joints* also form at depth before uplift and as a consequence of the influence of high fluid pressures but the cause of the high fluid pressures is sedimentary compaction. (3) *Unloading joints* are formed close to the Earth's surface and after uplift. The attitudes of unloading joints are controlled by the orientation of either tectonic or residual stresses at the time of their formation. The idea that some joints are caused by unloading after uplift is not a new one: Price (1959) had introduced the concept to explain the widespread occurrence of joints in otherwise 'undeformed' rocks. Exposed *neotectonic joints* (Hancock & Engelder 1989) are a special category of unloading joints that propagated in a stress field that has persisted with little or no change of orientation until the present day (see Chapter 18). (4) *Release joints* are also formed after uplift but their orientations and locations are determined by pre-existing structures such as cleavage planes. The above mechanisms of joint formation, which are important in areas of simple structure, such as the Appalachian Plateau, are also capable of giving rise to joints in more structurally complex terrains that have been buried and uplifted several times.

Joint sequence

The order of formation of neighbouring joints can be determined from three relationships. (1) Where a joint is offset across a fault, vein or stylolite, the joint is the older structure (Fig. 5.18a). (2) The trace of a younger joint segment abuts that of an older joint (Fig. 5.18b). This relationship arises when a joint propagating through intact rock intersects an unsealed crack and is unable to jump the gap. (3) Where the short traces of small sealed joints terminating in intact rock are cut by the long trace of a large joint, the shorter joints are the older joints (Fig. 5.18c). Where two joint traces cross each other, their relative ages cannot be determined because one of them must have been sealed when the other propagated across it (Fig. 5.18d).

Although the relative ages of neighbouring joints can be determined easily, absolute ages are difficult to acquire. Letouzey & Trémolières' (1980) method of dating assemblages of small faults, veins and stylolites by relating directions of shortening inferred from them to a well-dated stratigraphic sequence can, in principal, also be applied to populations of joints. However, because younger joint sets are not everywhere superimposed on older sets (Hancock 1991) the principal must be applied cautiously. Where a joint set is subject to later superimposed shear, *faulted joints* can result or distinctive patterns of *en échelon cracks, horsetail fractures* and *kinked joint ends* develop (Cruikshank *et al.* 1991)

Joint spacing, density and intensity

Measuring the 'abundance' of joints via parameters such as spacing (separation), frequency, periodicity, density and intensity has a long history. The motive for these studies is often connected with assessing either fracture porosity/permeability or rock-mass stability, but some workers have suggested a link between 'abundance' and stress intensity.

For each set and layer, joint spacing should be recorded and a note should be made of the thickness and lithology of the layer containing the joints. *Joint spacing* should be measured in plan or profile using a tape extended along a scan line oriented normal to the set. *Joint frequency* per metre for each set can be determined from spacings along several scan lines. Knowing the range of spacings and the mean spacing, joint periodicity can be assessed. A *joint-periodicity index (JPI)*, which expresses the evenness, or otherwise, of spacing of joints in a set, can be calculated from σ/S, where σ is the standard deviation of the spacing and S is the arithmetic mean spacing. Where fractures are perfectly periodic the *JPI index* will be zero and where they are completely non-periodic it will be one. *Joint zones* (Hodgson 1961), narrow, but widely separated, belts containing many joint planes replacing each other across stepovers, lead to a joint set being exceptionally non-periodic. Engelder (1987) has interpreted the observation that individual fractures in joint zones overlap each other as indicating that the initiation of one joint is not linked to that of the tip of its predecessor. By contrast, Olson & Pollard (1991) think that during the development of *en échelon* veins the location and orientation of an older crack influences that of a younger one.

The work of Harris *et al.* (1960), Price (1966), Hobbs (1967), Ladeira & Price (1981), Huang & Angelier (1989), Narr & Suppe (1991) and Rives *et al.* (1992), among others, has demonstrated that joint spacing in sedimentary rocks is largely a function of (1) layer thickness, (2) lithology, and (3) 'intensity' of deformation. Spacing is closest in thin beds, brittle lithologies and where stress levels have been greatest. The thicknesses of imcompetent layers enclosing competent ones may also influence the spacing of fractures in competent layers (Hobbs 1967).

A useful measure of the areal abundance of joints in a rock mass is *joint density*, that is, the total trace length of all joints in a unit area. *Joint intensity* is a measure of the total area of joint planes within a unit volume of rock, say 1 m³. It can be estimated knowing the joint density on three exposure surfaces that are mutually

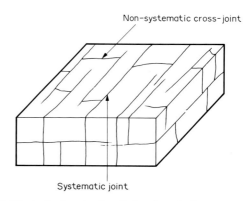

Fig. 5.19. A single set of well-developed planar systematic joints linked by small non-systematic joints of the type commonly known as cross-joints. Note that in plan there is a ladder-like pattern of joint traces.

orthogonal. Refined methods of assessing joint intensity are discussed by Wheeler & Dixon (1980).

Geometry and architecture of joint systems

A joint that is planar or approximately planar and belongs to a regularly oriented set (or angular continuum of surfaces) is called a *systematic joint* (Fig. 5.19) (Hodgson 1961). Systematic joints are interpreted as products of stress fields of deep-seated origin. By contrast, *non-systematic joints*, which are mostly irregular surfaces, not belonging to well-defined sets or continua, are of superficial origin. Non-systematic joints are generally small and abut systematic joints, indicating that they are younger than them. A common variety of non-systematic joint is a *cross-joint*, linking neighbouring systematic joints roughly at right angles, and defining a ladder-shaped pattern of traces.

Within a single exposure, a *joint set* comprises a family of parallel joints (Grout & Verbeek 1983), but the use of the words joint set in the context of a region does not always carry the implication that they are perfectly parallel throughout it, merely that the set maintains a constant angular relationship to some other structural trend. If the dispersion about the mean orientation of a group of joints exceeds about 10° the group may contain two sets enclosing a small dihedral angle, or it may define a *joint continuum* comprising an angular continuum of joint orientations that are coaxial about a common direction. A joint system consists of two or more sets symmetrically arranged with respect to each other or a reference direction. A *conjugate joint system* comprises two roughly coeval sets enclosing an acute angle. Although it is tempting to assign joints to conjugate sets on a statistical basis, and after fieldwork, this procedure is not recommended because if the dihedral angle between sets is small, the half-angle between the sets may exceed the angular dispersion in each one. Rather, joints should be assigned to sets in the field on the basis of joint style, including *joint-system architecture*. This latter term, introduced by Hancock (1985), describes the spatial relationship between neighbouring joints. By contrast, *joint-system geometry*

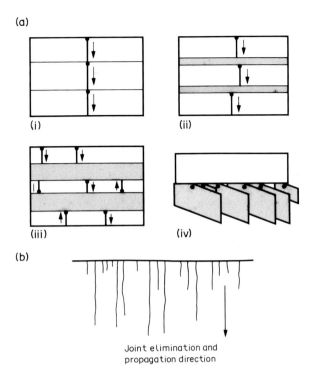

Fig. 5.20. (a) Joint communication styles between adjacent layers. (i) In-plane addition of joint segments, (ii) out-of-plane addition of joint segments, (iii) independent joint segments, (iv) non-planar addition of joint segments. Blank, competent jointed unit; stipple, inhibiting incompetent unit. After Helgeson & Aydin (1991). (b) Joint trace patterns illustrating the determination of joint propagation direction from the direction of joint elimination away from a surface on which the joints nucleated.

describes only the angular relationships between sets in a system. The complete representation of the architecture of a joint system is possible only when detailed plans and profiles of continuously exposed parts of an outcrop have been prepared. Such architectural plans are costly and time consuming to prepare unless drawn from photographs (e.g. Hancock 1986, Rawnsley et al. 1992). For some purposes it is only necessary to compare stylized trace patterns with the shapes of letters in the Latin alphabet.

Some joint sets are restricted to particular lithologies. Irrespective of whether *joint containment* by lithology exists, bedding planes commonly occlude individual joints. *Single-layer joints* are restricted to one bed whereas *multiple-layer joints* communicate across bedding planes. Helgeson & Aydin (1991) have argued on the basis of an analysis of fractographic features that relationships between joints and layer interfaces can be used to determine the propagation histories of joint segments. *In-plane addition of joint segments* occurs where layers of similar properties are adjacent to each other, *out-of-plane addition of joint segments* occurs where a thin or discontinuous inhibiting (generally incompetent) layer intervenes between the jointed layers, *independent joint segments* form where a thick inhibiting layer is intercalated in the sequence, and the *non-planar addition of joint segments* occurs where

younger joints have propagated in a former inhibiting layer from an older joint of different strike in an adjacent layer (Fig. 5.20a). The idea of a *joint elimination direction* (Pollard & Aydin 1988) permits the overall propagation direction of a set to be determined because joints are eliminated in their propagation direction, and away from the surface, such as a bedding plane, on which they nucleated (Fig. 5.20b).

Determination of genetic classes of joints and palaeostress axes

Relationships between joints and stress axes. Although a fracture initiated as a joint can be reactivated in another mode, or can coincide with an older surface, many joints are *neoformed fractures* in the sense of Angelier (1989 and Chapter 4). That is, they are fractures that propagated through formerly intact rock, or rock containing sealed fractures. In the following discussion of relationships between joints and stresses, the joints are assumed to be neoformed and the influence of anisotropy, which is of great importance during faulting (e.g. Peacock & Sanderson 1992, also see Chapter 4), is not considered.

The majority of joints are *extension fractures*, that is, fractures formed normal to the direction of σ_3 at the time of failure (Fig. 5.21a). Such joints are also *Mode 1 cracks* (Chapter 3) that propagated normal to the direction of opening of the crack. If σ_3 was tensile at the time of fracturing the joints are *tension joints*. In a general discussion of conjugate fractures, Angelier (1984) stated that they can be regarded as conjugate if they: (1) were formed in a uniformly oriented stress field, (2) are roughly coeval as inferred from mutual abutting/cutting relationships, (3) were formed within a volume of brittle rock that was behaving as if it was mechanically intact, and (4) propagated along planes whose orientations with respect to principal stress axes accord with the Coulomb–Mohr hypothesis of brittle shear failure. Use of the composite failure envelope to describe stress conditions at the time of fracturing

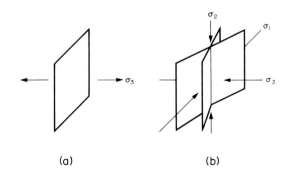

Fig. 5.21. (a) Relationship between a neoformed extension fracture and the orientation of the σ_3 axis at the time of failure. (b) Relationship between neoformed conjugate fractures and total principal stress axes at the time of failure. *Notation*: σ_1, σ_2 and σ_3, maximum, intermediate and minimum principal stresses, respectively.

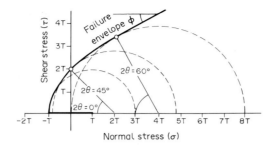

(a) Coulomb-Mohr diagram

(b) Extension fracture (c) Conjugate shear fractures

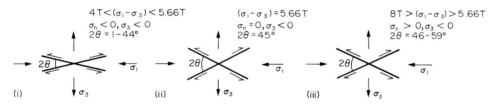

(d) Conjugate hybrid fractures

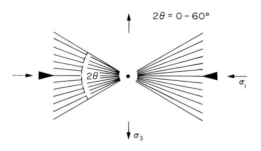

(e) Fracture spectrum

Fig. 5.22. Stress conditions during the formation of neoformed fractures in intact rock. (a) Composite failure envelope plotted on a Coulomb–Mohr stress diagram. (b) Extension fractures. (c) Conjugate shear fractures. (d) Three sub-classes of conjugate hybrid fractures. (e) A fracture spectrum comprising a coaxial angular continuum of extension, hybrid and shear fractures shown at 10° intervals and intersecting at a common node. The orientation of the symmetry axis of the spectrum is indicated by a solid arrowhead. *Notation:* σ_1 and σ_3, maximum and minimum principal stresses; σ_n, normal stress acting across fracture at time of failure; τ, shear stress; ϕ, angle of internal friction; 2θ, conjugate shear angle; T, tensile strength of intact rock. Stresses shown are total stresses and the intermediate principal stress (σ_2) is taken to be perpendicular to the page. Modified from Hancock (1986).

(Fig. 5.22a) carries with it, as argued by Price & Cosgrove (1990), the implication that there is a transition from single sets of *extension fractures*, through conjugate sets of *hybrid fractures* (synonyms — *oblique extension fractures* (Dennis 1972) and *extensional shear fractures* (Etheridge 1983)) enclosing small dihedral angles, to conjugate sets of *shear fractures* enclosing dihedral angles close to 60° (Figs 5.22b–d). Orientations of principal stress axes can be inferred from conjugate fractures because the attitude of the acute bisector between conjugate sets yields the σ_1 axis, the orientation of the obtuse bisector gives σ_3, and the attitude of the intersection direction between sets yields σ_2 (Fig. 5.21b).

Deciding whether conjugate fracture sets comprise *shear* or *hybrid fractures* is done by considering the magnitude of the *conjugate shear* or *dihedral angle* (2θ) between the sets, and making reasonable allowances for both angular despersion within a set, and variations of 2θ with lithology. The following ranges of 2θ values can

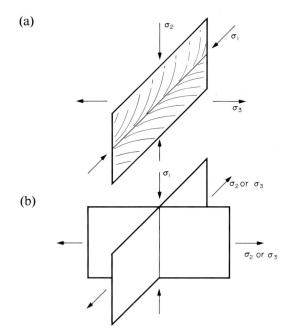

Fig. 5.23. Inferring directions of σ_1, σ_2 and σ_3 from: (a) a single extension fracture bearing a plumose marking with a well-developed axis, and (b) roughly coeval orthogonal extension joints. *Notation*: σ_1, σ_2 and σ_3, maximum, intermediate and minimum principal stresses, respectively.

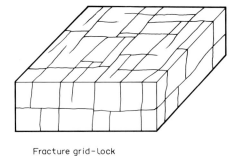

Fig. 5.24. A fracture grid-lock comprising coeval orthogonal joints. Note that abutting relationships indicate that some members of one set are younger than those in the other set, and vice versa.

be regarded as characterizing different genetic categories of fractures:

	2θ
extension fractures	$1-10°$
conjugate hybrid fractures	$11-50°$
conjugate shear fractures	$>50°$

Despite agreement that many joints are extension fractures, there is controversy about whether some are hybrid or shear fractures. Thus, Pollard & Aydin's (1988) recommendation that the name joint should only be applied to extension fractures could be interpreted as totally excluding both hybrid and shear fractures from being classed as joints. However, because: (1) many fractures that are morphologically indistinguishable from extension joints can be interpreted as hybrid fractures, and (2) across conjugate fractures enclosing a 2θ angle of less than 45° there is a component of dilation (Fig. 5.22d(i)), this sub-class of fractures can also be called joints, provided that no displacement along them is detectable in the field, and provided they are not veins. Surfaces resembling joints but interpreted as shears we call fractures without a descriptive qualifier.

The systematic joints in some outcrops do not belong to well-defined sets, but rather they belong to a coaxial angular continuum of fracture planes enclosing a maximum 2θ angle of about 45°. Hancock (1986) has interpreted some *joint continua* as comprising a spectrum of joint classes dominated by extension and hybrid (oblique extension) fractures. Theoretically, a *fracture spectrum* could enclose a maximum 2θ angle of 60° and comprise extension, hybrid and shear fractures (Fig. 5.22e), but a *joint spectrum* contains only the first two classes. To infer the orientations of principal stress axes from a joint spectrum, it is necessary to determine the attitude of the symmetry axis bisecting the continuum, which yields σ_1, the common coaxial direction which gives σ_2, and the perpendicualr to the plane containing these two axes which is parallel to σ_3. A joint spectrum is unlikely to comprise fractures intersecting at a common node, as shown in Fig. 5.22(e), but its presence will be reflected on a stereoplot by a girdle of poles that encloses a maximum angle of about 45°. From joints belonging to conjugate sets or defining a joint spectrum, not only can the orientations of all three principal stresses be determined, but an estimate can also be made from 2θ values of the likely differential stress ($\sigma_1 - \sigma_3$) in terms of multiples of the tensile strength (T) of the rock (Fig. 5.22a).

Although the orientations of σ_1 and σ_2 cannot be inferred from a single set of featureless extension joints, they can be determined in two ways from extension joints provided that either: (1) some of the surfaces display the linear axis of a plumose marking, or (2) the joints belong to a network of orthogonal sets that are coeval. Plume axes form locally parallel to the direction of σ_1 (Fig. 5.23a) (Kulander *et al.* 1979). Orthogonal extension joint sets known to be coeval on the basis of mutual abutting relationships define a *fracture grid-lock* (Hancock *et al.* 1987) in which some members of one set abut the other, and vice versa (Fig. 5.24). Caputo & Caputo (1988) presented the first attempt to estimate a complete stress tensor from such orthogonal extension joint sets, and Caputo (in review) has proposed that they are a product of the rapid 90° switching of σ_2 and σ_3 axes of roughly equal magnitude. The direction of σ_1 remains oriented parallel to the intersection direction between the sets (Fig. 5.23b).

Distinguishing between extension joints and conjugate joints. The genetic class of a joint must be identified in order to infer palaeostress axes from single or conjugate sets. Distinguishing between joints of different class can be carried out with the aid of several criteria, none of which is, by itself, diagnostic (Hancock 1985).

(1) *Microscopic attributes.* A thin section across a

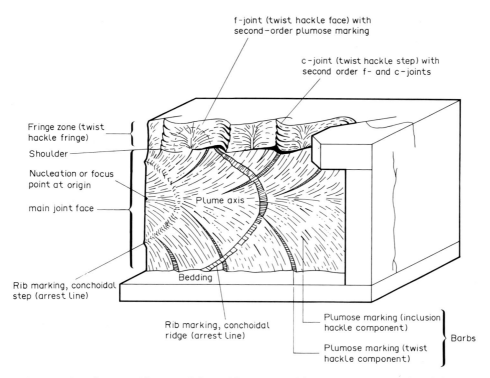

Fig. 5.25. Nomenclature of surface markings on joints. Plumose markings occur on extension joints, and plumes plus a fringe zone probably characterize hybrid joints. Modified after Kulander et al. (1979). See text for details.

joint can reveal whether at a scale of less than a millimetre there is evidence (e.g. offset markers, crack-margin matches) for shear or dilation (e.g. Engelder 1982). Evidence for shear offset does not, however, prove that the joint was initiated as a shear; it could be a reactivated extension fracture.

(2) *Fractography*. Some joints, especially in well-indurated fine-grained homogeneous rocks, display delicate surface markings that reflect their propagation history. Only one or two sets, out of a population of three or more, are likely to show such decoration. This outline of fractography principles is taken largely from reviews by Kulander et al. (1979), Engelder (1987), Ramsay & Huber (1987) and Price & Cosgrove (1990).

Two principal categories of marking occur on *main joint faces*: (1) rectilinear and curvilinear marks parallel to the direction of crack propagation (*plumose markings*), and (2) curvilinear marks perpendicular to the direction of crack propagation (*arrest lines*) (Fig. 5.25). The latter structures indicate where propagation was arrested or slowed during the punctuated growth of a joint. The principal components of a plumose marking (syn. *plume structure*) are: (1) a *nucleation* or *focus point* at the *fracture origin*, (2) the *plume axis* marking the leading tip of the joint front and indicating the overall propagation direction, and (3) *barbs* (syn. *hackle plumes, striations* and *feathers*). There are two types of barbs: (a) *inclusion hackles* and (b) *twist hackles*, both parallel to local joint propagation directions. Bahat & Engelder (1984) recognized two principal types of plumose markings: (1) *S-type plumes* (syn. *chevron* or *herringbone structure*) with straight axes on cracks that propagated a long distance at a constant velocity, and (2) *C-type plumes*, with several linked curving axes, reflecting a variable velocity of crack propagation along short segments.

The fracture origin is commonly an inhomogeneity, such as a clast or bedding plane perturbation, on which the joint nucleated. Close to the nucleation point the joint can be planar and smooth in a *mirror zone* lacking barbs and other markings. Plume axes, the best guides to the dominant propagation direction of a joint, are straight or curved, and in the centre of a layer or at its margins. Arrest lines are expressed by: *conchoidal steps, ribs* and *boundaries between areas of barbs*. They are everywhere convex in the direction of joint enlargement, concentric about fracture origins, and, if elliptical, possess long axes parallel to the dominant direction of joint propagation. Counting the number of arrest lines allows the number of propagation episodes to be estimated. Even on large, well-exposed joints there are rarely more than about ten arrest lines, each less that one or two metres apart. Thus, because most joints are both roughly planar and display only a small number of closely spaced arrest lines it is concluded that, in general, only a few short crack propagation events are involved in the enlargement of an individual joint and that, on the scale of an entire joint plane, principal stress axes do not rotate significantly during joint propagation. Where a plumose marking on a joint plane is cut by the trace of an intersecting joint, the fracture bearing the plume must be the older one.

Although markings on some joints are restricted to the main face, they continue on others within a *fringe zone* separated from the main face by a *shoulder* (Fig. 5.25). The principal structures within a fringe zone are a set of closely spaced en échelon cracks called *f-joints* (syn. *twist hackle faces*) linked by abutting *c-joints* (syn. *twist hackle steps*). On f-joints there may be second-order plumose markings and arrest lines indicating that f-joints propagated normal to the dominant direction of propagation of the main joint face. Whether plumose markings indicate failure in shear (e.g. Roberts 1961, Syme-Gash 1971) or tension (e.g. Kulander *et al.* 1979, Engelder 1982, Pollard & Aydin 1988) has been much debated: the balance of evidence favouring their formation on tensile fractures. However, the observations that: (1) f-joints are arranged en échelon, and (2) that the sense of stepping is generally the same where fringe zones occur on both sides of a main face, and on many joints in the same set, suggests that along an array of f-joints there could have been a small component of shear, in addition to dilation. Such shear could be related to the principal stress plane having locally twisted out of alignment with a main joint face, or it could be a product of the joint being a hybrid fracture. Thus, joints bearing plumose markings unaccompanied by a fringe zone are likely to be extension fractures, but those bearing both plumose markings and fringe zones might be either extension fractures or hybrid fractures.

(3) *Parallelism with a neighbouring kinematic indicator*. Joints parallel to nearby kinematic indicators might be of identical genetic class but without supporting criteria such an interpretation can be misleading. For example, conjugate joints are unlikely to be of the same age as parallel shear zones because the two types of structure evolve in rocks of different properties, and hence at different times.

(4) *Continuity with a kinematic indicator*. Interpreting a joint that is in continuity with, and parallel to, a kinematic indicator as belonging to the same class as the indicator is likewise tempting, but can also be misleading without supporting evidence.

(5) *Symmetry with respect to allied kinematic indicators*. Again, care must be exercised when applying this criterion because suites of associated structures may not be of precisely the same age.

(6) *Symmetry within a fold*. Joint sets that are symmetrically arranged within a fold are often (e.g. Stearns 1964, Hancock 1985, Price & Cosgrove 1990, Turner & Hancock 1990) interpreted as being extension or conjugate joints on the basis of whether they are normal or oblique to symmetry lines or planes, such as hinge lines, axial planes and layering (e.g. Fig. 5.26).

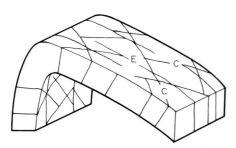

Fig. 5.26. A fold containing extension joints (E) normal to its hinge line, and two sets of conjugate (C) joints that are symmetrically arranged about the hinge line.

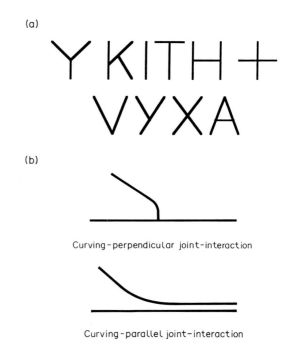

Fig. 5.27. (a) Description of joint-system architecture by comparing the stylized shapes of joint traces with letters in the Latin alphabet. (b) Joint trace patterns illustrating Dyer's (1988) curving-perpendicular and curving-parallel joint-interaction geometries. See text for details.

Not all joint sets symmetrically enclosing fold hinge lines are conjugate; some are unrelated paired sets of extension fractures (see e.g. Engelder & Geiser 1980).

(7) *Refraction*. The segments of a joint that is refracted at bedding planes separating layers of different lithology are generally of different genetic class. Refraction is commonest in sequences of alternating competent and incompetent rocks and where σ_1 acted perpendicular to layering (cf. Peacock & Sanderson 1992).

(8) *Curviplanar morphology*. Joints cutting a bed within which material properties vary continuously can be curviplanar. For example, within a normally graded bed, a vertical extension joint perpendicular to layering might characterize the lower coarse-grained unit, and this extension joint might curve via a hybrid joint to join a shear fracture at a moderate angle to the bedding in the upper muddy part of the layer. Other curved joints, especially those cutting a uniform lithology, probably formed when stess trajectories were markedly curved.

(9) *Joint-system architecture*. The idea of categorizing the architecture of a joint system by comparing the pattern made by its component traces with the shapes of letters in the Latin alphabet was introduced earlier (Fig. 5.27a). *Y*- and K-shapes characterize extension joints that propagated when all three principal stresses were of the same magnitude. I-shapes develop when the differential stress is small and σ_3 is uniformly oriented normal to the extension joint set. T- and +-shapes characterize orthogonal sets of extension joints, including those in fracture grid-locks within which the abutting arm of the T is sometimes one set, and sometimes the other (Fig. 5.24). Where younger joints

cut sealed older ones, +-shapes develop. When all the members of one set in an orthogonal network are younger than those in the other set, both T- and +-shapes also arise but the sense of abutment is consistent. H-shapes and ladder-like trace patterns arise where younger short non-systematic cross-joints abut neighbouring systematic joints (Fig. 5.19).

V-, Y-, and X-shapes characterize conjugate sets of hybrid and shear fractures (Fig. 5.27a). Y-shaped patterns are the commonest, Vs are fairly common, and Xs relatively rare. Patterns that are Y-shaped are formed when a joint propagating towards its conjugate neighbour abuts it. V-shapes are generally formed when two conjugate joints propagate away from an inhomogeneity, such as a perturbation at the base or top of a bed. Crossing conjugate joints, defining X-shapes, only form where a younger joint cuts a sealed one. A-shapes and Vs, Ys and Xs with cross bars characterize conjugate sets accompanied by younger non-systematic joints. Although V-, Y- and X-shapes are typical trace patterns associated with conjugate joints they can also develop when one extension set is obliquely superimposed on another.

According to Dyer (1988) two special trace patterns can be used to estimate stress ratios at the time of jointing. They are *curving-perpendicular* and *curving-parallel joint-interaction geometries* (Fig. 5.27b) which arise when the σ_2/σ_3 far-field stress ratio is either between about -0.3 and 1.0, or -3.0 and -0.3, respectively.

Individual examples of architectural style are of little significance but their repetition, particularly when combined with their symmetry about other structures, increases confidence in their meaning. Pattern recognition is easiest where fractures are closely spaced.

The above method of describing joint-system architecture is applicable where the joints in each set occur throughout an exposure. Associated with some faults, however, there are, in addition to Riedel shears and/or en échelon veins, locally developed joint sets restricted to narrow zones in either, or both, the hangingwall and footwall of the fault. For example, within broad shallow-formed shatter zones accompanying some neotectonic normal faults, such joints define an orthogonal fracture grid-lock (Fig. 5.28a) (Stewart & Hancock 1990) (Chapter 18). More commonly, the two principal types of fault-related joints are: (1) *pinnate joints*, enclosing an acute angle with the fault that closes in the direction of motion of the block containing the joints (Fig. 5.28b); and (2) *comb fractures* (Hancock & Barka 1987) that subtend an angle of about 90° with the fault (Fig. 5.28c). Both varieties of joint intersect the host fault normal to the inferred slip vector. A quadrantal architecture characterizes some pinnate joints accompanying small faults. Such a distribution allows the former locking-area on the fault to be located (Hancock 1985, Engelder 1989).

(10) *Dihedral angle.* Assuming that the composite failure envelope describes stress conditions at the time of failure, the occurrence of a single set of joints displaying a dispersion about its mean orientation of less than about 10° indicates that they are extension joints. Where two maxima reflect the presence of two sets, and the 2θ angle between the sets is about 10 to 45°, the surfaces are likely to hybrid joints. A coaxial continuum of orientations enclosing a 2θ angle between

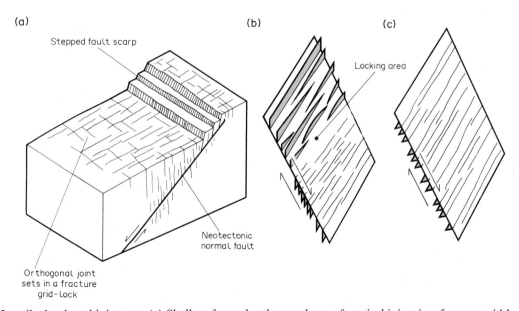

Fig. 5.28. Locally developed joint sets. (a) Shallow-formed orthogonal sets of vertical joints in a fracture grid-lock restricted to the immediate hangingwall and footwall of a neotectonic normal fault. Modified from Stewart & Hancock (1990). (b) Pinnate joints accompanying a normal fault. Note that the pinnate joints enclose an acute angle with the fault in the direction of motion of the block containing the joints, and that they intersect the fault normal to the slip vector. The joints display a quadrantal distribution, being in the hangingwall above the locking-area and in the footwall below the locking-area. The quadrants containing the pinnate joints will be those in which the first P-wave arrivals are dilational. (c) Comb fractures related to a normal fault. Note that the comb fractures are roughly perpendicular to the fault and intersect it normal to the slip vector. (b) and (c) are much modified from Hancock & Barka (1987).

about 10° and a maximum of about 50° reflects the presence of a joint spectrum comprising extension and hybrid joints.

CONCLUSIONS

(1) Small-scale brittle and semi-brittle structures, such as striated faults (see Chapter 4), fibrous veins, stylolites, brittle shear zones, fissures, sediment-filled fractures, and joints can be analyzed to yield sequences of palaeostress orientations. Conjugate sets of hybrid joints permit a rough estimate of differential stresses to be made.

(2) The value of small-scale structures as palaeostress indicators is greatest in regions of mild deformation and simple structure, such as the platform covers of cratons, where uniformly oriented sets of structures are related to remote (far-field) stresses. Small structures are products of local stresses in more structurally complicated settings. Brittle structures are more reliable paleostress indicators in beds of uniform thickness and in fine-grained homogeneous rocks, situations where regional stress fields are less perturbed.

(3) Although knowing the orientations of brittle mesostructures is central to most palaeostress investigations, it is also necessary to record their styles, for example, surface morphology, architecture, and abutting and cutting relationships. Understanding such style features permits discrimination between fractures of different genetic class, and allows the order in which palaeostress trajectories acted to be determined.

(4) In selected tectonic settings, such as platforms, analysis of small-scale brittle structures can yield a more detailed palaeostress history than the analysis of map-scale structures. Such histories are often best preserved in the forelands to mountain belts, rather than in the mountain belts themselves.

REFERENCES

Alvarez, W., Engelder, T. & Geiser, P. A. 1978. Classification of solution cleavage in pelagic limestones. *Geology* **6**, 263–266.

Angelier, J. 1984. Tectonic analysis of fault slip data sets. *J. geophys. Res.* **89**, 5835–5848.

Angelier, J. 1989. From orientation to magnitudes in paleostress determinations using fault slip data. *J. Struct. Geol.* **11**, 37–50.

Bahat, D. & Engelder, T. 1984. Surface morphology on cross-fold joints of the Appalachian Plateau, New York and Pennsylvania. *Tectonophysics* **104**, 299–313.

Brantley, S. L., Fisher, D. & Engelder, T. 1989. Calculated rates of formation of syntectonic quartz pressure shadows and quartz veins. *Geol. Soc. Am. Abstr. Prog.* **21**, A143.

Caputo, R. 1991. A comparison between joints and faults as brittle structures used for evaluating the stress field. *Annls Tecton.* **5**, 74–84.

Caputo, R. In review. Fracture grid-lock evolution as a result of stress swaps. *J. Struct. Geol.*

Caputo, M. & Caputo, R. 1988. Estimate of the regional stress field using joint systems. *Bull. geol. Soc. Greece* **23**, 101–118.

Cox, S. F. 1987. Antitaxial crack-seal vein microstructures and their relationship to displacement paths. *J. Struct. Geol.* **9**, 779–787.

Cruickshank, K. M., Zhao, G. & Johnson, A. M. 1991. Analysis of minor fractures associated with joints and faulted joints. *J. Struct. Geol.* **13**, 865–886.

De Paor, D. G., Simpson, C., Bailey, C. M., McCaffery, K. J. W., Beam, E., Gower, R. J. W. & Aziz, G. 1991. The role of solution in the formation of boudinage and transverse veins in carbonate rocks at Rheems, Pennsylvania. *Bull. geol. Soc. Am.* **103**, 1552–1563.

Dennis, J. G. 1972. *Structural Geology*. Ronald Press, New York.

Dunne, W. M. & North, C. P. 1990. Orthogonal fracture systems at the limits of thrusting: an example from southwestern Wales. *J. Struct. Geol.* **12**, 207–215.

Durney, D. W. & Ramsay, J. G. 1973. Incremental strains measured by syntectonic crystal growths. In: *Gravity and Tectonics* (edited by DeJong, K. A. & Scholten, R.). Wiley, New York, 67–95.

Dyer, R. 1988. Using joint interactions to estimate paleo-stress ratios. *J. Struct. Geol.* **10**, 685–699.

Elliott, D. 1976. The energy balance and deformation mechanisms of thrust sheets. *Phil. Trans. R. Soc.* **283A**, 289–312.

Ellis, M. A. 1986. The determination of progressive deformation histories from antitaxial syntectonic crystal fibres. *J. Struct. Geol.* **8**, 701–709.

Engelder, T. 1982. Is there a genetic relationship between selected regional joints and contemporary stress within the lithosphere of North America? *Tectonics* **1**, 161–177.

Engelder, T. 1985. Loading paths to joint propagation during a tectonic cycle: an example from the Appalachian Plateau, U.S.A. *J. Struct. Geol.* **7**, 459–476.

Engelder, T. 1987. Joints and shear fractures in rock. In: *Fracture Mechanics of Rock* (edited by Atkinson, B.). Academic Press, London, 27–69.

Engelder, T. 1989. Analysis of pinnate joints in the Mount Desert Island granite: implications for postintrusion kinematics in the coastal volcanic belt, Maine. *Geology* **17**, 564–567.

Engelder, T. & Geiser, P. 1980. On the use of regional joint sets as trajectories of paleostress fields during the development of the Appalachian Plateau, U.S.A. *J. Geophys. Res.* **85**, 6319–6341.

Engelder, T. & Marshak, S. 1985. Disjunctive cleavage formed at shallow depths in sedimentary rocks. *J. Struct. Geol.* **7**, 327–343.

Etchecopar, A. & Vasseur, G. 1987. A 3-D kinematic model of fabric development in polycrystalline aggregates: comparisons with experimental and natural examples. *J. Struct. Geol.* **9**, 705–717.

Etheridge, M. A. 1983. Differential stress magnitudes during regional deformation and metamorphism: upper bound imposed by tensile fracturing. *Geology* **11**, 231–234.

Eyal, Y. & Reches, Z. 1983. Tectonic analysis of the Dead Sea Rift regions since the late-Creaceous based on meso-structures. *Tectonics* **2**, 167–185.

Fisher, D. M. 1990. Orientation history and rheology in slates, Kodiak and Afognak Islands, Alaska. *J. Struct. Geol.* **12**, 483–498.

Fletcher, R. C. & Pollard, D. D. 1981. Anticrack model for pressure solution surfaces. *Geology* **9**, 419–424.

Foreman, J. L. & Dunne, W. M. 1991. Conditions of vein formation in the southern Appalachian foreland: constraints from vein geometries and fluid inclusions. *J. Struct. Geol.* **13**, 1173–1183.

Geiser, P. & Sansone, S. 1981. Joints, microfracturers, and the formation of solution cleavage in limestone. *Geology* **9**, 280–285.

Groshong, R. H. Jr 1975. 'Slip' cleavage caused by pressure solution in buckle fold. *Geology* 3, 411–413.
Grout, M. A. & Verbeek, E. R. 1983. Field studies of joints: insufficiencies and solutions, with examples from the Piceance Creek basin, Colorado. In: *Proc. 16th Oil Shale Symp. Golden, Colorado* (edited by Gary, J. H.). Colorado School of Mines Press, Golden, 68–80.
Hancock, P. L. 1985. Brittle microtectonics: principles and practice. *J. Struct. Geol.* 7, 437–457.
Hancock, P. L. 1986. Joint spectra. In: *Geology in the Real World — The Kingsley Dunham Volume* (edited by Nichol, I. & Nesbitt, R. W.). Institution of Mining and Metallurgy, London, 155–164.
Hancock. P. L. 1991. Determining contemporary stress directions from neotectonic joint systems. *Phil. Trans. R. Soc. Lond.* A337, 29–40.
Hancock, P. L. & Barka, A. A. 1987. Kinematic indicators on active normal faults in western Turkey. *J. Struct. Geol.* 9, 573–584.
Hancock, P. L. & Engelder, T. 1989. Neotectonic joints. *Bull. geol. Soc. Am.* 101, 1197–1208.
Hancock, P. L., Al-Kadhi, A., Barka, A. A. & Bevan, T. G. 1987. Aspects of analyzing brittle structures. *Annls Tecton.* 1, 5–19.
Harris, J. F., Taylor, G. L. & Walper, J. L. 1960. Relation of deformation fractures in sedimentary rocks to regional and local structures. *Bull. Am. Ass. Petrol. Geol.* 44, 1853–1873.
Helgeson, D. E. & Aydin, A. 1991. Characteristics of joint propagation across layer interfaces in sedimentary rocks. *J. Struct. Geol.* 13, 897–911.
Hobbs, D. W. 1967. The formation of tension joints in sedimentary rocks: an explanation. *Geol Mag.* 104, 550–556.
Hodgson, R. A. 1961. Regional study of jointing in Comb Ridge – Navajo Mountain area, Arizona and Utah. *Bull. geol. Soc. Am.* 45, 1–38.
Houseknecht, D. W. 1987. Intergranular pressure solution in four quartzose sandstones. *J. sedim. Pet.* 58, 228–246.
Huang, Q. & Angelier, J. 1989. Fracture spacing and its relation to bed thickness. *Geol. Mag.* 126, 355–362.
Kerrich, R., Beckinsale, R. D. & Durham, J. J. 1977. The transition between deforming regimes dominated by intercrystalline diffusion and intra-crystalline creep evaluated by oxygen isotope thermometry. *Tectonophysics* 38, 241–257.
Knipe, R. J. 1986. Microstructural evolution of vein arrays preserved in Deep Sea Drilling Project cores from the Japan Trench, Leg 57. In: *Structural Fabrics in Deep Sea Drilling Project Cores from Forearcs* (edited by Moore, J. C.) *Mem. geol. Soc. Am.* 166, 75–88.
Knipe, R. J. & White, S. H. 1979. Deformation in low grade shear zones in the Old Red Sandstone, S. W. Wales. *J. Struct. Geol.* 1, 53–65.
Kulander, B. R., Barton, C. C. & Dean, S. L. 1979. The application of fractography to core and outcrop fracture investigations. *Tech. Rep. U.S. Dept. Energy* METC/SP-79/3, Morgantown Energy Technology Center, 1–174.
Laderia, F. L. & Price, N. J. 1981. Relationship between fracture spacing and bed thickness. *J. Struct. Geol.* 3, 179–183.
Letouzey, J. & Trémolières, P. 1980. Paleo-stress fields around the Mediterranean since the Mesozoic derived from microtectonics: comparison with plate tectonic data. *Mem. Bur. Rech Geol. Min.* 115, 261–273.
Lundberg, N. & Moore, J. C. 1986. Macroscopic structural features in Deep Sea Drilling Project cores from forearc regions. In: *Structural Fabrics in Deep Sea Drilling Project Cores from Forearcs* (edited by Moore, J. C.). *Mem. geol. Soc. Am.* 166, 13–44.
McEwen, T. J. 1981. Brittle deformation in pitted pebble conglomerates. *J. Struct. Geol.* 3, 25–37.
Mullenax, A. C. & Gray, D. R. 1984. Interaction of bed-parallel stylolites and extension veins in boudinage. *J. Struct. Geol.* 6, 63–72.
Narr, W. & Suppe, J. 1991. Joint spacing in sedimentary rocks. *J. Struct. Geol.* 13, 1037–1048.
Nelson, R. A. 1981. Significance of fracture sets associated with stylolite zones. *Bull. Am. Ass. Petrol. Geol.* 65, 2417–2425.
Nicholson, R. & Ejiofor, I. B. 1987. The three-dimensional morphology of arrays of echelon and sigmoidal mineral-filled fractures: data from north Cornwall. *J. geol. Soc. Lond.* 144, 79–83.
Nicholson, R. & Pollard, D. D. 1985. Dilation and linkage of echelon cracks. *J. Struct. Geol.* 7, 583–590.
Nitecki, M. H. 1962. Observations on slickolites. *J. sedim. Pet.* 32, 435–439.
Olson, J. C. & Pollard, D. D. 1991. The initiation and growth of en échelon veins. *J. Struct. Geol.* 13, 595–608.
Park, W. C. & Schot, E. H. 1968. Stylolites: Their nature and origin. *J. sedim. Pet.* 38, 175–191.
Peacock, D. C. P. & Sanderson, D. J. 1992. Effects of layering and anisotropy on fault geometry. *J. geol. Soc. Lond.* 149, 793–802.
Pollard, D. D. & Aydin, A. 1988. Progress in understanding jointing over the past century. *Bull. geol. Soc. Am.* 100, 1181–1204.
Pollard, D. D. & Segall, P. 1987. Theoretical displacement and stresses near fractures in rocks with applications to faults, joints, veins, dikes, and solution surfaces. In: *Fracture Mechanics of Rock* (edited by Atkinson, B. K.). Academic Press, London, 277–350.
Pollard, D. D., Segall, P. & Delaney, P. T. 1982. Formation and interpretation of dilatant echelon cracks. *Bull. geol. Soc. Am.* 93, 1291–1303.
Price, N. J. 1959. Mechanics of jointing in rocks. *Geol. Mag.* 96, 149–167.
Price, N. J. 1966. *Fault and Joint Development in Brittle and Semi-Brittle Rock*. Pergamon Press, Oxford.
Price, N. J. & Cosgrove, J. W. 1990. *Analysis of Geological Structures*. Cambridge University Press, Cambridge.
Ramsay, J. G. 1977. Pressure solution — the field data. *J. geol. Soc. Lond.* 134, 72.
Ramsay, J. G. 1980a. The crack-seal mechanism of rock deformation. *Nature, Lond.* 284, 135–139.
Ramsay, J. G. 1980b. Shear zone geometry; a review. *J. Struct. Geol.* 2, 83–99.
Ramsay J. G. & Huber, M. I. 1983. *The Techniques of Modern Structural Geology, Volume 1: Strain Analysis*. Academic Press, London.
Ramsay, J. G. & Huber, M. 1987. *The Techniques of Modern Structural Geology, Volume. 2: Folds and Fractures*. Academic Press, London.
Rawnsley, K. D., Rives, T., Petit, J.-P., Hencher, S. R. & Lumsden, A. C. 1992. Joint development in perturbed stress fields near faults. *J. Struct. Geol.* 14, 939–951.
Rickard, M. J. & Rixon, L. K. 1983. Stress configurations in conjugate quartz-vein arrays. *J. Struct. Geol.* 5, 573–577.
Rispoli, R. 1981. Stress fields about strike-slip faults inferred from stylolites and tension gashes. *Tectonophysics* 75, T29–T36.
Rives, T., Razack, M., Petit, J.-P. & Rawnsley, K. D. 1992. Joint spacing: analogue and numerical simulations. *J. Struct. Geol.* 14, 925–937.
Roberts, J. C. 1961. Feather-fractures and the mechanics of rock jointing. *Am. J. Sci.* 259, 481–492.
Rutter, E. H. 1976. The kinetics of rock deformation by pressure solution. *Phil. Trans. R. Soc.* 283A, 203–219.
Schader, F. 1988. Symmetry of pebble-deformation involving solution pits and slip-lineations in the northern Alpine Molasse Basin. *J. Struct. Geol.* 10, 41–52.
Srivastava, D. C. & Engelder, T. 1990. Crack-propagation sequence and pore-fluid conditions during fault-bend

folding in the Appalachian Valley and Ridge, central Pennsylvania. *Bull. geol. Soc. Am.* **102**, 116–128.

Srivastava, D., Engelder, T. & Lacazette, A. 1990. Fluid inclusions as a stressmeter: overburden thickness during emplacement of the Yellow Springs thrust sheet, Appalachian fold-thrust belt, Pennsylvania. *Geol. Soc. Am. Abstr. Prog.* **22**, A94.

Stearns, D. W. 1964. Macrofracture patterns on Teton anticline, northwestern Montana. *Trans. Am. geophys. Un.* **45**, 107.

Stewart, I. S. & Hancock, P. L. 1990. Brecciation and fracturing within neotectonic normal fault zones in the Aegean region. In: *Deformation Mechanisms, Rheology and Tectonics* (edited by Knipe, R. J. & Rutter, E. H.). *Spec. Publs geol. Soc. Lond.* **54**, 105–119.

Stubbs, J. L. & Wheeler, R. L. 1975. Style elements of systematic joints. *Geol. Soc. Am. Abstr. Prog.* **7**, 1286.

Syme Gash, P. J. 1971. A study of surface features relating to brittle and semi-brittle fractures. *Tectonophysics* **12**, 349–391.

Tapp, B. & Cook, J. 1988. Pressure solution zone propagation in naturally deformed carbonate rocks. *Geology* **16**, 182–185.

Turner, J. P. & Hancock, P. L. 1990. Relationships between thrusting and joint systems in the Jaca thrust-top basin, Spanish Pyrenees. *J. Struct. Geol.* **12**, 217–226.

Urai, J. L., Williams, P. F. & van Roermund, H. L. M. 1991. Kinematics of crystal growth in syntectonic fibrous veins. *J. Struct. Geol.* **13**, 823–835.

Watts, N. L. 1983. Microfractures in chalks of Albuskjell Field, Norwegian sector, North Sea: possible origin and distribution. *Bull. Am. Ass. Petrol. Geol.* **67**, 201–234.

Wheeler, R. L. & Dixon, J. M. 1980. Intensity of systematic joints — methods and application. *Geology* **8**, 230–233.

Winslow, M. A. 1983. Clastic dike swarms and the structural evolution of the foreland fold and thrust belt of the southern Andes. *Bull. geol. Soc. Am.* **94**, 1073–1080.

Zhao, G. & Johnson A. M. 1992. A sequence of deformations recorded in joints and faults, Arches National Park, Utah. *J. Struct. Geol.* **14**, 225–236.

CHAPTER 6

Linked Fault Systems; Extensional, Strike-Slip and Contractional

I. DAVISON

INTRODUCTION

A *LINKED fault system* can be defined as a network of broadly contemporaneous branching faults, which link-up over a length-scale much greater than individual fault segments. Linked faults truly branch rather than cross-cut each other. Linked fault systems extend from the intracrystalline to global scale (e.g. Tchalenko & Berberian 1975), and are one of the most important features of upper-crustal deformation (Fig. 6.1a).

Hard-linkage occurs where faults directly link together, as opposed to *soft-linkage*, where faults may be linked by ductile highly-strained zones, without through-going faults developing (Walsh & Watterson 1991a). Faults produce strained zones which may extend up to 60 km either side of the fault plane during one large earthquake (Stein *et al.* 1988), hence there is a mechanical interaction between most faults, even if there is no direct linkage between them. Seismic evidence from active fault systems also indicates an interaction, as the seismicity pattern of a single fault appears different from the seismic behaviour of a fault system. Fault systems obey the *fractal frequency-magnitude Richter–Gutenberg relationship* whereas individual faults tend to produce constant magnitude events over short geological time periods (Wesnousky *et al.* 1983). Linked faults have been mapped at the Earth's surface since the nineteenth century (see references in Coward *et al.* 1986) and since 1960, seismic reflection profiling has provided good evidence that some major fault systems also link down-dip (e.g. Bally *et al.* 1966) (Fig. 11.20a in Chapter 11). However, many highly-strained and faulted terrains do not involve a fully-linked system. Even the largest faults in a system die out laterally and vertically within the deformed rock mass, with displacement taken up by various mechanisms which can produce a more distributed strain.

Most of the seismic deformation in the upper 10–15 km of the crust is taken up on large displacement faults which produce large earthquakes ($M_b \geqslant 5.5$, England & Jackson 1989, Scholz & Cowie 1990). However, in some areas such as the Mediterranean, the seismic strain rates constitute only a fraction of the relative plate motion and much of the upper crustal deformation is thought to be aseismic (e.g. Jackson & McKenzie 1988).

Many small active faults do not break to the surface because of incohesive material at the Earth's surface, which inhibits fault propagation (Marone & Scholz 1988). Therefore, most linked fault systems are initially blind, but large and intermediate-scale faults may grow to the surface with increasing deformation (see e.g. Stewart & Hancock 1991). Linkage at depth can be observed once the system is exhumed, or imaged seismically.

This review covers the mechanics of fault/shear zone growth and evolution towards a linked array. It subsequently describes and compares extensional, strike-slip and contractional fault systems in nature, theory, and experiment. Most of the geological data on linked fault systems are presented in either map or cross-sectional form, and these two observation planes are discussed separately as the patterns which develop are distinct for each mode of faulting. Natural examples of fault patterns are used to illustrate the principles of linked fault development. The importance of fault linkage to the tectonic evolution of regions experiencing large-scale extension, strike-slip motion or contraction is amplified in Chapters 11, 12, 13 and 14.

SHEAR ZONE LINKAGE

Although this chapter mainly considers brittle faults, linkage in the plastic shear regime is crucial to the understanding of linked faults (also see Chapter 1). Plastic *shear zones* in the lower/middle crust link together in more simple smoothly curved patterns than brittle faults (Fig. 6.1b). Usually shear zones widen with depth up to a maximum of 50 km in width (Grocott 1977). Hence, there is a greater likelihood of shear zone linkage rather than fault zone linkage because fault zones are much narrower in the upper crust (typically less than 100 m).

A well-preserved example of linked deep-level shear zones can be seen in Northeast Brazil (Fig. 6.1b). The major Precambrian strike-slip shear zones developed in amphibolite facies metamorphic conditions and affect an area over 300,000 km². They have a horizontal spacing which rarely exceeds 80 km, and the widest shear zones are up to 14 km. With such wide, but

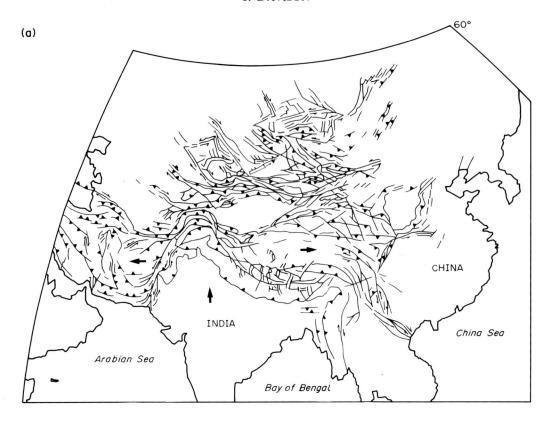

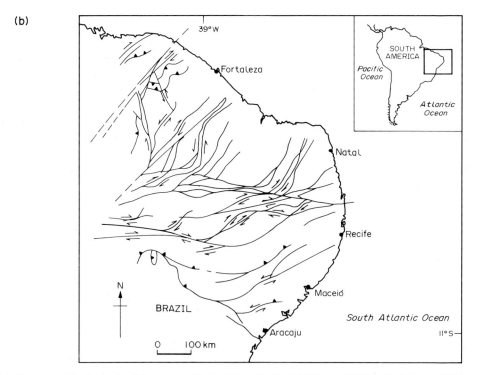

Fig. 6.1.(a) Pattern of faults in the Himalayan Chain (after Cobbold & Davy 1988). (b) Pattern of Precambrian linked strike-slip shear zones in NE Brazil (after Davison & Powell 1991). Thrusts shown with teeth in the hangingwall, strike-slip faults by arrow-couplets, normal faults and faults of unknown displacement are simple black lines.

closely-spaced curved shear zones, it is hardly surprising that all the shear zones link together.

Deep seismic reflection data, especially from extended offshore sedimentary basins, have also provided good evidence of tectonic linkage in the lower crust. The highly reflective lower crust has been interpreted to be caused by systems of extensional shear zones which link with faults in the upper crust (e.g. Reston 1988). Most lower crustal reflections are gently dipping events, which suggest that the steep upper-crustal faults flatten into shear zones (see Chapter 11 for more extended argument).

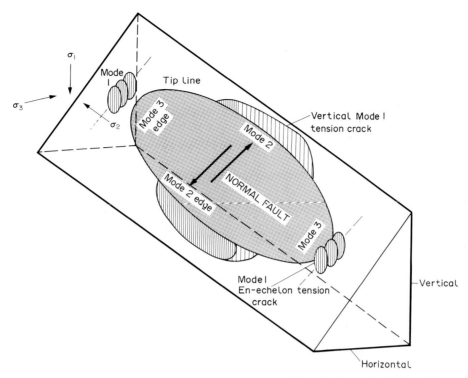

Fig. 6.2. Normal fault plane indicating Mode 2 and 3 crack edges and Mode 1 cracks (adapted from Scholz 1989). σ_1, maximum principal stress; σ_2, intermediate principal stress; σ_3, minimum principal stress.

FAULT PROPAGATION

This review deals with faults where there is relative movement of one block against the other, parallel to the fault plane, in contrast to the *Mode 1 cracks* discussed in Chapter 3. Many laboratory studies have confirmed that the Mohr – Coulomb criterion for brittle failure is a valid model for prediction of fault orientations if principal stress orientations are known (Anderson 1951) (see Chapter 5). The theory has also been applied to inhomogeneous rocks in the presence of microcracks (Brace 1960). However, more recent work on fracture mechanics analysis and observations of faults at widely differing scales indicate that the edges of Mode 2 and 3 shear cracks will not propagate simply within their own plane (Lawn & Wilshaw 1975, Scholz 1989, 1990), and only tensile Mode 1 cracks are capable of this (Fig. 6.2). This paradox has been investigated experimentally by Cox & Scholz (1988), who induced Mode 3 shear (fault) propagation in rock. The fault initiated as an unlinked en échelon tensile crack array (see Chapter 5), which became linked by later shear cracks forming a rubble zone that eventually became a continuous *Mode 3 brittle shear zone* (Fig. 6.3a). The dynamics of the crack linkage appeared too complex to analyze theoretically. Horii & Nemat-Nasser (1986) investigated *Mode 2 fracture propagation*, both in theory and experiments with resins which contained inserted open flaws. They found similar results, where linkage of randomly oriented flaws was due to Mode 1 fracture propagation, which produced an irregular fault plane (Fig. 6.3b). Lockner *et al.* (1991) carried out an interesting experiment on shear crack propagation using acoustic emission locations in a Westerly granite sample (Fig. 6.3c). They were able to show that microcracking was initially evenly distributed throughout the sample, and that cracking became localized at the point where fracture propagation took place. The fracture spread as a smoothly curving front through the specimen. They noted that the seismic *b*-value decreased, just before localized crack propagation took place. Their results suggest that the crack propagated more quickly parallel to the shear direction than at right angles to it.

In nature, it is very difficult to observe fault propagation, as no single fault has produced enough earthquake cycles since detailed fault studies began. However, some faults provide clues to large-scale propagation. The El-Asnam (Algeria) earthquake in 1980 produced an aftershock sequence which indicated that conjugate strike-slip faults are forming some thirty kilometres in advance of the propagating Mode 3 edge of the major thrust fault, which produced the main earthquake (King & Yielding 1984) (Fig. 6.4a). It would appear that the principal stress orientations (σ_2 and σ_3) may have switched ahead of a Mode 3 propagating fault tip. In contrast to this, en échelon tension fractures (Mode 1) in a way, have been found along the upper Mode 3 edge of a strike-slip fault cutting basalts in Iceland (Einarsson & Eiriksson 1981) (Fig. 6.4b). It is thought that in both these cases subsequent faulting would probably use these pre-existing faults and fractures and a rough through-going fault plane would result. The irregular surface geometry of faults from the scale of 10^{-5} up to 10^5 m (Scholz & Aviles 1986) is highly suggestive that many fault systems are produced by multiple fault and fracture linkage at all scales.

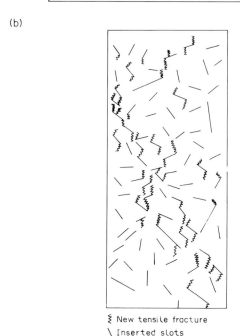

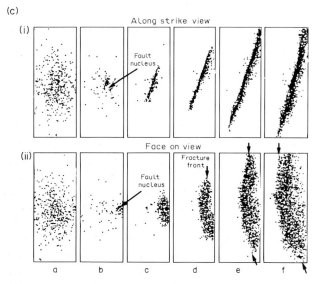

Fig. 6.3. (a) Propagation of a Mode 3 fault edge in Westerly granite. Note that the fault propagates across the gap between two pre-cut slots (after Cox & Scholz 1988). (b) Propagation of a Mode 2 fault edge through resin cut by randomly oriented flaws (after Horii & Nemat-Nasser 1986). (c) Micro-cracking acoustic emission locations showing nucleation and propagation of a shear crack in Westerly granite; (i) cross-sectional view, (ii) view at right angles to the developing shear crack (after Lockner et al. 1991).

FAULT SCALING AND DISPLACEMENT MODELS

Measurements on unlinked normal faults (up to approximately 2 km wide) in the British coalfields, and on seismic sections in the North Sea, indicate that displacement usually dies out gradually from a centrally located maximum (Barnett et al. 1987) (Fig. 6.5a) (see Chapter 11). Although few faults have been mapped with completely closed *tip lines*, it appears that displacement dies out more abruptly in the slip direction where dip-slip faults cut flat-lying sediments. Isolated blind normal faults have an elliptical shape with a typical width/length ratio between 2:1 and 3:1 (Fig. 6.5b). These observations strongly suggest that linkage of dip-slip fault systems is more likely along strike, as faults propagate more quickly in this direction, rather than down-dip. This has important implications for the *piggy-back thrusting* or the *in-sequence model of thrust fault propagation* which has become popular (e.g. Boyer & Elliot 1982) (also see Chapter 13). Most fault propagation models have been developed in two-dimensional cross-sections. However, there is great likelihood of out-of-sequence fault propagation which results from the more effective Mode 3 edge propagation along strike, compared to Mode 2 edge up-dip propagation in dip-slip faults. Barnett et al. (1987) suggest the ellipticity of fault tip-lines may be due to material anisotropy of flat-lying sediments. This is the most likely explanation, as the experiment of Lockner et al. (1991) described above (Fig. 6.3c) predicts that faults should grow more quickly in the slip direction if the material is homogeneous, which is the opposite of the behaviour inferred from observed tip lines in horizontally layered sediments.

Larger displacement faults should be longer and wider (Fig. 6.6). There is an increasing probability of linkage between high-displacement faults in a given population. The maximum total displacement on a fault increases with fault width, and the displacement/width ratio depends on material properties.

Although there are still few supporting data from larger faults, there appears to be a linear relationship between *mean seismic slip* (S) during a single earthquake event and the *longest dimension* (L) *of the ruptured area* so that $S = kL$, where k is a constant which depends principally on rock rheology (elastic and shear moduli) and perhaps on fault mode (strike-slip, thrust, normal) (Scholz 1982). For large faults, the rupture area may not cover the whole pre-existing fault zone, and there will be no direct correlation between seismic slip and the total length of a pre-existing fault (see also Szhwartz & Coppersmith 1984, Jackson & White 1989).

The remarkable observation that total fault slip is proportional to fault dimensions seems to ignore the presence of physical barriers in the fault plane. The importance of physical *fault barriers* may be reduced on larger displacement faults which produce big earth-

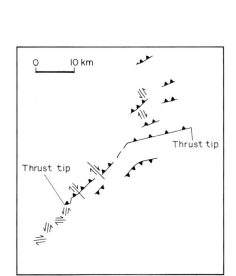
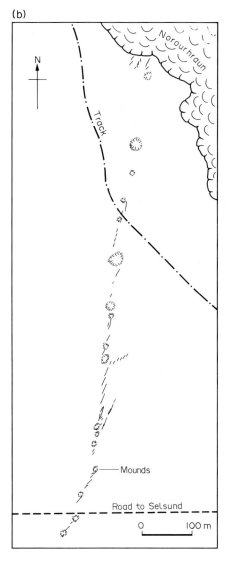

Fig. 6.4. (a) Map of major fault pattern associated with the 1980 El Asnam earthquake, showing conjugate strike-slip faults at the SW thrust tip (after King & Yielding 1984). (b) En échelon tension fractures caused by the underlying strike-slip earthquake in the Rangárvellir District, Iceland, 6th May, 1912, suggesting that Mode 1 fractures formed ahead of the propagating Mode 3 edge of the strike-slip fault (after Einarsson & Eiríksson 1981).

quakes. However, earthquake ruptures often die out at geometrical or strength barriers, and these may be so effective that they are left unbroken behind a *slip front* (Tchalenko & Berberian 1975). *Geometrical barriers* have been characterized as *jogs* (destructive) and *steps* (conservative) depending on their orientation with respect to the slip direction (Scholz 1990), but this nomenclature is not universally followed. Mode 2 fault edges encounter jogs and are more effective in arresting propagating ruptures than steps, which control Mode 3 fault propagation (see also King & Yielding 1984). As surface mapped fault irregularities on dip-slip faults tend to be conservative steps, they generally retain more irregularities than strike-slip faults in horizontal map view. Destructive jogs observed in map view of strike-slip fault will be progressively smoothed out with increasing displacement (Wesnousky 1988) (also see Chapters 11 and 12).

Mindful of geometrical barriers, Schwarz & Coppersmith (1984) suggested a *characteristic earthquake model* for fault displacements along individual fault segments of a larger fault over several seismic cycles (also see Chapter 18). Using data from the Wasatch (extensional) and San Andreas (strike-slip) fault zones, they suggested that individual fault segments between barriers regularly slip to produce constant-sized earthquakes. The segments are separated by barriers where fault slip is minimum, and deformation must be accommodated by other means. As displacements build up on a particular segment the strain in the barrier zone must increase until a through-going fault develops and the barrier is eventually destroyed. Thus, the *characteristic earthquake model* must eventually break down. As most barriers are not permanently fixed the fault will tend to distribute slip more evenly over time as barriers are destroyed. However, creation of new barriers is possible and ensures that the fault plane will not necessarily decrease in roughness with increasing displacement. Geometrical barriers appear to be separated by up to a maximum of

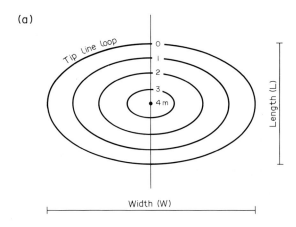

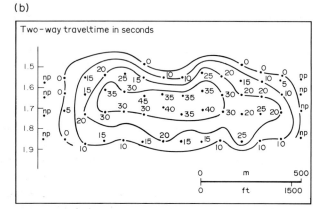

Fig. 6.5. (a) Theoretical displacement contour map of an isolated dip-slip fault. Contours in metres (after Barnett *et al.* 1987). (b) Displacement contour diagram of an extensional fault from the U.K. sector of the North Sea. Displacement values are measured as two way seismic wave travel times in millisecs, np denotes fault not present (after Barnett *et al.* 1987).

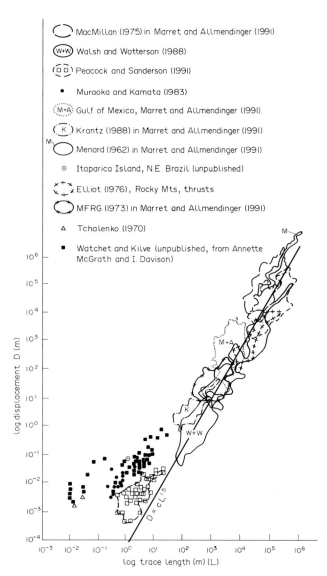

Fig. 6.6. Log displacement–log fault trace length relationship for faults cutting a wide range of lithologies in the three fault modes. The large variation in fault trace length for any one fault displacement is interpreted to be due to lithology, fault mode, barriers, strain rate variation, fault linkage, proximity to neighbouring faults and measurement error (unpublished data and data from Ellis & Dunlap 1987, Marrett & Allmendinger 1991, Peacock & Sanderson 1991, Walsh & Watterson 1991b).

15–20 km along active high-displacement faults, a value which coincides with the thickness of the seismogenic zone, although the significance of this is still not fully understood (Jackson & White 1989, Wallace 1989).

Walsh & Watterson (1988), Scholz & Cowie (1990) and Marret & Allmendinger (1991) have shown that faults increase their displacement (D) as they increase their surface trace length (L). However, the observed field data are very scattered with faults of the same displacement, having trace lengths which vary by three orders of magnitude, and it is unlikely whether one $D-L$ relationship can explain all the data. Scholz & Cowie (1990) have proposed a linear $D-L$ relationship; whereas Watterson (1986) and Walsh & Watterson (1988) proposed a $D = kL^2$, and Marrett & Allmendinger (1991) proposed a $D = kL^{1.5}$, relationship. The best fit to the observed data at high displacements (above 1 m) is the latter (Fig. 6.6). This relationship implies that the difference between each successive fault slip event increases linearly with the total number of slip events (Marrett & Allmendinger 1991). However, with single slip event faults, or faults with displacements less than 1 m, it appears that the faults have higher displacements than is predicted by the best-fit line through the faults with displacements greater than 1 m (Fig. 6.6). This suggests that smaller faults may not propagate in the same manner as higher displacement faults.

Once the upper and lower fault tips grow to reach the surface and the plastic deformation zone respectively, the fault will be constrained to propagate laterally as a brittle fault, and vertically as a shear zone. Thus, the brittle part of the fault will change from an elliptical shape to an elongate rectangular shape, and faults should lose their self-similar geometry once they propagate through the whole of the brittle upper crust. Seismic data indicate that the earthquake magnitude–frequency relationship, described by the *b*-value, changes for large earthquakes cutting the whole of the

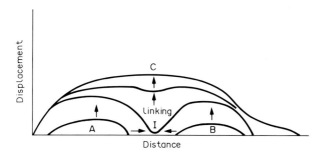

Fig. 6.7. Fault displacement plotted against distance along the fault for two individual faults which subsequently become linked.

seismogenic crust; which also supports this idea of there being a change in fault behaviour (Shimazaki 1986, Pacheco *et al.* 1992).

If a fault zone evolves to become part of a plate boundary (*interplate fault*) the intense deformation causes the major fault zones to weaken. This is evidenced by stress drops associated with interplate earthquakes which are approximately six times smaller than intraplate events (Scholz *et al.* 1986). Kanamori & Allen (1986) also observed how stress drop and earthquake repeat time are related to the fault plane area. Long fault repeat times and larger stress drops are believed to be associated with small-area stronger fault planes, where there has presumably only been a limited amount of strain softening. This observation partly explains why highly-strained tectonic-plate boundaries produce a worldwide linked fault network that is active over a 100 Ma time scale.

FAULT LINKAGE

The fundamental concept of a *linked fault system* is that it develops by growth and connection of individual *geometrical fault segments* as total displacement increases. If regional plate displacement rates are constant, the number of active faults will decrease with increasing strain and time as the large displacement linked faults begin to predominate. This behaviour has been convincingly demonstrated in extensional experiments on cohesive clay and cohesionless dry sand (e.g. Tchalenko 1970, Vendeville 1987). However, no experiments have been conducted on crystalline rock samples of sufficient size to be able to observe true fault propagation and linkage. Linkage will be suppressed if faults are widely-spaced, and the separation of faults is dependent upon the amount of strain softening of the fault planes. Stronger faults will tend to be more closely spaced, and hence linkage will be more likely.

Once faults link together, displacement should be smaller on linked faults compared to an individual fault of equivalent width; and as deformation increases the irregular displacement distribution associated with linkages will be smoothed-out (Peacock & Sanderson 1991) (Fig. 6.7).

Model experiments of strike-slip faults with tens of centimetres of displacement in cohesionless sand and clay show how *Riedel shear faults* progressively join together across late-developed *P shears*, to produce a *linked braided shear fault plane* (Fig. 6.8a) (e.g. Tchalenko 1970, Naylor *et al.* 1986) (Chapter 12). The theoretical stress concentrations between two interacting offset relay faults propagating along Mode 3 edges have been examined by Segall & Pollard (1980) and Rodgers (1980) (Fig. 6.9). However, it must be remembered that their stress analyses are only valid for small amounts of movement in a homogeneous elastic medium, when both fault strands are propagating instantaneously. They state that there should be significant interaction between strike-slip faults if they are separated by less than twice the depth of faulting. Computation of principal stresses around *fault bridges* indicates that *transtensional overlaps* (*jogs, steps*) produce a tensional area which extends well outside the area of overlap, and tensional fractures are predicted to form between the two faults (Fig. 6.9b) (also see Chapter 18). If tensional fractures link the two fault strands together, slippage may develop along this link. Tension fractures tend to turn away from the fault bridge and linkage is suppressed in *transpressional regimes* because of the higher frictional component which would be produced by high normal stresses across the linking fault (Fig. 6.9a), whereas transtensional bridges favour development of a linking extensional fault (Fig. 6.9b). Subsidiary shear fault orientations in the bridge area can also be predicted; these depend on fault overlap, spacing, rock rigidity and fault mode (Fig. 6.8b) (e.g. Woodcock & Fischer 1986). King & Sammis (in press) suggest that where overlapping faults occur, the intervening rock slab is subject to bending stresses which create new fractures at high angles to the initial fractures (Fig. 6.8c). This process can act at all scales, so that very complex fracture patterns can develop which lead to a *gouge zone* with *fractal grain-size distribution* (Fig. 6.8c).

An indication that separate co-planar faults may have linked together to produce a single fault can be inferred from plots of the fault displacement against distance from a fixed reference point on the fault (Eisenstadt & DePaor 1987, Ellis & Dunlap 1988, Peacock & Sanderson 1991). These authors show that faults of varying magnitudes show several displacement minima along the fault trace which have been interpreted as fault linkages, and fault displacement maxima appear to be located at more competent horizons where faulting is presumed to have nucleated.

THREE-DIMENSIONAL GEOMETRY OF LINKED FAULT SYSTEMS

There are many common features between linked fault systems in contractional, extensional and strike-slip regimes (cf. Gibbs 1984, Woodcock & Fischer 1986, and Chapters 11, 12 and 13). However, there are also important differences. These are in part due to the

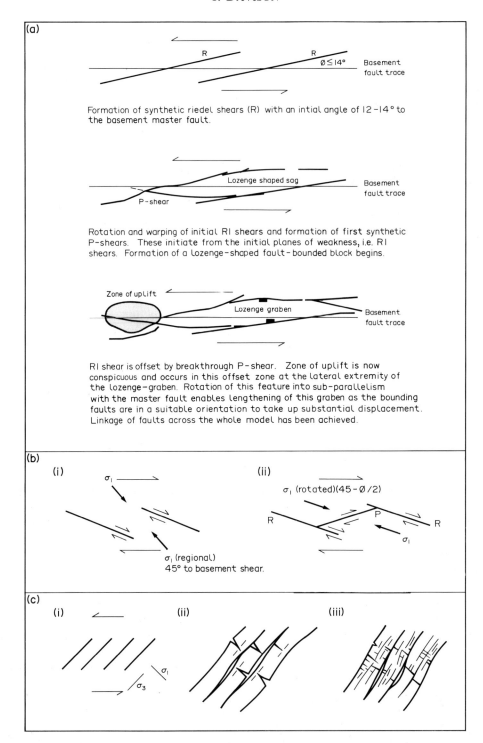

Fig. 6.8. (a) Diagram indicating the evolution of a principal strike-slip fault produced by linkage of isolated R shears and later P shears in a clay model (after T. Dooley, unpublished data). (b) Schematic representation of the stress refraction caused by R shear overlap causing subsequent formation of P shears (after Mandl 1988). (c) Development of tensional fractures across fault-bounded slabs produces a complex fractal fault gouge size distribution (after King & Sammis in press).

orientation of the principal stress axes in relation to weak bedding planes, which are assumed to be flat-lying in this discussion. In contractional systems, where the maximum principal stress is subhorizontal, development of bedding-parallel décollement zones in weak horizons is favoured, along with ramps cutting up at 30° dip through stronger layers. The fault ramps ensure the development of fault-bend folds and duplexes. This contrasts with extensional systems, where initial fault dips of 60° are most common, and subhorizontal décollement zones are suppressed due to the high normal stress across gently dipping planes. Hence, ramp flat-structures, fault bend folds and duplexes rarely appear in extensional regimes affecting flat-lying sediments. The majority of strike-slip faults dip steeply, between 65 and 90°, with dip direction often varying along the strike producing a *propeller geometry*, and linkages will be much more common in map view than in vertical sections of strike-slip faults.

Another important difference between the three

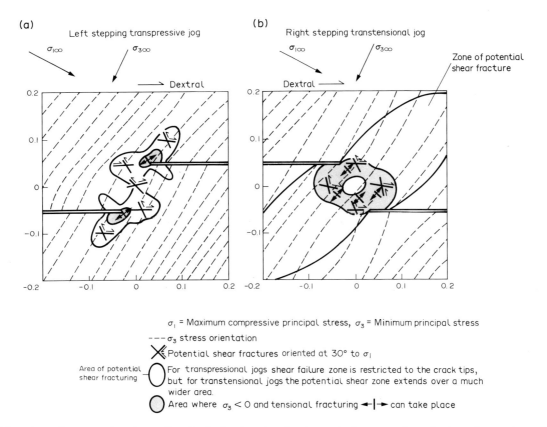

Fig. 6.9. Map view of two overlapping strike-slip faults showing the theoretical stress trajectories in an isotropic medium; (a) transtensional situation, (b) transpressional (after Segall & Pollard 1980).

faulting modes arises because of the angle that the faults meet the free surface of the Earth: vertical strike-slip faults tend to produce symmetrical flower structures; whereas normal faults (usually 50–80° dips) and thrust faults (10–60°) produce asymmetric deformation patterns which can develop above or below the original main fault.

The relative magnitude of the three principal stresses is important in determining the fault pattern and linkage as large differences in the three principal stresses give rise to a minimum of four fault plane orientations arranged in a symmetrical orthorhombic pattern, which favours fault linkage (Reches 1978, Krantz 1988).

Extensional fault systems

Extensional fault linkage in map view. Many outcropping examples of *linked extensional fault systems* have been published (e.g. Robson 1971, Illies

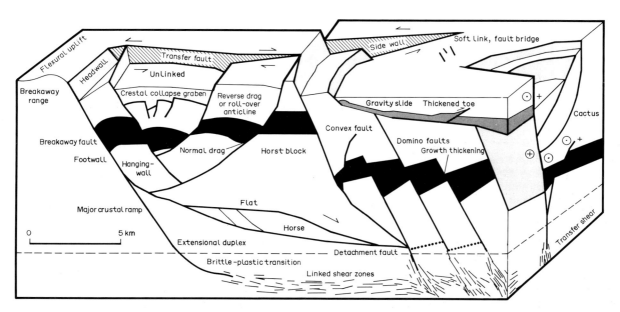

Fig. 6.10. Extensional fault nomenclature.

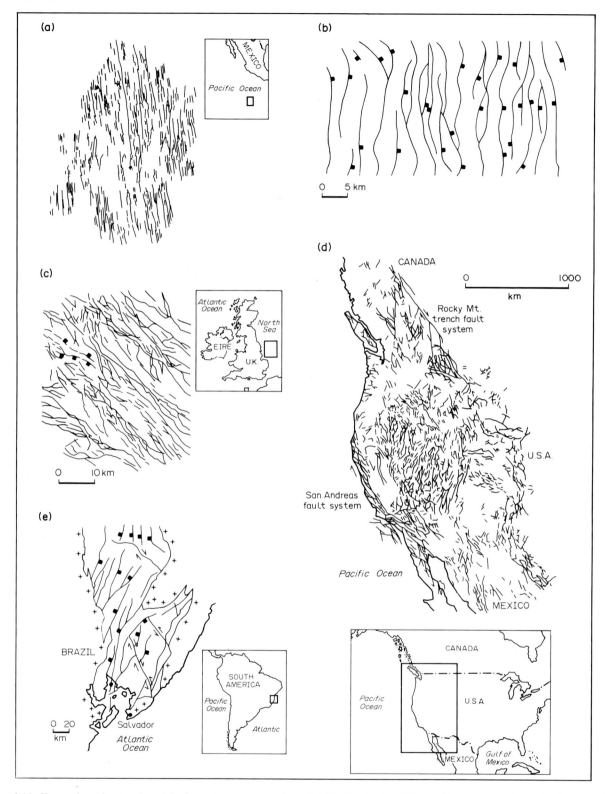

Fig. 6.11. Examples of extensional fault systems in map view. (a) Faults detected from side-scan sonar on the East Pacific Rise from 8°30′ to 9°50′N. These faults are mainly unlinked due to their parallelism (after Cabotte & MacDonald 1990). (b) Horizontal section through a sandbox model showing almost completely linked extensional faults (after McClay 1990). (c) Top Rotliegendes fault pattern map of the Sole Pit Basin, southern North Sea (after Walker & Cooper 1987). (d) Map of the main faults in the Basin and Range province (centre of map) showing highly variable fault orientations and mainly unlinked faults (after Eaton 1980). (e) Fault pattern map of the Recôncavo Basin, NE Brazil showing reticulate pattern with highly oblique transfer faults linking to the main extensional system (after Milani & Davison 1988).

1977, Proffett 1977, Eaton 1980, Miller *et al.* 1983) and recent advances in the acquisition and processing of seismic reflection profiles have provided an enormous impetus to our understanding of three-dimensional linked extensional fault geometry (e.g. Barr 1985, Etheridge *et al.* 1987, Rosendahl 1987, Tankard &

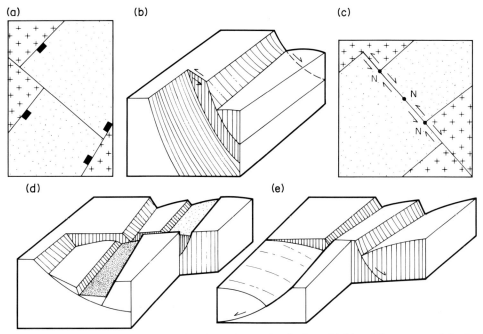

Fig. 6.12. Properties of transfer faults. (a) Termination of transfer fault at highly oblique normal faults (map view). (b) Change of throw and dip of normal fault either side of transfer fault. (c) Movement-sense change along strike of transfer fault (map view). (d) Offset depocentres by transfer which are parallel to rift opening. (e) Flip of depocentre across transfer fault (after Milani & Davison 1988).

Welsink 1987, Badley et al. 1988, Milani & Davison 1988, Colletta et al. 1991). This has produced many new descriptive terms. The terminology for the geometric features of linked extensional fault systems is summarized in Fig 6.10. The largest faults on maps of extended provinces often link up, but the smaller faults are isolated. However, even highly extended areas like the Basin and Range province (U.S.A.) appear to have dominantly isolated faults in map view (Fig. 6.11d). Maps of linked extensional fault systems using outcrop, seismic or experimental data show a gradation of patterns from anastamosing with smaller angular variations (Fig. 6.11a), through anastamosing variable orientations (Figs 6.11b–d), to a reticulate style with a dominant extensional trend, and highly oblique *transfer faults* (Fig. 6.11e).

These differences are probably due to variations in the ($\sigma_2 - \sigma_3$) horizontal principal stress differential and inherited basement weaknesses. Theoretically and experimentally, highly oblique *transfer faults* do not form in a homogeneous material, even when the horizontal principal stresses (σ_2 and σ_3) are markedly different (Cloos 1955, Reches 1978). This suggests that they are likely to form along favourably oriented basement weaknesses. Transfer faults act as a 'passive' response to 'active' normal faults which link with the transfer (Fig. 6.12). Hence, transfer faults can reverse their strike-slip sense in time and space and show apparent offsets opposite to the true movement sense (Fig. 6.12) (Milani & Davison 1988). Transfer faults may show a switch of basin asymmetry where the transfer acts as a *scissor fault*. They can cut right across, or die out, within an extended basin (Fig. 6.12a). In section, transfer faults can exhibit flower, cactus, and palm tree structures typical of strike-slip faults (Milani & Davison 1988). On seismic sections they commonly show a disrupted zone, where there are no reflections due to intense deformation and bedding rotation (e.g. Etheridge et al. 1987).

Complex soft-linked accommodation zones are more common than transfer faults between major fault systems of similar or opposite polarity (Bosworth 1985, Rosendahl 1987, Morley et al. 1990). The variation of accommodation zones in relation to linkage, fault overlap, displacement and spacing has been well-documented in the East African rift (Rosendahl 1987, Morley et al. 1990) (Fig. 6.13).

The reason for basins switching polarity across transfers or accommodation zones is commonly controlled by pre-existing basement structure (e.g. Chorowicz 1983, Rosendahl 1987, Milani & Davison 1988). However, Boswell (1985) has suggested that this may be due to the natural curvature of listric boundary faults curving towards the extension direction at their lateral tips, with polarity flipping to maintain a roughly constant direction of basin opening. This suggestion has still be be examined rigorously, as the boundary faults have been interpreted as being straighter by other workers (Morley 1989). Besides accommodation zones, simple relay ramp structures commonly occur between interfering faults which dip in similar directions (McGill & Stromquist 1979, Larsen 1988) (Fig. 6.14), or twist zones are developed between faults with opposing dips (Colletta et al. 1991).

Peacock & Sanderson (1991) and Peacock (1991) have examined *relay structures* between fault segments along normal and strike-slip faults, respectively. They indicate that the fault displacement decreases markedly at the relay ramp due to deformation being taken up by rotation of the zone between the two fault planes. They also show how the amount of overlap and displacement on the faults will have a marked effect on the shape of the relay ramp.

Extension fault linkage in cross-section. The most common type of major basement faulting, where faults are typically separated by 15–20 km, consists of *unlinked planar faults* in the brittle crust, although these faults may be softly-linked by ductile shear zones at greater depth (Fig. 6.10). One model of deformation in the lower crust involves *anastamosing low-angle shear zones* surrounding lozenge-shaped low-strain zones which produce the highly reflective lower crust below some extended terrains (e.g. Reston 1988), although many authors suggest that this is due to stratiform magmatic intrusions.

In section, extensional faults show two main styles: *domino faults* which are usually subparallel planar faults, where fault planes and bedding rotate simultaneously (Wernicke & Burchfiel 1982, Davison 1989); and linked *listric faults* which shallow out in a common detachment horizon, usually situated in the sedimentary basin fill along a weak horizon such as overpressured shale or salt. Linked fault systems, which involve the basement down to mid- and lower-crustal levels, have been inferred in many deep seismic reflection interpretations (e.g. Beach 1984, Enachescu 1987, Etheridge *et al.* 1987, Gibbs 1987); but direct evidence of their presence is limited due to the lack of outcropping examples and poor seismic resolution at these depths. However, linked detachments involving basement rocks have been carefully mapped out in the Basin and Range province (e.g. Miller *et al.* 1983).

Subsidiary faults can curve gradually into the basal detachment, or abut, to produce angular bends. The ramp-flat angle is probably related to the rock properties. Experimental studies on fine-grained halite gouge indicate that it can have an extremely low coefficient of sliding friction (<0.1), or it can behave plastically at high confining pressures of 200 MPa (Shimamoto & Logan 1981). Frictional sliding or plastic shearing may preferentially take place on low-angle faults, where high normal stresses are acting. Clay-rich fault gouge can be developed along the basal detachment, which is weaker than the surrounding rock, and aseismic sliding can occur. Weak fault gouge in low-angle basement-involved detachments may explain the apparent lack of seismic activity on faults with dips of less than 30° (Jackson 1987).

Ramp-flat geometry is uncommon in extensional regimes, but does exist, and can occur with extension of pre-existing staircase thrust planes, or because of fault refraction through varying lithologies (Bradshaw & Zoback 1986), and with varying compaction states (Davison 1987). Ramp-flat geometry can produce associated thrust faults and folds at convex-upwards kinks, and these can be used to infer the geometry of the detachment at depth (Gibbs 1984, McClay 1990, Dart 1991, McClay & Scott 1991).

The initiation of a detachment, before appreciable strain softening, probably requires a topographic dip, so that there is a gravitational component of shear stress acting on the potential detachment. Cross-section restoration shows that most extensional detachments dipped in the same direction as the surface topography when faulting was active, rather than against it, as in thrust wedges. Hence, the topography-extensional

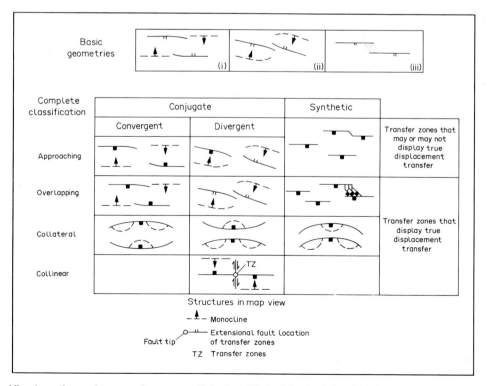

Fig. 6.13. Classification of transfer zones between unlinked and linked faults (after Morley *et al.* 1990). The three basic types of transfer show faults that dip in opposite directions (Conjugate, i and ii), and in the same direction (Synthetic, iii).

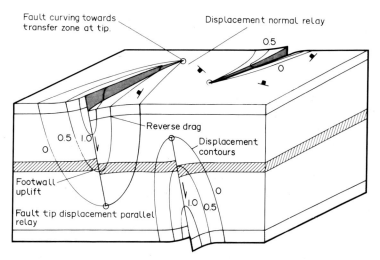

Fig. 6.14. Block diagram showing displacement-parallel and displacement-perpendicular relay ramps with displacement contours (after Peacock & Sanderson 1991, redrawn by Matt White).

detachment angle can be lower (<7°) than the critical Coulomb wedge angle in thrust belts 7–15° (see Chapter 13) (Dahlen 1990).

As major faults link up with the basal detachment, the amount of displacement accumulates at the down-dip end of the detachment, which must have a greater displacement than any of the individual fault displacements in the uppermost layer. The faults must rotate to lower angles as deformation increases, which produces more horizontal extension for the same fault displacement. The blocks at the down-dip end of the detachment must undergo large translations which can only be quantified by sequential restoration of the fault displacements.

Deformation below the detachment may be minimal, and if the crust below the detachment is also extending this can take place in another area. *Gravitationally-driven extensional detachments* in the sedimentary cover are linked to thrusts, fold belts, and erosional truncations close to the surface (Fig. 6.15), whereas *tectonically driven detachments* affecting basement may root down into the ductile lower crust.

Strike-slip fault systems

Strike-slip fault linkage in map view. In map view, strike-slip faulted terrains tend to be narrower and more continuous than either contractional or extensional terrains (England & Jackson 1989). Hence, there should be a higher tendency for *along-strike linkage* of strike-slip faults compared with dip-slip modes. In map view, *strike-slip faults* commonly take on a braided linked

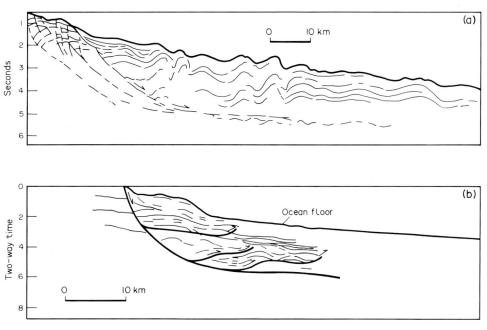

Fig. 6.15. Thin-skinned gravity driven slide showing deformation at the toe taken up by contractional structures. (a) Mexican Fold Ridges, Gulf of Mexico (after Buffler 1983). (b) Thrusting at the toe of a gravitational side in the Barreirinhas Basin continental slope, NE Brazil (after Azevedo 1991).

pattern of anastamosing contemporaneous faults (Fig. 6.16). These can be narrow discrete zones, or affect a large area where two conjugate sets are usually developed. In both cases a large degree of linkage is developed. Strike-slip faults generally have smoother, more continuous mapped traces than dip-slip faults, because jogs are more effectively smoothed-out by abrasion and *short-cut faults*, with increasing displacement (Wesnousky 1988).

Growth of Mode 3 edges of strike-slip faults is particularly amenable to investigation using physical analogue models (e.g. Tchalenko 1970, and references listed in Sylvester 1988). All models (independent of the modelling materials) show a general pattern of en échelon *Reidel shear faults* (R shears) (Chapter 5). These shears produce a reorientation of the stress field and *P shears* form at a low angle to the principal displacement zone, which then link with the R shears to produce a vertical braided continuous fault zone (see Chapter 12) (Fig. 6.8a). Similar patterns are observed on a larger scale with mature faults in nature, where Reidel shears are often observed on either side of the principal displacement zone, and their angular relation can be used to infer the sense of shear (Tchalenko 1970). En échelon fault arrays have not been observed as precursors to experimentally produced, or naturally occurring, Mode 2 edges of dip-slip faults (e.g. Cloos 1955), but unfortunately there is little information available on Mode 3 edge propagation of dip-slip faults to directly compare with Mode 3 edges of strike-slip faults.

Offsets and bifurcations in strike-slip faults cause either transpressional or transtensional areas (Chapter 12). Offset faults can remain unlinked until larger displacements have built up, resulting in a *pull-apart basin* or transpressional *push-up mountain range*. Subsequent linkage across two overlapping offset strike-slip faults leads to extinction of a pull-apart basin or push-up mountain range.

The reason for offsets in strike-slip faults may be due to propagation of separate faults which subsequently link, or may be imposed by a pre-existing curved block geometry. Aydin & Nur (1985) suggested that the curved North Anatolian fault has systematic right-stepping offsets, so that straight fault segments are produced between offsets, because this configuration would be

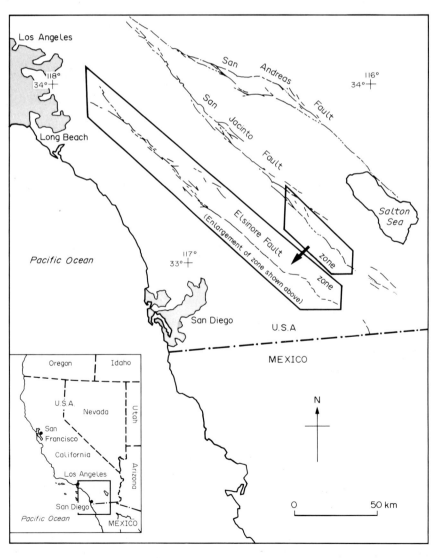

Fig. 6.16. Typical map patterns of Southern Californian strike-slip faults (Brown & Sibson 1989).

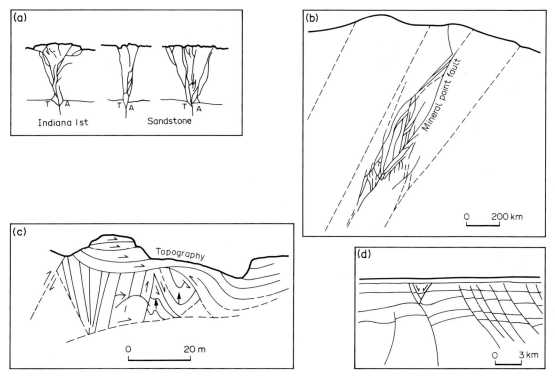

Fig. 6.17. Vertical sections through strike-slip faults. (a) Pure strike-slip mode faults in limestone (Bartlett *et al.* 1971). (b) Transpressional Mineral Point fault (after Wallace & Morris 1986). (c) Transpressional Painted Canyon Fault, California (after Sylvester 1988). (d) Transtensional fault, Andaman Sea (Harding 1983).

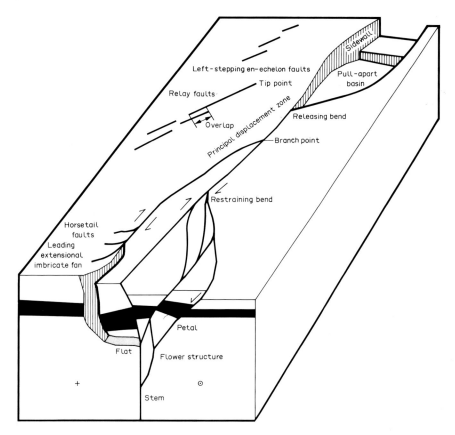

Fig. 6.18. Block diagram showing terminology of features associated with strike-slip faults.

energetically more efficient. The length of segments between offsets appears to be related to the curvature of the fault block boundary, with shorter spacing between offsets occurring along faults with larger curvature (Tim Dooley personal communication).

Strike-slip fault linkage in cross-section. In cross-section, strike-slip faults usually show a steeply dipping principal displacement zone which is linked to faults that branch outwards from the steep root towards the surface. Major strike-slip faults are usually steeply dipping complex zones and are difficult to image on seismic reflection profiles, and three-dimensional linkage is therefore difficult to see. In a purely strike-slip mode, Reidel shears will initially dominate. They have been shown to have a helicoidal geometry in experiments using homogeneous sand (Naylor *et al.* 1986), but a more complex geometry in layered rocks (Fig. 6.17). Individual faults may branch out, up to 20 km from the principal displacement zone, and entire sedimentary basins may be enclosed within a *flower structure* (e.g. Gibbs 1986). Faults may bend into weak subhorizontal layers to form upper detachments (convex-upwards faults defining *palm structures*), or may steepen to the surface controlled by an ambient stress regime (listric faults, *tulip structure* or *cactus structure*).

Strike-slip regimes may involve low-angle flats and be thin-skinned (e.g. Woodcock 1987). For example, major detachments below the Vienna Basin have been suggested on the basis of geophysical evidence (Royden 1985). However, it is most common for major strike-slip faults to be subvertical. A summary of the terminology of strike-slip structures is shown in Fig 6.18.

Contractional fault systems

Thrust fault systems have been mapped in more detail than the other modes of faulting because they generally occur in zones of good outcrop in mountainous terrain, except where erosion has denuded the upper brittle-thrust systems and plastic thrusts are exposed in flat-lying Precambrian terrains (e.g. Lucas 1989). Thrusts tend to link together more in vertical section than in horizontal map view where, commonly, flat-lying basal décollement thrusts are developed. However, some major thrusts are not linked and do not appear to flatten in the continental crust (e.g. Wind River thrust, Wyoming, U.S.A.). Thrust systems are especially well developed in accretionary prisms (Chapters 9 and 16) and regions of continental collision (Chapter 13).

Contractional fault linkage in map view. Most maps of *contractional fault systems* which have well-documented fault patterns come from Phanerozoic mountain belts. These are in *thin-skinned tectonic regimes* where flat-lying sedimentary sequences have been involved in the thrusting. There is a wide variation in fault patterns from regularly spaced subparallel systems (e.g. Rocky Mountains, Fig. 6.19a) through highly variable curved patterns such as those in the Himalayan *syntaxis* regions (*syntaxis*: sharp curve of mountain chains in plan) (Fig. 6.19b), to reticulate patterns such as the Pine Mountain thrust area in the Appalachians (Mitra 1988) (Fig. 6.19c). Thrusts can link together in smoothly bifurcating and anastamosing patterns or be joined by highly oblique faults termed *transfer faults*, or oblique and *lateral ramps* (Fig. 6.19c)

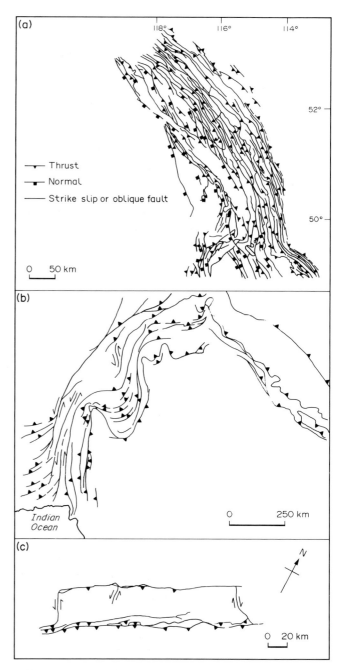

Fig. 6.19. Maps of thrust fault patterns. (a) Main thrusts in the southern Rocky Mountains (after McMehan & Thompson 1989). (b) Map of the southwestern frontal Himalayan chain showing three important syntaxes (after Butler *et al.* 1987). (c) The Pine Mountain thrust system, Southern Appalachians showing a reticulate pattern with lateral ramps (after Mitra 1988). In (a–c) teeth shown in hangingwalls of thrusts.

(Dahlstrom 1970, Butler 1982). Lateral ramps are likely to be induced by basement heterogeneity, or by lateral strength variations along the fault plane or in the surrounding rocks. Hence, linkage of faults in the horizontal surface will be enhanced by lateral strength variations, normally caused by sedimentary facies and thickness variations.

The contrasting lithology or thermal structure of colliding crust can produce rigid indentors which plough through softer crustal blocks and produce very complex linked fault systems, where both strike-slip and thrust faults are equally important (Fig. 6.1a). In such cases there is usually a high degree of linkage of the faults because they branch away from a common stress concentration nucleus at the point of the indentor.

Contractional fault linkage in cross-section. Contractional fault linkage in vertical sections parallel to the main transport direction is the most likely situation in which to observe fault linkage because of the tendency for contractional faults to develop along low-angle planes of weakness and to cut up through the section in stronger horizons. Where sedimentary cover rocks are faulted, evaporites or shales are generally the preferred lithologies along which décollement surfaces develop and the same weak zone may be identified over thousands of square kilometres in thrusted terrains (e.g. the Triassic cargneule in the Western Alps).

It is still hotly debated whether linked fault systems develop sequentially with faults propagating off the basal décollement surface upwards into the hangingwall block (*forward-propagating thrust faults*) (Fig. 6.20a), or whether the faults initiate in the most competent horizons in the hangingwall above the future décollement surface, and then propagate back and downwards to join it (*backward-propagating thrust faults*) (Eisendstadt & DePaor 1987) (Fig. 6.20b). Alternatively, development of folding in the hangingwall may serve to weaken the overturned highly sheared limbs of asymmetric folds developing above a décollement, and this serves to nucleate a fault which propagates down to

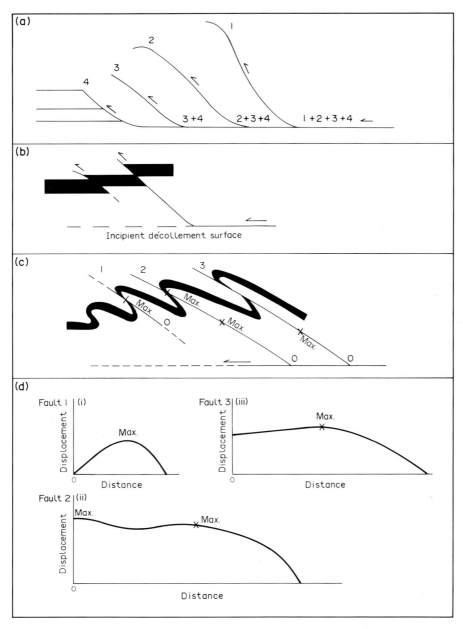

Fig. 6.20. (a) Forward-propagating thrust faults branching off a foreland propagating basal décollement surface. (b) Thrust fault development in competent layers which propagate downwards to an underlying décollement surface. (c) Thrust development from the weakened inverted limbs of asymmetric folds. (d) Displacement history of a fault which links to the basal décollement after initiation above it. (i) Displacement – distance diagram for fault 1 in (c). (ii) Displacement – distance diagram for fault 2. (iii) Displacement – distance diagram for fault 3.

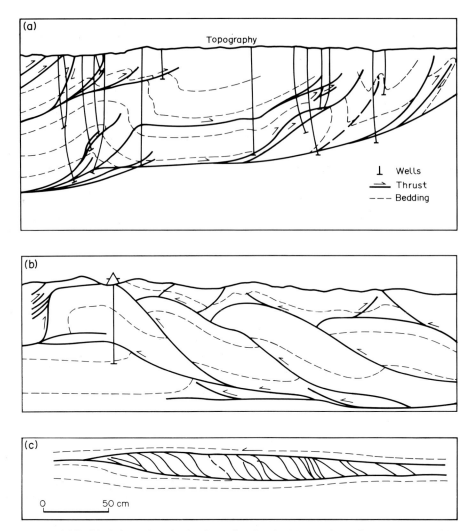

Fig. 6.21. (a) Emergent and blind imbricate fans in the Canadian Rocky Mountains (after Lamerson 1982). (b) Irregular roof duplex in the Appalachians (after Perry 1978). (c) Smooth roof duplex in sandstones and shales in the Hartland Quay area, SW England (after Tanner 1991).

the décollement (e.g. Dixon & Shumin 1991) (Fig. 6.20c). Displacement – distance analysis of some thrust faults indicates that thrust faults initiate in the most competent horizons (Eisenstadt & DePaor 1987, Ellis & Dunlap 1987). However, in situations where the thrust fault has propagated back downwards onto the décollement surface the displacement may build up from the branch point off the décollement surface, and the original displacement maximum at the point of nucleation will be masked by the gradual upward decrease of displacement away from the décollement surface (Fig. 6.20d).

The individual thrusts which link down onto a décollement surface can either die out upwards as a

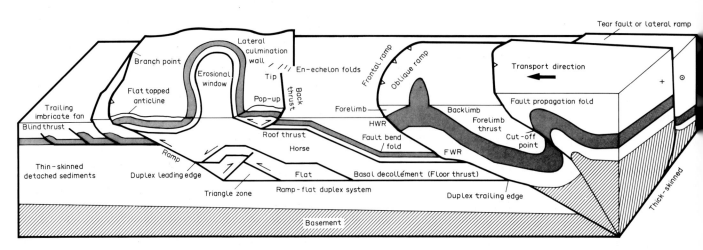

Fig. 6.22. Terminology of contractional fault related structures.

blind imbricate system, intersect the surface (*emergent imbricate system*), or link onto an upper detachment surface to produce a *duplex system* (Figs 6.21b & c) (Boyer & Elliot 1982). Duplex systems can have a *rough roof-thrust* which is produced by linking together different faults with different amounts and ages of displacement (Fig. 6.21b), or more commonly, they may be *smooth roof-thrusts* where the individual faults in the duplex are moving synchronously (Fig. 6.21c) (Tanner 1991). The individual link thrust faults in the smooth-roofed duplexes are often sheared at the contacts with both the roof and floor thrust to produce a smoothly curved sigmoidal shape, and the central thrusts of the duplex appear to have rotated more than the thrusts at the leading and trailing edges (Fig. 6.21c) (Tanner 1991).

Most thrust systems are *hinterland-dipping* with less *backthrusting* than *foreland-directed thrusting* developed. Experimental modelling suggests that backthrusts seem to be better developed when there is a low friction component on the basal décollement surface (Liu Huiqui *et al.* 1991). Once backthrusts develop, linkage between faults is increased due to intersections and branching between thrusts with opposing dips. A summary of the terminology of structures associated with contractional faults is shown in Fig. 6.22, and the reader is referred to a recent glossary of thrust tectonics for additional details (McClay 1991).

CONCLUSIONS

Linked fault systems commonly occur in nature in dip- and strike-slip modes of faulting from the crystalline to global scale, and their geometry appears to be scale independent, except perhaps at the smallest (< 1 mm) scale, where crystal structure controls fault patterns. However, it must be remembered that there are many apparently unlinked fault systems. Linked fault systems will predominate in highly-deformed zones at high-strain rates, and are equally developed at the margins of, and within, tectonic plates. Field mapping and seismic studies have greatly improved our understanding of three-dimensionally linked tectonics down to mantle level, and it is now generally agreed that brittle faults link to shear zones, and that linked highly strained zones can affect the whole crust.

A linked fault system has important implications for fault sequence and the dimensions of faults, which in turn can have a strong influence on strain, sedimentation and erosion patterns, and fluid and magma flow.

Acknowledgements — I would like to thank Chris Dart for a useful review of this manuscript. The Fault Dynamics Project at Royal Holloway gave financial help.

REFERENCES

Anderson, E. M. 1951. *The Dynamics of Faulting*. Oliver and Boyd, Edinburgh.

Aydin, A. & Nur, A. 1985. The types and role of stepovers in strike-slip tectonics. In: *Strike-slip Deformation, Basin Formation, and Sedimentation* (edited by Biddle, K. T. & Christie-Blick, N.). *Spec. Publs Soc. Econ. Palaeont. Mineral.* **37**, 35–44.

Azevedo, R. P. 1991. Tectonic evolution of Brazilian Equatorial continental margins. Ph.D. Thesis, University of London.

Badley, M. E., Price, J. D., Rambech Dahl, C. & Agdestein, T. 1988. The structural evolution of the northern Viking Graben and its bearing upon extensional modes of basin formation. *J. geol. Soc. Lond.* **145**, 455–472.

Bally, A. W., Gordy, P. L. & Stewart, G. A. 1966. Structure, seismic data, and orogenic evolution of southern Canadian Rocky Mountains. *Bull. Can. Petrol. Geol.* **14**, 337–381.

Barnett, J. A. M., Mortimer, J., Rippon, J. H., Walsh, J. J. & Watterson, J. 1987. Displacement geometry in the volume containing a single normal fault. *Am. Assoc. Petrol. Geol. Bull.* **71**, 925–938.

Barr, D. 1985. 3D palinspastic restoration of normal faults in the Inner Moray Firth. *Earth Planet. Sci. Lett.* **75**, 491–500.

Bartlett, W. L., Friedman, M. & Logan, J. M. 1981. Experimental folding and faulting of rocks under confining pressure. Part IX. Wrench faults in limestone layers. *Tectonophysics* **79**, 255–277.

Beach, A. 1984. The structural evolution of the Witch Ground Graben. *J. geol. Soc. Lond.* **141**, 621–628.

Bosworth, W. 1985. Geometry of propagating rifts. *Nature* **316**, 625–627.

Boyer, S. E. & Elliot, D. 1982. Thrust systems. *Am. Assoc. Petrol. Geol. Bull.* **66**, 1196–1230.

Brace, W. F. 1960. An extension of the Griffith theory of fracture of rocks. *J. geophys. Res.* **65**, 3477–3480.

Bradshaw, G. A. & Zoback, M. D. 1988. Listric normal faulting, stress refraction, and state of stress in the Gulf Coast Basin. *Geology* **16**, 271–274.

Brown, N. N. & Sibson, R. H. 1989. Structural geology of the Octillo Badlands antidilational fault jog, Southern California. Proceedings USGS Conference on Fault Segmentation and Controls of Rupture Initiation and Termination. *U.S. geol. Surv. Open File Rep.* 89–315, 94–109.

Buffler, R. T. 1983. Structure of the Mexican Ridges Foldbelt, Southwest Gulf of Mexico. In: *Seismic Expression of Structural Styles* (edited by Bally, A. W.). *Am. Assoc. Petrol. Geol.* **2**, 2.3.3-16–2.3.3-21.

Butler, R. W. H. 1982. Terminology of structures in thrust belts. *J. Struct. Geol.* **4**, 239–245.

Butler, R. W. H., Coward, M. P., Harwood, G. M. & Knipe, R. J. 1987. Salt control on thrust geometry, structural style and gravitational collapse along the Himalayan mountain front in the Salt Range of Northern Pakistan. In: *Dynamical Geology of Salt Related Structures* (edited by Lerche, I. & O'Brien, J. J.). Academic Press, London, 339–418.

Cabotte, S. M. & Macdonald, K. C. 1990. Causes and variation in fault-facing direction on the ocean floor. *Geology* **18**, 749–752.

Chorowicz, J. 1983. Le rift est-africain: début d'ouverture d'un océan. *Bull. Centre Rech. Explor. Prod. ELF Aquitaine* **7**, 155–162.

Cloos, E. 1955. Experimental analysis of fracture patterns. *Bull. geol. Soc. Am.* **66**, 241–256.

Cobbold, P. R. & Davy, P. 1988. Indentation tectonics in nature and experiment. 2. Central Asia. *Bull. geol. Instn Univ. Uppsala (New Ser.)* **14**, 143–162.

Colletta, B., Le Quellec, P., Letouzey, J. & Moretti, I. 1991. Longitudinal evolution of the Suez Rift structure (Egypt). *Tectonophysics* **153**, 221–233.

Coward, M. P., Deramond, J., Hossack, J. & Platt, J. 1986. Preface to thrusting and deformation. *J. Struct. Geol.* **8**, v–vi.

Cox, S. J. & Scholz, C. H. 1988. On the formation and growth of faults: an experimental study. *J. Struct. Geol.* **10**, 413–430.

Dahlen, F. A. 1990. Critical taper model of fold-and-thrust belts and accretionary wedges. *A. Rev. Earth Planet. Sci.* **18**, 55–99.

Dahlstrom, C. D. A. 1970. Structural geology in the eastern margin of the Canadian Rocky Mountains. *Bull. Can. Petrol. Geol.* **18**, 332–406.

Dart, C. 1991. Carbonate sedimentation and extensional tectonics in the Maltese Graben Systems. Ph.D. Thesis, University of London.

Davison, I. 1987. Normal fault geometry in relation to sediment compaction and burial. *J. Struct. Geol.* **9**, 393–401.

Davison, I. 1989. Extensional domino faulting: kinematics and geometrical constraints. *Ann. Tecton.* **3**, 12–24.

Davison, I. & Powell, D. 1991. Deformation along a mid-crustal continental strike-slip shear zone: the Pernambuco Lineament, NE Brazil. *Mitt. Geol. Inst. ETH Zurich Neue Folge* **239b**, 14–15.

Dixon, J. M. & Shumin Liu 1991. Centrifuge modelling of the propagation of thrust faults. *Thrust Tectonics* (edited by McClay, K. R.). Chapman and Hall, 53–71.

Eaton, G. P. 1980. Geophysical and geological characteristics of the crust of the Basin and Range Province. In: *Continental Tectonics*. National Academy of Sciences, Washington, D.C., 96–113.

Einarsson, P. & Eiriksson, J. 1981. Earthquake fractures in the Districts Land and Rangárvellir in the South Iceland Seismic Zone. *Jokull* **32**, 113–120.

Eisenstadt, G. & DePaor, D. C. 1987. Alternative model of thrust-fault propagation. *Geology* **15**, 630–633.

Ellis, M. A. & Dunlap, W. J. 1988. Displacement variation along thrust faults: implications for development of large faults. *J. Struct. Geol.* **10**, 183–192.

Ellis, P. G. & McClay, K. R. 1988. Listric extensional fault systems: results of analogue model experiments. *Basin Res.* **1**, 55–70.

Enachescu, M. E. 1987. Tectonic and structural framework of the Northeast Newfoundland continental margin. *Can. Soc. Petrol. Geol. Mem.* **12**, 117–146.

England, P. & Jackson, J. A. 1989. Active deformation of the continents. *A. Rev. Earth Planet. Sci.* **17**, 197–226.

Etheridge, M. A., Branson, J. C. & Stuart-Smith, P. G. 1987. The Bass, Gippsland and Ottoway Basins Southeast Australia: a branched rift system formed by continental extension. *Can. Soc. Petrol. Geol. Mem.* **12**, 147–162.

Gibbs, A. D. 1984. Structural evolution of extensional basin margins. *J. geol. Soc. Lond.* **142**, 609–620.

Gibbs, A. D. 1986. Strike-slip basins and inversion: a possible model for the southern North Sea gas areas. In: *Habitat of Paleozoic Gas in N.W. Europe* (edited by Brooks, J. *et al.*). *Spec. Publs geol. Soc. Lond.* **23**, 23–35.

Gibbs, A. D. 1987. Linked tectonics of the Northern North Sea Basins. *Can. Soc. Petrol. Geol. Mem.* **12**, 163–172.

Grocott, J. 1977. The relationship between Precambrian shear belts and modern fault systems. *J. geol. Soc. Lond.* **133**, 257–262.

Harding, T. P. 1983. Divergent wrench fault and negative flower structure, Andaman Sea. In: *Seismic Expression of Structural Styles* (edited by Bally, A.). **3**, 4.2-1–4.2-8.

Horrii, H. & Nemat-Nasser, S. 1986. Brittle failure in compression: splitting faulting and brittle-ductile transition. *Phil. Trans. R. Soc.* **A319**, 337–374.

Illies, J. H. 1977. Ancient and recent rifting in the Rhinegraben. *Geologie Mijnb.* **56**, 329–350.

Jackson, J. A. 1987. Active normal faulting and crustal extension. In: *Continental Extension Tectonics* (edited by Coward, M. P., Dewey, J. F. & Hancock, P. L.). *Spec. Publs. geol. Soc. Lond.* **28**, 3–18.

Jackson, J. A. & McKenzie, D. 1988. Rates of active deformation in the Aegean Sea and surrounding regions. *Basin Res.* **1**, 121–128.

Jackson, J. A. & White, N. J. 1989. Normal faulting in the upper continental crust: observations from regions of active extension. *J. Struct. Geol.* **11**, 15–36.

Kanamori, H. & Allen, C. R. 1986. Earthquake repeat time and average stress drop. *Earthquake Source Mechanics* (edited by Das, S., Boatwright, J. & Scholz, C. H.). *Geophys. Monogr. Washington DC* **37**, 227–235.

King, G. & Yielding, G. 1984. The evolution of a thrust fault system: processes of rupture initiation, propagation and termination in the 1980 El Asnam (Algeria) earthquake. *Geophys. J. R. astr. Soc.* **77**, 915–933.

King, G. C. P. & Sammis, C. In press. Mechanisms of finite brittle strain. *Pageoph.*

Krantz, R. W. 1988. Multiple fault sets and three dimensional strain: theory and application. *J. Struct. Geol.* **10**, 225–337.

Lamerson, P. R. 1982. The Fossil Basin area and its relationship to the Absaroka thrust fault system. In: *Geologic Studies of the Cordilleran Thrust Belt* (edited by Power, R. B.). Rocky Mountain Association of Petroleum Geologists, 279–340.

Larsen, P.-H. 1988. Relay structures in a Lower Permian basement-involved extensional system, East Greenland. *J. Struct. Geol.* **10**, 3–8.

Lawn, B. R. & Wilshaw, T. R. 1975. *Fracture of Brittle Solids*. Cambridge University Press, Cambridge.

Liu Huiqi, McClay, K. R. & Powell, D. Physical models of thrust wedges. In: *Thrust Tectonics* (edited by McClay K. R.). Chapman and Hall, 71–82.

Lockner, D. A., Byerlee, J. D., Kursenko, V., Ponomarev, A. & Sidorin, A. 1991. Quasi-static fault growth and shear fracture energy in granite. *Nature* **350**, 39–42.

Lucas, S. 1989. Structural evolution of the Cape Smith Fold Thrust Belt and the role of out-of-sequence faulting in the thickening of mountain belts. *Tectonics* **8**, 655–676.

Mandl, G. A. 1988. *Mechanics of Faulting*. Elsevier, Amsterdam.

Marone, C. & Scholz, C. H. 1988. The depth of seismic faulting and the upper transition from stable to unstable slip regimes. *Geophys. Res. Lett.* **15**, 621–624.

Marret, R. & Allmendinger, R. W. 1991. Estimates of strain due to brittle faulting: sampling of fault populations. *J. Struct. Geol.* **13**, 735–738.

McClay, K. R. 1990. Analogue modelling review. *J. Marine Petrol. Geol.* **7**, 206–233.

McClay, K. R. 1991. A glossary of thrust tectonics terms. In: *Thrust Tectonics* (edited by McClay, K. R.) Chapman and Hall, 419–433.

McClay, K. R. & Scott, A. 1991. Hangingwall deformation in ramp-flat listric extensional fault systems. *Tectonophysics* **188**, 85–96.

McGill, G. E. & Stromquist, A. W. 1979. The grabens of Canyonlands National Park, Utah, geometry, mechanics, and kinematics. *J. geophys. Res.* **84**, 4547–4563.

McMechan. M. E. & Thompson, R. I. 1989. Structural style and history of the Rocky Mountain Fold and Thrust Belt. In: *Western Canada Sedimentary Basin, A Case History* (edited by Ricketts, B. D.). *Can. Soc. Petrol. Geol.* 47–72.

Milani, E. J. & Davison, I. 1988. Basement control and transfer tectonics in the Recôncavo-Tucano Jatoba Basin, NE Brazil. *Tectonophysics* **154**, 41–70.

Miller, E. L., Gans, P. B. & Garing, J. 1983. The Snake Range Décollement: an exhumed Mid-Tertiary ductile-brittle transition. *Tectonics* **2**, 239–263.

Mitra, S. 1988. Three dimensional geometry and kinematic evolution of the Pine Mountain thrust system, southern Appalachians. *Bull. geol. Soc. Am.* **100**, 72–95.

Morley, C. K. 1989. Extension, detachments and sedimentation in continental rifts (with particular reference to the east Africa). *Tectonics* **8**, 1175–1192.

Morley, C. K., Nelson, R. A., Patton, T. L. & Munn, S. G. 1990. Transfer zones in the East African Rift System and their relevance to hydrocarbon exploration in rifts. *Bull. Am. Assoc. Petrol. Geol.* **74**, 1234–1253.

Naylor, M. A., Mandl, G. & Sijpestein, C. H. K. 1986. Fault geometries in basement-induced wrenching under different initial stress states. *J. Struct. Geol.* **7**, 737–752.

Pacheco, J. F., Scholz, C. H. & Sykes, L. R. 1992. Changes in frequency-size relationship from small to large earthquakes. *Nature* **355**, 71–73.

Peacock, D. C. P. 1991. Displacements and linkage in strike-slip fault zones. *J. Struct. Geol.* **13**, 1025–1035.

Peacock, D. C. P. & Sanderson, D. 1991. Displacements, segments linkage and relay ramps in normal fault zones. *J. Struct. Geol.* **13**, 721–733.

Perry, W. J. 1978. Sequential deformation in the Central Appalachians. *Am. J. Sci.* **278**, 518–542.

Pollard, D. D. & Aydin, A. 1984. Propagation and linkage of oceanic ridge segments. *J. geophys. Res.* **89**, 10,017–10,028.

Proffett, J. M. 1977. Cenozoic geology of the Yerington district Nevada, and its implications for the nature and origin of Basin and Range Faulting. *Bull. geol. Soc. Am.* **88**, 247–266.

Reches, Z. 1978. Analysis of faulting in a three-dimensional strain field. *Tectonophysics* **47**, 109–129.

Reston, T. J. 1988. Evidence for shear zones in the lower crust offshore Britain. *Tectonics* **7**, 929–945.

Robson, D. A. 1971. The structure of the Gulf of Suez (Clysmic) rift, with special reference to the eastern side. *J. geol. Soc. Lond.* **127**, 247–276.

Roberts, A., Yielding, G. & Freeman, B. 1990. Conference report, on the geometry of normal faults. *J. geol. Soc. Lond.* **147**, 185–187.

Rodgers, 1980. Analysis of pull-apart basin development produced by en-échelon strike-slip faults. In: *Sedimentation in Oblique-slip Mobile Zones* (edited by Ballance, P. F. & Reading, H. G.). International Association of Sedimentologists, Blackwell, Oxford, 27–42.

Ron, H. & Eyal, Y. 1985. Intraplate deformation by block rotation and mesostructures along the Dead Sea transform, northern Israel. *Tectonics* **4**, 85–105.

Rosendahl, B. R. 1987. Architecture of continental rifts with special reference to East Africa. *A. Rev. Earth Planet. Sci.* **15**, 445–503.

Royden, L. H. 1985. The Vienna Basin, a thin-skinned pull-apart basin. In: *Strike-slip Deformation, Basin Formation and Sedimentation* (edited by Biddle, K. T. & Christie Blick, N.). *Spec. Publs Soc. Econ. Palaeont. Mineral.* **37**, 319–338.

Segall, P. & Pollard, D. D. 1980. Mechanics of discontinuous faults. *J. geophys. Res.* **85**, 4337–4350.

Scholz, C. H. 1982. Scaling laws for large earthquakes: consequences for physical models. *Bull. seism. Soc. Am.* **72**, 1–14.

Scholz, C. H. 1989. Mechanics of faulting. *A. Rev. Earth Planet. Sci.* **17**, 309–334.

Scholz, C. H. 1990. *Mechanics of Earthquakes and Faulting*. Cambridge University Press, Cambridge.

Scholz, C. H. & Aviles, C. A. 1986. The fractal geometry of faults and faulting. In: *Earthquake Source Mechanics* (edited by Das, S., Boatwright, J. & Scholz, C. H.). *Geophys. Monogr. Washington DC* **37**, 147–156.

Scholz, C. & Cowie, P. 1990. Determination of total strain from faulting using slip measurements. *Nature* **346**, 837–839.

Scholz, C. H., Aviles, C. A. & Wesnousky, S. G. 1986. Scaling differences between large interplate and intraplate earthquakes. *Bull. seism. Soc. Am.* **76**, 65–71.

Schwartz, D. P. & Coppersmith, K. J. 1984. Fault behaviour and characteristic earthquakes: examples from the Wasatch and San Andreas fault zones. *J. geophys. Res.* **89**, 5681–5698.

Shimamoto, T. & Logan, J. 1981. Effects of simulated fault gouge on the sliding behaviour of Tennessee Sandstone, non-clay gouges. *J. geophys. Res.* **96**, 2902–2914.

Shimazaki, K. 1986. Small and large earthquakes: the effects of the thickness of the seismogenic layer and the free surface. In: *Earthquake Source Mechanics* (edited by Das, S., Boatwright, J. & Scholz, C. H.). *AGU geophys. Monogr.* **37**, 209–216.

Stein, R., King, G. C. P. & Rundle, J. B. 1988. The growth of geological structures by repeated earthquakes 2. Field examples of continental dip-slip faults. *J. geophys. Res.* **93**, 13,319–13,331.

Stewart, I. S. & Hancock, P. L. 1991. Scales of structural heterogeneity within neotectonic normal fault zones in the Aegean region. *J. Struct. Geol.* **13**, 191–204.

Sylvester, A. G. 1988. Strike-slip faults. *Bull. geol. Soc. Am.* **100**, 1666–1702.

Tchalenko, J. S. 1970. Similarities between shears of different magnitudes. *Bull. geol. Soc. Am.* **81**, 1625–1640.

Tankard, A. J. & Welsink, H. J. 1987. Extensional tectonics and stratigraphy of Hibernia Oil Field, Grand Banks, Newfoundland. *Am. Assoc. Petrol. Geol.* **71**, 1210–1232.

Tanner, P. W. G. 1991. The duplex model; implications from a study of flexural-slip duplexes. In: *Thrust Tectonics* (edited by McClay, K. R.). Chapman and Hall, 201–208.

Tchalenko, J. S. 1970. Similarities between shear zones of different magnitudes. *Bull. geol. Soc. Am.* **81**, 1625–1640.

Tchalenko, J. S. & Berberian, M. 1975. Dasht-e-Bayaz fault, Iran: earthquake and earlier related structures in bedrock. *Bull. geol. Soc. Am.* **86**, 703–709.

Vendeville, B. 1987. Champs de failles et tectonique en extension. Memoires et Documents du Centre Amoricaine d'etude structurale des socles No. 15. Doctorate thesis, University of Rennes, France.

Walker, I. M. & Cooper, W. G. 1987. The structural and stratigraphic evolution of the northeast margin of the Sole Pit Basin. In: *The Geology of the North European Margin* (edited by Brooks, J. & Glennie, K.). Graham and Trotman, London, 263–275.

Wallace, R. E. 1989. Fault plane segmentation in the brittle crust and anisotropy in loading system: Proceedings USGS Conference on Fault Segmentation and Controls of Rupture Initiation and Termination. *U.S. geol. Surv. Open File Rep.* 89–315, 400–408.

Wallace, R. E. & Morris, H. T. 1986. Characteristics of faults and shear zones in deep mines. *Pageoph* **124**, 107–125.

Walsh, J. J. & Watterson, J. 1988. Analysis of the relationship between displacements and dimensions of faults. *J. Struct. Geol.* **10**, 238–247.

Walsh, J. J. & Watterson, J. 1991a. Geometric and kinematic coherence and scale effects in normal fault systems. In: *The Geometry of Normal Faults* (edited by Roberts, A., Yielding, G. & Freeman, B.). *Spec. Publs geol. Soc. Lond.* **56**, 193–206.

Walsh, J. J. & Watterson, J. 1991b. New methods of fault projection for coalmine planning. *Proc. Yorkshire geol. Soc.* **48**, 209–219.

Watterson, J. 1986. Fault dimensions, displacements and growth. *Pure appl. Geophys.* **124**, 365–372.

Wernicke, B. & Burchfiel, C. 1982. Modes of extensional tectonics. *J. Struct. Geol.* **4**, 105–115.

Wesnousky, S. G. 1988. Seismological and structural evolution of strike-slip faults. *Nature* **335**, 340–342.

Wesnousky, S. G., Scholz, C. H., Shimazaki, K. & Matsuda,

T. 1983. Earthquake frequency distribution and the mechanics of faulting. *J. geophys. Res.* **88**, 9331–9340.

Woodcock, N. H. 1987. Kinematics of strike-slip faulting, Builth Inlier, Mid-Wales. *J. Struct. Geol.* **9**, 353–363.

Woodcock, N. H. & Fischer, M. 1986. Strike-slip duplexes. *J. Struct. Geol.* **8**, 725–735.

Zhang, P., Burchfiel, B. C., Chen, S. & Deng, Q. 1989. Extinction of pull-apart basins. *Geology* **17**, 814–817.

CHAPTER 7

Prelithification Deformation

ALEX MALTMAN

INTRODUCTION

TRADITIONALLY, tectonic studies have been confined to structures formed by deformation of deep-seated origin, thought only to affect rocks long after their burial and lithification. However, the subject has had to expand during the last twenty years to include a range of newly discovered shallow deformational conditions and processes, that can affect sediments long before they are turned into rock.

Modern oceanographic work has detected the gravitational deformation of sediments on an enormous scale (e.g. Bugge *et al.* 1988). Recognition of the widespread control of tectonics on sedimentation (e.g. Ingersoll 1988, Allen & Allen 1990) has prompted the realization that near-surface sediments must commonly be subjected to tectonism (see Chapter 16). Localized processes that can deform sediments, such as glacial, seismic, and igneous phenomena, have been investigated, disclosing astonishingly rock-like structures (e.g. Brodzikowski & van Loon 1983). It has been realized that sediments can persist to considerable depths of burial without lithification (e.g. Jones & Addis 1986). Environments have been discovered where stresses of deep origin are being transmitted up through sedimentary piles, to generate in the wet sediments a wide range of structures (e.g. Lundberg & Moore 1986). It is now clear that stress distributions in natural sediments can be highly localized, with the resulting deformation commonly being heterogeneous, diachronous, and even overlapping with sedimentation.

All this means that some old concepts have had to be discarded, including the idea that sedimentation is followed by slow burial and lithification before eventually a pulse of tectonism deforms all the materials simultaneously. Tectonic studies now have to incorporate a knowledge of how sediments deform. This entails some acquaintance with the vast body of knowledge that has been assembled by that branch of engineering known as soil mechanics, and, because water is so important in determining sediment behaviour, it requires some knowledge of hydrology. In ways such as these, the study of prelithification deformation is highly inter-disciplinary. Indeed, most of the concepts and terms to be discussed here are not new in themselves: it is their adaptation in a tectonic context that is new. This immediately introduces a complication, for some of the terms are already deeply entrenched in their respective disciplines, commonly with differing meanings.

The first parts of this chapter are therefore concerned with explaining terms in a way that is consistent and useful in geological studies of prelithification deformation. Later parts of the chapter indicate the kinds of geological situations in which sediments become deformed and the range of structures that are produced. Unfortunately, the structures now known to form in sediments are deceptively similar to those traditionally associated with rock deformation, and in the geological record their identification can be exceedingly difficult. Certainly, a pressing need is to extend the knowledge of prelithification deformation derived from modern environments back through geological time. The chapter concludes with some remarks on possible approaches to this problem.

TERMINOLOGY: LITHIFICATION, AND TECTONIC VERSUS GRAVITY DEFORMATION

Lithification is the means by which sediment becomes rock, but defining this change of state in a rigorous but useful way is difficult (Maltman 1984). For structural geological purposes, it is perhaps best done on the basis of the grain-scale movements (Table 7.1).

The grains are the strongest part of a sediment, which therefore deforms by *grain boundary sliding*. The effect is called in soil mechanics *particulate deformation*. With *progressive lithification*, the apparent inter-grain strength increases and hence the tendency grows for the grains themselves to participate in the deformation. This intermediate stage is *partial lithification*. Knipe (1986a) has discussed these progressive changes in the context of accretionary prisms. In this article, the term *sediment* is extended to all material that is not fully lithified.

Shortcomings of this approach to defining lithification include the following: (1) the nature of the grain-scale movements depend not only on the intrinsic condition of the material but on external physical factors such as burial pressure and strain rate, and these

Table 7.1. Glossary of some terms useful in prelithification deformation studies

Burial pressure: the load acting on a sediment that is induced by the overlying material, including any fluids. Its magnitude depends on the thickness and bulk density of the overlying sediment aggregate together with any water above the sediment, and the gravitational constant. Being a pressure, no direction is specified.

Burial stress: as burial pressure, but acting in the vertical direction, due to gravity, unless specified to be the lateral stress component.

Compaction: physical reductions in bulk sediment volume, by loss of pore space and/or the volume occupied by the mineral framework, resulting from burial or any other process.

Compression: the stress configuration that tends to shorten objects, but in the present context it is also used in the sense of reduction in volume of the mineral framework.

Consolidation: the time-dependent reduction of sediment porosity, normally the result of burial.

Diagenesis: the chemical, as opposed to mechanical, contribution to lithification. The diagenetic reactions that take place in sediments lead to increased particle bonding and sediment strength.

Effective stress: stress applied to a sediment, including burial and tectonic stresses, minus any fluid pressure in the sediment.

Fluid potential: a measure of the pressure in a fluid at a specified point, expressed as a length. It refers to the height of a column of the fluid that could be supported at that point, adjusted for its elevation.

Fluid potential gradient: change of fluid potential with distance: a measure of the tendency for the fluid to attempt to migrate (see *permeability*).

Fluid pressure: the pressure exerted by the fluid in a sediment due to the load of any overlying fluid. *Normal fluid pressure* depends solely on the average density and vertical height of the overlying fluid, and the gravitational constant. *Overpressured fluids* bear an additional load, such as a proportion of that arising from overlying sediment grains.

Hydrostatic: the stress configuration that exists in a static fluid, that is, of equal magnitude in all directions.

Lithification: the conversion of a sediment into a rock. In the present context, the mechanical and chemical processes that progressively transform an aggregate that deforms largely by *frictional grain boundary* sliding into one in which *intra-grain mechanisms* dominate.

Permeability: a measure of the rate at which fluid can pass through a sediment. Normally taken to be the constant that relates the flow rate to: cross-sectional area and length of a straight flow path through the sediment, the fluid potential gradient driving the flow, and the viscosity of the fluid.

Porosity: the percentage of pore space in the the total sediment aggregate.

Strength: a measure of the stress sustainable by a material. In sediment studies, the ***peak strength*** (maximum sustainable stress) and ***residual strength*** (a post-failure stress reasonably constant with progressive strain) are most often cited.

Tectonic: (1) as a noun, for the structure of an earth mass as it appears after some modification (as in the 'tectonics of a region'); (2) as an adjective, to describe forces of deep-seated origin, in contrast to other forces or the deformation structures they produce (as in 'tectonic folds versus slump folds').

---

can fluctuate. A sudden drop in pore water pressure, for example, can temporarily induce an aggregate to change shape by intra-grain deformation, in a kind of quasi-lithified state. (2) In ancient rocks, any record of the grain behaviour at the time of deformation may have been obscured by later events. The geologist will not be able to discern the role of the grains at the time of deformation.

The term *tectonic* is used today in two main ways (Table 7.1): as a noun, as in the 'tectonics of a particular region'; and as an adjective, to distinguish 'tectonic' from other kinds of stresses. In much of structural geology the two meanings are closely intertwined, with the structures, or tectonics, of an area resulting from the operation of deep or tectonic forces. However, there are exceptions, for example salt tectonics (see Chapter 8) and glacio-tectonics exclude the second meaning (e.g. Sharp *et al.* 1988). The term 'syn-sedimentary tectonics' (e.g. Thomas & Baars 1988) seems confusing and is discouraged here.

The first usage of tectonics commonly involves large-scale structures, and this meaning pervades much of the second part of this book. For the present chapter, however, the second sense is particularly useful, in order to contrast the products of tectonic forces with deformation related to localized processes and the effects of gravity. In general, gravitational forces are of most importance in the upper parts of a sedimentary pile and tectonic forces are more significant deeper down, but not exclusively so, and there is gradation and overlap. (I avoid here any discussion of the physical differences between gravity and tectonic force (e.g. Means 1976) and their ultimate connection (e.g. DeJong & Scholten 1972).)

MATERIAL CONDITIONS AND PROPERTIES

The following section outlines terminology, much of it drawn from the soil mechanics literature, useful in conceptualizing how sediments deform in geological circumstances (see also Jones & Preston 1987). *Cohesion* can be viewed both empirically and mathematically. As sedimentary grains are buried they are brought closer together and electrostatic bonding grows. The attraction is particularly marked in the clay minerals because of their large surface areas and numerous free bonds. The strength imparted to the aggregate by this bonding is qualitatively referred to as *cohesion*. Quantitatively, in the Coulomb–Mohr equation of the *shear strength* (τ) of an aggregate:

$$\tau = c + \sigma \tan \phi,$$

c is called the cohesion (sometimes symbolized as τ_0), σ is the normal stress, and ϕ the angle of internal friction.

Inter-grain friction provides the other intrinsic contribution of the sediment to its shear strength. Tan ϕ in the above equation is known as the *coefficient of internal friction*. Its magnitude varies with such factors as porosity, particle shape and size, and burial stress (e.g. Karig 1986). Because its value is normally greater in coarsely granular sediments than in clays, tan θ is commonly visualized as arising directly from the physical friction between the grains, although it is strictly a mathematical function. The inter-grain friction depends on the normal stresses acting across the particle boundaries, and these tend to increase with progressive burial. One geological repercussion of this is that near the surface, clayey sediments tend to be stronger than clastic materials because of their greater cohesion, but the latter soon become the stronger with burial, as the contribution of the frictional component becomes dominant.

Karig (1986) has emphasized the sparseness of quantitative data on these parameters for geological conditions, and only very approximate ranges are known. For example, Bryant *et al.* (1981) found surficial oceanic sediments to have shear strengths in the range 4–8 kPa, rapidly increasing with burial to around 100 kPa at 200 m depth. However, values were found as low as 20 kPa and as high as several hundred kPa, depending on lithology. Karig (1986) argued that the cohesion within some modern *accretionary-prism toes* probably varies between 0.1 MPa and less than 10 MPa, with friction coefficients ranging between 0.2 and 0.7, depending on porosity (see Chapters 9 and 16). Multiplication of the friction coefficients by the normal stresses arising through burial, following the Coulomb–Mohr equation, means that at most depths the contribution of friction to shear strength is much greater than that of cohesion. However, the situation is complicated by factors such as pore fluid pressures (see section on *Fluid Flow*).

Both components of shear strength increase with greater *burial pressure* (Table 7.1), as the grains are brought closer together (e.g. Sills & Been 1984) and the density of the aggregate increases (Fig. 7.1). The mechanical characteristics of shallowly buried marine sediments are now much investigated for seabed engineering purposes (e.g. Denness 1984). Eventually, burial reaches a point where the inter-grain strength fully exceeds that of the actual particles, and lithification is complete. The pressure due to burial is therefore critical to prelithification behaviour. It is variously termed *burial, overburden, lithostatic,* or *geostatic pressure*; the first is used here. In experimental work it is normally referred to as *confining pressure*. In nature it tends to be dominated by the downward component acting under gravity and, being directional, is better called a stress, though it can be *hydrostatic*, that is, acting equally in all directions. In the present chapter, pressure is used where no direction is implied, but where the vertical or downward component is meant, it is referred to as *burial stress*. Horizontal stresses depend on the *lateral stress ratio*, k. A common condition is the *lateral stress ratio at rest*, k_0, which allows no lateral strain (see Fig. 7.2). Common near-surface values of k_0 are 0.3–0.4 for loose sands and 0.6–0.8 for clays (e.g. Lambe & Whitman 1979). The depth at which k achieves unity, giving the hydrostatic burial conditions commonly assumed in structural geology, and the mechanisms by which this state is achieved, remain unspecified.

Progressive packing of sediment particles under increasing burial pressure reduces the *porosity* (Table 7.1) of the aggregate, but the efficiency of this depends critically on the pore fluids being able to dissipate. Porosity reduction is largely irreversible, although other factors have an influence (Scherer 1987) and there may be some rebound at low stresses (Karig, in Hill *et al.* in press). If water is occupying the pores, burial will prompt a concomitant decrease in the water content (Table 7.1) of the sediment. Inter-grain slip will be increasingly curbed.

The amount of porosity reduction commonly reflects the degree of lithification. Numerous porosity–depth curves have been compiled (e.g. Bayer & Wetzel 1989, Einsele 1989). Bryant *et al.* (1981) suggested that sediments buried to a depth of 1 km will normally have reduced their porosity to between 45 and 20%. Besides such summaries of the spatial variation of porosity, Karig (1990) and Bray & Karig (1985) have taken a time-based approach to porosity change, viewing sediment as it moves along flow trajectories through accretionary-prism toes.

Compaction is a general term useful in geology for all reductions in the *bulk volume* of an aggregate (Table 7.1) (Jones & Addis 1985), including adjustments to the grains themselves. *Consolidation* has commonly been used synonymously with compaction but in analyzing sediment deformation it is better restricted to mechanical reductions in porosity. Any rearrangement of the aggregate framework, independent of the pore water expulsion effects of consolidation, is termed *compression*. This is merely a

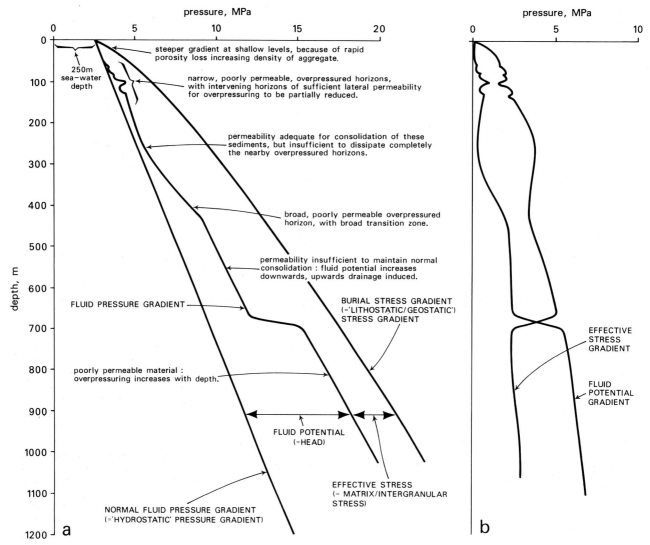

Fig. 7.1.(a) Rates of increase of pressure and stress due to burial. The normal fluid pressure gradient is based on an overlying sea-water depth of 250 m and a constant fluid density in the entire column of 1025 kg m^{-3}. The burial stress gradient is based on a grain density of 2650 kg m^{-3} and a porosity decrease with depth following the average values for sand – silt – clay of Einsele (1989). The hypothetical fluid pressure gradient shown here illustrates the kinds of effects that can arise in a sequence of varied sediments. (b) Effective stress gradient and fluid potential gradient derived from the curves in (a). Note that the fluid potential and its gradient should properly be expressed in units of metres rather than pascals, as the height that could be supported of a column of specified fluid, normally water or mercury.

more specialized usage, long-established in soil mechanics, of compression in the wider geological sense of the force that tends to shorten objects.

Therefore, in the usage encouraged here, consolidation and compression are both components of compaction. All are purely mechanical processes, and all help convert sediment into rock. They are therefore important contributors to lithification, but are not synonymous with that term, which also includes chemical effects (see *Fluid Flow* section).

Attempts to quantify the deformation behaviour of sediments, largely relying on laboratory testing, have led to the plotting of *stress paths* (Fig. 7.2) and *strain paths* (Lambe & Whitman 1979). The extent to which the pore fluids sustain pressure depends on whether the fluid is allowed to escape during deformation (*drained tests*) or not (*undrained tests*), and so the successive stress conditions are best plotted as *effective stress paths*, where effective stress = total stress − fluid pressure. Consideration of the pore water behaviour during burial and deformation is paramount, and is discussed further in the next section. The sediment mechanics also depend on the burial history of the sediment (Fig. 7.3). Apart from *normal consolidation*, where pore fluid loss has adjusted to the burial depth, *overconsolidated sediments* have the mechanical properties of once having been consolidated at some greater depth, whereas *underconsolidated sediments* have not yet reached a consolidation state in equilibrium with their present depth. Similar principles apply to the volume loss due to extraction from hydrocarbon reservoirs, some of which effectively change shape by particulate deformation and are therefore not fully lithified (Jones *et al.* 1987).

The Mohr – Coulomb failure criterion was developed for truly brittle materials but does seem to apply

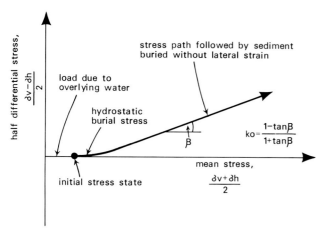

Fig. 7.2. A stress path. In this hypothetical example the principal stresses are vertical (σ_v) and horizontal (σ_h), and treated in two dimensions. The sediment is sub-aqueous, undergoes initial hydrostatic loading, and is then buried at a vertical stress – lateral stress ratio which allows no lateral strain: the k_0 condition. Fluid pressure is neglected. See text for details.

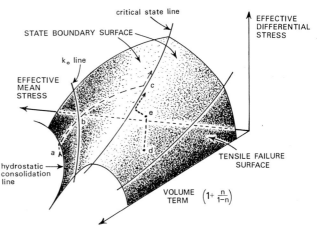

Fig. 7.4. Some aspects of the critical state concept. In the volume term, n = porosity. Stress – volume combinations outside the space bounded by the shaded surfaces are beyond the failure state and cannot be reached by a deforming sediment. Initial positions within the space depend on consolidation conditions. Stress paths rise on the addition of differential stress, and on meeting a boundary surface the stress paths follow that surface to the critical state line, which is then followed if stress continues to increase. Where drainage is curbed, the stress ratios will vary and constant volume will be maintained, but drainage allows volume change with a constant stress ratio. The dashed line illustrates: at (a) a hydrostatically consolidating sediment, decreasing in volume along the floor of the diagram, and addition of differential stress causing it to rise from the floor, following the state boundary surface; at (b) it crosses a possible position of a sediment undergoing consolidation in the k_0 condition (differential stress appropriate for no lateral strain); at (c) having reached failure, the sediment ascends along the critical state line. An overconsolidated sediment will originate towards the right of the diagram floor (e.g. at d), having a low mean stress – volume ratio, and rise steeply with added differential stress to meet the state boundary surface (e.g. at e). Sediments lacking cohesion will undergo tensile failure at the high differential – low mean effective stress conditions represented by the right-hand side of the diagram.

adequately for sediment undergoing bulk failure, that is, ceasing to support the applied stress. Many sediments, including dense sands and overconsolidated clays, show a *peak strength* at failure, followed by an overall *strain softening* as they drop to some value of *residual strength* (Fig. 7.3), largely governed by the frictional characteristics of the aggregate (e.g. Skempton 1985, Lupini *et al.* 1981). Other sediments show some degree of *strain hardening* (Fig. 7.3).

More recent soil mechanics thinking has synthesized possible relationships between applied stress, in particular the mean and differential effective stresses, and resulting volume changes into the *critical-state concept* (Fig. 7.4). Extensive laboratory testing, in both drained and undrained conditions, has suggested that the various stress paths followed by normal and overconsolidated clays define two *state-boundary surfaces*, when plotted in three-dimensional stress volume-change space. Similar principles apply also to sands.

The two state-boundary surfaces intersect to define the *critical-state line*. An individual stress path leads from the initial consolidation position to meet its boundary surface, which it then follows before joining and following the critical-state line. Two generalized examples are shown in Fig. 7.4. At the critical-state condition the sediment will be at failure and additional stress increments will induce large strains. It is here that the chief deformation occurs, and many of the various prelithification structures are produced. Stress magnitudes greater than the critical-state condition will be unsustainable. Thus, although any effective stress path within the volume enclosed by the boundary surfaces is possible, given appropriate stress – volume combinations, locations outside that volume are mechanically inachievable. Figure 7.4 is highly generalized: the ideas are carefully and systematically

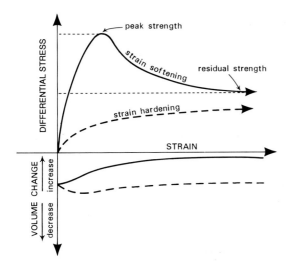

Fig. 7.3. Generalized stress – strain and volume change curves to show the overall difference in behaviour of dense sandy sediments and overconsolidated clays (solid lines) versus loose sandy sediments and normally consolidated clays (dashed lines). Some terms used in the text are illustrated.

explained in the textbook by Atkinson & Bransby (1978).

The value of the critical-state approach is in providing a unified conceptual framework for understanding sediment deformation and, for modern sediments, a quantitative and predictive tool (Jones & Addis 1986). For example, knowing the initial consolidation conditions and the stresses to be applied, the stress–volume response of the sediment during deformation can be predicted. Extensions to the model, such as the Cam-clay theory, allow sediment strains also to be anticipated for specified increments of stress or volume change.

The critical state concept has been used by Jones & Addis (1984) to explain the geometry and spacing of growth faults, by Brandon (1984) to elucidate the formation of mélanges, and by Karig (1990) to delimit the mechanical behaviour of sediments in modern accretionary prisms. These efforts illustrate the value of combining a knowledge of sediment properties with soil mechanics concepts in order to understand better how sediments deform.

A shortcoming of the critical-state approach for ancient sediment deformation is that the interpretation has to rely largely on analogy with modern, measurable sediments. Unlike, say, the Mohr–Coulomb criterion which incorporates an orientation factor that, in principle at least, can be measured in rocks, the parameters of the critical-state theory are not preserved through geological time. Moreover, the effects of the chemical changes that sediments almost invariably undergo during compaction are not included. Nevertheless, the power with which the theory unifies a wide variation in the mechanical behaviour of sediments makes this a compelling new concept in the understanding of prelithification deformation.

FLUID FLOW

The ability of a sediment to dissipate pore fluid depends upon its *permeability* (Table 7.1). This, following the classical investigations of Darcy, is strictly a mathematical constant, although it is commonly thought of as a material property (Freeze & Cherry 1979, Chapman 1981). In most hydrology, the fluid is assumed to be water at surface temperature, and therefore of constant viscosity, in which case the constant is properly termed *hydraulic conductivity*. This has units of rate, normally cm s^{-1}. In geology, however, the nature and temperature of the permeant can vary, so its viscosity has to be included, leading to the constant of permeability having units of area, normally m^2. Typical values range from 10^{-7} m^2 for loose sands to 10^{-19} m^2 or less for clayey sediments (Lambe & Whitman 1979). Hence, materials are commonly viewed as possessing different permeabilities, and being variously permeable.

The expulsion of pore water during laboratory tests is traditionally called *drainage*, and in natural sediment piles as *dewatering*, but the two terms are being increasingly interchanged. The process is central to consolidation but *tectonic dewatering* is being increasingly recognized in nature, for example, Carson & Berglund (1986).

Sediment pores are normally assumed to interconnect and allow expulsion of the fluid to keep pace with the increments of burial load, and thus maintain a *normal fluid pressure* (Table 7.1). This term is preferred to the commonly used *hydrostatic pressure*, which properly refers to a stress configuration and can exist in a medium that has no fluid at all. Normal fluid pressure results solely from the load of superjacent water and therefore typically has gradients between 9.7 and 10.9 kPa m^{-1}, depending on the fluid density profile. Any fluid on top of the sediment pile is commonly ignored, but its contribution to the magnitude of fluid pressure can be large, especially in deep-water sediments.

The fluid pressure can be regarded as one aspect of the burial pressure to which a sediment at depth is subjected. In this view, the other component is the load borne by the sediment grains themselves. This is of a magnitude equal to the burial stress minus the fluid pressure, and is therefore equivalent to the *effective stress* (Fig. 7.1). This term is preferable to *intergranular stress*, which, although having an average value equal to the effective stress, will vary in magnitude along irregular grain boundaries and could be critically high at point contacts. Hence, in detailed studies of grain–fluid junctions, the inter-granular and effective stresses will not equate. Note that the effective stress is independent of any water column above the sediment pile (Fig. 7.1).

If the permeability of sediment undergoing burial is insufficient to allow consolidation to proceed at the equilibrium rate, the pore fluid, being incompressible, begins to sustain some of the load usually taken by the grains. The fluid pressure thus exceeds the normal value, to give an *overpressure*. The effective stress component is reduced by the equivalent amount. Synonyms for overpressure include *geopressure, hydropressure* and *abnormal fluid pressure*, but the latter is imprecise because *underpressure* can also exist, albeit commonly as a result of man's interference (e.g. Belitz & Bredehoeft 1988).

Graphs of fluid pressure with depth are commonly presented as straight lines or lines with a sharp bend where overpressuring arises, below which the straight line resumes at the elevated values. However, in most natural piles of sediment the hydraulic properties will vary intricately, and the fluid pressure curves are likely to be complex. The fluid pressure curve in Fig. 7.1 shows some of the effects that are likely to arise, although this, too, is probably an oversimplification of most real systems. Also, the curve will change with time: Fig. 7.1 represents possible pressure curves at a single instant.

The fluid in a sedimentary pile is expelled upwards overall, in order to attain a lower *fluid potential*, also referred to as *head*. The rate at which this happens depends on permeability and the *fluid potential gradient*

(Table 7.1 and Fig. 7.1). There can be deviations from the overall upwards gradient. Some horizons shown in Fig. 7.1 (between 100 and 250 m depth in that hypothetical situation) are overpressured and overlie more permeable material. They are in a position of higher fluid potential than the materials immediately below, and this will induce some local downwards drainage. Horizontal gradients can also arise, and Bredehoeft et al. (1988) have documented lateral dewatering on the regional scale. However, even if a material has considerable lateral permeability, it will only be utilized if the flow paths to a site of significantly lower potential are short enough to be effective in the time available (Chapman 1981).

Burial stress has a paramount influence on sediment strength, so that if the effective stress is reduced by overpressuring, the sediment will be weakened. Any deformation of a sediment pile will tend to be concentrated at weak horizons, and hence material will fail preferentially at sites of overpressuring. Slump masses, for example, may detach along horizons of minimum effective stress, and the décollement horizon of accretionary prisms will tend to localize along overpressured zones (e.g. von Huene & Lee 1983). Deeply-buried but overpressured sediments will behave mechanically in the same way as near-surface sediments (e.g. Jones & Addis 1985). Thus the implications of overpressures for prelithification deformation, and the resulting structures, are enormous. However, any quantitative assessment of the real fluid pressure profiles remains extremely elusive. Interpretations from the geological record are difficult to make; even in the modern oceans, attempts to directly measure overpressures continue to meet frustration.

Overpressures arising in the way outlined above, with drainage unable to keep pace with increasing burial load, probably most commonly come about through rapid sedimentation (e.g. Gretener 1981). However, other mechanisms have been suggested for increasing fluid pressure above normal, for example, the rapid addition of deviatoric stress (Yassir 1990), mineral dehydration (Bruce 1984), *biopressuring* resulting from organic decay (Nelson & Lindsley-Griffen 1987), and *aquathermal pressuring* arising from temperature changes (Magara 1975). It has been supposed that additions of more highly-pressured fluids from greater depths may cause the local fluid pressure to actually exceed the burial stress (Fyfe et al. 1978). Horizons where overpressuring exceeds sediment strength may develop features such as *water sills* (Fyfe et al. 1978) and, if there is any slope, *detachment zones* (Fitches et al. 1986). If fluid pressures exceed sediment strength at a grain-to-grain scale, *liquefaction* and *fluidization* will be induced (e.g. Owen 1987), with the production of structures such as *sediment diapirs* (e.g. Brown & Westbrook 1988).

Permeability emerges as a fundamental control on sediment deformation behaviour, yet it is poorly investigated in geological circumstances. Vast numbers of on-land permeability measurements have been collected in engineering hydrology, where a single, scalar value for the sediment at rest is taken to be representative. The effects of inhomogeneities and deformation features have been little considered (Atkinson & Richardson 1987, Bosscher et al. 1987), apart from at the regional scale (Bethke 1989). Some permeabilities of deep-sea sediments have been assessed in onshore laboratory testing by Schultheiss & Gunn (1985) and Bryant et al. (1981), and values around 10^{-12} m^2 for silt and 10^{-13} m^2 for pelagic clay were obtained. Values from core retrieved from the Barbados forearc varied between 10^{-11} and 10^{-16} m^2 (Taylor & Leonard 1990).

However, deformed sediments will commonly contain microfabrics that give the material an *anisotropic permeability* (Arch & Maltman 1990). This comes about because the microstructures present flow paths of differing *tortuosity*. Moreover, the permeabilities of an actively deforming sediment may differ substantially from rest values, especially if the material is dilating to allow grain boundary sliding to take place. Some effects of *dilatancy*, an overall volume change, have been considered by Harp et al. (1990) and experimentally by Shimizu (1982). Preliminary investigations of *dynamic permeability*, the permeability of an actively deforming material (Maltman et al. in preparation), indicate that, in silty clays at least, it can be several orders of magnitude different from the value in the static state. Whether it increases or decreases may depend on the consolidation state of the sediment. Yet further complications to permeability arise where the permeant is multiphase, for example, with a water – organic liquid – gas mixture (Parker 1989).

The importance of overpressuring to aspects of tectonics, such as fracture mechanics and thrust displacements, has long been realized, but the mutual interplay between fluid pressures and deformation is a recent concept. For example, fluid pressure has emerged to be a significant factor in the supposedly stable abyssal-plain sediments that have been considered as possible radioactive-waste disposal sites. There, it is associated with faults arising from differential compaction structures that can variously act as channels for episodic pore fluid movement (Buckley 1989) and, in some circumstances, as seals (Williams 1987). Karig (1990) has invoked fluctuations of fluid pressure to explain how accretionary prisms can undergo local brittle deformation, roughly synchronous with overall ductile deformation.

The concept of *deformation – fluid flow interaction* is particularly topical in accretionary prism studies (Moore 1989, Langseth & Moore 1990) (see also Chapter 9). Here, fluid pressures influence the kinds of deformation that take place, and the resulting structures influence the flow and pressures of the fluids. Attempts to explore the hydrogeology of accreting sediments are involving seismic studies, direct observations from submersibles, geothermal studies and theoretical modelling (see references in Langseth & Moore 1990). These approaches have now documented the importance of discontinuities such as faults in the fluid flow regime. An additional line of evidence has been the discovery of *seep biological communities*, restricted to localities where fluids appear to be actively leaking to the sea floor along tectonic structures. Fujioka & Taira (1989) recognized four tectono-sedimentary settings that promoted seep communities: fault scarps; submarine canyons; fan valleys; and slide escarpments (also see Chapter 16). The great academic and applied importance of deformation – fluid flow interplay in a host of different environments in which sediments deform is likely to keep this concept topical for a considerable time to come.

CHEMICAL PROCESSES: DIAGENESIS

The previous two sections have treated the lithification of sediment in purely mechanical terms. The compaction of sediments, in which the grains are increasingly bonded electrostatically, simply because they are being brought closer together, is a fundamental process of lithification, but it is almost invariably accompanied by chemical changes, referred to here as *diagenesis*. Diagenesis covers the chemical reactions that take place in sediments, one result of which is an increase in particle bonding and hence the strength of the sediment. I believe that using this term as synonymous with lithification, and hence including compaction, consolidation, and compression, is less useful. These mechanical processes have been dealt with at some length here because they are inextricably bound up with deformation processes; the interaction between diagenesis and deformation is more subtle.

The diagenetic processes affect deforming on-land sediments (e.g. van der Meer 1985) but the reactions are much more vigorous in fluid-rich marine sediments (e.g. Marshall 1987). Diagenesis is dominated by *cementation,* that is, precipitation to fill pore-spaces, *recrystallization*, and *diffusion-mass-transfer*, all of which curb inter-grain slippage and hence promote lithification. The last process (see Chapter 1), often called *pressure solution*, can become significant in the later stages of lithification and may generate cementing material (e.g. Tada & Siever 1989).

Most diagenetic processes tend to decrease porosity, thus increasing the strength of the sediment independent of burial depth and prompting the effect of overconsolidation. However, burial pressure has some influence on the efficiency of diagenetic processes (e.g. Colten-Bradley 1987). At the same time, bedding-parallel fabrics formed by burial-induced consolidation can be enhanced by diagenesis (Maltman 1981, Craig *et al.* 1982).

Sample (1990) illustrated some effects of increasing carbonate cementation on deformation styles, and Oertel & Curtis (1972) documented the interaction between compaction and the diagenetic growth of concretions. Some authors have regarded cementation as a component of compaction (e.g. Jones & Addis 1985), and in practice the two processes are closely linked (e.g. Lasemi *et al.* 1990), but the usage of the term compaction encouraged here excludes chemical effects. Hydrothermal processes associated with igneous intrusion into sediments can reduce porosity and induce other diagenetic changes in the host material, as documented from the Gulf of California (Einsele *et al.* 1980).

That deformation structures influence fluid flow and hence diagenesis is being increasingly documented (e.g. Tribble 1990), but the extent to which diagenesis itself can induce significant deformation is less clear. It has, for example, been argued that some conditions of concretion growth can deform the host sediment (Dewers 1990), and Gray & Nickelsen (1989) have documented strain features such as slickensides that formed through the *in situ* expansion and contraction of clays. Maliva & Siever (1988) considered some force-inducing effects of diagenesis.

Diagenesis is a huge field of study with numerous ramifications. Together with the compactional processes, it is paramount in helping convert sediment into rock. However, so far, its role in directly influencing the mechanics of lithifying sediments has been little explored.

GEOLOGICAL SETTINGS AND CAUSES OF DEFORMATION

Compaction induces volume changes as outlined earlier, and may induce some *in situ* deformation such as shape fabrics (see later) and localized macroscopic structures (e.g. Allen 1983, Shanmugan *et al.* 1987). There has been much discussion of how strain analysis of sedimentary rocks should allow for any prelithification fabrics (e.g. Oertel & Curtis 1972) (also see Chapter 2). Compactional strains can change the geometry of early structures (Davison 1987), and differential compaction can produce significant faulting, even in submarine sediments that are tectonically stable (Buckley & Grant 1985).

Most prelithification deformation comes about either from gravity-driven bulk translation — *mass movement* — of the material, or from the addition of further stress, particularly in some non-vertical direction. The additional stress may be tectonic or it may arise from some local external agent such as igneous activity or glacial movement. The stress magnitude required for failure need only be small, because surficial sediments commonly have strengths considerably less than 1 MPa, much less if pore-fluid pressures are greater than normal.

Ice movement has long been known to induce structures in sediments, but the widespread extent to which sediments deform below glaciers has only recently been recognized (e.g. Alley *et al.* 1986, Blankenship *et al.* 1986). Indeed, the observation of sediment deformation under a modern glacier in Iceland (Boulton & Jones 1979), together with seismic results

(Blankenship et al. 1987) and direct sampling (Engelhardt et al. 1990) that indicated a water-saturated sediment layer below the west Antarctica ice sheet, have begun to reveal the substantial contribution that the deforming substrate makes to allowing the ice-sheet to move. The mechanical properties of deforming substrate sediments have been investigated both theoretically and experimentally by Murray (1990).

The major part of the ice-sheets that covered North America and Europe during the Quaternary are now thought to have moved on deforming unlithified substrates (Boulton et al. 1985; Boulton & Hindmarsh 1987). This new thinking and the resulting interest in how sediments deform has been called by Boulton (1986) 'a *paradigm shift in glaciology*'.

The wide range of structures that can be produced in association with ice has also only recently begun to be appreciated (e.g. Brodzikowski & van Loon 1983, van der Meer 1985, 1987, Croot 1988). Structures associated with glacial features such as drumlins (Menzies & Rose 1987, Menzies 1989) have much in common with those formed in other geological settings (Menzies & Maltman 1990). Moreover, ice-related deformation of sediments includes disturbances in the periglacial environment (Washburn 1978) and the results of isostatic readjustments to changing ice loads (e.g. Eyles & Clarke 1985). These, in turn, lead to differential consolidation and mass movements (e.g. Pederson 1987).

Igneous activity can deform and have marked local effects on the mechanical properties of sediments. Examples preserved in the geological record have been described by Grapes et al. (1974), Morris (1979) and Kokelaar et al. (1985). Spreading centres are one situation where the interaction of magma and wet sediments is likely to be common. Einsele (1982) has measured, in the Guayamas Basin of the Gulf of California, some mechanical effects of magma intruding the topmost three hundred metres of sediment. In this example, the activity is still in progress — the igneous rocks and the host wet sediments are still hot!

Downslope mass movement of sediment is now known to be a widespread phenomenon. It can come about where the *plastic limit* of the sediment is exceeded by its water content, giving rise to *sediment flows* (e.g. Einsele 1989), and where the external stresses exceed the shear strength of cohesive sediment. There are numerous approaches to classifying the mass movement of cohesive sediment, depending partly on whether the perspective is engineering (e.g. Bonnard 1989, Brabb & Harrod 1989), geophysical (e.g. Coleman & Prior 1988) or geological (e.g. Elliott & Williams 1987). *Slides* move largely along a basal detachment, and the bulk of the translating material remains intact apart from localized shear zones. They grade into *slumps*, where internal deformation is significant, typically producing folds. If slides and slumps cease to remain intact, they grade into *flows*. If the environment of movement is known, the prefix *land-* or *submarine-flows* can be used as appropriate. The synonymous use of landslide for slide, a practice growing in some quarters, is discouraged as it leads to 'submarine landslide', which is awkward and mutually contradictory. (Note that in this context the term 'slide' has a meaning entirely independent of its use, common in Caledonide studies, for high-strain zones that disrupt the stratigraphy of metamorphic rocks.)

Different morphologies of mass flows on the NW Atlantic continental slope have been documented by Booth et al. (1988). Martinsen & Bakke (1990) illustrated the compressional and extensional effects of sediment mass movements, as preserved in Carboniferous rocks in western Ireland. Problems remain, however, in recognizing in the rock record mass-movement features of the enormous scale now known on the modern ocean floor (e.g. Moore et al. 1989). Exposure restrictions are partly responsible for the discrepancy, but because many of the structures produced by slumping and sliding are so similar to those resulting from rock deformation, it is probable that the mass-movement origin of some structures in rocks is not being identified (Woodcock 1979).

Moreover, the distinction between such essentially gravity-induced deformation and that due to tectonic forces is commonly blurred. Gravitational mass movement is probably often prompted by seismic activity. On-land, unlithified cover can deform in response to gravitational instabilities caused by tectonic movements below (e.g. Brodzikowski et al. 1987a). Gravitational adjustments can accompany neotectonic deformation (Vita-Finzi 1985), which is now well known to induce various structures in on-land, poorly lithified sediment (see Chapter 18).

The marine setting in which the tectonic deformation of sediment is best documented is that of the accretionary prism, where sediments are offscraped at convergent plate margins (see Chapters 9 and 16). Also in this setting, tectonically induced dewatering of the sediments has been invoked (Carson & Berglund 1986). Bray & Karig (1986) have discussed the interplay of consolidation and tectonic stresses in prompting porosity loss across accretionary prisms.

Frequent reference is made in this chapter to the variety of tectonic structures that form in accretionary prisms, with the stresses having been transmitted through the poorly lithified sediments, even as far as the sediment surface. Faults propagate through weak sediments high in the Barbados prism (Moran & Christian 1990) and intersect the sea floor (Moore et al. 1988). In the Nankai prism, the frontal thrust passes through loose sediments still containing over 40% water. It cross-cuts other tectonic structures, which presumably formed when the sediment was even wetter (Hill et al. in press). Tectonically-induced microstructures are known in the Nankai prism at a depth of only 260 m below the sea floor, in sediments with porosities greater than 50%. Cores retrieved from these sediments are soft enough to be easily cut with a spatula, yet, while in place, they were able to undergo tectonic deformation. Observations such as these should dispel any remaining notions that tectonic deformation only occurs at substantial depths and only affects material which has completed lithification.

EXAMPLES OF DEFORMATION STRUCTURE IN SEDIMENTS

The following is merely an outline of the range of structures that are now recognized to form by prelithification deformation. The examples are mentioned roughly in order of increasing scale. *Microfabrics* can arise in sediments through the preferred dimensional alignment of grains during their consolidation (Moon & Hurst 1984). How shallow the levels of burial can be is unclear because of the difficulties of sampling very weak, near-surface material, but Faas & Crockett (1983) detected preferred alignments at 45 m sub-bottom depths and 250 kPa burial stress. Maltman (1987a) recorded deformation microstructures formed in association with near-surface mass-movement of sediments, and Hounslow (1990) documented magnetic grain fabrics from deep-sea cores that had no visible expression.

Some microstructures of deformed ice-related deposits closely resemble those of non-glacial sediments, but for others a series of specialized terms (e.g. *plasmic unistrial* and *sepic fabrics* (van der Meer 1985)) has evolved. Recognition of microstructures such as these can throw light on the dynamics of glaciation (e.g. van der Meer 1987, Mahang 1990). A number of tectonically-induced microstructures are now known from deformed sediments. Those recorded from modern marine forearcs were included in the review by Lundberg & Moore (1986).

Microscopic shear zones are common in sediments deformed in various environments (Maltman 1988), and have been investigated experimentally by Maltman (1987b) and Arch *et al.* (1988). The reorientation of phyllosilicates within restricted zones is the basis of other important microstructures. The abrupt deflection of bedding-parallel fabrics into semi-planar zones, termed *deformation bands* by Karig & Lundberg (1990), is known from several accretionary prisms and may be an important response to the bulk shortening of sedimentary piles. Certainly in the Nankai prism, such bands appear to reflect much of the early shortening strains before they were supplanted by discrete megascopic dislocations and other prelithification microstructures (Hill *et al.* in press). Lash (1989) documented the transition from localized shear zones affecting consolidation fabrics to scaly clays, as preserved in the Ordovician trench mudstones of the central Appalachians.

Scaly clays tend to acquire their characteristic fissile, polished aspect through phyllosilicate reorientation (Lundberg & Moore 1986), although Agar *et al.* (1988) have pointed out that this is not always so. Scaly clays are found in landslides (van den Berg 1987), *mud diapirs* (Barber *et al.* 1986), and are common in high strain zones within marine sediments, for example, the faults and décollement zones of accretionary prisms (Brown & Behrmann 1990). Moore *et al.* (1986) have postulated that oceanic scaly clays form at less than 25°C, pressures less than 4 MPa, and strain rates of about 10^{-13} s^{-1}, typically in underconsolidated sediment. The shear zones found in these oceanic dislocations are remarkably similar to the microstructures of some glacially-deformed clayey sediments (Menzies & Maltman 1990).

At the macroscopic scale, many of the structures long known to form in deforming rocks are now also known in sediments. Folds are known of various kinds (e.g. kink folds: van Loon *et al.* 1985; sheath folds: Agar 1988; recumbent folds: Brodzikowski *et al.* 1987a) and at various scales, from the centimetre size of small slump folds (Farrell & Eaton 1988) to kilometre-wavelength folds outboard of the Makkran accretionary complex (White 1982).

Similarly, macroscopic faults in sediments, not involving fracture of individual grains, can arise through various agents, including ice (Brodzikowski & van Loon 1983) and tectonism (Moore & Byrne 1987). Williams (1987) discussed faults arising from the differential compaction of abyssal plain sediments, which can produce structures with throws of hundreds of metres and both normal and reverse senses of movement.

Folds and faults can be associated in deforming sediments, both in extensional situations such as *rollover anticlines* (see Chapter 6) forming adjacent to normal faults (e.g. Mandl & Crans 1981) and in compressional regimes (e.g. Agar 1988). *Boudinage* occurs due to gravitational instabilities (e.g. Morris 1979) and in association with folds, and normal, reverse, and thrust faults (Pickering 1987). A range of volcanically induced structures in sediments is reported by Kokelaar *et al.* (1985) and Decker (1990).

Mineralized veins formed in partially-lithified sediments have been described, for example, by Fisher & Byrne (1990) and Brown & Behrmann (1990). Three-dimensional networks of *veinlets* (see Chapter 5) in clastic sediments have been termed *web structures* (Cowan 1982) but may require substantial lithification for their formation, as their development has been inferred to involve some cataclasis (Byrne 1984), and they have been recorded from basic igneous rocks. Remarkable breccias developed in existing on-land sediments are illustrated by Brodzikowski & van Loon (1985).

The similarity in appearance of all these structures to those formed by regional tectonism at depth can cause awkward problems of interpretation (e.g. Hendry & Stauffer 1977), as discussed in the next section. A famous example is the problem of the origin of *mélanges* (e.g. Raymond 1984, Brandon 1989).

There are some macroscopic structures that are thought to be restricted to the prelithification state, as they are largely related to the effects of fluid movement on the loose sediment. *Sedimentary dykes* are known from many environments but all are thought to require overpressuring to allow injection of fluidized sediment. They have been described from modern oceans (Lundberg & Moore 1986) and preserved in rocks on scales from the microscopic to hundreds of metres in length, in swarms of great intensity (Winslow 1983) (also see Chapter 5).

Vein-like structures retrieved from modern ocean sediments (Kemp in press) and preserved in on-land rocks (Pickering *et al.* 1990) have also been ascribed to dewatering processes (also see Chapter 5). Other water escape structures such as *flame structures* are less regular analogues, and are known from various settings, for example, sub-glacial (Brodzikowski & Haluszczak 1987). They grade up in scale to *diapirs* (see Chapter 8). In general, these larger intrusions, rooted deeper in a sediment pile than dykes, are composed of mud, which at depth is commonly weaker than sand. *Mud diapirs* in association with *slump sheets* have been described by Martinsen (1987) and in the accretionary prism setting by Brown (1990). Because most sedimentary dykes are inferred to form early in a burial history, they have been used as markers for consolidation strains. However, Taira (1984) has argued that dykes preserved in the Shimanto Belt of SW Japan formed in material of about 30% porosity and therefore at burial depths of at least several hundreds of metres.

Prelithification deformation also occurs on the regional scale, particularly where sediments are accreting at active plate margins. *Accretionary prisms* (see Chapter 9) have volumes measured in thousands of cubic kilometres, and include much actively deforming material that is not fully lithified. The results, as preserved in on-land rocks, are described, for instance, by Taira & Ogawa (1988), but most knowledge of structures and processes at this scale continues to be derived largely from marine research. For example, the structures and mechanics of accreting sediments are being investigated by geophysical methods (e.g. Davis & von Huene 1987), by theoretical modelling (e.g. Cloos & Shreve 1988a, b) and by a combination of theoretical, experimental, and observational approaches (see references in Langseth & Moore 1990). It is this kind of ocean-based work that has revealed, more than any other effort, the extent to which sediments are affected by deformation before they are lithified.

RECOGNITION OF PRELITHIFICATION DEFORMATION IN THE GEOLOGICAL RECORD

The extent of prelithification deformation occurring today is conspicuously more than that recognized in the rock record. This is partly because of on-land exposure restrictions and the full range of deformational conditions not always being considered. However, the discrepancy largely reflects the practical difficulties of distinguishing between structures formed before, during and after lithification. As explained earlier, the essential differences are at the grain scale and these are vulnerable to later overprinting.

At the more usual scales of observation of a hand specimen and larger, structures originating at different stages of lithification are indistinguishable morphologically. Even the recognition of surficial slump structures is problematic, as reviewed by Elliott & Williams (1988). Gravity slump structures have been thought of as being chaotic and disorganized or having some other hallmark, but while this may be true in some cases it is not generally so, and it does not apply to prelithification structures of tectonic origin. Often cited distinguishing features, such as faults, cleavages, veins, particular orientations, etc., do not hold up to scrutiny as they are now known in both sediments and rocks.

Strain geometry may help distinguish deformation due to compaction as it will be dominantly uniaxial and vertical. For example, Cowan (1982b) analyzed strains recorded by structures in Franciscan sediments, California, and argued that the bedding-parallel axially symmetric extensions reflected vertical shortening in upper structural levels rather than tectonism at depth. There are potential criteria for distinguishing between gravity and tectonic structures but as yet they remain uncertain. As an example, whether or not *axial-plane fabrics* can be generated by slump folding has been much discussed (e.g. Williams *et al.* 1969, Tobisch 1984, Elliott & Williams 1988). The structures expected in a theoretical free-flowing granular aggregate have been calculated (e.g. Drake 1990, Rothenburg & Bathurst 1989), but the stress systems and material properties in a natural, moving, surficial sheet are likely to be complex and constantly changing, and perhaps unlikely to produce a single systematic fabric. As yet, there is no good record of a slump fold with an axial-plane foliation that definitely formed through slumping rather than consolidation, but the question remains open.

The most useful general approach to identifying the stage of lithification at which a structure was produced is by noting its relationship to some feature whose position in the sequence is known. For example, *bioturbation* is restricted to unlithified material, features such as *septarian concretions* reflect early diagenesis, and *chlorite-mica stacks* commence growth in late diagenesis. *Sedimentary dykes* are universally regarded as forming well before lithification is complete. Hence in the Southern Andes, where sedimentary dykes were reported by Winslow (1983) to cross-cut joints, incipient cleavage, and some thrusting and buckling, these latter structures were interpreted as prelithification in origin. In SE France, sedimentary dykes post-date fracture sets, which are similarly interpreted (Huang 1988). At the other end of the lithification span, clearly any questionable structures that also affect igneous or metamorphic bodies are almost certainly of post-lithification origin. But even these are somewhat uncommon situations, and there can be exceptions. For instance, a diagenetic feature might not have formed *in situ* but have been derived from an earlier cycle. Magma injected into sediment could be deformed before it crystallized. A recognition criterion taken alone, therefore, is not diagnostic.

Despite all the rapid progress in understanding prelithification deformation, it still seems, as the writer remarked over a dozen years ago, that to identify it "the geologist has to assess the cumulative weight of several not infallible criteria" (Fitches & Maltman 1978). However, because prelithification deformation is now known to be a widespread phenomenon, the difficulty cannot be ignored. When observing deformation structures in sedimentary or even meta-sedimentary rocks, it cannot simply be assumed that they formed after lithification. The geologist has to have some familiarity with the extended range of circumstances in which structures can form, including the possibilities and the limitations of interpretation. A major prospect facing tectonic geologists is the application of the wealth of information being derived from modern prelithification structures to the recognition and understanding of those preserved in the geological record.

SUMMARY OF NEW CONCEPTS

The concepts that have been outlined here include the following.

(1) Sediments respond to stress by undergoing *particulate deformation*, which involves *grain boundary sliding*. During progressve lithification this mechanism is increasingly curbed and supplanted by *intra-grain processes*. When it is no longer dominant, the material can be regarded as lithified.

(2) *Lithification* comes about chiefly by two groups of processes: *compaction*, which is some combination of pore water expulsion (*consolidation*) and mechanical rearrangement of the aggregate framework (*compression*), and *diagenesis*, the chemical changes that fill pore spaces and increase grain bonding.

(3) The relationships between consolidation, volume, and applied stresses are synthesized in the *critical state theory*, which includes the prediction that above certain specified combinations of these factors, called the *critical state*, the bulk sediment will be mechanically unstable and undergo yielding. Many prelithification structures are produced while the sediment is in this state.

(4) The role of pore-water fluids is critical to prelithification deformation as, through the effective stress concept, it governs sediment strength and behaviour. *Permeability*, a dynamic and non-scalar variable, together with the fluid potential gradient, govern pore fluid dissipation, which if inadequate can produce overpressuring.

(5) Differential compaction can generate important structures even in tectonically stable marine environments. Sediment deformation can arise through localized geological circumstances, such as in association with volcanism. Beneath ice-sheets, sediment deformation can be intense, and largely responsible for the ice movement. Gravity-induced mass movements, including on-land and submarine sliding and slumping, are widespread at the surface of the earth today and can happen on an enormous scale — larger than so far recognized in the geological record.

(6) Tectonic deformation is the response to forces generated at depth in the earth, as opposed to gravity. Sediments can persist at depth through overpressuring and be subjected to tectonic forces, and tectonic forces can be transmitted through unlithified aggregates, and hence affect surficial material. Tectonic forces help induce dewatering, and can therefore play a part in generating overpressures.

(7) Sedimentation and tectonism commonly overlap. Deformation of a sedimentary pile can be diachronous and have complex spatial variations.

(8) A wide range of structures are now recognized from the different environments in which sediments deform, from the micro- to the mega-scale. Tectonic structures are particularly varied, and important in actively deforming sedimentary environments like accretionary prisms.

(9) Practicable criteria for distinguishing between pre-, syn-, and post-lithification structures are sparse and usually require subjective assessments on the part of the geologist. Nevertheless interpretations of structures preserved in sedimentary and metasedimentary rocks should consider the possibility of prelithification deformation having been involved.

(10) The knowledge of prelithification deformation being obtained from modern sediments should enable greater understanding of the resulting structures where preserved in the geological record.

REFERENCES

Agar, S. M. 1988. Shearing of partially consolidated sediments in a lower trench slope setting. Shimanto Belt, SW Japan. *J. Struct. Geol.* **10**, 21–32.

Agar, S. M., Prior, D. J. & Behrmann, J. H. 1988. Back-scattered electron imagery of the tectonic fabrics of some fine-grained sediments: implications for fabric nomenclature and deformation processes. *Geology* **17**, 901–904.

Allen, J. R. L. 1984. Sedimentary structures, their character and physical basis. *Developments in Sedimentology 30*. Elsevier, Amsterdam.

Allen, P. A. & Allen, J. R. 1990. *Basin Analysis*. Blackwell, Oxford.

Alley, R. B., Blankenship, D. D., Bentley, C. R. & Rooney, S. T. 1986. Deformation of till beneath Ice Stream 'B', West Antarctica. *Nature* **322**, 57–59.

Arch, J. & Maltman, A. J. 1990. Anisotropic permeability and tortuosity in deformed wet sediments. *J. geophys. Res.* **95**, 9035–9045.

Arch, J., Maltman, A. J. & Knipe, R. J. 1988. Shear-zone geometries in experimentally deformed clays: the influence of water content, strain rate and primary fabric. *J. Struct. Geol.* **10**, 91–99.

Atkinson, J. H. & Bransby, P. L. 1978. *The Mechanics of Soils: An Introduction to Critical State Soil Mechanics*. McGraw-Hill, London.

Atkinson, J. H. & Richardson, D. 1987. The effect of local

drainage in shear zones on the undrained strength of overconsolidated clay. *Geotechnique* **37**, 393–403.

Barber, A. J., Tjokrosapoetro, S. & Charlton, T. R. 1986. Mud volcanoes, shale diapirs, wrench faults, and melanges in accretionary complexes, eastern Indonesia. *Bull. Am. Ass. Petrol. Geol.* **70**, 1729–1741.

Bayer, U. & Wetzel, A. 1989. Compactional behaviour of fine-grained sediments — examples from Deep Sea Drilling Project cores. *Geol. Rundsch.* **78**, 807–819.

Belitz, K. & Bredehoeft, J. D. 1988. Hydrodynamics of Denver Basin: explanations of subnormal fluid pressures. *Bull. Am. Ass. Petrol. Geol.* **72**, 1334–1359.

van den Berg, L. 1987. Experimental redeformation of naturally deformed scaly clays. *Geologie Mijnb.* **65**, 309–315.

Blankenship, D. D., Bentley, C. R., Rooney, S. T. & Alley, R. B. 1986. Seismic measurements reveal a saturated porous layer beneath an active Antarctic ice stream. *Nature* **322**, 54–57.

Blankenship, D. D., Bentley, C. R., Rooney, S. T. & Alley, R. B. 1987. Till beneath Ice stream 'B' 1. Properties derived from seismic travel times. *J. geophys. Res.* **92**, 8903–8911.

Bonnard, C. (ed.) 1989. *Landslides: Proceedings of the 5th International Symposium, Lausanne.* A. A. Balkema, Rotterdam.

Booth, J. S., O'Leary, D. W., Poperoe, P., Robb, J. M. & McGregor, B. A. 1988. Map and tabulation of Quaternary mass movements along the United States – Canadian Atlantic continental shelf from 32°00' to 47°00'N latitude. *U.S. geol. Surv. Misc. Field Stud. Map* MF-2027.

Bosscher, P. J., Bruxvoort, G. P. & Kelley, T. 1987. Influence of discontinuous joints on permeability. *J. geotech. Eng.* **114** 1318–1331.

Boulton, G. S. 1986. A paradigm shift in glaciology. *Nature* **322**, 18.

Boulton, G. S. & Hindmarsh, R. C. A. 1987. Sediment deformation beneath glaciers: rheology and geological consequences. *J. geophys. Res.* **92**, 9059–9082.

Boulton, G. S. & Jones, A. S. 1979. Stability of temperate ice caps and ice sheets resting on beds of deformed sediment. *J. Glaciol.* **90**, 29–43.

Brabb, E. E. & Harrod, B. L. (eds) 1989. Landslides: extent and economic significance. In: *Proceedings of the 28th International Geological Congress Symposium on Landslides, Washington D. C.* A. A. Balkema, Rotterdam.

Brandon, M. T. 1984. Deformational processes affecting unlithified sediments at active margins: a field study and structural model. Ph. D. thesis, Seattle, University of Washington, 159 pp.

Bray, C. E. & Karig, D. E. 1985. Porosity of sediments in accretionary prisms and some implications for dewatering processes. *J. geophys. Res.* **90**, 768–778.

Bredehoeft, J. D., Djevanshir, R. D. & Belitz, K. R. 1988. Lateral fluid flow in a compacting sand–shale sequence: South Caspian Basin. *Bull. Am. Ass. Petrol Geol.* **72**, 416–424.

Brodzikowski, K. 1981. The role of dilatancy in the deformational process of unconsolidated sediments. *Ann. geol. Soc. Poland* **51**, 83–98.

Brodzikowski, K. & Cegla, J. 1981. Kink folding in unconsolidated Quaternary sediments. *Ann. geol. Soc. Poland* **51** 63–82.

Brodzikowski, K. & Haluszczak, A. 1987. Flame structures and associated deformations in Quaternary glaciolacustrine and glaciodeltaic deposits: examples from central Poland. In: *Deformation of Sediments and Sedimentary Rocks* (edited by Jones, M. E. & Preston, R. M. F.). *Spec. Publs geol. Soc. Lond.* **29**, 279–286.

Brodzikowski, K. & van Loon, A. 1983. Sedimentology and deformational history of unconsolidated Quaternary sediments of the Jaroszow zone (Sudetic foreland). *Geologia Sud.* **18** 123–237.

Brodzikowski, K. & van Loon, A. J. 1985. Penecontemporaneous non-tectonic brecciation of unconsolidated silts and muds. *Sediment. Geol.* **41**, 269–282.

Brodzikowski, K., Gotowala, R., Haluszczak, A., Krzyszkowski, D. & van Loon, A. 1987. Soft-sediment deformations from glaciodeltaic, glaciolacustrine and fluviolacustrine sediments in the Kleszczow Graben (central Poland). In: *Deformation of Sediments and Sedimentary Rocks* (edited by Jones, M. E. and Preston, R. M. F.). *Spec. Publs geol. Soc. Lond.* **29**, 255–267.

Brodzikowski, K., Krzyszkowski, D. & van Loon, A. 1987b. Endogenic processes as a cause of penecontemporaneous soft-sediment deformations in the fluviolacustrine Czyzow Series (Kleszczow Graben, central Poland). In: *Deformation of Sediments and Sedimentary Rocks* (edited by Jones, M. E. and Preston, R. M. F.). *Spec. Publs geol. Soc. Lond.* **29**, 269–278.

Brown, K. M. 1990. The nature and hydrogeological significance of mud diapirs and diatremes for accretionary systems. *J. geophys. Res.* **95**, 8969–8982.

Brown, K. M. & Behrmann, J. 1990. Genesis and evolution of small-scale structures in the toe of the Barbados accretionary wedge. *Proc. Ocean Drilling Prog. Scient. Res.* **110**, 229–241.

Bruce, C. H. 1984. Smectite dehydration — its relation to structural development and hydrocarbon accumulation in northern Gulf of Mexico Basin. *Bull. Am. Ass. Geol.* **68**, 673–683.

Bryant, W. R., Bennett, R. H. & Katherman, C. E. 1981. Shear strength, consolidation, porosity, and permeability of oceanic sediments. In: *The Sea, Volume 7* (edited by Emiliani, C.). Wiley, New York.

Buckley, D. E. 1989. Small fractures in deep sea sediments: indications of pore fluid migration along compactional faults. In: *Advances in Underwater Technology, Ocean Science and Offshore Engineering, Volume* 18 (edited by Freeman, T. J.). Graham and Trotman, London, 115–135.

Buckley, D. E. & Grant, A. C. 1985. Fault-like features in abyssal plain sediments: possible dewatering structures. *J. geophys. Res.* **90**, 9173–9180.

Bugge, T., Belderson, R. H. & Kenyon, N. H. 1988. The Storegga slide. *Phil. Trans. R. Soc. Lond.* **325**, 357–388.

Byrne, T. 1984. Structural geology of melange terranes in the Ghost Rocks Formation. In: *Melanges: their Nature, Origin and Significance. Spec. Pap. geol. Soc. Am.* **198**, 21–51.

Byrne, T. & Fisher, D. 1990. Evidence for a weak and overpressured decollement beneath sediment-dominated accretionary prisms. *J. geophys. Res.* **95**, 9081–9097.

Carson, B. 1977. Tectonically induced deformation of deep-sea sediments off Washington and northern Oregon: mechanical consolidation. *Mar. Geol.* **24**, 289–307.

Carson, B. & Berglund, P. L. 1986. Sediment deformation and dewatering under horizontal compression: experimental results. *Mem. geol. Soc. Am.* **166**, 135–150.

Carson, B., von Huene, R. & Arthur, M. 1982. Small-scale deformation structures and physical properties related to convergence in Japan slope sediments. *Tectonics* **1**, 277–302.

Chapman, R. E. 1981. *Geology and Water. An Introduction to Fluid Mechanics for Geologists.* Martinus Nijhoff/Dr W. Junk Publishers, The Hague.

Cloos, M. & Shreve, R. L. 1988a. Subduction-channel model of prism accretion, melange formation, sediment subduction, and subduction erosion at convergent plate margins: 1. Background and description. *Pure appl. Geophys.* **128**, 455–500.

Cloos, M. & Shreve, R. L. 1988b. Subduction channel model of prism accretion, melange formation, sediment subduction, and subduction erosion at convergent plate margins: 2. Implications and discussion. *Pure appl. Geophys.* **128**, 501–545.

Coleman, J. M. & Prior, D. B. 1988. Mass wasting on continental margins. *A. Rev. Earth Planet. Sci.* **16**, 101–119.

Colten-Bradley, V. A. 1987. Role of pressure in smectite dehydration — effects on geopressure and smectite-illite transformation. *Bull. Am. Ass. Petrol. Geol.* **71**, 1414–1427.

Cowan, D. S. 1982a. Origin of 'vein structure' in slope sediments on the inner slope of the Middle America Trench off Guatemala. In: *Initial Reports of the Deep Sea Drilling Project* (edited by Auboin, J., von Huene, R. et al.). **67**, 645–650.

Cowan, D. S. 1982b. Deformation of partly dewatered and consolidated Franciscan sediments near Piedras Blancas Point, California. In: *Trench-forearc Geology: Sedimentation and Tectonics on Modern and Ancient Plate Margins. Spec. Publs geol. Soc. Lond.* **10**, 439–457.

Craig, J., Fitches, W. R. & Maltman, A. J. 1982. Chlorite-mica stacks in low-strain rocks from central Wales. *Geol. Mag.* **119**, 243–256.

Croot, D. G. (ed.) 1988. *Glaciotectonics: Proceedings of the Glaciotectonics Working Group.* Outtawa, A. A. Balkema, Rotterdam.

Davis, D. M. & von Huene, R. 1987. Inferences on sediment strength and fault friction from structures at the Aleutian Trench. *Geology* **15**, 517–522.

Davison, I. 1987. Normal fault geometry related to sediment compaction and burial. *J. Struct. Geol.* **9**, 393–402.

Decker, P. L. (ed.) 1990. Structural style and mechanics of liquefaction-related deformation in the Lower Absaroka Volcanic Supergroup (Eocene), east-central Absaroka Range, Wyoming. *Spec. Pap. geol. Soc. Am.* **240**.

Denness, B. (ed.) 1984. *Seabed Mechanics.* Graham and Trotman, London.

DeJong, K. A. & Scholten, R. (eds) 1972. *Gravity and Tectonics.* Wiley, New York.

Dewars, T. 1990. Force of crystallization during the growth of siliceous concretions. *Geology* **18**, 204–207.

Drake, T. G. 1990. Structural features in granular flows. *J. geophys. Res.* **95**, 8681–8696.

Einsele, G. 1982. Mechanism of sill intrusion into soft sediment and expulsion of pore water. In: *Initial Reports of the Deep Sea Drilling Project* (edited by Curray, J. R., Moore, D. G. et al.). **64**, 1169–1176.

Einsele, G. 1989. In-situ water contents, liquid limits, and submarine mass flows due to a high liquefaction of slope sediments (results from DSDP and subaerial counterparts). *Geol. Rundsch.* **78**, 821–840.

Einsele, G. et al. 1980. Intrusion of basaltic sills into highly porous sediments, and resulting hydrothermal activity. *Nature* **283**, 441–445.

Elliott, C. G. & Williams, P. F. 1988. Sediment slump structures: a review of diagnostic criteria and application to an example from Newfoundland. *J. Struct. Geol.* **10**, 171–182.

Engelhardt, H., Humphrey, N., Kamb, B. & Fahnestock, M. 1990. Physical conditions at the base of a fast moving Antarctic ice stream. *Science* **248**, 57–59.

Eyles, N. & Clark, B. M. 1985. Gravity-induced soft-sediment deformation in glaciomarine sequences of the Upper Proterozoic Port Askaig Formation, Scotland. *Sedimentology* **32**, 789–814.

Faas, R. W. & Crockett, D. S. 1983. Clay fabric development in a deep-sea core. Site 515. *Deep Sea Drilling Proj. Leg 72, Initial Rep. DSDP* **72**, 519–525.

Farrell, S. G. & Eaton, S. 1988. Foliations developed during slump deformations of Miocene marine sediments, Cyprus. *J. Struct. Geol.* **10**, 567–576.

Fisher, D. & Byrne, T. 1990. The character and distribution of mineralized fractures in the Kodiak Formation, Alaska: implications for fluid flow in an underthrust sequence. *J. geophys. Res.* **95**, 9069–9080.

Fitches, W. R. & Maltman, A. J. 1978. Deformation of soft sediments, Conference report. *J. geol. Soc. Lond.* **135**, 245–251.

Fitches, W. R., Cave, R., Craig, J. & Maltman, A. J. 1986. Early veins as evidence of detachment in the Lower Palaeozoic rocks of the Welsh Basin. *J. Struct. Geol.* **8**, 607–620.

Freeze, R. A. & Cherry, J. A. 1979. *Groundwater.* Prentice-Hall, Englewood Cliffs, New Jersey.

Fujioka, K. & Taira, A. 1989. Tectono-sedimentary settings of seep biological communities — a synthesis from the Japanese subduction zones. In: *Sedimentary Facies in the Active Plate Margin, Terrapub, Tokyo* (edited by Taira, A. & Masuda, F.). 577–602.

Fyfe, W. S., Price, N. J. & Thompson, A. B. 1978. *Fluids in the Earth's Crust.* Elsevier, Amsterdam.

Gray, M. B. & Nickelsen, R. P. 1989. Pedogenic slickensides, indicators of strain and deformation processes in redbed sequences of the Appalachian foreland. *Geology* **17**, 72–75.

Gretener, P. E. 1981. Pore pressure: fundamentals, general ramifications, and implications for structural geology (revised). *Am. Ass. Petrol. Geol. Ed. Course Note Ser* **4**.

Harp, E. L., Wells, W. G. & Sarmiento, J. G. 1990. Pore pressure response during failure in soils. *Bull. geol. Soc. Am.* **102**, 428–438.

Hendry, H. E. & Stauffer, M. R. 1977. Penecontemporaneous folds in cross-bedding: inversion of facing criteria and mimicry of tectonic folds. *J. geol. Soc. Lond.* **88**, 809–812.

Hill, I., Taira, A. et al. In press. *Proc. Ocean Drilling Prog.* **131**.

Hounslow, M. W. 1990. Grain fabric measured using magnetic susceptibility anisotropy in deformed sediments of the Barbados accretionary prism. *Leg 110, Proc. Ocean Drilling Prog. Scient. Res.* **110**, 257–275.

Huang, Q. 1988. Geometry and tectonic significance of Albian sedimentary dykes in the Sisteron area, SE France. *J. Struct. Geol.* **10**, 453–462.

Von Huene, R. & Lee, H. J. 1983. The possible significance of pore fluid pressures in subduction zones. In: *Studies in Continental Marine Geology* (edited by Watkins, J. S. & Drake, C. L.). *Mem. Am. Ass. Petrol. Geol.* **34**, 781–789.

Ingersoll, R. V. 1988. Tectonics of sedimentary basins. *Bull. geol. Soc. Am.* **100**, 1704–1719.

Jones, M. E. & Addis, M. A. 1984. Volume change during sediment diagenesis and the development of growth faults. *Mar. Petrol. Geol.* **1**, 118–122.

Jones, M. E. & Addis, M. A. 1985. On the changes in porosity and volume during a burial of argillaceous sediments. *Mar. Petrol. Geol.* **2**, 247–252.

Jones, M. E. & Addis, M. A. 1986. The application of stress path and critical state analysis to sediment deformation. *J. Struct. Geol.* **8**, 575–580.

Jones, M. E. & Preston, R. M. F. (eds) 1987. Deformation of sediments and sedimentary rocks. *Spec. Publs. geol. Soc. Lond.* **29**.

Jones, M. E., Leddra, M. J. & Addis, M. A. 1987. Resevoir compaction and subsurface subsidence due to hydrocarbon extraction. *Offshore Technol. Rep. OTH 87 276.* HMSO, London.

Karig, D. E. 1986. Physical properties and mechanical state of accreted sediments in the Nankai Trough, Southwest Japan Arc. *Mem. geol. Soc Am.* **166**, 117–133.

Karig, D. E. 1990. Experimental and observational constraints on the mechanical behaviour in the toes of accretionary prisms. In: *Deformation Mechanisms, Rheology and Tectonics* (edited by Knipe, R. J. & Rutter, E. H.). *Spec. Publs geol. Soc. Lond.* **54**, 383–393.

Karig, D. E. & Lundberg. N. 1990. Deformation bands from

the toe of the Nankai accretionary prism. *J. geophys. Res.* **95**, 9099–9109.

Kemp, A. E. S. In press. Fluid flow in 'vein structures' in Peru forearc basins: evidence from back scattered electron microscope studies. *Proc. Ocean Drilling Program Sci. Res.* **112**.

Knipe, R. J. 1986a. Deformation mechanism path diagrams for sediments undergoing lithification. *Mem. geol. Soc. Am.* **166**, 151–160.

Knipe, R. J. 1986b. Faulting mechanisms in slope sediments: examples from Deep Sea Drilling Project cores. *Mem. geol. Soc. Am.* **166**, 45–54.

Kokelaar, B. P., Bevins, R. E. & Roach, R. A. 1985. Submarine silicic volcanism and associated sedimentary and tectonic processes, Ramsey Island, SW Wales. *J. geol. Soc. Lond.* **142**, 591–613.

Lambe, T. W. & Whitman, R. V. 1979. *Soil Mechanics* (S. I. version, 2nd edn). Wiley, New York.

Langseth, M. G. & Moore, J. C. 1990. Introduction to special section on the role of fluids in sediment accretion, deformations, diagenesis, and metamorphism in subduction zones. *J. geophys. Res.* **95**, 8737–8741.

Lasemi, Z., Sandberg, P. A. & Boardman, M. R. 1990. New microstructural criteria for differentiation of compaction and early cementation of fine-grained limestones. *Geology* **18**, 370–373.

Lash, G. 1989. Documentation and significance of progressive microfabric changes in middle Ordovician trench mudstones. *Bull. geol. Soc. Am.* **101**, 1268–1279.

Van Loon, A. J., Brodzikowski, K. & Gotowala, R. 1985. Kink structures in unconsolidated fine-grained sediments. *Sediment Geol.* **41**, 283–300.

Lundberg, N. & Moore, J. C. 1986. Macroscopic structural features in Deep Sea Drilling Project cores from forearc regions. *Mem. geol. Soc. Am.* **166**, 13–44.

Lupini, J. F., Skinner, A. E. & Vaughan, P. R. 1981. The drained residual strength of cohesive soils. *Geotechnique* **31**, 181–213.

Magara, K. 1975. Importance of aquathermal pressuring effect in Gulf Coast. *Bull. Am. Ass. Petrol. Geol.* **59**, 2037–2045.

Mahaney, W. C. 1990. Macrofabrics and quartz microstructures confirm glacial origin of Sunnybrook drift in the Lake Ontario basin. *Geology* **18**, 145–148.

Maliva, R. G. & Siever, R. 1988. Diagenetic replacement controlled by force of crystallization. *Geology* **16**, 688–691.

Maltman, A. J. 1981. Primary bedding-parallel fabrics in structural geology. *J. geol. Soc. Lond.* **138**, 475–483.

Maltman, A. J. 1984. On the term 'soft-sediment deformation'. *J. Struct. Geol.* **6**, 589–592.

Maltman, A. J. 1987a. Microstructures in deformed sediments, Denbigh Moors, North Wales. *Geol. J.* **22**, 87–94.

Maltman, A. J. 1987b. Shear zones in argillaceous sediments — an experimental study. In: *Deformation of Sediments and Sediementary Rocks* (edited by Jones, M. E. & Preston, R. M. F.). *Spec. Publs geol. Soc. Lond.* **29**, 77–87.

Maltman, A. J., 1988. The importance of shear zones in naturally deformed wet sediments. *Tectonophysics* **145**, 163–175.

Mandl, G. & Crans, W. 1981. Gravitational gliding in deltas. In: *Thrust and Nappe Tectonics* (edited by McClay, K. R. & Price, N. J.). *Spec. Publs geol. Soc. Lond.* **9**, 41–54.

Marshall, J. D. (ed.) 1987. *Diagenesis of Sedimentary Sequences. Spec. Publs geol. Soc. Lond.* **36**, 1–360.

Martinsen, O. J. 1987. Styles of soft-sediment deformation on a Namurian (Carboniferous) delta slope, Western Irish Namurian Basin, Ireland. In: *Deltas, Sites and Traps for Fossil Fuels* (edited by Whateley, M. K. G. & Pickering, K. T.). *Spec. Publs geol. Soc. Lond.* **41**, 167–178.

Martinsen, O. J. & Bakke, B. 1990. Extensional and compressional zones in slumps and slides in the Namurian of County Clare, Ireland. *J. geol. Soc. Lond.* **147**, 153–164.

Means, W. D. 1976. *Stress and Strain. Basic Concepts of Continuum Mechanics for Geologists.* Springer, New York.

van der Meer, J. J. M. 1985. Sedimentology and genesis of glacial deposits in the Goudsberg, Central Netherlands. *Meded. Rijks geol. Dienst* **39**, 1–29.

van der Meer, J. J. M. 1987a. Micromorphology of glacial sediments as a tool in distinguishing genetic varieties of till. In: *INQUA Till Symposium, Finland* (edited by Kujanso, R. & Saarnisto, M.). *Spec. Pap. geol. Surv. Finland* **3**, 77–89.

van der Meer, J. J. M. (ed.) 1987b. Tills and glaciotectonics. *Proc. INQUA Meet. the Genesis and Lithology of Glacial Deposits, Amsterdam.* A. A. Balkema, Rotterdam.

Menzies, J. 1989. Subglacial hydraulic conditions and their possible impact upon subglacial bed formation. *Sediment. Geol.* **63**, 125–150.

Menzies, J. & Maltman, A. J. 1990. Microstructures in diamictons — evidence of subglacial bed conditions. *3rd Drumlin Symp. Oulu, Finland.* (Abstr.).

Menzies, J. & Rose, J. (eds) 1987. *Drumlin Symp.: First Int. Conf. Geomorphology, Manchester.* A. A. Balkema, Rotterdam.

Miyata, T. 1990. Slump strain indicative of paleoslope in Cretaceaous Izumi sedimentary basin along Median Tectonic Line, southwest Japan. *Geology* **18**, 392–394.

Moon, C. F. & Hurst, C. W. 1984. Fabric of muds and shales: an overview. In: *Fine-Grained Sediments: Deep-Water Processes and Facies* (edited by Stow, D. A. V. & Piper, D. J. W.). *Spec. Publs geol. Soc. Lond.* **15**, 579–594.

Moore, J. C. 1989. Tectonics and hydrogeology of accretionary prisms: role of the decollement zone. *J. Struct. Geol.* **11**, 95–106.

Moore, J. C. & Byrne, T. 1987. Thickening of fault zones: a mechanism of melange formation in accreting sediments. *Geology* **15**, 1040–1043.

Moore, J. C., Roeske, S., Lundberg, N., Schoonmaker, J., Cowan, D. S., Gonzales, E. & Lucas, S. 1986. Scaly fabrics from Deep Sea Drilling Project cores from forearcs. *Mem. geol. Soc. Am.* **166**, 55–74.

Moore, J. G., Clague, D. A., Holcomb, R. T., Lipman, P. W., Normark, W. R. & Torresam, M. E. 1989. Prodigious submarine landslides on the Hawaiian Ridge. *J. geophys. Res.* **94**, 17,465–17,000.

Moran, K. & Christian, H. A. 1990. Strength and deformation behaviour of sediment from the Lesser Antilles forearc accretionary prism. *Proc. Ocean Drilling Program Scient. Res.* **110**, 279–288.

Morris, J. H. 1979. Lower Palaeozoic soft-sediment deformation structures in the western end of the Longford-Down inlier, Ireland. In: *The Caledonides of the British Isles — Reviewed* (edited by Harris, A. L., Holland, C. H. & Leake, B. E.). *Spec. Publs geol. Soc. Lond.* **8**, 513–516.

Murray, T. 1990. Deformable glacier beds: measurement and modelling. Ph. D. Thesis, Aberystwyth, University of Wales, 321 pp.

Nelson, R. B. & Lindsley-Griffin, N. 1987. Biopressured carbonate turbidite sediments: a mechanism for submarine slumping. *Geology* **15**, 817–820.

Oertel, G. & Curtis, C. D. 1972. Clay-ironstone concretion preserving fabrics due to progressive compaction. *Bull. geol. Soc. Am.* **83**, 2597–2606.

Owen, G. 1987. Deformation processes in unconsolidated sands. In: *Deformation of Sediments and Sedimentary Rocks* (edited by Jones, M. E. & Preston, R. M. F.). *Spec. Publs geol. Soc. Lond.* **29**, 11–24.

Parker, J. C. 1989. Multiphase flow and transport in porous media. *Rev. Geophys.* **27**, 311–328.

Pederson, S. A. S. 1987. Comparative studies of gravity tectonics in Quaternary sediments and sedimentary rocks related to fold belts. In: *Deformation of Sediments and Sedimentary Rocks* (edited by Jones, M. E. & Preston, R. M. F.). *Spec. Publs geol. Soc. Lond.* **29**, 165–180.

Pickering, K. T. 1987. Wet-sediment deformation in the Upper Ordovician Point Leamington Formation: an active thrust-imbricate system during sedimentation. Notre Dame Bay, north-central Newfoundland. In: *Deformation of Sediments and Sedimentary Rocks* (edited by Jones, M. E. & Preston, R. M. F.). *Spec. Publs geol. Soc. Lond.* **29**, 213–239.

Pickering, K. T., Agar, S. M. & Prior, D. J. 1990. Vein structure and the role of pore fluids in early wet-sediment deformation; Late Miocene volcaniclastic rocks, Miura Group, SE Japan. In: *Deformation Mechanisms, Rheology and Tectonics* (edited by Knipe, R. J. & Rutter, E.). *Spec. Publs geol. Soc. Lond.* **54**, 417–430.

Raymond, L. A. & Terranova, T. 1984. The melange problem — a review. *Spec. Pap. geol. Soc. Am.* **198**, 1–5.

Rothenburg, L. & Bathurst, R. J. 1989. Analytical study of induced anisotropy in idealized granular materials. *Geotechnique* **39**, 601–614.

Sample, J. C. 1990. The effect of carbonate cementation of underthrust sediments on deformation styles during underplating. *J. geophys. Res.* **95**, 9111–9121.

Scherer, M. 1987. Parameters influencing porosity in sandstones: a model for sandstone porosity. *Bull. Am. Ass. Petrol. Geol.* **71**, 485–491.

Schulthesis, P. J. & Gunn, D. E. 1985. The permeability and consolidation of deep-sea sediments. *Inst. Oceanogr. Sci. Rep.* 201.

Schultz, A. P. 1986. Ancient, giant rockslides, Sinking Creek Mountain, southern Appalachians, Virginia. *Geology* **14**, 11–14.

Shanmugan, G., Moiola, R. J. & Sales, J. K. 1988. Duplex-like structures in submarine fan channels, Ouachita Mountains, Arkansas. *Geology* **16**, 229–232.

Sharp, M., Lawson, W. & Anderson, R. S. 1988. Tectonic processes in a surge-type glacier. *J. Struct Geol.* **10**, 499–516.

Shimizu, M. 1982. Effect of overconsolidation on dilatancy of a cohesive soil. *Soils Found.* **22**, 121–135.

Sills, G. C. & Been, K. 1984. Escape of pore fluid from consolidating sediment. In: *Transfer Processes in Cohesive Sediment Systems* (edited by Parker, W. R. *et al.*). Plenum Press, New York, 109–125.

Skempton, A. W. 1985. Residual strength of clays in landslides. *Geotechnique* **35**, 3–18.

Tada, R. & Siever, R. 1989. Pressure solution during diagenesis. *A. Rev. Earth Planet. Sci.* **17**, 89–118.

Taira, A. 1984. Tectonic significance of sandstone dikes in the Eocene Murohanto Group, Shimanto Belt (subduction complex). *Shikoku Res. Rep. Kochi Univ.* **32**, 161–189.

Taira, A & Ogawa, Y. (eds) 1988. The Shimanto belt, southwest Japan — studies on the evolution of an accretionary prism. *Mod. Geol.* **12** (special issue).

Taylor, E. & Leonard, J. 1990. Sediment consolidation and permeability at the Barbados forearc. *Proc. Ocean Drilling Prog. Scient. Res.* **110**, 2898–3080.

Thomas, W. A. & Baars, D. L. (convenors) 1988. Synsedimentary tectonics, Penrose Conference report. *Geology* **16**, 190–191.

Tobisch, O. T. 1984. Development of foliation and fold interference patterns produced by sedimentary processes. *Geology* **12**, 441–444.

Tribble, J. S. 1990. Clay diagenesis in the Barbados accretionary complex: potential impact on hydrology and subduction dynamics. *Proc. Ocean Drilling Prog. Scient. Res.* **110**, 97–110.

Vita-Finzi, C. 1986. *Recent Earth Movements. An Introduction to Neotectonics.* Academic Press, London.

Washburn, A. L. 1978. *Geocryology. A Survey of Periglacial Processes and Environments.* Arnold, London.

White, R. S. 1982. Deformation of the Makran accretionary sedimentary prism in the Gulf of Oman (north-west Indian Ocean). In: *Trench-forearc Geology: Sedimentation and Tectonics on Modern and Ancient Plate Margins* (edited by Legget, J. K.). *Spec. Publs Geol. Soc. Lond.* **10**, 357–372.

Williams, P. F., Collins, A. R. & Wiltshire, R. G. 1969. Cleavage and penecontemporaneous deformation structures in sedimentary rocks. *J. Geol.* **77**, 415–425.

Williams, S. J. 1987. Faulting in abyssal-plain sediments, Great Meteor East, Mafeira abyssal plain. In: *Geology and Geochemistry of Abyssal Plains* (edited by Weaver, P. P. E. & Thomson, J.). *Spec. Publs geol. Soc. Lond.* **31**, 87–104.

Winslow, M. A. 1983. Clastic dike swarms and the structural evolution of the foreland fold and thrust belt of the southern Andes. *Bull. geol. Soc. Am.* **94**, 1073–1080.

Woodcock, N. H. 1979. Sizes of submarine slides and their significance. *J. Struct. Geol.* **1**, 137–142.

Yassir, N. A. 1990. The undrained shear behaviour of fine grained sediments. In: *Deformation Mechanisms, Rheology and Tectonics* (edited by Knipe, R. J. & Rutter, E.). *Spec. Publs geol. Soc. Lond.* **54**, 399–404.

CHAPTER 8

Advances in Salt Tectonics

M. P. A. JACKSON and C. J. TALBOT

BACKGROUND

THE last major benchmark publications in salt tectonics were in 1968, with the simultaneous appearance of *Diapirism and Diapirs* as a Memoir of the American Association of Petroleum Geologists and *Saline Deposits* as a Special Paper of the Geological Society of America. Since then, there were few major advances in salt tectonics until the early 1980s, when the topic gained momentum to the present high level in both academia and industry. Virtually all the new concepts summarized in this chapter are from the 1980s.

The most important reason for the advances in salt tectonics during the 1980s was the significant improvement in seismic acquisition and processing. The most dramatic harvest from this technological leap is the recognition of the huge size and abundance of allochthonous salt sheets in the Gulf of Mexico. The leading edges of these sheets along the abyssal Sigsbee Escarpment had been seismically imaged in the late 1960s. Their potential scale and importance were perceptively deduced from shallow sparker lines in 1978, but were not fully revealed and publicly recognized until the middle 1980s. Newly recognized processes, such as the emplacement and reactivation of salt sheets and the formation of salt welds by removal of salt, were revealed by the new seismic data. Salt-cored, gravity-induced fold-and-thrust belts were discovered under the base of the continental slope of the Gulf of Mexico, where they are being overridden by allochthonous salt sheets. Quantitative restorations of cross-sections reemphasized the importance of differential loading in promoting diapirs and of downbuilding as the dominant mode of diapiric growth. Reflection-seismic surveys revealed remarkable structures resulting from diapirism combined with prodigious extension on the west-central African margin. Other new geophysical advances benefitting salt tectonics include the GLORIA (*Geological LOng-Range Inclined Asdic*) sidescan sonography and multibeam bathymetric swath mapping of the deep Gulf of Mexico. In contrast, the reinterpretation of 1950s data yielded new ideas on mushroom diapirs and salt canopies in one of the world's least-known areas of salt diapirism, the Great Kavir of Central Iran.

Experimental rock mechanics cast doubt on the suitability of salt as a geological repository for high-level nuclear wastes. Rock salt was found to soften to a virtually Newtonian viscous fluid when deformed with only 0.01% water. Dynamic recrystallization is achieved as brine inclusions smear into thin films that allow highly mobile grain boundaries. This solution-transfer creep appears to be the dominant deformation mechanism in rock salt at geological stresses and strain rates. Investigation of the rheology of damp salt was triggered by the surprising observation in the late 1970s that an Iranian salt glacier flowed only after rainfall, but at rates 100,000 times faster than predicted by experiments on dry salt. Many other salt glaciers in southern Iran provide potential large-scale natural experiments in flow folding and shear zone development, in tectonites deforming on a human time scale.

Physical modelling of salt tectonics was marked by a resurgence of centrifuge modelling, using viscous overburdens to elucidate downbuilding, the effects of progadation, mushroom diapirs and salt canopies. At the same time, new experimental techniques under normal-gravity used sand as a brittle overburden over a viscous substratum to investigate the relationships between faulting and diapirism driven by density inversion, differential loading, and gravity spreading or gliding. Numerical modelling, still restricted to two dimensions, investigated downbuilding, diapirism affected by superimposed multiple wavelengths and thermal convection, and the effects of brittle overburdens.

READING THIS CHAPTER

The chapter is designed for reading through in a sequence in which ideas and nomenclature are introduced progressively. The trend is thus from simple to complex, from qualitative to quantitative, and from geometric to dynamic, ending with three short appendices. Some older terms are defined, but, as implied by the title, most terms are new words coined to express new discoveries and advances in thinking. Papers cited are those in which terms were originally used or concepts first recorded, based on our knowledge

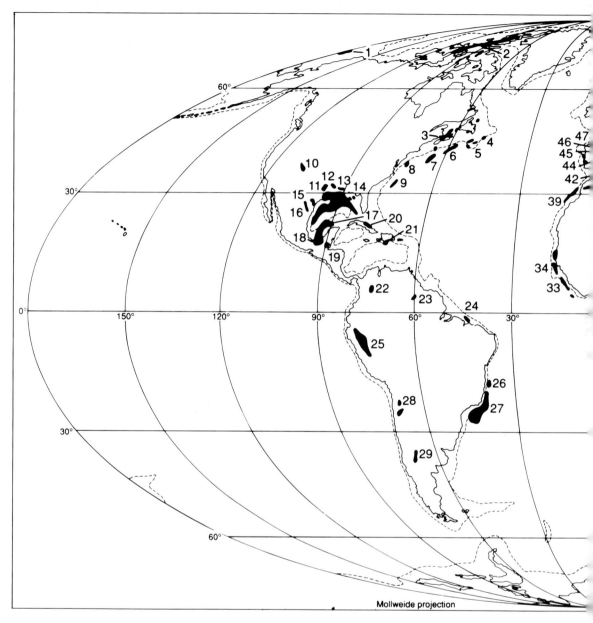

Fig. 8.1. Equal-area projection showing the global distribution of salt structures (black). The map depicts only those parts where deformation is sufficiently great that salt structures are prominent on reflection seismic profiles; the map does not show areas of undeformed salt. Dashed line represents the continental shelf break. Salt provinces are categorized in terms of petroleum production, based on the American Association of Petroleum Geologists classification of sedimentary provinces of the world (St. John *et al.* 1984). *Giant*: (2) Sverdrup; (4) Jeanne d'Arc; (5) Grand Banks; (10) Paradox; (11) East Texas; (12) North Louisiana; (14) Gulf Coast; (18) Salinas; (22) Zipaquira; (31) Cabinda; (32) Gabon; (47) Aquitaine; (55) Southern North Sea; (56) Northwest German; (57) Northern North Sea; (58) Tromsø; (61) Dnepr-Donetz; (62) North Caspian; (64) Tadjik; (68) Zagros; (70) Suez; (71) Arabian; (72) Red Sea East; (74) Oman. *Subgiant*: (6) Scotian;

of the literature. This is not an authoritative etymology, and readers may well know of earlier citations. Three appendices contain information too specialized or quantitative to be incorporated into the main body of text without disruption.

COMPONENTS OF SALT TECTONICS

Figure 8.1 reflects our current knowledge of the global distribution of salt structures. Compared with compilations of 20 years ago, the most striking additions are the many discoveries of diapir provinces offshore. Enough of these provinces are now known that a general pattern emerges. Salt diapirs are common in divergent continental margins, rifts, aulacogens and pull-apart basins. Salt diapirs are virtually absent from continental margins characterized by present-day subduction of oceanic crust. The widespread discoveries of salt diapirs offshore have revealed a large range of salt structures formed under highly variable rates of sedimentation and extension. All these structures arise from the same basic sedimentary components, which are as follows.

In salt tectonics, *substratum* refers to the ductile salt layer below a brittle overburden and above the subsalt

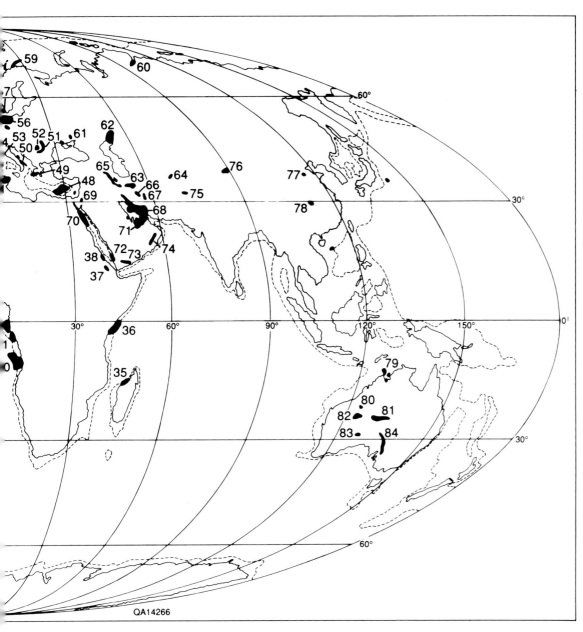

(13) Mississippi; (16) Sabinas; (19) Petenchiapas; (23) Takutu; (25) Oriente-Ucayali; (26) Esperito Santo; (27) Campos-Santos; (30) Kwanza; (39) Essaouira; (40) Atlas; (41) Pelagian; (46) Cantabrian; (49) Ionian; (50) South Adriatic; (51) Carpathian; (52) Transylvanian; (63) Great Kavir; (77) Bohai Bay; (79) Bonaparte; (80) Canning; (81) Amadeus. *Nonproductive*: (1) Chukchi; (3) Moncton; (15) South Texas; (7) Georges Bank; (8) Baltimore Canyon; (9) Carolina; (17) Sigsbee; (20) Cuban; (21) Haitian; (24) Barreirinhas; (28) Atacama; (29) Neuquen; (33) Liberia; (34) Senegal; (35) Comoros; (36) West Somali; (37) Danakil; (38) Red Sea West; (42) Algerian-Alboran; (43) Balearic; (44) Ebro; (45) Jaca; (48) Levantine; (53) Ligurian; (54) Rhodanian; (59) Nordkapp; (60) Yenisey-Khatanga; (65) Tabriz; (66) Yazd-Kalut; (67) North Kerman; (69) Dead Sea; (73) Hadhramaut; (75) Salt Range; (76) Qaidam; (78) Jianghan; (82) Woolnough; (83) Officer; (84) Flinders.

strata or basement. If the substratum supplies salt for the growth of salt structures, it is termed a *source layer* (syn. *mother salt*), a specific type of substratum. Overburden is generally used as a stratigraphic term; for example, *allochthonous salt* structurally overlies part of its (stratigraphically younger) overburden. The entire sedimentary pile above the basement and including both substratum (salt) and overburden is known as the *cover*.

SHAPE OF SALT STRUCTURES

Several terms describe the geometry of the interface between the overburden and the source layer or substratum. These can be arranged in approximate order of structural maturity. A *salt anticline* (syn. *salt welt*) is an elongated upwelling of salt with concordant overburden (Fig. 8.2) (DeGolyer 1925, Harrison & Bally 1988), whereas a *salt pillow* is a subcircular upwelling of salt, also with concordant overburden (Fig. 8.2) (Trusheim 1960). The imprecise, general term, *salt dome*, refers to a domal upwelling comprising a salt core and its envelope of deformed overburden (Harris & Veatch 1899). The salt core may or may not be diapiric (i.e. discordant). A *salt diapir* comprises a mass of salt that has flowed in a ductile manner and appears to have disconcordantly pierced or intruded the overburden

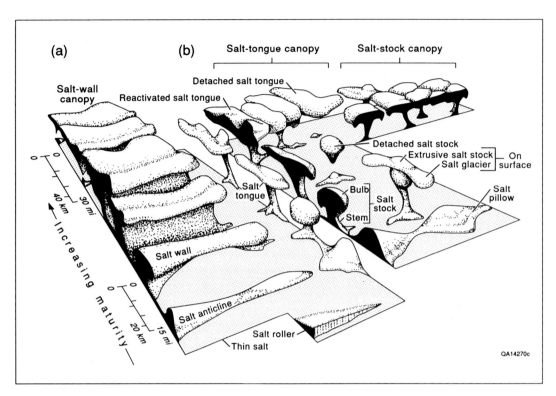

Fig. 8.2. Block diagrams showing the schematic shapes of known classes of salt structures. Structural maturity and size increase towards the composite, coalesced structures in the rear. The left panel shows elongated structures rising from line sources. The right panel shows structures rising from point sources. (Updated from Jackson & Talbot 1989b.)

(Mrazec 1907). In its broadest sense (which we recommend), diapir includes either lateral or vertical intrusion of any symmetry, upwelling of either buoyant or non-buoyant rock, migma, or magma, or emplacement by downbuilding or by faulting of prekinematic overburden.

A *salt roller* (Bally 1981) is a low-amplitude, asymmetric salt structure comprising two flanks: a gently dipping flank in conformable stratigraphic contact with the overburden; and a more-steeply dipping flank in normal-faulted contact with the overburden (Fig. 8.2). Salt rollers are an unequivocal sign of regional thin-skinned extension (see Chapters 6 and 11) perpendicular to the strike of the salt rollers. A plug-like salt diapir with subcircular planform (Fig. 8.2) is called a *salt stock* (syn. *salt plug*) (Trusheim 1960). An elongated upwelling of diapiric (discordant) salt, commonly forming sinuous, parallel rows, is a *salt wall* (Fig. 8.2) (Trusheim 1960). A *bulb* is the swollen, crestal portion of a salt diapir (Fig. 8.2) (Jackson & Talbot 1986); its overhanging periphery constitutes an *overhang*. Extremely broad bulbs grade into salt sheets. The comparatively slender part of a salt diapir below the bulb is the *stem* (Fig. 8.2), whose base is the *root*.

A *salt canopy* is a composite diapiric structure formed by partial or complete coalescence of diapir bulbs or salt sheets (Figs 8.2 and 8.3) (Jackson *et al.* 1987). These coalescing bodies may or may not be connected to their source layer by feeder stems. If necessary, canopies can be further differentiated by their components: *salt-stock canopies, salt-wall canopies* and *salt-tongue canopies* form by coalescence of stocks, walls and tongues, respectively (Figs 8.2 and 8.3). The junction between salt structures that have coalesced laterally to form a canopy is a *salt suture* (Fig. 8.3). Incomplete sutures are recognized by lensoid basins of overburden along them. Complete sutures are recognized by folds in the overburden above them: either the suture is below a subsided, cuspate synform (Nelson & Fairchild 1989); or it is the junction of a pair of appressed, raised salt antiforms (Lee *et al.* 1989).

Sheet-like salt bodies emplaced at stratigraphic levels above the autochthonous source layer are known as *allochthonous salt* (allochthonous, Naumann 1858, Wilckens 1912; allochthonous salt, Bally 1981). Allochthonous salt lies on stratigraphically younger strata (theoretically, allochthonous salt could overlie older strata, but such examples have not yet been reported). We suggest that the term allochthonous salt can be applied even if the salt sheet remains attached to its source layer. There are two reasons for this liberal definition. (1) The accepted definition of a *tectonic allochthon* is a sheet transported from its original position and resting on a different substratum, regardless of whether its roots remain attached. (2) It is generally impracticable to determine whether a salt sheet is detached or connected by only a thin stem; assigning a name should not be delayed until geophysical technology can readily distinguish diapiric stems. Where possible, it is useful to distinguish between allochthonous salt that is attached to or detached from its source layer. Another aspect affecting terminology is whether an allochthonous sheet emplaced beneath a thin, weak roof is intrusive (in a geological context) or extrusive (in a mechanical context). The terminology for allochthonous salt structures is still extremely 'fluid'. Several overlapping terms are in currency, and the definitions cited below are still provisional.

Allochthonous salt whose breadth is several times greater than its maximum thickness is a *salt sheet* (O'Brien & Lerche 1988), a broad term that includes salt tongue, salt laccolith and salt sill. A *salt laccolith* (new term) is an intrusive salt sheet whose ratio of maximum width to maximum thickness is (arbitrarily and tentatively) less than 20. The upper contact of a salt laccolith is typically concordant, whereas its lower contact is commonly discordant. A *salt sill* is a thinner, intrusive salt sheet (Watkins *et al*. 1978, Nelson & Fairchild 1989) whose ratio of maximum width to maximum thickness is (arbitrarily and tentatively) greater than 20. A salt sill is typically intruded at depths of a few hundred metres or less. Contact relations are similar to those of a salt laccolith.

A highly asymmetric variety of salt sheet or salt laccolith fed by a single stem is called a *salt tongue* (Figs 8.2 and 8.3). Individual salt tongues are as large as 80 km long and 7 km thick. The name is typically applied to wedge-like bodies with large taper angles and which do not resemble a salt sill (Watkins *et al*. 1978, D'Onfro 1988, Wang 1988). Structures rising by reactivation of allochthonous salt are known as *second-cycle structures*. Even *third-cycle structures* rising from salt sheets derived from still older salt sheets have been inferred in the Gulf of Mexico (Diegel and Schuster 1990, Hossack & McGuinness 1990); compare diapir family.

A *salt glacier* (syn. *namakier*) flows as a sheet-like extrusion of salt from an exposed diapir and spreads beneath air or water (Fig. 8.2) (De Böckh *et al*. 1929, Talbot & Jarvis 1984).

INTERNAL STRUCTURES

Minor structures within deformed salt bodies are spectacular, but few are unique to salt. We describe those peculiar to or well developed in salt and other evaporites. Three types of *ductile shear zones* (also see Chapter 1) associated with salt diapirs have been recognized. An *external shear zone* is a sheath-like collar around the contacts of a diapir; it is formed by the contact strain during rise of the diapir relative to the surrounding overburden. The inner part of the external shear zone typically comprises ductile evaporites. Depending on the rheology of the overburden, the external part of this shear zone can be a ductile shear zone, a brittle shear zone, a fault, or a growth fault. A *boundary shear zone* is enclosed within the diapir and contains screens or clast trains of country rock. Some boundary shear zones could result from external shear zones becoming entrained within mushroom structures.

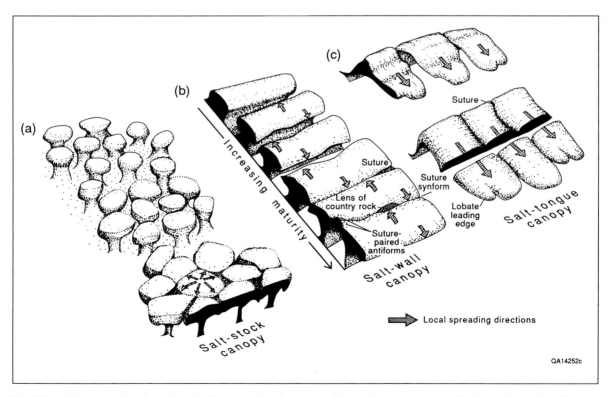

Fig. 8.3. Block diagrams showing three basic types of salt canopy formed by coalescence of salt stocks, salt walls and salt tongues, respectively. The degree of coalescence increases towards the front of each group. The sutures between coalesced salt structures can be recognized by lenses of country rock between salt, by cuspate synforms in the overburden above, and by pairs of raised, appressed antiforms in the margins of the coalesced salt sheets. Sutures of salt-wall canopies are perpendicular to the main spreading directions (arrowed), whereas sutures of salt tongue canopies are parallel to the main spreading direction and to the lobate leading edge of salt intrusion.

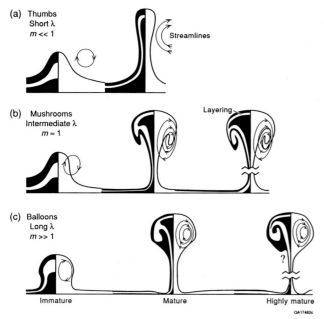

Fig. 8.4. Vertical sections through fluid diapirs rising through fluid overburdens, showing the effects of viscosity contrast on shapes and characteristic wavelength of diapirs. Streamlines are toroidal (doughnut-shaped stream surfaces) in three dimensions. The variable m is the effective viscosity ratio between overburden and source layer. Immature domes show only subtle differences as m varies, but these differences increase with maturity. At maturity, $m \sim 1$ bulbs have simple external mushroom structure; $m \gg 1$ bulbs look like balloons on strings and may contain simple internal mushroom structure. By the highly mature stage, both these types of bulb can coil into vortex mushroom structures. The highly mature shape of $m \ll 1$ bulbs is qualitatively similar to that shown for the mature stage. Differences in wavelength have been reduced to keep the diagram compact. (After Jackson & Talbot 1989a.)

An *internal shear zone* is also enclosed within the diapir, but comprises only sheared evaporites (Kupfer 1976). A *spine* is a steep-sided mass of salt within a larger salt body that moves or has moved faster or slower than neighbouring spines, from which it is separated by boundary shear zones or internal shear zones (Kupfer 1976).

The internal structure and shape of bulbs rising through fluid overburdens is controlled by viscosity and by structural maturity (Fig. 8.4). Two types of major fold can form in diapirs. A *curtain fold* is a cylindrical fold with radial axial trace and steeply plunging hinge within a diapir, possibly incorporating sheath folds originally in the source layer (Fig. 8.5) (Stier 1915). This type of fold is relatively common in diapirs.

The second type of major fold is restricted to *mushroom diapirs* having a broad bulb fringed by one or more laterally flattened, pendant *skirts* of deformed evaporites that envelop the stem of the diapir (Fig. 8.6a) (Jackson & Talbot 1989a). Both the skirts and the *infolds* they surround are downward-facing folds (Fig. 8.6c). With *external mushroom structure*, skirts surround antiformal, infolded synclinal infolds of overburden (Fig. 8.6a). With *internal mushroom structure*, the skirts are contained entirely within the diapir and thus envelop antiformal, infolded synclinal infolds of evaporite (Fig. 8.6a) (Jackson & Talbot 1989a). Both the skirt and infold are a form of *crescentic fold* having concentric axial trace and gently plunging fold hinge and are likely within the bulb of mushroom diapirs (Figs 8.6b & c). Crescentic folds whose lowest parts are rolled or coiled inwards about fold hinges concentric to the diapir core define a *vortex structure* (Figs. 8.6b & c, left side of diapir) (Jackson & Talbot 1989a).

Common in both diapiric and non-diapiric salt is the intestine-like *enterolithic fold*, variously ascribed to: (1) volume changes, such as the hydration of anhydrite to gypsum, which involves a 61% increase in volume; (2) diagenesis; (3) minor folding by gravity; and (4) larger-scale folding.

Glacial structures are not unique to salt glaciers, but they provide an unrivalled display of progressive strain in salt (shown from left to right in Fig. 8.7).

SEDIMENTARY RECORD OF SALT TECTONICS

Major sedimentary structures record the timing and character of the flow of underlying, and even overlying, salt. Correct interpretation of sedimentary structures is vital to reconstruct the history of salt flow, especially after much of the salt has vanished. Both salt flow and episodic local sedimentation influence each other. The sedimentary sequence can be divided into two or more major units, based on their age compared with the deformation. A *prekinematic layer* (syn. *isopachous layer*) is an interval of strata whose initial stratigraphic thickness is constant (or varying no more than typical for the region) above a salt structure or its adjacent rim syncline (but below any synkinematic layer) (Fig. 8.8). The prekinematic layer records sedimentation before salt movement, or any other deformation, began. In contrast, the *synkinematic layer* records sedimentation during salt flow or during any other type of deformation and typically overlies the prekinematic layer. The synkinematic layer shows local stratigraphic thickening (above structures, or parts of structures, such as withdrawal basins that subside faster than their surroundings), or local thinning (above relatively rising structures) (Fig. 8.8). Changes in thickness can also be recorded by onlap or truncation at all levels of the synkinematic layer. The interval of strata recording sedimentation after salt flow, or any other deformation, has ceased is the *postkinematic layer* (Fig. 8.8). Basal postkinematic strata can onlap an underlying, uneven, deformed surface but show no thickness changes ascribed to local deformation; the postkinematic layer typically overlies the synkinematic layer.

Truncations are particularly useful in distinguishing the three main units described above. Removal and termination of dipping deformed strata by subaerial or subaqueous erosion forms the truncation, representing

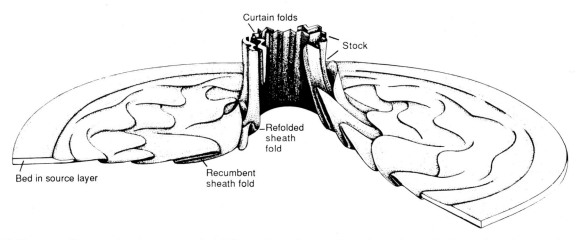

Fig. 8.5. Cutaway diagram showing progressive deformation of a representative bed near the base of a salt source layer. As the bed flows towards the central stock, recumbent sheath folds become folded, and perhaps refolded, above steps in the basement (which are omitted for simplicity). The sheath folds then rotate into the base of the salt stock where they are refolded into steeply plunging curtain folds. (Based on Talbot & Jackson 1987a, b.)

a sequence boundary. Truncations cutting across anticlines or fault blocks indicate that these structures had surface relief at the time of erosion. Undeformed truncations mark the base of the postkinematic layer. Where a truncation is itself folded or faulted, it marks an episode of spasmodic deformation in the synkinematic layer.

Structural inversion is the reversal of vertical motion recorded by stratigraphic thickening or thinning of synkinematic strata (Fig. 8.9) (Zeigler 1978, obsolete syn. transformation of structural relief, Trusheim 1960). Commonly viewed in the context of changes in regional tectonics (see Chapter 14), local structural inversion provides a valuable diagnostic criterion in salt tectonics. It is recognized by local changes in thickness or onlap directions and truncations. For example, a

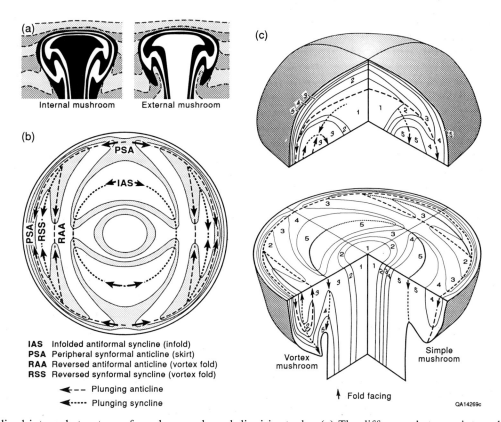

Fig. 8.6. Stylized internal structure of mushroom-shaped diapiric stocks. (a) The difference between internal and external mushroom structures. (b) Horizontal section through a vortex mushroom diapir, showing the names and axial traces of the internal crescentic folds; arrows indicate fold axial plunge; stippled layers mark alternating lithologic units. (c) Isometric, cutaway diagram of the mushroom diapir in (b); its vortex structure is displayed along the southwest – northeast line (left vertical section). In adjoining quadrants (right vertical section), its structure is that of a simple mushroom diapir. Numbers refer to stratigraphic sequence of layers, and arrows indicate direction of fold facing. (Based on Jackson & Talbot 1989a.)

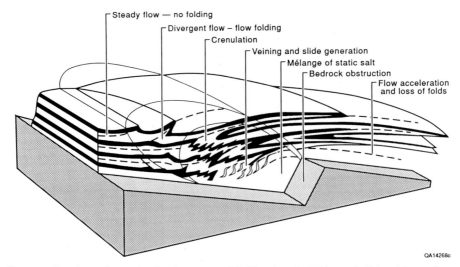

Fig. 8.7. Block diagram showing schematic development of folds, sheath folds and slides (sheared out limbs of folds) in glacial salt flowing from left to right across a bedrock step. Where flow decelerates upstream of the step, streamlines (dashed lines) diverge and deflect layering in the salt to form flow folds on the left. Flow then accelerates above the step and streamlines converge, thereby tightening and elongating folds into sheath folds (large recumbent folds on right). (After Talbot 1981, Talbot & Jackson 1987a.)

thickened interval (with outward onlap) overlying a thinned interval indicates local uplift followed by local subsidence (Fig. 8.9). In contrast, a thinned interval of synkinematic strata (with inward onlap) overlying a thickened interval indicates local subsidence followed by local uplift (Fig. 8.9). Salt-tectonic examples of inversion are a diapir rising in a graben then sagging under continuing extension, or the decay of a dynamic bulge after the underlying stem pinches off.

Several terms describe the basins formed by evacuating flow of underlying salt. *Rim syncline*, a purely structural term, refers to a fold having an arcuate or subcircular axial trace on the outer margin of a salt upwelling, typically resulting from salt withdrawal in the source layer (Nettleton 1934). The term *marginal syncline* is more appropriate for withdrawal basins flanking elongated salt walls, which lack an arcuate rim. Synkinematic *peripheral sinks* of sediments accumulate within rim synclines as a result of salt withdrawal (Trusheim 1960). Peripheral sinks can have any plan shape but are differentiated according to thickness variations within them. The *primary peripheral sink* accumulates on the flanks of growing salt pillows and comprises strata that thin towards the salt structure. The *secondary peripheral sink* accumulates around diapirs withdrawing salt from their precursor pillow and contains strata that thicken towards the salt diapir. The *tertiary peripheral sink* accumulates around mature diapirs that have exhausted the salt supply from their precursor pillow and are rising by necking or thinning of

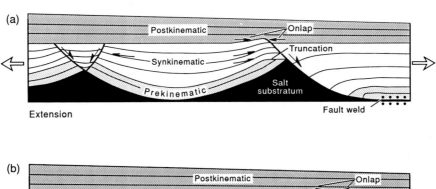

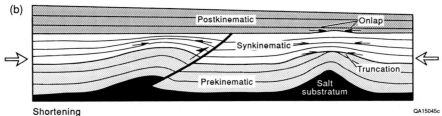

Fig. 8.8. Sedimentary record of salt flow during extension (above) and shortening (below). The prekinematic layer was deposited before salt flow began. The synkinematic layer accumulated during salt flow and may include internal onlap or truncation. The postkinematic layer was deposited after salt flow ceased; its base may onlap or truncate originally opencast (exposed) structures in the synkinematic layer.

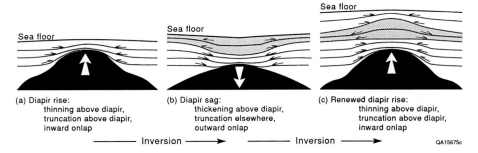

Fig. 8.9. Two structural inversions caused by subsidence and renewed uplift of a diapir. The inversions are recorded by changes in local thickness and onlap directions (barbed arrows). Changes in the sites of local truncation can also indicate inversion but are omitted here for simplicity. (Inspired by Nelson & Fairchild 1989.)

the diapir's stem; coeval strata thin over the crest of the salt diapir. The mechanical significance of peripheral sinks is questioned by some workers. Mounded strata between salt diapirs and having a flat base and rounded crest over a sedimentary thickness wedge constitute a *turtle structure anticline* (Trusheim 1960). A turtle structure is formed by structural inversion of a primary peripheral sink when salt is withdrawn from the margins of the sink by growing diapirs. The plan shape of turtle structures is typically highly irregular, depending on the number, location and relative vigour of the diapirs flanking it. An analogous mound flanked by a single diapir is a *half-turtle structure*.

Newer terms have also been coined for a synkinematic *intrasalt basin* (syn. *minibasin*) subsiding into relatively thick, allochthonous or autochthonous salt. These basins may express spoke circulation or may reflect merely random patterns of differential loading. Some intrasalt basins form upslope of large salt tongues, which impede transport of sediments downslope. A *depotrough* is an intrasalt basin with deep-water, mass-flow or turbiditic deposits uplapping the salt diapirs (Spindler 1977). A *depopod* is an intrasalt basin with shallow-water, deltaic deposits draping over or against the flanking salt diapirs (Spindler 1977).

A *family* of adjoining salt walls or stocks contains more than one generation of diapirs. This concept, originally proposed by Sannemann (1968), is currently being reevaluated in Germany. A mother diapir is flanked by daughter diapirs, which, in turn, can be flanked farther outward by granddaughter diapirs. The secondary peripheral sink of the mother diapir laterally merges with the primary peripheral sink of the daughter diapir; the same relationship holds for succeeding offspring. Thus the secondary peripheral sinks of each generation overlap imbricately. The family comprises diapirs of different age rooted in the same source layer. Contrast this relationship with generations of salt structures distributed vertically, with each generation of salt sheet providing a new source layer for the next (second-cycle and third-cycle structures).

The surface or zone joining strata originally separated by autochthonous or allochthonous salt is a *salt weld* (Jackson & Cramez 1989). A weld is a negative salt structure resulting from the complete, or near-complete, removal of intervening salt. The weld can consist of a brecciated, insoluble residue or salt too thin to be resolved in reflection seismic data. The weld is usually but not invariably marked by a structural discordance. Another distinctive feature of welds is a structural inversion above them. *Primary welds* join strata originally separated by gently dipping, autochthonous, bedded salt. *Secondary welds* join strata originally separated by steep-sided salt diapirs (walls or stocks). *Tertiary welds* join strata originally separated by gently dipping, allochthonous salt sheets (Jackson & Cramez 1989). A *fault weld* is a fault surface or fault zone joining strata originally separated by autochthonous or allochthonous salt (Hossack & McGuinness 1990). A fault weld is thus equivalent to a salt weld along which there has been faulting or shearing.

PROCESSES OF SALT TECTONICS

Salt tectonics (syn. *halotectonics*) refers to any tectonic deformation involving salt, or other evaporites, as a substratum or a source layer. *Halokinesis* is a form of salt tectonics in which flow of salt is powered entirely by gravity, that is, by release of gravity potential energy alone without significant lateral tectonic forces (Trusheim 1957).

Several geologic processes reflect or affect salt tectonics. They are seen in their purest form where gravity operates alone, the maximum principal stress (σ_1) is vertical, and all lateral stresses are equal to lithostatic pressure. Any extension, including normal faulting, represents only local stretching related to diapiric intrusion. The following rates (dimensions LT^{-1}) are the most important (the actual symbols used are a personal choice): $\dot{A}$, aggradation rate of overburden; $\dot{D}$, dissolution rate of salt structure; $\dot{E}$, extension rate of overburden or salt; $\dot{I}$, lateral injection rate of salt structure; $\dot{P}$, progradation rate of overburden; $\dot{R}$, rate of increase in relief of salt structure; $\dot{S}$, shortening rate of overburden or salt.

Downbuilding (syn. *passive piercement*) describes syndepositional diapir growth: the diapir increases in relief by growing downward relative to the sedimentary surface (Fig. 8.10b) (Barton 1933). The base of the diapir subsides, together with surrounding strata, as the basin fills with sediment. The diapir's crest remains at

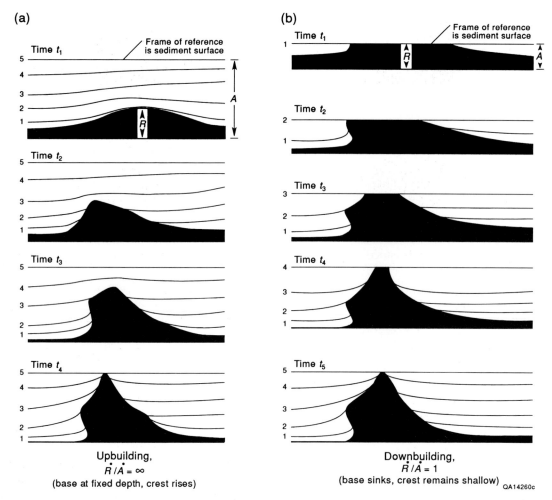

Fig. 8.10. Two end-member mechanisms of diapir growth from early (time t_1) to the same final stage (time t_4 or t_5). The upbuilding mechanism portrayed is intrinsically unlikely because (1) overburden area must be reduced to create space to accommodate the intruding salt, and (2) a considerable thickness and strength of overburden must be penetrated and displaced by the relatively weak buoyancy forces of diapirism. These two problems are relieved only by regional extension or by the downbuilding mechanism shown on the right.

or just below a thin roof, which is continually thickened by sedimentation and thinned by erosion or extensional thinning. R is the height of the crest of the salt structure above its base. A is the depth from the sediment surface to the base of the salt source layer. Pure downbuilding is characterized by equal rates of diapiric rise, $\dot{R}$, and aggradation, $\dot{A}$. However, most diapirs do not grow with perfect equilibrium between these two rates. Many diapirs grow syndepositionally. Their crests rise to the surface, and their bases sink ($\dot{R}/\dot{A}>1$); these diapirs can eventually extrude. Conversely, where $\dot{R}/\dot{A}<1$, the diapir continues to increase in relief but is nevertheless buried progressively deeper. Where $\dot{R}/\dot{A} = 0$, the diapir is dormant and is merely buried without growing. The shape of diapirs surrounded by non-fluid overburdens can depend strongly on the relative rates of $\dot{R}$ and $\dot{A}$. Passive piercement refers to the continued growth of a diapir during downbuilding after its crest has reached the surface (Nelson 1989).

Upbuilding (syn. *active piercement*, Nelson 1989; obsolete syn. *upthrusting*, Barton 1933) in its most extreme or ideal form refers to postdepositional diapir growth through prekinematic overburden, where $\dot{R}/\dot{A}$ = ∞ (Fig. 8.10a) (Jackson *et al.* 1988). Thus, as the diapir increases in relief, its base remains at a constant depth below the sedimentary surface, while its crest rises towards the surface. This kind of upbuilding, known as active piercement, is probably only possible in relatively tall diapirs overlain by relatively thin overburden, unless the overburden is being extended or is unusually weak, or is fluid.

More usually, diapirs grow by a combination of the ideal end members of upbuilding and downbuilding. Sediments accumulate around the diapir as it works its way to the surface.

Diapiric growth vertically or laterally restricted by the strength of its overburden is *constrained growth* (Bishop 1978). An example is forceful intrusion at moderate depths, characteristic of upbuilding. By contrast, *unconstrained growth* includes overturns of Rayleigh–Taylor instabilities, downbuilding, shallow injection of salt sheets into weak sediments, very shallow vertical intrusion by upbuilding, extrusive protuberance of diapirs and glacial flow. *Asymmetric intrusion* is upwelling of salt diapirs with two contrasting flanks (Trusheim 1960, Nelson 1989): (1) a

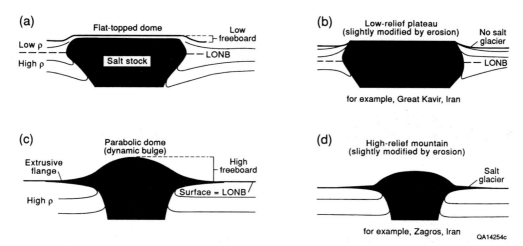

Fig. 8.11. Relation between relief and level of neutral buoyancy (LONB). Flat-topped dome (a) modified by erosion (b) to form a pancake-like geomorphology. If the overburden is sufficiently yielding, a ductile strain collar surrounds the diapir. The diapir spreads most widely at the LONB rather than at the surface. The freeboard is low because the density contrast is also low, like an iceberg. Because of the low freeboard, there is little tendency for the salt to extrude laterally. Compare with the parabolic bulge of salt (c), whose dynamic pressure has raised its crest higher than the LONB at the surface because the overburden (e.g. carbonates) is denser than the salt, even at the surface. Ductile yielding of the country rocks is negligible. Despite erosion (d), such diapirs can form rounded mountains, whose high freeboard is maintained by large density contrasts or lithostatic pressures from below. The high freeboard ensures that glacial salt extrudes as long as the salt supply is adequate, and the climate is arid or semiarid. (Based on Jackson et al. 1990.)

steep, discordant flank — typically a growth fault or shear zone — separating salt from thick, gently dipping strata; and (2) a gently-dipping, concordant flank overlain by a tilted, trapdoor-like interval of sediments, that are commonly erosionally truncated and which can preserve prekinematic or synkinematic strata (Fig. 8.8). During *mature piercement*, a diapir grows by necking of its stem after the adjoining source layer has been exhausted (Nelson 1989) during the formation of a tertiary peripheral sink. Because salt supply is restricted, growth is slow, and the diapir can easily become buried if sedimentation continues.

Differential loading creates lateral pressure gradients on salt due to lateral variation in thickness, density, or strength of the overburden (Hollingworth et al. 1945). Such variations may be sedimentary (e.g. fans, deltas or lobes) or structural (e.g. thinning by rifting or thickening in growth-fault hangingwalls) (Fig. 8.8). Unlike the buoyancy mechanism, differential loading requires no density inversion or burial of salt below the level of neutral buoyancy to operate; it effectively initiates halokinesis and is particularly efficient near the surface. Differential loading is augmented by buoyancy after the overburden sediments flanking the salt structure have compacted to densities greater than that of salt.

Buoyancy is an instability caused by an overburden sinking into a less-dense source layer (Arrhenius 1913). Buoyancy is driven by lateral pressure gradients caused by relief in the upper surface of the salt source layer. The buoyancy concept is generally linked with Rayleigh–Taylor instability, which was originally defined for two viscous layers, but can be applicable to multilayers with other properties. The *density inversion*, or superposition of a denser layer above a less-dense layer, is a primary one due to compositional differences.

Gravity overturning refers to the resulting exchange in position of the source layer and overburden such that the less-dense layer is eventually uppermost; the process is rarely completed in nature.

Lateral spreading of salt (and commonly its overburden) above the level of neutral buoyancy is *gravity spreading* (Fig. 8.11) (Ramberg 1981). A dipping upper surface allows a small component of the gravitational body force to be resolved into a downslope shear stress to drive spreading. Examples are the spread of salt glaciers or lateral injection of salt sheets.

Thermal convection is the subsolidus rise of hotter salt and the sinking of cooler salt because of a thermal gradient between the top and bottom of the convecting layer (Talbot 1978). A temperature-induced density inversion provides the necessary instability. Thermal convection is a cyclic process, unlike the single overturn of systems with a compositional density inversion. Convection is promoted by increase of layer thickness, density, temperature gradient, thermal expansivity and by decrease in viscosity (hence, grain size too) and thermal diffusivity. Thermal convection is theoretically possible in a salt layer thicker than 2.9 km and having a viscosity of $<10^{16}$ Pa in a geothermal gradient of $30°C$ km^{-1} (see Rayleigh number, Appendix 3).

The depth (or range of depth) at which the bulk density of the overburden is equal to that of salt, about 2200 kg m^{-3}, is the *level of neutral buoyancy* (new term). At this depth a diapir will spread most rapidly if surrounded by a yielding, fluid overburden (including air or water) (Fig. 8.11). Terrigenous clastics must be buried before they compact, dehydrate, or cement to reach the density of salt. This burial depth varies from basin to basin but is approximately 450–900 m under normal compaction gradients (hydrostatic pore pressure) and around 1500 m where shales are undercompacted (pore pressure elevated between hydrostatic and lithostatic). A *dynamic bulge* is caused by pressure of flowing fluid against its overburden; for example, the viscous flow of salt upward from a diapir stem; an analogy is a fountain (Fig. 8.11) (Talbot & Jarvis 1984). Such bulges can extend above the level of neutral buoyancy, the most striking examples being the mountainous extrusive salt domes of southern Iran. Dynamic bulges may subside after the underlying stem pinches off or the source layer is exhausted; the record of a decaying bulge is preserved as structural inversion in any synkinematic overburden. Dynamic bulges can dome strata above the sediment surface, deformation that is recorded by onlapping strata or erosional truncations, which may themselves be deformed later. The height of the crest of a salt diapir or sheet above the surrounding sediment surface is its *freeboard* (Fig. 8.11) (new term). Bulbs of Tertiary diapirs in Central Iran have low freeboards because their level of neutral buoyancy is at least 500 m below the surface. They are analogous to ductile icebergs floating while they spread slowly within soft, clay-rich, halite-bearing overburden. In contrast, diapirs of Precambrian Hormuz salt in southern Iran have high freeboards. They rise in viscous fountains with parabolic profiles because their level of neutral buoyancy is the land surface above carbonates denser and stronger than salt. Where the rate of extrusion exceeds that of erosional wasting, salt glaciers spread from such domes.

Two related terms describe progressive loss of salt. *Salt reduction* is the mass transfer of salt over time, resulting in an obvious change in area of salt in cross-section, by (1) volume loss due to dissolution, (2) isochoric (constant volume) flow out of the plane of section, including smearing along décollement faults, or (3) isochoric flow within the plane of section but beyond the ends of the cross-section (Jackson & Cramez 1989). In contrast, the redistribution of salt without obvious change in its area in cross-section is known as *salt withdrawal*. Examples are the migration of salt from the flanks of a salt pillow into its core as it evolves into a diapir, or the flow of salt along a wall into local culminations that evolve into stocks.

AUTOCHTHONOUS GROWTH DURING REGIONAL EXTENSION

Because salt is weak as well as buoyant, its presence promotes thin-skinned regional deformation and profoundly influences the overall structural style. Extensional salt tectonics results where σ_1 is vertical, and σ_3 is horizontal (or slope parallel) and less than the lithostatic stress. This type of salt tectonics is generally

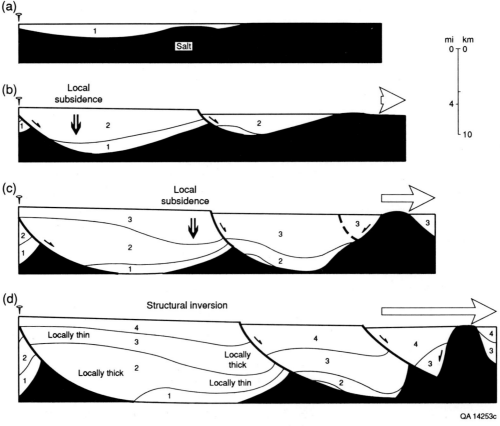

Fig. 8.12. Formation of deep growth faults on thick salt during regional thin-skinned extension (indicated by progressive lengthening of horizontal arrows) over flowing salt (black). Within each growth-fault block, strata initially thicken to the left as the block rotates counterclockwise. Overlying strata thicken to the right as the block rotation reverses because salt continues to flow to the right after the left end of the fault block has grounded on subsalt strata. (After Worrall & Snelson 1989.)

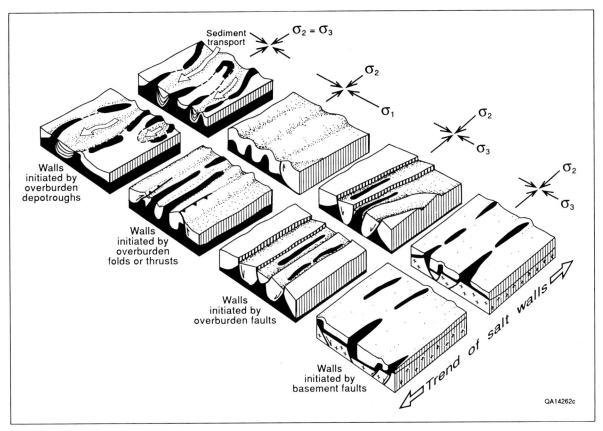

Fig. 8.13. Initiation and localization of parallel salt walls by four types of boundary conditions, each of which is illustrated by two dip-parallel sections. Three of these involve regional tectonics deforming basement or overburden; the fourth is controlled by the pattern of sedimentation. σ_1, σ_2, and σ_3 are the maximum, intermediate and minimum principal stresses. (Based on numerous experiments at the Universities of Texas, Uppsala and Rennes.)

associated with regionally developed normal faults in the overburden (also see Chapter 11).

Thin-skinned extension by gravity spreading or gravity gliding, where the base of the salt remains undeformed, is common on continental slopes and prograding deltas; for example, the Kwanza basin and Gabon basin (West Africa), the Campos basin (Brazil) and the Gulf of Mexico. Characteristic faults are *listric normal growth faults* (see Chapter 6) in synkinematic strata where the hangingwall is stratigraphically thickened with respect to the footwall. The footwall is typically underlain by diapiric salt or shale, whose presence appears to be required for listric growth faulting (Figs 8.12 and 8.13). In contrast, during crustal extension, for example, rifting, the base of the source layer is offset by faulting (Fig. 8.13); examples are the central North Sea, northwest German basin, Paradox basin (Utah), Saudi Arabia and the inverted Atlas basin (Morocco). Experiments suggest that for faults of a given throw, overburden faulting has a stronger control on the location and evolution of salt diapirs than does basement faulting. Uniform thinning of salt and overburden — if this were possible — would retard upwelling of salt. In contrast, where extension is non-uniform, salt diapirism is promoted because local structural thinning creates differential loading (Fig. 8.14). The response of the salt to extension depends critically on the rate of diapir rise, $\dot{R}$, to that of extension, $\dot{E}$, a topic that cannot be fully explored in this summary.

Parallel salt walls are now thought to result from linear boundary effects (Fig. 8.13), rather than extra-thick source layers, as was previously supposed. The strongest boundary effects are exerted by overburden faulting (Vendeville & Jackson 1992a); less so by folding and by the pattern of differential loading in the overburden, especially during progradation or downslope mass flow. Other boundary effects include basement-fault offsets in the base of the source layer and tilted or wedge-shaped source layers or overburdens.

Salt rollers are the first response of the salt to non-uniform extension of overburden (Fig. 8.14). A roller with a single crest forms at the base of the footwall of a half-graben. A pair of rollers forms below a graben, one in each footwall. With increased extension, the salt pierces the graben, forming a salt wall. With the most extreme thin-skinned extension, the overburden ruptures into rafts which can be translated tens of kilometres (Fig. 8.15).

Raft tectonics is advanced extension, where deep, syndepositional grabens open and the intervening overburden separates into rafts, which slide like block-glide landslides downslope on a décollement of thin salt (Figs 8.14 and 8.15) (Burollet 1975, Duval et al. 1992). A *raft*, itself, is a fault block of allochthonous

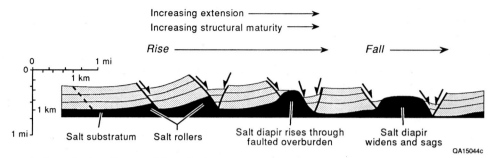

Fig. 8.14. The evolution of salt rollers to rising diapirs to falling diapirs during regional thin-skinned extension. (Based on Vendeville & Jackson 1992a, b, Jackson & Vendeville 1990.)

overburden that has separated so far that it no longer rests on its original footwall (the adjoining fault block); instead, the raft lies entirely on a décollement layer, which typically consists of thin salt (Figs 8.14 and 8.15). Generally the length of a raft exceeds its thickness. Rafts may comprise both prekinematic and synkinematic strata and may themselves consist of smaller, older rafts, which became yoked by later sedimentation before being ruptured again, so that they moved as a single, large raft. Rafts are separated by trough-like, typically asymmetric, depocentres of younger, synkinematic strata (Vendeville & Jackson 1992b). The underlying décollement faults may commonly be cryptic salt welds if all the salt is removed or thinned beyond seismic recognition. The type area for raft tectonics is the continental margin of Angola (Duval et al. 1992).

AUTOCHTHONOUS GROWTH DURING REGIONAL SHORTENING

Contractional salt tectonics requires either the minimum principal stress (σ_3) to be vertical (creating shortening) or the intermediate principal stress (σ_2) to be vertical (resulting in transpression). As with regional extension, regional shortening can be thin-skinned (only the cover contracts) or it can involve the basement. This tectonics is well covered in other chapters (6 and 13), so we focus here only on modifications induced by salt, which affect at least twelve major orogenic belts worldwide. The pressure of a salt décollement horizon promotes decoupling of the overburden, which has the following effects (Fig. 8.16): (1) folding is more common than thrusting; (2) folds and thrusts lack consistent vergence, being commonly expressed as box folds and backthrusts: (3) low-angle cross-sectional taper, which results in extremely broad fold-and-thrust belts; and (4) complex three-dimensional edge effects formed at the lateral boundary of the salt substratum. Angular folds (box folds, chevron folds and mitre folds) predominate, reflecting internal deformation of anisotropic multilayers above a thin, ductile substratum. If the ductile substratum is thick, the multilayers tend to buckle as a single unit into large, sinusoidal folds, with consequent loss of fold angularity (see critical Coulomb taper, Appendix 2).

Lateral shortening encourages upwelling by thickening the substratum and creating the structural relief necessary for buoyancy. Where anticlines are preferentially thinned by erosion, this would encourage diapirs to pierce them. Salt diapirs that grow in fluid overburdens before, after and during regional shortening differ in shape and location, as shown by centrifuge modelling (Koyi 1988). Most preshortening diapirs are elongated by folding parallel to the trend of regional synclines, whose folding distorts the diapirs; the diapirs are local upwellings that tend to lie in

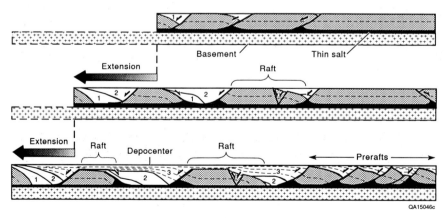

Fig. 8.15. Raft tectonics during extreme thin-skinned extension over a décollement layer. Prerafts evolve into rafts when the hangingwall blocks no longer overlie their original footwalls. (Based on Burollet 1975, Jackson & Cramez 1989; Duval et al. 1992.)

regional synclines nucleated by withdrawal basins. During regional folding, undeformed salt becomes rucked into ridges that spawn new synshortening diapirs in the cores of regional anticlines, whose folding elongates the planform of diapirs normal to anticline hinges. Postshortening diapirs are also concentrated in regional anticlines but tend to be circular and larger than synshortening diapirs.

ALLOCHTHONOUS GROWTH

Perhaps the single most important discovery in salt tectonics within the last decade has been the recognition that irregular masses of salt hundreds of square kilometres in area are relatively thin, allochthonous sheets rather than thick, autochthonous massifs. These salt bodies do not extend down several kilometres to the source layer; rather they are underlain by the source layer's overburden. This realization opens vast volumes of subsalt potential reservoirs for hydrocarbon exploration, particularly in the Gulf of Mexico. Because knowledge on this topic is advancing rapidly and little has been published so far, the terminolgy is unsettled at this stage.

Salt laccoliths and sills are emplaced as thin sheets between overburden strata a few hundred metres or less below the sediment surface by *sheet injection* (syn *sheet spreading*) (Fig. 8.17). Injection is driven by gravity spreading (typically, but not invariably, down local dips) through weak, unconsolidated, mud-rich, less dense overburdens under low confining pressures (Nelson & Fairchild 1989). *Sheet thickening* is the vertical thickening of a laterally injected salt laccolith or sill. *Reactivation* is a general term for the deformation of a salt sheet, which thus acts as a second-generation diapiric source layer by differential loading or burial below the level of neutral buoyancy. Deformation involves local upwelling of salt and sheet segmentation by partitioning the salt sheet into separate salt structures by subsidence of intrasalt basins or by growth faulting (Fig. 8.18).

Two stratigraphic markers elucidate the timing or direction of emplacement of allochthonous salt. Oblique truncation of strata against the lower contact of a salt tongue forms a *basal cut-off* (Figs 8.17 and 8.18) (new term, after fault cut-off). Each cut-off marks the advancing position of the leading edge of a spreading tongue at successive stratigraphc times. The basal cut-offs of typical laccoliths and sills suggest that their injection rate, $\dot{I}$, is about 100 times greater than the aggradation rate, $\dot{A}$, during sheet injection. *Salt ramps* and *salt flats* are the steeply inclined and gently inclined segments, respectively, of the stairstep basal contact of a salt tongue (Figs 8.17 and 8.18) (J. R. Hossack, written communication). Salt ramps cut up stratigraphic section in the direction of emplacement; they dip in a direction opposite to the spreading direction of a tongue. Ramps form where the ratio of aggradation to salt spreading is high; conversely, flats form where this ratio is low.

APPENDIX 1: RHEOLOGY, MICROSTRUCTURE AND STRAIN RATES

Steady-state flow of halite results from the combined effect of solution transfer creep (dominant at low deviatoric stresses, low strain rates and fine grain size) and dislocation creep (dominant at high deviatoric stresses, high strain rates and coarse grain size). The strain rate can be approximated by the following composite creep law (supplied by Janos Urai 1989 and based on Carter & Hansen 1983, Wawersik & Zeuch 1986, Urai *et al.* 1986, 1987, and Spiers *et al.* 1988):

$$\dot{e} = (7.6 \times 10^{-4} \exp(-66500/RT) \sigma^{4.5}) + (4.2 \times 10^{-3} \exp(-26959/RT) \sigma/Td^3)$$

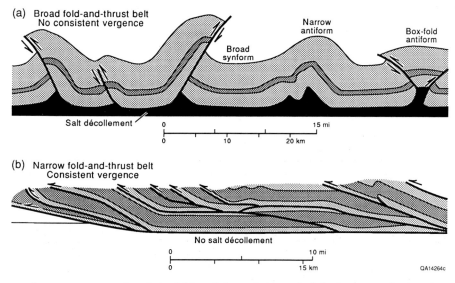

Fig. 8.16. The contrast between structural style in fold-and-thrust belts underlain (upper section) and not underlain (lower section) by a décollement layer of thin salt. (Based on Laubscher 1961, Roeder *et al.* 1978, Davis & Engelder 1987.)

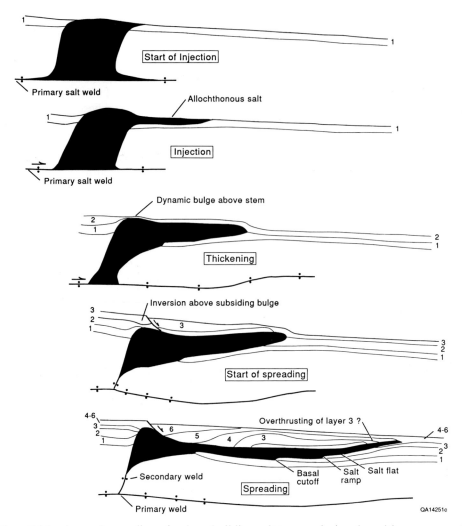

Fig. 8.17. Injection, thickening and spreading of a downbuilding salt tongue during deposition on a continental slope. The stem of the diapir closes as strata to its left move rightward along a primary fault weld. (After Wu 1989, Nelson & Fairchild 1989, unpublished seismic data.)

where: $\dot{e}$ = strain rate (s^{-1}); R = universal gas constant (8.315 J K^{-1} mol^{-1}); T = temperature (K); σ = deviatoric stress (MPa, typically <3); d = grain size (mm, typically 1–10 but smaller in extrusions).

Solution-transfer creep is approximately Newtonian flow in slightly damp salt under typically geologic conditions of low strain rate and low deviatoric stress under confining pressure. Dynamic recrystallization is achieved by migration of grain boundaries: grains with higher internal strain energy dissolve, liberating ions that diffuse across a thin film of brine and precipitate on a neighbouring grain with lower internal strain energy.

Dislocation creep is intracrystalline plastic flow involving a combination of glide and climb of dislocations through the crystal lattice to produce change in grain shape. At lower temperatures, gliding can be obstructed by grain boundaries and inclusions, resulting in dislocation tangles and work hardening. At higher temperatures, recovery takes place by cross-slip and climb past obstacles; this annihilates dislocation tangles or orders them into dislocation walls by polygonization.

Rheology of overburden — overburden is commonly modelled as one of two ideal rheologies: fluid and brittle. Both these approximations treat sediments as a continuum. Fluid behaviour is modelled as a power-law, time-dependent flow in which the shear viscosity decreases as rate of shear increases. The logarithm of shear stress is proportional to the logarithm of shear-strain rate, as approximated by:

$$\dot{e} = A_0 \, \sigma^n \exp[-Q / RT]$$

where: $\dot{e}$ = strain rate; A_0 = material parameter; Q = apparent activation energy for creep (J mol^{-1}); R = universal gas constant (8.315 J K^{-1} mol^{-1}); T = temperature (K); σ = deviatoric stress (Pa); $1 < n < 10$ = stress exponent.

Brittle behaviour of the shallow crust (less than, say 8 km deep) has been simulated by experimental rock mechanics. This is typically modelled by a time-independent, elastic–plastic, Prandtl body (represented by a solid block pulled by a spring across a rigid plate). Low deviatoric stresses induce an elastic (Hookean), recoverable response in which shear stress, τ, is linearly proportional to shear strength, γ:

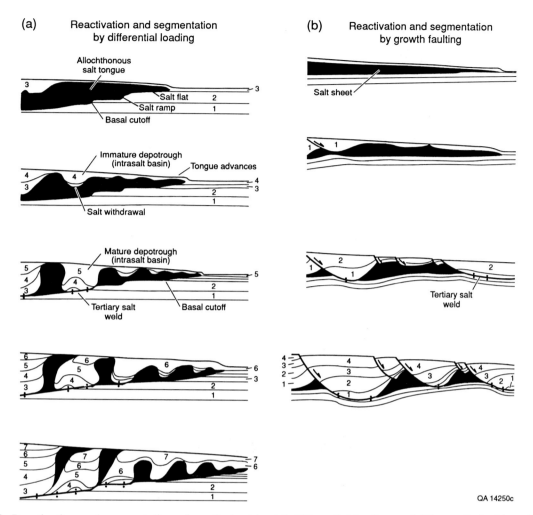

Fig. 8.18. Reactivation and segmentation of a salt sheet by (a) differential loading and (b) growth faulting. (Inspired by Humphris 1979, Worrall & Snelson 1989, West 1989.)

$$\tau = G\gamma$$

where G is the shear modulus. Higher stresses above the yield strength, τ_y, induce non-recoverable, frictional plastic deformation. For most geological strains, the elastic component is much less than the non-recoverable strain. Failure during frictional plastic deformation follows the *Mohr–Coulomb fracture criterion*:

$$|\tau| = \tau_0 + \sigma_n \tan \phi = \tau_0 + \sigma_n \mu$$

where: τ = shear strength; τ_0 = cohesive shear strength; σ_n = normal stress; ϕ = angle of internal friction, or slope of Mohr envelope; μ = Coulomb coefficient.

Strain rates in salt tectonics — bulk strain rates based on overall change in length per time can be estimated (the length change is vertical in most salt structures, but horizontal for salt sheets, including glaciers). Actual strain rates may greatly exceed such values locally because of strain inhomogeneity. Salt values listed are representative ranges, not extreme limits. Diapiric rates are based on the well-explored Gulf of Mexico and show wide variations between young, fast diapirs on the

Type of flow	Strain rate	Velocity	Velocity
	(s^{-1})	$(mm\ a^{-1})$	
Lava flow	10^{-5} to 10^{-4}	5×10^{11} to 3×10^{13}	1 to 60 km h^{-1}
Ice glacier	10^{-10} to 5×10^{-8}	3×10^5 to 2×10^7	1 to 60 m day^{-1}
Salt glacier	1×10^{-11} to 2×10^{-9}	2×10^3 to 2×10^6	10 to 100 km Ma^{-1}
Mantle currents	10^{-14} to 10^{-15}	10 to 1×10^3	2 m a^{-1} to 5 m day^{-1}
Salt tongue injection (<30 km wide)	8×10^{-15} to 1×10^{-11}	2 to 20	2 to 20 km Ma^{-1}
Salt tongue injection (>30 km wide)	3×10^{-16} to 1×10^{-15}	0.5 to 3	0.5 to 3 km Ma^{-1}
Salt diapir rise	2×10^{-16} to 8×10^{-11}	1×10^{-2} to 2	10 m to 2 km Ma^{-1}

continental slope and old, slow ones onshore. Comparative rates are given for other types of active geologic flow. Conventional strain rates and velocities for structures of representative size are given; flows are listed in decreasing order of velocity in both constant (left) and characteristic (right) units.

APPENDIX 2: MECHANICS OF WEDGE TAPER

Critical Coulomb taper — the taper angle of an accretionary wedge where compressive forces at any point are balanced by frictional resistance of the wedge ahead of that point. The critical taper is the sum of α, the mean local topographic slope, and β, the mean local dip of the décollement surface. The critical taper is thought to be as low as a few tenths of a degree for wedges shortened over a salt layer (Davis & Engelder 1987). The taper may be estimated for a cohesionless, time-independent, homogeneous, isotropic wedge deforming over a salt décollement surface (Davis & Engelder 1987, adapted from Dahlen et al. 1984):

$$\alpha + \beta = \frac{\beta \rho g z + (1 - \lambda)\tau_0 - [2S_0 \cot\phi\,(\alpha + \beta)/(\csc\phi - 1)]}{\rho g z\,[1 + 2(1 - \lambda)/(\csc\phi - 1)]}$$

where: ρ = bulk density of overburden; g = gravitational acceleration; z = depth; λ = pore-pressure coefficient (quotient of pore pressure, P_p and lithostatic pressure, $\rho g z$); τ_0 = shear strength of basal detachment in salt; S_0 = cohesive strength of overburden; ϕ = internal friction angle of wedge.

APPENDIX 3: FLUID MECHANICS

Characteristic wavelength, λ_c — theoretical periodic spacing of the fastest mode of growth of an instability, represented by the principal eigenvalue of growth rate, κ^*_m. Amplitude of the displacement at time t on the ith interface is

$$y_i(t) = y_i(0)\exp[\kappa^*_m q t]$$

where: y_i, amplitude of perturbation on layer i; q, scaling factor (s^{-1}) (Ramberg 1981).

Dominant wavelength, λ_d — observed periodic spacing of an instability; it may differ from the characteristic wavelength if the interface is initially deformed or if Rayleigh–Taylor instability is influenced by other effects, such as differential loading, regional strain, thermal convection or faulting (Schmeling 1987).

Streamlines — laminar flow lines for which the tangent at any point is the local, instantaneous velocity vector of steady flow; decreasing spacing of streamlines indicates increasing velocities of flow, and vice versa.

Non-uniform flow — non-parallel streamlines; converging streamlines represent accelerating flows, diverging streamlines represent decelerating flows.

Unsteady flow — flow pattern varying with time, so that streamlines cross layering and deflect it as flow folds.

Rayleigh number, Ra — non-dimensional number to measure the vigour of thermal convection. For convection to occur, Ra must be larger than the *critical value*, Ra_c, which depends on boundary conditions.

$$Ra = \frac{\rho g \alpha \beta h^4}{\kappa \eta}$$

where: ρ = density (kg m^{-3}); g = gravitational acceleration (9.8 m s^{-2}); α = thermal expansivity (K^{-1}); β = temperature gradient (K m^{-1}); h = total thickness of layer (m); κ = thermal diffusivity (nominally 10^{-6} m^2 s^{-1}); η = dynamic shear viscosity (Pa s).

Rayleigh–Taylor instability — a layer of viscous fluid of uniform density overlying a compositionally less-dense layer is unstable. Small perturbations in the horizontal interface become amplified at a rate represented by an eigenvalue (the special value of a variable parameter for which the solution of an equation is non-trivial). The amplification rate depends on the thickness, density and viscosity of every layer, size of initial perturbation and time elapsed.

Rayleigh–Taylor number Rt — non-dimensional number to measure the vigour of Rayleigh–Taylor instability and its capacity to advect heat. There is no critical Rt number below which the system is stable.

$$Rt = \frac{\Delta\rho g h^3}{\kappa \eta}$$

where: $\Delta\rho$ = density contrast at the interface; g = gravitational acceleration; h = total thickness of the layer; κ = thermal diffusivity; η = dynamic shear viscosity (Schmeling 1988).

The relative dominance of Rayleigh–Taylor instability or thermal convection can be expressed by a non-dimensional number representing the following ratio (R. Weijermars, written communication):

$$Rt/Ra = \frac{\Delta\rho}{\rho \alpha \beta h}$$

Where this ratio is greater than one, the overturn of the system is dominated by compositional density and by Rayleigh–Taylor instability. Where this ratio is less than one, the overturn is dominated by heat-induced density contrasts and by thermal convection.

Movement cell — volume containing closed loops of streamlines or stream surfaces in an overturning system of source layer and overburden. *First-order cells* contain several diapirs and may be as large as an entire sedimentary basin. *Second-order cells* are on the scale of a single diapir; the centre of the cell is the diapir where the steamlines rise, its lateral margins are defined by the outer limits of the sinking streamlines. *Third-order cells* affect only part of a single diapir, as in the growth of mushroom bulbs (Talbot & Jackson 1987).

Spoke circulation — second-order cellular pattern

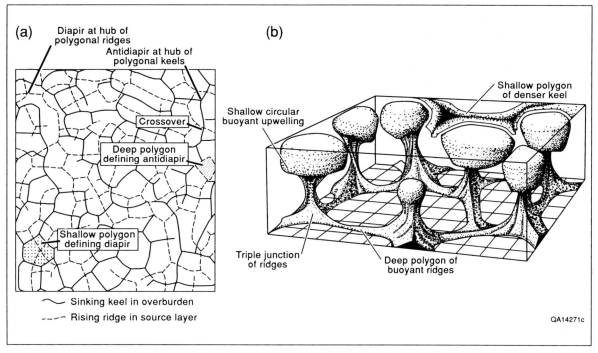

Fig. 8.19. Spoke circulation, the basic pattern of Rayleigh−Taylor gravitational overturn of a two-layer system of fluids. (a) Spoke pattern in experiment. (b) Perspective view of a source layer undergoing spoke circulation; the overburden has been removed for clarity except for a small piece of one horizon, illustrating a sinking keel. (After Talbot et al. 1991.)

similar to that of the most vigorous mode of thermal convection replicated in Rayleigh−Taylor overturn (Fig. 8.19) (Talbot et al. 1991). The pattern is best developed where viscosity contrasts are small; it comprises (1) downwelling, circular basins and radiating keels of overburden fluid, and (2) upwelling circular domes and radiating ridges of buoyant fluid. Downwelling basins are surrounded by upwelling polygons; upwelling domes are surrounded by downwelling polygons (Fig. 8.19). Shallow polygonal keels are laterally offset about half a wavelength relative to the deep polygonal ridges.

Acknowledgements — We thank R. Weijermars, D. D. Schultz-Ela, B. C. Vendeville, S. J. Seni, H. Koyi, L. Fairchild, J. R. Hossack, R. Evans, D. M. Worrall, R. G. Martin and W. Shengyu for their helpful suggestions, but we alone are responsible for any remaining errors. Figures were drafted by Joel L. Lardon, Kerza Prewitt, Wade Kolb, Patrice A. Porter and Margaret L. Evans under the direction of Richard L. Dillon, and by the authors. Bruno Vendeville contributed much to several new concepts described here, and he and Shing-Tzong Lin helped a lot with initial computer drafting of some of the figures.

This research was funded by the Texas Advanced Research Program, the Texas Advanced Technology Program, the Swedish Natural Science Foundation and by the following oil companies: Agip SpA, Amoco Production Company, ARCO Oil and Gas Company, BP Exploration Inc., Chevron OFR and USA Inc., Conoco Inc., Elf Exploration Inc., Exxon Production Research Company, Marathon Oil Company, Mobil Research and Development Corporation, Petroleo Brasileiro SA, Phillips Petroleum Company, Texaco Inc. and Total Minatome Corporation.

REFERENCES

Arrhenius, S. 1913. Zur Physik der Salzlagerstätten. *Meddelanden Vetensskapsakademiens Nobelinstitut* **2**, 1−25.

Bally, A. W. 1981. Thoughts on the tectonics of folded belts. In: *Thrust and Nappe Tectonics* (edited by McClay, K. R. & Price, N. J.). Spec. Publs geol. Soc. Lond. **9**, 13−32.

Barton, D. C. 1933. Mechanics of formation of salt domes with special reference to Gulf Coast salt domes of Texas and Louisiana. *Bull. Am. Ass. Petrol. Geol.* **17**, 1025−1083.

Bishop, R. S. 1978. Mechanism for emplacement of piercement diapirs. *Bull. Am. Ass. Petrol. Geol.* **62**, 1561−1583.

Braunstein, J. & O'Brien, G. D. 1968. Diapirism and diapirs — a symposium. *Mem. Am. Ass. Petrol. Geol.* **8**, 444 pp.

Burollet, P. 1975. Tectonique en radeaux en Angola (Raft tectonics in Angola). *Bull. Soc. géol. Fr.* **17**, 503−504.

Carter, N. L & Hansen, F. D. 1983. Creep of rocksalt. *Tectonophysics* **92**, 275−333.

Dahlen, F. A., Suppe, J. & Davis, D. 1984. Mechanics of fold-and-thrust belts and accretionary wedges: cohesive Coulomb theory. *J. geophys. Res.* **89**, 10087−10101.

Davis, D. M. & Engelder, T. 1987. Thin-skinned deformation over salt. In: *Dynamical Geology of Salt and Related Structures* (edited by Lerche, I. & O'Brien, J. J.). Academic Press, Orlando, Florida, 301−337.

De Bockh, H., Lees, G. M. & Richardson, F. D. S. (eds) 1929. *Contribution to the Stratigraphy and Tectonics of the Iranian Ranges*. Methuen, London.

DeGolyer, E. L. 1925. Origin of North American salt domes. *Bull. Am. Ass. Petrol. Geol.* **9**, 831−874.

Diegel, F. A. & Cook, R. W. 1990. Palinspastic reconstruction of salt-withdrawal growth-fault systems, northern Gulf of Mexico. *Geol. Soc. Am. Abstr. Prog.* **22** (7), A48.

Diegel, F. A. & Schuster, D. C. 1990. Regional cross sections and palinspastic reconstructions, northern Gulf of Mexico. *Geol. Soc. Am. Abstr. Prog.* **22** (7), A66.

D'Onfro, P. 1988. Mechanics of salt tongue formation with examples from Louisiana slope. *Bull. Am. Ass. Petrol. Geol.* **72**, 175 (Abstr.)

Duval, B., Cramez, C. & Jackson, M. P. A. 1992. Raft tectonics in the Kwanza basin, Angola. *Mar. Petrol. Geol.* **9**, 389−404.

Harris, G. D. & Veatch, A. C. 1899. A preliminary report on the geology of Louisiana. *Geol. Surv. Louisiana Rep. Baton Rouge*, 9–138.

Harrison, J. C. & Bally, A. W. 1988. Cross-section of the Parry Islands fold belt on Melville Island, Canadian Arctic Islands: implications for the timing and kinematic history of some thin-skinned décollement systems. *Bull. Can. Petrol. Geol.* **36**, 311–332.

Hollingworth, S. E., Taylor, J. M. & Kellaway, G. A. 1945. Large scale superficial structures in the Northampton ironstone field. *Q. J. geol. Soc. Lond.* **100**, 1–34.

Hossack, J. R. & McGuinness, D. B. 1990. Balanced sections and the development of fault and salt structures in the Gulf of Mexico. *Geol. Soc. Am. Abstr. Prog.* **22** (7), A48.

Humphris, C. C. Jr. 1978. Salt movement on continental slope, northern Gulf of Mexico. In: *Framework, Facies, and Oil-Trapping Characteristics of the Upper Continental Margin* (edited by Bouma, A. H., Moore, G. T. & Coleman. J. M.). *Am. Ass. Petrol. Geol. Stud. Geol.* **7**, 69–85.

Jackson, M. P. A. & Cornelius, R. R. 1987. Stepwise centrifuge modeling of the effects of differential sediment loading on the formation of salt structures. In: *Dynamical Geology of Salt and Related Structures* (edited by Lerche, I. & O'Brien, J. J.). Academic Press, Orlando, Florida, 163–259.

Jackson, M. P. A. & Cramez, C. 1989. Seismic recognition of salt welds in salt tectonics regimes. *GCSSEPM Found. 10th A. Res. Conf. Prog. Extended Abstr., Houston, Texas*, 66–89.

*Jackson, M. P. A. & Talbot, C. J. 1986. External shapes, strain rates, and dynamics of salt structures. *Bull. geol. Soc. Am.* **97**, 305–323.

Jackson, M. P. A. & Talbot, C. J. 1989a. Anatomy of mushroom-shaped diapirs. *J. Struct. Geol.* **11**, 211–230.

Jackson, M. P. A. & Talbot, C. J. 1989b. Salt canopies. *GCSSEPM Found. 10th A. Res. Conf. Prog. Extended Abstr., Houston, Texas*, 72–78.

Jackson, M. P. A. & Vendeville, B. C. 1990. The rise and fall of diapirs during thin-skinned extension. *Bull. Am. Ass. Petrol. Geol.* **74**, 683 (Abstr.).

Jackson, M. P. A., Cornelius, R. R., Craig, C. H. & Talbot, C. J. 1987. The Great Kavir salt canopy: a major new class of salt structure. *Geol. Soc. Am. Abstr. Prog.* **19**, 714.

Jackson, M. P. A., Talbot, C. J. & Cornelius, R. R. 1988. Centrifuge modeling of the effects of aggradation and progradation on syndepositional salt structures. *University of Texas at Austin, Bureau of Economic Geology Report of Investigations* 173, 93 pp.

Jackson, M. P. A., Cornelius, R. R., Craig, C. H., Gansser, A., Stöcklin, J. & Talbot, C. J. 1990. Salt diapirs of the Great Kavir, Central Iran. *Mem. geol. Soc. Am.* **177**, 139 pp.

*Jenyon, M. K. 1986. *Salt Tectonics*. Elsevier, London.

Kupfer, D. H. 1976. Shear zones inside Gulf Coast stocks help delineate spines of movement. *Bull. Am. Ass. Petrol. Geol.* **60**, 1434–1447.

Laubscher, H. P. 1961. Die Fernschubhypothese der Jurafaltung. *Eclog. geol. Helv.* **54**, 221–282.

Lee, G. H., Bryant, W. R. & Watkins, J. S. 1989. Salt structures and sedimentary basins in the Keathley Canyon area, northwestern Gulf of Mexico: their development and tectonic implications. *GCSSEPM Found. 10th A. Res. Conf. Prog. Extended Abstr., Houston, Texas*, 90–93.

Mattox, R. B., Holser, W. T. & Odé, H (eds). 1968. Saline deposits. *Spec. Pap. geol. Soc. Am.* **88**, 1–701.

Mrazec, L. 1907. Despre cute cu simbure de strapungere (On folds with piercing cores). *Bull. Soc. Stiite Romania* **16**, 6–8.

---

*Asterisks indicate reviews suitable to introduce salt tectonics.

Naumann, C. F. 1858. *Lahrbuch der Geognesie* (vol. 1, 2nd edn). Wilhelm Engelmann, Liepzig.

Nelson, T. H. 1989. Style of salt diapirs as a function of the stage of evolution and the nature of the encasing sediments. *GCSSEPM Found. 10th A. Res. Conf. Prog. Extended Abstr., Houston, Texas*, 109–110.

Nelson, T. H. & Fairchild, L. H. 1989. Emplacement and evolution of salt sills in northern Gulf of Mexico. *Bull. Am. Ass. Petrol. Geol.* **73**, 395 (Abstr.).

Nettleton, L. L. 1934. Fluid mechanics of salt domes. *Bull. Am. Ass. Petrol. Geol.* **18**, 1175–1204.

O'Brien, J. J. & Lerche, I. 1988. Heat flow through and around salt sheets. *Bull. Am. Ass. Petrol. Geol.* **72**, 230 (Abstr.).

*Ramberg, H. 1981a. *Gravity, Deformation and the Earth's Crust in Theory, Experiments and Geological Application* (2nd edn). Academic Press, London.

Ramberg, H. 1981b. The role of gravity in orogenic belts. *Thrust and Nappe Tectonics* (edited by McClay, K. R. & Price, N. J.). *Spec. Publs geol. Soc. Lond.* **9**, 125–140.

Roeder, D., Gilbert, O. E. Jr & Witherspoon, W. D. 1978. Evolution and macroscopic structure of Valley and Ridge thrust belt, Tennessee and Virginia. *Univ. Tennessee Dept. geol. Sci. Stud. Geol.* **2**, 1–25.

Schmeling, H. 1987. On the relation between initial conditions and late stages of Rayleigh–Taylor instabilities. *Tectonophysics* **133**, 16–31.

Schmeling, H. 1988. Numerical models of Rayleigh–Taylor instabilities superimposed upon convection. *Bull. geol. Inst. Univ. Uppsala N. S.* **14**, 95–109.

Spiers, C. J., Urai, J. L., Lister, G. S., Boland, J. N. & Zwart, H. J. 1986. The influence of fluid–rock interaction on the rheology of salt rock. *Nucl. Sci. Technol. EUR 10399 EN*, 131 pp.

Spindler, W. M. 1977. Structure and stratigraphy of a small Plio-Pleistocene depocenter, Louisiana continental shelf. *Gulf Coast Ass. geol. Soc. Trans.* **27**, 180–196.

St. John, B., Bally, A. W. & Klemme, H. D. 1984. Sedimentary provinces of the world — hydrocarbon productive and nonproductive. *Am. Ass. Petrol. Geol. Map. Ser.*, 35 pp.

Stier, K. 1915. Strukturbild des Benther Salzgebirges. *Jahresber. niedersächs. geolog. Ver.* **8**.

Talbot, C. J. 1978. Halokinesis and thermal convection. *Nature* **273**, 739–741.

Talbot, C. J. 1981. Sliding and other deformation mechanisms in a glacier of salt, S. Iran. In: *Thrust and Nappe Tectonics* (edited by McClay, K. R. & Price, N. J.). *Spec. Publs geol. Soc. Lond.* **9**, 13–32.

Talbot, C. J. & Jackson, M. P. A. 1987a. Internal kinematics of salt diapirs. *Bull. Am. Ass. Petrol. Geol.* **71**, 1068–1093.

*Talbot, C. J. & Jackson, M. P. A. 1987b. Salt tectonics. *Sci. Am.* **256**, 70–79.

Talbot, C. J. & Jarvis, R. J. 1984. Age, budget and dynamics of an active salt extrusion in Iran. *J. Struct. Geol.* **6**, 521–533.

Talbot, C. J., Rönnlund, P., Schmeling, H., Koyi, H. & Jackson, M. P. A. 1991. Diapiric spoke patterns. *Tectonophysics* **188**, 187–201.

Trusheim, F. 1957. Über Halokinese und ihre Bedeutung für die strukturelle Entwicklung Norddeutschlands. *Zeitschr. Deutsch. Geol. Ges.* **109**, 111–151.

Trusheim, F. 1960. Mechanism of salt migration in northern Germany. *Bull. Am. Ass. Petrol. Geol.* **44**, 1519–1540.

Urai, J. L., Spiers, C. J., Zwart, H. J. & Lister, G. S. 1986. Weakening of rock salt by water during long-term creep. *Nature* **324**, 554–557.

Urai, J. L., Spiers, C. J., Peach, C. J., Franssen, R. C. M. W. & Liezenberg, J. L. 1987. Deformation mechanism operating in naturally deformed halite rocks as deduced from microstructural investigations. *Geologie Mijnb.* **66**, 165–176.

*Vendeville, B. C. & Jackson, M. P. A. 1992a. The rise of diapirs during thin-skinned extension. *Mar. Petrol. Geol.* **9**, 331–353.

Vendeville, B. C. & Jackson, M. P. A. 1992b. The fall of diapirs during thin-skinned extension. *Mar. Petrol. Geol.* **9**, 354–371.

Wang, Y. F. 1988. Salt tongues in northern Gulf of Mexico. *Bull. Am. Ass. Petrol. Geol.* **72**, 256 (Abstr.).

Watkins, J. S., Ladd, J. W., Buffler, R. T., Shaub, F. J., Houston, M. H. & Worzel, J. L. 1978. Occurrence and evolution of salt in deep Gulf of Mexico. *Am. Ass. Petrol. Geol. Stud. Geol.* **7**, 43–65.

Wawersik, W. R. & Zeuch, D. H. 1986. Modelling and mechanistic interpretation of creep of rock salt below 200°C. *Tectonophysics* **121**, 125–152.

West, D. B. 1989. Model for salt deformation on deep margin of central Gulf of Mexico basin. *Bull. Am. Ass. Petrol. Geol.* **74**, 1472–1482.

Wilckens, O. 1912. *Grundzüge der Tektonischen Geologie.* Fischer, Jena.

*Worrall, D. M. & Snelson, S. 1989. Evolution of the northern Gulf of Mexico, with emphasis on Cenozoic growth faulting and the role of salt. In: *The Geology of North America — An Overview* (edited by Bally, A. W. & Palmer, A. R.). *Geol. Soc. Am.* 97–138.

Wu, S. 1989. Allochthonous salt, structure and stratigraphy of the northeastern Gulf of Mexico. Unpublished M. A. thesis, Rice University, Houston, Texas, 219 pp.

Ziegler, P. A. 1978. Northwestern Europe: tectonics and basin development. *Geologie Mijnb.* **57**, 589–626.

Part 2
Deformation Styles and Tectonic Settings

CHAPTER 9

Arc-Trench Tectonics

J. CHARVET and Y. OGAWA

GENERAL SETTING

ARC systems are linked to the sinking or *subduction* of oceanic lithospheric plates beneath *overriding plates* that can comprise continental, oceanic or transitional lithosphere. The descending part of the *underthrusting plate* that is disappearing beneath the overlying plate is the *downgoing* or *subducting slab* (Fig. 9.1).

Seismicity

Subduction is mainly recorded at depth by the famous Wadati or Benioff plane; a double Wadati–Benioff plane being definable in several places (e.g. Yoshii 1979, Uyeda 1982). One is along the top surface of the slab, and shows compressive focal mechanisms, the other is inside the slab, with extensional focal mechanisms. This relationship is tentatively interpreted as a result of the elastic response of the down-bent slab, that tends to unbend (Uyeda 1986). The shallow seismicity allows us to define an *aseismic front* in the overriding plate (Yoshii 1979), at some distance outward (*outboard*), from the *volcanic front*. The former corresponds to an abrupt thickness decrease of the rigid (seismically active) part of the overriding plate (Fig. 9.1). On average, focal mechanisms of earthquakes indicate: (1) compression in the overriding plate, (2) thrust shear at the interface between the rigid upper plate and the slab, and (3) extension in the shallowest part of the downgoing plate, in the bending area.

Material versus mechanical boundaries between convergent plates

On maps of small scales, the superficial boundary between convergent plates is usually drawn at trench axes, where the sediments of the oceanic plate begin to be buried beneath the trench fill. This is the *material boundary*, bordering the landward (or *inboard*) edge of the downgoing plate. But the *mechanical boundary*, where the relative motion between plates is accommodated, corresponds to a *megathrust plane*, regarded as the fault zone along which the relative motion occurs during great underthrusting earthquakes; in other words, the shallow expression of the Benioff plane. This seismogenic and tsunamigenic megathrust does not reach the trench but propagates upwards into the trench wedge and reaches the surface between the trench and the trench-slope break (e.g. Fukao 1979, Nakamura *et al.* 1984, Kagami 1986). The part of the *subduction complex* or *accretionary prism* located oceanward from this zone corresponds to *non-seismogenic structures* (also see Chapter 16).

Driving forces and stress distribution

If we eliminate mantle convection, not yet very well understood and sometimes considered as a product, not a major cause, of plate motion (Alvarez 1982, Hamilton 1988), the *driving forces of lithospheric plates* are anomalous pressures acting on the edges of the plates (Forsyth & Uyeda 1975). They develop opposing resistances, such as *drag forces* between the lithosphere and the asthenosphere. The horizontal component results from the lateral distribution of different buoyancies. Three main driving forces are known. *Ridge push* (Lister 1975, Dahlen 1981) is the slide of plates away from ridges, due to the inclined boundary between the base of the cooling oceanic lithosphere and the underlying less-dense asthenosphere (Fig. 9.2). *Slab pull* is the force acting on the subducting plate at the subduction boundary, mainly due to the negative buoyancy of the dense slab (Fig. 9.2). On average, slab pull is 2.5 times as important as ridge-push in moving plates (Carlson 1981). *Trench suction* (Chase 1978) is the plate boundary force which acts on the overriding plate at a subduction plate boundary, pulling the plate towards the trench. Models show that slab pull and trench suction are higher when the subduction fault is free, unlocked, without shearing stress within the fault plane, and much smaller when the fault is locked (Bott *et al.* 1989). The former case corresponds to the *decoupling* of convergent plates; the latter one to a strong, or tight, *coupling*, allowing better transmission of horizontal compressive stress between the subducting and the overlying plates.

The stress distribution in trench-arc–back-arc systems is variable in space and time. Considering the general regime one can recognize two opposite types of arc systems and subduction (Uyeda & Kanamori 1979,

Uyeda 1982). The *Mariana type* is dominated by extension, with back-arc opening, and the *Chilean type* is dominated by compression (Fig. 9.3), without a back-arc basin. There are also many intermediate arc types. Among others, the dip of the slab can be an important factor controlling the distinction (Bott *et al.* 1989). In a single system, there is a systematic *stress gradient* within the upper plate, from the trench to the back-arc region, inferred from observations and calculations (Nakamura & Uyeda 1980, von Huene 1984, Froidevaux *et al.* 1988, Bott *et al.* 1989). Compressive stress, which always exists to some extent in the trench area, transmitted from the subducting oceanic plate, attenuates landward and becomes tectonically insignificant in many fore-arc basins (Nakamura & Uyeda 1980, von Huene 1984). It combines with the other main source of tectonic stress: the vertical one linked to lateral density variations corresponding to topography and its deep compensation (Froidevaux *et al.* 1988). The tectonic style across the system depends upon the difference between these horizontal and vertical stresses. The most extensional feature is expected in the highest part of the volcanic arc (Fig. 9.4). An additional local back-arc tension is produced by an anomalously low density mantle beneath it (Bott *et al.* 1989).

In general, the most common regime in an overriding plate is one of extension, not shortening. However, this regime can change through time. A classic example is the northern Honshu arc: extension dominated during the Miocene; but compression prevails today as in the eastern border of the Japan Sea. Alternating locking and unlocking of the subduction fault may be an important controlling factor, and the abundance of sediments at the trench is another (see later).

Tectonic structures of the fore-arc: compressional versus extensional convergent margins

The first models of arc tectonics (e.g. Karig 1971, 1972, 1982, Karig & Sharman 1975, Dickinson & Seely 1979) assumed that there is a well-developed wedge of thrust sheets between the trench and the arc. Results of oceanographic cruises demonstrated that some trench inner-slopes are devoid of any compressive structure at the surface but show normal faults and collapse structures, even near the trench. One can also contrast the *convergent – compressional margins* (Barbados, SW Japan) with *convergent – extensional margins* (Guatemala, Mariana, Peru), in other words *extensional* or *collapsing active margins* (Aubouin *et al.* 1984, Bourgois *et al.* 1988).

Location of the plate boundary: migration of the trench

In many cases of subduction of *normal* (*mature*) oceanic lithosphere, with a moderate to steep inclination of the slab, the slab is pulled down more rapidly than

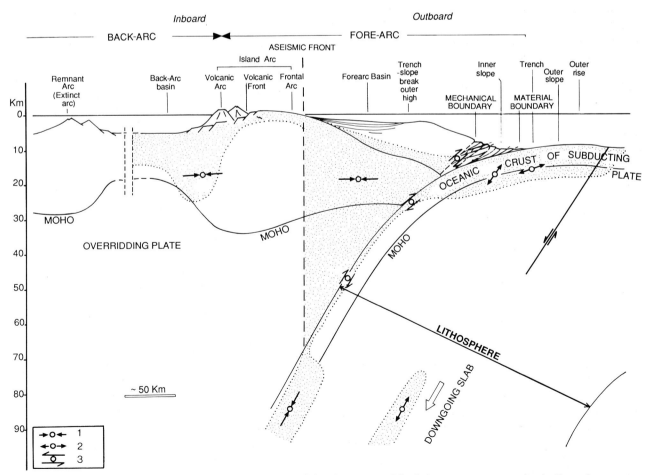

Fig. 9.1. Idealized cross-section of a subduction zone, of island-arc type. Stippled areas represent seismically active zones. Focal mechanisms of earthquakes: (1) compression, (2) extension and (3) thrust shear.

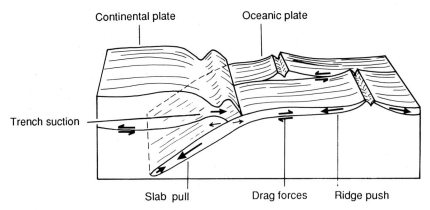

Fig. 9.2. Forces acting on plates (simplified after Forsyth & Uyeda 1975).

the rate of convergence of plates (Molnar & Atwater 1978, Hynes & Mott 1985, Hamilton 1988), and it sinks more steeply than the inclination of the Benioff zone. This is called the *roll back* phenomenon (or *subduction roll back*) (Dewey 1980) which implies a migration of the trench axis, that is, *trench retreat*, and of the hinge (bending area) of the downgoing plate away from the arc, that is, outboard or oceanward (Fig. 9.5a). Another cause of trench retreat and migration of the arc oceanward is related to an *anchored slab* in the mantle flowing away from the landward plate (Carlson & Melia 1984) (Fig. 9.5b).

COMPRESSIVE DEFORMATION: TECTONIC ACCRETION

In the fore-arc area, the tectonic transfer of material from the subducting plate to the overriding plate, or *tectonic accretion* (*sensu stricto*), leads to the growth of an *accretionary prism* or *accretionary wedge* (Fig. 9.6). An accretionary wedge is a volume of rocks (initially sediments) bounded at depth by a rather flat *décollement horizon* separating an accreted part from a subducted part, and a *backstop* at the rear. The latter is a buttress made of the older part of an overlying plate,

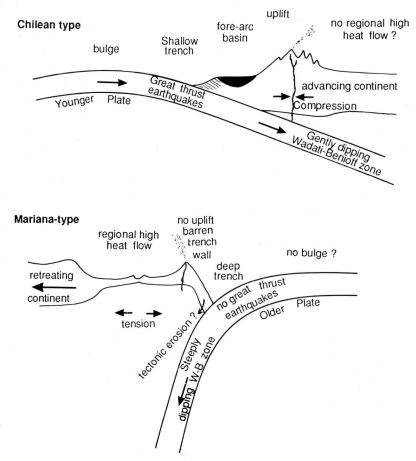

Fig. 9.3. Two models of subduction, possible implications and causes (modified after Uyeda 1982).

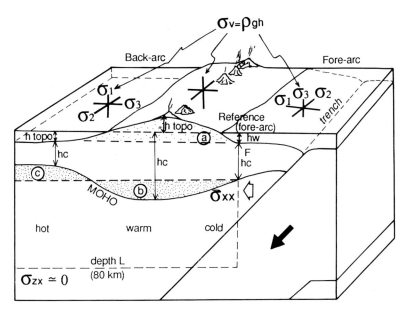

Fig. 9.4. Schematic block diagram of model island-arc structures (modified after Nakamura & Uyeda 1980, and Froidevaux *et al.* 1988). Reference density structure taken at fore-arc: water depth h_w = 4 km; crustal thickness h_c^F = 15 km; linear geotherm increasing to 1300°C at 80 km depth. Dotted area: (a) topographic load, (b) crustal root (buoyancy), (c) mantle root. Lateral variations of topography assumed to be compensated at 80 km depth. $\sigma_1 > \sigma_2 > \sigma_3$: principal stresses.

dipping either trenchward or arcward (e.g. Westbrook 1982, Mascle *et al.* 1986, Hamilton 1988, Westbrook *et al.* 1988). On the floor of the sea, the boundary of the prism goes from the *deformation front* at the toe of the prism, through a more or less complex transition to the fore-arc basin.

In principle, compressive deformation and tectonic shortening, and growth of the wedge, can occur by incorporation of new material at every place (Fig. 9.6): at the toe, the top, the back, and the base of the prism. Actually, it rarely extends to the fore-arc basin area, as in the Barbados. In detail, the shortening rate varies greatly inside the prism, corresponding to a complex deformation distribution accounting for the tectonic *convergence accommodation*.

The existence of an accretionary prism is not obligatory in the fore-arc of every arc-trench system. Well developed in the Barbados or at Nankai, it is lacking, or very small, in the Mariana or Japan trench inner slopes. It is clear that an important factor controlling the development of the prism is the thickness of the sedimentary pile of the subducting oceanic plate, especially the trench fill; in other words, the proximity of the detrital source supplying turbidites to the trench.

This has been well verified, for the bulk geometry (width of the prism) as well as its detailed structure, for the Barbados and Nankai prisms (for instance Biju-Duval *et al.* 1984, Mascle *et al.* 1985, Le Pichon *et al.* 1987a, b).

Accretion at the toe of the prism: frontal accretion

Frontal accretion is typically due to the offscraping of sediments of the downgoing plate and their incorporation in the prism by underthrusting at the deformation front, and additional deformation and shortening inside the lower slope area, that is, the non-seismogenic part of the fore-arc.

Accretion is selective and only the upper part of the sedimentary pile is added to the prism, forming the *offscraped sequence* or *accreted sequence* of *offscraped sediments* separated from the underlying *underthrust* or *subducted sequence* of *subducted sediments* by a *décollement horizon* or *level*, which corresponds to a discrete stratigraphic level, that is generally porous and water-rich (Fig. 9.7).

Seaward of the *frontal thrust*, that is the lowest thrust reaching the surface at the arcward side of the trench,

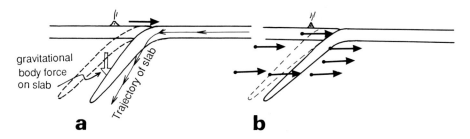

Fig. 9.5. Two causes of trench retreat: roll back relative to an inert asthenosphere (a) and a slab anchored to oceanward asthenospheric flow (b).

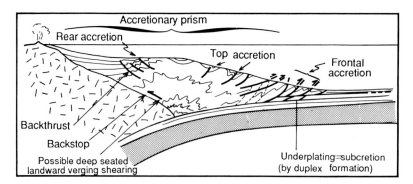

Fig. 9.6. Schematic section across an accretionary prism (modified after Mascle *et al.* 1986).

blind thrusts develop upward from the décollement level in the *proto-thrust zone*. The décollement level propagates into a *potential* or *incipient décollement*. Thus, the lower part of the prism possesses an *imbricate structure* corresponding to the *imbricate thrust zone*, made of a stack of elementary thrust sheets (Fig. 9.7) bounded by *in sequence thrusts or bedding-plane stepthrusts*. An anticline tends to develop in the hangingwall of the thrust, a syncline in the footwall.

In some cases, like at Nankai, the first surface deformation appearing at the toe, the deformation front, is expressed by these folds (often box folds) developed above blind thrusts. As a result of this mechanism the imbricate thrust zone shows a typical morphology of anticlinal ridges and intervening troughs. In detail, several tectonic styles can be seen at the deformation front (Fig. 9.8) and the décollement level must transfer to deeper stratigraphic horizons,

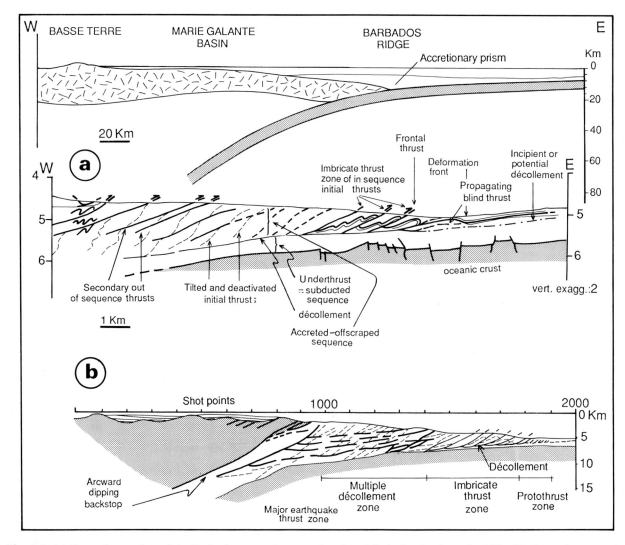

Fig. 9.7. (a) Frontal accretion of the Barbados accretionary prism (modified after Beck *et al.* 1988). (b) Tectonic zones of the Nankai accretionary prism (after Kagami 1986).

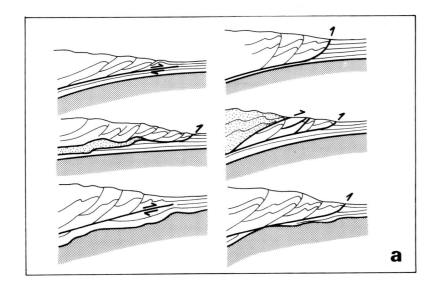

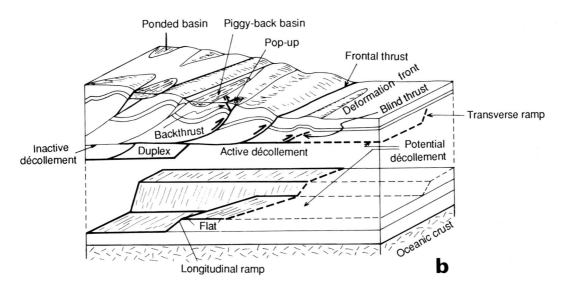

Fig. 9.8. (a) Tectonic styles at the toe of an accretionary prism (after Mascle *et al.* 1986). (b) Schematic diagram of thrust systems operating below a prism and their expression as a surface morphology (modified after Mascle *et al.* 1986).

according to the *flat* and *ramp* geometry of thrust systems (Westbrook & Smith, 1983) (also see Chapters 6, 13 and 16 for general discussions of such faulting).

Observations (e.g. Moore *et al.* 1983, 1987, Karig 1985, Le Pichon *et al.* 1987a, b, Beck *et al.* 1988) and modelling (Malavieille 1984) indicate that only the 2–3 frontal thrusts, especially the first one, are active in accommodating the convergence by tectonic shortening. They are quickly deactivated. This *thrust deactivation*, or *fault abandonment*, is most likely due to the rotation of stress during motion, and to strain hardening during water loss, leading to the propagation of thrusts seaward (Figs 9.7 and 9.8), and the landward tilting of initial thrusts, now inactive, or abandoned, by addition of new wedges of accreted sediments.

Then, landward from the imbricate zone, the growth of the prism is related to another mechanism: the initially deactivated and tilted thrusts are cut by active *secondary thrusts, out-of-sequence thrusts*, thus creating a pile of duplex structures in the *multiple décollement zone* (Fig. 9.7). These late thrusts were first interpreted from seismic records (Kagami 1986); they were documented by drilling in the Barbados by ODP Leg 110 (Moore *et al.* 1987, Beck *et al.* 1988). The late thrusts use what some authors call *secondary décollements*.

Although seaward-verging structures are predominant in the frontal part, some arcward-verging thrusts occur in several places (Silver 1972, Seely 1977). They are either restricted antithetic *backthrusts* which delimit a *pop-up* structure with conjugate seaward main thrusts (Fig. 9.8b); or, like off Washington and Oregon, initial *landward-verging thrusts*.

The above shortening represents only part of the total deformation due to plate convergence. The thickening of the prism, above the décollement level, implies, together with secondary stacking, a certain amount of diffuse deformation (Karig 1985) and addition of older material at the bottom.

Accretion below the prism: underplating or subcretion

Addition of sediments at the base of the prism, or *underplating* (*subcretion*), is commonly inferred, as, for example, in the Barbados, Makran, and Kodiak Island (Platt *et al.* 1985, Mascle *et al.* 1986, Sample & Fisher 1986, Byrne 1986). The basic mechanism is the landward down-stepping of the décollement level, along ramps. This *changing décollement level* leads to the formation of duplexes, or *duplex accretion* (Figs 9.7 and 9.8).

Accretion at the back of the prism: rear accretion

Comprehensive deformation may reach the inner boundary of the accretionary prism, leading to folding of fore-arc sediments and *backthrusting* of the wedge over the fore-arc basin (Fig. 9.6). This phenomenon is well documented and seismically imaged, in the case of the Barbados (Fig. 9.7), where it occurs above a seaward-dipping backstop (Westbrook 1982, Mascle *et al.* 1986, Silver & Reed 1988, Westbrook *et al.* 1988). It has also been reported from the Mediterranean ridge, where the system is now actually collisional (Le Pichon *et al.* 1982), and from the Sunda arc (Karig *et al.* 1980); for this latter case, the backstop dip is debated (Hamilton 1988, Silver & Reed 1988).

Such backthrusting is predicted by physical modelling, over a seaward-dipping backstop (Malavieille 1984, Davis *et al.* 1986). Then landward-verging shearing taking place in the deeper part of the prism (Fig. 9.6) is likely to explain some *synmetamorphic landward-verging structures* found in former accretionary prisms like the Shimanto belt of Japan (Fabbri *et al.* 1987, 1990, Charvet *et al.* 1990).

Accretion above the prism

Slope sediments are deposited in the *syntectonic interfold basins* or *piggy-back basins* developed in the synclinal troughs of the imbricated structure (Fig. 9.8b). Due to general arcward tilting, these basins evolve into asymmetrical *tilted basins*. Later, as with the flat sediments of some *ponded basins* formed upslope, this material, that accumulated over the offscraped deposits, can be incorporated into the prism, as a result of the activity of secondary thrusts (Fig. 9.7) (also see Chapter 16).

Small-scale structures

Small-scale structures, revealed by DSDP and ODP cores from fore-arc regions where accretion is occurring, and from on-land observations of old accreted sequences, include: microfaults, stratal disruption, cataclastic fabrics, scaly fabrics, vein structures, folds, kink bands (also see Chapters 5, 7 and 16). Compressional structures are generally restricted to lower-slope sites and to the accreted sequence. In the trench, and in the underthrust sequence, deformation is dominated by layer-parallel extension. This implies that the maximum stress is gently dipping above the décollement level, and steeply dipping below it (Moore & Byrne 1987, Byrne & Fisher 1990, Karig *et al.* 1990) (Fig. 9.9).

Stratal disruption refers to pervasive bedding discontinuities characterized by dismemberment of sand-rich bodies in a matrix of mudstone to siltstone

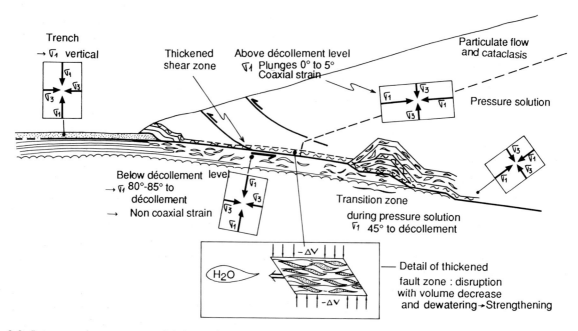

Fig. 9.9. Interpretative summary of deformation regimes and stresses within an accretionary prism (after Byrne & Fischer 1990 and Moore & Byrne 1987), showing some causes of mélange formation by stratal disruption. Principal stresses — $\sigma_1 > \sigma_2 > \sigma_3$.

(Lundberg & Moore 1986). This disruption may be accompanied by grain breakage or *cataclastic fabrics* (Lucas & Moore 1986). *Scaly fabrics* (*scaly foliation*) are defined as fabrics of anastomosing, curviplanar polished and lineated fracture surfaces, pervasive on a scale of millimetres, that define phacoids (Moore *et al.* 1986). They develop especially in weak smectitic mudstones along main faults, or even at the décollement level. *Striated foliation* and *scaly cleavage* are synonyms for scaly fabric.

Vein structure comprises parallel sets of small (<2 mm) planar to curviplanar, dark, mud-filled seams cross-cutting bedding without any offset. *Veins* display several morphological types (Carson *et al.* 1982, Knipe 1986, Lundberg & Moore 1986, Lindsley-Griffin *et al.* 1990) but are basically *dewatering structures*, linked to water loss from sediments. When closely spaced, they form a *spaced foliation*. Besides *kink bands* and *recumbent folds, crenulation folds* develop: asymmetrical microfolds with parallel axial surfaces. Folding can be non-cylindrical and may produce non-coaxial strain as *sheath-like folds*, like in the Shimanto prism (Hibbard & Karig 1987).

Reverse microfaults, conjugate or otherwise, are encountered in the accreted sequence. In the trench, in rare cases (e.g. Nankai), *normal microfaults* are recorded; they also occur on the upper slope, where they are associated with strike-slip faults.

In general, structures that record layer-parallel compression or shear dominate sites on lowermost slopes; whereas cores from upper slopes, trench and underthrust sequences show structures characteristic of layer-parallel extension.

EXTENSIONAL DEFORMATION, TECTONIC EROSION

General vertical movement of fore-arcs

With the exception of Barbados, most of the fore-arc basins display only normal faulting and subsidence. Surprisingly, several modern convergent margins: Japan, Mariana, Guatemala, Peru, underwent mainly subsidence during the subduction process (Aubouin *et al.* 1982a, b, 1984, Uyeda 1982, von Huene *et al.* 1982, von Huene *et al.* 1988, von Huene & Lallemand 1990), and a tectonic regime dominated by extensional faulting, even in the lower part of the slope. Moreover, the width of the margin may have diminished, by a landward retreat of the arc and the trench (Shipboard Scientific Party 1980).

Those extensional convergent margins with no or very small accretionary prism show normal faults of limited size linked to slope failure driven by gravity tectonics, leading to *mass wasting* (*collapse tectonics*) by gravity sliding (Fig. 9.10).

The subsidence of the margin, which cannot be explained by crustal thinning through major listric faulting or the systematic change of slab dip, is best explained by erosion of the upper plate (von Huene & Culotta 1989, von Huene & Lallemand 1990).

Subduction erosion (tectonic erosion)

Subduction erosion (Scholl *et al.* 1980) or *tectonic erosion* (Karig 1974) is subtracting material from the upper plate. *Frontal erosion* occurs at the leading edge, where the debris produced by collapse tectonics (slumping) is mainly subducted together with the lower plate sedimentary pile (Figs 9.10a & b). The *horst and graben chain-saw model* (Fig. 9.10c) (Hilde 1983) is inadequate to explain this (von Huene & Lallemand 1990). A mechanism derived from the *Coulomb wedge model* is proposed for the Japan (Fig. 9.11) and Peru trenches (Lallemand & Le Pichon 1987, von Huene & Lallemand 1990). It underscores the role of wedging and break-up of the upper plate by the subduction of a step, to produce debris transported away by sediment subduction. *Basal erosion* is illustrated by truncation of older rocks at the base of the upper plate (von Huene *et al.* 1988, von Huene & Culotta 1989) (Fig. 9.10). This basal erosion, or *attrition* (Moberly *et al.* 1982), is a principal explanation for the Neogene subsidence of the Japan and Peru margins, although the precise mechanism is unclear (von Huene & Lallemand 1990). Quantitative estimates indicate rates of erosion comparable to known rates of accretion (von Huene & Lallemand 1990).

Accretion versus subduction, and erosion and uplift versus subsidence

Tectonic regimes vary through space and time along the same convergent margin. General uplift or subsidence of the fore-arc is linked to accretion or erosion, which mainly depends on the thickness of the sedimentary pile at the trench. They are independent of seismic coupling and of whether there is Mariana or Chilean-type subduction.

Rifting and magmatism in fore-arcs

Besides the previously reported small-scale extensional fractures, fore-arc basins frequently show, on seismic records, a graben structure bounded by *fore-arc horsts* determining the outer-arc high. Stratigraphic evidence shows that one or several extensional phases can occur in a true fore-arc position, that is, on the trench side of a working arc, whether the arc is built on continental lithosphere, like Japan (e.g. Geological Survey of Japan 1982), or is an intraoceanic arc, like Izu-Bonin (Taylor *et al.* 1990). The rifting and subsidence phases of the fore-arc seem to be synchronous with the rifting of the arc and back-arc opening (Letouzey & Kimura 1985, Lagabrielle *et al.* 1990).

The term *fore-arc magmatism* is most often misleading, as the rocks considered are usually an older part of the arc, now isolated in front of the present volcanic chain. The first examples of real fore-arc

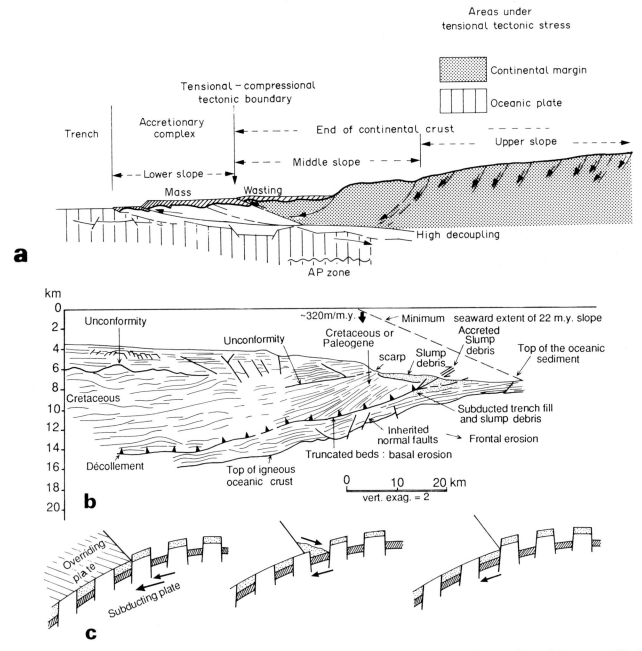

Fig. 9.10. (a) Diagrammatic sketch illustrating the convergent extensional margin or collapsing margin model, as exemplified by the Andean margin off Peru (after Bourgois *et al.* 1988). (b) Frontal and basal tectonic erosion at the Japan trench (after von Huene & Lallemand 1990). (c) Horst and graben chain-saw model (after Hilde 1983).

magmatic activity come from the Mariana and Izu-Bonin fore-arc basins (Fryer *et al.* 1990a, Lapierre personal communication). Similarly, the terms *fore-arc ophiolites* and *fore-arc ophiolitic emplacement* refer to dredged or drilled rocks which constitute the basement of intraoceanic arcs; older than the initiation of subduction.

Specific features of the Mariana and Izu-Bonin 200-km-wide fore-arcs are *serpentinite seamounts*, aligned along the outer high and upper slope. Circular, 10–20 km wide and about 1000 m high, they are made of detritic serpentinite including blocks of ultramafites and mafites. Their origin, as *serpentinite diapirs* (Hussong & Fryer 1985, Fryer *et al.* 1990b) or *serpentinite domes* (Lagabrielle *et al.* 1992) is still debated.

Rifting, arc-splitting and back-arc basin opening

Statistically, the opening of back-arc basins starts with an initial rifting splitting the arc and leaving behind a *remnant arc* driven away from the still active arc by the opening of the *back-arc basin* (Karig 1971). The location of rifting within the volcanic arc is assigned to several causes; mainly the weakness of the hot volcanic zone (Dewey 1980), and the topographic effect inducing the highest vertical stress (Froideveaux *et al.* 1988). Arc rifting may be of *single-rift* type or *multi-rift* type (Fig. 9.12) (Tamaki 1988).

The origin of back-arc basins is still a matter for debate (e.g. Uyeda 1986). If we except, on the one hand, the process of *entrapment* of old oceanic crust (e.g. Hilde & Lee 1984), which predates the back-arc

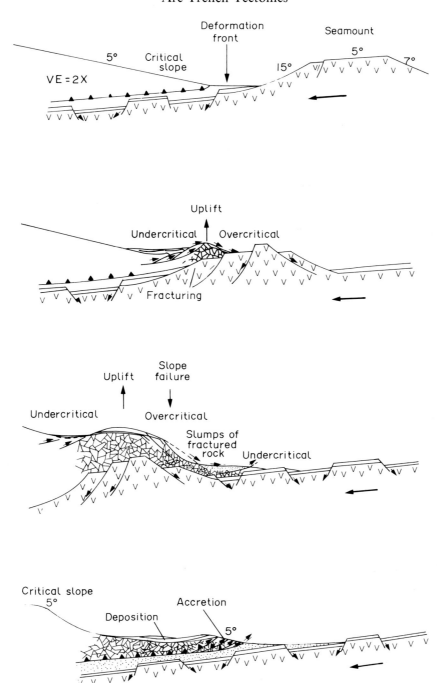

Fig. 9.11. Tectonic erosion; effect of subduction of an asperity (seamount) according to the Coulomb wedge model (after von Huene & Lallemand 1990). Inspired from the Kashima seamount and Japan–Kuril trench junction areas. Two times vertical exaggeration.

environment, and, on the other hand, the abandoned concept of *in situ* oceanization-basification (Beloussov & Ruditch 1968), two main basic mechanisms are advocated for back-arc spreading: *active* and *passive spreading*. Active spreading is induced from below by an asthenospheric flow: it includes the models of *mantle diapir rise* (Karig 1971), *asthenospheric injection* (Tatsumi *et al.* 1989), induced *wedge flow* (Toksöz & Bird 1977, Jurdy & Stefanick 1983) and the influence of a hypothetical deep *'hot region'* (Miyashiro 1986). Passive spreading results from extensional forces acting on the plates, that is, at the lithospheric level. They may result from the global interaction of the major plates:

(1) absolute *landward retreat* of the landward plate (Chase 1978, Uyeda & Kanamori 1979) with respect to a stable slab 'anchored' to stationary mantle; (2) *trench retreat* due to gravitational roll back of a collapsing slab (Elsasser 1971, Molnar & Atwater 1978); (3) *slab anchorage* in an oceanward mantle flow (Alvarez 1982); and (4) *spherical shell buckling* (Uyeda 1986). For continent-based island arcs, at least, a substitute or additional mechanism of *marginal sea opening* is the action of stress propagating within the continent, and generating strike-slip motions (Otsuki & Ehiro 1978). The *indentation model* (Fig. 9.13) is advocated for the opening of the South China Sea, due to the India-

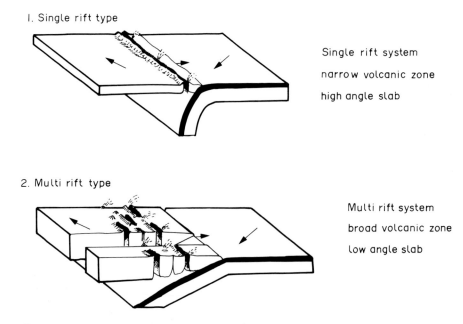

Fig. 9.12. Single-rift and multi-rift types of arc rifting and back-arc spreading (after Tamaki 1988).

Himalaya collision (Molnar & Tapponnier 1975, Tapponnier et al. 1982), as well as for the opening of the Okinawa trough, due to Taiwan collision (Letouzey & Kimura 1985). Alternatively, strike-slip extension, with an en échelon pattern of grabens, as in the Okinawa trough (and Andaman Sea?) is induced by the oblique convergence of the downgoing plate with respect to the arc and trench trends (Sibuet et al. 1987) (Fig. 9.14). Similarly, strike-slip motion works for the pull-apart basin model (Lallemand & Jolivet 1986, Jolivet et al. 1989) applied to the Sea of Japan and the Western Pacific marginal basins in general. In those cases, the initial splitting and rifting does not follow the arc trend but is oblique to it. Recently, the post-collisional thinning of a continental crust, previously thickened, has been proposed as a possible factor for intracontinental marginal basins (Faure & Charvet 1990). Most likely, a combination of several different mechanisms occurs. For instance, in addition to extensional strike-slip, the Okinawa trough opening requires a trench retreat component with lateral anchoring due to the Taiwan and Palau-Kyushu ridge indentations (Viallon et al. 1986).

Closure of back-arc basins

Marginal seas seem to have a limited spreading lifetime (Jurdy & Stefanick 1983). The Sea of Japan is an example of a back-arc basin which began to close about 2 Ma ago (Nakamura 1983, Lallemand et al. 1985, Tamaki & Honza 1985, Tamaki 1988, Tamaki et al. 1990), as documented by compressive structures on its eastern border.

ROLE OF FLUIDS

Fluids play a very important role in the tectonic style of fore-arcs. The *pore-fluid pressure*, reducing normal stress (see Chapters 7 and 16), facilitates shear failure. Deep-sea drilling results show that décollement levels and main thrusts in the prism correspond to a zone of *pore-fluid overpressure*, in underconsolidated stratigraphic levels, when the pore pressure is almost equivalent to the lithostatic pressure (Aubouin et al. 1982, Moore et al. 1983, von Huene 1984, Moore et al. 1987, Beck et al. 1988). Observations on land also support the existence of an overpressured décollement

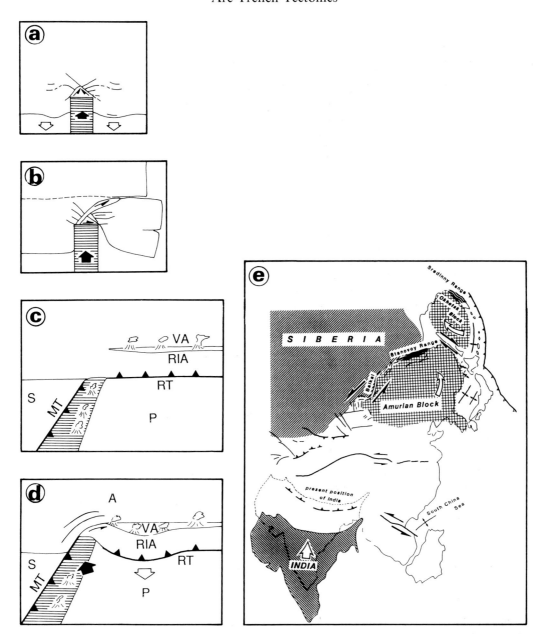

Fig. 9.13. Indentation model of back-arc basin opening. (a) and (b) Physical modelling (after Tapponnier *et al.* 1982). (c) and (d) Okinawa trough opening in response to Taiwan collision (after Letouzey & Kimura 1985). (e) Indo-Eurasian collision and spreading of the Sea of Japan, Kuril Basin and South China Sea (after Tamaki 1988, modified from Tapponnier *et al.* 1982).

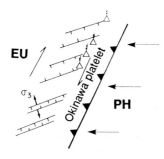

Fig. 9.14. Schematic diagram showing the extensional strike-slip motion of the Okinawa platelet with respect to Eurasia (EU), opening the Okinawa trough with an en échelon graben pattern. PH: Philippine Sea plate (after Sibuet *et al.* 1987).

level beneath accretionary prisms, as for instance in Alaska (Byrne & Fisher 1990). Water loss and fluid pressure release along faults is likely to be responsible for strain hardening and fault deactivation. Fluids circulate along those specific surfaces leading to *fluid advection, fluid drainage* and then *fluid venting* (Fig. 9.15) of methane- and H_2S-bearing water at the surface, allowing the life of benthic biological communities associated with chemosynthetic processes (Kulm *et al.* 1986, Suess *et al.* 1986, Boulègue *et al.* 1987, Cadet *et al.* 1987, Le Pichon *et al.* 1987, Pautot *et al.* 1987).

Fluid overpressure could be a cause of decoupling between plates at depth below the fore-arc (von Huene 1984). Other manifestations of pore-fluid pressure are *mud volcanism, mud diapirism* and *shale diapirism* (Fig. 9.15). Mud volcanoes are very common in subduction zones, on the accretionary prism as well as

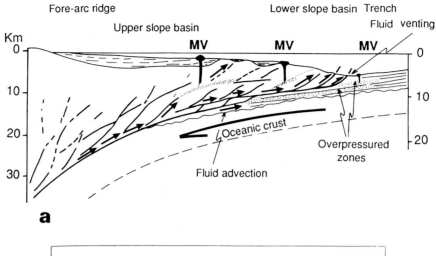

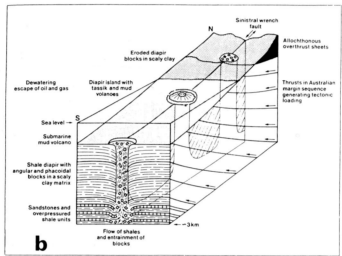

Fig. 9.15. (a) Schematic model of fluid advection, distribution of overpressured zones and mud volcanoes (MV) (after Barber & Brown 1988). (b) Conceptual model of relationship between shear zones, shale diapirs and mud volcanoes (after Barber *et al.* 1986).

in front of it and on the subducting plate (Biju-Duval *et al.* 1982, Westbrook & Smith 1983, Barber *et al.* 1986, Barber & Brown 1988, Brown & Westbrook 1988). They are the surface expression of mud diapirism, that is, the rising of buoyant and fluid-overpressured shales coming from depths of 1.5–3 km (Yamagata & Ogawa 1989), in response, at least partly, to tectonic loading (Fig. 9.15). Pressure release can be due to faulting, as indicated by the concentration of mud volcanoes near transverse fractures (Fig. 9.15b), while rising shales incorporate various country rocks (Fig. 9.15b). Shale diapirism could be a main mechanism for chaotic *mélange* formation (Barber *et al.* 1986, Barber & Brown 1988, Pickering *et al.* 1988, Yamagata & Ogawa 1989).

ROLE OF SUBDUCTING SLAB GEOMETRY AND CHARACTER

Tectonic inheritance, subduction of asperities

The structural grain of the subducting plate, that is, the distribution of fractures initially created at the ridge, parallel to magnetic anomalies, influences the tectonics of the inner slope. These faults are reactivated at the outer slope of the trench, as normal faults if they are subparallel to the trench, or as strike-slip faults if they are oblique (Lallemand *et al.* 1986). In this latter case, new normal faults are also created by bending, parallel to the trench. This pattern may be detected through the lower slope morphostructure, as in the Japan trench (Lallemand *et al.* 1986). This implies that oceanic faults remain active beneath the toe of the prism (Fig. 9.16).

Subduction of oceanic highs, like seamounts, first develops additional local compressional deformation at the toe when the seamount enters the subduction zone; then, it increases the erosional process when the high passes beneath the toe. The Kashima seamount, cut into two blocks by normal faults achieving nearly 1500 m of total offset (Fig. 9.17), and the Japan–Kuril trench junction are illustrative examples (Fig. 9.11) (Kobayashi *et al.* 1987, Lallemand *et al.* 1989, von Huene & Lallemand 1990).

Subduction versus collision: 'B' versus 'A' subduction

Statistically, when normal *oceanic subduction*, or *'B' subduction* (Bally 1981), is working, the compressional deformation, if any, is clearly limited to a narrow fore-arc zone (e.g. von Huene 1984, Hamilton 1988). By

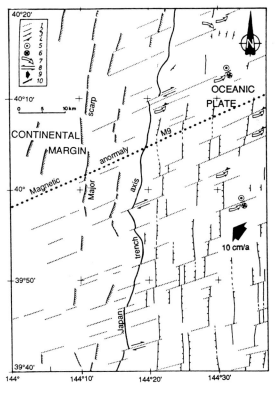

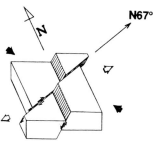

Fig. 9.16. Structural map of a part of the Japan trench, and schematic block diagram showing reactivation of old faults parallel to magnetic anomalies, new normal faults parallel to the trench, and their influence on the margin morphostructure. The block diagram shows how N–S trending normal faults are offset by reactivated vertical (or reverse) faults (after Lallemand et al. 1986). (1) Old fault (lineament) subparallel to magnetic anomalies, or structural grain. (2) Vertical fault. (3) New normal fault, triangles on the down-thrown side. (4) Apparent reverse fault. (5) Upfaulted block. (6) Downfaulted block. (7) Intersection point between N65°E lineaments and an E–W seismic profile. (8) Fault with at least an important right-lateral strike-slip motion component. (9) Plate convergence vector: Pacific plate motion relative to the Japanese plate (N55°W). (10) Topographic scarp of the margin, facing the trench axis.

contrast, subduction of buoyant lithosphere, or collision, extends dramatically the horizontal stress gradient and the compressionally deformed area. Spatial transition from subduction to collision is best illustrated by the Luzon and Sunda-Banda arcs, which collided with the Chinese margin in Taiwan (Fig. 9.18), and the Australian shelf along the Timor trough, respectively (e.g. Hamilton 1979, Suppe 1981, Karig et al. 1987). Continental subduction, or 'A' subduction (Bally 1981), is the most typical case of collision (see Chapter 13) and corresponds to the initial definition of

subduction (Amstutz 1951). The compressional deformation then extends to the back-arc area, and is reflected by *back-arc thrusting* (Fig. 9.19) and the creation of a *back-arc accretionary prism* (Silver & Reed 1988). Back-arc thrusting, east and west verging, is occurring along the eastern border of the Sea of Japan (Lallemand et al. 1985, Tamaki & Honza 1985, Tamaki 1988), although no buoyant feature is subducting at the Japan trench. However, the case is not typical because the Honshu arc is now dependent on the Okhotsk or North American plate (Nakamura 1983, Seno 1985).

Effect of large transverse faults and subducting ridges

Besides the tectonic effects of the newly formed and reactivated faults at the toe of the margin, subduction of major ridges is supposed to be responsible for *transverse faults*, segmenting the arc and allowing a specific Mg-rich volcanism (Andreieff et al. 1987). The arc may be dissected by linear depressions, corresponding to strike-slip faults, as at the Kerama gap or the Miyako depression, cutting the Ryukyu arc and offsetting left-laterally the structural zones of Japan (Charvet & Fabbri 1987, Kizaki 1978).

SOME PROBLEMS RELATED TO ARC-TRENCH TECTONICS

Formation of mélanges in subduction environments

Broken formations, chaotic rocks and mélanges with a pelitic matrix are very frequent in assumed fossil accretionary wedges (e.g. Suzuki 1986). Two main basic origins are advocated. The *olistostrome type* implies a sedimentary process of reworking of sedimentary and/or igneous blocks, due to gravity sliding or slumping, on the oceanic plate or at the trench, before subduction of the resulting mélange (also see Chapters 7 and 16). An actualistic model can be seen in the volcanic and sedimentary detritus at the foot of the Kashima seamount (Pautot et al. 1987) and the Sunda submarine slide (Moore et al. 1976). The *tectonic mixing* may involve several mechanisms during accretion: (1) *thickening of fault zones* (Moore & Byrne 1987) due to simple shearing at the base of the accreted sequence (Fig. 9.9), strain hardening, a fluid pressure drop, and fault orientation during stratal disruption; (2) *hydrofracturing* (see Chapter 3) and shearing leading to stratal disruption within the subducted sequence (Fisher & Byrne 1987, Byrne & Fisher 1990) (Fig. 9.9); and (3) *flow mélange* linked to mud diapirism and mud injections, related to fluid overpressure (Williams et al. 1984, Barber et al. 1986, 1989, Yamagata & Ogawa 1987, Pickering et al. 1988) (Figs 9.15 and 9.20).

Accretion versus collision: palaeo-accretionary prisms and palaeo-subduction complexes

Extending the present style of tectonics in arc-trench systems to the past was achieved by some authors by considering the structure of classical examples, such as

194   J. CHARVET and Y. OGAWA

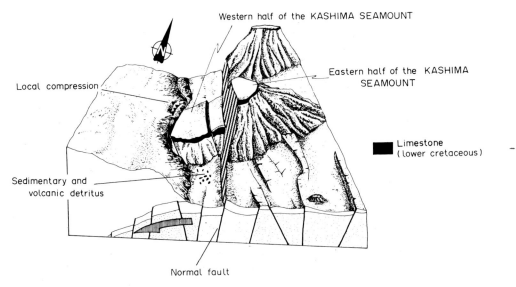

Fig. 9.17. Tectonic diagram of the Daiichi–Kashima seamount (after Cadet *et al.* 1987). Normal faulting of the seamount, local compressional deformation at the margin toe and volcanic debris in a sedimentary mélange are all shown.

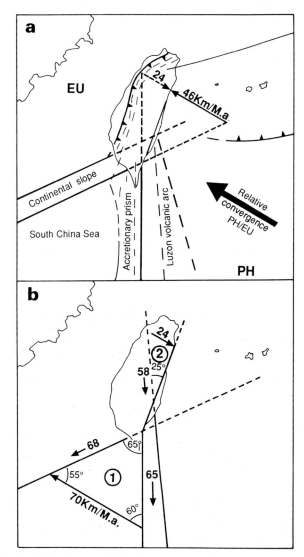

Fig. 9.18. From subduction to collision: the example of Taiwan. (a) Main units affected. Solid arrow: relative motion of collision boundary/Eurasia. Dashed arrow: relative motion of collision boundary/Philippine plate. (b) Velocity triangles. (1) PH/EU collision, (2) internal deformation of PH; all speeds in km/Ma. PH, Phillipine sea plate. EU, Eurasia plate (after Suppe 1981).

Japan or California, as simple *palaeo-accretionary wedges* or subduction complexes related to a steady landward subduction of oceanic crust since Palaeozoic time. This model corresponds to the *Pacific-type orogeny* model (Matsuda & Uyeda 1971). Recent *fold-and-thrust belts* parallel to a trench and bordering the coast, like the Cretaceous-Lower Tertiary Shimanto belt of SW Japan, are even more often simply regarded as the emerged part of an accretionary prism, still active at the present trench, such as the Nankai trough. However, for the former case (e.g. Charvet *et al.* 1990), as well as for the latter one (Shimanto), a careful study of: (1) structural style, timing and development; (2) facies and unconformities; (3) uplift; (4) post-folding magmatism; and (5) modelling, argues for a succession of subduction episodes ending with collision events during the building of these belts (Charvet & Fabbri 1987). Collision can occur with young and hot back-arc lithosphere (Hibbard & Karig 1990) or, more likely, with buoyant continent or arc-type lithosphere (Charvet & Fabbri 1987, Charvet *et al.* 1990). Several assumed palaeo-accretionary prisms may, alternatively, actually be polyphase collision-related belts (also see Chapter 13).

Geothermal gradient and metamorphism in arc systems

Sinking of a cold lithospheric slab into the mantle corresponds to a low geothermal gradient (Honda 1985); it is thus considered as a likely setting for *high-pressure–low-temperature metamorphism* (Ernst 1972, Miyashiro 1972). On the contrary, a high geothermal gradient below the volcanic arc may correspond to the location of *high-temperature–low-pressure metamorphism*. The assumption of contemporaneous development of these metamorphic rocks led to the concept of *paired metamorphic belts*, first using the example of Japan (Miyashiro 1973). In reality, respective low- and high-geothermal gradients are predicted by modelling

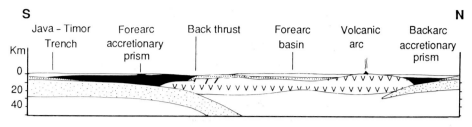

Fig. 9.19. Schematic cross-section of the Sunda arc (after Silver & Reed 1983), corresponding to a collisional stage.

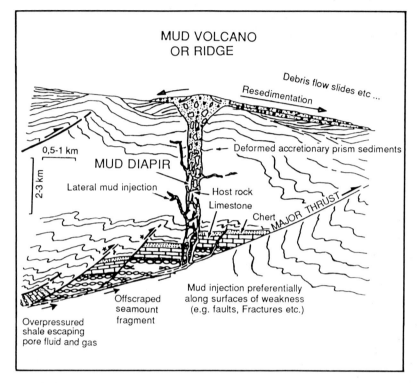

Fig. 9.20. Speculative model of wet sediment injection (diapirism) for genesis of muddy chaotic mélanges that contain mixed basalt – chert – limestone associations (after Pickering et al. 1988). Accretion of basalt, by offscraping, has not been observed.

and documented by heat flow distribution (e.g. Uyeda 1986). However, the structural development of these belts suggests that the high-temperature metamorphism post-dates, and may overprint, the high-pressure one. The metamorphic belts are geographically, but not chronologically or genetically, paired (e.g. Charvet et al. 1990).

REFERENCES

Alvarez, W. 1982. Geological evidence for the geographical pattern of mantle return flow and driving mechanism of plate tectonics. *J. geophys. Res.* **87**, 6697–6710.

Amstutz, A. 1951. Sur l'évolution des structures alpines. *Arch. Sci. Genève* **4**, 323–329.

Andreieff, P., Bouysse, P. & Westercamp. D. 1987. Géologie de l'arc insulaire des petites Antilles et évolution géodynamique de l'Est-Caraïbe. *Thèse Doc. Sci. Bordeaux* **921**, 359 p.

Aubouin, J., von Huene, R., Baltuck, M., Arnott, R., Bourgois, J., Filewicz, M., Helm, R., Kvenvolden, K., Lienert, B., McDonald, T., McDougall, K., Ogawa, Y., Taylor, E. & Winsborough, B. 1982a. Leg 84 of the Deep Sea Drilling Project subduction without accretion: Middle America Trench off Guatemala. *Nature* **297**, 458–460.

Aubouin, J., von Huene, R., Coulbourn, W. et al. 1982b. *Init. Rep. DSDP* **67**. U.S. Govt. Printing Office, Washington.

Aubouin, J., Bourgois, J. & Azéma, J. 1984. A new type of active margin: the convergent-extensional margin, as exemplified by the Middle America Trench off Guatemala. *Earth Planet. Sci. Lett.* **67**, 311–218.

Bally, A. W. 1981. Thoughts on the tectonics of folded belts. In: *Thrust and Nappe Tectonics* (edited by McClay, K. & Price, N. J.). *Spec. Publs geol. Soc. Lond.* **9**, 13–32.

Barber, A. J., Tjokrosapoetro, S. & Charlton, T. R. 1986. Mud volcanoes, shale diapirs, wrench faults and melanges in accretionary complexes, Eastern Indonesia. *Am. Ass. Petrol. Geol. Bull.* **70**, 1729–1741.

Barber, T. & Brown, K. 1988. Mud diapirism: the origin of melanges in accretionary complexes? *Geology Today* May–June 88, 89–94.

Beck, C., Blanc, G., Mascle, A., Moore, J. C., Taylor, E., Alvarez, F., Andreieff, P., Barnes, R., Behrmann, J., Brown, K., Clark, M., Dolan, J., Fisher, A., Gieskes, J., Hounslow, M., McLellan, P., Moran, K., Ogawa, Y., Sakai, T., Schoonmaker, J., Vrolijk, P., Wilkens, R. & Williams, C. 1988. Anatomie et physiologie d'un prisme d'accrétion: premiers résultats des forages du complexe de la ride de la Barbade, Leg O.D.P. 110. *Bull. Soc. geol. Fr.* **8**, 129–140.

Beloussov, V. V. & Ruditch, E. M. 1961. Island arcs in the development of the earth's structure (especially in the region of Japan and Sea of Okhotsk). *J. Geol.* **69**, 647–658.

Biju-Duval, B., Le Quellec, P., Mascle, A., Renard, V. & Valéry, P. 1982. Multibeam bathymetric survey and high resolution seismic investigations on the Barbados Ridge complex (eastern Caribbean): a key to the knowledge and interpretation of an accretionary wedge. *Tectonophysics* **86**, 275–304.

Biju-Duval, B., Moore, J. C. et al. 1984. *Init. Rep. DSDP* **78A**. U.S. Govt. Printing Office, Washington, 848 pp.

Boulègue, J., Benedetti, E. L., Dron, D., Mariotti, A. & Létolle, R. 1987. Geochemical and biogeochemical observations on the biological communities associated with fluid venting in Nankai Trough and Japan Trench subduction zones. *Earth Planet. Sci. Lett.* **83**, 343–355.

Bourgois, J., Pautot, G., Bandy, W., Boinet, T., Chotin, P., Huchon, P., Mercier de Lépinay, B., Monge, F., Monlau, J., Pelletier, B., Sosson, M. & von Huene, R. 1988. Sea Beam and seismic-reflection imaging of the tectonic regime of the Andean continental margin off Peru (4°S to 10°S). *Earth Planet. Sci. Lett.* **87**, 111–126.

Bott, M. H. P. Waghorn, G. D. & Whittaker, A. 1989. Plate boundary forces at subduction zones and trench-arc compression. *Tectonophysics* **170**, 1–15.

Brown, K. M. & Westbrook, G. K. 1988. Mud diapirism and subcretion in the Barbados Ridge accretionary complex: the role of fluids in accretionary processes. *Tectonics* **7**, 613–640.

Byrne, T. 1986. Eocene underplating along the Kodiak shelf, Alaska: implications and regional correlations. *Tectonics* **5**, 403–421.

Byrne, T. & Fisher, D. 1990. Evidence for a weak and overpressured décollement beneath sediment-dominated accretionary prisms. *J. geophys. Res.* **95**, 9081–9097.

Cadet, J. P., Kobayashi, K., Lallemand, S., Jolivet, L., Aubouin, J., Boulègue, J., Dubois, J., Hotta, H., Ishii, T., Konishi, K., Niitsuma, N. & Shimamura, H. 1987. Deep scientific dives in the Japan and Kuril Trenches. *Earth Planet. Sci. Lett.* **83**, 313–328.

Carlson, R. L. 1981. Boundary forces and plate tectonics. *Geophys. Res. Lett.* **8**, 958–961.

Carlson, R. L. & Melia, P. J. 1984. Subduction hinge migration. *Tectonophysics* **102**, 399–411.

Carson, B., von Huene, R. & Arthur, M. 1982. Small-scale deformation structures and physical properties related to convergence in Japan Trench slope sediments. *Tectonics* **1**, 277–302.

Charvet, J. & Fabbri, O. 1987. Vue générale sur l'orogenèse Shimanto et l'évolution tertiaire du Japon sud-ouest. *Bull. Soc. geol. Fr.* **8**, 1171–1188.

Charvet, J., Faure, M., Fabbri, O., Cluzel, D. & Lapierre, H. 1990. Accretion and collision during east-Asiatic margin building. A new insight on the peri-Pacific orogenies. In: *Terrane Analysis of China and the Pacific Rim* (edited by Wiley, T. J., Howell, D. G. & Wong, F. L.). *Circum-Pacific Council for Energy and Mineral Resources Earth Sciences Series* **13**, 161–185. Houston, Texas.

Chase, C. G. 1978. Extension behind island arcs and motions relative to hot spots. *J. geophys. Res.* **83**, 5385–5387.

Dahlen, F. A. 1981. Isostasy and the ambient state of stress in the oceanic lithosphere. *J. geophys. Res.* **86**, 7801–7807.

Davis, D. M., Suppe, J. & Dahlen, F. A. 1983. Mechanics of fold-and-thrust belts and accretionary wedges. *J. geophys. Res.* **88**, 1153–1172.

Dewey, J. F. 1980. Episodicity, sequence and style at convergent plate boundaries. *Geol. Ass. Can. Spec. Pap.* **20**, 553–573.

Dickinson, W. R. & Seely, D. R. 1979. Structure and stratigraphy of forearc regions. *Am. Ass. Petrol. Geol. Bull.* **63**, 31.

Elsasser, W. M. 1971. Sea-floor spreading as thermal convection. *J. geophys. Res.* **76**, 1101–1112.

Ernst, W. G. 1971. Metamorphic zonations on presumably subducted lithospheric plates from Japan, California and the Alps. *Contrib. Mineral. Petrol.* **34**, 43–59.

Ernst, W. G. 1972. Occurrence and mineralogic evolution of blueschist belts with time. *Am. J. Sci.* **272**, 657–668.

Fabbri, O., Charvet, J. & Faure, M. 1987. Phase ductile à vergence nord dans la zone Shimanto de Kyushu (Japon SW). *C.R. Acad. Sci. Paris* **304**, 923–927.

Fabbri, O., Faure, M. & Charvet, J. 1990. Back-thrusting in accretionary prisms: microtectonic evidence from the Cretaceous–Tertiary Shimanto Belt of Southern Japan. *J. Southeast Asian Earth Sci.* **4**, 195–201.

Faure, M. & Charvet, M. 1991. L'ouverture de mers marginales périasiatiques est-elle initiée par amincissement d'une croûte préalablement épaissie? *C.R. Acad. Sci. Paris* **313**, 1367–1372.

Fisher, D. & Byrne. T. 1987. Structural evolution of underthrusted sediments: evidence from Kodiak Island, Alaska. *Tectonics* **6**, 775–793.

Forsyth, D. & Uyeda, S. 1975. On the relative importance of the driving forces of plate motion. *Geophys. J. R. astr. Soc.* **43**, 163–200.

Froidevaux, C., Uyeda, S. & Uyeshima, M. 1988. Island arc tectonics. *Tectonophysics* **148**, 1–19.

Fryer, P., Pearce, J. A., Stokking, L. B. et al. 1990a. *Proc. ODP Init. Rep.* **125**. Ocean Drilling Program, College Station, TX, 1092 pp.

Fryer, P., Saboda, K. L., Johnson, L. E., Mackay, M. E., Moore, G. F. & Stoffers, P. 1990b. Conical seamount. Sea MARC II, *Alvin* submersible and seismic-reflection studies. In: Fryer, P., Pearce, J. C., Stokking, C. B. et al. *Proc. O.D.P. Init. Rep.* **125**. Ocean Drilling Program, College Station, TX, 69–80.

Fukao, Y. 1979. Tsunami earthquakes and subduction processes near deep sea trenches. *J. geophys. Res.* **84**, 2303–2314.

Geological Survey of Japan, 1982. Geological Atlas of Japan, 119 pp.

Hamilton, W. P. 1979. Tectonics of the Indonesian region. *U.S. Geol. Surv. Prof. Pap.* **1078**, 1–345.

Hamilton, W. B. 1988. Plate tectonics in island arcs. *Bull. geol. Soc. Am.* **100**, 1503–1527.

Hibbard, J. & Karig, D. E. 1987. Sheath-like folds and progressive fold deformation in Tertiary sedimentary rocks of the Shimanto accretionary complex, Japan. *J. Struct. Geol.* **9**, 845–857.

Hibbard, J. & Karig, D. E. 1990. Alternative plate model for the early Miocene evolution of the southwest Japan margin. *Geology* **18**, 170–174.

Hilde, T. W. C. 1983. Sediment subduction versus accretion around the Pacific. *Tectonophysics* **99**, 381–397.

Hilde, T. W. C. & Lee, C. S. 1984. Origin and evolution of the West Philippine basin: a new interpretation. In: *Geodynamics of Back-arc Regions* (edited by Carson, R. L. & Kobayashi, K.). *Tectonophysics* **102**, 85–104.

Honda, S. 1985. Thermal structure beneath Tohoku, Northeast Japan — a case study for understanding the detailed thermal structure of the subduction zone. *Tectonophysics* **112**, 69–102.

Hussong, D. M. & Fryer, P. 1985. Fore-arc tectonics in the northern Mariana arc. In: *Formation of Active Ocean Margins* (edited by Nasu, N. et al.). Terrapub, Tokyo, 273–290.

Hynes, A. & Mott, J. 1985. On the causes of back-arc spreading. *Geology* **13**, 387–389.

Jolivet, L., Huchon, P. & Rangin, C. 1989. Tectonic setting of western Pacific marginal basins. *Tectonophysics* **160**, 23–48.

Jurdy, D. M. & Stefanick, M. 1983. Flow models for back-arc spreading. *Tectonophysics* **99**, 191–206.

Kagami, H. 1986. The accretionary prism of the Nankai Trough off Shikoku, southwestern Japan. In: Kagami, H., Karig, D. E., Coulbourn, W. T. *et al*. *Init. Rep. DSDP* **87**, U.S. Govt. Printing Office, Washington, 941–953.

Karig, D. E. 1971. Origin and development of marginal basins in the western Pacific. *J. geophys. Res.* **76**, 2542–2561.

Karig, D. E. 1972. Remnant arcs. *Bull. geol. Soc. Am.* **83**, 1057–1068.

Karig, D. E. 1974. Evolution of arc systems in the western Pacific. *A. Rev. Earth Planet. Sci.* **2**, 51–75.

Karig, D. E. 1982. Deformation in the forearc: implications for mountain belts. In: *Mountain Building Processes* (edited by Hsü, K. J.). Academic Press, London, 59–71.

Karig, D. E. 1985. Kinematics and mechanics of deformation across accreting forearcs. In: *Formation of Active Ocean Margins* (edited by Nasu, N. *et al*.). Terrapub, Tokyo, 155–178.

Karig, D. E. & Sharman, G. F. 1975. Subduction and accretion in trenches. *Bull. geol. Soc. Am.* **89**, 265–276.

Karig, D. E., Lawrence, M. B., Moore, G. F. & Curray, J. 1980. Structural framework of the fore-arc basin, NW Sumatra. *J. geol. Soc. Lond.* **137**, 77–91.

Karig, D. E., Barber, A. J., Charlton, T. R., Klemperer, S. & Hussong, D. M. 1987. Nature and distribution of deformation across the Banda Arc-Australian collision zone in Timor. *Bull. geol. Soc. Am.* **98**, 18–32.

Karig, D. E., Moran, K. & Leg 131 Scientific Party. 1990. A dynamically sealed decollement: Nankai prism. In: *Fluids in Subduction Zones, Abstracts Volume* (edited by Cadet, J. P. & Le Pichon, X.). Nov. 1990, Paris.

Kizaki, K. 1978. Tectonics of the Ryukyu island arc. *J. Phys. Earth* **26**, 301–307 (Suppl.).

Knipe, R. J. 1986. Microstructural evolution of vein arrays preserved in Deep Sea Drilling Project cores from the Japan Trench, Leg 57. In: *Structural Fabrics in Deep Sea Drilling Project Cores from Forearcs* (edited by Moore, J. C.). *Mem. geol. Soc. Am.* **166**, 75–88.

Kobayashi, K., Cadet, J. P., Aubouin, J., Boulègue, J., Dubois, J., von Huene, R., Jolivet, L., Kanazawa, T., Kasahara, J., Koizumi, K., Lallemand, S., Nakamura, Y., Pautot, G., Suyehiro, K., Tani, S., Tokuyama, S. & Yamazaki, T. 1987. Normal faulting of the Daiichi-Kashima seamount in the Japan Trench revealed by the Kaiko I cruise, Leg 3. *Earth Planet. Sci. Lett.* **83**, 257–266.

Kulm, L. D., Suess, E., Moore, J. C., Lewis, B. T., Ritger, S. D., Kadko, D. C., Thornburg, T. M., Embley, R. W., Rugh, W. D., Massoth, G. J., Langseth, M. G., Cochrane, G. R. & Scamman, R. L. 1986. Oregon subduction zone: venting, fauna and carbonates. *Science* **231**, 561–566.

Lagabrielle, Y., Karpoff, A. M., & Cotten, J. 1992. Mineralogical and geochemical analyses of sedimentary serpentinites from Conical Seamount (Hole 778A): implications for the evolution of serpentinite seamounts. *Proc. ODP Sci. Res.* **125**, Ocean Drilling Program, College Station, TX, 325–342.

Lallemand, S. & Jolivet, L. 1986. Japan Sea, a pull-apart basin? *Earth Planet. Sci. Lett.* **76**, 375–389.

Lallemand, S. & Le Pichon, X. 1987. Coulomb wedge model applied to subduction of seamounts in the Japan Trench. *Geology* **15**, 1065–1069.

Lallemand, S., Okada, H., Otsuka, K. & Labeyrie, L. 1985. Tectonique en compression sur la marge est de la mer du Japon: mise en évidence de chevauchements à vergence orientale. *C.R. Acad. Sci. Paris* **301**, 201–206.

Lallemand, S., Cadet, J. P. & Jolivet, L. 1986. Mécanisme de tectogenèse à la base du mur interne de la fosse du Japon (au large de Sanriku, Japon NE) rejeu des failles océaniques sous la marge. *C.R. Acad. Sci. Paris* **302**, 319–324.

Lallemand, S., Culotta, R. & von Huene, R. 1989. Subduction of the Daiichi-Kashima seamount in the Japan Trench. *Tectonophysics* **160**, 231–247.

Le Pichon, X., Lyberis, N., Angelier, J. & Renard, V. 1982. Strain distribution over the east Mediterranean ridge, a synthesis incorporating new sea-beam data. *Tectonophysics* **86**, 243–274.

Le Pichon, X., Iiyama, T., Boulègue, J., Charvet, J., Faure, M., Kano, K., Lallemant, S., Okada, H., Rangin, C., Taira, A., Urabe, T. & Uyeda, S. 1987. Nankai Trough and Zenisu Ridge: a deep-sea submersible survey. *Earth Planet. Sci. Lett.* **83**, 285–299.

Le Pichon, X., Iiyama, T., Chamley, H., Charvet, J., Faure, M., Fujimoto, H., Furuta, T., Ida, Y., Kagami, H., Lallemant, S., Leggett, J., Murata, A., Okada, H., Rangin, C., Renard, V., Taira, A. & Tokuyama, H. 1987a. Nankai Trough and the fossil Shikoku Ridge: results of Box 6 Kaiko survey. *Earth Planet. Sci. Lett.* **83**, 186–198.

Le Pichon, X., Iiyama, T., Chamley, H., Charvet, J., Faure, M., Fujimoto, H., Furuta, T., Ida, Y., Kagami, H., Lallemant, S., Leggett, J., Murata, A., Okada, H., Rangin, C., Renard, V., Taira, A. & Tokuyama, H. 1987b. The eastern and western ends of Nankai Trough: results of Box 5 and Box 7 Kaiko survey. *Earth Planet. Sci. Lett.* **83**, 199–213.

Letouzey, J. & Kimura, M. 1985. Okinawa Trough genesis: structure and evolution of a back-arc basin developed in a continent. *Mar. Petrol. Geol.* **2**, 111–130.

Lindsley-Griffin, N., Kemp, A. & Swartz, F. 1990. Vein structures of the Peru margin, Leg 112. In: Suess, E., von Huene, R. *et al*., *Proc. ODP, Sci. Results* **112**, Ocean Drilling Program, College Station, TX, 3–16.

Lister, C. R. B. 1975. Gravitational drive on oceanic plates caused by thermal contraction. *Nature* **257**, 663–665.

Lucas, S. E. & Moore, J. C. 1986. Cataclastic deformation in accretionary wedges: Deep Sea Drilling Project Leg 66, southern Mexico, and onland examples from Barbados and Kodiak Islands. In: *Structural Fabrics in Deep Sea Drilling Project Cores from Forearcs* (edited by Moore, J. C.). *Mem. geol. Soc. Am.* **166**, 89–103.

Lundberg, N. & Moore, J. C. 1986. Macroscopic structural features in Deep Sea Drilling Project cores from forearc regions. In: *Structural Fabrics in Deep Sea Drilling Project Cores from Forearcs* (edited by Moore, J. C.). *Mem. geol. Soc. Am.* **166**, 13–44.

Malavieille, J. 1984. Modélisation expérimentale des chevauchements imbriqués: application aux chaînes de montagnes. *Bull. Soc. geol. Fr.* **7**, 129–138.

Mascle, A., Cazes, M. & Le Quellec, P. 1985. Structures des marges et bassins caraïbes, une revue. In: *Symposium Géodynamique des Caraïbes* (edited by Mascle, A.). Ed. Technip, Paris, 1–20.

Mascle, A., Biju-Duval, B., de Clarens, P. & Munsch, H. 1986. Growth of accretionary prisms: tectonic processes from Caribbean examples. In: *The Origin of Arcs* (edited by Wezel, F. C.). Elsevier, Amsterdam, 375–400.

Matsuda, T. & Uyeda, S. 1971. On the Pacific type orogeny and its model extension of the paired belts concept and possible origin of marginal seas. *Tectonophysics* **11**, 5–27.

Miyashiro, A. 1972. Metamorphism and related magmatism in plate tectonics. *Am. J. Sci.* **272**, 629–656.

Miyashiro, A. 1973. Paired and unpaired metamorphic belts. *Tectonophysics* **17**, 241–254.

Miyashiro, A. 1986. Hot regions and the origin of marginal basins in the Western Pacific. *Tectonophysics* **122**, 195–216.

Moberly, R., Shepard, G. L. & Coulbourn, W. C. 1982. Forearc and other basins, continental margin of northern and southern Peru and adjacent Ecuador and Chile. In:

Trench-Forearc Geology (edited by Legett, J. K.). *Spec. Publs geol. Soc. Lond.* **10**, 171–189.

Molnar, P. & Atwater, T. 1978. Interarc spreading and cordilleran tectonics as alternates related to the age of subducted oceanic lithosphere. *Earth Planet. Sci. Lett.* **41**, 330–340.

Molnar, P. & Tapponnier, P. 1975. Cenozoic tectonics of Asia: effects of a continental collision. *Science* **189**, 419–426.

Moore, D. G., Curray, J. R. & Emmel, F. J. 1976. Large submarine slide (olistostrome) associated with Sunda Arc subduction zone, northeast Indian ocean. *Mar. Geol.* **21**, 211–226.

Moore, J. C. & Byrne, T. 1987. Thickening of fault zones: a mechanism of melange formation in accreting sediments. *Geology* **15**, 1040–1043.

Moore, J. C., Biju-Duval, B. *et al.* 1983. Offscraping and underthrusting of sediment at the deformation front of the Barbados Ridge: results from Leg 78A, DSDP. *Bull. geol. Soc. Am.* **93**, 1065–1077.

Moore, J. C., Roeske, S., Lundberg, N., Schoonmaker, J., Cowan, D. S., Gonzales, E. & Lucas, S. E. 1986. Scaly fabrics from Deep Sea Drilling Project cores from forearcs. In: *Structural Fabrics in Deep Sea Drilling Project Cores from Forearcs* (edited by Moore, J. C.). *Mem. geol. Soc. Am.* **166**, 55–73.

Moore, J. C., Mascle, A. & ODP Leg 110 Scientific Party. 1987. Expulsion of fluids from depth along a subduction zone decollement horizon. *Nature* **326**, 785–788.

Nakamura, K. 1983. Possible nascent trench along the eastern Japan Sea as the convergent boundary between Eurasian and North American plates. *Bull. Earthq. Res. Inst. Univ. Tokyo* **58**, 711–722.

Nakamura, K. & Uyeda, S. 1980. Stress gradient in arc-backarc regions and plate subduction. *J. geophys. Res.* **85**, 6419–6428.

Nakamura, K., Shimazaki, K. & Yonekura, N. 1984. Subduction, bending and eduction. Present and Quaternary tectonics of the northern border of the Philippine Sea plate. *Bull. Soc. geol. Fr.* **7**, 221–253.

Otsuki, K. & Ehiro, M. 1978. Major strike-slip faults and their bearing on spreading in the Japan Sea. *J. Phys. Earth* **26**, 537–555 (Suppl.).

Pautot, G., Nakamura, K., Huchon, P., Angelier, J., Bourgois, J., Fujioka, K., Kanazawa, K., Nakamura, Y., Ogawa, Y., Séguret, M. & Takeuchi, A. 1987. Deep-sea submersible survey in the Suruga, Sagami and Japan trenches, preliminary results of the 1985 Kaiko cruise, Leg 2. *Earth Planet. Sci. Lett.* **83**, 300–312.

Pickering, K. T., Agar, S. M. & Ogawa, Y. 1988. Genesis and deformation of mud injections containing chaotic basalt-limestone-chert associations: examples from the southwest Japan forearc. *Geology* **16**, 881–885.

Platt, J. P., Leggett, J. K., Young, J., Raza, H. & Alam, S. 1985. Large-scale sediment underplating in the Makran accretionary prism, southwest Pakistan. *Geology* **13**, 507–511.

Sample, J. C. & Fisher, D. M. 1986. Duplex accretion and underplating in an ancient accretionary complex, Kodiak Islands, Alaska. *Geology* **14**, 160–163.

Scholl, D., von Huene, R., Vallier, T. & Howell, D. 1980. Sedimentary masses and concepts about tectonic processes at underthrust ocean margins. *Geology* **8**, 564–568.

Seely, D. R. 1977. The significance of landward vergence and oblique structural trends on trench inner slopes. In: *Island Arcs, Deep Sea Trenches and Back-arc Basins* (edited by Talwani, M. & Pitman, W.C. III). *Maurice Ewing Series* **1**, A.G.U., 187–198.

Seno, T. 1985. Is northern Honshu a microplate? *Tectonophysics* **115**, 177–196.

Shipboard Scientific Party. 1980. Sites 438 and 439, Japan deep-sea terrace. In: von Huene, R., Nasu, N. *et al., Init. Rep. DSDP* **56–57**. U.S. Govt. Printing Office, Washington, Part 1, 23–191.

Sibuet, J. C., Letouzey, J., Barbier, F., Charvet, J., Foucher, J. P., Hilde, T. W. C., Kimura, M., Chiao, L. Y., Marsset, B., Muller, C. & Stephan, J. F. 1987. Back arc extension in the Okinawa Trough. *J. geophys. Res.* **92**, 14041–14063.

Silver, E. A. 1972. Pleistocene tectonic accretion of the continental slope off Washington. *Mar Geol.* **13**, 239–249.

Silver, E. A. & Reed, D. L. 1988. Backthrusting in accretionary wedges. *J. geophys. Res.* **93**, 3116–3126.

Suess, E., Carson, B., Ritger, S. D., Moore, J. C., Jones, M. J., Kulm, L. D. & Cochrane, G. R. 1986. Biological communities at vent sites along the subduction zone off Oregon. *Bull. biol. Soc. Wash.* **6**, 475–484.

Suppe, J. 1981. Mechanics of mountain building and metamorphism in Taiwan. *Mem. geol. Soc. China* **4**, 67–89.

Suzuki, T. 1986. Melange problems of convergent plate margins in the circum-Pacific regions. *Mem. Fac. Sci. Kochi Univ. Ser. E Geol.* **7**, 23–48.

Tamaki, K. 1988. Geological structure of the Japan Sea and its tectonic implications. *Bull. geol. Surv. Jap.* **39**, 269–365.

Tamaki, K. & Honza, E. 1985. Incipient subduction and obduction along the eastern margin of the Japan Sea. *Tectonophysics* **119**, 381–406.

Tamaki, K., Pisciotto, K. A., Allan, J. *et al.* 1990. *Proc. ODP Init. Rep.* **127**. Ocean Drilling Program, College Station, TX, 844 pp.

Tapponnier, P., Peltzer, G., Le Dain, A. Y., Armijo, R. & Cobbold, P. 1982. Propagating extrusion tectonics in Asia: new insights from simple experiments with plasticine. *Geology* **10**, 611–616.

Tatsumi, Y., Otofuji, Y. I., Matsuda, T. & Nohda, S. 1989. Opening of the Sea of Japan back-arc basin by asthenospheric injection. *Tectonophyics* **166**, 317–329.

Taylor, B., Fujioka, K. *et al.* 1990. *Proc. ODP Init. Rep.* **126**. Ocean Drilling Program, College Station, TX, 1002 pp.

Toksöz, M. N. & Bird, P. 1977. Formation and evolution of marginal basins and continental plateaus. In: *Island Arcs, Deep Sea Trenches and Back-arc Basins* (edited by Talwani, M. & Pitman, W. C. III). *Maurice Ewing Series* **1**, A.G.U., 379–393.

Uyeda, S. 1982. Subduction zones: an introduction to comparative subductology. *Tectonophysics* **81**, 133–159.

Uyeda, S. 1986. Facts, ideas, and open problems on trench-arc-backarc systems. In: *The Origin of Arcs* (edited by Wezel, F.C.). Elsevier, Amsterdam, 435–460.

Uyeda, S. & Kanamori, K. 1979. Back-arc opening and the mode of subduction. *J. geophys. Res.* **84**, 1049–1061.

Viallon, C., Huchon, P. & Barrier, E. 1986. Opening of the Okinawa basin and collision in Taiwan. A retreating trench model with lateral anchoring. *Earth Planet. Sci. Lett.* **80**, 145–155.

von Huene, R. 1984. Observed strain and the stress gradient across some forearc areas of modern convergent margins. In: *Origin and History of Marginal and Inland Seas. Proc. 27th Int. geol. Congr.* **23**, 155–188.

von Huene, R. & Culotta, R. 1989. Tectonic erosion at the front of the Japan Trench convergent margin. *Tectonophysics* **160**, 75–90.

von Huene, R. & Lallemand, S. 1990. Tectonic erosion along the Japan and Peru convergent margins. *Bull. geol. Soc. Am.* **102**, 704–720.

von Huene, R., Langseth, M., Nasu, N. & Okada, H. 1982. A summary of Cenozoic tectonic history along the IPOD Japan Trench transect. *Bull. geol. Soc. Am.* **93**, 829–846.

von Huene, R., Suess, E. & Leg 112 Shipboard Scientists. 1988. Results of Leg 112 drilling, Peru continental margin, Part 1, tectonic history. *Geology* **16**, 934–938.

Westbrook, G. K. 1982. The Barbados Ridge complex. Tectonics of a mature forearc system. In: *Trench and Forearc Geology* (edited by Leggett, J. K.). *Spec. Publs geol. Soc. Lond.* **10**, 275–290.

Westbrook, G. K. & Smith, M. J. 1983. Long decollements and mud volcanoes: evidence from the Barbados Ridge complex for the role of high pore-fluid pressures in the development of an accretionary complex. *Geology* **11**, 279–283.

Westbrook, G. K., Ladd, J. W., Buhl, P., Bangs, N. & Tiley, G. J. 1988. Cross section of an accretionary wedge: Barbados Ridge complex. *Geology* **16**, 631–635.

Williams, P. R., Pigram, L. J. & Dow, D. B. 1984. Melange production and the importance of shale diapirism in accretionary terrains. *Nature* **309**, 145–146.

Yamagata, T. & Ogawa, Y. 1989. Role of mud diapirism for the formation of chaotic rocks. *J. geol. Soc. Jap.* **95**, 297–310.

Yoshii, T. 1979. A detailed cross-section of the deep seismic zone beneath northeastern Honshu, Japan. *Tectonophysics* **55**, 349–360.

CHAPTER 10

Craton Tectonics, Stress and Seismicity

R. G. PARK and W. JAROSZEWSKI

CRATONS, STABILITY AND INTRAPLATE TECTONIC REGIMES

WHAT is a craton? A *craton* may be defined as the relatively stable part of a continent, or the interior of a continental plate. Prior to the development of plate tectonic theory, the territory of the present-day continents was regarded as divisable, for any given geological period, into cratons and mobile or orogenic belts. In plate tectonic terms, the craton represents that part of a continental plate not affected by contemporaneous plate boundary activity. The term is more frequently used for Precambrian regions; for example, the 'Tanzanian craton', the 'Superior craton', etc. A piece of crust that forms part of a mobile belt in one period may become part of a craton in a subsequent period, after stabilization or *cratonization* has occurred; that is, when the tectono-thermal activity characteristic of the mobile belt has ceased and the area has become stable.

The definition of craton embodies the concept of relative tectonic stability. A *stable tectonic zone* implies a lack of tectonic activity. However the term is purely relative, since all parts of the crust undergo more or less continuous activity on a geological time scale, particularly in terms of vertical movement. What distinguishes 'stable' from 'unstable' tectonic zones is, essentially, their comparatively slow rate of movement over the time interval in question.

A *shield* is that part of a craton where crystalline basement is exposed. Such areas must have experienced net uplift. The related term *platform* has been used to represent that part of the craton where sedimentary cover is exposed. Thus for much of the Phanerozoic, the European craton could be said to contain the Baltic shield on its northwest side and the Russian platform to the southeast. However, in the classical Russian tradition, 'platform', or more strictly 'continental platform' involves two parts, a strongly folded, generally crystalline basement and a mostly flat-lying, although locally deformed cover.

Like the term craton, both shield and platform (and indeed basement and cover) are age-specific: yesterday's orogen may become today's shield or platform. Moreover, the boundaries of shield and platform are continually changing: for example the Baltic shield as presently exposed was covered by marine sediments in the Silurian, and was part of the platform. The *foreland* is the zone of undeformed craton bordering an orogenic belt, and towards which overfolds and thrusts are directed (see Chapter 13). For example, the Alpine foreland during the Mesozoic and Cenozoic is the relatively stable part of the European plate immediately west and north of the Alpine front. Forelands typically exhibit deep linear basins that are genetically related to the adjoining orogenic belts. These are termed *foreland basins*. Although in one sense part of the craton, foreland basins are clearly related to plate-boundary processes. Sengör (1984) has suggested distinguishing between a *foreland* and a *hinterland* on the basis of whether a terrain was, or is, part of a subducting or overriding plate, respectively.

Cratons thus represent the continental parts of plate interiors, or *continental intraplate regions*. According to classical plate tectonic theory, plates are essentially stable internally, and tectonic effects are concentrated at their boundaries. However, it has always been recognized that this was true only to a first approximation, and that all regions of the Earth's surface experience tectonic effects to some degree. The most common type of intraplate tectonic activity is vertical movements leading to uplift and depression of parts of the continental surface with respect to its mean level. Accurate geodetic measurements have shown that most parts of the continental crust are experiencing slow uplift or depression with respect to present-day mean sea level. However, the measured rates of such movements are typically at least an order of magnitude slower than those at plate boundaries. Lateral variations in these movements create a system of depressed and uplifted areas (basins and uplifts) which are the characteristic structures of cratons.

A widespread type of intraplate tectonic activity is *strike-slip movement* — probably the main mechanism for the release of intraplate horizontal stress. Such movements take place along faults and shear zones, and some attain rates of displacement comparable with those at plate margins, for instance certain strike-slip faults in Turkey and Central Asia. The frequency and distribution of intraplate earthquakes is another indication of widespread tectonic activity, albeit to a lesser degree than at plate boundaries.

CRATONIC STRUCTURES

Types of cratonic structure

Any piece of cratonic crust is composed of rocks and structures ranging in age from its date of initial formation up to the present. In terms of craton tectonics, however, we are concerned only with those structures formed since the time of cratonization. The most widespread cratonic structures are *basins* and *uplifts* (or *swells, arches*), and most of the surface area of a craton can be divided into one or other of these two basic types. Other important types of structure are extensional *rift systems* (see Chapter 11) and other linear fault zones, many of which have a long and complex history of movement, and constitute significant topographic features. Such structures are often referred to as *crustal lineaments*. Localization of deformation of younger cover strata above a basement lineament is a characteristic feature of cratonic regions and accounts for many examples of localized folding. Some lineaments represent old crustal sutures which mark the site of former collision zones between two continental plates or terranes.

A much less common type of structure is the compressional fold/thrust belt. For reasons that will be discussed below, extensional structures far exceed compressional in importance in the intraplate regime.

Basins

The term *basin* is used geologically in two different senses. In a strictly structural sense, it describes a synformal structure with a near-circular outcrop, and whose limb dips are radial. In the present context, the term is used to describe a topographic depression that either contains sediments or is capable of receiving them. The term thus has a topographic/sedimentological or stratigraphic connotation. Such basins may be either marine or continental, or may change from one to the other through time, and exhibit a wide variation in size and shape. Most basins are tectonic in origin, and are produced by active depression of the surface of the crust by gravitational or extensional forces. Numerous types of basins have been distinguished, which can be broadly subdivided into those formed on oceanic or continental crust, respectively. A further important

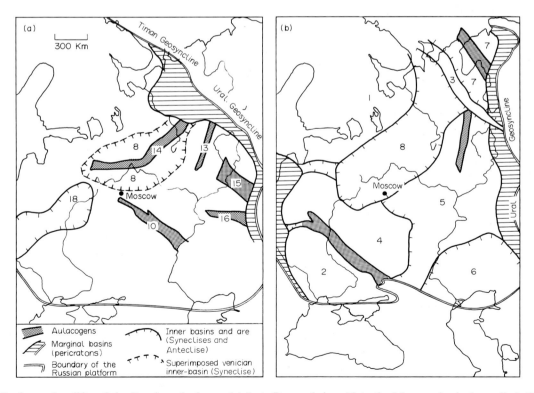

Fig. 10.1. Basins and uplifts of the Russian platform. (a) Late Precambrian. Note the Moscow basin (syneclise) (8) and the Orshansk basin (18); the Moscow basin is situated over the mid-Russian aulacogen (14) which is one of a set of late Precambrian aulacogens which project into the craton from its margin. (b) Upper Palaeozoic. Note the much enlarged Moscow basin, the Peri-Caspian syneclise (6) and the large uplift comprising the Ukraine (2) and Voronezh (4) anteclises separated by the Dnieper–Donetz aulacogen (11). After Aleinikov *et al.* (1980).

subdivision may be made into basins of intracratonic (or intraplate) origin and those whose origin is directly atttributable to processes associated with plate boundaries.

On the Russian platform, for example (Fig. 10.1), major tectonic basins can be recognized which are characterized by continued depression over periods of 100–1000 Ma. These basins (termed *syneclises* in the Russian literature) are large features with dimensions of the order of 500–1000 km across, and represent depressions in the basement of, in some cases, over 5 km depth. They are therefore major crustal structures and form, with the associated uplifts (or *anteclises*), the most important and widespread type of cratonic structure. Over a long time-scale, the vertical movements responsible for these structures constitute a significant departure from intraplate stability.

The term *aulacogen* (Shatsky 1955) is used, particularly by Eastern European workers, for a special category of linear, trough-like basins. Such structures form an intermediate stage between rifts and basins, in the sense that many basins appear to evolve out of rifts through an aulacogen stage into much larger basins centered over the aulacogen. For example, the Moscow basin (Fig. 10.1) is situated over the earlier mid-Russian aulacogen. We shall now describe briefly several well-studied examples of basins from different continents and with varied tectonic settings: the Paris basin in Western Europe, the Michigan basin in the northern U.S.A., the Taoudeni basin in West Africa and the currently active North Sea basin.

The Paris basin, situated in northern France, is about 600 km in diameter and is one of the best-defined basin structures in Europe (see Megnien & Pomerol 1980). It rests on Palaeozoic basement to the west (Massif Armoricain), south (Massif Central) and east (Ardennes massif), but extends to the north into the English Channel, and northeastwards towards the North Sea. The basin probably originated in the Late Triassic but existed as a well-defined structure only during the Jurassic and Cretaceous, during which time a total maximum thickness of 2900 m of sediments accumulated. However the area of maximum subsidence migrated southwards during this period, so that the total thickness at any one point is rather less than this. Megnien & Pomerol (1980) demonstrated, from a study of the variation of cumulative thickness with time, a fairly steady rate of accumulation of sediments of around 0.2 mm/yr. They attributed the growth of the basin over this period to flexure of the crust induced by sediment loading, but do not comment on the mechanism of initiation.

The Michigan basin (Fig. 10.2) has been intensively studied by means of numerous boreholes, and by gravity and magnetic surveys (see Walcott 1970). The basin was formed in mid-Ordovician times and lasted at least until the late Carboniferous. The basal sediments rest unconformably on Precambrian basement. Structural contours on the basement surface indicate an unusually regular, almost perfectly circular,

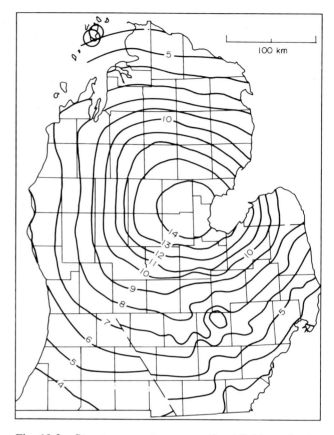

Fig. 10.2. Structure contours on the Michigan basin (1000 ft intervals) showing its near-circular plan. Note the steepening dip towards the centre. After Sleep & Sloss (1980).

shape with a gradual increase in depth towards a well-defined central point. Most Palaeozoic units thicken towards the centre of the basin, and facies variations indicate consistently deeper-water conditions there. It can be concluded that subsidence has continued throughout most of the history of the basin, and that the depression was achieved by downward flexure rather than by faulting. The total thickness in the centre is about 3 km, from which an average rate of subsidence of 0.3 mm/yr is obtained — very similar to that of the Paris basin.

The Taoudeni basin (Fig. 10.3) is one of the most prominent structures of the African craton (see Bronner *et al.* 1980). It lies in the western Sahara region of West Africa, mainly in Mauritania and Mali, and is about 1300 km across, with an area of 2×10^6 km^2. The sedimentary thickness varies from 1000 to 1500 m. The basin rests on Archaean and Early Proterozoic basement which forms shield areas to the north and south. It is bounded on the west by the Palaeozoic Mauretanide belt and on the east by the late Precambrian to early Cambrian Pharuside belt.

The sedimentary fill ranges in age from mid-Proterozoic to Carboniferous. The central part of the basin is covered by a thin veneer of Mesozoic and Cenozoic sediments. The basin sequence is divided into four supergroups, separated by unconformities or

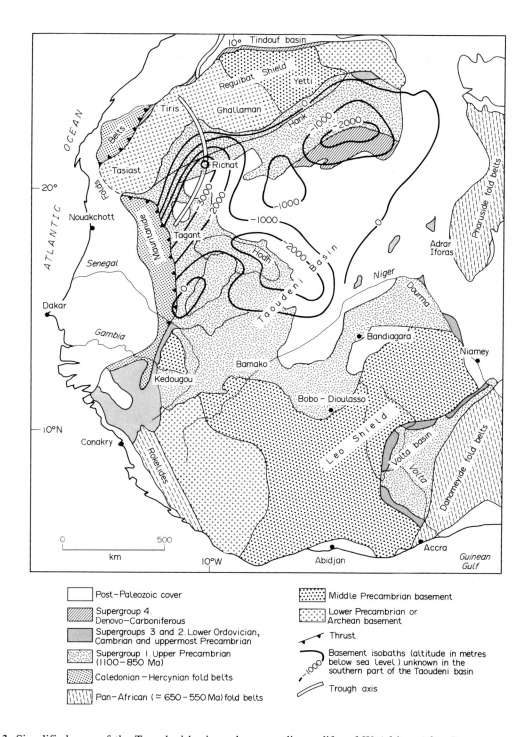

Fig. 10.3. Simplified map of the Taoudeni basin and surrounding uplifts of W Africa. After Bronner et al. (1980).

disconformities, which show quite different subsidence rates, varying from 0.04 mm/yr for supergroup 1 (late Precambrian) to 1.6 mm/yr for supergroup 4 (Devonian). The mean subsidence rate overall is 0.05 mm/yr. Within the main basin is a curved linear trough, or aulacogen, with a maximum depth of 5–6 km which appears to have existed as a major rift zone in the late Precambrian. This rift does not extend into the Phanerozoic since the basal sediments of supergroup 2 rest unconformably on the rift-fill sediments.

The North Sea basin (Fig. 10.4) is a basin situated on continental crust of the northwest Eurasian plate between Britain and Norway, opening into the Atlantic Ocean in the north. It is lozenze-shaped, measuring 1000 km along its long axis and about 500 km across. Apart from a narrow strip along the Norwegian coast, it is never more than 200 m deep at present.

The history of the basin is summarized by Ziegler (1982). The basin probably originated in the Devonian as one of a number of extensional rifts developed in the region of the Caledonian orogenic belt in Britain and Norway. During the Permian, two separate basins existed in the northern and southern parts of the North the rift fault system. During the mid- and late Cretaceous, active extension appears to have ceased and been replaced by subsidence over the whole area of the basin, resulting in up to 2 km of Cretaceous sediments which blanket the rift topography. Further subsidence took place during the Cenozoic, when an additional 3.5 km of sediment was deposited in the central area of the basin. It is thought that the region of maximum Cenozoic sedimentation corresponds to the zone of maximum crustal thinning during the preceding extensional phase (Donato & Tully 1981).

Gravity profiles indicate a mass excess beneath the Viking and Central grabens which is consistent with seismic refraction evidence of a much shallower Moho beneath the graben. Normal crustal thicknesses of 30–35 km beneath Norway and Shetland decrease to about 20 km in the Viking graben, where there is 8–10 km of sediment fill (Fig. 10.4c). Seismic reflection profiles across the graben show an asymmetric half-graben structure formed by a series of tilted fault blocks, thought to rest on a low-angle detachment zone (e.g. Beach 1986).

The four examples briefly described above range in duration from 150 to about 700 Ma and also in thickness of sediment fill. There is no simple relationship between duration and thickness, and average sedimentation rates vary from 0.2–0.4 mm/yr for the relatively short-lived basins to only 0.002 for the longer-lived Taoudeni basin. However, these average rates conceal periods of non-sedimentation and even erosion and do not give a reliable guide to the true rate of tectonic depression. The longer the life of a basin, in general, the smaller will be the apparent average rate of sedimentation. It seems likely that long continued basins are the locus of several distinct episodes of tectonic depression separated by periods of low tectonic activity. The origin of basins will be considered below but at this point it is worth noting that in many basins the existence of a rift or graben system in the early stages of evolution of the basin can either be proved or reasonably inferred (also see Chapter 11).

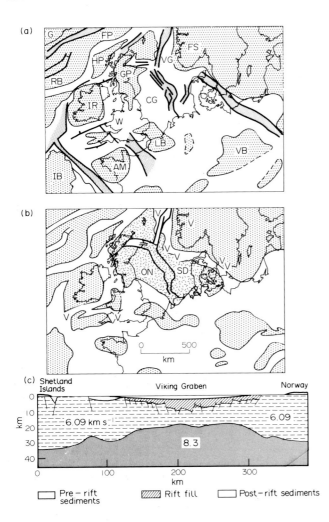

Fig. 10.4. The North Sea basin. (a) Reconstruction of land (open stipple) and sea (blank) areas during the early Jurassic; major rifts (close stipple) with boundary faults are shown. (b) Reconstruction of land and sea areas during the mid-Jurassic. The large area indicated by dense stipple is the central North Sea dome. Areas of contemporary vulcanicity are indicated by the symbol 'v'. G, Greenland; RB, Rockall bank; FP, Faraoes plateau; HP, Hebridean platform; FS, Fennoscandia; IR, Irish massif; GP, Grampian–Pennine massif; AM, Armorican massif; LB, London–Brabant massif; VB, Vindelician–Bohemian massif. (c) Crustal profile across the Viking graben showing crustal thinning and sediment thickening towards the rift. After Ziegler (1982).

Sea respectively, separated by the mid-North Sea high. The present North Sea basin was initiated in the Triassic period as part of an extensive rift system connecting with the major Arctic–Atlantic rift (Fig. 10.4a). Two major rifts underlie the basin, the Viking and Central graben, lying midway between Norway and Shetland, and between Norway and mainland Britain respectively. These two rifts meet at a triple junction with a graben which extends westwards into the Moray Firth. In mid-Jurassic times, a large rifted dome existed in the central North Sea with volcanic centres situated at the triple junction and at several other points along the rift system (Fig. 10.4b). In the late Jurassic and early Cretaceous, deep water conditions were re-established and sedimentation continued under renewed extension on

Uplifts

An *uplift* is a structure caused by the upward movement of part of the Earth's surface within a stable zone or craton. The terms *swell*, *dome*, *arch* and *anteclise* have been used synonymously. Uplifts are typically of the order of 500–1000 km across and may exist for long periods of geological time, undergoing slow and perhaps spasmodic upwards movements and, through erosion, providing a supply of sediment to surrounding basins. They are as important as basins in the tectonic history of the cratons and occupy similar surface areas. Volumetrically, they must approximately balance basins over long periods of time in order to maintain the sediment supply of the latter. Uplifts are more difficult to study than basins because, in many cases, the stratigraphic record is either incomplete or

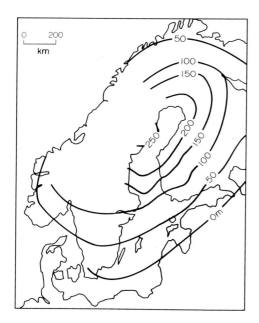

Fig. 10.5. The Fennoscandian post-glacial uplift. Contours in m of uplift of the Baltic area since 6800 BC. After Vita-Finzi (1986) and Zeuner (1958).

totally missing. For example, the major Precambrian shield regions are exposed because they form uplifts; but the detailed history of their uplift has not, in general, been recorded.

Those regions of the northern hemisphere covered by the Quaternary ice sheets show high recent uplift rates that have been attributed to post-glacial isostatic response to the removal of the ice load. However, these same regions have acted as uplifts over much longer periods and the Quaternary movements merely accentuate a long-continued tectonic trend.

The Fennoscandian uplift is a familiar example of recent post-glacial uplift, as is Laurentia. Both these uplifts, which have been studied in detail, involve vertical movements ranging up to 100 mm/yr over periods of the order of 10,000 yr. The presently glaciated regions of Greenland and Franz Josef Land yield very similar uplift rates.

The shape of the Fennoscandian uplift (Fig. 10.5) has been determined from raised post-glacial shore lines. The contours on these show an oval uplift, 1800 km long and about 1000 km across. The central portion has been elevated by over 250 m relative to present sea level since 6800 BC, corresponding to an average uplift rate of 28 mm/yr. A very similar recent uplift pattern has been determined by precise re-levelling since 1835. The uplift corresponds to a marked negative Bouguer gravity anomaly. Zones of active seismicity and other evidence of recent fault movements suggest that the uplift has been at least partly accomplished by faulting (Vita-Finzi 1986) (also see Chapter 18). The post-glacial Fennoscandian uplift corresponds spatially to the Precambrian Baltic shield which has constituted an uplift throughout most of the Phanerozoic. Since cratonization at the end of the Sveconorwegian orogeny, about 1000 Ma ago, erosion has removed perhaps several kilometres of crust. Uplift rates cannot be calculated but the overall average rate must be very much lower, by at least three orders of magnitude, than the recent rate.

Many continental uplifts are associated spatially with rift zones. Among such structures are the East African and Ethiopian plateaus, the Rhenish shield of Western Europe (see ahead to Fig. 10.8) and the Voronezh – Ukraine uplift of the Russian platform (see Fig. 10.1b). Other uplifts are associated with present-day plate boundaries, for example the Tibetan plateau, or with former rifted margins, as in the case of the Deccan plateau.

The Colorado plateau is a frequently cited example of a currently active uplift (e.g. McGetchin et al. 1980). It lies immediately east of the Basin and Range province of the western U.S.A. The plateau has a mean elevation of 1800 m and is about 700 km across. Undeformed Mesozoic and Cenozoic strata are exposed on the highest surfaces, and the uplift is considered to have taken place during Cenozoic time, principally during the Miocene, when most of the 1.5 – 2 km of uplift took place, at an average rate of up to 20 mm/yr. The crust beneath the uplift is unusually thick (45 km) and exhibits shield-like properties on the basis of the behaviour of seismic waves. However, the upper mantle below the uplift is characterized by an anomalously low P_n wave velocity and high heat flow of about 70 mWm^{-2}. The plateau lies within a large region of Cenozoic igneous activity and its uplift mechanism is presumed to be a related thermal effect.

Lineaments and fault zones

In the broadest sense, following O'Leary et al. (1976), the term *lineament* refers to a linear feature of the Earth's surface, which can be recognized on maps or on aerial or satellite images, and must be at least a few kilometres long. A stricter definition (Hobbs 1912) refers to 'significant lines of landscape which reveal the hidden architecture of the rock basement'. A *megalineament* has a transregional extent, such that it crosses boundaries between different cratonic domains. Real lineaments (i.e. in the sense of Hobbs) are the surface expression of faults or fault zones, together with associated structures. Because the depth of faults is generally proportional to their length (e.g. see Sankov 1989), the larger lineaments, and particularly megalineaments, represent deep faults or deep-seated faults, which penetrate to the base of the crust and in many cases into the lithospheric mantle (Fig. 10.6). However, the reverse is not the case, since many deep faults covered by younger strata are not visible as surface lineaments or, if they are, their surface expression may not be commensurate with their structural significance. The idea of deep faults is historically associated with the rhegmatic concept (see

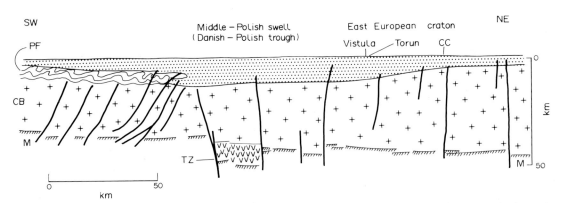

Fig. 10.6. Deep faults in the boundary zone between the East European craton and the West European platform, interpreted from international seismic profile VII (Pozaryski 1975). M, Moho; TZ, transitional crust/mantle zone; CB, crystalline basement; PF, Palaeozoic folds; CC, cratonic cover (Palaeozoic – Quaternary). The boundary corresponds to the Teisseyre-Tornquist line.

later) but in practice has a usefulness that is quite independent of the validity of rhegmatic tectonics.

Active cratonic lineaments are faults or fault zones on which movement has taken place during the lifetime of the craton. In many cases such faults are older structures, dating from before the period of cratonization, which have been re-activated. Primarily extensional structures (grabens, rifts and aulacogens), compressional structures (reverse faults and thrusts) and strike-slip structures can be distinguished, but many are composite, showing variation in sense of motion with time.

Deep faults are in fact fault zones, with widths of up to tens of kilometres, displaying many structural features, of which three are essential (Peyve 1990): (1) great length; (2) depth extent at least to the base of the crust; and (3) long-lived and multiphase development, often with reversals of motion. Many deep faults mark sharp differences in depth of up to 20 km of the Moho and other major reflectors (Fig. 10.6), as well as contrasts in other geophysical characteristics, such as seismic wave dispersion, high gravity gradients, magnetic anomalies, and increased heat flow. Magmatism and hydrothermal mineralization are also important manifestations.

Many deep faults have been active throughout much of the Phanerozoic, for example the Tyrnau fault along the north slope of the Caucasus, the Talassa – Fergana fault in the Tien Shan, and the trans-African fault which runs from the Gulf of Guinea to the Gulf of Aden (Kozerenko & Lartzev 1987). Such faults have a significant influence on the sedimentary record, constituting important palaeogeographical and tectonic boundaries, renewed many times in their history, and expressed, for example, as belts of barrier reefs or olistholiths. They may separate areas of different metamorphic, magmatic or tectonic development, and can often be characterized by enhanced metamorphism and deformation. The character of the fault zone changes and widens with depth, under the increasing influence of mylonitization and plastic flow, to become eventually a wide zone of concentrated displacement, rather than a series of discrete faults. Production of thermomechanical heat may locally be sufficient to generate anatectic melts (Patakha *et al.* 1978). Such zones exhibit markedly reduced viscosities compared with the surrounding craton; and Kucay *et al.* (1981) have demonstrated that viscosity increases exponentially with distance from the axial region of certain deep fault zones.

The steep attitude of most well-known cratonic deep faults suggests that they may be primarily strike-slip in origin, although many exhibit episodes of dip-slip movement during their period of activity. Many deep fault zones exhibit significant economic potential. Near-surface endogenic ore mineralization in the Ukrainian shield shows an asymmetric distribution, being concentrated in the hangingwall above the region where the fault lies within the crust (Cekunov & Kucma 1979).

Extensional structures. A *graben* is a down-faulted block, and in this context a *rift* is a major linear graben, consisting of a complex of faults exending laterally for many hundreds of kilometres (see Chapter 11). The major currently active rift zones correspond to constructive plate boundaries, and are either situated on oceanic crust (e.g. the mid-Atlantic rift) or contain newly created oceanic crust (e.g. the Red Sea rift). Cratonic rifts that do not presently constitute active plate boundaries, such as the East African rift system, may connect with adjacent plate boundaries via a triple junction (Fig. 10.7a), but others (e.g. the Baikal rift) have no obvious link to the active plate boundary network. Rifts are usually associated with magmatism, although the nature and extent of the magmatic effects vary widely, both within a single rift and between one rift and another. The sedimentary cover along the flank regions of rifts is frequently affected by asymmetric monoclinal folds or inclined or tilted fault blocks (Fig. 10.7b).

The failed arms of rift triple junctions are an important class of rift defined by Burke & Dewey (1973), who suggested that continental splitting took place by joining pairs of rifts from adjoining triple junctions to form a continuous but irregular

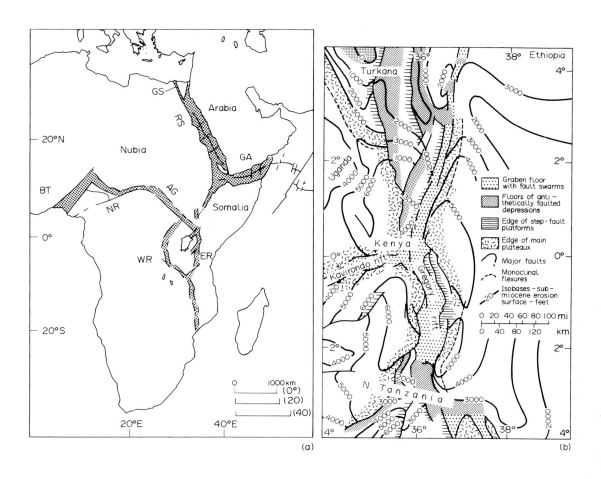
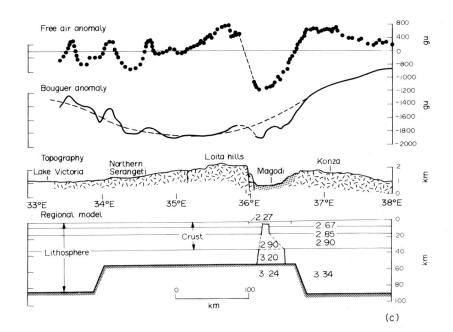

Fig. 10.7. (a) Major Phanerozoic rifts of Africa. BT, Benue trough; NR, Ngaoundere rift; AG, Abu Gabra; WR, western E African; ER, eastern E African; GA, Gulf of Aden; RS, Red Sea; GS, Gulf of Suez. After Girdler & Darracott (1972). (b) Structure of the Kenya (Eastern) rift system. Note the irregular fault pattern and the monoclinal flexures following the rift 'shoulders'. Contours (in feet) indicate amount of uplift since the mid-Cenozoic. From Baker & Wohlenberg (1971). (c) Free-air and Bouger gravity profiles across the Kenya rift together with crustal section and gravity model. Note the wide region of thin lithosphere underlying the regional dome, and the narrow zone of high-density material immediately underlying the rift. From Darracott *et al.* (1973).

constructive boundary. The rifts that did not develop into the subsequent plate boundary were the *failed arm graben*. A good example of a failed-arm graben is the Benue trough of West Africa (Fig. 10.7a), which is part of the Niger triple junction developed in the Cretaceous, prior to the opening of the South Atlantic Ocean. The other two rifts of the junction developed into the ocean while the Benue trough remained as a failed arm.

Active rifts exhibit anomalous crustal and upper mantle profiles, usually interpreted in terms of lithosphere thinning due to extension or to asthenospheric diapirism (see later). The geophysical characteristics of rifts are similar to those of ocean ridges, and are explained in terms of a region of low-density mantle material often termed the *rift pillow* which correlates with a zone of high heat flow. This rift pillow often supports an uplifted plateau or dome in the flanking regions. Many rifts show evidence of partial control by lines of structural weakness. For example, the East African rift commonly parallels the trend of Precambrian mobile belts, and follows the boundaries of older cratons. Such structural control (or *inherited structure*) is never complete however, and rifts also locally crosscut previous structures. We shall consider the East African and Rhine rifts in some detail. They are both well studied examples and exhibit several contrasting properties.

The East African rift system is part of a complex network of rifts within the African continent linking the Atlantic margin in the west with the Red Sea and Gulf of Aden in the east (Fig. 10.7a). The Kenya or Eastern rift (e.g. Baker & Wohlenberg 1971) extends from Tanzania in the south, where it joins the Western rift, to the Red Sea in the north, at the Afar triple junction in Ethiopia. It is about 2800 km long and crosses a broad elliptical dome in the Kenya sector which is 1000 km in length. In this sector (Fig. 10.7b), the main graben is 60 – 70 km wide and 750 km long, and is bounded by normal faults arranged in an en échelon pattern. Between the ends of adjacent en échelon faults, sloping ramps descend to the rift floor from the flanks, which are elevated by up to 2000 m above the floor. From the thickness of the rift fill, the total throw on the bounding normal fault system is estimated to be around 3 – 4 km. The graben floor is cut by a swarm of minor faults and fissures.

The evolution of the Eastern rift commenced with a monoclinal flexure in the mid-Cenozoic along the western rift flank. This was succeeded by voluminous vulcanicity during the Miocene. Extensive rift faulting accompanied by additional vulcanicity developed in the early Pliocene and continued to the present day. The crustal extension for this rift is estimated at 5 – 10 km.

The rift closely follows the axis of a prominent negative Bouguer anomaly interpreted as a zone of thin lithosphere (Fig. 10.7c). The gravity data also suggest the presence of a narrow shallow crustal body of dense material thought to represent a basic magma chamber. Focal mechanisms of earthquakes along the rift indicate

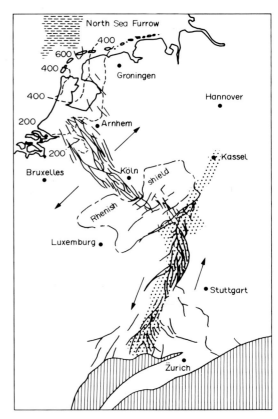

Fig. 10.8. The Rhine – Ruhr rift system. The rift extends from the Alpine front in the south (vertical ornament) to the edge of the North Sea basin in the north, crossing the Rhenish shield uplift in the centre. Note the differences in fault pattern between the northern (extensional) and southern (sinistral transpressional) sectors. Stipple indicates extent of rift before the late Miocene. After Illies (1981).

WNW – ESE extensional stress across the structure.

The Rhine-Ruhr rift system (Fig. 10.8) is another well known currently active system (Illies & Greiner 1978) extending from the Alpine front in the south near Basel, to the coastal plain of the Netherlands, west of Arnhem. The southern sector of the rift, from Basel to Frankfurt, has an overall NNE – SSW trend and is characterized by a complex network of branching faults which tend to curve from NNE, parallel to the trend of the rift, into a N – S orientation. Both dip-slip and strike-slip movements are recorded on the faults, and at many localities former dip-slip movements have been overprinted by strike-slip striations. *In situ* stress determinations indicate a general NW – SE maximum compressive stress axis across this part of Western Europe and this is consistent with fault-plane solutions of earthquakes along the rift which indicate oblique sinistral shear with an overthrust component. The northern sector of the system, from Frankfurt to Arnhem, trends NW – SE and is characterized by abundant normal faults, parallel to the trend of the rift, showing dip-slip movements. Both sectors of the system are consistent with NE – SW extension in the present stress field.

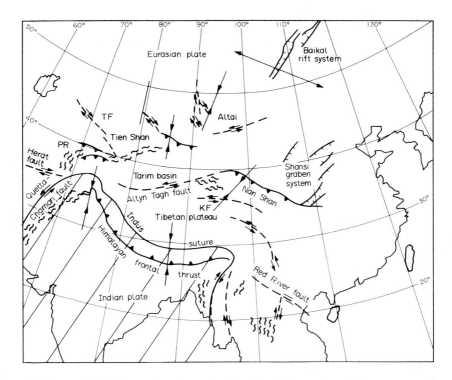

Fig. 10.9. Intracratonic deformation in the Asian plate related to the India–Asia collision. Note compressional fold-thrust belts (toothed lines); extensional rifts and graben systems (ticked lines); and strike-slip zones (dashed lines with half-arrows showing sense of movement). Compression and extension directions derived from earthquake focal-mechanism solutions are indicated by double arrows. TF, Talasso-Fergana fault; PR, Pamir range; KF, Kunlun fault. After Molnar & Tapponnier (1975).

Tectonic activity in the rift system appears to have commenced in the late Cretaceous, about 80 Ma ago, coinciding with the initiation of compressive deformation in the Alps. Rifting and subsidence continued through the Eocene and Oligocene but appear to have ceased in the mid-Miocene when the whole region became uplifted. In the Pliocene, tectonic activity was resumed but with a change in orientation of the principal stress axes, from WNW–ESE extension prior to the Miocene, to NW–SE compression, NE–SW extension from the late Pliocene onwards. This change resulted in the southern rift sector changing from extension to sinistral strike-slip.

Vertical movements on the rift are considerable. In the Pleistocene alone, depression of the rift floor reached a maximum of 380 m. The flanks of the rift are elevated as part of a regional upwarp associated with the rift structure. The geophysical evidence indicates a

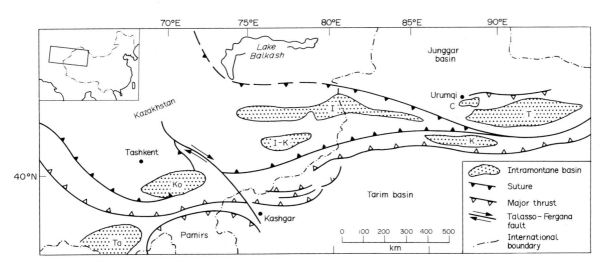

Fig. 10.10. The Tien Shan mountain range in E Kazakhstan and NW China, showing positions of major sutures, thrusts and intramontane basins. Ta, Tadjik; Ko, Kokland; I-K, Issyk-Kul; I-Ili; K, Korla; C, Chai Wo Pu; T, Turfar. From Windley *et al.* (1990).

thinned crust (c. 24 km) beneath the rift associated with high heat flow.

Compressional structures. In contrast to extensional structures, which are widespread, compressional structures are comparatively uncommon in cratonic regions. This is a consequence of the fact that the lithosphere is much stronger in compression than it is under extension and much higher stresses would be required to produce failure in normal intraplate lithosphere. It is to be expected therefore, that only under exceptional circumstances — for example, in a zone of particularly weak lithosphere — would compressional structures be formed.

Kamaletdikov *et al.* (1987) analyze many cases of intracratonic compressional structures and conclude that zones of overthrusting, some reaching to the base of the crust and re-activated many times, are an important feature of cratonic basement regions, and are generally responsible for zones of intracratonic cover folding.

The Eurasian plate north of the India − Asia collision zone (see Fig. 10.9 and Chapters 6, 12 and 13) contains several *intra-cratonic fold/thrust belts,* one of which is the Tien Shan belt (Fig. 10.10). This belt, which lies partly in Kazakhstan and partly in the Sinkiang region of northwest China, extends for about 2500 km in an E − W direction and is about 250 km in width (e.g. see Windley *et al.* 1990). It is separated from the Himalayas by the stable Tarim block which is 600 km across. Cenozoic deformation, linked to the post-collisional shortening of Central Asia, appears to be related mainly to movements on several major thrust zones. Deformation commenced in Oligocene times but appears to have intensified in the Pliocene.

A major S-directed thrust system marks the southern margin of the range, and the amount of shortening on this system increases westwards, corresponding to a decrease in distance between the Tien Shan range and the Himalayas. The reason for the occurrence of Cenozoic shortening here appears to lie in the previous history of the belt, which contains two Palaeozoic sutures representing fundamental planes of weakness within the craton. The intervening Tarim basin is believed to overlie a stable cratonic block during the Phanerozoic. It is probable that the shortening displacement across the Tien Shan range, which may amount to perhaps 300 km (e.g. see Molnar & Tapponnier 1975) may be connected to the plate boundary suture via the northern front of the Pamir range, which connects in the southwest with the northwestern corner of the Indian plate (Fig. 10.9).

A Carboniferous example of an intracratonic thrust belt is described by Teyssier (1985). The belt lies along the northern margin of the Amadeus basin in central Australia and separates it from the adjoining Arunta block. The sedimentary sequence in the basin is affected by a south-directed thin-skinned thrust system, which also extends down into the basement rocks to produce recumbent folds and nappes. Minimum estimates of horizontal shortening are 50 − 100 km. The belt can be regarded as a crustal-scale duplex, resting on a sole thrust, which has resulted in isostatic downbending of the footwall slab. In the absence of evidence of contemporary plate-boundary orogens, either north or south of the belt, the origin of the belt is ascribed by Teyssier to general N − S shortening of the craton. However it is conceivable that the displacement may be transferred laterally along a basal crustal detachment to the contemporary plate boundary in eastern Australia.

Strike-slip zones. Active strike-slip faults and fault zones are comparatively common in cratonic regions. Several well-known examples of strike-slip zones are situated in Central Asia north of the Himalayas (Fig. 10.9), the Altyn Tagh and Kunlun faults being two of the most important (Molnar & Tapponnier 1975). In Europe, the Tornquist line is another well known lineament with a long history of movement (Jaroszewski 1972, Brockwicz-Lewinski *et al.* 1981).

Many sectors of extensional or compressional fault zones are oblique to the general trend and exhibit a strike-slip component of movement. The example of the southern part of the Rhine rift system has already been described. Under conditions of general compression or extension of a craton, old fault zones may become reactivated and produce strike-slip movements if they are suitably oriented.

Intracratonic folds

In the early part of this century, when the notions of platform and craton were created by Stille (1913) and others, the cratons were thought to be the exclusive domain of germano-type (fault-block) tectonics, and folds of the craton cover were interpreted mostly as passive adjustments of that cover to fault movements in the underlying basement. This approach has particularly dominated views on the East European (Russian) platform even up to the present, and the advent of plate tectonics merely confirmed many geologists in their views on the essential cohesiveness, or lack of lateral strain, shown by the cratons.

An examination of the actual distribution of folds on cratonic areas however, reveals several categories.

(1) *Drape folds* include both monoclinal folds generated over single dip-slip faults in basement and symmetrical folds above graben or horst blocks (Fig. 10.11a). In the case of reverse faults (Fig. 10.11b), the drape fold may accommodate some horizontal shortening. In normal faults, there is always extension.

(2) *Folds developed over strike-slip faults* result from the action of a horizontal shear couple within the sedimentary cover, transmitted upwards from transcurrent movement on a basement fault (also see Chapter 12). The folds are typically arranged en échelon, oblique to the fault strike. Folds of this type are well known from the San Andreas fault zone in California (Harding 1973) and have been described, in a cratonic setting, from the Donbass Coal basin (Fig. 10.12) by Korcemagin & Ryabostan (1987). Larger

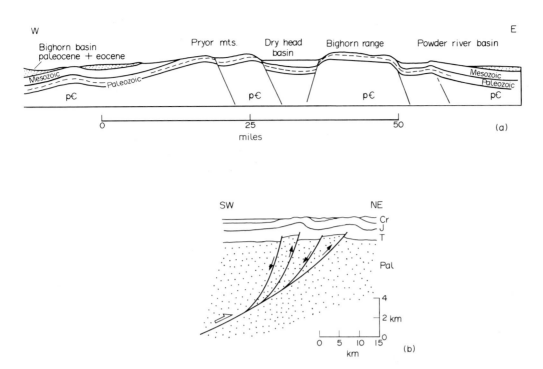

Fig. 10.11. (a) Drape folds in cover over basement blocks in the Bighorn Range, Wyoming, after Thom (1923) p€, Precambrian. (b) Asymmetric folds in Mesozoic cover over reverse faults in basement, near Radomsko, Central Poland, after Czubla (1988) Pal, Palaeozoic; T, Triassic; J, Jurassic; Cr, Cretaceous.

folded zones form in areas of fault overlap, for example the Nepa zone in the Siberian craton (Riazanov 1973).

(3) *Folds resulting from inversion* over dip-slip basement faults which have changed their sense of movement (e.g. from normal to reverse) exhibit a quite different geometry to the drape folds discussed above

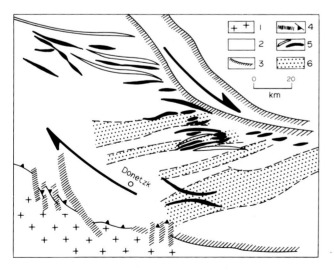

Fig. 10.12. Oblique folds and uplifted blocks related to Alpine strike-slip reactivation of the Donbass Coal basin, after Korcemagin & Ryabostan (1987). 1, crystalline basement; 2, Palaeozoic cover; 3, fault zones bordering the Donbass graben; 4, flexures and normal faults; 5, en échelon folds; 6, en échelon geomorphological structures; arrows give sense of strike-slip displacement.

(see Chapter 14). They are markedly disharmonic in relation to the basement structure, the disharmony being compensated by considerable lateral changes in layer thickness.

(4) *Folds produced by diapiric movement* resulting from the spontaneous gravitational displacement of salt and other buoyant masses are common in cratonic regions. Folds resulting from this process occur in regions of former salt basins such as the Gulf of Mexico, North Germany, North Poland and the peri-Caspian depression. Such folds are formed without any bulk shortening of the basement. A detailed account of salt tectonics is given in Chapter 8.

(5) *Rootless folds* form above a detachment surface entirely within the craton cover (Kamaletdinov *et al.* 1980). Structures of this kind have been described from the Turan craton south of the Urals (Popkov 1985) and a number of cases in the East European, Siberian, N American and African cratons are analyzed by Kamaletdinov *et al.* (1981) who ascribe cratonic folds in general to this mechanism.

Bulk strain within cratons

Since the work of Molnar & Tapponnier (1975) the idea of bulk shortening and extension within cratonic regions such as Central Asia has been much more generally accepted than before. However, in the Central Asian model, the bulk strain is accomplished by discrete displacements on faults, whereas Goguel (1983) recognized the importance of creep and elastic strain in the bulk deformation of cratons. The existence of bulk

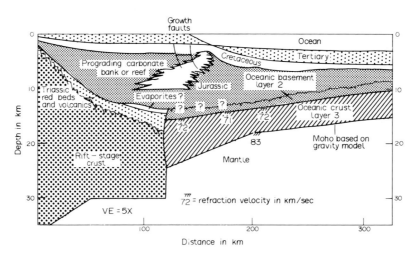

Fig. 10.13. Passive-margin basin of the Eastern U.S.A., off New Jersey. Interpretative profile based on gravity, seismic and borehole data. Note asymmetric nature of the basin in the Jurassic. In the Cenozoic, there are two separate basins, one on the continental shelf (cratonic), the other oceanic. After Sawyer et al. (1982).

strain is required by the high measured stresses within cratons (see below). In order to store such stress, the craton has to be in a state of strain. Indications of bulk penetrative shortening within cratonic areas are provided by the widespread occurrence of stylolites (see Chapter 5). In the platform region of southern Germany, according to Wagner (1967), the loss of rock volume due to stylolite formation under horizontal compression is at least 0.3 – 0.7%. Another example has been reported from the Transjordan block (Mikbel & Zacher 1986), where stylolite formation is associated with intracratonic folding. The mechanism for basement shortening may be pressure solution in crystalline rocks (see Cox & Etheridge 1989, O'Hara 1990) or shear along numerous dispersed discontinuities.

General penetrative bulk strain in the horizontal plane is probably common, if not universal, in cratonic regions but is quantitatively too small to be readily noticeable. Moreover, the strain rate is likely to be very much lower than that involved in plate boundary deformation. Slow strain rates would encourage pressure solution and stylolite formation rather than folding.

Passive continental margins

Passive continental margins represent former plate boundaries that now lie within plates and either are tectonically inactive or are the sites of intraplate basins and subject to continuing subsidence. Such structures are marginal to the cratons, because they are developed partly on continental and partly on oceanic crust. They typically show a lower faulted section corresponding to an early rifting stage in the evolution of the basin. At this stage, the basin is the product of a divergent plate boundary. However during the later stages of its evolution, the basin becomes part of the stable plate interior, as the boundary migrates oceanwards with continued ocean spreading.

The Atlantic coastal basins provide good examples of such structures. The continental shelf and slope bordering the Atlantic coast of the northern U.S.A. has been intensively studied using borehole, seismic reflection and gravity data. A section across the New Jersey coastline, crossing the Baltimore trough (Fig. 10.13), shows 18 km of sediment deposited since the late Triassic, an average rate of deposition of 0.08 mm/yr. However, the period of maximum subsidence occurred during the Jurassic when the initial extension in the Central Atlantic took place. Above the Jurassic sediments, a much thinner sequence of Cretaceous strata is draped over the edge of the continental shelf without showing any marked thickness variation. During the Cenozoic, the basin is divided into two separate structures: a continental shelf wedge, which constitutes a cratonic margin structure, and an oceanic basin which thins out along the continental slope (also see Chapter 11).

PHYSICAL PROPERTIES OF CRATONIC LITHOSPHERE

Physical nature of cratonic lithosphere

The *lithosphere* is defined on seismological criteria as the strong outer layer of the Earth; it can also be regarded as the cool surface layer of the Earth's convective system. The lithosphere cannot be considered in isolation, but must be taken together with the asthenosphere and mesosphere as part of a system where the boundaries are transitional, and continuously changing with time.

The base of the lithosphere is usualy defined on the

Table 10.1. Heat flow variation in cratons.*

Region	Heat flow (mWm^{-2})	in HFU
(A) Shield		
Superior Province[2]	34 ± 8	0.85
West Australia[2]	39 ± 8	0.98
West Africa (Niger)[2]	20 ± 8	0.50
South India[2]	49 ± 8	1.23
Mean Archaean and older Proterozoic[5]	41 ± 10	1.03
(B) Intermediate		
Eastern U.S.A.[2]	57 ± 17	1.43
England and Wales[2]	59 ± 23	1.48
Central Europe (Bohemian massif)[2]	73 ± 18	1.83
Northern China[1]	75 ± 15	1.89
Mean younger Proterozoic[5]	50 ± 5	1.25
Mean Palaeozoic[5]	62 ± 20	1.55
(C) Thermally active		
Rhine graben[3]	107 ± 35	2.68
(flanks—Rhenish massif)[3]	(73 ± 20)	(1.83)
Baikal rift[3]	97 ± 22	2.43
(S.E. flank—older Palaeozoic)[3]	(55 ± 10)	(1.38)
East African rift[3]	105 ± 51	2.63
(flanks)[3]	(52 ± 17)	(1.30)
Basin and Range province[2]	92 ± 33	2.30
(E. flank—Colorado plateau)[4]	(60)	(1.50)

*Divided into *shield* (i.e. Archaean and early Proterozoic craton, ranging from 20 to 50 mWm^{-2}); *intermediate* (i.e. late Proterozoic to Phanerozoic mobile belts — 50–75 mWm^{-2}); and *thermally active* (i.e. active rifts — 90–110 mWm^{-2}) regimes.

Heat flow data from Pollack and Chapman (1977)[1], Vitorello and Pollack (1980)[2], Morgan (1982[3], 1983[4], in press[5]).

From Kusznir & Park (1984).

basis of a relatively rapid change in seismic wave velocity (of both P and S waves), and specifically, of a decrease in the rate of increase of V_P and V_S, which takes place at depths of around 150 km beneath the continents. This change in velocity is related to changes in density and rheology that in turn reflect the temperature variation downwards through the crust and upper mantle. Although there may be compositional differences between the lithospheric and asthenospheric mantle, these are not the main factor in differentiating the two layers, which are most conveniently distinguished in terms of their viscosity. The viscosity of the *asthenosphere* is usually estimated to be around an order of magnitude lower than that in the lower part of the lithosphere (McKenzie 1967). The relative plasticity of the asthenosphere is due therefore to the effect of elevated temperature on the rheology of the material and enables flow to take place at geologically significant rates. It is this property, of course, that enables the plate tectonic process to operate.

Short-term mechanical properties

The formation of cratonic structures is ultimately dependent on the mechanical properties of the cratonic lithosphere. The important properties that control deformation are elasticity, viscosity, fracture strength and yield strength (also see Chapters 1 and 3). These properties vary with rock composition, depth and temperature. Information provided by laboratory experiments on rock materials yields estimates for these mechanical parameters under a limited range of conditions. Likewise, various indirect geophysical methods, particularly the study of seismic waves, provides data on the short-term mechanical behaviour of the cratonic lithosphere. The values of elasticity and rigidity derived from seismic wave velocities define the elastic properties of the lithosphere over very short time periods (0.1 s to 1 h). However, the strength of cratonic lithosphere calculated in this way (often called the *instantaneous strength*) is very much greater than its strength when subjected to forces applied for periods of tens of Ma, and is of little value in considering the origin of cratonic structures.

However, useful information about the structure and strength distribution on the cratons can be obtained by studying lateral variations in certain physical parameters. Several independent geophysical methods indicate zones of anomalous physical properties corresponding to those exhibited by the asthenosphere. Zones following the active continental rifts exhibit low P_n wave velocity, greater surface wave dispersion, low Q (or high degree of attenuation), inefficient propagation of S_n waves and low electrical conductivity. It may be

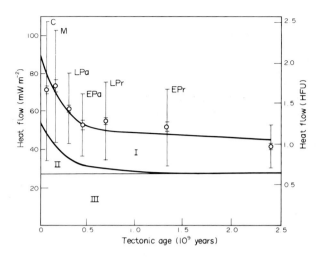

Fig. 10.14. Graph showing decrease of continental heat flow with increase in tectonic age. The upper curve is interpreted as made up of three components: (I) radiogenic heat from the crust; (II) heat from a transient thermal perturbation associated with a tectonothermal event; and (III) background heat flow from a deeper source. C, Cenozoic; M, Mesozoic; LPa and EPa, late and early Palaeozoic, respectively; LPr and EPr, late and early Proterozoic, respectively. After Vitorello & Pollack (1980).

concluded that such zones exhibit abnormally low instantaneous strength indicating anomalously thin and weak lithosphere. They may be compared with other active tectonic zones such as ocean ridges and volcanic arcs that follow current plate boundaries.

Thermal structure and viscosity

Mean values of *heat flow* per unit surface area for the different continents are shown in Table 10.1. These values do not differ significantly from the oceanic values. More than 99% of the Earth's surface has a normal heat flow of around 1.5 HFU (= 60 mWm^{-2}). Anomalous zones of higher heat flow on the cratons correspond mostly with areas of current volcanic activity along the major continental rifts. The close correspondence between heat flow and other geophysical anomalies confirms the view that the zones of anomalous physical properties are thermal in origin.

A closer examination of the variation in heat flow within cratonic regions indicates a tendency for heat flow to show an inverse correlation with age of a tectonic province, that is, the heat flow increases with decrease in age of the last major tectono-thermal event to have affected the region (Table 10.1). Vitorello & Pollack (1980) show that this relationship can be approximately fitted to an exponential curve (Fig. 10.14) representing the gradual decay of a transient thermal perturbation together with a component of heat loss due to gradual erosion of radiogenic crust. This variation has important implications for cratonic strength, older cooler Precambrian shield regions being significantly stronger then younger warmer regions involved in Phanerozoic thermal events.

Long-term strength

In tectonically stable cratonic lithosphere subjected to an applied force over geologically significant periods of time, the *strength* of the piece of *lithosphere* as a whole is dependent on the distribution of strength with depth. *Brittle strength* is controlled primarily by lithostatic pressure and increases with depth, whereas *ductile strength* is controlled by temperature and decreases with depth because of the geothermal gradient (also see Chapter 1). The net effect is that the initial stress decays with time in the lower ductile part due to viscous creep, and becomes concentrated in the strong competent elastic region in the *upper crust* (Kusznir & Park 1982). Significant deformation within the piece of lithosphere will only occur when the level of accumulated stress overcomes the strength of this strong layer. This process, termed *whole-lithosphere failure* by Kusznir and Park, will result in a rapid increase in strain across the whole thickness of the lithosphere, allowing geologically significant levels of deformation to take place. The process is critically dependent on the geothermal gradient because of the control of temperature over ductile strength. In this way, the relatively small stresses calculated for the main plate boundary effects (e.g. ridge-push and slab-pull) may be amplified to the point where they are sufficient to overcome the much larger strengths of crustal rocks.

Kusznir & Park (1984) use a mathematical model to calculate the stress distributions with time, assuming Maxwell visco-elastic properties for the lithosphere. They show that for levels of initial applied stress comparable with those predicted for the main plate boundary stress sources (i.e. around 20–30 MPa) (Table 10.2) extensional failure is predicted after periods of 1 Ma only for continental lithosphere with higher than average heat flow of about 70 mWm^{-2} (Fig. 10.15). For normal cratonic lithosphere with heat flows of around 60 mWm^{-2}, extensional failure is predicted only for the very large applied stresses such as those associated with plateau uplifts. For Precambrian shield regions with heat flows generally below 50 mWm^{-2}, failure should not occur, according to the model. If this analysis is correct, it suggests that extensional structures within cratons may be generated only under abnormal thermal conditions or an unusually high applied stress. Plateau uplifts, particularly if they are also associated with thermal anomalies, are the most likely source of extensional failure. Alternatively, for normal thermal conditions, an unusually large net extensional plate boundary stress would be required to generate extensional failure. It has been suggested by Bott & Kusznir (1984) that such conditions might have accompanied the break-up of Pangaea and other earlier supercontinents, if they were surrounded by subduction zones.

Table 10.2. Levels of calculated stress for the principal sources of renewable stress acting on cratonic lithosphere*

Mechanism	Compression or tension	Approximate level of stress differences	Stresses subject to amplification effects	Stresses significant in tectonics
subduction slab pull (plate boundary force)	tension (normally)	0.50 MPa†	yes	yes
subduction trench suction (plate boundary force)	tension (normally)	0–30 MPa†	yes	yes
ridge push (plate boundary force)	compression	20–30 MPa	yes	yes
mantle convection or asthenosphere drag	both	1–50 MPa	yes	probably not
lithosphere loading (uncompensated)	both (mainly tension)	35 MPa‡	no	locally perhaps
lithosphere loading (compensated)	both (mainly tension)	50 MPa‡	yes	locally yes

*Note that these fall in the range 0–50 MPa (0–500 bars) but that most are capable of amplification and are therefore tectonically significant.
†Highly variable in space and time because of variation in resistances, subduction velocity, length of slab, etc.
‡For 2 km elevation.
After Bott & Kusznir (1984).

Because rocks are much stronger under compression than under tension, the compressive strength of the lithosphere as a whole is considerably greater than its extensional strength. The model predicts that compressional failure would only occur either for abnormally high heat flows or for unusually large compressive stresses, certainly larger than would be generated by ridge-push alone. It is likely that the rather limited examples of compressional cratonic failure are confined to zones of abnormally weak lithosphere. For example, the compressional fold/thrust belts of Central Asia, such as the Tien Shan, occur along Mesozoic or Palaeozoic sutures which represent fundamental weaknesses within the craton, and might also be expected to exhibit above average heat flows.

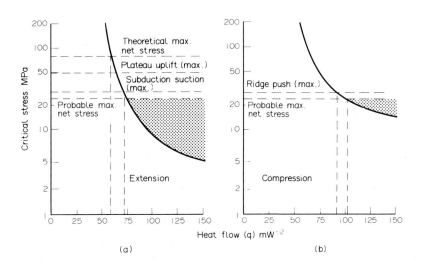

Fig. 10.15. Relationship between lithosphere strength and heat flow. Curves of variation of strength (stress required to produce whole-lithosphere failure after 1 Ma) with heat flow in extension (a) and compression (b) compared with likely levels from major sources of stress. Region of potential deformation shown stippled. Numerical model, after Kusznir & Park (1986).

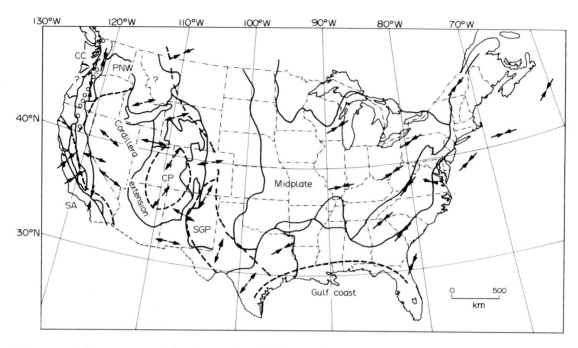

Fig. 10.16. Generalised stress map of the (conterminous) U.S.A. The greatest horizontal principal stress is indicated by inward-directed arrows where compressional or outward-directed arrows where extensional. Note the uniformity of the ENE–WSW to NE–SW maximum compressive stress in the cratonic (mid-plate) region of the central and eastern U.S.A. PNW, Pacific NW province; SA, San Andreas; CP, Colorado Plateau; SPG, S Great Plains; and CC, Cascades. After Zoback et al. (1989).

Seismicity and stress

Seismicity. Although most of the Earth's seismicity is concentrated along zones following the plate boundaries, a significant number of shallow earthquakes occur in intraplate regions, particularly in the cratons (also see Chapter 18). There is a marked concentration of seismicity along active continental rifts such as the African rift system, but there are also a considerable number of seismic events distributed throughout cratonic areas. These are typically of low magnitude although there are also infrequent events of greater magnitude, some of which have caused significant damage, for example the 1990 earthquake in Eastern Australia and the New Madrid earthquakes of 1811–1812 in Missouri. Many cratonic events can be assigned to movements on known faults or major lineaments but the source is often difficult to locate. It would appear that, unlike the oceanic intraplate regions, which are for the most part seismically quiescent, the cratonic crust contains numerous weaknesses that are exploited under stress to form localized zones of movement.

The distribution of intraplate seismicity in Central Asia was used by Molnar & Tapponnier (1975) in their influential account of the recent deformation of Asia in response to the India–Asia collision (Fig. 10.9). They showed that much of the seismicity could be divided into three simple categories: (1) thrust-sense events associated with active fold/thrust belts (e.g. along the Pamir and Tien Shan ranges; (2) normal-slip events associated with active graben and rift zones (e.g. along

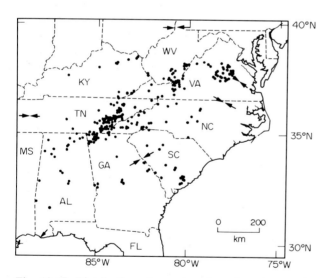

Fig. 10.17. Distribution of recent earthquake epicentres in the SE U.S.A. from 1977 to 1985. Note the concentration along zones marking major intracratonic lineaments: that is, the southern Appalachian zone and the South Carolina–Georgia zone. Arrows show directions of greatest horizontal stress. Standard abbreviations used to indicate location of states. After Kuang et al. (1989).

the Baikal rift, the Shansi graben system and on the Tibetan plateau); and (3) strike-slip events of both sinistral and dextral sense associated with a number of major fault zones and lineaments oblique to the N–S convergence direction between the two plates (e.g. the Altyn Tagh, Kunlun and Talasso-Fergana fault zones). Thus the sense of movement on many of the faults associated with the recent seismicity of Asia is consistent with a state of general N–S compression and E–W extension inferred for this cratonic region.

Zoback & Zoback (1989) analyze the state of stress in the conterminous U.S.A. using seismic data as well as *in situ* stress determinations (Fig. 10.16). For the cratonic part of the U.S.A., that is, the Mid-Continent east of the Cordilleran belt, the rather sparse seismic events yield fault-plane solutions in close agreement with the stress orientations determined from *in situ* measurements (see below and Chapter 18).

In an analysis of intraplate seismicity and stress in the southeastern U.S.A., Kuang *et al.* (1989) showed that this region has experienced few earthquakes with magnitudes greater than 4.0, the largest reported event being the 1886 Charleston earthquake with an estimated magnitude of 6.8–7.1. Most of the epicentres of recent earthquakes (Fig. 10.17) are concentrated in two broad zones: the Southern Appalachian seismic zone and the Charleston–Summerville area. In the former zone, more than 30 earthquakes with magnitudes greater than 3 were recorded between 1928 and 1981. Both zones are interpreted as linear belts of weakness within the otherwise largely aseismic craton.

Stress. Estimates of the current state of stress in the cratonic lithosphere are derived from a variety of sources. In addition to earthquake focal mechanism studies, direct measurement of both magnitude and orientation of the stress field can be made by various *in situ* techniques, of which the most reliable are the borehole break-out, hydraulic fracturing and overcoring techniques used to determine the stress at depth in wells or boreholes (see Chapter 18 for details). Young geological structures, such as neotectonic joints, can also be employed to estimate comtemporary horizontal stress directions in cratons according to Hancock & Engelder (1989). Data are now available for many cratonic regions. Zoback *et al.* (1989) discuss global patterns of tectonic stress on the basis of the International Lithosphere World Stress Map database. Their analysis uses 3142 measurements, but these are distributed very unevenly. The quality of the data is also highly variable and the authors recognize four categories of quality, A, B, C and D, of which only categories A, B, and C are plotted. Of their A (highest) quality data points, the North American plate contains 450 and Eurasia 100 (of which 66 are in Europe, almost all in W Europe). In contrast the Indo-Australian plate contains only 16 points, South America 7, and Africa and Antarctica none.

It must be realized therefore, that generalisations about cratonic stress are currently based on only three well known areas: North America, Western Europe and Central Asia. In these regions, it is clear that both horizontal stress orientations and stress magnitudes are uniform over large areas. Moreover, the cratons are in a state of general compressive horizontal stress, with the exception of zones of localized extension associated with rifts and recent uplifted plateaus.

Zoback & Zoback (1989) in their analysis of stress in the conterminous U.S.A. (Fig. 10.16), show that most of the central and eastern U.S.A. is part of a large cratonic region (which also includes the Canadian shield), east of the Cordilleran belt, characterized by uniform NE–SW to ENE–WSW–oriented maximum horizontal compressive stress. West of this region is a wide zone, including the Basin and Range province, characterized by extensional horizontal stress and recent thermal uplift. The stress regime in the western coastal province is clearly related to strike-slip movement on the San Andreas fault. The stress state in the Cordilleran and western provinces is therefore related to current or relatively recent plate boundary activity, and is not considered to be genuinely intraplate.

The overall uniformity in the midplate stress province, which continues into the western half of the Atlantic ocean floor to within about 250 km of the mid-Atlantic ridge, is interpreted by Zoback & Zoback (1980) as evidence that the ridge-push force is the source of the stress field. The possibility of drag-related compression as a mechanism is discounted by the authors on the grounds that, since the craton is moving southwest in a lower mantle reference frame, NE–SW compression should be most pronounced along the leading edge of the craton instead of the N–S extension which currently exists there. Moreover, seismic activity increases from west to east within the craton, whereas a decrease east might be expected with the basal drag model.

The relatively large number of data points for the western part of the Eurasian craton indicates a nearly uniform NW–SE oriented maximum horizontal compressive stress. Zoback *et al.* (1989) point out that this direction is significantly clockwise of the WNW–directed absolute velocity of the plate. They prefer to relate the stress field to the convergence velocity between the Eurasian and African plates, which correlates rather better with the data. However, the extensive U.K. data set indicates a more N–S direction for the maximum horizontal compression than that predicted either from the continent convergence or from the North Atlantic ridge push force. The authors point out that more sophisticated modelling is required to predict accurately the stress field arising from a combination of plate boundary forces. Nevertheless it is clear that, Worldwide, there is a correlation between stress and absolute plate velocity that is too good to be fortuitous, prompting Zoback *et al.* (1989) to conclude that the net plate boundary forces responsible for moving the plates also dominate the stress distribution in the plate interiors.

Regions of extension on the continents occur mainly

in areas of anomalously high elevation. Such areas include, for example, the intracratonic rift systems of Africa, plateau uplifts associated with recent collision (e.g. Tibet) and thermally active extensional regions (e.g. western North America).

PROCESSES: ORIGIN OF CRATONIC STRUCTURES

Types of cratonic process

It is convenient to divide possible modes of origin of cratonic structures into those relating to horizontal plate movements, and those that are not directly related to such movements. Those structures that can be explained by horizontal compression or extension have already been discussed in the context of present-day stress distribution. It has been argued that forces arising from plate boundary processes are most probably responsible for the intraplate stress distribution and, by inference, these forces must also be responsible to a significant extent for the structures themselves. The close correspondence between the earthquake focal mechanism data and the *in situ* stress determinations indicates a direct relationship between the regional stress field and individual fault movements.

The origin of vertical intraplate movements, however, has been the subject of considerable uncertainty and debate. Plate boundary processes can produce vertical movements at considerable distances from the plate boundary, as in the case of Central Asia. Loading of the lithosphere by thrust sheets, leading to crustal thickening, produces an isostatically controlled uplift in the region of the load, and a flexural depression beneath, which extends to the region beyond the load, producing a foreland basin. Similar processes undoubtedly operate in the case of intraplate fold/thrust belts, such as the Tien Shan, where the horizontal displacement may be transferred eventually to a nearby plate boundary.

Other processes that have been considered are unrelated, or only indirectly related, to plate movements or plate boundary effects. The most important of such processes is the isostatic response to changes in the density structure or thickness of the lithosphere, or to the removal of a load, as in post-glacial rebound. The most important problems in the interpretation of cratonic structures arise in considering the origins of basins, uplifts, rifts and lineaments, and these will now be considered.

Origin of basins and rifts

Intraplate cratonic basins are divided into three categories by Bally (1980, 1982): those located on earlier rifts, those located in transform zones and those located on former back-arc spreading basins. In both cases, the origin of the basins is regarded as extension, leading to crustal thinning and depression of the surface. A further category recognized by Bally is the passive-margin basin, which originates as a plate-boundary structure but evolves into an intraplate structure with time.

The implication of Bally's classification of basins is that the origin of the basin is related to an inherited crustal structure, that is, one resulting from a former period of tectonic activity that caused the surface depression. There is considerable controversy over the origin of the depressions required to create cratonic basins, and the establishment of a generally agreed model has been hampered by lack of evidence as to the deep structure of many basins. A number of possible mechanisms that have been put forward are listed by Bally: (1) sediment loading; (2) isostatic response to cooling and density increase; (3) lithosphere stretching and thinning; (4) emplacement of dense, mantle-derived, igneous material within the crust; (5) density increase of lower-crustal rocks due to a gabbro-eclogite phase change; (6) basification or 'oceanization' of the continental crust by the introduction of ultramafic material from the mantle; and (7) creep of ductile lower-crustal material towards the ocean at a passive margin. Most of these suggested mechanisms are variants of two basic processes: extensional thinning and 'basification' by introduction of mantle-derived mafic/ultramafic material into the crust. Both processes result in an isostatic response to the change in density distribution in the lithosphere.

Extensional thinning is now widely referred to as the '*McKenzie model*' (McKenzie 1978) (also see Chapter 11). This model initially envisaged 'instantaneous' lithosphere extension, producing a surface depression and a corresponding upwards bulge in the base of the lithosphere. The thinned lithosphere generates a thermal anomaly, since the warm asthenosphere is now closer to the surface in the stretched section. As the initial basin fills with sediment, the thermal anomaly gradually decays, and the base of the lithosphere returns to its original level. This slow cooling results in gradual isostatic subsidence over a period of about 60 Ma. Jarvis & McKenzie (1980) showed that the instantaneous model predicts the subsidence geometry accurately for many basins, provided that the duration of the initial stretching is less than $60/\beta$, where β, the stretching factor, is less than 2 (i.e. 100% extension) and $60/(1 - 1/\beta)$ if $\beta \geq 2$. The model has been applied, for example, to the Michigan basin (Klein & Hsui 1987) and the North Sea basin (Christie & Sclater 1980, Barton & Wood 1984).

The second basic process that must be considered is the isostatic response to a change in the density structure of the lithosphere occurring in the absence of an initial stretching. This is the mechanism associated with an initial thermal anomaly, such as that produced by a *mantle plume* or *hot spot* (e.g. see Burke & Dewey 1973). Such a process produces an initial decrease in density to which the response is an isostatic uplift. This in turn may lead to extensional rifting and magma emplacement. On cooling, the thermal response is similar to that of the McKenzie model, with the cooled denser lithosphere sinking isostastically to form a surface depression.

There has been considerable discussion on the origin of rifts between proponents of the '*plume-generated*' or '*mantle-activated*' *mechanism*, and those who support an extensional or '*lithosphere-generated*' *origin* (e.g. see Condie 1982). It seems probable that both mechanisms operate. It is accepted that plumes exist and in many cases underlie lithosphere in a state of general compression, as in Africa at the present time. In such cases, clearly, the extensional structures present within the craton are a secondary effect of the thermal anomaly generated by the plume. It is equally valid to point out that there are periods in earth history, such as that attending the break-up of Pangaea and earlier supercontinents, when a general state of extensional stress appears to have existed throughout large regions of the contemporary craton. At such times, numerous coeval rifts and extensional basins were formed (e.g. see Figs. 10.1a & b), many of which display no evidence of prior thermal elevation, or igneous effects that could be attributed to the existence of a mantle plume. Indeed, given the inferred distribution of plumes during the Pangaea break-up, for instance, it would not be possible to explain all the extensional structures on that basis. Regardless of the actual mechanism of initiation of the rift, it is clear that many basins that have been studied in detail are located on early rift structures, and that there is a marked concentration of these structures at particular periods of geological history, in fact precisely those periods when supercontinent break-up occurred (e.g. see Klein & Hsui 1987).

Subsidence due to sediment loading is a subsidiary mechanism which is dependent on the gravitational response to imposing an extra mass on the surface (Dietz 1963, Walcott 1972). This mechanism cannot initiate subsidence, since an initial depression must exist to enable sufficient sediment to accumulate; however, it must operate in conjunction with other mechanisms to produce an additional downward force. Bott (1980), in an adaptation of the flexure model developed by Walcott (1972) shows that, in a 200 km wide transition zone at a passive continental margin, a thick pile of sediment at a major delta, such as that of the Niger, will produce a downwarp extending about 200 km beyond the area of the initial load, and may eventually produce a sediment pile twice to three times the original water depth in thickness. The sediment loading mechanism provides a means of explaining the continued evolution of a basin long after the initial causal mechanism has ceased to operate and may explain the very long life (100–700 Ma) of many cratonic basins.

Flexure is not considered by Bally (1980) as a mechanism for producing basins. However it is now accepted that *foreland basins*, which are marginal cratonic structures, may be produced by the flexural response of the lithosphere to the loading effect of thrust sheets at the leading edge of a collisional orogenic belt such as the Alps or Himalayas. The *flexural depression* extends much further into the craton than the margin of the load, forming a basin in advance of the thrust front. Intracratonic basins can form in the same way, marginal to intracratonic fold/thrust belts. Flexural basins will be underlain by crust of normal thickness, in contrast to the thinned crust underlying extensional basins.

Origin of uplifts

Three plausible mechanisms have been proposed to explain intraplate uplifts: (1) the isostatic response to a reduction in density; (2) the imposition of a superincumbent load, resulting in lithosphere thickening; and (3) upward flexure of the lithosphere. Isostatic uplift may be a response to a reduction in mass (e.g. due to erosion of a topographic high) or to a reduction in mean density, due to the thermal effect of a mantle plume or other local heat source. The production of thermally generated uplifts, and their relationship to rifting and subsequent basin formations has already been discussed. Examples of such structures are the East African and Rhenish uplifts (see above). The Colorado plateau is an example of a thermally generated uplift that has not been affected by extensional rifting, although it is bounded by regions (the Basin and Range province and the Rio Grande rift) that have experienced extension.

The erosion of material from a thermally generated uplift will produce a further isostatic response which will tend to renew and perpetuate the uplift until such time as cooling of the thermal anomaly produces an isostatic response in the reverse direction.

Loading of the lithosphere by compressional thrusting is a relatively unimportant mechanism in the formation of cratonic uplifts simply because such structures are comparatively rare, for reasons that have already been explored. Again, erosion will provoke an isostatic response that will act to renew the uplift until the over-thickened lithosphere returns to normal.

Upward flexure of the lithosphere will take place as a consequence of its lateral strength. Flexural downbending in one place, due to loading or isostatic depression (or subduction, in the case of oceanic lithosphere) is accompanied by a tilting of the lithosphere towards the area of the flexural depression, causing a corresponding, but smaller, *flexural upwarp*. The uplift will be compensated by the flow of asthenospheric material into the area of the bulge producing a small positive thermal anomaly which will tend to accentuate the uplift. Erosion and thermal decay will counteract and eventually reverse the uplift process.

The processes of uplift and depression are complementary, and probably balance out, broadly, over long periods of time. Both processes appear to be self-limiting, indeed self-reversing, with time because of the tendency of any gravitational (isostatic) or thermal anomaly to reduce and eventually be eliminated due respectively to ductile creep and thermal decay. Such reversals or inversions are characteristic of the evolution of cratonic structure over periods of the order of hundreds of Ma, as exemplified for example in the history of the Russian platform (Aleinikov *et al.* 1980).

Origin of lineaments: rhegmatic or planetary faulting

The idea of deep faults is historically associated with the *rhegmatic concept*. Rhegmatic or planetary tectonics is a global-scale fracture system ascribed to forces arising from the behaviour of the planet as a whole. The classical concept of rhegmatic tectonics (e.g. Hobbs 1911, Sonder 1947) is inconsistent with plate tectonic theory, although attempts have been made to reconcile the two (e.g. Kogan & Kitarov 1982). It has been suggested that the long-term permanence of the rhegmatic pattern in certain cratonic regions (e.g. Kvet 1983) may be explained by the inheritance of old directions in younger epochs influenced by possibly quite different stress fields (e.g. Bryan 1986, Rathore & Hospers 1986). The concept of long-lived deep faults or lineaments, however, has a usefulness that is quite independent of the validity of rhegmatic tectonics.

REFERENCES

Aleinikov, A. L., Bellavin, O. V., Bulashevich, Yu. P., Tavrin, I. F., Maksimov, I. M., Rudkevich, M. Ya., Nalivkin, V. D., Shablinskaya, N. V. & Surkov, V. S. 1980. Dynamics of the Russian and West Siberian platforms. In: *Dynamics of Plate Interiors* (edited by Bally, A. W., Bender, P. L., McGetchin, T. R. & Walcott, R. I.). *Am. Geophys, U./Geol. Soc. Am. geodynam. Ser.* **1**, 53–71.
Baker, B. H. & Wohlenberg, J. 1971. Structure and evolution of the Kenya rift valley. *Nature, Lond.* **229**, 538–542.
Bally, A. W. 1980. Basins and subsidence — a summary. In: *Dynamics of Plate Interiors* (edited by Bally, A. W., Bender, P. L., McGetchin, T. R. & Walcott, R. I.). *Am. Geophys. U./Geol. Soc. Am. Geodynam. Ser.* **1**, 5–20.
Bally, A. W. 1982. Musings over sedimentary basin evolution. *Phil. Trans. R. Soc. Lond.* **A305**, 325–338.
Barton, P. & Wood, R. 1984. *Geophys. J. R. astr. Soc.* **79**, 987–1022.
Beach, A. 1986. A deep seismic reflection profile across the northern North Sea. *Nature, Lond.* **323**, 53–55.
Bott, M. P. 1980. Mechanisms of subsidence at passive continental margins. In: *Dynamics of Plate Interiors*. (edited by Bally, A. W., Bender, P. L., McGetchin, T. R. & Walcott, R. I.). *Am. Geophys. U./Geol. Soc. Am. Geodynam. Ser.* **1**, 27–35.
Bott, M. P. & Kusznir, N. J. 1984. Origins of tectonic stress in the lithosphere. *Tectonophysics* **105**, 1–14.

Brockwicz-Lewiński, W., Pożaryski, W. & Tomczyk, H. 1981. Large-scale strike-slip movements along SW margin of the East-European Platform in the Early Paleozoic (in Polish). *Przeglad Geologiczny* **8**, 385–397.
Bronner, G., Roussel, J. & Trompette, R. 1980. Genesis and geodynamic evolution of the Taoudeni cratonic basin (upper Precambrian and Paleozoic), western Africa. In: *Dynamics of Plate Interiors* (edited by Bally, A. W., Bender, P. L., McGetchin, T. R. & Walcott, R. I.). *Am. Geophys, U./Geol. Soc. Am. Geodynam. Ser.* **4**, 81–90.
Bryan, W. B. 1986. Tectonic controls on initial continental rifting and the evolution of young ocean basins — a planetary perspective. *Tectonophysics* **132**, 103–115.
Burke, K. & Dewey, J. F. 1973. Plume-generated triple junctions: key indicators in applying plate tectonics to old rocks. *J. Geol.* **81**, 406–433.
Christie, P. A. & Sclater, J. C. 1980. An extensional origin for the Buchan and Witchground graben in the North Sea. *Nature, Lond.* **283**, 729–732.
Condie, K. C. 1982. *Plate Tectonics and Continental Drift.* Pergamon, Oxford.
Cox, S. F. & Etheridge, M. A. 1989. Coupled grain-scale dilatancy and mass transfer during deformation at high fluid pressures: examples from Mount Lyell, Tasmania. *J. Struct. Geol.* **11**, 147–162.
Cekunov, A. V. & Kucma, V. G. 1979. The deep structure of faults (in Russian). *Geotektonika* **5**, 24–37.
Czubla, P. 1988. Tectonics of the Radomsko Elevation on the basis of mesostructural methods (in Polish). *Przeglad Geologiczny* **10**, 560–566.
Daracott, B. W., Fairhead, J. D., Girdler, R. W. & Hall, S. A. 1973. The East African rift system. In: *Implications of Continental Drift to the Earth Sciences* (edited by Tarling, D. H. & Runcorn, S. K.). Academic Press, London.
Dietz, R. S. 1963. Collapsing continental rises: an actualistic concept of geosynclines and mountain building. *J. Geol.* **71**, 314–333.
Donato, J. A. & Tully, M. C. 1981. A regional interpretation of North Sea gravity data. In: *Petroleum Geology of the Continental Shelf of Northwest Europe* (edited by Illing, L. V. & Hobson, G. D.). Heyden, London, 65–75.
Girdler, R. W. & Darracott, B. W. 1972. African poles of rotation. *Comments Earth Sci. Geophys.* **2**, 7–15.
Goguel, J. 1983. A short note on continuity or discontinuity in the global tectonic plate velocities. *Tectonophysics* **100**, 1–4.
Hancock, P. L. & Engelder, T. 1989. Neotectonic joints. *Bull. geol. Soc. Am.* **101**, 1197–1208.
Harding, T. P. 1973. Newport-Inglewood trend, California — an example of wrenching style of deformation. *Bull. Am. Ass. Petrol. Geol.* **57**, 97–116.
Hobbs, W. H. 1911. Repeating patterns in the relief and in the structure of the land. *Bull. geol. Soc. Am.* **22**, 123–176.
Hobbs, W. H. 1912. *Earth Features and their Meaning.* Macmillan, New York.
Illies, J. H. 1981. Mechanism of graben formation. *Tectonophysics* **73**, 249–266.
Illies, J. H. & Greiner, G. 1978. Rhinegraben and the Alpine system. *Bull. geol. Soc. Am.* **89**, 770–782.
Jarozsewski, W. 1972. Mesoscopic structural criteria of tectonics of non-orogenic areas: an example from the north-eastern margin of the Swietokrzyskie Mountains (in Polish). *Studia Geol. Pol.* **38**. Wydawnicta Geologiczne, Warszawa.
Jarvis, G. T. & McKenzie, D. P. 1980. Sedimentary basin formation with finite extension rates. *Earth Planet. Sci. Lett.* **48**, 42–52.
Kamaletdinov, M. A., Kazancev, Y. V. & Kazanceva, T. T. 1980. Origin of the rootless platform structures (in Russian). *Doklady AN SSSR* **250**, 1204–1208.
Kamaletdinov, M. A., Kazancev, Y. V., Kazanceva, T. T. & Postnikov, D. V. 1981. *Charriage and Overthrust*

Structures in Platform Basements (in Russian). Nauka, Moscow.
Kameletdinov, M. A., Kazancev, Y. V., Kazanceva, T. T. & Postnikov, D. V. 1987. *Charriage and Overthrust Structures in Platform Basements* (in Russian). Nauka, Moscow.
Klein, G. de V. & Hsui, A. T. 1987. Origin of cratonic basins. *Geology* 15, 1094–1098.
Kogan, A. B. & Kitarov, Y. N. 1982. Main regularities of the fault network on platform areas of USSR (in Russian). *Geotektonika* 6, 80–87.
Korcemagin, V. A. & Ryabostan, Y. S. 1987. Tectonics and stress fields of Donbass. In: *Stress Fields and Deformations in the Earth's Crust* (in Russian). Nauka, Moscow, 164–170.
Kozerenko, V. N. & Lartzev, V. S. 1987. On the trans-African zone of deep faults (as a part of the global fault net) (in Russian). *Bull. Mosk. Obsc. Isp. Prir.* 62(2), 18–27.
Kuang, J., Long, L. T. & Mareschal, J.-C. 1989. Intraplate seismicity and stress in the southeastern United States. *Tectonophysics* 170, 29–42.
Kucay, V. K., Guseva, T. V. & Ulasina, S. A. 1981. A contribution to the geodynamics of faults (in Russian). *Izvestya AN SSSR* 10, 45–56.
Kusznir, N. J. & Park, R. G. 1982. Intraplate lithosphere strength and heat flow. *Nature* 299, 540–542.
Kusznir, N. J. & Park, R. G. 1984. Intraplate lithosphere deformation and the strength of the lithosphere. *Geophys. J. R. astr. Soc.* 79, 513–538.
Kusznir, N. J. & Park, R. G. 1986. The extensional strength of the continental lithosphere: its dependence on geothermal gradient, and crustal composition and thickness. In: *Continental Extensional Tectonics* (edited by Coward, M. P., Dewey, J. F. & Hancock, P. L.). *Spec. Publs geol. Soc. Lond.* 28, 35–52.
Kvet, R. 1983. *Ruptures of the Earth's Crust and Regularities in their Orientation* (in Czechoslovakian). Ustav, Brno.
McGetchin. T. R., Burke, K. C., Thompson, G. A. & Young, R. A. 1980. Mode and mechanisms of plateau uplifts. In: *Dynamics of Plate Interiors* (edited by Bally, A. W., Bender, P. L., McGetchin, T. R. & Walcott, R. I.). *Am. Geophys. U./Geol. Soc. Am. Geodynam. Ser.* 1, 99–110.
McKenzie, D. P. 1967. The viscosity of the mantle. *Geol. J. R. astr. Soc.* 14, 297–305.
McKenzie, D. P. 1978. Some remarks on the devolopment of sedimentary basins. *Earth Planet. Sci. Lett.* 40, 25–32.
Megnien, C. & Pomerol, C. 1980. Subsidence of the Paris basin from the Lias to the late Cretaceous. In: *Dynamics of Plate Interiors* (edited by Bally, A. W., Bender, P. L. McGetchin, T. R. & Walcott, R. I.). *Am. Geophys. U./Geol. Soc. Am. Geodynam. Ser.* 1, 91–92.
Mikbel, S. & Zacher, W. 1986. Fold structures in northern Jordan. *Neues Jb. Geol. Palaont. Mh.* 4, 248–256.
Molnar, P. & Tapponnier, P. 1975. Cenozoic tectonics of Asia: effects of a continental collision. *Science* 189, 419–426.
Morgan, P. 1982. Heat flow in rift zones. In: *Continental and Oceanic Rifts* (edited by Palmason, G.). *Am. geophys. U. Geodym. Ser.* 8, 107-122.
Morgan, P. 1983. Uplift of the Colorado plateau and its relationship to volcanism and rifting in the adjacent Basin- and- Range and Rio Grande rift. *Proc. XVIII General Assembly Int. U. Geodesy Geophys.* Hamburg (Abs), 576.
Morgan, P. 1984. The thermal structure and thermal evolution of the continental lithosphere. *Phys. Chem. Earth.* 15, 107–193.
O'Hara, K. 1990. State of strain in mylonites from the western Blue Ridge province, southern Appalachians: the role of volume loss. *J. Struct. Geol.* 12, 419–430.
O'Leary, D. W., Friedman, J. D. & Pohn, H. A. 1976. Lineament, linear, lineation: some proposed new standards for old terms. *Bull. geol. Soc. Am.* 87, 1463–1469.
Patakha, E. I., Polyakov, A. I. & Sevrugin, N. N. 1978. The role of mechanical factors in thermal regimes of great fault zones (in Russian). *Geotektonika* 4 79–90.
Peyve, A. V. 1990. *Deep-seated Faults and their Role in the Structure and Evolution of the Earth's Crust* (selected works) (in Russian). Nauka, Moscow.
Pollack, H. N. & Chapman, D. S. 1977. On the regional variation of heat flow, geotherms and lithosphere thickness. *Tectonophysics* 38, 279–296.
Popkov, V. I. 1985. Traces of tangential compression in the platform structure of the western regions of the Turan plate (in Russian). *Doklady AN SSSR* 284, 674–678.
Pożaryski, W. 1975. Geological interpretation of DSS international profile VII (in Polish). *Przegląd Geologiczny* 4, 163–171.
Rathore, J. S. & Hospers, J. 1986. A lineaments study of southern Norway and adjacent offshore areas. *Tectonophysics* 131, 257–285.
Riazanov, G. V. 1973. *Morphology and Origin of the Folds in the Nepa Zone (Southern Part of the Siberian Platform)* (in Russian). Nauka, Novosibirsk.
Sankov, V. A. 1989. *The Depth of Fault Penetration.* (in Russian). Nauka, Novosibirsk.
Sawyer, D. S., Swift, B. A., Sclater, J. G. & Toksoz, M. N. 1982. Extensional model for the subsidence of the northern United States Atlantic continental margin. *Geology* 10, 134-140.
Sengör, A. M. C. 1984. The Cimmeridge orogenic system and the tectonics of Eurasia. *Spec. Publs geol. Soc. Am.* 195, 1–82.
Shatsky, N. S. 1955. The origin of the Pachelma trench: comparative tectonics of ancient platforms. *Bull. Moscow Soc. Nat. Geol. Sec.* 30, 5–26.
Sleep, N. H. & Sloss, L. L. 1980. The Michigan basin. In: *Dynamics of Plate Interiors* (edited by Bally, A. W., Bender, P. L., McGetchin, T. R. & Walcott, R. I.). *Am. Geophys. U./Geol. Soc. Am. Geodynam. Ser.* 1, 93–98.
Sonder, R. A. 1947. *Mechanik der Erde.* Schweizerbart, Stuttgart.
Stille, H. 1913. *Tektonische Evolutionen und Revolutionen in der Erdrinde.* Veit, Leipzig.
Teyssier, C. 1985. A crustal thrust system in an intracratonic tectonic environment. *J. Struct. Geol.* 7, 689-700.
Thom, W. T. Jr 1923. The relation of deep-seated faults to the surface structural features of central Montana. *Am. Ass. Petrol. Geol. Bull.* 7, 1–13.
Vita-Finzi, C. 1986. *Recent Earth Movements.* Acadamic Press, London.
Vitorello, I. & Pollack, H. N. 1980. On the variation of heat flow with age and the thermal evolution of continents. *J. geophys. Res.* 85, 983–995.
Wagner, G. H. 1967. Druckspannungsindeizien in den Sedimenttafeln des Rheinischen Schildes. *Geol. Rdsch.* 56, 906–913.
Walcott, R. I. 1970. Flexural rigidity, thickness, and viscosity of the lithosphere. *J. geophys. Res.* 75, 3941–3954.
Walcott, R. I. 1972. Gravity, flexure, and the growth of sedimentary basins at a continental edge. *Bull. geol. Soc. Am.* 83, 1845–1848.
Windley, B. F., Allen, M. B., Zhang, C., Zhao, Z.-Y. & Wang, G. R. 1990. Paleozoic accretion and Cenozoic redeformation of the Chinese Tien Shan range, Central Asia. *Geology* 18, 128–131.
Zeuner, F. E. 1958. *Dating the Past* (4th edn). Methuen, London.
Ziegler, P. A. 1982. Faulting and graben formation in western and central Europe. *Phil. Trans. R. Soc. Lond.* A305, 113–143.
Zoback, M. L. & Zoback, M. 1980. State of stress in the

conterminous United States. *J. geophys. Res.* **85**, 6113–6156.

Zoback, M. L. & Zoback, M. 1989. Tectonic stress field of the continental United States. In: *Geophysical Framework of the Continental United States.* (edited by Pakiser, L. C. & Mooney, W. D.) *Mem. geol. Soc. Am.* **172**, 523–539.

Zoback, M. L., Zoback, M. D. *et al.* 1989. Global patterns of tectonic stress. *Nature, Lond.* **341**, 291–298.

CHAPTER 11

Continental Extensional Tectonics

ALAN ROBERTS and GRAHAM YIELDING

INTRODUCTION

IN this chapter we review current understanding of how the continental lithosphere (crust and uppermost mantle) accommodates extensional strain. The majority of the chapter is devoted to a discussion of stretching processes that affect the whole lithosphere and are driven by the relative motion of lithospheric plates. Towards the end of the chapter, however, we contrast the processes and resulting structures of whole-lithosphere deformation with those produced purely by 'thin-skinned' deformation of the uppermost part of the continental crust. Such 'thin-skinned' extension is fundamentally different from whole-lithosphere extension. It is generally driven by 'down-slope' gravity processes or the flow of low-viscosity material, such as salt (see Chapter 8), and is decoupled from the underlying basement. Thin-skinned extension can be regarded merely as the translation of material across the surface of the underlying lithospheric plate.

An appropriate sub-title to this chapter might be "Processes controlling the formation of extensional sedimentary basins". We attempt to review the processes at three different scales. Firstly, we look at stretching models, developed to explain whole-basin geometry. Secondly, we come down in scale and review knowledge of the major faults which partition whole basins into discrete structural elements (also see Chapter 6). Such faults are believed to penetrate the entire thickness of the seismogenic layer of the upper crust and transfer their displacement into the aseismic lower crust. Thirdly, we come down in scale again, and look at 'small' faults contained entirely within the seismogenic upper crust. These are the types of fault most commonly observed in surface outcrop.

The increased understanding, over the last 10–15 years, of the processes which control the formation of extensional basins, derives from a number of sources. Notable among these are the increasing availability of reflection seismic data, the undertaking of deep seismic-reflection profiling, an increased understanding of the physical processes that are likely to operate in the lower crust and upper mantle, and perhaps also the beginning of collaboration between academic and industrial geologists and geophysicists working on similar problems. Our review, however, takes one particular publication as a landmark (McKenzie 1978), and we begin our discussion from there. Much subsequent and important work has its origins in the original 'McKenzie model'.

STRETCHING MODELS

Instantaneous stretching

McKenzie (1978) was the first to describe quantitatively the implications of extension of the continental lithosphere. He proposed a simple model in which the continental lithosphere is uniformly (within any given vertical column) and rapidly stretched (Fig. 11.1). Stretching is accomplished by lithospheric thinning which allows passive upwelling of hot asthenosphere below the stretched lithosphere. The amount of vertical thinning accommodated by the lithosphere is referred to as the *stretching factor*, denoted by β, the length ratio of a deformed line to its undeformed equivalent. Isostatic compensation is assumed to accompany stretching. Provided the unstretched crust was initially greater than 18 km in thickness, isostatic compensation results in subsidence of the top surface of the lithosphere and a rise of the Moho. The amount of this initial subsidence (S_i), that is the depth of the newly formed basin, is given by:

$$S_i = d\left(1 - \frac{1}{\beta}\right) \qquad (1)$$

where d is a factor which incorporates the initial thicknesses of crust and lithosphere, the densities of mantle, crust and new basin infill, the temperature at the base of the lithosphere, and the coefficient of thermal expansion for mantle and crust (see also Le Pichon & Sibuet 1981). Barr (1987a) has subsequently shown that, for an initial lithosphere in isostatic equilibrium, most of these parameters are not independent. The value of d is largely controlled by the density of the new basin infill, within the range c. 2.5 (for an air-filled basin) to c. 7.3 (for a sediment-filled basin).

In the basic *McKenzie model* the initial stretch and

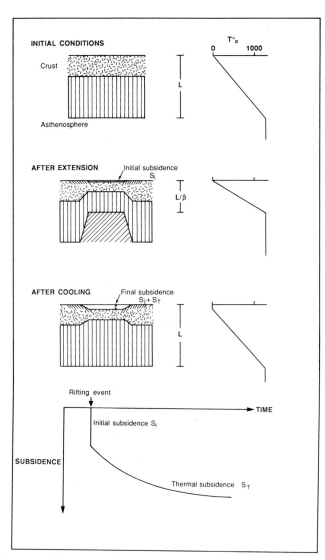

Fig. 11.1. Illustration of the lithospheric stretching model of McKenzie (1978). The crust (stippled) and mantle lithosphere (vertical lines) are stretched instantaneously by a factor β. Below the stretched basin, lithosphere is replaced by hotter asthenosphere (diagonal lines), while the geothermal gradient within the lithosphere is raised. Extension causes an initial subsidence of the basin floor, S_i. Cooling of the thermal anomaly causes further subsidence, which decays exponentially, S_T, and converts uprisen asthenosphere to mantle lithosphere.

subsidence are assumed to be geologically instantaneous. Subsequent to this *instantaneous stretching event* a thermal anomaly exists below the newly formed basin, as a result of the uprisen asthenosphere (Fig. 11.1). The lithosphere/asthenosphere boundary is simply an isotherm (generally taken to be at 1333°C) within the mantle, not a material boundary. With time the thermal anomaly decays, replacing relatively low-density asthenosphere with relatively high-density mantle lithosphere. This increase in bulk density loads the overlying basin from below, and causes a period of *time-dependent, thermal subsidence*, whose magnitude decays exponentially. In a simplified form, thermal subsidence at a time t is given by:

$$S_T(t) \approx Er(1 - \exp\frac{t}{\tau}) \quad (2)$$

where E depends on properties of the lithosphere, r depends on the stretching factor β, and τ is the thermal time-constant of the lithosphere.

Thus the basic McKenzie model is a two-stage model involving:

(i) an initial, rapid, isostatically-controlled subsidence, accompanying geologically-instantaneous stretching of the lithosphere;

(ii) a slow, exponential subsidence caused by the maintenance of isostatic equilibrium during the cooling of upwelled asthenosphere. No lithospheric stretching accompanies this period of thermal subsidence.

The basic McKenzie model is elegant in its simplicity, and has provided an explanation for the common occurrence of relatively unfaulted *sag-basins* overlying rifts or areas of tilted fault-blocks. Many subsequent workers have developed the model further, in order to incorporate second-order effects that may be geologically important.

Time-dependent (finite-rate) stretching

A fundamental assumption of the McKenzie (1978) model is that stretching occurs geologically instantaneously, such that all thermal cooling effects occur after the initial subsidence of the basin. A reasonable question to ask would be whether this model remains valid if stretching is not instantaneous but occurs over a finite interval of time.

This question has been addressed by Jarvis & McKenzie (1980) and Cochran (1983). Their work has shown that, at relatively fast stretching rates, the partitioning of total basin subsidence into *initial (syn-stretching)* and *thermal (post-stretching)* components approximates closely to instantaneous stretching. If, however, stretching rates are relatively slow, then significant thermal subsidence may accompany stretching. This results in an amplification of S_i, the calculated initial subsidence, and a reduction in S_T, the post-stretch thermal subsidence. Total basin subsidence (nominally at infinite time) remains the same for both instantaneous and time-dependent models.

Whether stretching is 'fast' or 'slow' depends on the average strain rate relative to the rate at which the stretched lithosphere can lose heat by conduction. Jarvis & McKenzie (1980) showed that, in general, if stretching occurs in a time less than $60/\beta^2$ Ma, then the time-dependent model gives results very similar to the instantaneous-stretching model. When stretching lasts longer than $60/\beta^2$ Ma, however, then subsidence predictions from the instantaneous model begin to deviate significantly from the more complete treatment of the time-dependent model. In practice this means that in a weakly stretched basin (e.g. $\beta<1.5$), providing stretching lasted less than c. 30 Ma, the basic McKenzie model is a good approximation. In a highly-stretched basin (e.g. $\beta>2.5$), however, instantaneous stretching is only a reliable approximation if stretching lasted less than c.10 Ma.

Lateral heat-flow models

An assumption of both the instantaneous and time-dependent stretching models outlined above is that the thermal anomaly generated below the stretched lithosphere decays by vertical conduction only. There is no heat loss to the rift margins in these models. Jarvis (1984) and Alvarez *et al.* (1984) examined the effects of lateral heat flow at basin margins, that is, conduction from the thermal anomaly below the basin into the cooler, unstretched lithosphere flanking the rift. The models predict that this will have a number of effects.

While the basin itself undergoes rapid initial subsidence, conduction of heat into the basin margins lowers their bulk density and causes the surface of the lithosphere to rise isostatically. Such uplift may cause erosion of the marginal areas, preserved as an unconformity in the geological record.

The *marginal-uplift*, however, being of thermal rather than mechanical origin, is entirely recoverable. During the *post-stretch phase of thermal subsidence* the basin margins subside, together with the basin itself. If no erosion occurred, the basin margins will return to their initial elevation, perhaps leaving no geological record of their temporary elevation. If, however, erosion accompanied initial uplift, total thermal recovery will cause the basin margins to subside below their initial, pre-stretch elevation.

Lateral heat flow also affects the values of both S_i and S_T within the basin itself. As an approximate rule, however, the equations of McKenzie (1978) remain valid at the centre of a basin if the basin width is greater than the thickness of the lithosphere (*c.* 125 km).

Depth-dependent stretching

Another assumption of the basic McKenzie model that has been the subject of subsequent modification is that of uniform stretching of the lithosphere within any given vertical column. The stretching factor (β) in the McKenzie model is depth-independent. A number of subsequent models have been published in which β varies with depth.

The first depth-dependent stretching model was developed by Royden & Keen (1980). From a study of wells, offshore of eastern Canada, they found that the initial and thermal subsidence components, recorded by the stratigraphy in the wells, were not in the proportions predicted by the McKenzie model. In general they found that the thermal-subsidence component was too great for both the initial-subsidence component and the calculated crustal extension. They inferred that a greater thermal anomaly than that predicted by the McKenzie model had developed below the basin in question. To explain the presence of this anomaly they suggested that the mantle lithosphere below the basin had been stretched more than the crust. The resulting increased

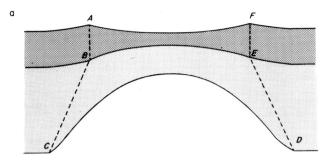

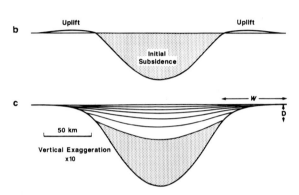

Fig. 11.2. Depth-dependent stretching. (a) Stretching in the lithospheric mantle (light stipple) distributed over a wider area than stretching in the crust (dark stipple). This causes (b) syn-rift uplift of the rift flanks and enhanced initial subsidence of the basin, followed by (c) reduced thermal subsidence within the basin, by comparison with the McKenzie model, and collapse of the rift flanks. From: (a) Rowley & Sahagian (1986), (b) & (c) White & McKenzie (1988).

thermal anomaly would thus act both to buffer initial subsidence, by reducing bulk density, and to amplify thermal subsidence.

The problem with Royden & Keen's model is that it invokes both β (stretching factor) and e (the actual amount of extension) to be greater in the mantle lithosphere than in the crust. The model therefore requires a physical decoupling of the crust and lithospheric mantle and implies that mass is not preserved during stretching.

In order to circumvent this problem, Hellinger & Sclater (1983), Rowley & Sahagian (1986), Coward (1986) and White & McKenzie (1988) have each proposed similar models in which a gradation in β exists between crust and mantle lithosphere, but in which total lithosphere extension is constant at all levels of the lithosphere (Fig. 11.2). Physical decoupling within the lithosphere is not a prerequisite of these models, while they all preserve mass during extension. A mass-balanced two-layer stretching model, which still invokes decoupling within the lithosphere, has been advocated by Weissel & Karner (1989).

In each of these models the total subsidence at any particular point in the basin, at infinite time, depends only on β_c (crustal stretching), as the total lithosphere

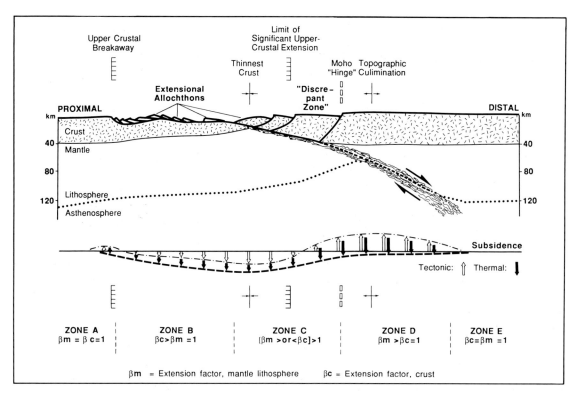

Fig. 11.3. The lithospheric 'simple shear' model of Wernicke (1985). Extension occurs by movement on an asymmetric shear zone which cuts the entire lithosphere. (Reprinted from *Canadian Journal of Earth Sciences*, Vol. 22, Wernicke B., Uniform, normal-sense simple shear of the continental lithosphere, pp. 108–125, © 1985 with permission from National Research Council.)

will eventually re-equilibrate to its pre-stretching thickness. If, however, the stretching of the mantle lithosphere (β_m) is greater than crustal stretching at a given position, then initial subsidence is buffered and thermal subsidence increased, by comparison with the instantaneous-stretching model. Conversely, if $\beta_c > \beta_m$ then the initial subsidence is increased at the expense of thermal subsidence. If the zone of mantle lithosphere extension extends beneath the flanks of the basin (i.e. at the flanks $\beta_m \gg \beta_c$ such that beneath the basin $\beta_c > \beta_m$), the basin flanks will be initially uplifted and then subside, in a manner similar to that predicted by models of lateral heat-flow (Fig. 11.2).

Many records exist of deeply-eroded basin margins (e.g. Hellinger & Sclater 1983) and a common explanation of this observation has been to invoke thermal uplift of rift margins as a result of depth-dependent stretching. A number of authors, however, have suggested that flank uplift is a mechanical process associated with faulting at the basin margin, rather than a transient thermal effect. Such fault-related processes are discussed later in this chapter.

Lithospheric simple shear

Each of the models discussed so far assumes that the gross extension of the lithosphere approximates to a pure shear deformation, with the faulted upper surface of the basin directly overlying the site of upwelled asthenosphere. These *pure shear models* impart a gross symmetry to the resulting basin, although it is not required that the distribution of β be uniform across the basin axis.

An alternative model of gross lithosphere deformation has been advanced by Wernicke (1981, 1985), in which it is envisaged that lithospheric extension is accommodated on a major, gently-dipping zone of simple shear, which cuts through from the free surface to the base of the lithosphere (Fig. 11.3). Extension of the faulted upper crust is therefore spatially remote from the upwelling asthenosphere which accommodates extension of the lower lithosphere.

Isostatic compensation is again assumed to operate in this model. Thus in the 'proximal' area of upper crustal faulting there will be an initial subsidence proportional to the extension, accompanied by a rise of, but no extension at, the base of the lithosphere. At the opposite 'distal' end of the shear zone, however, there will be a significant, isostatic uplift caused by thinning of the lower lithosphere beneath unthinned crust.

Following the extensional movement on the shear zone the 'proximal' area of thinned crust undergoes a modest thermal subsidence as the lithosphere re-equilibrates. The 'distal' uplift will also undergo thermal subsidence, which in the absence of erosion, would return the surface of the lithosphere to its pre-uplift elevation. The behaviour of the lithosphere at the distal part of the shear zone is therefore similar to that of the rift margins in some of the depth-dependent stretching models (e.g. Rowley & Sahagian 1986).

Wernicke (1981, 1985) described the *simple shear*

model in qualitative terms only, much of it based on observations in the Basin and Range province of the western U.S.A. More recent numerical studies by Buck *et al.* (1988) and Voorhoeve & Houseman (1988) have quantified the uplift and subsidence patterns of the model. Buck *et al.* (1988) have, in addition, contrasted the results of simple shear modelling with those of a pure shear extensional model. Significant results from these studies are as follows. (1) Maximum net uplift associated with the shear zone occurs in its hangingwall, above the point at which it leaves the base of the crust. (2) Uplift in the footwall to the shear zone is small, and is produced by lateral heat conduction, recoverable on cooling. (3) The total subsidence at infinite time, for any point in the resulting basin, is the same in the simple shear model as in a pure shear model. Total subsidence depends on the amount of crustal thinning only. (4) For the same values of β_c the simple shear model predicts less thermal subsidence than the pure shear model, with a corresponding increase in initial subsidence.

Wernicke (1981, 1985) applied the simple shear model to the evolution of the Basin and Range province, U.S.A., where its applicability is still debated (see later section on Metamorphic Core Complexes). Buck *et al.* (1988) showed that a simple shear model cannot account for the observed heat flow in the northern Red Sea, where asymmetric topography has been cited as evidence for simple shear.

The lithospheric simple shear model has also been applied, in a qualitative sense, to the evolution of the Viking graben in the northern North Sea (Beach 1986, Beach *et al.* 1987). There is, however, much debate over its applicability in this area, with Klemperer (1988), White (1989, 1990) and Klemperer & White (1989) arguing that deep seismic data and subsidence calculations favour an origin for the Viking graben by pure-shear stretching.

We would thus conclude our discussion of the simple shear model by pointing out that while variants on the pure shear model have now been successfully applied to many basins World-wide, a quantitative test of the applicability of the simple shear model has yet to be found.

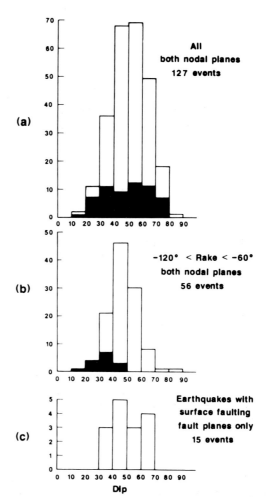

Fig. 11.4. Histograms of normal fault-plane dip compiled from seismological fault-plane solutions. (a) includes both nodal planes, either of which could be the fault, from 127 events, (b) is a subset of (a) in which the 56 events have near dip-slip displacement, (c) is a subset of (a) for cases in which nodal plane and surface fault break can be matched. All three histograms show the most common dip for the nodal planes to be in the range 30–60°. From Jackson & White (1989). (Reprinted from *Journal of Structural Geology*, Vol. 11, Jackson, J. A. and White, N. J. Normal faulting in the upper continental crust, pp. 15–36, © 1989 with permission from Pergamon Press Ltd, Headington Hill Hall, Oxford OX3 0BW, U. K.)

LARGE EXTENSIONAL FAULTS

The seismogenic layer

We define *large extensional faults* as those whose dimensions are sufficient to cut the entire seismogenic layer. The *seismogenic layer* is that part of the crust which responds to stress by brittle failure, that is, fault slip (cf. p. 378, Chapter 18). A significant proportion of fault slip occurs as earthquakes. On the time-scale of a single earthquake (tens of seconds), with very fast strain rates, the whole crust and mantle may behave as an elastic solid. Earthquakes within extending continental crust are, however, typically observed to nucleate only within its uppermost 15 km. Thus the range of recorded earthquake focal depths defines the seismogenic layer, in extensional areas, as being of approximately 15 km thickness (e.g. Chen & Molnar 1983, Jackson 1987).

The faults we consider within this section are therefore those which can be reasonably assumed to penetrate the crust to a depth of about 15 km. The reason for distinguishing these large faults from more minor faults (confined within the seismogenic layer) involves the way in which the rock volume around the faults deforms in response to slip on the fault surface. As discussed below, different controls act depending on fault size. The thickness of the seismogenic layer acts as an upper bound to the growth of large faults (Jackson & White 1989).

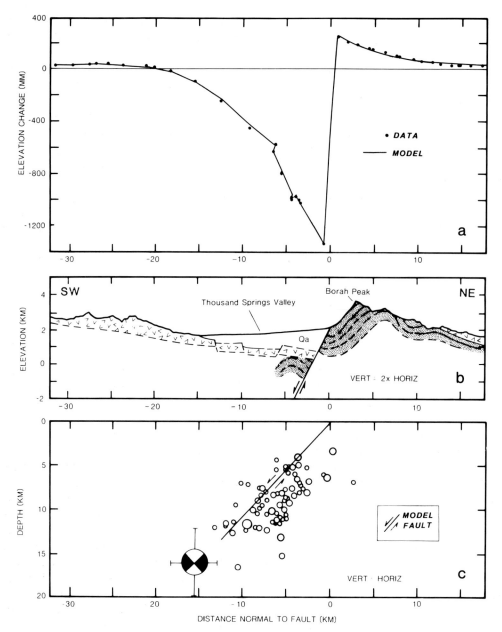

Fig. 11.5. Geodetic and geological data from the Lost River Fault, on which occurred the 1983 Borah Peak earthquake (from Stein *et al.* 1988). (a) Co-seismic deformation (dots) associated with the earthquake, measured by resurveying a levelling line across the fault. The model fit to the data (solid line) was calculated assuming a planar fault in an elastic medium. (b) Geological cross-section across the Lost River Fault. Qa = Quaternary alluvium, V = Tertiary volcanics, stipple = Palaeozoic and older. (c) Seismological data for the 1983 earthquake sequence. Error bars bracket the mainshock location, with the focal mechanism shown at the hypocentre. Small circles indicate after-shock locations. The 'model fault' is that used to generate the geodetic curve (solid) in (a). (Reprinted from *Journal of Geophysical Research*, Vol. 93, Stein, R. S., King, G. C. P. and Rundle, J. B. The growth of geological structures by repeated earthquakes, pp. 13319–13331, © 1988 American Geophysical Union.)

The geometry of major normal faults

Throughout the 1980s there was considerable debate about whether the major faults that accommodate crustal extension are essentially planar structures throughout the seismogenic layer or whether they form linked networks of 'listric' faults within the upper crust (also see Chapter 6).

We cannot directly observe the geometry of major sub-surface faults. Seismological observations of active faults can, however, provide a wealth of data on their geometry and kinematics. The overwhelming evidence of the earthquake data indicates that active, large normal faults are essentially *planar faults* (straight in cross-section) throughout the seismogenic layer, with dips in the range 30–60° (e.g. Stein & Barrientos 1985, Jackson 1987, Jackson *et al.* 1988, Jackson & White 1989) (Figs 11.4 and 11.5). There is no evidence for the existence of seismically-active, low-angle or markedly *listric faults* (concave upwards profile) provided by earthquake data.

The problem of determining the geometry at depth of now-inactive, large normal faults is probably best addressed using seismic reflection data. Conclusions about fault geometry drawn from reflection data can, however, be more subjective than conclusions derived from earthquake seismology. Seismic reflection data may be used to image fault surfaces directly, to infer their presence from cut-offs of imaged strata, or to constrain geological cross-sections from which fault geometry may be inferred by geophysical/geological modelling. From a purely interpretational viewpoint many examples can be found in the literature where both major listric faults and major planar faults have purportedly been recognized, the geometry of the interpreted faults depending largely on the bias of the interpretation. Studies of fault-plane reflections on seismic data (e.g. McGeary & Warner 1985, Kusznir & Matthews 1988, Young et al. 1990, Cartwright 1991, Yielding et al. 1991) are, however, beginning to suggest that now-inactive major faults present around the Northwest European and North Atlantic margin are planar structures, similar in geometry to the active normal faults described by earthquake data. Similarly, geophysical modelling of structural geometries observed on seismic data around major normal faults also supports the conclusion that such structures are predominantly planar (Kusznir & Egan 1989, Marsden et al. 1990, Roberts & Yielding 1991, Kusznir & Ziegler 1992).

In the remainder of this section we accept the view that large extensional faults accommodate slip as near-planar structures and expand the discussion from this point. We have suggested that the definition of a large extensional fault is one which cuts through the entire (c. 15 km) seismogenic layer. Two pertinent further questions could be asked about the gross geometry of these structures. Do they have a finite lateral extent, and how is displacement distributed below the seismogenic layer?

No modelling studies have yet addressed the subject of the maximum theoretical strike-length of a normal fault surface. Jackson & White (1989) and Roberts & Jackson (1991) have, however, noted that no recorded examples of active, normal fault surfaces with strike-length greater than c. 25 km exist, and suggest that this may be approximately the maximum size of a normal fault surface capable of accommodating slip in a single rupture event. Illustration of individual fault traces extending for about 100 km or more along-strike is an oversimplification caused by the difficulty in representing detail in a regional map, or indeed the inability to define detail in a regional mapping exercise (see discussion by Walsh & Watterson 1991). Large fault traces are invariably a composite of discrete slip surfaces defining *fault segments*.

It is generally assumed that plastic flow in the lower crust (Kusznir & Park 1987) accommodates brittle extension within the seismogenic layer (Jackson et al. 1988, Jackson & White 1989, Kusznir & Egan 1989). The way in which plastic deformation of the lower crust accommodates the slip from individual earthquakes is perhaps not yet completely understood, because being aseismic no direct record can be obtained of displacement within this zone. On a geological time-scale, the evidence from exhumed metamorphic terranes suggests that displacement is accommodated on broad, mylonitic (perhaps now blastomylonitic) shear zones (e.g. Escher & Watterson 1974, Barr et al. 1986). Note, however, that most of our understanding of deep-crustal structure has been derived from eroded, compressional mountain belts, rather than from the lower crustal 'roots' of extensional basins. Studies of deep seismic reflection data are beginning to suggest that broad shear zones may exist in the lower crust beneath extensional basins (e.g. Reston 1988, Klemperer 1988). The success of modelling studies which assume bulk pure-shear to act below the seismogenic layer (or indeed throughout the lithosphere) suggests that the gross geometry of such deep-crustal zones of simple shear may in fact approximate a bulk pure-shear deformation.

Deformation in the volume around major normal faults

Deformation in response to a single earthquake. When considering the deformation of the earth's surface produced by large normal faults we must consider the effects on two time-scales: (a) the co-seismic/post-seismic deformation which occurs during a single earthquake, and (b) the longer-term, isostatic response to repeated increments of such deformation (e.g. King et al. 1988, Stein et al. 1988). We define *co-seismic* as the initial rapid part of the earthquake cycle during which rupture occurs and seismic waves are generated (also see Chapter 18). *Post-seismic* refers to that part of the earthquake cycle after the initial rupture. The most accurate method of measuring co-seismic deformation around an active fault is to survey geodetically the earth's surface before and after a major earthquake. To date, four geodetic profiles have been re-surveyed across normal faults that have moved in earthquakes, namely, 1928 Bulgaria (Jankhof 1945, Richter 1958); 1954 Fairview Peak, U.S.A. (Whitten 1957); 1959 Hebgen Lake, U.S.A. (Myers & Hamilton 1964), and 1983 Borah Peak, U.S.A. (Stein & Barrientos 1985). No footwall measurements were made at Hebgen Lake, but in the other three examples fault slip was accomplished by both *hangingwall subsidence* (vertical lowering of hangingwall) and *footwall uplift* (vertical elevation of footwall). In the Bulgarian and Borah Peak examples footwall uplift constitutes about 20% of the fault throw, whereas at Fairview Peak it was about 5%. Using sea-level as a reference, rather than geodetic surveying, Vita-Finzi & King (1985) reported a footwall uplift proportion of about 10% for the 1981 Corinth earthquakes.

The elevation changes associated with major earthquakes are generally modelled using the elastic half-space solutions of Mansinha & Smylie (1971). In such elastic models the proportion of co-seismic footwall uplift to hangingwall subsidence is critically dependent on fault dip, steeper faults producing greater

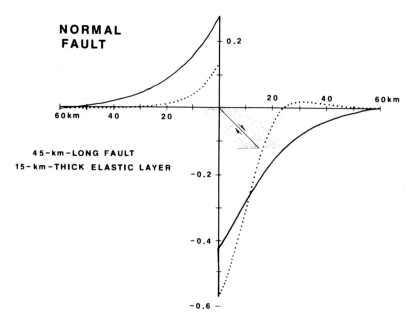

Fig. 11.6. Model profile showing the co-seismic (dotted) and co-seismic plus post-seismic (solid) deformation associated with a normal fault dipping at 45°, cutting on elastic layer of 15 km thickness. Note the post-seismic relaxation causes an uplift of both hangingwall and footwall. (Reprinted from *Journal of Geophysical Research*, Vol. 93, King, G. C. P., Stein, R. S. and Rundle, J. B. The growth of geological structures by repeated earthquakes, pp. 13307–13319, © 1988 American Geophysical Union.)

footwall uplift. A vertical fault would produce equal amounts of uplift and subsidence.

The analysis of the 1983 Borah Peak earthquake by Stein & Barrientos (1985) has become the classic reference for this type of elastic modelling (Fig. 11.5). They showed that the co-seismic elevation changes across the Lost River fault are excellently matched by a model comprising a planar, rectangular fault surface, dimensions 23 × 18 km, dipping at 47° with 2 m of slip. The co-seismic deformation around the Lost River fault produced a hangingwall rollover anticline (see Chapter 6), extending *c.* 20 km away from the fault and a footwall syncline, extending across about 15 km.

Thus we can conclude that the co-seismic deformation around major normal faults can be effectively modelled as a purely elastic response to slip on the fault surface. Co-seismic footwall uplift and hangingwall subsidence are not, however, the only ground surface deformations which occur in response to a single earthquake. Following initial rupture, plastic deformation within the aseismic zone acts to relieve the *elastic strain* (non-permanent strain) field accumulated during instantaneous fault slip and in so doing restores isostatic equilibrium within the lithosphere. This results in the phenomenon of *'post-seismic rebound'*. For a non-vertical fault, post-seismic rebound acts to equalize the asymmetric, co-seismic movements of hangingwall and footwall. Thus, for a normal fault, post-seismic rebound acts as a broad uplift centred on the fault (Fig. 11.6).

Modelling of the post-seismic movements associated with dip-slip faults has been published by many authors (e.g. Koseluk & Bischke 1981, Melosh 1983, Cohen 1984, Lang *et al.* 1984, Reilinger 1986). In such models an elastic layer (equivalent to the seismogenic layer) overlies a viscoelastic half-space (equivalent to the lower crust and uppermost mantle). The duration of post-seismic rebound is controlled by the viscosity of the underlying material, and is expected to occur at an exponentially-decreasing rate over a period of about a thousand years, following the earthquake. The earthquake recurrence rate on major normal faults is thought to be at least this long and possibly ten times as long (Wallace 1984). Thus the post-seismic rebound (i.e. restoration of isostatic equilibrium) following one earthquake should be essentially complete before the next occurs.

Post-seismic movements associated with reverse faults are well documented (e.g. Thatcher 1984). However, only one observational record exists of the rebound associated with a normal fault earthquake. Re-levelling data show that at Hebgen Lake (U.S.A.) up to 30 cm of uplift occurred in the 24 years following the 1959 earthquake, which had a co-seismic throw of 7–8 m (Reilinger 1986). Post-seismic rebound at this fault is presumably far from complete.

Documentation of post-seismic rebound indicates that the deformation accumulated around a normal fault is not simply the sum of strains associated with (elastic) slip events. Areally distributed vertical movements, between earthquakes, must also be considered.

Flexural/isostatic models of long-term deformation. In this section we describe models for long-term fault behaviour which approximate to a summation of repeated co-seismic and post-seismic deformation around a major normal fault. On a geological time-scale (millions of years) a more appropriate model for the continental lithosphere is that of an elastic layer (equivalent to the seismogenic layer) overlying a viscous, fluid substratum (equivalent to the lower crust

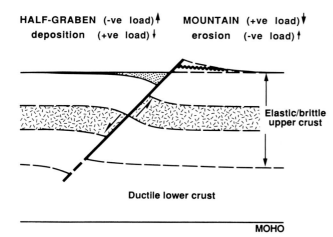

Fig. 11.7. Schematic diagram of the isostatic loads generated by a large normal fault. Uplift of the footwall and subsidence of the hangingwall generate positive and negative loads respectively. Erosion of the footwall and sediment deposition in the hangingwall act to reduce these initial loads. The isostatic loads are supported by the flexural rigidity of the elastic/brittle upper crust. After Yielding & Roberts (1992), modified from King et al. (1988).

and mantle lithosphere). Over such a time-scale, thermally activated creep processes in the aseismic layer are capable of relieving stress, thus allowing its long-term treatment as a viscous fluid.

The term *'flexure'* is used here to describe the bending of a floating elastic beam or layer in response to imposed loads. In geological terms flexure describes the response of the elastic upper crust to loads such as mountain ranges and sedimentary basins, balancing the isostatic response to loading against the strength of the upper crust (Fig. 11.7).

In a simple mathematical *flexural model* a downwards-acting point load on an elastic beam will produce a surface depression around the load and a much smaller *peripheral uplift or bulge*. If the beam is cut (i.e. faulted) and then loaded downwards the flexural response becomes asymmetric but still incorporates a peripheral bulge. The resistance to bending of a beam, or layer, is known as its flexural rigidity, D:

$$D = \frac{ET_e^3}{12(1-v^2)}$$

where T_e is the effective elastic thickness of the beam, E is Young's modulus and v is Poisson's ratio. Thus small changes in T_e produce significant variations in D.

Studies of earthquake focal mechanisms show that the elastic, seismogenic layer of the continental crust has a thickness of about 15 km (the so-called schizosphere, see p. 378, Chapter 18). Yet detailed gravity and flexural modelling studies of extended (continental) sedimentary basins show T_e typically to have a value of 5 km or less (Stein *et al.* 1988, Watts 1988, White & McKenzie 1988, Kusznir & Egan 1989, Fowler & McKenzie 1989, Barton & Wood 1984, Marsden *et al.* 1990, Roberts & Yielding 1991, Kusznir *et al.* 1991,

Roberts *et al.* 1993b). How can we reconcile the elastic/brittle response shown by earthquakes with the flexural response to loading? One possible answer is that the seismogenic layer, on a geological time-scale, is an imperfect elastic solid, and is cut by non-elastic imperfections such as fault and fracture surfaces. These serve to reduce the *effective elastic thickness* of the seismogenic layer to the values of *apparent elastic thickness* yielded by basin modelling studies (Buck 1989, Kusznir *et al.* 1991). Thus, on geological time-scales the 15 km seismogenic layer has similar properties to a perfectly elastic beam 5 km, or less, in thickness.

The role of isostasy and flexure in normal faulting was first considered by Vening Meinesz (1950) and Heiskanen & Vening Meinesz (1958). Recent renewed interest in flexure as a mechanism is probably largely the result of work by Jackson & McKenzie (1983), although others (e.g. Bott 1976, Zandt & Owens 1980) had previously adapted the Vening Meinesz hypothesis, while Weissel & Karner (1989) have subsequently adapted it further.

Flexural models of isolated normal faults show that imposing a fault displacement on the elastic layer causes subsidence of the hangingwall and uplift of the footwall (Fig. 11.7), with a much reduced peripheral bulge in the hangingwall and *peripheral sink* on the footwall. Thus on a scaleless cartoon the flexural response to long-term faulting might look similar to the co-seismic elastic response to individual earthquakes (Fig. 11.5). Controls on the deformation are, however, different. In the co-seismic elastic models the radius of the deformed volume around the fault, that is, the lateral extent of hangingwall subsidence and footwall uplift, is controlled by the size of the fault surface. In the long-term flexural models the radius of deformation is instead controlled by the flexural parameter, α, which is a function of the *flexural rigidity* (resistance of an elastic layer to bending) of the elastic layer and the densities of the overlying and underlying material (see review by Weissel & Karner 1989). The flexural parameter is independent of fault slip. The flexural response deforms a significantly larger volume than the co-seismic elastic deformation.

The flexural response to faulting maintains isostatic equilibrium within the lithosphere and is therefore sensitive to external loads imposed on the elastic layer. When the elastic layer is overlain by a homogeneous overburden, be it air, water or sediment, footwall uplift and hangingwall subsidence are partitioned approximately equally across the fault. If, however, the overlying load is asymmetric, that is, sediment/water in the hangingwall and air in the footwall, then the flexural response to faulting is asymmetric. Preferential loading of the hangingwall leads to increased hangingwall subsidence and reduced footwall uplift (King *et al.* 1988, Yielding & Roberts 1992). In geological terms we can see that under conditions of unequal loading the sedimentary response to tectonics (i.e. sediment load) may feed back and further control the tectonics. Jackson & McKenzie (1983) suggested that in the case of asymmetric loading footwall uplift might in general comprise approximately 10% of the total fault throw. Given the sensitivity of the flexural response to *half-graben* (subsidence zone bounded on one side by a normal fault) loading a more cautious estimate might quote the range 5–25%.

In the case of asymmetric loading, a common circumstance when fault blocks lie close to sea-level, uplift:subsidence ratios predicted by flexural models are coincidentally in the same range as uplift:subsidence ratios predicted for moderately dipping elastic dislocations (see earlier). This explains why elastic and flexural/isostatic models have both been applied, with success, to the Gulf of Corinth half-graben (Jackson *et al.* 1982, Jackson & McKenzie 1983). It is useful to bear in mind, however, that for given amounts of hangingwall subsidence and footwall uplift, flexural models will predict deformation across a much wider area than the corresponding elastic model, that is, both half-graben and footwall uplands will be wider in the flexural model.

The 'flexural cantilever' model

The flexural models described above consider the flexural response to a single, major normal fault. To place such models in a more meaningful geological context two additional factors must be considered. (1) Single fault flexural models calculate deformation relative to an undeformed datum lying beyond the deformed volume around the fault. Large normal faults, however, are normally found in extensional basins, and consequently are involved in the overall isostatic subsidence of the basin floor (see Stretching Models). Thus the undeformed datum may be involved in a basin-wide subsidence relative to a truly-fixed datum such as sea-level. Any measurements of absolute, rather than relative, uplift and subsidence must therefore take basin-floor subsidence into acount, in addition to more local deformation around the fault. (2) Because of the thickness of the elastic layer, the flexural response to a single, major normal fault is typically distributed across a zone several tens of kilometres in width. Large normal faults spaced at less than the width of the flexural response will therefore possess overlapping zones of deformation. A typical fault spacing of 5–30 km will make this the rule rather than the exception. Hence in modelling the deformation associated with any one particular fault, the interference effect of deformation around neighbouring faults should also be considered.

These two points are addressed by the flexural cantilever model of continental lithosphere extension, developed by Kusznir and co-workers (Fig. 11.8). The *flexural cantilever model* combines simple-shear (fault-controlled) deformation of the upper crust with bulk pure-shear deformation of the lower crust and mantle lithosphere.

Kusznir *et al.* (1987) developed the initial model, in which lithospheric extension is achieved by major listric faults in the upper crust and by pure shear below a specified level of horizontal detachment (Fig. 11.8). In order to bring the model into accordance with both earthquake-seismic and reflection-seismic evidence for

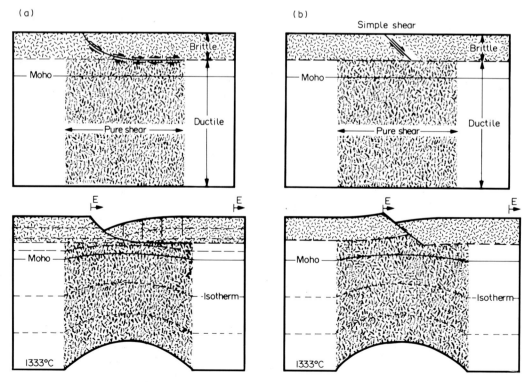

Fig. 11.8. A schematic representation of lithosphere extension by faulting in the upper crust and pure shear in the lower crust and mantle, showing crustal thinning, perturbation of the temperature field and the geometric effects of faulting. (a) Construction for listric faults. (b) Construction for planar faults, the flexural cantilever model. After Kusznir *et al.* (1991).

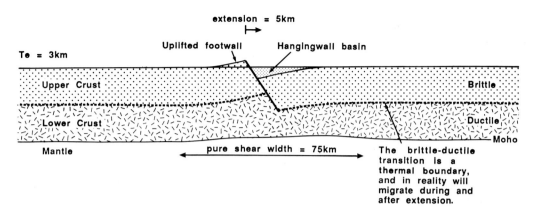

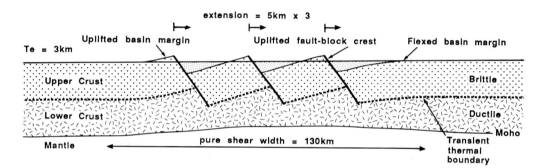

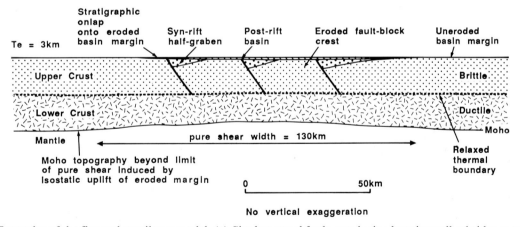

Fig. 11.9. Examples of the flexural cantilever model. (a) Single normal fault, producing hangingwall subsidence and footwall uplift. (b) Multi-faulted model (no erosion), producing domino-style fault blocks within the basin, while the basin margin and fault-block crests are uplifted above sea level. (c) Multi-faulted model, incorporating erosion of the basin margin and fault-block crests, followed by post-rift thermal relaxation. From Roberts & Yielding (1991).

the geometry of upper crustal faults (e.g. Jackson 1987, Kusznir & Matthews 1988) the model was modified by Kusznir & Egan (1989) and Kusznir et al. (1991) to allow upper crustal deformation to occur on an array of planar faults, separating fault blocks, which behave mechanically as *interacting flexural cantilevers* (Fig. 11.8). The planar fault model does not require a detachment separating zones of pure and simple shear. Instead this transition is a function of the rheological and thermal properties of the lithosphere.

Like the more simple flexural models the flexural cantilever model considers the lithosphere as comprising a thin elastic/brittle layer, deformed by faulting, overlying a viscous fluid. Motion on upper crustal faults thus initiates an isostatic response and causes a flexural subsidence of hangingwall blocks and flexural uplift of footwall blocks (Fig. 11.9). Zones of deformation around closely-spaced faults interact with each other, such that a single fault block may deform as a footwall to one fault and a hangingwall to another. In addition, overall crustal thinning induces a general basin-floor subsidence superimposed on the more local fault-related deformation.

As in simple flexural models, the width of the deformed zone around an individual fault is controlled primarily by the apparent elastic thickness of the crust. Low values of T_e produce tight, steep-gradient deformation around the faults, while high values of T_e distribute the deformation over a wide area. Uplift and subsidence values for a particular fault are controlled on the local scale principally by fault-dip and density of sediment load, but in addition have a regional basin-floor subsidence superimposed on them.

As a result of the response of the elastic layer to isostatic loads the fault surfaces themselves are deformed during extension, in addition to the adjacent fault blocks. In general, fault-surface deformation approximates a rotation to a gentler dip. Following large displacements the fault becomes non-planar, with a gentler dip on the exposed footwall.

By explicitly including whole-lithosphere extension in the model, the evolution of a basin into its post-stretch, thermal-subsidence phase may be considered (Fig. 11.9). The progressive burial of fault-block crests may be studied in this way, or alternatively emergent fault-block crests may be allowed to erode and the basin to readjust isostatically prior to thermal subsidence.

The flexural cantilever model has been applied to a number of geographically disparate extensional basins with a considerable degree of success, for example, the Jeanne d'Arc, Canada (Kusznir & Egan 1989, Kusznir et al. 1991), the Viking graben, U.K./Norway (Marsden et al. 1990, Roberts et al. 1993b), the East African rift (Kusznir & Ziegler 1992), Rhine graben and western Aegean (Kusznir personal communication), and the basin margins to the North Sea rift (Roberts & Yielding 1991).

At basin margins the flexural cantilever model predicts that the footwalls to the basin itself will be elevated above their initial datum and perhaps subject to erosion (Fig. 11.9). Such flank uplift is also predicted by stretching models incorporating lateral-heat-flow or depth-dependent-stretching. In these latter two types of model, uplift is produced entirely by temporarily elevated geothermal gradients and is completely recovered on cooling. Marginal uplift in the flexural cantilever model is largely a mechanical response to faulting, with perhaps a small thermal component involved, and therefore remains largely unrecovered during thermal cooling of the basin. A mechanical model for rift flank uplift, similar to the flexural cantilever model, has also been advocated by Weissel & Karner (1989).

The advantage that the flexural cantilever model enjoys over other basin models is that it explicitly includes fault-block motions, and incorporates many processes and parameters that affect uplift and subsidence within an extensional basin. Within the interior of a basin a convenient, geometric simplification of the flexural cantilever model can be made, in order to allow rapid study of individual large fault blocks. This simplification is the domino model of faulting.

The domino model

It has long been recognized that the geometry of many extensional fault systems approximates to that of a set of 'rotating dominoes', or tilted books on a bookshelf (Emmons & Garrey 1910, Thompson 1960, Morton & Black 1975, Wernicke & Burchfiel 1982, Mandl 1987, Davison 1989). In its simplest form the *domino model* of faulting involves an array of faults with the same dip and same displacement separated by rigid fault-blocks of the same size. Thus as extension on the array occurs, all faults achieve extension together by the same amount, rotating the intervening fault blocks through the same angle. A geometric consequence of the domino model is *passive rotation*, to gentler dips, of the fault planes.

The flexural cantilever model produces rotation of faults and fault-blocks as a consequence of interacting hangingwall and footwall flexure. There are, however, circumstances in which the geometric response to the flexural cantilever model approaches closely that of the domino model. When faults are closely and evenly spaced, even at a T_e as low as c. 5 km, overlap of deformation fields around the faults results in a series of uniformly and equally dipping fault blocks, comparable to a domino array (Fig. 11.9). Alternatively, at high values of T_e ($\sim$>15 km), the radius of deformation around individual faults is sufficient to achieve uniform tilt cross a large area, again mimicking the geometry of domino fault-blocks.

Thus we can view the domino model as a geometric simplification of the more complete flexural cantilever model, in which, for a given fault spacing, the T_e is sufficiently large to produce a homogeneous, rather than heterogeneous, deformation of the fault blocks.

Most descriptions of the domino model simply describe the geometric consequences of block rotation. Barr (1987a, b) and Jackson et al. (1988) incorporated rotating-domino fault-blocks within the context of a subsiding basin-floor (Fig. 11.10). Both used the simple McKenzie model to describe overall basin-floor subsidence and superimposed on this the geometric consequences of rotating fault blocks, in order to define the magnitude of footwall uplift and hangingwall subsidence associated with each fault. If there were no basin-floor subsidence the fulcrum separating uplift from subsidence on a rotating-domino block would lie precisely half-way along its length. Incorporating basin-floor subsidence, however, causes the fulcrum to move towards the fault block crest, its position being controlled by the magnitude of block rotation, the size of the rotating block, and the density of the sediment load.

In broad terms, for any given combination of initial fault angle and sediment load, Barr's models recognize

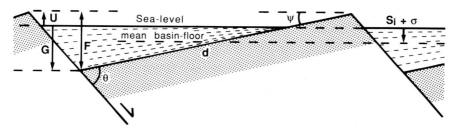

Fig. 11.10. The domino model of continental extension. All fault blocks and faults rotate simultaneously as extension proceeds. The stretching factor $\beta = \sin\theta/[\sin(\theta - \psi)]$, and the fault throw $F = d\sin\psi$. The average level of the basin floor undergoes a tectonic subsidence S_i that is a function of β, and there is a further subsidence σ caused by sediment loading. Footwall uplift $U = F/2 - S_i - \sigma$. From Yielding (1990).

three categories of fault-block, defined by their size (across strike) and motion of the footwall crest relative to an initial starting point (such as sea-level). (1) In a domino array with small fault-blocks, footwall crests immediately submerge below sea-level and continue to subside with increased rotation/extension. (2) Large fault-blocks are characterized by uplift of footwall crests above sea-level upon the initiation of extension. Footwall crests remain emergent throughout the duration of extension. (3) In between these extremes the crests of medium-sized fault-blocks are initially uplifted above sea-level and then subside below sea-level as extension increases. In any individual case the amounts of footwall uplift and hangingwall subsidence are dependent upon fault-block size (most importantly), initial fault-dip, density of sediment load, and degree of erosion experienced by emergent crests (Fig. 11.10).

From his calculations, Barr constructed a series of half-graben models with sediment infill. These showed that, given the ideal circumstances described by his models, steady onlap towards emergent fault-block crests will always occur as extension increases, that is, emergent 'footwall islands' become narrower (Fig. 11.11). The only unconformity predicted upon the block crests lies at the base of the syn-rift fill. There is no post-rift unconformity predicted by the simple domino model. Incorporating time-dependent stretching and/or time-varying sediment loading into the domino model at the expense of steady state conditions will, however, alter the detail of the predicted half-graben sequence (Roberts et al. 1993a).

Many authors have successfully applied the domino model to natural examples of rotated fault blocks. These applications include the northern North Sea (Barr 1987a, 1991, White 1990, Yielding 1990, Roberts et al. 1993a), the Aegean Sea (Barr 1987a), the Armorican margin (Barr 1987a) and the Gulf of Suez (Jackson et al. 1988).

The domino model appears consistent with observations made within many extended basins and its advantages are its simplicity and ease of application. The model breaks down, however, when applied to basin margins, because clearly the basin margin cannot rotate freely as a rigid block (cf. Sclater & Celerier 1988, Davison 1989). Basin margins cannot be analyzed by using the domino model, and a more appropriate way in

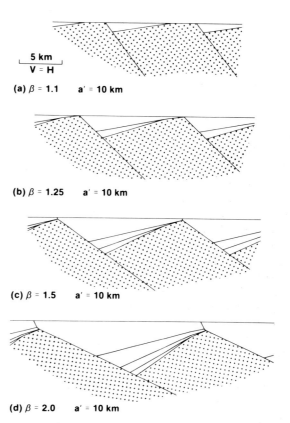

Fig. 11.11. Model set of half-grabens and their stratigraphic fills constructed by the domino model. The fault blocks have an initial width of 10 km and the initial fault dip is 60°. The horizontal line is sea-level, with older, rotated sea-level markers below. Pre-rift basement is stippled. Note initial uplift and erosion of fault-block crests, followed by their burial below the onlapping syn-rift sequence. After Barr (1987b). (Reprinted from *Journal of Structural Geology*, Vol. 9, Barr, D., Structural/stratigraphic models for extensional basins, pp. 491–500, © 1987 with permission from Pergamon Press Ltd, Headington Hill Hall, Oxford OX3 0BW, U. K.)

which to model major faults at basin margins is by the application of a flexural or flexural cantilever model (e.g. Jackson & McKenzie 1983, King et al. 1988, Weissel & Karner 1989, Roberts & Yielding 1991). Similarly, if fault blocks of radically varying size are juxtaposed within a basin then again the domino model is inapplicable.

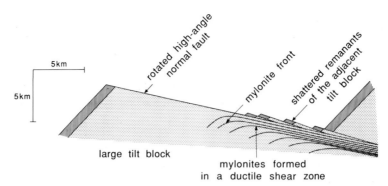

Fig. 11.12. A model for the formation of low-angle detachment faults and metamorphic core complexes by the rotation of domino-style fault blocks. The model supposes the detachment fault to be a rotated, high-angle normal fault, with the core complex in its footwall. After Lister & Davis (1989), who did not favour this model.

Metamorphic core complexes (the Basin and Range province)

Perhaps the most extensively studied extensional province in the world, by virtue of its exposure at the surface, is the Basin and Range province in the western U.S.A. Despite the intensity of investigation into this area, however, the kinematics of the major extensional faults within the Basin and Range, and indeed even the dynamics of whole-lithosphere extension in the province, remain subjects of unresolved controversy.

The geometry of normal faults in the Basin and Range presents an apparent paradox. Those faults active at the present day are delineated by earthquake data to be planar structures, penetrating to about 15 km depth, dipping at about 45° (e.g. Stein & Barrientos 1985, Stein *et al.* 1988, Westaway 1989, 1991). Extensive fieldwork within the province has, however, identified many examples of now-inactive normal faults which are regionally flat or very gently inclined (e.g. Wernicke 1981, 1985, Hamilton 1987, John 1987), and have accommodated several tens of kilometres of displacement. These so-called *detachment faults* appear to be fundamentally different structures from those currently active faults accommodating smaller extensions.

A feature characteristic of the low-angle detachment faults is the presence of a strong contrast in metamorphic grade between the hangingwall and footwall blocks. The hangingwalls typically expose low-grade upper-crustal rocks, while the footwalls expose rocks at greenschist or amphibolite facies. The footwalls are interpreted as mid-crustal rocks, transported during extension to the earth's surface. The surface outcrops of intermediate/high-grade metamorphic rocks in the footwalls of some low-angle normal faults are commonly referred to as a *metamorphic core complex*. An excellent review of metamorphic core complexes and detachment faults is provided by Lister & Davis (1989). In the review they discuss, in more detail, the apparent paradox of Basin and Range extension, and very clearly highlight that although large low-angle normal faults are observed in the Basin and Range, there is no seismological evidence from there, or elsewhere, that continental extension occurs along such surfaces. Lister & Davis (1989) considered, and rejected, the possibility that core complexes may be the highly-rotated footwalls of domino-style fault blocks, with the low-angle faults bounding the core complexes having rotated from an initially-high angle (Fig. 11.12). Instead they considered the bulk of geological evidence to indicate that 'detachment faults' initiate with, and remain at, a low dip. A consequence of this is that such faults must be considered aseismic and fundamentally different from those major faults described earlier in this section.

Lister & Davis's model for the formation of core complexes incorporates both the steep, planar faults evidenced by seismological data and the large, low-angle faults observed in the field (Fig. 11.13). They suggested that small extensions are accommodated by steep, seismically active faults which detach into an aseismic shear zone close to the level of the brittle/ductile transiton. At higher extensions aseismic low-angle faults are fired from this shear zone, cutting up through the older steep faults. Crustal attenuation above the shear zone causes isostatic uplift of the shear zone itself and its mid/lower-crustal footwall. This doming of the shear zone renders it inactive and younger, higher structures take over as the active, basal fault. Ultimately, the domed shear zone is exposed at the surface as a low-angle *detachment fault*. Its footwall comprises a metamorphic core complex. This model is comprehensive and well considered, incorporating both geological and geophysical evidence about fault geometry, plus an acknowledgement of the role of isostasy in the evolution of a large fault system. The model, however, is not quantitative, it is purely descriptive.

The importance of isostasy (flexure) in the evolution of metamorphic core complexes has been modelled quantitatively by Buck (1988). As the starting point for his model Buck took a flexural/isostatic treatment of fault-related deformation, similar to those described above (Fig. 11.7) (see Flexural/isostatic models of long-term deformation). On the basis of seismological data the faults in Buck's model initiate with a dip of 60°. Most flexural models have considered the flexural response to a normal fault accommodating only a few kilometres of displacement. In such instances, footwall uplift is observed and a rotation of the fault plane through 10–15° may occur. Buck allowed his model to accommodate much larger extensions, several tens of kilometres, and investigated the flexural consequences (Fig. 11.14). Buck's mathematical treatment demonstrated that at large extensions it is possible to rotate a fault, with an initial dip of 60°, into an areally extensive, flat-lying structure, still with an active root inclined at 60° (Fig. 11.14). The rotation of the fault-plane through 60° is an example of extreme footwall uplift. As a consequence of this uplift and rotation, Buck's model also carries mid-crustal rocks near to the surface, where they would form a metamorphic core complex beneath the flat-lying fault.

Buck's model for the formation of large low-angle faults and metamorphic core complexes is very appealing. It is mathematically based and can make quantitative predictions. It also illustrates a method by which both the inactive, low-angle detachment faults and the active, seismic, steeply dipping faults of the Basin and Range may be produced as part of a continuum, rather than by two separate styles of crustal extension. At present the controversy over the formation of low-angle faults in the Basin and Range remains. It would be ill-considered therefore to regard either the *Lister & Davis detachment-fault model* or the *Buck flexural model* as proven. Both models, however, illustrate an elegant solution to the Basin and Range paradox.

SMALL EXTENSIONAL FAULTS

Small extensional (normal) faults are defined here as those which can be reasonably considered to be

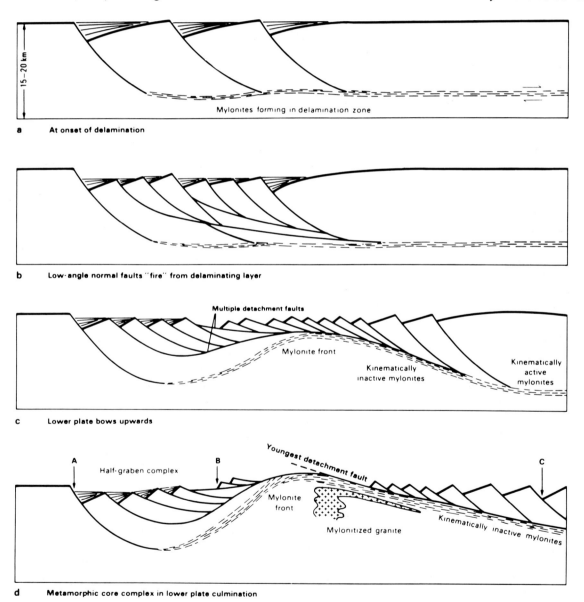

Fig. 11.13. Model for the evolution of metamorphic core complexes below an initially flat-lying detachment fault. Extension begins (a) on steeply-dipping, detached normal faults. Subsequently, (b) low-angle normal faults propagate upwards from the detachment, breaking up the older steep faults. Upper crustal thinning (c) causes isostatic uplift of the early-formed detachment fault, and ultimately, (d) its footwall is exposed at the surface as a metamorphic core complex. After Lister & Davis (1989). (Reprinted from *Journal of Structural Geology*, Vol. 11, Lister, G. S. and Davis, G. A., The origin of metamorphic core complexes, pp. 65–94, © 1989 with permission from Pergamon Press Ltd, Headington Hill Hall, Oxford OX3 0BW, U. K.)

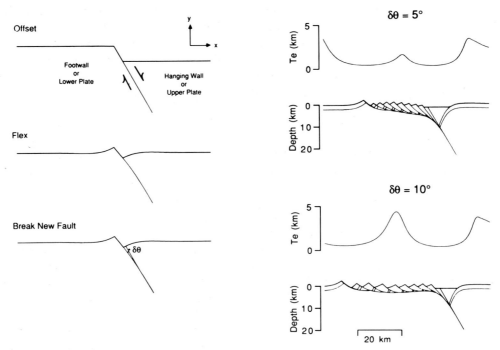

Fig. 11.14. Buck's flexural model for the formation of detachment faults and metamorphic core complexes. The left-hand column shows the basic flexural model (cf. Figs 11.7, 11.8 and 11.9), including the formation of a new fault break in the hangingwall after a rotation through $\delta\theta$ by the upper part of the initial fault. The right-hand column shows two models, incorporating spatially varying T_e (effective elastic thickness) and sediment loading, in which the initially steep (60°) fault has rotated in its upper part to a flat dip. The flat portion of the fault lies near the surface, where its footwall will constitute a metamorphic core complex. After Buck (1988).

confined within the seismogenic layer. Such faults do not generate significant topographic loads and thus in their treatment the effects of isostasy can be excluded.

The geometry of small normal faults

Knowledge of small normal faults has significantly increased in recent years, largely as a result of the work of Watterson and co-workers. From detailed analysis of coal-mine surveys and seismic reflection data they have demonstrated systematic relationships between the displacements and dimensions of small normal faults (e.g. Watterson 1986, Barnett *et al.* 1987, Walsh & Watterson 1987, 1988, 1989, 1991, Gibson *et al.* 1989, Childs *et al.* 1990) (also see Chapter 6).

The observations on many *small isolated normal*

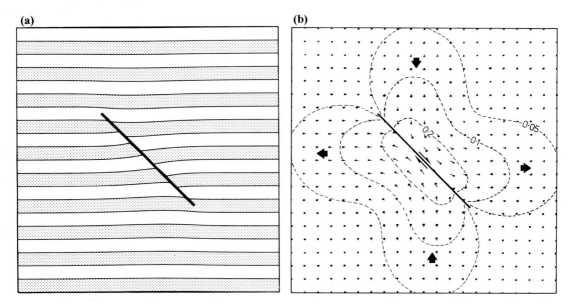

Fig. 11.15. (a) Model blind normal fault, confined within an elastic medium. The fault has a maximum displacement at its centre decreasing gradually to a zero displacement tip. The along-strike fault width is the same as its down-dip length. Note that strain around the fault is partitioned equally and oppositely into footwall uplift and hangingwall subsidence. (b) Particle trajectories (tadpoles) and gross displacement directions (arrows) in the deformed volume around the blind fault. The contours of displacement are calibrated relative to the displacement at the fault centre. Note the symmetry in the strain field across the fault. (Yielding & Freeman, previously unpublished. Derived from the solutions of Mansinha & Smylie 1971.)

faults show them to be structures with finite dimensions. An individual fault surface, confined totally within a rock volume, is typically near-planar and has an approximately elliptical outline. The maximum displacement on the fault lies at the centre of the ellipse, with displacement decreasing in all directions, including downwards, away from the central maximum. Elliptical contours of equal displacement can be constructed on the fault surface. The outermost (zero displacement) contour defines the limit of the fault surface and is termed the *tip-line loop*. Deformation associated with the fault is not restricted to the slip surface but extends in three dimensions around the fault. This deformation takes the form of *reverse drag* (folding round a fault of opposite polarity to that resulting from frictional drag on the fault), with footwalls exhibiting synformal uplift and hangingwalls exhibiting *antiformal rollover* (Fig. 11.15). Local extensional and contractional strains must develop on both sides of the fault, in order to accommodate this deformation, as a confined, isolated fault intersects no free surfaces.

The maximum displacement on an isolated fault is found to be systematically related to the dimensions of the fault. More specifically displacement is related to the along-strike dimension of the fault through a power-law relationship, having an exponent of about 1.5 (Walsh & Watterson 1988, Marrett & Allmendinger 1991). Thus a small fault with a strike-length of 100 m may have a maximum displacement of about 0.1 m, whereas a fault with a strike-length of ten kilometres will have a maximum displacement of a few hundred metres. Strike-length/displacement ratio is also affected by the material properties of the surrounding rock.

A fault growth model

Although the total, finite displacement on a fault is related through a power law to the along-strike fault dimensions, studies of slip during single earthquakes show individual slip-increments to be directly proportional to fault size. Watterson (1986) demonstrated that these two observations can be reconciled by a simple model of fault growth. The model is based on the accumulation of permanent strains on and around the fault as the result of successive increments of elastic movement.

In *Watterson's fault growth model*, the fault increases in size by a small but constant amount each time it slips in an earthquake. The slip increment (in each earthquake) is directly proportional to fault size, and so it too increases by a constant amount each earthquake. At any given time, however, the total displacement is the sum of all previous slip events. Because fault-slip increases at a steady rate, total fault-displacement must continually increase by larger and larger slip-increments, at a rate faster than that at which the fault is growing along-strike. This results in total displacement growing faster than the along-strike dimension. From observational evidence, Walsh & Watterson (1988) have suggested that this fault growth model may be applicable not just to small normal faults, but also to structures such as large-displacement thrusts and ocean-ridge transform faults.

Deformation around small normal faults

The form of the three-dimensional strain field which develops around faults is complex, and in theory only decreases to zero at infinite distance from the fault (Mansinha & Smylie 1971). The shape and magnitude of the strain field is, however, symmetric across the fault if the fault is remote from the Earth's surface. An effective limit can be defined at which the amount of deformation becomes vanishingly small. Barnett *et al.* (1987) and Gibson *et al.* (1989) take this *maximum reverse drag radius*, situated perpendicular to the fault centre, to be equal to half the strike extent of the fault. Thus the size of the deformed zone around the fault is not constant, as in the flexural models, but is directly proportional to the dimensions of the fault itself.

A *blind fault* surface is one entirely confined within a rock volume, intersecting no free surfaces. The discussion above, however, has shown that the deformation associated with small normal faults is not two-dimensional and affects a three-dimensional rock volume. A *blind fault volume* is therefore one in which the 'effective limit' of deformation around a fault is entirely remote from any free surface (Fig. 11.15).

Within a blind fault volume, strain is partitioned symmetrically across the fault (Fig. 11.15). Thus at the fault surface, displacement of any given marker is equal about its original intersection point with the fault. The shape of a reverse drag profile across the fault is, however, dependent upon the location of the marker horizon within the deformation volume, and on the steepness of the fault relative to bedding. In the general case of a normal fault intersecting bedding at an angle other than 90°, hangingwall rollover is wider and more open, within the upper half of the fault volume, than is the corresponding footwall-uplift syncline. Conversely, in the lower half of the same fault volume the footwall syncline is wider and more open than the hangingwall rollover (Barnett *et al.* 1987, Gibson *et al.* 1989).

When an active fault intersects the Earth's surface, or is close to it, the deformation around the fault is modified. The deformation volume is not only truncated by the ground surface, but the hangingwall subsidence is significantly enhanced (Fig. 11.16). This happens because the surface is no longer 'held up' by overlying material of finite strength. As a result, fault slip is partitioned into a greater component of hangingwall rollover than footwall uplift. An empirically-derived expression describing this partitioning across a *truncated fault volume* is given as (Gibson *et al.* 1989):

$$HW\% = 110 - \frac{2}{3}\theta$$

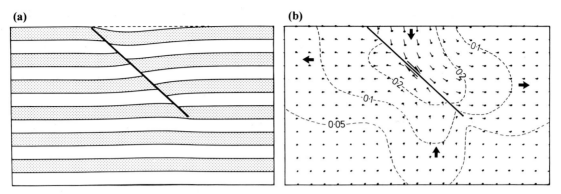

Fig. 11.16. (a) Model normal fault similar to that in Fig. 11.15, but whose upper tip extends to the contemporary free surface. Hangingwall subsidence is here very much greater than footwall uplift. (b) Particle trajectories (tadpoles) and gross displacement directions (arrows) in the deformed volume around the fault in (a). The contours of displacement are calibrated relative to the maximum displacement at the fault centre. Note the larger area of high strain in the hangingwall. (Yielding & Freeman, previously unpublished. Derived from the solutions of Mansinha & Smylie 1971.)

where HW% is the percentage contribution of hangingwall displacement and θ is the fault dip (>30°).

This equation returns us to our earlier discussion in which we considered the deformation in response to a single earthquake around a large normal fault. There, the modelling of the short-term co-seismic deformation of the earth's surface was discussed as an elastic response to fault slip. This is the same approach as the above treatment of the deformation around small faults intersecting the surface. Although the dimensions of the faults considered are different, we are able to neglect isostasy in both cases and thus arrive at comparable solutions. Thus application of the above equation of hangingwall displacement to the Lost River fault, inclined at 47° (Stein & Barrientos 1985), predicts a hangingwall displacement equal to 79% of the total co-seismic fault slip. Displacement in the 1983 earthquake was partitioned about 80:20 (hangingwall:footwall).

Fault populations

Having made a distinction between large and small normal faults, on the basis of whether or not they cut the entire thickness of the seismogenic layer, it is of interest to consider the relative contribution they make to regional extension. Do the large faults account for almost all the extension or are small ones significant?

One approach to this problem is to examine the amounts of slip (or strictly seismic movement) on a population of faults in a seismically active area. The seismicity represents an increment of the total geological deformation accumulated by the area. It has been known for many decades that the frequency of earthquakes is related to their size by the relationship

$$\log N = a - bM$$

where N is the number of earthquakes of magnitude greater than magnitude M, and a and b are constants (Gutenberg & Richter 1944). The constant a depends on the sample size, but the constant b determines the relative portions of large and small earthquakes. Since large earthquakes require large fault surfaces (e.g. Wesnousky et al. 1983, Schwartz & Coppersmith 1984) the *Gutenberg–Richter relationship* defines the size distribution of active faults in an area. Earthquake magnitude is a logarithmic measure of fault size (Kanamori & Anderson 1975, Sibson 1989), and so this size distribution is linear when plotted on log–log scales. This implies that the active fault population is *fractal* (scale invariant) in form (Aki 1981, King 1983).

The value of the constant b in the Gutenberg–Richter relationship is typically about 1. It can be shown (Wyss 1973, Caputo 1987) that the largest earthquakes therefore account for an overwhelming proportion of the seismic slip. This is in spite of the fact that there are many more small faults than large ones. The same conclusion has been reached from direct comparison of seismic moments and plate motions in the Aegean (Jackson & McKenzie 1988). Thus, on the basis of seismically active normal faults, it would seem that only the largest faults are significant in terms of regional extension.

However, recent studies have suggested that the finite geological deformation is not distributed in the same way as the instantaneous deformation described by a seismicity sample. During continuing extension over geological time, faults will grow and may also become inactive. These processes act to shift the distribution of the total population of faults present in the sample volume (as opposed to the population active at any one time). Walsh & Watterson (1992) argued that the resulting 'dead' population at the end of the faulting episode will comprise a greater proportion of small faults than that predicted by the Gutenberg–Richter relationship with $b = 1$. Childs et al. (1990) showed from analysis of coal-mine plans and seismic reflection sections that this is indeed the case. Kautz & Sclater (1988) found the same result from studies of analogue models deformed by extension. Thus in terms of the accumulated geological extension, a significant proportion may occur on the smaller-scale structures.

This approach may reconcile extension estimates made from measurements on faults with those derived from other sources. Typically, extension estimates obtained by analyzing faults on seismic reflection profiles are lower than those obtained from seismic refraction or subsidence calculations (e.g. Ziegler 1983, Barton & Wood 1984, Hellinger et al. 1989). However, seismic reflection data have only a limited resolution, and smaller faults will not be imaged. Walsh et al. (1991) suggested that the seismically observable extension may only constitute 50–75% of the total extension accommodated by the full fault population.

STRAIN PARTITIONING IN THREE DIMENSIONS

Fault linkage (also see Chapter 6)

In previous sections we have described the finite nature of individual fault surfaces, both in terms of the maximum observed size of slip surface active during individual earthquakes, and in terms of the way in which total displacement is distributed on individual fault surfaces. Somehow, however, the strains associated with individual faults must combine to accommodate a bulk upper-crustal strain. In this section we describe how this may occur, with the description being applicable in general terms to both large and small faults.

Conceptually, the simplest way to accommodate linkage of fault strain within an evolving basin is to transfer slip on individual fault surfaces downwards onto a detachment horizon (e.g. Gibbs 1983, 1984, 1990) (Fig. 11.17). Such models assume that individual fault surfaces and fault-slip do not terminate downwards. In the case of small normal faults at least, such models are clearly at variance with observations (e.g. Rippon 1985, Watterson 1986, Barnett et al. 1987). It is becoming increasingly apparent that the transference of slip onto discrete detachment faults is not the way in which the upper crust accommodates bulk stretching. Instead a number of authors have suggested that what is required is not necessarily physical linkage of fault surfaces, but a three-dimensional linkage/overlap of fault deformation volumes. The characteristic of strain linkage in such a manner must be that it approximates a vertically-homogeneous stretch at any given point in the basin, assuming a bulk McKenzie-type stretching mechanism to operate.

Strain linkage in three dimensions does not rule out fault linkage, either vertically or laterally. Observational evidence, both at outcrop and in the subsurface, shows that faults may link to form interconnected networks (Chapter 6). Linkage of faults is, however, not a geometrical necessity of three-dimensional strain. An excellent review of this subject is provided by Walsh & Watterson (1991), and is also covered more briefly by Jackson & White (1989).

Within a three-dimensional fault array individual faults will initiate as small-displacement fractures and will grow into larger structures (Watterson 1986). As individual faults grow they may intersect other, either active or dormant, fault surfaces. Direct physical linkage of faults can occur in this way, and the linked faults may now behave as a single, larger fault surface or fault network. Large faults many tens of kilometres long possibly form by such coalescence of smaller faults, while individual slip events on such fault traces have not been observed to exceed a lateral extent of about 30 km (e.g. Stein & Barrientos 1985, Jackson & White 1989, Roberts & Jackson 1991).

Transfer zones. The finite lateral extent of individual fault traces, both at the large and small scale, requires that the strain associated with any one fault must ultimately be transferred laterally onto other structures. Direct linkage of near-co-planar fault surfaces, as described above, can accommodate this. Commonly, however, fault traces are offset from each other in an en échelon fashion. The concept of physical or *hard-*

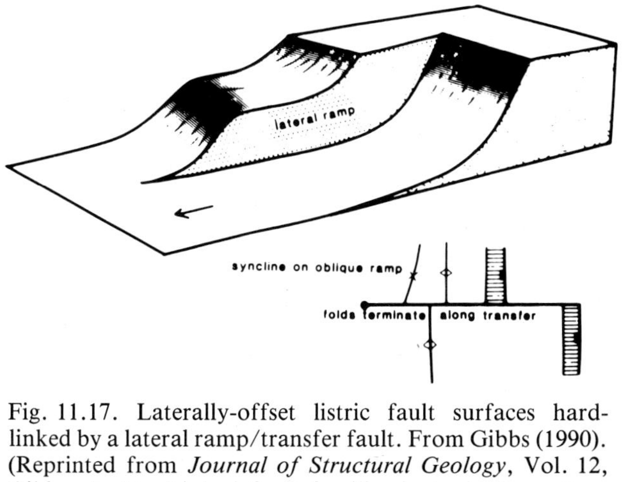

Fig. 11.17. Laterally-offset listric fault surfaces hard-linked by a lateral ramp/transfer fault. From Gibbs (1990). (Reprinted from *Journal of Structural Geology*, Vol. 12, Gibbs, A. D., Linked fault families in basin formation, pp. 795–803, © 1990 with permission from Pergamon Press Ltd, Headington Hill Hall, Oxford OX3 0BW, U. K.)

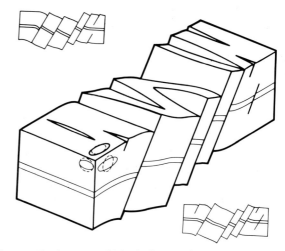

Fig. 11.18. Conceptual block diagram illustrating the soft-domino model and the concept of soft-linkage between offset normal faults by ductile strain. Insets show cross-sections on the block sides. Strain ellipses show the generalized strain (of arbitrary amount) in each plane. After Walsh & Watterson (1991).

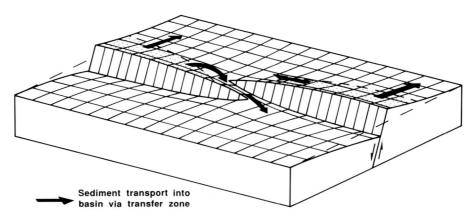

Fig. 11.19. Cartoon illustrating an en échelon offset between two major normal faults. The relay ramp or transfer zone between the two faults necessarily has a lower footwall elevation than the central parts of each fault. It therefore acts as a sediment transport route from the footwalls to the hangingwall basin. Syn-rift sands will be preferentially deposited near these transfer zones. After Yielding & Roberts (1992).

linkage, as displayed by detachment-fault models, demands linkage of offset normal faults by oblique-slip *transfer faults*, that is, faults whose strikes lie close to their slip directions, which transfer displacement between laterally offset faults (e.g. Gibbs 1983, 1984, 1990) (Fig. 11.17). In this way the slip on one fault is entirely conserved and transferred to its neighbour. Although transfer faults undoubtedly exist, increasing evidence has come to light that such faults are only an end-member mechanism of offset displacement transfer. The more general mechanism involves en échelon *soft-linkage* (Larsen 1988, Morley *et al.* 1990, Roberts & Jackson 1991, Roberts *et al.* 1991, Stewart & Hancock 1991, Walsh & Watterson 1991, Yielding & Roberts 1992), in which displacement transfer is accommodated by ductile strain rather than physical linkage (Figs 11.18 and 11.19). Such soft-linked faults are separated from each other by a *fault bridge* or *relay ramp* which dips sub-parallel with fault-strike and orthogonal to the regional dip of the basin. *Fault bridges* may be zones of ductile deformation or intense small-scale fracturing, and they may accommodate very high displacement gradients on their bounding faults.

It is likely that at high extensions fault bridges eventually lose their coherence and break through into discrete transfer faults, in this way achieving hard-linkage (e.g. Stewart & Hancock 1991, Roberts *et al.* 1991). A common misconception about such transfer faults is that they are by definition strike-slip structures. This is not so. Transfer faults accommodate both the vertical and horizontal components of displacement exhibited by the faults which they link. In the general case therefore they are steeply inclined, oblique-slip structures.

Three-dimensional controls on sediment dispersal. We have discussed earlier that uplift of footwalls during extensional faulting is the rule rather than the exception. Elevation of footwall blocks, including basin margins, above sea-level provides a mechanism for the formation and continued rejuvenation of sediment source areas during basin formation. Following derivation by erosion, the dispersal of clastic detritus within, or into, the basin will be largely controlled by the three-dimensional topography of the deforming basin floor and basin margin. Generalized models for sediment and facies distribution within evolving half-graben have been produced by Leeder & Gawthorpe (1987) and applied by Gawthorpe (1987). Roberts & Jackson (1991) have illustrated the three-dimensional patterns of sediment dispersal associated with active fault zones in central Greece.

In general, it is expected that sediment will disperse from topographic highs into topographic lows, via slopes in the basin floor (Fig. 11.19). It is often considered that erosion of footwall highs primarily results in sediment transport directly across the fault scarp into the immediate hangingwall low. It has become apparent, however, that while hangingwall sediment fans do occur a significant, perhaps the larger, volume of sediment is transported back down the more gently-sloping dip-surface of the fault block, resulting in the formation of footwall fans.

Within the North Sea both types of fan can be recognized. The Brae complex (e.g. Turner *et al.* 1987) is a series of hangingwall fans derived from the uplifted adjacent platform. Conversely, the Magnus sand (De'Ath & Schuleyman 1981) is a footwall fan deposited on the dip slope of a large fault block.

Transfer zones, and in particular fault bridges, may be critical in controlling the distribution of dip-slope detritus, because they form a physical link from the footwall of one fault scarp into the hangingwall of another (Fig. 11.19). In this way sediment shed back from an emergent basin-margin footwall, onto the adjacent platform, may still ultimately find a route into a hangingwall-low within the basin. Jackson & White (1989) have suggested that many of the Brae complex fans owe their siting not necessarily to direct derivation from the adjacent platform but to dispersal into the basin via offsets in the marginal fault zone. A similar model has been used in the North Sea Central graben by Roberts *et al.* (1991).

Continental Extensional Tectonics

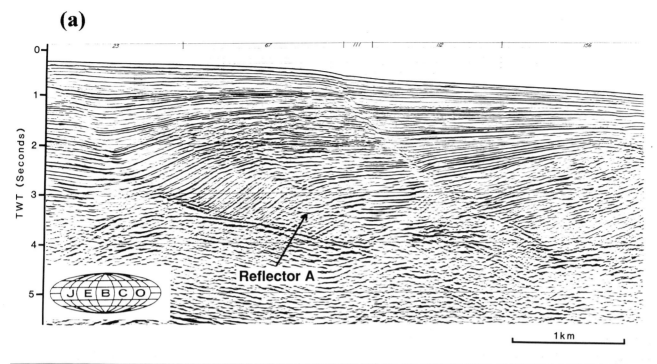

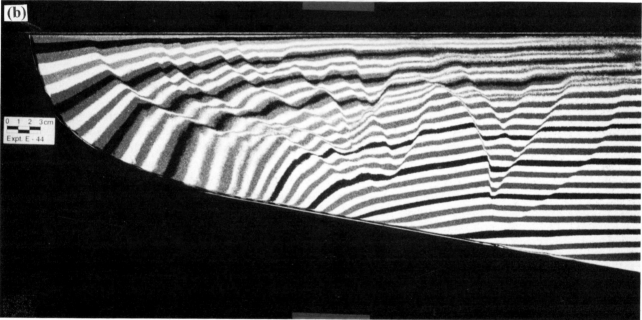

Fig. 11.20. (a) Seismic section showing gravity-driven, listric normal faults from the Gulf of Mexico. (b) Analogue sandbox model, with imposed rigid footwall, exhibiting hangingwall geometries, notably a pronounced rollover anticline, comparable with those observed in (a). After McClay et al. (1991).

GRAVITY-DRIVEN EXTENSIONAL FAULTS

In this section we briefly review extensional structures which result from gravity-induced slope failure (also see Chapters 6 and 7). All too frequently in the literature insufficient distinction is made between such structures and the crustal-stretching faults reviewed in the bulk of this chapter. We repeat here our earlier distinction, that crustal-stretching faults accommodate the upper-crustal part of a whole lithosphere strain, whereas gravity-driven structures may be unrelated to lithospheric extension and are simply a translation of material across the surface of lithospheric plates. It is therefore entirely wrong to use one class of structure as an analogue for the other, although this has frequently been done (e.g. Wernicke & Burchfiel 1982, Gibbs 1983, 1984, Shelton 1984).

Gravity-driven extensional faults are typically found within the 'unbuttressed' sediment piles of (a) continental margins (e.g. the Gulf coast of America (Wernicke & Burchfiel 1982, fig. 15, Jackson & Galloway 1984) (Fig. 11.20) or the west African coast), (b) progradational deltas (e.g. the Nile (Beach & Trayner 1991) and the Niger) and (c) fjords (e.g. Syvitski & Farrow 1989). Similar structures, on a smaller scale, may also develop on the degrading crests of uplifted fault blocks (e.g. Speksnijder 1987, Livera & Gdula 1990) and within sub-aerial landslides (e.g. Brunsden & Jones 1976, Bishop & Norris 1986).

The mechanical development of such structures has been reviewed by Price (1977), Mandl & Crans (1981) and Mandl (1988). Here, we are more concerned with the geometry of such fault systems. It should be borne in mind that while we discuss such structures as extensional, strictly speaking they are translational. Extension occurs at the upslope head of the load, but must be balanced by an equal amount of contraction or spreading at the downslope toe.

Fault geometry

It is a geometric necessity of gravity-driven systems that extensional faults in the head area should pass their displacement into a downslope translation. This requires the development of the classic listric fault geometry, in which steeply/moderately dipping faults flatten markedly with depth (Figs 11.17 and 11.20a). A series of listric faults will commonly link into one sole fault or detachment, which ultimately accommodates the slip on the whole fault system (e.g. Gibbs 1983, 1984, 1990).

Gravity-driven structures typically form in unconsolidated, or at best partially-consolidated, sediments (also see Chapters 6 and 7). In mechanical terms such rocks have little strength and will not exhibit the elastic properties of consolidated sediment or crystalline basement. This is borne out by the aseismic nature of major gravity-driven extensional provinces, such as the Gulf Coast. Because the rocks have little strength the development and growth of gravity-driven faults is not controlled by elastic failure and we cannot successfully model their slip surfaces as elastic dislocations. Fault-slip is accomplished entirely by downslope movement of the hangingwall while the footwall remains undeformed (unless it too is rotating against another fault in the array at the same time).

Analogue models. It is difficult to construct scaled physical models which simulate the brittle failure and isostatic loading associated with crustal-stretching faults. Modelling such structures is therefore best performed numerically. It is relatively simple, however, to induce gravitational failure in a scaled model and many analogue models simulating gravity-driven extensional faults have been made.

Some of the earliest modelling of normal faulting was that by Cloos (e.g. 1968). Similar experiments have subsequently been performed by McClay & Ellis (1987), Ellis & McClay (1988), McClay et al. (1991) (Fig. 11.20b), Vendeville et al. (1987), Vendeville & Cobbold (1988) and Vendeville (1991). These models illustrate the patterns of faulting and related ductile strain that can be expected in gravitational systems, and here we choose to highlight two particular results of the modelling.

First, it is very difficult to simulate the development of a listric fault system which propagates by footwall-collapse. Such idealized fault systems, postulated by Gibbs (1983, 1984), form by sequential movement on listric faults which propagate back into previously-undeformed footwall. Analogue models tend to show the synchronous development of a fault array, which approximates a detached domino geometry. The only way in which deforming-hangingwall against unrotating-footwall can be modelled is by forcing this geometry into the model with a rigid footwall (Fig. 11.20). Similarly, ramp-and-flat geometries do not initiate of their own accord in experimental models, but must also be forced into the models with an imposed footwall geometry.

Second, when a rigid footwall is imposed, and all deformation is confined to hangingwall movement above a listric fault, complex hangingwall strains are developed (Fig. 11.20). In such circumstances hangingwall deformation occurs by both folding and faulting. The result of this deformation is always to impose a *rollover anticline* within the hangingwall, whose precise geometry is controlled by fault shape. Rollover anticlines clearly also occur in nature above listric faults (e.g. Wernicke & Burchfiel 1982, fig. 15) (Fig. 11.20). Many geometric techniques exist for the analysis of rollover/fault geometry; these are briefly reviewed below.

Geometric section-balancing

The purpose of geometric section-balancing is to relate hangingwall shape to fault geometry, and thus construct a cross-section that will restore to a geologically realistic template. The prime assumption built into all methods of extensional section-balancing is that deformation in response to fault-slip is confined entirely to the hangingwall, although rigid-body rotation of the footwall can be accommodated. A plethora of techniques now exists for attempting section-balance, each of which makes different assumptions about how the hangingwall deforms.

The earliest construction is known as the Chevron construction (Verrall 1981) (Fig. 11.21). This assumes that hangingwall deformation is accomplished entirely by vertical simple shear and that horizontal translation of the hangingwall (heave) is constant and equal to the magnitude of extension. As a result of this, fault-slip is required to diminish with decreasing fault-dip.

A modification of the vertical shear assumption was proposed by White et al. (1986), who suggested a more general solution in which hangingwall strain is accommodated by arbitrarily inclined simple shear. In such circumstances, shear antithetic to the main fault results in heave less than extension, while synthetic shear results in heave greater than extension (Fig. 11.22). Observations of natural and analogue examples suggest that hangingwall shear on antithetic, rather than synthetic, slip surfaces is most likely.

Groshong (1989) has produced a model similar to that of White et al. (1986), but which will allow ramp-and-flat structures to be accommodated by variable

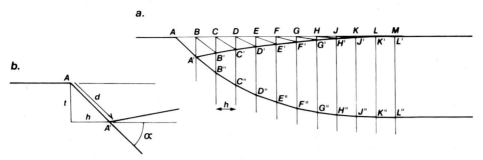

Fig. 11.21. The constant-heave Chevron construction, which relates hangingwall rollover geometry directly to fault shape. The hangingwall is assumed to collapse by vertical simple shear during extension. In this way displacement trajectories B–B', C–C' etc. can be assumed to parallel exactly the underlying fault plane. The fault plane may thus be constructed from rollover geometry or vice versa. After Williams & Vann (1987). (Reprinted from *Journal of Structural Geology*, The geometry of listric normal faults. Vol. 9, Williams G. and Vann, I., pp. 789–795, © 1987 with permission from Pergamon Press Ltd, Headington Hill Hall, Oxford OX3 0BW, U. K.)

directions of simple shear within the hangingwall. A further modification of the inclined simple shear technique is that of White (1987) and White & Yielding (1991). This technique allows for inclined simple shear and compaction in the hangingwall and can be used iteratively to test several beds within the hangingwall, with the aim of identifying the single fault geometry most likely to have produced the variable shape of the various beds.

A construction involving constant fault-slip, with resulting variable hangingwall heave, was introduced by Gibbs (1985). Like the constant heave models this involves elongation of bed-length within the rollover. As an alternative, Davison (1986) suggested that in some cases bed-length might be preserved while slip and heave are allowed to vary. Unfortunately neither of these constructions, nor the slip-line construction of Williams & Vann (1987), preserves hangingwall area during deformation, making them geometrically, and probably also geologically, invalid in their simplest form (Wheeler 1987).

All balancing techniques can be applied in two ways. Given a known fault geometry, forward modelling of the likely hangingwall deformation style can be established (Fig. 11.22), or, perhaps of more practical use, is the inverse model in which bed geometry can be used to predict fault shape at depth (Fig. 11.21). The inverse model is readily applicable to seismic data on which hangingwall folding may be defined but the fault itself remains unimaged.

It has frequently been stated, or inferred, in the literature that hangingwall folding above an extensional fault must indicate the presence of a curved fault surface, the geometry of which can be inferred from the aforementioned geometric reconstructions. We cannot stress strongly enough that this is not so. Hangingwall folding results from deformation above curved fault surfaces, from the imposition of elastic strain and from the imposition of isostatic loads. In any particular geological example the first analytical step towards understanding fault-related geometry should be the considered choice of the correct rheological and geometric model in the correct circumstances.

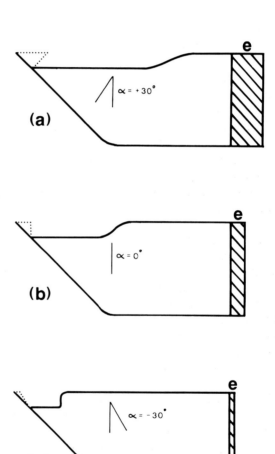

Fig. 11.22. Examples of variable simple shear in the hangingwall of a listric-shaped fault. In (a) the shear direction is 30° from vertical and antithetic to the main fault, extension is greater than fault heave. In (b) the shear direction is vertical (cf. Fig. 11.21), extension equals fault heave. In (c) the shear direction is 30° from vertical and synthetic to the main fault, extension is less than fault heave. After White (1987). (Reprinted from "Constraints on the measurement of extension in the brittle upper crust" by White, N. J., from *Norsk Geologisk Tidsskrift*, Vol. 67, pp. 269–279, © 1987 by permission of Scandinavian University Press, Olso, Norway.)

CONCLUSIONS

Within this chapter we have attempted to synthesize current understanding of continental extensional tectonics from the scale of the whole basin down to that of isolated small faults and superficial gravity-driven structures. In order to help in the choice of the correct model for the correct circumstances we highlight the following points as conclusions.

Stretching models

(1) The McKenzie uniform-stretching model, together with its derivatives, has been able to explain quantitatively the observed gross geometry of a number of extensional basins.

(2) The lithospheric simple shear model cannot be applied to most basins with the same success as the McKenzie-type derivatives.

(3) An understanding of the concept of isostasy is critical to understanding the evolution of sedimentary basins.

Large extensional faults

(1) Large extensional faults are those which completely cut the seismogenic layer. Their geometry is generally planar.

(2) The long-term deformation associated with large normal faults is a combination of co-seismic fault-displacement and a time-dependent, isostatic uplift.

(3) Co-seismic deformation is well accounted for by elastic dislocation models. Modelling of long-term deformation must include the response to isostatic loading (flexure).

(4) Both footwall uplift and hangingwall subsidence will always occur adjacent to large faults, the relative magnitude of each depending critically on the density of sediment load.

(5) The most complete treatment of the deformation and subsidence associated with crustal stretching on large faults is given by the flexural cantilever model.

(6) In circumstances where the spacing of major faults is approximately constant, the domino model is a useful geometric simplification of the more complete flexural solution.

Small extensional faults

(1) Small faults are those confined within the seismogenic layer.

(2) Deformation around small faults appears to be well described by dislocation models. The flexural response to loading can be ignored.

(3) Blind small faults produce equal magnitudes of footwall uplift and hangingwall rollover. Emergent small faults produce greater hangingwall rollover than footwall uplift.

(4) Fault population statistics imply that, in a given sample of faults, small faults will be more numerous than large faults, with the specific population corresponding to a logarithmic relationship between population density and fault size.

Gravity-driven extensional faults

(1) Gravity-driven faults are confined to the sediment pile and typically have a listric profile.

(2) The geometry of hangingwall deformation can be modelled by use of any one of a number of section-balancing techniques.

(3) Geometric section-balancing techniques which assume a rigid footwall and deforming hangingwall geometry are, however, inapplicable to the study of crustal-stretching faults.

Acknowledgements — Many of the concepts in this chapter have been made clear to us by patient discussion with a number of our colleagues. In particular we thank Brett Freeman, Mike Badley (Badley Earth Sciences), Nick Kusznir, Juan Watterson, John Walsh (Liverpool), James Jackson, Nicky White (Cambridge), Dave Barr (BP) and Geoff King.

REFERENCES

Aki, K. 1981. A probabilistic synthesis of precursory phenomena. In: *Earthquake Prediction: An International Review* (edited by Simpson, D. W. & Richards, P. G.). *Maurice Ewing Ser.* **4**, Am. geophys. U., 566–574.

Alvarez, F., Virieux, J. & Le Pichon, X. 1984. Thermal consequences of lithosphere extension over continental margins: the initial stretching phase. *Geophys. J. R. astr. Soc.* **78**, 389–411.

Barnett, J. A. M., Mortimer, J., Rippon, J. H., Walsh, J. J. & Watterson, J. 1987. Displacement geometry in the volume containing a single normal fault. *Bull. Am. Ass. Petrol. Geol.* **71**, 925–937.

Barr, D. 1987a. Lithospheric stretching, detached normal faulting and footwall uplift. In: *Continental Extensional Tectonics* (edited by Coward, M. P., Dewey, J. F. & Hancock, P. L.). *Spec. Publs geol. Soc. Lond.* **28**, 75–94.

Barr, D. 1987b. Structural/stratigraphic models for extensional basins of half-graben type. *J. Struct. Geol.* **9**, 491–500.

Barr, D. 1991. Subsidence and sedimentation in semi-starved half-graben: a model based on North Sea data. In: *The Geometry of Normal Faults* (edited by Roberts, A. M., Yielding, G. & Freeman, B.). *Spec. Publs geol. Soc. Lond.* **56**, 17–28.

Barr, D., Holdsworth, R. E. & Roberts, A. M. 1986. Caledonian ductile thrusting in a Precambrian metamorphic complex: the Moine of northwestern Scotland. *Bull. geol. Soc. Am.* **97**, 754–764.

Barton, P. & Wood, R. 1984. Tectonic evolution of the North Sea basin: crustal stretching and subsidence. *Geophys J. R. astr. Soc.* **79**, 987–1022.

Beach, A. 1986. A deep seismic reflection profile across the northern North Sea. *Nature* **323**, 53–55.

Beach, A. & Trayner, P. 1991. The geometry of normal faults in a sector of the offshore Nile Delta, Egypt. In: *The Geometry of Normal Faults* (edited by Roberts, A. M., Yielding, G. & Freeman, B.). *Spec. Publs geol. Soc. Lond.* **56**, 173–182.

Beach, A., Bird, T. & Gibbs, A. 1987. Extensional tectonics and crustal structure: deep seismic reflection data from the northern North Sea Viking Graben. In: *Continental Extensional Tectonics* (edited by Coward, M. P., Dewey,

J. F. & Hancock, P. L.). *Spec. Publs geol. Soc. Lond.* **28**, 467–476.

Bishop, D. G. & Norris, R. J. 1986. Rift and thrust tectonics associated with a translational block slide, Abbotsford, New Zealand. *Geol. Mag.* **123**, 13–25.

Bott, M. H. P. 1976. Formation of sedimentary basins of graben type by extension of the continental crust. *Tectonophysics* **36**, 77–86.

Brundsden, D. & Jones, D. K. C. 1976. The evolution of landslide slopes in Dorset. *Phil. Trans. R. Soc. Lond.* **A283**, 605–631.

Buck, W. R. 1988. Flexural rotation of normal faults. *Tectonics* **7**, 959–993.

Buck, W. R., Martinez, F., Steckler, M. S. & Cochran, J. R. 1988. Thermal consequences of lithospheric extension: pure and simple. *Tectonics* **7**, 213–234.

Caputo, M. 1987. The interpretation of the b and b_o values and its implications on the regional deformation of the crust. *Geophys. J. R. astr. Soc.* **90**, 551–573.

Cartwright, J. A. 1991. The kinematic evolution of the Coffee Soil Fault. In: *The Geometry of Normal Faults* (edited by Roberts, A. M., Yielding, G. & Freeman, B.). *Spec. Publs geol. Soc. Lond.* **56**, 29–40.

Chen, W. P. & Molnar, P. 1983. Focal depths of intracontinental and intraplate earthquakes and their implications for the thermal and mechanical properties of the lithosphere. *J. geophys. Res.* **88**, 4183–4214.

Childs, C., Walsh, J. J. & Watterson, J. 1990. A method for estimation of the density of fault displacements below the limits of seismic resolution in reservoir formations. In: *North Sea Oil and Gas Reservoirs II*. Norwegian Institute of Technology, Trondheim, Graham & Trotman, 309–318.

Cochran, J. R. 1983. Effects of finite rifting times on the development of sedimentary basins. *Earth Planet. Sci. Lett.* **66**, 289–302.

Cloos, E. 1968. Experimental analysis of Gulf Coast fracture patterns. *Bull. Am. Ass. Petrol. Geol.* **52**, 420–444.

Cohen, S. C. 1984. Postseismic deformation due to subcrustal viscoelastic relaxation following dip-slip earthquakes. *J. geophys. Res.* **89**, 4538–4544.

Coward, M. P. 1986. Heterogeneous stretching, simple shear and basin development. *Earth Planet. Sci. Lett.* **80**, 325–336.

Davision, I. 1986. Listric normal fault profiles: calculation using bed-length balance and fault displacement. *J. Struct. Geol.* **8**, 209–210.

Davison, I. 1989. Extensional domino fault tectonics: kinematics and geometrical constraints. *Ann. Tecton.* **3**, 12–24.

De'Ath, N. G. & Schuyleman, S. F. 1981. The geology of the Magnus Oilfield. In: *Petroleum Geology of the Continental Shelf of North-West Europe* (edited by Illing, L. V. & Hobson, G. D.). Heyden, 342–351.

Ellis, P. G. & McClay, K. R. 1988. Listric extensional fault systems — results of analogue model experiments. *Basin Res.* **1**, 55–70.

Emmons, W. H. & Garrey, G. H. 1910. General geology. In: *Geology and Ore Deposits of the Bullfrog District, Nevada* (edited by Ransome, F. L., Emmons, W. H. & Garrey, G. H.). *Bull. U.S. geol. Surv.* **407**, 19–89.

Escher, A. & Watterson, J. 1974. Stretching fabrics, folds and crustal shortening. *Tectonophysics* **2**, 223–231.

Fowler, S. & McKenzie, D. P. 1989. Gravity studies of the Rockall and Exmouth Plateaux using SEASAT altimetry. *Basin Res.* **2**, 27–34.

Gawthorpe, R. L. 1987. Tectono-sedimentary evolution of the Bowland Basin, northern England, during the Dinantian. *J. geol. Soc. Lond.* **144**, 59–72.

Gibbs, A. D. 1983. Balanced cross-section construction from seismic lines in areas of extensional tectonics. *J. Struct. Geol.* **5**, 153–160.

Gibbs, A. D. 1984. Structural evolution of extensional basin margins. *J. geol. Soc. Lond.* **141**, 609–620.

Gibbs, A. D. 1985. Discussion on the structural evolution of extensional basin margins. *J. geol. Soc. Lond.* **142**, 941–942.

Gibbs, A. D. 1990. Linked fault families in basin formation. *J. Struct. Geol.* **12**, 795–803.

Gibson, J. R., Walsh, J. J. & Watterson, J. 1989. Modelling of bed contours and cross-sections adjacent to planar normal faults. *J. Struct. Geol.* **11**, 317–328.

Groshong, R. H. 1989. Half-graben structures: balanced models of extensional fault-bend folds. *Bull. geol. Soc. Am.* **101**, 96–105.

Gutenberg, B. & Richter, C. F. 1944. Frequency of earthquakes in California. *Bull. seism. Soc. Am.* **34**, 185–188.

Hamilton, W. 1987. Crustal extension in the Basin and Range Province, southwestern United States. In: *Continental Extensional Tectonics* (edited by Coward, M. P., Dewey, J. F. & Hancock, P. L.). *Spec. Publs geol. Soc.* **28**, 155–176.

Heiskanen, W. A. & Vening Meinesz, F. A. 1958. *The Earth and its Gravity Field*. McGraw-Hill, New York.

Hellinger, S. J. & Sclater, J. G. 1983. Some comments on two-layer extensional models for the evolution of sedimentary basins. *J. geophys. Res.* **88**, 8251–8269.

Hellinger, S. J., Sclater, J. G. & Giltner, J. 1989. Mid-Jurassic through mid-Cretaceous extension in the Central Graben of the North Sea — part 1: estimates from subsidence. *Basin Res.* **1**, 191–200.

Jackson, J. A. 1987. Active normal faulting and crustal extension. In: *Continental Extensional Tectonics* (edited by Coward, M. P., Dewey, J. F. & Hancock, P. L.). *Spec. Publs geol. Soc. Lond.* **28**, 3–18.

Jackson, M. P. A. & Galloway, W. E. 1984. *Structural and Depositional Styles of Gulf Coast Tertiary Continental Margins: Application to Hydrocarbon Exploration*. AAPG Continuing Ed. Course Note Ser. No. 25. AAPG, Tulsa, 1–226.

Jackson, J. A. & McKenzie, D. P. 1983. The geometrical evolution of normal fault systems. *J. Struct. Geol.* **5**, 471–482.

Jackson, J. A. & McKenzie, D. P. 1988. The relationship between plate motions and seismic moment tensors, and the rates of active deformation in the Mediterranean and Middle East. *Geophys. J.* **93**, 45–73.

Jackson, J. A. & White, N. J. 1989. Normal faulting in the upper continental crust: observations from regions of active extension. *J. Struct. Geol.* **11**, 15–36.

Jackson, J. A., King, G. & Vita-Finzi, C. 1982. The neotectonics of the Aegean: an alternative view. *Earth Planet. Sci. Lett.* **61**, 303–318.

Jackson, J. A., White, N. J., Garfunkel, Z. & Anderson, A. 1988. Relations between normal-fault geometry, tilting and vertical motions in extensional terrains, an example from the southern Gulf of Suez. *J. Struct. Geol.* **10**, 155–170.

Jankhof, K. 1945. Changes in ground level produced by the earthquakes of April 14 to 18, 1928 in southern Bulgaria. In: *Tremblements de Terre en Bulgarie*, Nos. 29–31. Institut Meterologique Central de Bulgaria, Sofia, 131–136 (in Bulgarian).

Jarvis, G. T. 1984. An extensional model of graben subsidence — the first stage of basin evolution. *Sediment. Geol.* **40**, 13–31.

Jarvis, G. T. & McKenzie, D. P. 1980. Sedimentary basin formation with finite extension rates. *Earth Planet. Sci. Lett.* **48**, 42–52.

John, B. E. 1987. Geometry and evolution of a mid-crustal extensional fault system: Chemehueui Mountains, southeastern California. In: *Continental Extensional Tectonics* (edited by Coward, M. P., Dewey, J. F. &

Hancock, P. L.). *Spec. Publs geol. Soc. Lond.* **28**, 313–336.

Kanamori, H. & Anderson, D. L. 1975. Theoretical basis of some empirical relations in seismology. *Bull. seism. Soc. Am.* **65**, 1073–1095.

Kautz, S. A. & Sclater, J. G. 1988. Internal deformation in clay models of extension by block faulting. *Tectonics* **7**, 823–832.

King, G. C. P. 1983. The accommodation of large strains in the upper lithosphere of the earth and other solids by self-similar fault systems: the geometrical origin of b-value. *Pure appl. Geophys.* **121**, 761–815.

King, G. C. P., Stein, R. S. & Rundle, J. B. 1988. The growth of geological structures by repeated earthquakes, 1, conceptual framework. *J. geophys. Res.* **93**, 13307–13319.

Klemperer, S. 1988. Crustal thinning and nature of extension in the northern North Sea from deep seismic reflection profiling. *Tectonics* **7**, 803–822.

Klemperer, S. L. & White, N. J. 1989. Coaxial stretching or lithospheric simple shear in the North Sea? Evidence from deep seismic profiling and subsidence. In: *Extensional Tectonics and Stratigraphy of the North Atlantic Margins* (edited by Tankard, A. J. & Balkwill, H. R.). *Mem. Am. Ass. Petrol. Geol.* **46**, 511–521.

Koseluk, R. A. & Bischke, R. E. 1981. An elastic rebound model for normal fault earthquakes. *J. geophys. Res.* **86**, 1081–1090.

Kusznir, N. J. & Egan, S. S. 1989. Simple-shear and pure-shear models of extensional sedimentary basin formation: application to the Jeanne d'Arc Basin, Grand Banks of Newfoundland. In: *Extensional Tectonics and Stratigraphy of the North Atlantic Margins* (edited by Tankard, A. J. & Balkwill, H. R.). *Mem. Am. Ass. Petrol. Geol.* **46**, 305–322.

Kusznir, N. J. & Matthews, D. H. 1988. Deep seismic reflections and the deformational mechanisms of the continental lithosphere. *J. Petrol. (Special Lithosphere Issue)*, 66–87.

Kusznir, N. J. & Park, R. G. 1987. The extensional strength of the continental lithosphere: its dependence on geothermal gradient, and crustal composition and thickness. In: *Continental Extensional Tectonics* (edited by Coward, M. P., Dewey, J. F. & Hancock, P. L.). *Spec. Publs geol. Soc. Lond.* **28**, 35–52.

Kusznir, N. J. & Ziegler, P. A. 1992. The mechanics of continental extension and sedimentary basin formation: a simple-shear/pure-shear flexural cantilever model. *Tectonophysics* **215**, 117–131.

Kusznir, N. J., Karner, G. D. & Egan, S. 1987. Geometric, thermal and isostatic consequences of detachments in continental lithosphere extension and basin formation. In: *Sedimentary Basins and Basin Forming Mechanisms* (edited by Beaumont, C. & Tankard, A. J.). *Mem. Can. Soc. Petrol. Geol.* **12**, 185–203.

Kusznir, N. J., Marsden, G. & Egan, S. S. 1991. A flexural cantilever simple-shear/pure-shear model of continental lithosphere extension: application to the Jeanne d'Arc Basin, Grand Banks and Viking Graben, North Sea. In: *The Geometry of Normal Faults* (edited by Roberts, A. M., Yielding, G. & Freeman, B.). *Spec. Publs geol. Soc. Lond.* **56**, 41–60.

Lang, G., Vilotte, J. P. & Neugebauer, H. J. 1984. Relaxation of the Earth after a dip slip earthquake: dependence on rheology and geometry. *Phys. Earth Planet. Interiors* **36**, 260–275.

Larsen, P. H. 1988. Relay structures in a Lower Permian basement-involved extension system, East Greenland. *J. Struct. Geol.* **10**, 3–8.

Leeder, M. R. & Gawthorpe, R. L. 1987. Sedimentary models for extensional tilt-block/half-graben basins. In: *Continental Extensional Tectonics* (edited by Coward, M. P., Dewey, J. F. & Hancock, P. L.). *Spec. Publs geol. Soc. Lond.* **28**, 139–152.

Le Pichon, X. & Sibuet, J. C. 1981. Passive margins: a model of formation. *J. geophys. Res.* **86**, 3708–3720.

Lister, G. S. & Davis, G. A. 1989. The origin of metamorphic core complexes and detachment faults formed during Tertiary continental extension in the northern Colorado River region, U.S.A. *J. Struct. Geol.* **11**, 65–94.

Livera, S. E. & Gdula, J. E. 1990. Brent Oil Field. In: *Atlas of Oil and Gas Fields, Structural Traps II, Traps Associated with Tectonic Faulting* (edited by Beaumont, E. A. & Foster, N. H.). *Am. Ass. Petrol. Geol.* 21–63.

Mandl, G. 1987. Tectonic deformation by rotating parallel faults — the 'bookshelf' mechanism. *Tectonophysics* **141**, 277–316.

Mandl, G. 1988. *Mechanics of Tectonic Faulting*. Elsevier, Amsterdam, 407 pp.

Mandl, G. & Crans, W. 1981. Gravitational gliding in deltas. In: *Thrust and Nappe Tectonics* (edited by McClay, K. R. & Price, N. J.). *Spec. Publs geol. Soc. Lond.* **9**, 41–54.

Mansinha, L. & Smylie, D. E. 1971. The displacement fields of inclined faults. *Bull. seism. Soc. Am.* **61**, 1433–1440.

Marrett, R. & Allmendinger, R. W. 1991. Estimates of strain due to brittle faulting: sampling of fault populations. *J. Struct. Geol.* **13**, 735–738.

Marsden, G., Yielding, G., Roberts, A. M. & Kusznir, N. J. 1990. Application of a flexural cantilever simple-shear/pure-shear model of continental lithosphere extension to the formation of the northern North Sea Basin. In: *Tectonic Evolution of the North Sea Rifts* (edited by Blundell, D. J. & Gibbs, A. D.). Oxford University Press, Oxford.

McClay, K. R. & Ellis, P. G. 1987. Analogue models of extensional fault geometries. In: *Continental Extensional Tectonics* (edited by Coward, M. P., Dewey, J. F. & Hancock, P. L.). *Spec. Publs geol. Soc. Lond.* **28**, 109–125.

McClay, K. R., Waltham, D. A., Scott, A. D. & Abousetta, A. 1991. Physical and seismic modelling of listric normal fault geometries. In: *The Geometry of Normal Faults* (edited by Roberts, A. M., Yielding, G. & Freeman, B.). *Spec. Publs geol. Soc. Lond.* **56**, 231–239.

McGeary, S. & Warner, M. R. 1985. Seismic profiling the continental lithosphere. *Nature* **317**, 795–797.

McKenzie, D. P. 1978. Some remarks on the development of sedimentary basins. *Earth Planet. Sci. Lett.* **40**, 25–32.

Melosh, H. J. 1983. Vertical movements following a dip-slip earthquake. *Geophys. Res. Lett.* **10**, 47–50.

Morley, C. K., Nelson, R. A., Patton, T. L. & Munn, S. G. 1990. Transfer zones in the East African rift system and their relevance to hydrocarbon exploration in rifts. *Bull. Am. Ass. Petrol. Geol.* **74**, 1234–1253.

Morton, W. H. & Black, R. 1975. Crustal attenuation in Afar. In: *Afar Depression of Ethiopia* (edited by Pilgar, A. & Rosler, A.). Inter-Union Commission on Geodynamics, *Proc. Int. Symp. Afar Region Related Rift Problems*. E. Schweizerbart'sche Verlagsbuchhandlung, Stuttgart. Scientific Report No. 14, 55–65.

Myers, W. B. & Hamilton, W. 1964. Deformation accompanying the Hebgen Lake earthquake of August 17, 1959. *U.S. geol. Surv. Prof. Pap.* **435**, 55–98.

Price, N. J. 1977. Aspects of gravity tectonics and the development of listric faults. *J. geol. Soc. Lond.* **133**, 311–327.

Reilinger, R. 1986. Evidence for postseismic viscoelastic relaxation following the 1959 M = 7.5 Hebgen Lake, Montana, Earthquake. *J. geophys. Res.* **91**, 9488–9494.

Reston, T. J. 1988. Evidence for shear zones in the lower crust offshore Britain. *Tectonics* **7**, 929–945.

Richter, C. F. 1958. *Elementary Seismology*. W.H. Freeman, San Francisco.

Rippon, J. H. 1985. Contoured patterns of the throw and

hade of normal faults in the Coal Measures (Westphalian) of north-east Derbyshire. *Proc. Yorkshire geol. Soc.* **45**, 147–161.

Roberts, A. M. & Yielding, G. 1991. Deformation around basin-margin faults in the North Sea/Norwegian rift. In: *The Geometry of Normal Faults* (edited by Roberts, A. M., Yielding, G. & Freeman, B.). *Spec. Publs geol. Soc. Lond.* **56**, 61–78.

Roberts, A. M., Price, J. D. & Olsen, T. S. 1991. Late Jurassic half-graben control on the siting and structure of hydrocarbon accumulations: UK/Norwegian Central Graben. In: *Tectonic Events Responsible for Britain's Oil and Gas Reserves* (edited by Hardman, R. F. P. & Brooks, J.). *Spec. Publs geol. Soc. Lond.* **55**, 229–257.

Roberts, A. M., Yielding, G. & Badley, M. E. 1993a. Tectonic and bathymetric controls on stratigraphic sequences within evolving half-graben. In: *Tectonics and Seismic Sequence Stratigraphy* (edited by Williams, G. D. & Dobb, A.). *Spec. Publs geol. Soc. Lond.* **71**, 87–121.

Roberts, A. M., Yielding, G., Kusznir, N. J., Walker, I. M. & Dorn-Lopez, D. 1993b. Mesozoic extension in the North Sea : constraints from flexural backstripping, forward modelling and fault populations. In: *Petroleum Geology of Northwest Europe, Proceedings of the 4th Conference* (edited by Parker, J. R.). *Geol. Soc. Lond.*

Roberts, S. & Jackson, J. A. 1991. Active normal faulting in Central Greece: an overview. In: *The Geometry of Normal Faults* (edited by Roberts, A. M., Yielding, G. & Freeman, B.). *Spec. Publs geol. Soc. Lond.* **56**, 125–142.

Rowley, D. B. & Sahagian, D. 1986. Depth-dependent stretching: a different approach. *Geology* **14**, 32–35.

Royden, L. & Keen, C. E. 1980. Rifting process and thermal evolution of the continental margin of eastern Canada determined from subsidence curves. *Earth Planet. Sci. Lett.* **51**, 343–361.

Schwartz, D. P. & Coppersmith, K. J. 1984. Fault behaviour and characteristic earthquakes: examples from the Wasatch and San Andreas fault zones. *J. geophys. Res.* **89**, 5681–5698.

Sclater, J. G. & Celerier, B. 1989. Errors in extension measurements from planar faults observed on seismic reflection lines. *Basin Res.* **1**, 217–221.

Shelton, J. W. 1984. Listric normal faults: an illustrated summary. *Bull. Am. Ass. Petrol. Geol.* **68**, 801–815.

Sibson, R. H. 1989. Earthquake faulting as a structural process. *J. Struct. Geol.* **11**, 1–14.

Speksnijder, A. 1987. The structural configuration of Cormorant Block IV in context of the northern Viking Graben structural framework. *Geologie Mijnb.* **65**, 357–379.

Stein, R. S. & Barrientos, S. E. 1985. Planar high-angle faulting in the Basin and Range: geodetic analysis of the 1983 Borah Peak, Idaho, earthquake. *J. geophys. Res.* **90**, 11355–11366.

Stein, R. S., King, G. C. P. & Rundle, J. B. 1988. The growth of geological structures by repeated earthquakes, 2, field examples of continental dip-slip faults. *J. geophys. Res.* **93**, 13319–13331.

Stewart, I. S. & Hancock, P. L. 1991. Scales of structural heterogeneity within neotectonic normal fault zones in the Aegean region. *J. Struct. Geol.* **13**, 191–204.

Syvitski, J. P. M. & Farrow, G. E. 1989. Fjord sedimentation as an analogue for small hydrocarbon-bearing fan deltas. In: *Deltas: Sites and Traps for Fossil Fuels* (edited by Whately, M. K. G. & Pickering, K. T.). *Spec. Publs geol. Soc. Lond.* **41**, 21–43.

Thatcher, W. 1984. The earthquake deformation cycle at the Nankai Trough, Southwest Japan. *J. geophys. Res.* **89**, 3087–3101.

Thompson, G. A. 1960. Problem of late Cenozoic structure of the Basin Ranges. *Proc. 21st Int. Geol. Congr. Copenhagen* **18**, 62–68.

Turner, C. C., Cohen, J. M., Connell, E. R. & Cooper, D. M. 1987. A depositional model for the South Brae oilfield. In: *Petroleum Geology of North West Europe* (edited by Brooks, J. & Glennie, K. W.). Graham & Trotman, 853–864.

Vendeville, B. 1991. Mechanisms generating normal fault curvature: a review illustrated by physical models. In: *The Geometry of Normal Faults* (edited by Roberts, A. M., Yielding, G. & Freeman, B.). *Spec. Publs geol. Soc. Lond.* **56**, 241–249.

Vendeville, B. & Cobbold, P. R. 1988. How normal faulting and sedimentation interact to produce listric fault profiles and stratigraphic wedges. *J. Struct. Geol.* **10**, 649–659.

Vendeville, B., Cobbold, P. R., Davy, P., Brun, J. P. & Choukroune, P. 1987. Physical models of extensional tectonics at various scales. In: *Continental Extensional Tectonics* (edited by Coward, M. P., Dewey, J. F. & Hancock, P. L.). *Spec. Publs geol. Soc. Lond.* **28**, 95–107.

Vening Meinesz, F. A. 1950. Les grabens africains, resultat de compression ou de tension dans le croute terrestre? *Bull. Inst. R. Colonial Belge* **21**, 539–552.

Verrall, P. 1981. Structural interpretation with application to North Sea problems. *Course Notes No. 3*. JAPEC, U.K.

Vita-Finzi, C. & King, G. C. P. 1985. The seismicity, geomorphology and structural evolution of the Corinth area of Greece. *Phil. Trans. R. Soc. Lond.* **A314**, 379–407.

Voorhoeve, H. & Houseman, G. 1988. The thermal evolution of a lithosphere extending on a low-angle detachment zone. *Basin Res.* **1**, 1–9.

Wallace, R. E. 1984. Patterns and timing of late Quaternary faulting in the Great Basin Province and relation to some regional tectonic features. *J. geophys. Res.* **89**, 5763–5769.

Walsh, J. J. & Watterson, J. 1987. Distributions of cumulative displacement and seismic slip on a single normal fault surface. *J. Struct. Geol.* **9**, 1039–1046.

Walsh, J. J. & Watterson, J. 1988. Analysis of the relationship between displacements and dimensions of faults. *J. Struct. Geol.* **10**, 239–247.

Walsh, J. J. & Watterson, J. 1989. Displacement gradients on fault surfaces. *J. Struct. Geol.* **11**, 307–316.

Walsh, J. J. & Watterson, J. 1991. Geometric and kinematic coherence and scale effects in normal fault systems. In: *The Geometry of Normal Faults* (edited by Roberts, A. M., Yielding, G. & Freeman, B.). *Spec. Publs geol. Soc. Lond.* **56**, 193–203.

Walsh, J. J. & Watterson, J. 1992. Populations of faults and fault displacements and their effects on estimates of fault-related regional extension. *J. Struct. Geol.* **14**, 701–712.

Walsh, J. J., Watterson, J. & Yielding, G. 1991. The importance of small-scale faulting in regional extension. *Nature* **351**, 391–393.

Watterson, J. 1986. Fault dimensions, displacements and growth. *Pure appl. Geophys.* **124**, 365–372.

Watts, A. B. 1988. Gravity anomalies, crustal structure and flexure of the lithosphere at the Baltimore Canyon Trough. *Earth Planet. Sci. Lett.* **89**, 221–238.

Weissel, J. K. & Karner, G. D. 1989. Flexural uplift of rift flanks due to mechanical unloading of the lithosphere during extension. *J. geophys. Res.* **94**, 13919–13950.

Wernicke, B. 1981. Low-angle normal faults in the Basin and Range province: nappe tectonics in an extending orogen. *Nature* **291**, 645–648.

Wernicke, B. 1985. Uniform normal-sense simple shear of the continental lithosphere. *Can. J. Earth Sci.* **22**, 108–125.

Wernicke, B. & Burchfiel, B. C. 1982. Modes of extensional tectonics. *J. Struct. Geol.* **4**, 105–115.

Wesnousky, S. G., Scholz, C. H., Shimazaki, K. & Matsuda, T. 1983. Earthquake frequency distribution and mechanisms of faulting. *J. geophys. Res.* **88**, 9331–9340.

Westaway, R. 1989. Northeast Basin and Range province active tectonics: an alternative view. *Geology* **17**, 779–783.

Westaway, R. 1991. Continental extension on sets of parallel faults: observational evidence and theoretical models. In: *The Geometry of Normal Faults* (edited by Roberts, A. M., Yielding, G. & Freeman, B.). *Spec. Publs geol. Soc. Lond.* **56**, 143–169.

Wheeler, J. 1987. Variable-heave models of deformation above listric normal faults, the importance of area conservation. *J. Struct. Geol.* **9**, 1047–1050.

White, N. J. 1987. Constraints on the measurement of extension in the brittle upper crust. *Norsk geol. Tidsskr.* **67**, 269–279.

White, N. J. 1989. The nature of lithospheric extension in the North Sea. *Geology* **17**, 111–114.

White, N. J. 1990. Does the uniform stretching model work in the North Sea? In: *Tectonic Evolution of the North Sea Rifts* (edited by Blundell, D. J. & Gibbs, A. D.). Oxford University Press, Oxford.

White, N. J. & McKenzie, D. 1988. Formation of the "steer's head" geometry of sedimentary basins by differential stretching of the crust and mantle. *Geology* **16**, 250–253.

White, N. J. & Yielding, G. 1991. Calculating normal fault geometries at depth: theory and examples. In: *The Geometry of Normal Faults* (edited by Roberts, A. M., Yielding, G. & Freeman, B.). *Spec. Publs geol. Soc. Lond.* **56**, 251–260.

White, N. J., Jackson, J. A. & McKenzie, D. P. 1986. The relationship between the geometry of normal faults and that of the sedimentary layers in their hangingwalls. *J. Struct. Geol.* **8**, 897–909.

Whitten, C. A. 1957. Geodetic measurements in the Dixie Valley area. *Bull. seism. Soc. Am.* **47**, 321–325.

Williams, G. & Vann, I. 1987. The geometry of listric normal faults and deformation in their hangingwalls. *J. Struct. Geol.* **9**, 789–795.

Wyss, M. 1973. Towards a physical understanding of the earthquake frequency distribution. *Geophys. J. R. astr. Soc.* **31**, 341–359.

Yielding, G. 1990. Footwall uplift associated with Late Jurassic normal faulting in the northern North Sea. *J. geol. Soc.* **147**, 219–222.

Yielding, G. & Roberts, A. M. 1992. Footwall uplift during normal faulting — implications for structural geometries in the North Sea. In: *Structural and Tectonic Modelling and its Application to Petroleum Geology* (edited by Larsen, R. M. *et al.*). NPF *Special Publication* **1**, 289–304.

Yielding, G., Badley, M. E. & Freeman, B. 1991. Seismic reflections from normal faults in the northern North Sea. In: *The Geometry of Normal Faults* (edited by Roberts, A. M., Yielding, G. & Freeman, B.). *Spec. Publs geol. Soc. Lond.* **56**, 79–89.

Young, R. A., Stewart, S. C., Seaman, M. R. & Evans, B. J. 1990. Fault-plane reflection processing and 3D display: the Darling Fault, Western Australia. *Tectonophysics* **173**, 107–117.

Zandt, G. & Owens, T. J. 1980. Crustal flexure associated with normal faulting and implications for seismicity along the Wasatch Front, Utah. *Bull. seism. Soc. Am.* **70**, 1501–1520.

Ziegler, P.A. 1983. Discussion on: crustal thinning and subsidence in the North Sea. *Nature* **304**, 561.

CHAPTER 12

Continental Strike-Slip Tectonics

N. H. WOODCOCK and C. SCHUBERT

STRIKE-SLIP TECTONICS DEFINED

A STRIKE-SLIP deformation zone is one with a large component of its net slip parallel to the strike of the zone (Fig. 12.1a). The direction of slip can be specified by its pitch within the zone (Fig. 12.1a). Strike-slip zones have small pitch angles, *dip-slip zones* have pitch angles close to 90°, and *oblique-slip zones* have intermediate values. There are no agreed quantitative bounds to these classes. A strike-slip (deformation) zone is a rock volume deforming in strike-slip mode by ductile or brittle processes in any proportion. *Strike-slip fault* is used for a discrete brittle fracture and *strike-slip fault system* for a number of kinematically related strike-slip faults. *Strike-slip tectonics* describes the deformation process in a region on any scale dominated by strike-slip kinematics.

The term *wrench fault* is often used synonymously with strike-slip fault. However, as developed by Anderson (1942), wrench has a dynamic rather than a kinematic significance, and implies that the intermediate principal stress (σ_2) was vertical during deformation (Fig. 12.1b). More recent usages of the term wrench fault (Moody & Hill 1956, Wilcox *et al.* 1973, Biddle & Christie-Blick 1985) require that it cuts basement rocks, and thus their definition approaches synonymy with transcurrent fault. We favour preserving Anderson's dynamic definition of wrench, rather than allowing it to lose any rigorous meaning (Sylvester 1988).

Wrench faults in mechanically isotropic rocks theoretically comprise vertical strike-slip faults (Fig. 12.1b). Strike-slip faults in general need not be vertical. They may form parallel to a pre-existing bedding or fabric anisotropy, or may reactivate dipping faults formed in thrust or gravity dynamic regimes.

TECTONIC SETTING OF STRIKE-SLIP ZONES

A tectonic classification

This section assesses the location and prevalence of strike-slip faults in a global plate tectonic perspective. The classification used here (Fig. 12.2) depends on the siting of strike-slip faults with respect to plate boundaries (Fig. 12.3). Woodcock (1986) and Sylvester (1988) list recent and ancient examples of each class.

Most large-displacement strike-slip faults either themselves form a discrete plate boundary or occur in the distributed deformation zone that forms most plate boundaries involving continental crust. In the first case the fault is a *transform fault*, a strike-slip fault that cuts the whole lithosphere and fully accommodates the displacement across the plate boundary. Some such faults near convergent boundaries delimit small lithosphere fragments. It is a semantic matter whether these faults are regarded as transforms bounding 'microplates' or as faults in a zone of distributed plate boundary deformation (compare Woodcock 1986 and Sylvester 1988).

Large-displacement strike-slip faults that cut continental basement as well as sedimentary cover have come to be termed *transcurrent faults* (Moody & Hill 1956, Freund 1974). Sylvester (1988) suggested that the term be restricted to strike-slip faults confined within the lithosphere; that is, those that are not transforms. This is a helpful usage, excepting only the difficulty of estimating the effective depth of an ancient fault.

Strike-slip faults with small displacements occur both in deformation belts bordering plate boundaries, particularly as elements of linked fault systems (see Chapter 6), and in intra-plate settings (see Chapter 10).

Character of the classes of strike-slip fault

A *ridge transform fault* is a transform fault that joins two oceanic ridge segments with similar spreading vectors, and which therefore maintains a constant length through time (Gilliland & Meyer 1976) (Fig. 12.3). They are preserved in some ophiolite fragments that have been emplaced onto or against continental crust.

Boundary transform faults separate plates moving parallel to their mutual boundary over lengths of hundreds of kilometres (Fig. 12.3). In contrast to ridge transforms, boundary transforms usually link unlike plate boundaries (Gilliland & Meyer 1976), and often change length through time and involve continental crust on one or both bounding plates. Boundary transforms may persist for tens of millions of years and accumulate displacements of many hundreds of kilometres.

Trench-linked transforms or strike-slip faults form at obliquely convergent plate boundaries. Here the oblique displacement is commonly resolved kinematically into two components: dip-slip on the subduction zone itself and strike-slip on a major fault parallel to and some 100 km inboard from the trench (Fitch 1972, Woodcock 1986). This trench-linked strike-slip fault may properly be termed a *trench-linked transform* if it is a discrete, whole-lithosphere fracture which isolates a sizeable coherent 'buffer' plate between itself and the trench. Trench-linked transforms commonly border or bisect the volcanic arc (see Chapter 9). Ancient examples might therefore both cut and localize sub-arc intrusives, arc volcanics and derived volcaniclastics.

Indent-linked strike-slip faults form where a convergent plate boundary juxtaposes continental or arc crust on both plates (Woodcock 1986). The relative buoyancy of such crust causes collision and indentation of crust across the convergence zone (also see Chapters 6, 10 and 13). The kinematic consequences depend on the relative shapes and rheologies of opposing continents, but conjugate indent-linked strike-slip faults are an important element (Fig. 12.3) (Molnar & Tapponnier 1975). They allow both crustal shortening ahead of an indenting promontory and lateral *tectonic escape* (Burke & Şengör 1986) into uncollided segments of the convergence zone. They are necessarily restricted to continental crust, and might be diagnosed by their localization of non-marine sedimentation, and their association with limited silicic volcanics and S-type granites. Some indent-linked strike-slip faults may cut the whole lithosphere and may qualify as transforms if they link other plate boundaries and define coherent plates. Most are probably confined within the lithosphere and may root in low-dip detachments below or within the crust.

Tear and transfer faults are types of small-scale strike-slip faults restricted to the upper crust or to its sedimentary cover. They are common as parts of linked fault systems in the contractional and extensional deformation belts associated with active plate boundaries (see Chapter 6). A *tear fault* is any strike-slip or oblique-slip fault that runs across the strike of a contractional or extensional belt. It accommodates differential displacements between two segments of an allochthon, or between allochthon and autochthon. It may link unlike structural elements, for instance a thrust with a normal fault. A *transfer fault* is a sub-class of tear fault that transfers displacement between two similar structural elements, for instance two normal faults (also see Chapter 11). Transfer faults linking thrust faults are commonly known as *lateral ramps*, in contrast to the *frontal ramps* formed by the thrusts themselves (see Chapter 6). Both tear and transfer faults usually link downwards into a low-dip detachment. The term transfer fault is also used for the faults which connect overstepping segments of strike-slip faults (Sylvester 1988).

Intra-plate strike-slip faults occur remote from plate boundaries, often forced by intra-plate stresses reactivating old faults and propagating them laterally and vertically into unfaulted rock (see Chapter 10). Reactivation in strike-slip mode is favoured in a regional wrench regime (intermediate principal stress vertical) on faults which have moderate to steep dips. Although displacements are small, they may produce fault rocks and fault-surface features which obliterate earlier movement indicators.

Global frequency of strike-slip tectonics

Most strike-slip tectonics is related to plate boundaries with boundary-parallel or oblique displacement vectors. An estmate of its past global frequency is given by the proportion of different boundary types in the present-day plate system (Woodcock 1986). About 14% of boundaries show boundary-sub-parallel displacements, with a further 45% showing markedly oblique displacements. So a component of strike-slip tectonics is probably occurring along nearly 60% of present-day plate boundaries. This is a minimum

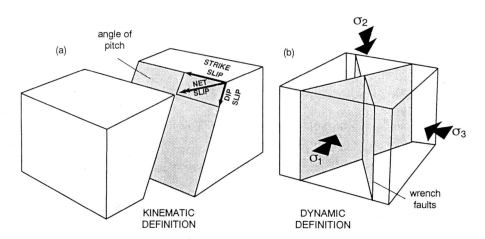

Fig. 12.1. Definition diagrams for strike-slip and wrench faults. (a) Kinematics. (b) Dynamics; σ_1, σ_2, σ_3, maximum, intermediate and minimun principal stresses.

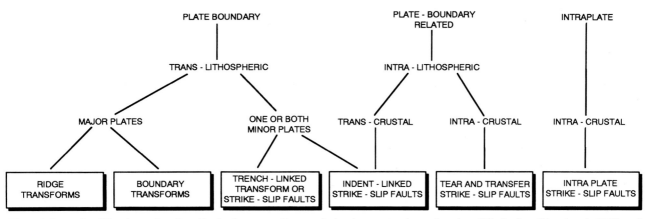

Fig. 12.2. A classification of strike-slip faults according to their plate tectonic setting (modified after Woodcock 1986 and Sylvester 1988).

estimate, ignoring as it does indent-linked strike-slip faults produced at head-on collision zones. Certainly, some strike-slip tectonics would be expected in the majority of old orogenic belts.

THEORETICAL AND EXPERIMENTAL BASIS OF STRIKE-SLIP TECTONICS

Theory, experiment and reality

Theory and experiment are more successful in explaining structures in real strike-slip zones than in dip-slip settings. This success arises from the tendency of rock layering to be gently-dipping and of strike-slip faults to be steeply-dipping. Structural geometries are therefore less influenced by rock anisotropy. This section summarizes the theoretical treatment first of simple strike-slip zones in rocks without steep anisotropy, and then of additional components of shortening and extension perpendicular to the margins of the deforming zone. Finally some additional concepts deriving from physical experiments are summarized.

Theoretical models of simple strike-slip

Most major continental strike-slip zones occur in domains of crustal-scale simple shear rather than pure shear because they are related to oblique convergence or divergence at plate boundaries. Sylvester (1988) reviewed the history of the pure versus simple shear debate. The theory of simple shear is reviewed by Ramsay & Huber (1983) (also see Chapters 1 and 2). It applies to shear zones in any orientation and of any scale. Being a plane strain process it is best viewed along the direction of no length change, that is on the section normal to the shear zone and containing the slip direction. This is a horizontal section for a vertical strike-slip zone (Fig. 12.4). The shear sense in Fig. 12.4 is *dextral* (or *right-lateral*), that is the far side of the zone is displaced to the right with respect to the near side, in downwards map view. The alternative is *sinistral* (or *left-lateral*) displacement.

The relevant kinematic rules of simple shear in isotropic rocks (Fig. 12.4) can be summarized as follows. (1) The first increment of displacement can be thought of as transforming an initially circular strain marker into a strain ellipse with axes at 45° to the shear zone boundaries. (2) Further displacements increase the axial ratio of the strain ellipse and cause its long axis to rotate towards parallelism with the shear zone boundaries. (3) Material lines rotate (clockwise under dextral shear), quickly if at a high angle to the shear zone, slowly if at a low angle and not at all if parallel to it.

The particular structures that develop depend on the bulk rheology and microscropic deformation mech-

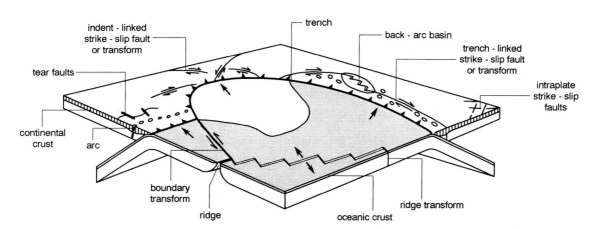

Fig. 12.3. Major classes of strike-slip fault in their plate tectonic setting (after Woodcock 1986).

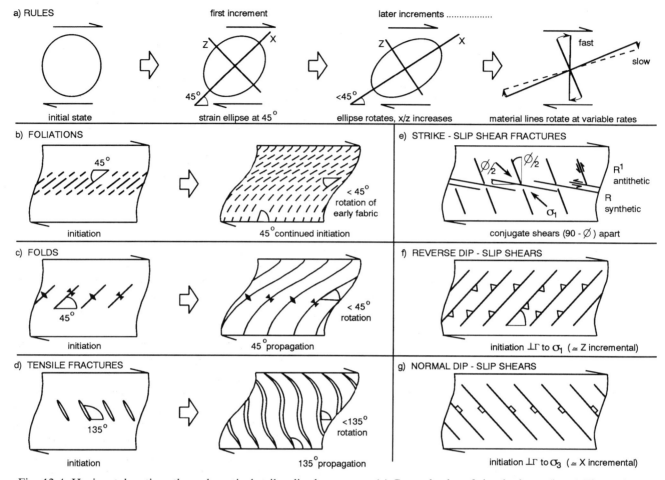

Fig. 12.4. Horizontal sections through vertical strike-slip shear zones. (a) General rules of simple shear. (b – g) Theoretically predictable geometric patterns formed by specific structures.

anisms in the zone. The structural geometries in Figs 12.4(b – g) are direct deductions from simple shear theory for a variety of real structures. In each case the initiation geometry can be predicted from the strain orientation after small simple shear across the zone. Progressive changes in geometry are predictable in zones with an element of continuous ductile deformation (Figs 12.4b – d) but less reliable for discontinuous brittle deformation (Figs 12.4e – g).

Foliations, cleavages and schistosities are taken to develop perpendicular to the short axis (Z) of the finite strain ellipse (Fig. 12.4b). In a typical zone, with higher shear strain in the centre than at the margins, foliation develops a sigmoidal pattern, making a 45° angle at the zone boundary but a lower angle in its centre. Sigmoidal foliations are common in ductile shear zones in high-grade rocks on a variety of scales (see review by Ramsay 1980).

Folds tend to develop at lower metamorphic grades and where layering, often nearly horizontal, is more effective. Fold hinges tend to nucleate perpendicular to the short axis (Z) of the strain ellipse, but to rotate as material lines with additional strain (Fig. 12.4c). They develop a sigmoidal pattern in heterogeneous shear zones, but do not quite parallel foliation (see transected folds later).

Tensile fractures develop in small-scale shear zones in low-grade rocks. They are particularly favoured at low differential stresses and high pore fluid pressures. The resulting *T fractures* initiate perpendicular to the minimum compressive stress, the long axis (X) of the incremental strain ellipse (Fig. 12.4d) (also see Chapter 5). Any ductile component rotates early formed segments in the centre of the zone, but new segments propagate at 135° towards the margin to give a sigmoidal geometry.

Strike-slip shear fractures are the main response of brittle zones. Two sets (termed *Riedel shears*) form, oriented so that their acute bisector is the direction of the maximum compressive stress (Fig. 12.4e) (also see Chapter 5). Brittle failure theory predicts that the acute angle between sets is $90° - \phi$ (where ϕ is the angle of internal friction, commonly about 30°) (Tchalenko & Ambrayses 1970). The maximum principal stress coincides with the short axis (Z) of the incremental strain ellipse at 45° to the shear zone boundary. One shear set (*R shears*) therefore forms at a low angle ($\phi/2$) to the zone, with a displacement sense synthetic to that of the whole zone (Fig. 4e). The other set (*R′ shears*) forms at a high angle ($90° - \phi/2$) and has an antithetic shear sense. More complex patterns of natural shear fractures develop at higher strains. These are best understood with respect to experimental studies described later.

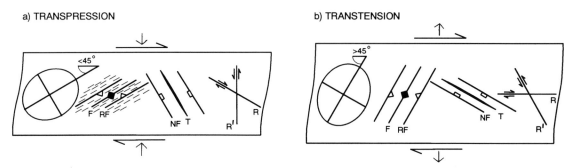

Fig. 12.5. Horizontal sections through vertical strike-slip zones undergoing (a) transpression and (b) transtension, to show surface traces of folds and foliation (F), reverse faults (RF), normal faults (NF), tension fractures (T) and Riedel shears (R, R').

Dip-slip shear fractures may also form with predictable orientations. Reverse dip-slip faults will strike perpendicular to the maximum compression (Fig. 12.4f) and normal dip-slip faults will strike perpendicular to the minimum compression (Fig. 12.4g). Both modes necessarily involve non-plane strain deformation, unless they mutually compensate each other. They are therefore more common when there is a component of shortening or extension across a strike-slip zone. These departures from simple shear are the subject of the next section.

Theory of transpression and transtension

By contrast with simple strike-slip shear zones, many real zones have a component of either shortening or extension across them. These situations of oblique shear are most commonly termed *transpression* and *transtension*, respectively (Fig. 12.5) (Harland 1971). *Convergent and divergent strike-slip or obliquely convergent and obliquely divergent* are alternative terms (Wilcox et al. 1973). Transpression has less commonly been applied to both modes (Sanderson & Marchini 1984), with a shortening component indicated by a positive sign and extension being negative. Nomenclature apart, both modes involve non-plane strain, shortening or extension across the zone being compensated by vertical thinning or thickening.

Transpression and transtension occur both on a regional scale, typically at plate boundaries with oblique displacement vectors, and more locally in any strike-slip zone which is not perfectly planar. In any case the deformation can be analyzed as a pure shear superimposed across a simple shear zone (Sanderson & Marchini 1984). The resulting strain magnitudes and orientations can be computed and the implications for the geometry of real structures deduced (Fig. 12.5). The suite of structures illustrated are generally not all developed in any one zone, nor do they develop synchronously; brittle structures will tend to post-date ductile structures.

Transpression (Fig. 12.5a) induces a horizontal strain ellipse with its long axis at a lower angle to the zone boundaries than in simple strike-slip. Foliations, folds and reverse faults therefore form at low angles to the zone, and tensile fractures and normal faults at high angles. Riedel shears form at higher angles than in simple shear, such that the antithetic set may develop at or beyond the zone normal. Because the zone thickens vertically, reverse displacements tend to dominate over normal displacements.

Transtension (Fig. 12.5b) causes the long axis of the horizontal strain ellipse to develop at a higher angle to the zone boundaries than in simple strike-slip. Foliations, folds and reverse faults form at high angles to the zone, and tensile fractures and normal faults at low angles. Riedel shears form at lower angles than in simple shear, such that the synthetic set may develop sub-parallel to the zone boundaries. Because the zone thins vertically, normal displacements tend to dominate over reverse displacements.

Experimental studies

Numerous laboratory experiments simulating strike-slip shear zones (reviewed by Sylvester 1988) have successfully reproduced the structural patterns predicted by theory. Some have produced more complex patterns which match naturally observed structures. Examples are the brittle fracture patterns which progressively develop after the first increments of shear, and the three-dimensional architecture of en échelon fracture and fold patterns.

Most simulations of brittle shear zones involve a basement of two stiff boards allowed to slip past each other, overlain by a cover of layered wax, clay or sand. Although details vary according to experimental material and conditions, some recurrent features of the progressive surface fracture pattern can be observed (Fig. 12.6a).

The first formed structures are en échelon *R shears*. They have a component of synthetic strike-slip, but also a small dip-slip component which changes downthrow sense at the mid-point of the shear. Each R shear is therefore a *scissor fault*. *Splay faults*, linking to a genetically related major fault, form at the tips of the R shears, at a higher angle to the zone. Antithetic Riedel shears (R') form only occasionally, in the overlap zones between R shears. At higher displacements the R shears

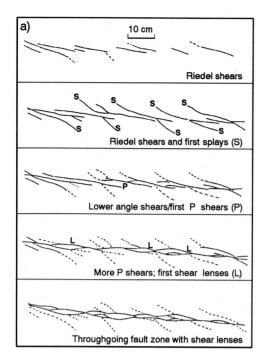

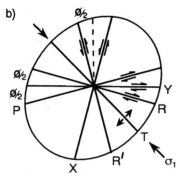

Fig. 12.6. (a) Sequence of shear fracture development in plan view of a sand box experiment above a dextral strike slip-fault (Naylor *et al.* 1986). (b) Compilation of theoretically and experimentally predicted fractures during simple shear (after Bartlett *et al.* 1978).

become linked by synthetic low-angle shears, particularly *P shears* oriented at $-\phi/2$ to the zone (Tchalenko 1970). These form in response to the locally rotated stresses within the array of R shears. Yet further displacement links segments of R and P shears into a throughgoing synthetic *principal displacement zone*, a narrow zone that accounts for most of the displacement (Tchalenko & Ambrayses 1970). This comprises an anastomosing array of faults defining shear lenses, with displacement concentrated on a central throughgoing *Y shear* (Morgenstern & Tchalenko 1967). It is most commonly applied to active faults (see Chapter 18).

The full suite of experimentally observed vertical fractures (Fig. 12.6b) also includes antithetic *X shears* at $90° + \phi/2$ to the principal displacement direction (Bartlett *et al.* 1981). The experiments by Naylor *et al.* (1986) in transpression and transtension also verify the angular shifts in Riedel shear orientation predicted by Sanderson & Marchini (1984) and summarized in Fig. 12.5. A further complication is the *Riedel-within-Riedel structure* where minor Riedel shears form

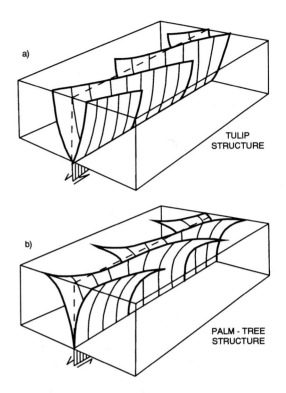

Fig. 12.7. Helicoidal forms of (a) Riedel shears in sandbox experiment (Naylor *et al.* 1976) and (b) axial surfaces of en échelon folds (Sylvester 1988).

oblique to a fault zone, itself a Riedel shear within a larger system (Tchalenko & Ambrayses 1970).

In three dimensions (Fig. 12.7a) the Riedel shear and related fractures are seen not to be vertical, but to have a helicoidal geometry, twisting to merge with the basement fault at depth. In the best documented experiment (Naylor *et al.* 1986) they are concave upwards, forming a *tulip structure* in cross-section. This geometry contrasts with the convex upwards palm-tree structures (Sylvester 1984) described from en échelon fold/fault zones (Fig. 12.7b) (Gamond & Odonne 1984), which may be characteristic of transpression rather than simple shear or transtension (Naylor *et al.* 1986).

REAL STRUCTURAL GEOMETRIES IN STRIKE-SLIP ZONES

Complications in the real world

The theoretical and experimental models of strike-slip zones fail to explain all structures naturally observed in such zones. There are two main reasons for this. First, real rocks at low to moderate metamorphic grades are strongly anisotropic. Should their pre-deformational structures not be so, then the fractures, folds and foliations produced by the strike-slip shear themselves form anisotropies which affect later deformation. Second, any long-lived strike-slip zone has a progressive deformation history which may superimpose structures of brittle and ductile rheologies, from newly initiated

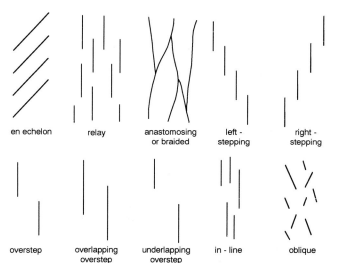

Fig. 12.8. Map views of structural geometries within planar strike-slip zones.

and strongly rotated generations, and from varying obliquities and senses of displacement vector across the zone. The following sections outline some commonly observed structural geometries in real strike-slip zones.

En échelon and relay patterns

Structures in strike-slip zones may be discrete, with either an *en échelon* or *relay arrangement* in map view, or may be interconnected into *anastomosing or braided patterns* (Fig. 12.8). These terms are defined as follows. *En échelon* describes a consistently overstepping and overlapping arrangement of structures, sub-parallel to each other but oblique to the planar zone in which they occur (Campbell 1958). By contrast a *relay pattern* comprises a shingled arrangement of sub-parallel faults with no consistent sense of overstep, each structure sub-parallel to the planar zone in which they occur (Harding & Lowell 1979) (Fig. 12.8). *Anastomosing (braided)* refers to branching and rejoining of a network of surfaces or surface traces, particularly faults (Fig. 12.8). En échelon patterns are common due to the monoclinic symmetry of simple strike-slip shear. The sense of *overstep* (discontinuity between two parallel faults, commonly in map view (Aydin & Nur 1982) from one structure to the next) is described as *right-* or *left-stepping* (Fig. 12.8). The overstep sense, where it is consistent along a zone of faults, can be used to tell the displacement sense on the zone (Fig. 12.4). Overstepping faults are commonly *overlapping*, that is, the normal to one fault tip intersects the other fault. The rock volume between overlapping strike-slip faults is called a *bridge* (Gamond 1987). *Underlapping* is the opposite of overlapping (Pollard & Aydin 1984).

In-line structures

In-line structures, that is, structures parallel to strike-slip zones, are more common than the theoretical models imply. In some cases these are directly related to the principal displacement zone; either shear lenses along it or *forced folds* (see Chapter 10) generated by draping of the cover over the faulted basement. However, in-line structures away from or between major faults can result from partitioning of oblique displacements (Fig. 12.9). *Partitioning* is the physical resolution of displacements or strain in to several components, reflected in several different structures. Oblique shortening is physically resolved into strike-slip on in-line faults and shortening normal to intervening in-line folds and thrusts. This partitioning presently operates in zones containing faults too weak to support much shear stress (Mount & Suppe 1987, Zoback *et al.* 1987).

Cleavage-transected folds

Some transpressively deformed areas show folds with contemporaneous non-axial planar cleavage (e.g. Powell 1974, Borradaile 1978, Murphy 1985, Soper *et al.* 1987, Woodcock *et al.* 1988). The characteristic relationship is *hinge transection*, where the cleavage does not contain the fold hinge (Borradaile 1978, Johnson 1991). In map views of sub-horizontal, upright folds the cleavage typically strikes anticlockwise of the fold hinges in dextral transpression (Fig. 12.10a).

Current explanations for fold transection invoke the rotational components in simple strike-slip or transpression, in particular the rotation of the finite strain ellipse and passive marker lines with respect both to each other and to the incremental strain ellipse (Fig. 12.10b). Three specific models (Fig. 12.10c–e) all assume that folds initiate at 45° and then rotate as passive markers. The '*strain fabric*' *model* (Fig. 12.10c) (e.g. Sanderson *et al.* 1980) assumes that the cleavage tracks the *XY* plane of the finite strain ellipsoid and rotates more slowly than the folds. The '*stress fabric*' *model* (Fig. 12.10d) (Soper 1986) postulates that pressure solution seams initiate perpendicular to the maximum principal stress, that is, in the *XY* plane of the incremental strain ellipsoid at 45°. The seams rotate as passive markers, but new seams continually develop

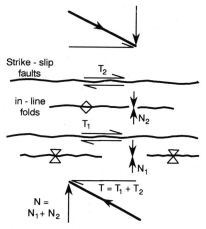

Fig. 12.9. Displacement partitioning in a transpressive zone to give in-line rather than en échelon structures.

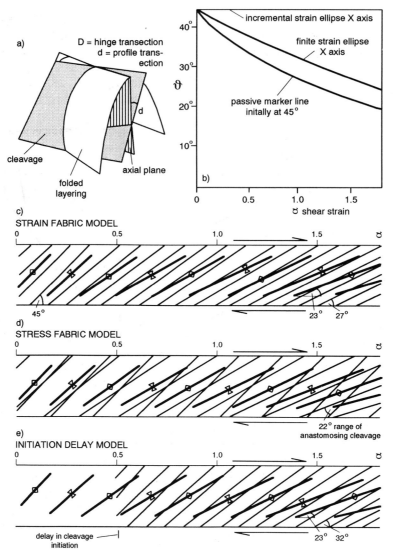

Fig. 12.10. (a) Geometry of a cleavage-transected fold. (b) Orientation of strain ellipse and passive markers during simple shear. (c–e) Map views of three models for formation of transected folds (see text for explanation).

at 45° to form an anatomosing spaced fabric with an average orientation transecting the rotated folds. The *'initiation delay'* model (Fig. 12.10e) enhances transection angles by postulating delayed cleavage initiation with respect to folds. This is effective with both stress-controlled and strain-controlled fabrics.

Fault stepovers, bends and straights

A segment of a strike-slip fault that is sub-parallel to the regional slip vector is a *straight* (Woodcock & Fischer 1986). However, strike-slip fault traces are never perfectly straight, for reasons of wall rock heterogeneity and linkage of non-colinear segments. Non-linearity produces numerous complexities (Fig. 12.11). Irregularities in fault traces can be divided into continuous bends and discontinuous stepovers or jogs, although these structures may link into each other at depth (also see Chapter 6 for a discussion of linked strike-slip faults). A *bend* is a deflection in the trace of a continuous fault. By contrast, a *stepover* is a discontinuity in a fault trace (Aydin & Nur 1982). Segall & Pollard (1980) and Sibson (1985) call a stepover a *jog*. Both may be either releasing or restraining depending on their sense of overstep with respect to the overall shear sense (Fig. 12.12). There are numerous synonyms for releasing and restraining geometries: *opening and closing, extensional and compressional, divergent and convergent, dilational and anti-dilational*. Releasing stepovers or bends tend to subside vertically. If faults

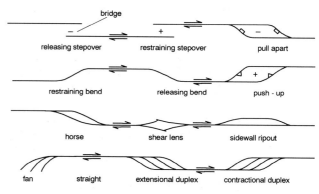

Fig. 12.11. Map views of structural geometries along strike-slip faults.

isolate the subsiding area it becomes a *rhomb graben* or *pull apart*, an important mode of basin formation (see later). The equivalent at a restraining stepover or bend is a *rhomb horst* or *push-up*.

Horses, duplexes and fans

Any fault-bounded sliver in a fault zone can be termed a *horse* (Fig. 12.11). If subject to near-surface uplift it is a *pressure ridge* or *fault-slice ridge*, that is, a linear topographic high (Crowell 1974b). A horse within the principal displacement zone is termed a *shear lens*, whereas one that is asymmetric, doubly tapered and was carved out of an adjacent wall is a *sidewall ripout* (Swanson 1989). In practice it may be difficult to tell whether a particular horse originated at a bend or on a straight segment of fault, particularly if it has been shunted along the system. A horse may still be in contact on one side with the wall rock from which it was detached (a *cognate horse*), or have become isolated within contrasting wall rocks as an *exotic horse* (Woodcock & Fischer 1986).

An imbricate array of two or more horses between two or more major, bounding faults is termed a duplex (Fig. 12.11). *Strike-slip duplexes* may form in a variety of ways (Woodcock & Fischer 1986), but commonly result from successive fault imbrication at bends or stepovers. If the bend is releasing, an *extensional duplex* forms, dominated by normal-oblique faults. *Contractional duplexes* form at restraining bends and comprise mainly reverse-oblique faults.

The ends of strike-slip faults may show *imbricate fans* or *horsetail splays*, imbricate arrays of sub-parallel curved oblique-slip faults each linked to the main strike-slip fault at one end but losing displacement to a fault tip at the other (Fig. 12.11) (Woodcock & Fischer 1986).

Flower structures

Vertical cross-sections through strike-slip zones, particularly from seismic data, commonly show diverging from a steep fault system (the *stem*) a pattern of upward-diverging faults termed a *flower structure* (Fig. 12.13) (Harding & Lowell 1939). *Positive flower structures* are dominated by reverse-oblique faults, and *negative flower structures* by normal-oblique faults

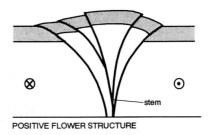

POSITIVE FLOWER STRUCTURE

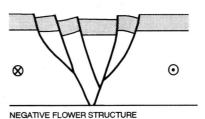

NEGATIVE FLOWER STRUCTURE

Fig. 12.13. Vertical cross-sections of flower structures.

(Harding 1985). The terms *tulip structure* and *palm tree structure* have been used for flower structures with concave-up and convex-up fault geometries respectively. Whilst tulip structures are common in transtension and palm tree structures in transpression, there is no consensus on how general is this tendency (compare Naylor *et al.* 1986, Sylvester 1988). A low-angle fault-bounded element in a flower structure is called a *leaf* (Gibbs 1986), a *flake* being a volume of rock along a low-dipping fault (Dewey 1982).

Not all upward diverging fault patterns on cross-sections indicate strike-slip tectonics. This difficulty of inverse interpretation is part of the general problem of diagnosing strike-slip components from vertical sections.

Block rotations and domainal structure

Palaeomagnetic studies (reviewed by Christie-Blick & Biddle 1985) have demonstrated widespread rotations of crustal blocks in strike-slip zones. Clockwise rotations

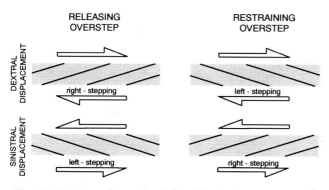

Fig. 12.12. Map views of relationships between senses of displacement and stepping of strike-slip faults and their restraining or releasing tendencies.

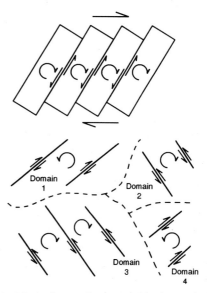

Fig. 12.14. Map views of rotated blocks and domainal structure.

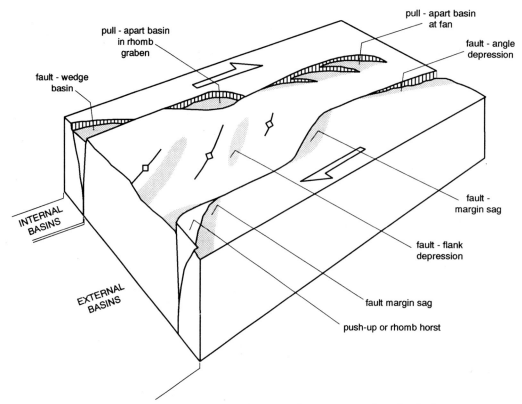

Fig. 12.15. Modes of basin formation in and adjacent to a strike-slip fault zone.

predominate in major dextral systems but the kinematics of rotation are complex. Present models follow Freund (1970) in favouring elongate blocks separated by antithetic sinistral faults (Fig. 12.14a). Some strike-slip zones and their adjacent forelands have a clear *domainal structure*, with discrete areas of sinistral faults abutting areas of dextral faults and anticlockwise-rotating blocks (e.g. Freund 1974, Ron *et al.* 1984). As a further complication, *nested domains* on a small-scale may occur within larger strike-slip domains, producing opposing rotations which may damp the finite palaeomagnetic rotation (Woodcock 1987).

BASIN FORMATION IN STRIKE-SLIP BELTS

Formation mechanisms and resulting geometries

Basin formation in strike-slip settings has been reviewed by Reading (1980) and Christie-Blick & Biddle (1985), and many pertinent examples are provided by the papers associated with these reviews.

Two sites of basin formation in strike-slip belts can be distinguished (Fig. 12.15). Basins internal to the strike-slip zone form in local transtensional areas along active fault strands. Most are fault controlled, for instance at releasing bends or offsets, or controlled by extensional fans. Basins external to the zone are either due to folding of adjacent cover rocks or to flexural loading of adjacent lithosphere by blocks thrust out from transpressional areas or by sediment eroded off internal push-ups.

Any segment of a strike-slip fault that is oblique to the regional slip vector will contain a component of dip-slip. Whether this component is normal or reverse, sediments may accumulate in the topographic depression on the downthrown side of the fault in a *fault-angle depression* (Ballance 1980). The upthrow side may provide a convenient local source. Special types of fault-angle depression occur at restraining bends. One slab may tend to override the opposite slab which is depressed by loading to form a *fault-margin sag* (Crowell 1976). Alternatively, faults in the restraining bend may define a compressional uplift called a *push-up block* or *rhomb horst* which can form fault-margin sags on both sides of the main fault zone (Mann *et al.* 1983, Aydin & Nur 1982). The sediment fill to these sags commonly becomes deformed and may eventually be buried, sliced, or uplifted as part of the push-up block. The push-ups themselves provide fast-rising local sources of sediment in continental settings (e.g. Mann *et al.* 1985).

The other class of external basins is the *fault-flank depressions* formed by the synclines in fold systems adjacent to a strike-slip fault belt (Crowell 1976). These folds will usually be oblique to the main fault zone (Fig. 12.15). Examples were described by Harding (1974). In transpressive situations, these depressions may deepen by the weight of the overriding slab and eventually be buried beneath it (Harding & Tuminas 1988).

The first class of internal basin forms at branches in

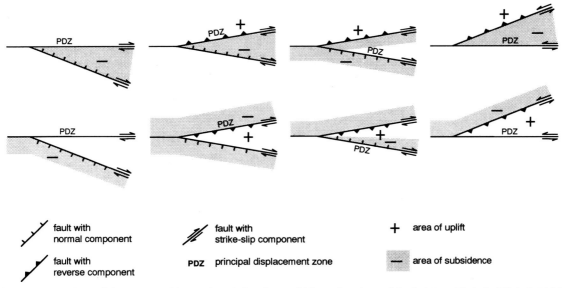

Fig. 12.16. Range of possible topographic results of slip along a bifurcating dextral fault (after Christie-Blick & Biddle 1985). Only some situations give fault-wedge basins *sensu stricto*.

strike-slip faults. These *fault-wedge basins* (*wedge grabens*) (Crowell 1974b) were originally assigned to sites where faults 'diverge', with uplift occurring at sites where faults converge. However, Christie-Blick & Biddle (1985) point out the paradox that faults which converge in one direction diverge in the opposite direction. The tendency of a fault wedge to subside depends in fact on the orientation of both the branch faults to the slip vector and to their relative displacement magnitudes. A range of topographic effects may result (Fig. 12.16).

The other common mode of internal basin formation occurs at releasing bends or stepovers. A *pull-apart basin* (Burchfiel & Stewart 1966) and the *rhomb graben* which is its host structure (Freund 1971) are defined by strike-slip strands parallel to the slip vector and by normal oblique-slip faults across the ends of the basin (Fig. 12.15). Basins may also form above imbricate fans or horsetails, and can be regarded as hybrid between pull-aparts and fault-wedge basins. The shape of pull-aparts varies with their progressive development (Fig. 12.17) (Mann *et al.* 1983) from spindle shape, through 'lazy-S' and rhomb shapes to complex multi-rhomb shapes.

Basin size and shape

Strike-slip related basins tend to be smaller, more elongate and deeper in relation to their width than basins formed by thermal subsidence or flexural loading (see Chapters 10, 11 and 13). Aydin & Nur (1982) found a fairly constant length to width ratio of 3:1 from rhomb-shaped pull-aparts over a scale range of tens of metres to tens of kilometres. Some basins are as deep as they are wide (Sylvester 1988). Model studies have tried to quantify the relationship between the geometry and depth of pull-apart basins, the master fault overlap and the strike-slip displacement (Rodgers 1980, Aydin & Nur 1982, Mann *et al.* 1983). Although some basins conform to these models (Schubert 1982) there are many exceptions (Mann *et al.* 1983). One generally accepted rule is that the length of a pull-apart basin is roughly equivalent to the strike-slip displacement needed to produce it.

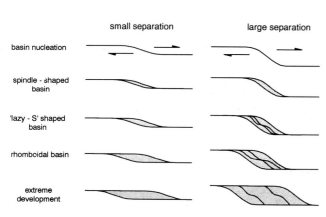

Fig. 12.17. Proposed development (Mann *et al.* 1983) of pull-apart basins in stepovers with small and large separation.

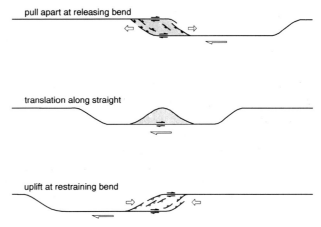

Fig. 12.18. Scheme for the inversion of a pull-apart basin by its translation from a releasing bend to a restraining bend.

Subsidence history

Observational data and theoretical models show that small strike-slip basins subside very rapidly. This is because subsidence on steep faults can be localized, and because lateral heat loss from a narrow subsiding sliver allows thermally driven subsidence to accompany, rather than post-date, the fault-compensated component. The absence of a superimposed thermal sag basin (see Chapter 11) is characteristic of strike-slip fault basins, though it can arise in larger stretched basins by lithosphere scale asymmetry of an extensional system. Zones with pure strike-slip displacement ideally involve no lithosphere thickness change. All subsidence and uplift can be compensated in the crust, and no long-wavelength thermal effects need occur.

Depositional environment

Characteristic fill patterns of strike-slip basins have been suggested, but are almost entirely based on continental environments. Because strike-slip tectonics can operate in all environments from abyssal to continental, a wide range of fill patterns is to be expected. Sediment starvation is common, due to subsidence rates exceeding sediment supply rates. Fault controlled facies patterns are common, but deciding whether the contemporary faults were dominantly strike-slip or dip-slip is usually impossible from the sedimentary record alone.

BASIN INVERSION IN STRIKE-SLIP BELTS

Inversion (shortening and uplift, see Chapter 14) of extensional basins requires a major switch in regional stress directions, if both extensional and contractional processes are envisaged as primarily dip-slip. In strike-slip dominated systems, basin inversion may arise under similarly oriented crustal stresses, by one of two processes. (1) A small rotation in regional stresses can put a previously transtensile zone into transpression. (2) *Heterogeneous flow* within a strike-slip zone means that at any one time some volumes are under contraction and some under extension. As the zone evolves, one volume can be transferred from an extensional to a contractional zone, giving inversion (Fig. 12.18). With these strike-slip involved mechanisms operating, local basin inversion need not involve regional adjustment of lithospheric stresses. The alternate uplift and subsidence of crustal blocks as they move along a strike-slip zone has been called *porpoising*, irrespective of whether basin inversion occurs (Crowell & Sylvester 1979).

REFERENCES

Anderson, E. M. 1942. *The Dynamics of Faulting and Dyke Formation with Application to Britain*. Oliver & Boyd, Edinburgh.

Aydin, A. & Nur, A. 1982. Evolution of pull-apart basins and their scale independence. *Tectonics* **1**, 91–105.

Ballance, P. F. 1980. Models of sediment distribution in non-marine and shallow marine environments in oblique-slip fault zones. In: *Sedimentation in Oblique-slip Mobile Zones* (edited by Ballance, P. F. & Reading, H. G.). *Spec. Publs Int. Ass. Sediment.* **4**, 229–236.

Bartlett, W. L., Friedman, M. & Logan, J. M. 1981. Experimental folding and faulting of rocks under confining pressure, Part IX. Wrench faults in limestone layers. *Tectonics* **79**, 255–277.

Biddle, K. T. & Christie-Blick, N. 1985. Glossary — strike-slip deformation, basin formation, and sedimentation. In: *Strike-slip Deformation, Basin Formation and Sedimentation* (edited by Biddle, K. T. & Christie-Blick, N.). *Spec. Publs Soc. econ. Palaeont. Miner. Tulsa* **37**, 375–385.

Borradaile, G. J. 1978. Transected folds: a study illustrated with examples from Canada and Scotland. *Bull. geol. Soc. Am.* **89**, 481–493.

Burchfiel, B. C. & Stewart, J. H. 1966. "Pull-apart" origin of the central segment of Death Valley, California. *Bull. geol. Soc. Am.* **77**, 439–442.

Burke, K. & Şengör, A. M. C. 1986. Tectonic escape in the evolution of continental crust. In: *Reflection Seismology: The Continental Crust* (edited by Barazangi, M. & Brown, L. D.). *Geodyn. Ser. Am. geophys. U.* **14**, 41–53.

Campbell, J. D. 1958. En échelon folding. *Econ. Geol.* **53**, 448–472.

Christie-Blick, N. & Biddle, K. T. 1985. Deformation and basin formation along strike-slip faults. In: *Strike-slip Deformation, Basin Formation and Sedimentation* (edited by Biddle, K. T. & Christie-Blick, N.). *Spec. Publs Soc. econ. Palaeont. Miner. Tulsa* **37**, 1–34.

Crowell, J. C. 1974a. Origin of the late Cenozoic basins in southern California. In: *Tectonics and Sedimentation* (edited by Dickinson, W. R.). *Spec. Publs Soc. econ. Palaeont. Miner. Tulsa* **22**, 190–204.

Crowell, J. C. 1974b. Sedimentation along the San Andreas Fault. In: *Modern and Ancient Geosynclinal Sedimentation* (edited by Dott, R. H. Jr). *Spec. Publs Soc. econ. Paleont. Miner. Tulsa* **19**, 292–303.

Crowell, J. C. 1976. Implications of crustal stretching and shortening of coastal Ventura Basin, California. In: *Aspects of the Geologic History of the California Continental Borderland* (edited by Howell, D. G.). *Misc. Publs Am. Ass. Petrol. Geol. Pacific. Sect.* **24**, 365–382.

Crowell, J. C. & Sylvester, A. G. 1979. Introduction to the San Andreas–Salton Trough juncture. In: *Tectonics of the Juncture Between the San Andreas Fault System and the Salton Trough, South-Eastern California. A Guidebook* (edited by Crowell, J. C. & Sylvester, A. G.). University of California, Santa Barbara, 1–13.

Dewey, J. F. 1982. Plate tectonics and the evolution of the British Isles. *J. geol. Soc. Lond.* **139**, 371–412.

Fitch, T. J. 1972. Plate convergence, transcurrent faults and internal deformation adjacent to Southeast Asia and the Western Pacific. *J. geophys. Res.* **77**, 4432–4460.

Freund, R. 1970. Rotation of strike-slip faults in Sistan, southeast Iran. *J. Geol.* **78**, 188–200.

Freund, R. 1971. The Hope Fault, a strike-slip fault in New Zealand. *Bull. geol. Surv. N. Z.* **86**, 1–49.

Freund, R. 1974. Kinematics of transform and transcurrent faults. *Tectonophysics* **21**, 93–134.

Gamond, J. F. 1983. Displacement features associated with fault zones: a comparison between observed examples and experimental models. *J. Struct. Geol.* **5**, 33–45.

Gamond, J. F. & Odonne, F. 1984. Criteres d'identification des plis induits par un decroachment profond: modelisation analogique et donnees de terrain. *Bull. Soc. geol. Fr.* **7**, 115–128.

Gibbs, A. D. 1986. Strike-slip basins and inversion: a possible model for the southern North Sea gas areas. In: *Habitat of Palaeozoic Gas in N.W. Europe* (edited by Brooks, J., Goff, J. C. & Van Hoorn, B.). *Spec. Publs geol. Soc. Lond.* **23**, 23–35.

Gilliland, W. N. & Meyer, G. P. 1976. Two classes of transform faults. *Bull. geol. Soc. Am.* **87**, 1127–1130.

Harding, T. P. 1983. Divergent wrench fault and negative flower structure, Andaman Sea. In: *Seismic Expression of Structural Styles* (edited by Bally, A. W.). *Am. Ass. Petrol. Geol.* **4.2**, 1–8.

Harding, T. P. 1984. Petroleum traps associated with wrench faults. *Bull. Am. Ass. Petrol. Geol.* **58**, 1290–1304.

Harding, T. P. 1985. Seismic characteristics and identification of negative flower structures, positive flower structures and positive structural inversion. *Bull. Am. Ass. Petrol. Geol.* **69**, 582–600.

Harding, T. P. & Lowell, J. D. 1979. Structural styles, their plate tectonic habitats and hydrocarbon traps in petroleum provinces. *Bull. Am. Ass. Petrol. Geol.* **83**, 1–24.

Harding, T. P. & Tuminas, A. C. 1988. Interpretation of footwall (lowside) fault traps sealed by reverse faults and convergent wrench faults. *Am. Ass. Petrol. Geol.* **72**, 738–757.

Harland, W. B. 1971. Tectonic transpression in Caledonian Spitsbergen. *Geol. Mag.* **108**, 27–42.

Johnson, T. E. 1991. Nomenclature and geometric classification of cleavage-transected folds. *J. Struct. Geol.* **12**, 261–274.

Lamb, S. H. 1988. Tectonic rotations about vertical axes during the last 4Ma in part of the New Zealand plate-boundary zone. *J. Struct. Geol.* **10**, 875–893.

Mann, P., Hempton, M. R., Bradley, D. C. & Burke, K. 1983. Development of pull-apart basins. *J. Geol.* **91**, 529–554.

Mann, P., Draper, G. & Burke, K. 1985. Neotectonics of a strike-slip restraining bend system, Jamaica. In: *Strike-slip Deformation, Basin Formation and Sedimentation* (edited by Biddle, K. T. & Christie-Blick, N.). *Spec. Publs Soc. econ. Palaeont. Miner. Tulsa* **37**, 211–226.

Molnar, P. & Tapponnier, P. 1975. Cenozoic tectonics of Asia: effects of a continental collision. *Science, N.Y.* **189**, 419–426.

Moody, J. D. & Hill, M. J. 1956. Wrench-fault tectonics. *Bull. geol. Soc. Am.* **67**, 1207–1246.

Morgenstern, N. R. & Tchalenko, J. S. 1967. Microscopic structures in kaolin subjected to direct shear. *Geotechnique* **17**, 309–328.

Mount, V. S. & Suppe, J. 1987. State of stress near the San Andreas fault: implications for wrench tectonics. *Geology* **15**, 1143–1146.

Murphy, F. C. 1985. Non-axial planar cleavage and Caledonian sinistral transpression in eastern Ireland. *Geol. J.* **20**, 257–279.

Naylor, M. A., Mandl, G. & Sijpesteijn, C. H. K. 1986. Fault geometries in basement-induced wrench faulting under different initial stress states. *J. Struct. Geol.* **8**, 737–752.

Pollard, D. D. & Aydin, A. 1984. Propagation and linkage of oceanic ridge segments. *J. geophys. Res.* **89**, 10017–10028.

Powell, C. McA. 1974. Timing of slaty cleavage during folding of Precambrian rocks, northwest Tasmania. *Bull. geol. Soc. Am.* **85**, 1043–1060.

Ramsay, J. G. 1980. Shear zone geometry: a review. *J. Struct. Geol.* **2**, 83–89.

Ramsay, J. G. & Huber, M. I. 1983. *The Techniques of Modern Structural Geology. Vol. 1*. Academic Press, London.

Reading, H. G. 1980. Characteristics and recognition of strike-slip fault systems. In: *Sedimentation in Oblique-slip Mobile Zones* (edited by Ballance, P. F. & Reading, H. G.). *Spec. Publs Int. Ass. Sediment.* **4**, 7–26.

Rodgers, D. A. 1980. Analysis of pull-apart basin development produced by en échelon and strike-slip faults. In: *Sedimentation in Oblique-slip Mobile Zones* (edited by Ballance, P. F. & Reading, H. G.). *Spec. Publs Int. Ass. Sediment.* **4**, 27–41.

Ron, H., Freund, R., Garfunkel, Z. & Nur, A. 1984. Block rotation by strike-slip faulting: structural and paleomagnetic evidence. *J. geophys. Res.* **89**, 6256–6270.

Sanderson, D. J. & Marchini, W. R. D. 1984. Transpression. *J. Struct. Geol.* **6**, 449–458.

Sanderson, D. J., Anderson, J. R., Phillips, W. E. A. & Hutton, D. H. W. 1980. Deformation studies in the Irish Caledonides. *J. geol. Soc. Lond.* **137**, 289–302.

Schubert, C. 1982. Origin of Cariaco Basin, southern Caribbean Sea. *Mar. Geol.* **47**, 345–360.

Segall, P. & Pollard, D. D. 1980. Mechanics of discontinuous faults. *J. geophys. Res.* **85**, 4337–4350.

Sibson, R. H. 1985. Stopping of earthquake ruptures at dilational fault jogs. *Nature* **316**, 248–251.

Soper, N. J. 1986. Geometry of transecting, anastomosing solution cleavage in transpression zones. *J. Struct. Geol.* **8**, 937–940.

Soper, N. J., Webb, B. C. & Woodcock, N. H. 1987. Late Caledonian (Acadian) transpression in North West England: timing, geometry and geotectonic significance. *Proc. Yorkshire geol. Soc.* **46**, 175–192.

Swanson, M. T. 1988. Pseudotachylyte-bearing strike-slip duplex structures in the Fort Foster Brittle Zone, S. Maine. *J. Struct. Geol.* **10**, 813–828.

Swanson, M. T. 1989. Sidewall ripouts in strike-slip faults. *J. Struct. Geol.* **11**, 933–948.

Sylvester, A. G. (ed.) 1984. Wrench fault tectonics. *Am. Ass. Petrol. Geol. Reprint Ser.* **28**, 1–374.

Sylvester, A. G. 1988. Strike-slip faults. *Bull. geol. Soc. Am.* **100**, 1666–1703.

Tchalenko, J. S. 1970. Similarities between shear zones of different magnitudes. *Bull. geol. Soc. Am.* **81**, 1625–1640.

Tchalenko, J. S. & Ambrayses, N. N. 1970. Structural analysis of the Dasht-e-Bayaz (Iran) earthquake fractures. *Bull. geol. Soc. Am.* **81**, 41–60.

Willis, B. 1928. Dead Sea problem: rift valley or ramp valley? *Bull. geol. Soc. Am.* **39**, 490–542.

Woodcock, N. H. 1986. The role of strike-slip fault systems at plate boundaries. *Phil. Trans. R. Soc. Lond.* **A317**, 13–29.

Woodcock, N. H. 1987. Kinematics of strike-slip faulting, Builth Inlier, Mid-Wales. *J. Struct. Geol.* **9**, 353–363.

Woodcock, N. H. & Fischer, M. 1986. Strike-slip duplexes. *J. Struct. Geol.* **8**, 725–735.

Woodcock, N. H., Awan, M. A., Johnson, T. E., Mackie, A. H. & Smith, R. D. A. 1988. Acadian tectonics in Wales during Avalonia/Laurentia convergence. *Tectonics* **7**, 483–495.

Zoback, M. D., Zoback, M. L., Mount, V. S., Suppe, J., Eaton, J. P., Healy, J. H., Oppenheimer, D., Resenberg, P., Jones, L., Raleigh, C. B., Wong, I. G., Scotti, O. & Wentworth, C. 1987. New evidence on the state of stress of the San Andreas Fault system. *Science* **238**, 1105–1111.

CHAPTER 13

Continental Collision

MIKE COWARD

INTRODUCTION

OCEANIC crust may be subducted along destructive plate margins but less dense continental crust is buoyant and remains on the upper surface of the lithosphere. Where two continental masses collide, the lower lithosphere may be subducted but the upper lithospheric mantle and crust is thickened, resulting in isostatic uplift. Hence a mountain belt is generally aligned along a zone of continental collision.

The most dramatic of the present day mountain belts is that which extends from the Pyrenees, through the Alps, Turkey and the Middle East, to the high mountains of the Himalayas and Tibet. This mountain belt resulted from the collision of a number of southern plates, which made up Gondwanaland, with Eurasia to the north. During the late Palaeozoic older mountain belts developed in Central and Western Europe forming the Variscan or Hercynian Chain. This fold belt continues into the southern Appalachians where it overlaps and merges with the older Taconic and Acadian mountain chains to the south, forming a continuation of the Caledonides of Britain and Scandinavia.

Bally & Snelson (1980) used the term *A-subduction* to describe the deep burial of continental crust beneath these mountain belts. Alpine geologists had long recognized that, on their cross-sections, the width of the sedimentary cover sequence was much greater than the width of the available basement section. Bally & Snelson (1980) suggested the term *B-subduction* for the removal of oceanic lithosphere and the development of an associated Benioff zone.

Tectonic processes associated with continental collision may be grouped according to their timing relative to ocean closure: (i) *pre-collisional tectonics*: the subduction of oceanic material and local obduction of ophiolites; (ii) *collisional tectonics*: the thickening of the crust and lithospheric mantle, associated with continental collision; and (iii) *post-collisional tectonics*: the effects of continued continental plate indentation and deformation associated with the gravitational instability formed by the thickened lithosphere. This chapter reviews collision tectonics according to the above classification.

PRE-COLLISIONAL TECTONICS

Crustal deformation associated with plate conversion

Figure 13.1 shows the initial stages of plate collision associated with plate convergence. The oceanic crust of one plate is subducted beneath a second plate, here shown as continental crust and lithosphere. The topmost deep-sea sediments of the oceanic crust, and sediments that have been washed on to the oceanic crust from the adjacent continent, may be scraped up in small thrust slices (see Chapter 9).

Where the subducting plate has a steep dip there is relatively little deformation of the overriding plate. However where the subducting plate has a gentle dip there is often large scale crustal shortening of the overriding plate, possibly due to viscous drag at the base of the plate. Thus, in Colombia and Venezuela, the Cordillera Orientale and Venezuelan Andes have been uplifted and show considerable crustal shortening above the shallowly subducting East Pacific plate. The crustal shortening involves thick-skinned deformation, often reworking earlier extensional structures. A similar mechanism has been invoked for the origin of the Laramide uplifts of the western U.S.A. It is often difficult to separate this Laramide style deformation from deformation produced by plate collision.

At deeper levels in the subducting slab, oceanic crust and its thin cover of sediments may melt and this melt will rise to *underplate* and intrude the overriding plate and locally may manifest itself on the surface as a volcanic arc, as in Japan or the west coast of South America. The addition of this volcanic and intrusive material will thicken the overriding continental crust, resulting in plateau uplift as in the Altiplana of Bolivia and Peru.

An *accretionary prism* of sediment accumulates ahead of the overthrusting plate and consists of ocean-floor sediments and some ocean-floor basalts, together with sediments washed into the trench from the adjacent overthrusting plate. The removal of trench sediments by a thrust and recumbent fold mechanism, followed by progressive shortening, thickening and locally polyphase deformation of the accretionary prism can also occur. These foredeep sediments and their associated

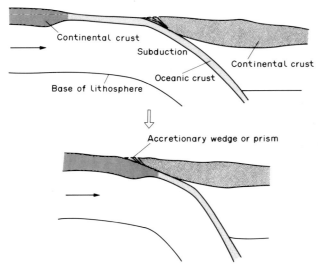

Fig. 13.1. Convergence of two plates with subduction of oceanic crust and the build-up of an accretionary wedge beneath one plate.

structures are described in more detail in Chapters 9 and 16.

The process of subduction causes *flexure of the overthrust plate,* resulting in stretching of the outer part of the flexed plate and compression in the inner part. The resultant bending strains have a pattern similar to that of a buckle fold (Fig. 13.2.) (see also Ramsay 1967, p. 397) in that the depth to the neutral surface will be approximately half that of the thickness of the bent plate and the strain intensity will increase with increasing curvature and distance away from the neutral surface. The upper part of the subducting plate often shows extensional faults related to bending (Fig. 13.2) (Dewey 1982).

Not all subducting slabs show extensional faults in the sedimentary cover to the subducting crust. The presence or absence of pre-foredeep extension depends on the curvature of the bent plate, which in turn depends on the lithospheric strength and also the position of the zone of maximum curvature relative to the accretionary prism.

A subducting slab consists largely of relatively cool lithospheric mantle which is denser than the adjacent hot asthenosphere. Thus the slab will be affected by a

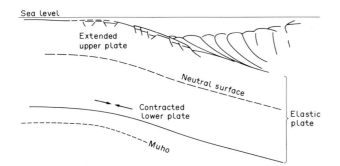

Fig. 13.2. The flexure of an oceanic lithospheric plate causing extension around the outer arc and compression in the inner arc.

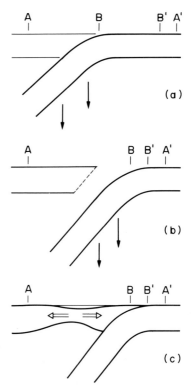

Fig. 13.3. Schematic diagrams showing (a) subduction in which the amount of plate convergence (rate of change of line A – A') is less than the subduction rate (rate of change of line B – B'). (b) The space between the upper and lower plate that would result if no extension occurred as the trench B retreats towards the foreland during subduction, due to the effects of gravity on the subducting slab. (c) The accommodation of different rates of convergence and subduction by extension of the overriding plate. From Burchfiel & Royden (1982).

vertically downwards gravitational force which may cause the hinge line of the bent plate to roll back (Fig. 13.3). The rate at which *subduction roll-back* occurs will partially depend on the strength of the subducting plate. If this rate is greater than the rate at which the overriding plate advances relative to a fixed point in the Earth, then a space will occur which must be accommodated by lithospheric spreading (Fig. 13.3). The zone of spreading usually occurs above or behind the volcanic arc, where the geothermal gradient is high as the result of the rise of melt from the subduction zone. A back-arc basin is generated (Fig. 13.3), causing continental extension and/or ocean crust development, depending on the type of overthrusting lithosphere.

Thus during plate convergence and prior to continental collision, the underthrusting plate may show renewed extension of oceanic crust or a passive continental margin. A *foredeep* will develop on the overthrusting plate, infilled on one side by an accretionary prism of imbricated and folded sediments behind which will lie a magmatic arc, resulting in a volcanic island chain, and plateau uplift in oceanic and continental environments, respectively. Along and behind the magmatic arc there may be large-scale extension. During continental collision similar processes

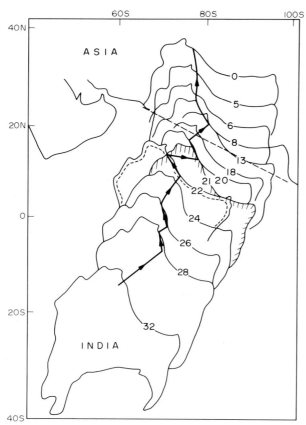

Fig. 13.4. Relative position of India with respect to Asia kept arbitrarily fixed in its present position. Ornamented lines (anomalies 21 and 22) mark the position of the northern margin of India at a time of discontinuous movement which may be the time of initial collision with the southern margin of Eurasia (Patriat & Achache 1984). Dashed line is the suggested margin of Eurasia according to Tapponnier *et al.* (1986). Note that Greater India, that is India with its thrust belts restored, extended north of the western Himalayas for *c.* 500–1000 km. This northern promontory collided with Asia before the rest of the Indian margin (Treloar & Coward 1991).

apply except that the subducting plate carries continental crust.

The exact timing of continental collision is often difficult to determine. During the early stages of collision the plates may be bordered by stretched continental crust of the old passive margin and hence deformation will take place below sea level. Early collision-related sediments will be subducted below the overthrusting plate and hence the earliest preserved sediments are those which were deposited a long time after the onset of the collisional process. For example the timing of collision between India and Asia is generally estimated from plate restorations using magnetic anomaly correlations (Achache *et al.* 1984, Patriat & Achache 1984, Tapponnier *et al.* 1986, Besse & Courtillot 1988, Dewey *et al.* 1989). Based on the successive positions of India relative to Eurasia (Fig. 13.4), three criteria have been used to date the onset of continental collision. (1) The shape of Greater India has been estimated from an approximate restoration of thrust structures in the Himalayas and the time at which the edge of Greater India intersected the projected edge of Asia, drawn approximately as a continuation of the Makran coast, i.e. *c.* 55 Ma (Tapponnier *et al.* 1986). (2) From 84–45 Ma, India moved approximately northeastwards in the Eurasian reference frame at an average velocity of *c.* 10 cm/Ma. From 45–0 Ma, India moved approximately northwards at a slower rate of *c.* 5 cm/Ma. This change in plate motion at 45 Ma marks the time of continental collision, when simple subduction of the Tethyan oceanic lithosphere, north of Greater India was replaced by continental shortening and thickening and the indentation of India into Eurasia (Dewey *et al.* 1989). (3) Some motion paths of India relative to Eurasia (e.g. Patriat & Achache 1984) show large zig-zag deviations during the early Tertiary (Fig. 13.4) which may indicate the onset of collision at approximately 50 Ma.

The timing of *obduction* of ophiolitic material cannot be used to date continental collision. For example, in southern Tibet and northern Pakistan the Spontang ophiolites were obducted onto the northern edge of India during the Maastrichtian/Danian, *c.* 20 Ma before continental collision (Searle *et al.* 1987). Precollisional ophiolite obduction of a similar age occurred along much of the Neo-Tethyan margin from southern Turkey through Iran and Oman, into western Pakistan. Along the margin of Arabia and India the ophiolites and associated mélanges, derived from the ophiolites, are overlain by lower Eocene carbonates.

Similar giant ophiolite nappes occur in central Tibet, where Jurassic ophiolites are preserved across a width of greater than 200 km of the Lhasa Block, and in the Ordovician Bay of Islands Complex in Newfoundland (Dewey & Bird 1971). *Ophiolitic obduction* is considered not to be related to collision but to the subduction of a thin and rifted continental margin below relatively thin oceanic lithosphere (Dewey *et al.* 1988). Across the northern margin of the Indian continent, ophiolite obduction occurred soon after the Late Cretaceous break-up of Gondwanaland, when the Tethyan Ocean began to close by the northward movement of Arabia and Africa. There is evidence that both the Arabian and Indian ophiolites were part of a thin oceanic lithosphere. The ophiolitic rocks, obducted onto the Arabian margin in Oman, are considered to have originated from a relatively young mid-oceanic ridge. As Greater India moved northwards during the Palaeocene–Eocene, it passed over a mantle hot spot resulting in uplift and the extrusion of large quantities of volcanic rock (Deccan Traps). The ophiolitic rocks obducted over the western and northern margin of Greater India may have been affected by the early stages of hot spot growth, causing local uplift and thinning of the oceanic lithosphere.

COLLISIONAL TECTONICS

Large-scale thrust kinematics

Where two slabs of continental crust collide, one slab will override the other (Fig. 13.5). Some continental crust may be subducted into the lithospheric mantle,

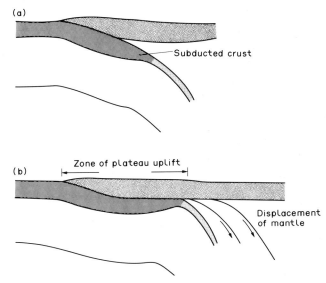

Fig. 13.5. Collision of plates carrying continental crust. (a) Limited subduction. (b) Generation of minor subduction zones. See text for details.

where it is affected by high temperature and high pressure metamorphism. Generally, however, such a slab will resist subduction and will *underplate* the overriding slab of crust, causing a zone of *plateau uplift* (see also Argand 1924, Bird 1978). This zone of uplift should roll back away from the collision suture zone with time. Mantle material will be scraped off the base of the overriding plate, possibly generating *minor subduction zones* as shown in Fig. 13.5(b).

As plate movement continues, either the overriding or overridden plate may break and stack up *crustal-scale thrust sheets* (Fig. 13.6). The thrusts will be linked at depth by a flat ductile detachment in the weak zone at the base of the crust or within the deeper parts of the lithospheric mantle (also see Chapter 6). Deep seismic reflection profiling has shown the positions of some of these major structures in older orogenic belts (e.g. in Scotland, see Cheadle *et al.* 1987). The seismic data show a zone of gently dipping reflectors near the base of the crust and moderately dipping reflectors, which branch off this zone, through the middle and upper crust. The moderately dipping reflectors are now considered to be the seismic signatures of the Outer Isles and Moine thrusts, together with other thrusts within the Moines. The thrusts have been reworked by later extensional movements.

The width and thickness of the thrust slabs depend on lateral variations in strength profiles through the crust and upper mantle. Where the lithosphere has been stretched to form a passive margin prior to collision, the stretched crust will be thin and will lack the weak low velocity zone near the base of the crust and hence the stretched lithosphere will be relatively strong. A passive margin forms a relatively *rigid peel* to a continent and hence the early lithospheric breaks often occur on the inner side of this peel. In the Western Alps, some of the earliest thrusts associated with Alpine collision occur in the outer zones of the Alps, deforming rocks of the Dauphinois Zone.

In Northwest Scotland the suture zone between the collided plates lies somewhere to the southeast of Scotland and the exact relationship between the Moine–Flannan thrust systems and this suture is obscured by subsequent strike-slip deformation. According to Coward (1990), after restoration of the Great Glen fault, which is the main strike-slip structure, the thrusts of Scotland are part of a major crustal scale Caledonidian thrust system whose continuation, but with opposing thrust vergence, can be seen in western Norway. The thrusts have opposing dips and vergence on either side of the suture, producing an approximately symmetrical mountain belt.

However, not all mountain belts show this symmetry. In the Western Alps of France and Switzerland, the majority of thrusts are NW-vergent and lie in the overridden plate. However, to the east in Austria and northern Italy the Alpine-age thrusts are divergent about a central axial zone.

Figure 13.7 shows a simplified section through Tibet, based on a preliminary seismic reflection survey of Hirn *et al.* (1984) and the work of the Sino-British team (Chang Chengfa *et al.* 1986, Coward *et al.* 1988, Dewey *et al.* 1988). The suture zone between the Indian and Asian plates lies along the Indus and Tsangpo valleys, immediately north of the High Himalayas. The main thrusts are S-vergent but affect both the overriding plate, thickening the crust to produce the Tibetan uplift, and also the overridden plate, thickening the crust to produce the main Himalayan mountains. The majority of thrusts dip to the north parallel to the suture. However, beneath the Indian plate there are offsets of the Moho which suggest local N-directed thrusts, of opposed dip to the main set. These N-directed thrusts may be due to either the presence of earlier crustal-scale fabrics, which control the dip of subsequent structures, or to the presence of a regional shear couple in the lithosphere. Experiments show (e.g. Vendeville *et al.* 1987), that if a dominant shear couple is applied at the base of a slab, this couple will control the sense of shear shown by the main faults within the slab. The shear couple, as shown in Fig. 13.7, is of Indian lithospheric mantle and/or lower crust moving northwards beneath Tibet, causing the main thrusts within Tibet to have the same S-directed shear couple. The lack of this shear

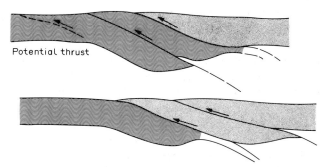

Fig. 13.6. The development of new crustal-scale thrust sheets in the footwall (top) or hangingwall (bottom) of the main suture.

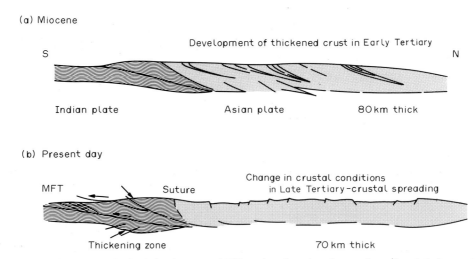

Fig. 13.7. Simplified section through the Himalayas and Tibet showing the change in style of deformation both in space (across Tibet) and with time (a) Miocene (b) present day.

couple at the base of the Indian plate allows the development of thrusts with opposing dips.

Fault plane solutions of earthquakes beneath the Himalayas indicate thrust faulting on planes dipping 25° NE (Ni & Barazangi 1984). These planes occur c. 100 km northeast of the mountain front and may indicate a steepening of the fault plane along which India underthrusts the Himalayas (Coward 1983,

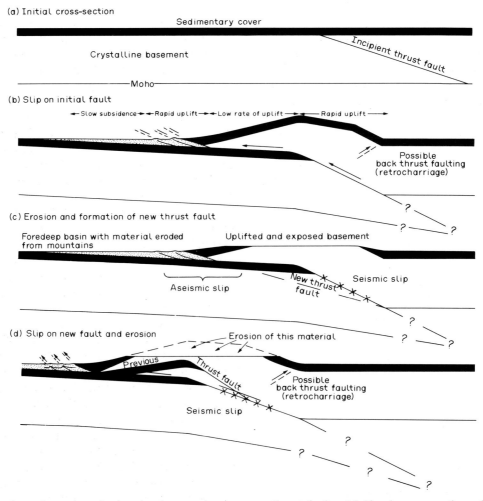

Fig. 13.8. Sketches of a series of migrating large-scale ramp overthrust faults. (a) Simple cross-section of basement and sedimentary cover. (b) Section after 30–50 km of slip — the underthrust block is flexed down to produce the foredeep basin. (c) Profile after erosion reduces the elevations. (d) Section after a new fault produces another 30 km displacement followed by erosion. From Molnar & Lyon-Caen (1988).

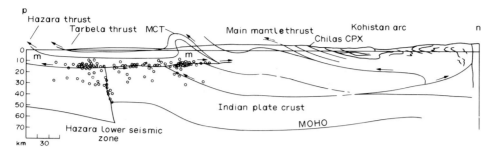

Fig. 13.9. Cross-section through the Himalayas, east of the Nanga Parbat syntaxis. Earthquake hypocentres are shown by circles (after Seeber *et al.* 1981). m — Miocene rocks overlying an unknown thickness of deformed Phanerozoic sediments. Fine stipple — Indian plate crust; coarse stipple — Mesozoic sediments above the Main Boundary thrust and the Main Mantle thrust. After Coward (1983).

Molnar & Lyon-Caen 1988). Beneath the frontal regions of the mountain belt the slip is probably dominantly aseismic along bedding-plane detachments and small thrust ramps which characterize the foreland fold and thrust belt. Hence the crustal-scale faults probably have a ramp/flat geometry on a crustal scale and the Himalayas can be considered partly as a consequence of uplift on these steep ramps (Fig. 13.8). The latter can be considered as *hangingwall ramp anticlines* to the crustal-scale thrusts.

In some places the intersection between deep and shallow thrusts can be seen from earthquake data. In the western Himalayas there is a region of active uplift between the Murree Hills and Nanga Parbat in northern Pakistan. Shallow seismic data define an active detachment zone at *c.* 10–30 km depth (Seeber & Armbruster 1979, Seeber *et al.* 1981) which probably connects with the gently dipping thrusts recognized in the Salt Ranges to the south (Fig. 13.9). This detachment is intersected by a deeper-level zone of seismic activity dipping 30° to the northeast. The dipping seismic zone underlies, and is probably responsible for, the region of recent uplift. The most dramatic region of fault-produced uplift occurs at the northeastern end of this zone, where the Nanga Parbat mountain is formed by uplift of a crustal-scale hangingwall anticline above an oblique thrust on its western edge. The fold has a half-wavelength of 50 km and can be mapped from the changes in dip of gneissic foliation and lithological banding (Coward 1986, Treloar *et al.* 1991). The core of the fold gives the youngest cooling ages (Ar/Ar and fission track, e.g. Zeitler 1985) suggesting that this structure has been growing at an uplift rate of 6 mm/yr (Treloar *et al.* 1991).

Not all continental collisions are head-on; many involve oblique or strike slip movements. Strike-slip faulting has played a significant role in the Mesozoic–Cenozoic structural history of the North American Cordillera, particularly in Canada and Alaska.

The stacking of large thrust sheets produces major thermal effects, resulting in new higher temperature assemblages in the overridden sheets (England & Thompson 1984). Figure 13.10 shows pressure–temperature–time paths modelled for a particular slab of crust, assuming purely conductive heat transport during thermal equilibration (Platt & Lister 1985). The overridden slab is originally rigid but with time and thermal equilibration it becomes more ductile relative to the overriding slab. The lower slab may then be deformed by ductile folding and thickening (Fig. 13.11). Rapid overthrust rates allow considerable underplating by the lower slab, but slower rates allow the lower plate to heat up and deform internally. The thermal maximum generally occurs *c.* 30–60 Ma after initial overthrusting (e.g. Fig. 13.10), often after much of the initial deformation in the lower plate. The metamorphic mineral assemblages therefore are late- to post-tectonic; it is often difficult to ascertain accurate metamorphic conditions during the main overthrusting.

In the Pakistan Himalayas the peak of high-temperature metamorphism occurred *c.* 15 Ma after initial overthrusting (Treloar *et al.* 1992). Purely conductive models of thermal evolution may be inappropriate in zones where there have been large amounts of ductile shearing and some form of strain or shear heating may be invoked to explain the short time scales involved.

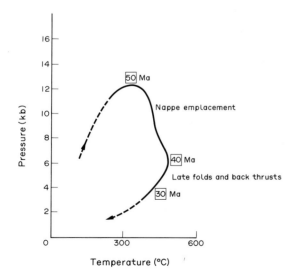

Fig. 13.10. Typical *P–T–t* path for rocks involved in thrust tectonics. This example is taken from interpretations of the metamorphic history of the Vanoise Massif in the Western Alps (Platt & Lister 1985). Boxed numbers indicate approximate ages in millions of years.

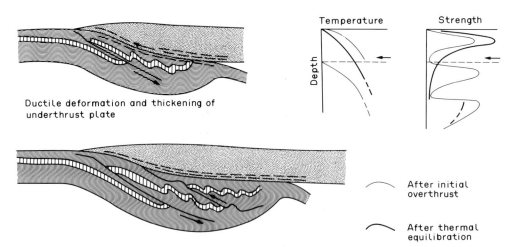

Fig. 13.11. Schematic profiles through thickened crust to show the development of structures in the more ductile overthrust plate and typical temperature and strength profiles after initial overthrusting and thermal equilibration.

Hence it is unlikely that rigid crust can underplate the overriding slab for any great distance (cf. Bird 1978). In the Indian and Pakistan Himalayas the slab has probably been underridden for only 500–750 km, a small distance compared to the width of the entire Tibetan-Himalayan mountain belt.

Locally, the earlier thrusts can be refolded or thrust, so that rocks, heated up following the early phase of overthrusting, are uplifted over rocks which were initially at a high level in the thrust pile. Thus in the Pakistan Himalayas there was reworking of the early metamorphic pile, the reimbrication of cover and basement sequences and the final stacking of the metamorphic rocks into a pile of internally imbricated S-verging thrust nappes. This stacking post-dates the metamorphism as shown by reworking of primary ductile shear assemblages by lower temperature, locally brittle, shear fabrics and by sharp metamorphic breaks across the shears between individual nappes. These later shears normally have higher-grade rocks on the hangingwall relative to those on the footwall. This inverted metamorphic sequence is generated by post-metamorphic stacking and not by syn-metamorphic emplacement of hot rocks upon cooler rocks.

As the crust thickens, its lower part becomes hotter and weaker and eventually may be too weak to support the load of the overlying thick crust. Sonder *et al.* (1987) calculated the time needed for the lower crust to become weak enough to spread, given that there are local boundary conditions enabling spreading to take place. They consider that reasonable spreading rates will occur when the lower crustal rocks reach a temperature of *c.* 700°C. The time needed for the lower crust to reach this temperature depends on the initial thermal gradient, that is the temperature of the Moho before thickening and the amount of crustal thickening. It may take up to 100 Ma between the end of compression and the onset of extension if the Moho temperature was low, for example 450°C. However, extension may follow immediately after release of compression if the initial Moho temperature exceeds 700°C.

Results of the above calculation are plotted graphically in Fig. 13.12. For the Himalayas and Tibet, which have present crustal thicknesses of 60–70 km (Barazangi & Ni 1982), that is 2–2.5 times the normal crustal thickness, thickening became widespread *c.* 50 Ma ago while extensional faulting, related to the spreading of the thickened crust, became important in only the past 5–10 Ma. Earthquake studies show that extensional faulting is the main deformation mechanism in Tibet at the present day. The data in Fig. 13.12 suggest that the initial Moho temperature was in the region of 500–600°C.

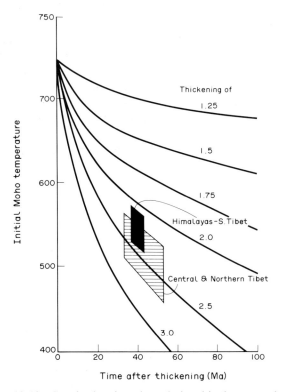

Fig. 13.12. Graph showing the relationship between the initial Moho temperature, the thickening and the time lapse after thickening, before the lower crust becomes too weak to support the thickened crust and spreads at geologically realistic strain rates. After Sonder *et al.* (1987).

The calculations of Sonder *et al.* (1987) also agree

with observations of the Tertiary extensional history in western America. Late Mesozoic to Early Tertiary compression in the north was accompanied by extensive calc-alkaline magmatic activity and followed almost immediately by extension. However in the south, where there is little or no evidence for syn-compressional igneous activity and hence there was probably a lower geothermal gradient, the time between compression and extension was 70 Ma.

Calculation of shortening during collision

In young orogenic belts the amounts and rates of shortening can be estimated from regional plate tectonic considerations, that is, from an analysis of the widths and ages of the magnetic stripes in the sea floor. Thus if the location of the relative pole of rotation and the angular rate of rotation (Ω) are known between two plates, the rate of surface shortening is given by:

$$\hat{e} = \Omega R^2 \sin\Phi$$

where Φ is the eulerian latitude and R the radius of the Earth. Where sea-floor magnetic data are unavailable, or where the shortening needs to be estimated across particular crustal segments, it is necessary to look for an offset crustal-scale structure (e.g. Shackleton 1969) or to produce *balanced and restored cross-sections*. Recently there has been a tendency to believe implicitly the accuracy of section restoration. However, section balancing techniques and section restoration are only as good as the input data and the assumptions used. Many thrust belts are *thin-skinned deformation belts* and hence area-balancing techniques can be used to limit amounts of thrust belt shortening. However, other belts involve *thick-skinned deformation* and sometimes the inclusion of *syn-rift sediments*. From surface data alone it is difficult to tell which thrust model applies. The use of an incorrect model can produce huge errors. For example in the Sulaiman Ranges of western Pakistan, Jadoon *et al.* (in press) assume thin-skinned deformation and calculate greater than 300 km shortening from area-balancing techniques, assuming a regional decoupling horizon. For the same area the present author has estimated only 25 km shortening based on line-length restoration techniques, assuming that the beds were uplifted on thick-skinned basement-involved faults.

Other errors arise from an assumption of plane strain, that is, that the material has remained on the same material plane throughout the deformation. This is clearly impossible in many thrust belts where the fault kinematics indicate variations in overthrust directions and often *rotation of thrust sheets about a vertical axis*. This rotation is due to pinning of the thrusts at their lateral tips. Treloar *et al.* (1992) propose greater than 50° rotation of parts of the Salt Ranges in the Pakistan Himalayas.

Section restoration is important for determining the deep structure of mountain belts. If section restoration indicates that the upper crust has undergone, for example, 100 km of shortening and the deformation is considered to be thin-skinned, then the basement to the thin-skinned thrust zone must be shortened the same amount beneath the mountain belt, or be subducted into the mantle (Boyer & Elliott 1982, Coward & Butler 1985, Butler 1986, Butler & Coward 1989, Le Pichon *et al.* in press).

Uplift and thickening

The elevation of a mountain belt, or any other segment of crust, can be considered in terms of its isostatic balance with reference to the mid-ocean ridge column. The elevation is given by:

$$e = S(\ell - \mu) - L(\mu - \alpha - x(\alpha - \omega) - h(\alpha - \beta b) \cdot 1/\alpha$$

where x is the depth of sea water (density ω) at the mid-ocean ridge, h is the thickness of ocean crust with a density β, above the asthenosphere with a density α. This is balanced by the mountain belt of lithospheric thickness L, crustal thickness S, average crustal density ℓ, and average mantle density μ. Appropriate values would be $x = 2.5$ km, $h = 5$ km, $\omega = 1.03$ g/cc, $\ell = 2.8$ g/cc at 0°C, $\beta = 2.9$ g/cc at 0°C, the average mantle density, μ or $\alpha = 3.33$ g/cc at 0°C, with the coefficient of thermal expansion = 0.000328 and the temperature at the top of asthenosphere at 1333°C. Thus this equation simplifies to:

$$e = \ell S(/-\mu) - L(\mu - 3.184) - 7.9\} / 3.184.$$

An alternative and to some extent oversimplified method for describing uplift or subsidence of a mountain belt is to relate elevation or subduction to the level of the mantle geoid, that is, the level to which hot asthenosphere at a temperature of 1333°C would rise if continental and oceanic lithosphere were removed (Fig. 13.13).

Assuming homogeneous shortening of the whole

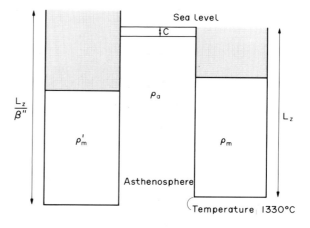

Fig. 13.13. The isostatic balance causing uplift or subsidence relative to the mantle geoid at a depth c km below sea level. ρ_c and ρ_m are the average crustal and mantle densities.

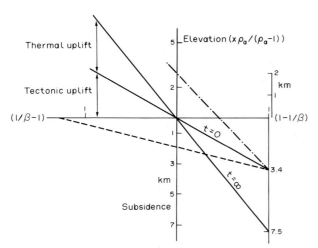

Fig. 13.14. Graph showing subsidence or elevation associated with changes in lithospheric thickness where $t = 0$ and $t = $ infinity. Note the scale above sea-level is different to that below sea-level to compensate for the loading of sea water. Dashed line-subsidence/uplift curve for crust initially 2 km below sea-level; dot-dashed line-subsidence associated with stretching of crust initially 2 km above sea-level. After Le Pichon et al. (1982).

lithosphere and that β is the amount of lithosphere shortening so that the thickening is given by $1/\beta$, then the uplift approximates to:

$$c(1/\beta - 1).$$

The constant c is equivalent to the depth of the mantle geoid below sea-level. This relationship is shown graphically in Fig. 13.14.

This equation assumes that the lithosphere thickened homogeneously to maintain the pre-collisional relationship between temperature and depth. If the crust was thin initially, then assuming normal densities, the top of the lithosphere would lie below sea-level, at a depth e_0 depending on the buoyancy of the local lithospheric column. The uplift can then be plotted as a simple straight line. A change in scale is used above sea-level to convert virtual water depth into actual altitude. The slope of the line is given by:

$$e_i = e_0 + (c - e_0)(1 - 1/\beta)$$

where e_i or e_0 are converted to values of minus water depth above sea-level. e is positive below sea-level, negative above sea-level. This conversion involves the multiplication of the height above sea-level by a factor of $\rho_a/(\rho_a - 1)$. The value of c is generally estimated to be c. 3.4 km below sea-level. To examine the uplift after thermal equilibration, a similar equation can be derived where:

$$s_{tot} = c'(1 - 1/\beta)$$

and c' equals approximately 7.5. This *thermal phase of uplift* will take place over several tens of million years. A simple exponential relationship between rate of uplift and cooling time, similar to the relationship between subsidence and time in a stretched basin (see Chapter 11), does not apply to rising mountain belts because of the effects of erosion and non-uniform radiogenic heating within the crust.

Figure 13.14 shows the straight line plots for different initial crustal thicknesses, based on Le Pichon et al. (1982). These simple plots show that it would take a homogeneous shortening of c. 65% to produce an elevation of 5 km from a lithosphere originally at sea-level, and a shortening of 84% to produce a similar mountain belt from a lithosphere whose top was originally 2 km below sea-level. This relationship has obvious implications when considering the shortening of previously thinned crust. Shortening factors of up to 50% may be required to elevate passive margins with thin crust above sea-level.

There is a maximum thickness sustainable by most segments of crust and this is governed by the strength of the thickened crust. Gravitational potential energy is stored in a mountain belt produced by thickened crust. As the crust/lithosphere thickens to produce the mountain belt an increasing amount of work is required to counteract this gravitational energy. The mountain range will reach a maximum mean elevation related to the force at which the plates are pushed together (Molnar & Lyon-Caen 1988). When the maximum elevation is reached, crustal shortening continues on the flanks of the range. Thus, where one mass of crust collides with, and indents another, the thickening will not be infinite but will reach a maximum value, after which the deformation will spread over a wide area. Crustal shortening will occur where the mountains are low and the thickness of the crust is normal until the elevations reach those of the adjacent highlands. Thus mountain ranges and high plateaux tend to grow outwards while the forces causing crustal shortening remain essentially the same (Molnar et al. 1977, Molnar & Tapponnier 1978, Molnar & Lyon-Caen 1988). Hence mountain belts must grow dominantly by the accretion of new thrust material to the already deformed zone. The mountain belts must grow by *footwall collapse*, that is, by accretion of material into the footwall of the

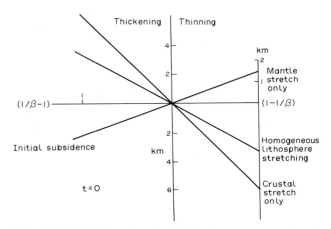

Fig. 13.15. Variations in uplift/subsidence rates associated with heterogeneous strains within the lithosphere.

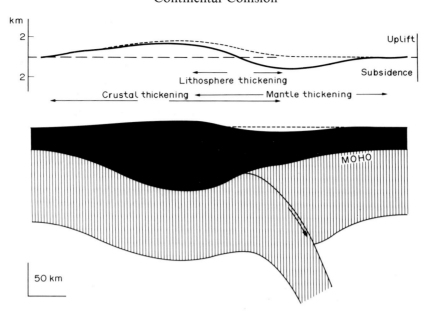

Fig. 13.16. Asymmetric thickening of a mountain belt involving simple shear deformation, equivalent to a two-layer model. The crust and the mantle are thickened in different places across the mountain belt, causing variable uplift and subsidence rates. A similar section has been drawn across the Alps (Butler 1986) to explain subsidence beneath the Po Basin. In the upper section the solid line shows the initial uplift or subsidence, the dotted line shows the uplift or subsidence following thermal equilibration.

thrust zones, and the thrusts must propagate outwards towards the foreland. *This pattern of foreland propagation*, or *piggy-back thrusting*, where earlier thrusts are carried on the backs of later thrusts is common for almost all mountain belts (e.g. Bally *et al*. 1966, Price 1981, Coward 1983, Butler 1985) (also see Chapter 6).

The growth in area of a mountain belt has been examined semi-quantitatively by England & McKenzie (1982), England & Houseman (1986) and Molnar & Lyon-Caen (1988). England & McKenzie (1982) and England & Houseman (1986) defined a parameter, the *Argand number* (Ar), which is a measure of the average strength of the lithosphere in a vertical direction. The stress-strain rate relationships can be defined as:

$$\tilde{\tau}_{ij} = B\dot{E}^{(1/n-1)}\dot{E}_{ij}$$

where $\tilde{\tau}_{ij}$ are the vertically averaged components of the deviatoric stress tensor, E_{ij} the strain rates, n is a constant that defines the vertically averaged power law rheology of the lithosphere and B is a constant that depends on the vertically integrated strength, that is, it depends on the Argand number. In reality the rheology will depend on the temperature distribution in the lithosphere and the laws which govern the constituent materials. The value B does, however, give a good approximation when investigating lithospheric behaviour. The Argand number is related to the constant B by the relationship:

$$Ar = g\frac{\rho_c(1-\rho_c/\rho_m)L}{B(V_0/L)^{1/n}}$$

where g is the gravitational acceleration, L the lithosphere thickness, V_0 the indentation velocity and ρ_c and ρ_m the densities of crust and lithospheric mantle.

The rates of deformation and the distribution of deformation intensities around an indentor depend on Ar and n. Where $Ar = 0$, then deformation is concentrated around the indentor, more so as n increases. Where $Ar>1$, then gravity acts to inhibit crustal thickening once a critical crustal thickness is reached and hence transfers the active thickening away from the indentor to the outer edge of the thickened plateau. Thus, according to this theory (England & McKenzie 1982, England & Houseman 1986) there is a progressive migration away from the indentor of the zone in which the rate of thickening is at a maximum. England & Houseman (1988) claim they can model the approximate thickening of the Tibet plateau as a result of the collision and indentation by India.

The strength of materials plays a key role in the migration of thrust faults; the simple migration of thrust faulting can be prevented, or at least modified, if the strength of the material in which the new fault forms is large (Molnar & Lyon-Caen 1988). Thus the Tarim basin in western China, which formed by early Mesozoic crustal extension behind the Kun Lun magmatic arc, was stronger than adjacent Tibetan and Tien Shan terrains. This region was not thickened during Tertiary collision tectonics and remained as a depressed basin between Tibet and the Tien Shan.

The above discussion on elevation assumes homogeneous thickening of the lithosphere. However, thickening of the crust involves displacement on large-scale gently dipping shear zones with a thrust sense. Thus the upper crust may thicken by thin-skinned tectonics above the main shear zones in the frontal regions of a mountain belt, while the lower crust and lithospheric mantle may thicken or subduct towards the *hinterland*. Thickening of the crustal part of the lithosphere will increase the topographic elevation, while thickening of the lithospheric mantle will depress the elevation (Fig. 13.15). Assuming a collisional model, as in Fig. 13.16, where the crust thickens in one region while the lithospheric mantle subducts, then there should be pronounced elevation of the frontal regions of the mountain belts. However, if the lithospheric mantle does not subduct, but instead underplates a region in the hinterland of a mountain belt, then this region will not elevate but instead will subside (Fig. 13.16). This may be one mechanism for generating basins on the hinterland side of mountain belts, on the side opposite to that of the advancing thrusts. The model in Fig. 13.16 is very simplistic; more realistic models should involve variable patterns of thickening of crust and lithospheric mantle across a mountain belt with time, and hence the patterns of uplift and subsidence may be complex.

The crustal thickening beneath Tibet, for example, does not account for the large area of uplift. Estimates of crustal shortening across a section line between Lhasa and Golmud suggest that there has been only 300 km shortening during the Tertiary (Coward *et al.* 1988). No high-grade rocks have been uplifted during Tertiary deformation, most of Tibet is underlain by low-grade metamorphic rocks or by the relicts of earlier foreland basins. N to NW-dipping thrusts of Tertiary age have been recognized, particularly in three zones, near Lhasa, in central Tibet, near Amdo and in the north, near the southern edge of the Kun Lun. In each of these zones, Tertiary thrusting is superimposed on older structures; presumably the older thrusts, related to the original accretion of Tibetan terranes (Coward *et al.* 1988, Dewey *et al.* 1988). These caused the upper crust to be locally weaker and hence more prone to thickening. However, the cumulative shortening on these thrusts is small, especially when compared to the large crustal-scale thrusts of the Himalayas, with cumulative displacements of greater than 500 km. Some other mechanism is required to produce crustal thickening in Tibet. Some thickening might have been associated with magmatic arc accretion during the Late Cretaceous and early Tertiary subduction of Tethys beneath the southern edge of Asia. Some thickening may have been associated with earlier subduction and collision during the Triassic and Jurassic. However, during the early Tertiary, marine carbonates were deposited across much of southern Tibet, suggesting that much of the Tibetan crust had not been thickened by this time.

The lack of upper crustal shortening across Tibet suggests a *two-layer thickening model*, where the lower crust thickens far in excess of the upper crust. The two layers of the crust must have been detached by ductile shear at mid-crustal levels. Zhao & Morgan (1985) suggested that excess crust may have been injected beneath the upper Tibetan crust by some process of lateral extrusion. If this is correct, then lateral extrusion must account for a minimum of 33% and up to 50% of the crustal shortening between India and Asia since the beginning of the Tertiary (Le Pichon *et al.* in press). Zhao & Morgan (1985) envisage that the lower crust was injected into the lower Tibetan crust beneath a head formed by the elevated Himalayas. This injected lower crust would then essentially jack-up the Tibetan crust. In older orogenic belts this injected material may be recognizable as large sheet-like intrusive bodies of granitic or dioritic rock.

A simpler model may involve two-layer shortening, where the Tibetan crust is partially underplated mechanically by thickened lower crust of the Indian plate. Unless the Tibetan lower crust is removed then space problems require that this lower crust is shortened far more than the upper crust. This model also requires thickening of the Tibetan lithospheric mantle.

Medium-scale structures and collision

Thrusts occur on all scales and many medium- to small-scale structures in an orogenic belt develop because of the bulk shape changes which are necessary to allow the large-scale thrust slabs to move. One of the problems which puzzled structural geologists for much of this century involved the driving mechanisms of large thrust sheets. From observations, these sheets were relatively thin and yet often had large displacements of many tens of kilometres. Studies of rock mechanics have shown that it is impossible to push thin slabs of rocks large distances without the slabs breaking apart. Alternative explanations were developed which involved, for example, lubricated bases to the slabs and sliding slabs down slopes. However, recently it was realized that the surface slope of the thrust wedge is an important controlling factor to thrust movement (Elliott 1976, Chapple 1978, Davis *et al.* 1983, Platt 1986). In the same way that glaciers can move uphill as long as there is a gravitational potential defined by the surface slope of the glacier, so relatively thin thrust wedges can be pushed up a basal slope as long as there is sufficient surface slope to the thrust wedge.

The crustal taper of the *thrust wedge* depends on the basal shear strength. Davis *et al.* (1983) envisaged the shape of the thrust wedge to be similar to that of sand being pushed by a mechanical shovel. Thrust wedges with low basal shear strength, for example, wedges moving over sediments with a high fluid pressure or over salt or gypsum deposits, will require only a low surface slope. However, thrust wedges moving over rocks with a high shear strength, such as crystalline basement gneisses or granites, will require a steeper surface slope. The surface slope can be changed by deformation within the wedge, by accreting material as thrust slices to the front of the wedge or by accretion of fault bounded packages of material from beneath the wedge (Fig. 13.17). As erosion removes material from the surface, so more internal deformation is needed or more material must be added to the wedge to maintain its critical taper (Davis *et al.* 1983).

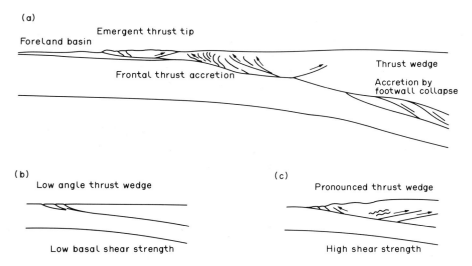

Fig. 13.17. Structures developed within a thrust wedge (a) and the different shapes of thrust wedge for low (b) and high (c) basal shear strength.

As the thrust wedge gradually climbs through the crust, it will pass over rocks with different shear strengths. A more pronounced wedge shape with a steeper surface slope will be required where the basal thrust passes through the strong middle to upper crust. This region is characterized by internal deformation involving folding and the development of local backthrusts (Fig. 13.18). At the surface it is marked by the main mountain front; the frontal parts of the High Himalayas overlie this part of the thrust zone. However, as the wedge overthrusts the higher level, and often still uncompacted weak sediments, its surface slope is too steep. The slope may be reduced by erosion or by the addition of *frontal thrust packages* but sometimes it collapses in a series *foreland-directed extensional faults*, similar to large landslides. These extensional faults have been recognized in the frontal regions of the Pakistan Himalayas as well as in older thrust belts, such as parts of the Moine thrust zone of Northwest Scotland (Coward 1982).

Davis *et al.* (1983) developed an analytical model for thrusting involving a sub-aerial wedge (Fig. 13.19), close to Coulomb failure:

$$\alpha + \beta = \frac{\beta + (1-\lambda_b)\mu_b}{1 + (1-\lambda)K}$$

where α is the forward topographic slope of the wedge, β the regional hinterland-directed dip of the basal thrust, λ and λ_b the ratio of pore fluid pressures to vertical normal stresses exerted by the lithostatic pressure within and at the base of the wedge respectively, μ_b the coefficient of sliding friction on the thrust and K a dimensionless quantity related to the direction of the maximum compression and dependent on μ_b and on the coefficient of internal friction (μ) within the wedge.

Davis *et al.* (1983) tested their model on the active thrust belt of western Taiwan; especially the accretionary prism of sediments in front of the Phillipine Sea plate. For a prism with a pore fluid pressure to confining pressure ratio (λ) of 0.675 and a

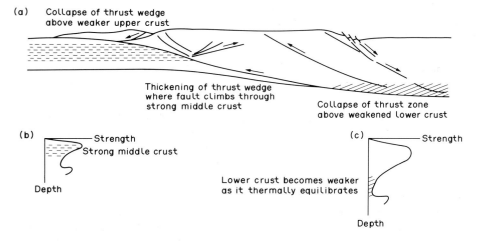

Fig. 13.18. (a) The change in shape of a thrust wedge as it moves over rocks with different basal shear strength — see text for discussion. (b) and (c) Strength profiles through the thrust wedge at the frontal tip and some distance back into the thrust zone.

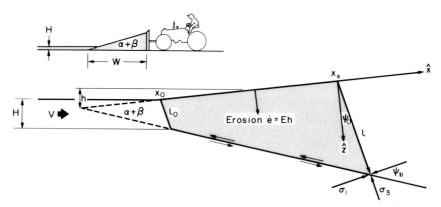

Fig. 13.19. Cross-section of a critically tapered wedge, illustrating the parameters employed in deriving the kinematic model. Cartoon illustrates the similar growth of a bulldozer wedge. From Davis *et al.* (1983) and Dahlen & Suppe (1987). See text for details.

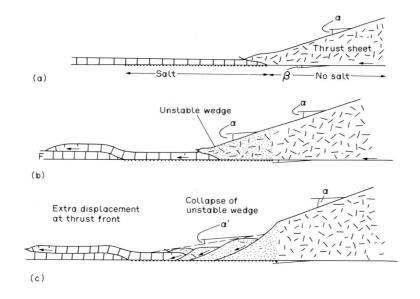

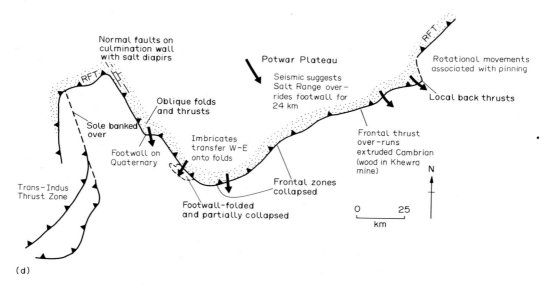

Fig. 13.20. (a–c) Simplified sections illustrating the growth of the frontal ranges of the Pakistan Himalayas and collapse of the thrust wedge as the basal thrust overrode the Cambrian salt. (d) Simplified map of the Salt Ranges (after Butler *et al.* 1987) showing the variations in fault kinematics (arrows show fault movement directions). Variations in displacement direction are due to pinning of the thrust tips at the eastern end of the Salt Ranges where the salt is thin and basal slip more difficult than in the west, and where the Salt Range thrusts interfere with thrusts from the main Himalayan Range (also see Fig. 13.32) and gravitational collapse of the thickened salt in the western part of the Salt Range thrust zone.

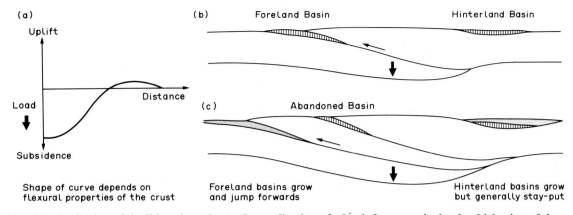

Fig. 13.21. (a) The flexing of the lithosphere due to the application of a load, for example, by the thickening of the crust. (b) and (c) The development of foreland and hinterland basins.

coefficient of basal sliding friction of 0.85, the average coefficient of internal friction would be c. 1.03.

Davis & Engelder (1985) applied the model of Davis et al. 1983) to thrust belts in low-strength evaporites. They pointed out that such thrust belts have a very small taper, usually with a flat topography and a zone of overthrusting which extends far out over the foreland. These observations can be applied to the gross form of the Himalayan frontal thrusts in northern Pakistan (Fig. 13.20). In the Salt Ranges – Potwar Plateau region the thrusts detach on Cambrian salt for c. 200 km onto the Indian plate (Fig. 13.20). Their topographic slope is near zero, the ratio of basal pore fluid pressure to vertical normal stress is locally measured as 0.93 ± 0.01 (Hubbert & Rubey 1959). For most of this region no internal deformation is required within the thrust sheet to increase the critical taper. However, in the eastern part of the Salt Ranges the basal slope is less than 1° (based on seismic data) but the amount of Cambrian salt decreases. Here, the thrust belt shows a very different structure, with internal deformation producing several large thrusts.

Davis & Engelder (1985) suggested that salt-based thrust belts need only maintain a foreland wedge taper of c. 1° to continue movement, compared to 8–12° for belts with relatively stronger basal detachments. The frontal slopes of the Salt Ranges are generally c. 30° with a present-day topographic relief of up to 1500 m above the foreland. Thus the potential is available to drive gravitational collapse, especially if there was a weakening of the salt layer. The presence of brine leads to a substantial loss of strength of salt at low temperatures.

Large wedges which involve rocks undergoing prograde metamorphism are unlikely to exhibit Coulomb-failure criteria. The deeper-level rocks will have negligible yield strength and have a non-linear viscous rheology. However, these wedges will behave in essentially the same way as Coulomb wedges, in that they will deform internally until they reach a stable wedge configuration, in which the gravitational forces exherted by the surface slope balance the resistance to movement on the underside. Accretion of material at the front of the wedge will lengthen the wedge, decreasing its taper and hence the wedge will shorten internally to maintain stability. The internal deformation will involve folding, the generation of *back-thrusts* and *back-folds* and *break-back fore-thrust systems*.

Platt (1986) derives a stability criterion for thick orogenic wedges:

$$\alpha = \frac{\tau}{\rho g \lambda}.$$

This equation is identical to the glacier sliding term already discussed and states that in the special case of a wedge that is neither shortening or extending, the surface slope is such that the gravity sliding stress balances the shearing stress at the base.

An increase in basal shear stress caused, for example, by an increase in subduction/overthrust rate, will lead to shortening and thickening of the thrust wedge. Variations in rate of plate movement or variations in partitioning of the plate movement may lead to alterations of contractional and extensional deformation (Dahlen 1984, Platt 1986). Platt (1986) uses this thrust wedge theory to explain the development of polyphase deformation in the Alps and the uplift of middle to lower crustal rocks.

Similarly, when subduction ceases the orogenic wedge may undergo extensional collapse because the basal drag exerted by the subduction no longer exists. In the deeper parts of a thrust wedge, and hence in the more internal parts of a mountain belt, the basal shear strength may decrease as the thickened crust becomes hotter. This region also collapses on *large-scale extensional faults* which generally dip in the same direction as the thrusts, and may link with the thrusts at depth. Large-scale extensional faults have been recognized around much of the Himalayan belt; the region to the north of Everest has dropped more than 10 km to the north on one such N-dipping fault (Burg & Chen 1984, Royden & Burchfiel 1987).

Flexural basins

The mass of the thickened lithosphere isostatically depresses the Moho. However, gravity anomalies show that mountain belts and adjacent basins are generally not in isostatic equilibrium. The basins are characterized by negative isostatic anomalies of 100 mgal or more suggesting large deficits in mass, while the mountain belts have positive anomalies and are thus underlain by excess mass. The 4–6 km high Himalayan-Tibet region is underlain by a c. 70-km-thick crust and the subsidence associated with this additional load flexes the adjacent unthickened crust. The crust is depressed on the foreland and hinterland sides of the thickened zone (Fig. 13.21) and the depressions, known as *foreland and hinterland basins*, are generally infilled with debris eroded from the mountain belts (also see Chapter 10). The depth of these basins increases gradually and generally smoothly towards the mountain front, so that the maximum depths are close to the mountain range. The shape of the flexure, that is, the width of the basin and the size of the small foreland bulge, depends on the flexural properties of the crust (Beaumont *et al.* 1982). Young, hotter and weaker crust has a low *flexural rigidity* and forms a narrow but often deep basin while older, cooler and stronger crust will form a wide basin and a correspondingly more distinct foreland bulge.

A *foreland (foredeep) basin* is characterized by sediments which feather out on to the foreland and may reach a maximum depth of up to 6 km. Figure 13.22 illustrates a schematic cross-section through the Alberta basin which typifies many of the structural feature of foreland basins. The basin is wedge shaped and reaches a maximum depth adjacent to the mountains. The stratigraphy consists of units which thin laterally and may be truncated by erosion away from the mountain belt, or may appear to thin towards a common point, or overstep each other. This 'pinch-out' is important in determining the foreland basin evolution. The sediments near the mountains may be folded and involved in basin-directed thrusts, which detach in the sediments or in the underlying basement. Often the lateral advance of the thrust belt has driven the foreland subsidence in front of it.

The youngest sediments are coarse grained and proximal near the mountain front, but finer grained and distal where they feather out against the foreland. They are mostly derived from the mountain belt, as the foreland side of the basin has only low relief. The clastic influx is governed by the rate of uplift of the mountains; major slumps and megaturbidites can indicate important phases of uplift. Under certain climatic conditions, foredeep basins may be filled by carbonates. Much of the drainage is axial, parallel to the mountain front; typical examples of this pattern are seen in the Ganges basin of India and the Jaca basin of the Spanish Pyrenees.

Much of the data for defining flexural parameters come from studies of the lithosphere response to loading by ice (e.g. Walcott 1970). These studies show that the response can be modelled as either an elastic or a visco-elastic plate overlying a weak substratum (Turcotte & Schubert 1982). In a simple elastic model the applied load, consisting of ice, volcanic seamount or overthrust mass, applies a bending moment to the underlying plate. The bent plate takes a shape as shown in Fig. 13.23 and the subsidence fits the equation:

$$w = w_0 e^{x/\alpha} \cos x/\alpha$$

where w is the subsidence or uplift, w_0 the maximum deflection beneath the applied load, x the distance away from the applied load and α the '*flexural parameter*'. A slight *peripheral bulge* develops, whose amplitude is given by:

$$w_b = w_0 e^{-\frac{3}{4}\pi} \cos \frac{2\pi}{4}$$

which simplifies to:

$$w_b = 0.0670 \, w_0.$$

The bulge is situated at a distance $x = 0.75 \, \alpha\pi$, and the hinge, where there is neither subsidence nor uplift, is at:

$$x = 0.5 \, \alpha\pi.$$

α depends on a parameter known as the *flexural rigidity* of the bent plate, where:

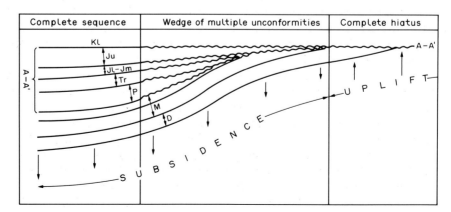

Fig. 13.22. Simplified section showing the stratigraphic pinch-outs (A–A') in the Alberta Basin, southern Canadian Rocky Mountains. D = Devonian, M = Mississipian, P = Permian, Tr = Triassic, J = Jurassic, K = Cretaceous. After Bally *et al.* (1966).

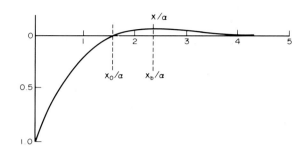

Fig. 13.23. The deflection of elastic lithosphere under load. See text for details.

$$\alpha = \frac{4D}{(\rho_m - \rho_w)g}^{1/4}$$

for submarine basins and:

$$\frac{4D}{(\rho_m - \rho_s)g}^{1/4}$$

for sedimentary basins, where the mass of sediment replaces water. D is related to the elastic moduli by:

$$D = ET^3/12(1 - v^2)$$

where E is Youngs modulus and v is Poissons ratio. T is the '*effective elastic thickness*' of the plate, and it is T which determines the amplitude and wavelength of the deflection. Thus the distance between the load and the peripheral bulge can be used to predict the flexural rigidity and estimate the effective elastic thickness (also see Chapter 10). However, beneath a mountain belt it is often difficult to determine the exact position of the load; generally the flexural rigidity has to be obtained from a curve fitted to the deflected lithosphere. In some areas, the position of an ancient peripheral bulge is difficult to locate as the position of the pinch-outs depends partly on changes in sea-level and subsidence due to sediment load, as well as that due to the mountain belt. Hence it is best not to use the feathering out of sediments to estimate the flexural parameters but to use the total basin shape.

The lithosphere is probably not continuous and uniform beneath the orogenic front; it may vary with age along the belt, as older cratonic material and young crust are variously involved in the collisional margin. The initial collision may have shortened a previously attenuated margin, where the crust was thin and the geothermal gradient high. Thus flexural rigidities can vary considerably along and across strike (Molnar 1988) and the width of a foreland basin can vary with time, as different parts of the lithosphere are overridden.

The northern margin of the Alpine chain in Europe has flexural basins with variable width (Homewood *et al.* 1986). In France, where the crust had been extended to form the Rhine–Bresse–Rhone graben prior to Alpine shortening, there is only a narrow flexural basin. Presumably, the flexural strength of the lithosphere was low because of Cenozoic faulting. In Germany and Austria however where the basement consists of Variscan crust covered by Mesozoic sediments, the flexural basin is wide. The Austro-German molasse basin widens towards the edge of the Bohemian Massif, where it ends abruptly against the continuation of a line of Variscan transcurrent faults. To the south and east of the Bohemian Massif, the flexural basin is very narrow in Slovakia, but then widens in Poland. The response of the European crust to Alpine loading is therefore very variable and depends on the age and structure of the basement.

As the underthrusting plate bends, the outer arc of the elastic portion of the plate will extend; the amount of extension will depend on the real elastic thickness of the plate and the depth of the neutral surface (see also Fig. 13.2). Note that this *real elastic thickness* may be very different from the effective elastic thickness (T) governing the shape of the flexural basin. The effective elastic thickness (T) is a parameter averaging the rigidity of the whole lithosphere, and not only the elastic lid of the crust. Extensional faults characterize the deeper parts of flexural basins, for example the Austrian molasse basin, where they form effective hydrocarbon seals.

A plate can be bent by other mechanisms (Fig. 13.24). A subducting slab of oceanic lithosphere can produce an additional load other than that expected from the weight of the mountains alone (Royden & Karner 1984). The occurrence of intermediate- to deep-earthquakes beneath the Carpathians suggest that oceanic lithosphere was subducted beneath these arcs in the last 10 Ma. Gravity acting on this subducted slab of oceanic lithosphere may have produced the force necessary for the production of the Carpathian foredeep basin (Royden & Karner 1984). Similarly, in the western part of Pakistan the Indus basin, which is wide and locally more than 7 km deep, is too large considering the mass of the adjacent Sulaiman and Kirthar mountain ranges. As the Indus basin was originally part of the much larger Indus–Katawaz basin, infilled with an 11 km thickness of Katawaz sediments, west of the Sulaiman Ranges, it is obvious that the load of the mountain belt is insufficient to flex the lithosphere to produce the basin. An additional load is required, probably provided by Eocene–Pliocene subduction of Indian oceanic crust beneath Afghanistan. The load provided by the Sulaiman and Kirthar Ranges is a late and relatively superficial load.

A bending moment can be applied to the end of a plate which can flex the plate up or down depending on its sign (Karner & Watts 1983). Lyon-Caen & Molnar (1983, 1985) argue that gravity acting on a detached slab of mantle lithosphere beneath the India–Asia collision zone applies a bending moment to the Indian plate that not only flexes it up but also contributes to the support of the Himalayas (Fig. 13.24).

From an examination of the different ages of sediments in foreland basins and their gradual incorporation into the thrust regime, the rates of advance and rates of displacement for the major

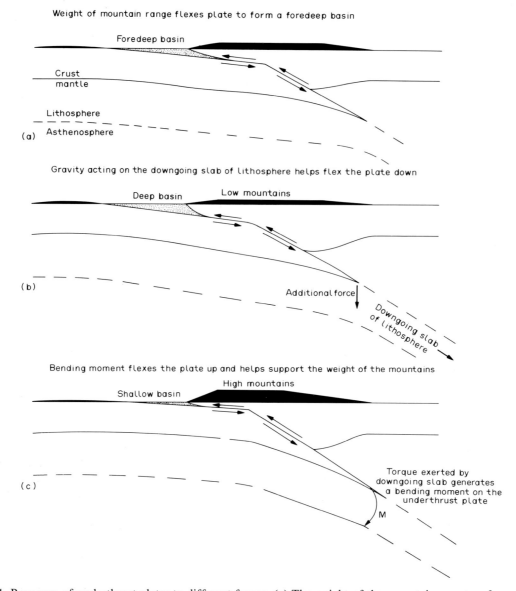

Fig. 13.24. Response of underthrust plates to different forces. (a) The weight of the mountains exerts a force that causes a foredeep basin. The depth of the basin and the average height of the mountains should be proportional, depending on the flexural parameter. (b) A downgoing slab of oceanic lithosphere can contribute an additional force that pulls the plate down. In such a case the basin should be deeper (or the mountains lower) than in the case where no buried load exists. (c) The mass in the downgoing slab could exert a bending moment on the lithosphere, locally flexing the lithosphere up. After Molnar & Lyon-Caen (1988).

thrusts, which bound the thickened zone, may be obtained. Subsidence curves can be plotted to separate the *flexural subsidence phase* from any earlier phase related to the development of the original passive margin. Thus the subsidence curves plotted for the Montana and Utah foreland basins show a gradual subsidence history, possibly related to earlier stretching, followed by rapid subsidence in the Middle Cretaceous during the growth of the Rocky Mountain overthrust belt (Cross 1986).

The successive positions of pinch-outs or depocentres across a basin may be plotted to give a rough estimate of the rate of migration of the mountain belt. However, the position of these pinch-outs is governed by other parameters, for example, the sediment loading, sea-level changes and the effect of visco-elastic flexure.

Figure 13.25 shows the progressive migration of pinch-outs for part of the Ganges basin, south of the Himalayas. This map illustrates the episodic growth of the Himalayan foreland basin during the Cenozoic (see also Treloar *et al.* 1992). A graph can be plotted showing the approximate age of the pinch-out relative to distance from the edge of the Ganges basin (Fig. 13.26). This plot suggests a migration rate for the basin of 1–1.5 cm/yr (Lyon-Caen & Molnar 1985) which is considered to be the relative convergence rate for the Himalayan mountain belt and the underthrusting Indian plate.

Similar constructions can be made for the Western Alps and molasse foreland basin in Switzerland (Homewood *et al.* 1986). The oldest basin deposits are the Tertiary flysch comprising calcareous and/or

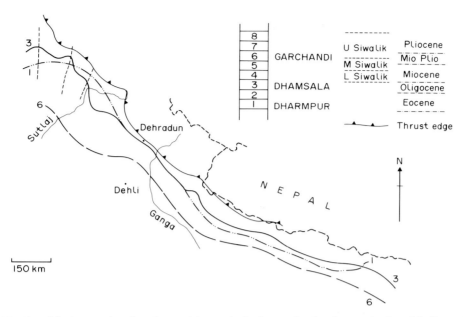

Fig. 13.25. Simplified map showing the positions of pinch-outs in the Ganges basin of India and Nepal.

siliceous turbidites which infill the marine foreland basin. During the middle Tertiary, terrestial deposits began to be deposited, although there are several fluctuations between terrestial and marine basin fill. The youngest sediments form the Recent molasse deposits of the Swiss plain. The earlier foreland basin sediments have been involved in the thrust tectonics and need to be restored to find the original positions of the pinch-outs. Restoration of the sections and analysis of the relative positions of pinch-outs suggests a convergence rate for the Alps of c. 0.7 – 0.8 km/yr.

The convergence rates shown by the migration of pinch-outs or depocentres need not be the same as the rate at which the frontal tips of thrusts migrate forwards. The zone of thrusting may locally migrate rapidly if the decoupling zone lies in an easy slip horizon such as shale or salt, where the thrust taper need not be large. In Northwest Pakistan the position of the thrust front has been estimated for different times using magnetostratigraphy (Fig. 13.27) and hence the rate of thrust migration and an idea of the critical thrust taper can be obtained through time.

Sometimes new large thrusts develop which carry forward the older foreland basin. This uplifted and transported basin is known as a *piggy-back basin*. Excellent examples occur in the northern Appennines. For piggy-back basins to survive, the thrust taper must be low; if thrust mechanics require a high critical taper, then the piggy-back basins will be destroyed and the sediments reworked into the new frontal foreland basin.

Hinterland basins are essentially similar to foreland basins but as the major thrusts are generally directed towards the foreland, the hinterland basins are less deformed and may remain stable throughout the history of the adjacent mountain belt. In China, the Tsaidam and Tarim basins form hinterland basins to the thickened Tibetan crust and show no migration of depocentres with time.

POST-COLLISIONAL STRUCTURES

Indentation and continental escape

In many mature mountain belts the thrusts and folds associated with approximately plane strain deformation are crossed by strike-slip faults and shear zones which make a low angle to the strike of the thrust belts. Molnar & Tapponnier (1975, 1977, 1978, 1979) Tapponnier & Molnar (1977, 1979) and Tapponnier *et al.* (1986) explained the pattern of strike-slip faults in Central Asia as being analogous to *slip lines* developed experimentally in plastic materials when indented by a

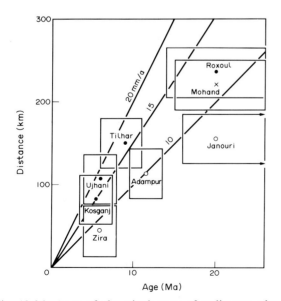

Fig. 13.26. Ages of the pinch-outs of sediments plotted against distance from an arbitrary line on the southern side of the Ganges basin, to show rate of migration of the basin. From Lyon-Caen & Molnar (1985).

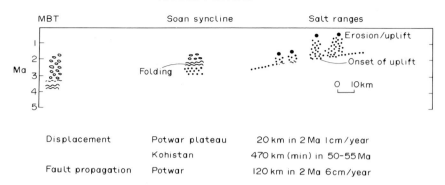

Fig. 13.27. Diagram to show the timing of folding (coarse wavy lines), local uplift (dotted lines), coarse sedimentation and final uplift and erosion (circles) on a section through the foreland basin of Pakistan. MBT, Main Boundary Thrust. From data by Burbank & Raynolds (1984).

rigid die (Fig. 13.28). The metal flows out and around the indentor and is crossed by a series of slip lines which are used as analogues for the major strike-slip faults.

The general tectonic framework of the Himalayas and Tibet and the extent of strike-slip faulting in China is shown in Fig. 13.29. Major strike-slip faults are left-lateral, east of India and right-lateral, west of India. The asymmetry is carried far into Asia, Mongolia and possibly into Siberia, where Lake Baikal exists along the axis marking the northern extension of the division between the left-lateral and right-lateral fields. The kinematic role of the strike-slip faults is to accommodate flow away from the penetrating indentor (hence *indentation tectonics*), that is, to allow *tectonic escape* of continental material (also see Chapters 6, 10 and 12).

According to Tapponnier et al. (1986) the geological record of Southeast Asia illustrates a polyphase extrusional model, with displacements greater than 1500 km, during which India has successively pushed Sundaland, Tibet and south China towards the ESE. These authors argue that the middle Tertiary movements mostly occurred along the Red River – Ailao Shan fault zone. Most of Sundaland lay in a frontal position with respect to impinging India.

However Dewey et al. (1988) argue that *lateral expulsion (extrusion)* occurred only late in the collisional process. According to Dewey et al. (1988) and Coward et al. (1988) the northern part of Tibet was thickened during the early stages of Himalayan collision tectonics in the Paleogene and there is no comparable region east of the eastern continuation of the Indian indentor. Hence, the degree of lateral expulsion appears less than that favoured by Tapponnier et al. (1986). Furthermore, Tertiary subduction-related volcanics and intrusives occur in the Lhasa terrain, relatively close to the suture and hence it is unlikely that a large mass of continental crust, the size of Sundaland, was expelled from the frontal collisional region, between the Indian and Lhasa terrains, during the early Tertiary.

The way in which the northward motion of India into Asia is *partitioned* into different styles and ages of structures is summarized in Fig. 13.30, which shows the neotectonic (Chapter 18) displacements in mm/yr across and within the convergent zone. A displacement of $18 +/- 7$ mm/yr is absorbed by thrusting in the Himalayas, $13 +/- 7$ mm/yr by thrusting in the Tien Shan, $6 +/- 4$ mm/yr by thrusting in the Kun Lun and Altyn Tagh and $11 +/- 7$ mm/yr as the N–S convergent component of strike-slip faulting (Molnar et al. 1987). The total displacement gives a N–S convergence rate of $48 +/- 25$ mm/yr which is consistent with the 56 mm/yr deduced from palaeomagnetic data (Dewey et al. 1988).

The tectonic history of Tibet can be summarized as follows. During the Eocene and Oligocene, c. 66 mm/yr convergence between India and Asia was accommodated in Tibet by N–S crustal shortening involving almost plane strain and S-directed thrusts (Coward et al. 1988, Dewey et al. 1988). The latter was the main collisional phase during which almost the

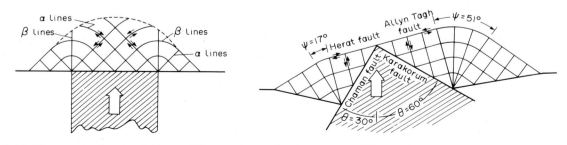

Fig. 13.28. Plane strain slip-line fields for different shapes of indentors and different shapes of indented rigid plastic material (after Tapponnier & Molnar 1976, Molnar & Tapponnier 1977, Tapponnier et al. 1986).

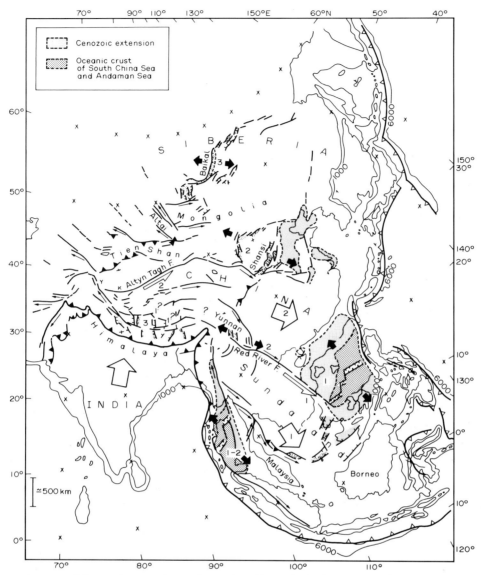

Fig. 13.29. Schematic map of Cenozoic extrusion and associated large faults in eastern Asia. Heavy lines — major faults or plate boundaries. Open barbs — present subduction of oceanic crust. Solid barbs — thrusts. Large open arrows — major block motions with respect to Siberia since the Eocene. Numbers refer to extrusion phases, 50–20 Ma, 20–0 Ma, and 0 to future, respectively. From Tapponnier et al. (1986).

entire Tibetan crust doubled in thickness to c. 65 km, close to its maximum sustainable limit (England & Houseman 1986). During the Eocene there was considerable overthrusting in the Himalayas, producing the major crustal-scale basement-cored nappes and the subsequent burial metamorphism, while during the Oligocene thrust tectonics in the Himalayas was much less; presumably the displacement of Indian into Asia had been partitioned so that more shortening occurred in Tibet. By c. 30 Ma, further shortening across a wider region of Central Asia, north of Tibet was blocked by the stronger lithosphere of the Tarim and Tsaidam basins and deformation spread farther south into the Himalayas. This reworking of the Himalayas (30–5 Ma) resulted in the development of a second mountain front, a new influx of sediments into the Siwalik foreland basin and shortening in the Tien Shan and Altyn Tagh. In Tibet, there was slow horizontal plane strain associated with strike-slip faulting during which time a small amount of E–W extrusion occurred. Much of this extrusion may have been accommodated by thrusting in the Long Men Shan, near Chengdu in Eastern Tibet. According to Dewey et al. (1988) there is no evidence for the very large strains and eastward extension proposed by Tapponnier et al. (1986). During phase 3, in the last 5 Ma, N–S shortening propagated into Asia north of the Tarim Basin and eastward extrusion of Tibet was accompanied by E–W extension.

Thus the lateral extrusion of Tibet appears to have developed from the following, (1) resistance to further thickening of the Tibetan lithosphere, and (2) resistance to lateral growth of the zone of thickening by adjacent stronger lithosphere. Such a change in deformation style may occur at different times in the collisional history of different orogenic belts. Sometimes it is possible to identify the buffer zones which inhibit plane strain styles of deformation. In the Alps, for example, the

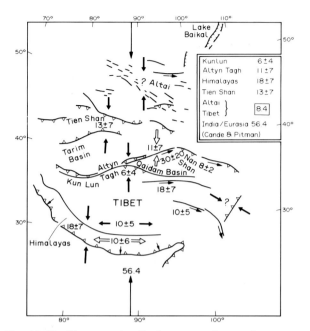

Fig. 13.30. Neotectonic displacements in mm/yr across and within the India–Eurasia convergent zone. Black arrows — shortening and extension values from thrust and graben zones, open arrows — displacements from strike slip faults. After Dewey et al. (1988).

northwest edge of the Valais zone presumably formed a rigid buffer to NW-directed plane strain shortening, and hence shows complex three-dimensional strains and strike-slip zones. Some of the strike-slip zones responsible for the West Alpine arc originate in this zone.

Strike-slip tectonics (also see Chapter 12) characterizes the mature stages of older orogenic belts. During Caledonian–Acadian–Variscan collision in eastern North America and Western Europe, there was lateral escape of a triangular-shaped fragment of Northwest Europe, away from the North African indentor. Along the eastern margin of the Appalachian thrust zone, the major shear zones are dominantly left lateral in the north but right lateral in the south.

Extensional *pull-apart basins* (see Chapter 12) have formed along side-stepped portions of the strike-slip faults, producing the Baikal rift system in Central Asia (Fig. 13.29) and the Devonian proto-Viking graben in the North Sea (Coward 1990).

Sections through older mountain belts

The deep seismic studies carried out by, for example, the American, British and French research groups (COCORP, BIRPS and ECORS, respectively) show clearly the variations in structure through the older collisional mountain belts of Europe and North America. These belts include the Caledonian and Variscan fold belts of Europe and the Appalachian–Ouachita fold belts of North America, all of which formed during Palaeozoic collision processes. The advantage of studying older orogenic belts is that they show a long history of collisional tectonics, whereas younger belts such as the Himalayan/Tibet belts are still tectonically active and all the post-collisional events have not yet occurred. In the former, the shear belts have been uplifted during post-collisional extension and thermal equilibrium, or by a subsequent phase of collision tectonics. There are some common features.

(1) The old orogenic belts show a relatively flat Moho, at c. 27–35 km depth, identified by a zone of moderate to strong semicontinuous reflectors which occur at approximately the same depth as seismic velocity changes determined from refraction data. The Moho is clearer where the belts have been reactivated by subsequent extension. There are no zones of anomalously thick crust observed on the seismic reflection data, nor are any 'roots' to a mountain belt observed. Where the Moho is irregular, it commonly rises beneath the internal parts of the orogenic belt as shown by the southern Appalachian transect (Fig. 13.31). The Moho has risen to a generally flat level by uplift and spreading of the thickened crust.

(2) The inner parts of the orogenic belts show a *reflective lower crust* and seismic sections showing numerous short and discontinuous reflectors. These reflectors indicate strongly *sheared lower crust* in which the shear zones form discontinuous planar fabrics, with slightly variable orientations, often wrapping round pods of less deformed material. A similar network of shear zones can be seen from surface data in regions of uplifted Precambrian rocks, such as Northwest Scotland. Similarly, in the internal parts of the Caledonian fold belt of Scotland, the deformation intensity varies in the Precambrian metasediments; the more intense foliation wraps round fold hinges or less intensely deformed pods of sediments.

(3) The zone of reflective lower crust sometimes underlies a zone of *less reflective upper crust*; the latter represents a less deformed overthrust slab which may be part of a different plate.

(4) There is generally a wide zone of reflectors which dips at c. 35° ± 10° throughout the *middle crust*. These reflectors represent *crustal-scale shear zones*, or ductile thrusts, which link the lower-crustal reflectors to the upper-crustal thin-skinned zone. The dips must have been modified by post- or late-orogenic uplift; the shears would have a lower dip relative to datum during the collisional processes.

(5) The shears link to a high-level relatively *flat thrust zone* on the foreland side of the orogenic belt, above undeformed basement. This zone is well preserved and clearly visible on the northern Appalachian seismic traverse (Fig. 13.31) and the ECORS line, north of the Paris Basin (Bois et al. 1988). It is preserved on-shore in northern Scotland and parts of the Variscan zone of southwest Britain but has been removed by erosion from above the section line shown by the MOIST traverse, offshore Northwest Scotland.

In some internal zones, prominant reflectors dip towards the hinterland and the foreland. This is shown by interpretations through parts of the Caledonides and

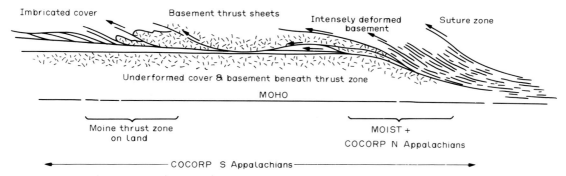

Fig. 13.31. Simplified model for the Caledonide–Appalachian structure, from BIRPS and COCORP data (modified from Brewer & Smythe 1984).

Eastern Alps/Carpathians (Tomek *et al.* 1987). Here, the crust has probably deformed as a series of *interthrusting wedges,* where slices of crust interfinger. Some orogenic belts appear symmetric in that they show overthrusting on both margins. The northern Caledonides is one such belt; to the southeast there is

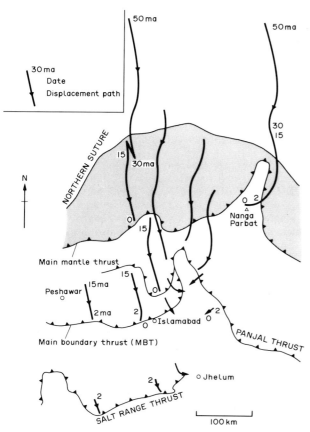

Fig. 13.32. Displacement paths of the thrusts from the western part of the Himalayas in northern Pakistan. The displacements are complex due to: (1) pinning of the main Himalayan thrusts in the Kohistan region, causing clockwise rotation of the main Himalayan thrust sheets and (2) local pinning of the Salt Ranges thrust and the MBT, causing local anti-clockwise rotation. To this pattern must be added the regional variations in thrust directions due to gravitational collapse of the thickened Tibetan crust and the local gravitational collapse of individual thrust wedges such as the Salt Ranges. No one section through this part of the western Himalayas has suffered plane strain and any two-dimensional balanced sections and palinspastic restorations can only give approximate figures for displacement.

c. 400–500 km of overthrusting onto the Scandinavian foreland and to the northwest there is more than 100 km of overthrusting onto the foreland of Greenland and Northwest Scotland.

The Caledonidian orogenic belt is similar to the recent Himalayan–Tibetan collisional zone of Asia. There are, however, some important differences between the older and younger belts.

(1) The younger belts show thickened crust. The depth of the Moho beneath Europe clearly demonstrates thickened crust beneath the Tertiary orogenic belts of the Alps and Pyrenees but shows the absence of thickened crust beneath older orogenic belts.

(2) In detail the Moho may be extremely irregular beneath the younger belts. The seismic work of Hirn *et al.* (1984) suggests a stepped Moho beneath Tibet, varying in level between *c.* 50–70 km. A similar step is observed from the ECORS seismic line across the northern Pyrenees. The Moho may act as a passive marker during collision which becomes offset by thrust or strike-slip faults affecting the lithospheric mantle. Thus during the later stages of collision the Moho must become smoothed out, either by the introduction of new material or by large-scale extensional flow.

As discussed earlier, the mean elevations of high plateaux are crude pressure gauges for the average compressive stress pushing on the margins of the plateau. Changes in the driving forces can lead to situations in which a high plateau can undergo crustal extension, while the flanks of the plateau undergo crustal shortening. The removal of compressive stresses by cessation of convergence, or the weakening of the lithosphere surrounding an elevated region can lead to extension of the mountain belt (England & Houseman 1988, Molnar & Lyon-Caen 1988). A third mechanism may be important in some belts, that of increasing the potential energy of the plateau. Thermal equilibration of thickened lithosphere will lead to uplift and hence to an increase in potential energy, but this takes place over a long time period. An alternative mechanism is suggested by Houseman *et al.* (1981) and England & Houseman (1988). Houseman *et al.* (1981) showed that the thickened lower lithospheric mantle became gravitationally unstable and sank into the convective mantle below. The descent of the lower mantle occurred some time after initial thickening; the duration of the delay depending on the Rayleigh number of the convecting mantle and the degree of thickening of the lower lithospheric mantle. Replacement of the lower lithosphere by hot asthenosphere results in an increase in surface elevation and hence in gravitational potential energy.

The loss of lower lithosphere would have the effect of raising the surface height of the region of thickened crust by 1–3 km. Shackleton & Cheng (1988) argued that the Tibetan plateau was affected by relatively rapid uplift associated with crustal extension in the Pliocene, so that the topography of Tibet is now a dissected Neogene plateau. Sediment accumulation rates in the adjacent Tarim and Tsaidam Basins show a rapid influx of sediment in the Pliocene and Pleistocene, suggesting rapid uplift of the Tibetan Plateau at this time (Wang & Coward 1990). Thus the most important structures forming the topography of Tibet at the present day are extensional. Likewise, the dominant structures seen on seismic sections through older mountain belts are the late extensional faults which collapse the orogenic belt and smooth out thickness variations in the lower crust.

CONCLUDING COMMENTS

The strength of the lithosphere controls the spacing and distribution of crustal-scale thrusts during continental collision, as well as the size and depth of associated flexural basins. The strength of the basal or 'sole' thrust to the overriding thrust wedge controls the distribution of small-scale thrusts, folds and normal faults within the overthrust wedge. In general, the volume of thickened crust grows with continued plate convergence and hence the dominant structures must involve footwall collapse, that is, foreland propagation. However, new thrusts could develop within the thrust wedge to maintain critical wedge taper, and normal faults form from changes in plate convergence rate and direction, as well as changes to the basal shear strength of the thrust wedge. Thrust propagation or movement may be locally hindered causing rotations of the overlying thrust sheets. See for example the suggested transport paths for the western Himalayas in Pakistan (Fig. 13.32). Thus it is highly unlikely that thrust and fold kinematics within a small zone of continental collision will bear a close resemblance to regional plate kinematics. Continental collision needs to be examined and synthesized on a large scale and requires an integration of stratigraphic, structural and geophysical techniques; no one discipline can give the full solution.

REFERENCES

Achache, J., Courtillot, V. & Xiu, Y. 1984. Paleomagnetic and tectonic evolution of southern Tibet since the middle Cretaceous time: new paleomagnetic data and synthesis. *J. geophys. Res.* **89**, 10311–10339.

Ampferer, O. 1906. Uber das Bewegungbild von Faltengebirgen, Austria. *Jahrb. geol. Bundesanst* **56**, 539–622.

Argand, E. 1924. La tectonique de l'Aisie. *13th Int. Geol. Congr.* **1**, 170–372.

Bally, A. W. & Snelson, S. 1980. Realms of subsidence. In: *Facts and Principles of World Oil Occurrence. Mem. Can. Soc. Petrol. Geol.* **6**.

Bally, A. W., Gordy, P. L. & Stewart, G. A. 1966. Structure, seismic data, and orogenic evolution of Southern Canadian Rocky Mountains. *Bull. Can. Petrol. Geol.* **14**, 337–381.

Barazangi, M. & Ni, J. 1982. Velocities and propagation characteristics on Pn and Sn beneath the Himalayan arc and the Tibetan plateau: possible evidence of overthrusting of Indian continental lithosphere beneath Tibet. *Geology* **10**, 179–185.

Beaumont, C., Keen, C. A. & Boutilier, R. 1982. A comparison of foreland and rift margin sedimentary basins. *Phil. Trans. R. Soc. Lond.* **A305**, 295–318.

Besse, J. & Courtillot, V. 1988. Paleogeographic maps of the continents bordering the Indian Ocean since the Early Jurassic. *J. geophys. Res.* **93**, 11791–11808.

Bird, P. 1978. Initiation of intracontinental subduction in the Himalaya. *J. geophys. Res.* **83**, 4975–4987.

Bois, C., Cazes, M., Hirn, A., Mascle, A., Matte, P., Montadert, L. & Pinet, B. 1988. Contribution of deep seismic profiling to the knowledge of the lower crust in France and neighbouring areas. *Tectonophysics* **145**, 253–275.

Boyer, S. & Elliott, D. 1982. Thrust systems. *Bull. Am. Ass. Petrol. Geol.* **66**, 1196–1230.

Brewer, J. A. & Smythe, D. K. 1984. MOIST and the continuity of crustal reflector geometry along the Caledonian-Appalachian orogeny. *J. geol. Soc. Lond.* **141**, 105–120.

Burbank, D. W. & Raynolds, R. G. H. 1984. Sequential late Cenozoic disruption of the northern Himalayan foredeep. *Nature* **311**, 114–118.

Burchfiel, B. C. & Royden, L. 1982. Carpathian fold and thrust belt and its relation to the Pannonian and other basins. *Bull. Am. Ass. Petrol. Geol.* **66**, 1179–1195.

Burg, J. P. & Chen, G. M. 1984. Tectonics and structural zonation of southern Tibet, China. *Nature* **311**, 219–223.

Butler, R. W. H. 1985. The restoration of thrust systems and displacement continuity around the Mount Blanc massif, NW external Alpine thrust belt. *J. Struct. Geol.* **7**, 569–582.

Butler, R. W. H. 1986. Thrust tectonics, deep structure and crustal subductions in the Alps and Himalaya. *J. geol. Soc. Lond.* **143**, 857–873.

Butler, R. W. H. & Coward, M. P. 1989. Crustal scale thrusting and continental subduction during Himalayan collision tectonics of the NW Indian Plate. In: *Tectonic Evolution of the Tethyan Region* (edited by Sengör, A. M. C.). *NATO ASI Ser.* C **259**, 387–413.

Butler, R. W. H., Coward, M. P., Harwood, G. M. & Knipe, R. J. 1987. Salt control on thrust geometry, structural style and gravitational collapse along the Himalayan Mountain Front in the Salt Ranges of Northern Pakistan. In: *Dynamic Geology of Salt and Related Structures* (edited by Lerche, I. & O'Brien, J. J.). Academic Press, London, 399–417.

Chang Chengfa et al. 1986. Preliminary conclusions of the Royal Society and Academia Sinica 1985 geotraverse of Tibet. *Nature* **323**, 501–507.

Chapple, W. M. 1978. Mechanics of thin-skinned fold and thrust belts. *Bull. geol. Soc. Am.* **89**, 1189–1198.

Cheadle, M. J., McGeary, S., Warner, M. R. & Matthews, D. H. 1987. Extensional structures on the western UK continental shelf: a review of evidence from deep seismic profiling. In: *Continental Extensional Tectonics* (edited by Coward, M. P., Dewey, J. F. & Hancock, P. L.). *Spec. Publs geol. Soc. Lond.* **28**, 445–466.

Coward, M. P. 1982. Surge zones in the Moine thrust zone of NW Scotland. *J. Struct. Geol.* **4**, 247–256.

Coward, M. P. 1983. Thrust tectonics, thin skinned or thick skinned and the continuation of thrusts to deep in the crust. *J. Struct. Geol.* **5**, 113–123.

Coward, M. P. 1986. Heterogeneous stretching and basin development. *Earth Planet. Sci. Lett.* **80**, 325–336.

Coward, M. P. 1990. The Precambrian, Caledonian and Variscan framework to NW Europe. In: *Tectonic Events Responsible for Britain's Oil and Gas Reserves* (edited by Hardman, R. F. P. & Brooks, J.). *Spec. Publs geol. Soc. Lond.* **55**, 1–34.

Coward, M. P. & Butler, R. W. H. 1985. Thrust tectonics and the deep structure of the Pakistan Himalaya. *Geology* **13**, 417–420.

Coward, M. P., Butler, R. W. H., Chambers, A. F., Graham, R. H., Izatt, C. N., Khan, M. A., Knipe, R. J., Prior, D. J., Treloar, P. J. & Williams, M. P. 1988. Folding and imbrication of the Indian crust during Himalayan collision. *Phil. Trans. R. Soc. Lond.* **A326**, 89–116.

Coward, M. P., Kidd, W. S. F., Pan Yung, Shackleton, R. M. & Zhang Hu. 1988. The structure of the Tibet Geotraverse, Lhasa to Golmud. *Phil. Trans. R. Soc. Lond.* **A327**, 307–336.

Cross, T. A. 1986. Tectonic controls of foreland basin subsidence and Laramide style deformation, western United States. In: *Foreland Basins* (edited by Allen, P. A. & Homewood, P.). *Spec. Publs int. Ass. Sediment* **8**, 15–39.

Dahlen, F. A. 1984. Noncohesive critical Coulomb wedges: an exact solution. *J. geophys. Res.* **89**, 10125–10133.

Dahlen, F. A. & Suppe, J. 1988. Mechanics, growth and erosion of mountain belts. *Spec. Pap. geol. Soc. Am.* **218**, 161–178.

Davis, D. & Engelder, T. 1985. The role of salt in fold and thrust belts. *Tectonophysics* **119**, 67–88.

Davis, D., Suppe, J. & Dahlen, F. A. 1983. Mechanics of fold and thrust belts and accretionary wedges. *J. geophys. Res.* **88**, 1153–1172.

Dewey, J. F. 1982. Plate tectonics and the evolution of the British Isles. *J. geol. Soc. Lond.* **139**, 371–412.

Dewey, J. F. & Bird, J. M. 1971. Origin and emplacement of the ophiolite suite: Appalachian ophiolites in Newfoundland. *J. geophys. Res.* **76**, 3179–3206.

Dewey, J. F., Shackleton, R. M., Chang Cheng Fa & Sun Yujin. 1988. The tectonic evolution of the Tibetan plateau. *Phil. Trans. R. Soc. Lond. Ser.* **A327**, 379–413.

Dewey, J. F., Cande, S. & Pitman, W. C. 1989. Tectonic evolution of the India-Eurasia collision zone. *Ecol. geol. Helv.* **82**, 717–734.

Elliott, D. 1976. The motion of thrust sheets. *J. geophys. Res.* **81**, 949–963.

England, P. & Houseman, G. 1986. Finite strain calculations of continental deformation. 2. Comparison with the India–Asia collision zone. *J. geophys. Res.* **91**, 3664–3676.

England, P. C. & Houseman, G. A. 1988. The mechanics of the Tibetan Plateau. *Phil. Trans. R. Soc. Lond.* **A326**, 301–320.

England, P. & McKenzie, D. 1982. A thin viscous sheet model for continental deformation. *Geophys. J. R. astron. Soc.* **70**, 295–296.

England, P. & Thompson, A. B. 1984. Pressure–temperature–time paths of regional metamorphism. 1. Heat transfer during the evolution of regions of thickened continental crust. *J. Petrol.* **25**, 894–928.

Hirn, A. et al. 1984. Lhasa block and bordering sutures: a continuation of a 500 km Moho traverse through Tibet. *Nature* **307**, 25–27.

Homewood, P., Allen, P. A. & Williams, G. D. 1986. Dynamics of the Molasse Basin of western Switzerland. In: *Foreland Basins* (edited by Allen, P. A. & Homewood, G. D). *Spec. Publs int. Ass. Sediment.* **8**, 199–217.

Houseman, G. A., McKenzie, D. P. & Molnar, P. 1981. Convective instability of a thickened boundary layer and its relevance for the thermal evolution of continental convergent belts. *J. geophys. Res.* **86**, 6115–6135.

Hubbert, M. K. & Rubey, W. W. 1959. Role of fluid pressure in mechanics of overthrust faulting. *Bull. geol. Soc. Am.* **70**, 115–166.

Jadoon, I. A. K., Lawrence, R. D. & Lillie, R. J. In press. The Sulaiman Lobe, Pakistan: Geometry, evolution, and shortening of the active fold-and-thrust belt over transitional crust west of the Himalaya. *Bull. Am. Ass. Petrol. Geol.*

Karner, G. D. & Watts, A. B. 1983. Gravity anomalies and flexure of the lithosphere at mountain ranges. *J. geophys Res.* **88**, 10449–10477.

Le Pichon, X., Angelier, J. & Sibuet, J. C. 1982. Plate boundaries and extensional tectonics. *Tectonophysics* **81**, 239–256.

Le Pichon, X., Fournier, M. & Jolvet, L. In press. Kinematics, topography, shortening and extrusion in the India–Asia collision. *Tectonics*.

Lyon-Caen, H. & Molnar, P. 1983. Constraints on the structure of the Himalaya from an analysis of gravity anomalies and a flexural model of the lithosphere. *J. geophys. Res.* **88**, 8171–8191.

Lyon-Caen, H. & Molnar, P. 1985. Gravity anomalies, flexure of the Indian plate, and the structure, support and evolution of the Himalaya and Ganga basins. *Tectonics* **4**, 513–538.

Molnar, P. 1988. A review of the geophysical constraints on the deep structure of the Tibetan plateau, the Himalaya and the Karakorum and their tectonic implications. *Phil. Trans. R. Soc. Lond.* **A326**, 33–88.

Molnar, P. & Lyon-Caen, H. 1988. Some simple physical aspects of the support, structure, and evolution of mountain belts. *Spec. Pap. geol. Soc. Am.* **218**, 179–207.

Molnar, P. & Tapponnier, P. 1975. Cenozoic tectonics of Asia: effects of a continental collision. *Science* **189**, 419–426.

Molnar, P. & Tapponnier, P. 1977. The relation of the tectonics of eastern Asia to the India–Eurasia collision: an application of slip line field theory to large scale continental tectonics. *Geology* **5**, 212–216.

Molnar, P. & Tapponnier, P. 1978. Active tectonics of Tibet. *J. geophys. Res.* **83**, 5361–5375.

Molnar, P. & Tapponnier, P. 1979. Active faulting and Cenozoic tectonics of the Tien Shan, Mongolia and Bayka regions. *J. geophys. Res.* **84**, 3425–3459.

Molnar, P., Chen, W.-P., Fitch, T. J., Tapponnier, P., Warsi, W. E. K. & Wu, F. T. 1977. Structure and tectonics of the Himalaya; a brief summary of the relevant geophysical observations. *Edit. Centre Nat. Rech. Scient. Sci. Terre Paris*, 267–294.

Ni, J. & Barazangi, M. 1984. Seismotectonics of the Himalayan collision zone: geometry of the underthrusting Indian Plate beneath the Himalaya. *J. geophys. Res.* **89**, 1147–1163.

Patriat, P. & Achache, J. 1984. India–Asia collision chronology has implications for crustal shortening and driving mechanisms of plates. *Nature* **311**, 615–621.

Platt, J. P. 1986. Dynamics of orogenic wedges and the uplift of high-pressure metamorphic rocks. *Bull. geol. Soc. Am.* **97**, 1037–1053.

Platt, J. P. & Lister, G. S. 1985. Structural history of high pressure metamorphic rocks in the Vanoise Massif, French Alps, and their relation to Alpine tectonic events. *J. Struct. Geol.* **7**, 19–36.

Price, R. A. 1981. The Cordilleran foreland thrust and fold belt in the southern Canadian Rocky Mountains. In:

Thrust and Nappe Tectonics (edited by McClay, K. R. & Price, N. J.). *Spec. Publs geol. Soc. Lond.* **9**, 427–448.

Ramsay, J. G. 1967. *Folding and Fracturing of Rocks.* McGraw Hill, New York.

Roeder, D. 1989. South-Alpine thrusting and trans-Alpine convergence. In: *Alpine Tectonics* (edited by Coward, M. P., Dietrich, D. & Park, R. G.). *Spec. Publs geol. Soc. Lond.* **45**, 211–228.

Royden, L. & Burchfiel, B. C. 1987. Thin-skinned N–S extension within the convergent Himalayan region: gravitational collapse of a Miocene topographic front. In: *Continental Extensional Tectonics* (edited by Coward, M. P., Dewey, J. F. & Hancock, P. L.). *Spec. Publs geol. Soc. Lond.* **28**, 611–619.

Royden, L. & Karner, G. D. 1984. Flexure of the lithosphere beneath Appalachian and Carpathian foredeep basins; evidence for an insufficient topographic load. *Bull. Am. Ass. Petrol. Geol.* **68**, 704–712.

Searle, M. P., Windley, B. F., Coward, M. P., Cooper, D. J. W., Rex, A. J., Rex, D. C., Tindong, L., Xuchang, X., Jan, M. Q., Thakur, V. C. & Kumar, S. 1987. The closing of Tethys and the tectonics of the Himalaya. *Bull. geol. Soc. Am.* **98**, 678–701.

Seeber, L. & Armbruster, J. 1979. Seismicity of the Hazara arc in northern Pakistan: decollement vs. basement faulting. In: *Geodynamics of Pakistan* (edited by Farah, A. & De Jong, K. A.). *Geol. Surv. Pak.* 131–142.

Seeber, L., Armbruster, J. & Quittmeyer, R. C. 1981. Seismicity and continental subduction in the Himalayan Arc. In: *Zagros, Hindu Kush, Himalaya, Geodynamic Evolution. Geodynamics Series 3* (edited by Gupta, H. K. & Delany, F. M.). *Am. geophys. U. Washington*, 215–242.

Shackleton, R. M. 1969. Displacement within continents: introductory remarks. In: *Time and Place in Orogeny* (edited by Kent, P. E., Satterthwaite, G. E. & Spencer, A. M.). *Spec. Publs geol. Soc. Lond.* **3**, 1–7.

Shackleton, R. M. & Chang Chengfa, 1988. Cenozoic uplift and deformation of the Tibetan Plateau: the geomorphic evidence. *Phil. Trans. R. Soc. Lond.* **A327**, 365–377.

Sonder, L. J., England, P. C., Wernicke, B. P. & Christiansen, R. L. 1987. A physical model for Cenozoic extension of western North America. In: *Continental Extensional Tectonics* (edited by Coward, M. P., Dewey, J. F. & Hancock, P. L.). *Spec. Publs geol. Soc. Lond.* **28**, 187–202.

Tapponnier, P. & Molnar, P. 1976. Slip-line field theory and large scale continental tectonics. *Nature* **264**, 319–324.

Tapponnier, P. & Molnar, P. 1977. Active faulting and tectonics in China. *J. geophys. Res.* **82**, 2905–2930.

Tapponnier, P. & Molnar, P. 1979. Active faulting and Cenozoic tectonics of the Tien Shan, Mongolia and Baikal regions. *J. geophys. Res.* **84**, 3425–3459.

Tapponnier, P., Pelzer, G. & Armijo, R. 1986. On the mechanism of collision between India and Asia. In: *Collision Tectonics* (edited by Coward, M. P. & Ries, A. C.). *Spec. Publs geol. Soc. Lond.* **19**, 115–157.

Tomek, C., Dvorakova, L., Imbrmajer, I., Jiricek, R. & Korab, T. 1987. Crustal profiles of active continental collisional belt: Czeckoslovak deep seismic reflection profiling in the West Carpathians. *Geophys. J.R. astron. Soc.* **89**, 383–388.

Treloar, P. J. & Coward, M. P. 1991. Indian plate motion and shape: constraints on the geometry of the Himalayan orogen. *Tectonophysics* **191**, 189–198.

Treloar, P. J., Potts, G. J., Wheeler, J. & Rex, D. C. 1991. Structural evolution and asymmetric uplift of the Nanga Parbat syntaxis, Pakistan Himalaya. *Geol. Rdsch.* **80**, 411–428.

Treloar, P. J., Coward, M. P., Chambers, A. F., Izatt, C. N. & Jackson, K. C. 1992. Thrust geometries, interferences and rotations in the Northwest Himalaya. In: *Thrust Tectonics* (edited by McClay, K. R.). Chapman & Hall, London, 325–342.

Turcotte, D. L. & Schubert, G. 1982. *Geodynamics; Applications of Continuum Mechanics to Geological Problems.* Wiley, New York.

Vendeville, B., Cobbold, P. R., Davy, P., Brun, J. P. & Choukroune, P. 1987. Physical models of extensional tectonics at various scales. In: *Continental Extensional Tectonics* (edited by Coward, M. P., Dewey, J. F. & Hancock, P. L.). *Spec. Publs geol. Soc. Lond.* **28**, 95–107.

Walcott, R. I. 1970. Flexural rigidity, thickness and viscosity of the lithosphere. *J. geophys. Res.* **75**, 3941–3954.

Wang, Q. M. & Coward, M. P. 1990. The Chaidam Basin (NW China): Formation and hydrocarbon potential. *J. Petrol. Geol.* **13**, 93–112.

Zeitler, P. K. 1985. Cooling history of the NW Himalaya, Pakistan. *Tectonophysics* **4**, 127–151.

Zhao, W. L. & Morgan, W. J. 1985. Uplift of Tibetan Plateau. *Tectonics* **4**, 359–369.

CHAPTER 14

Inversion Tectonics

MIKE COWARD

INTRODUCTION *

THE presence of early synsedimentary faults is now being recognized in many orogenic belts and it is known that much intra-continental deformation within the crust is accommodated by the reactivation of pre-existing structures. This is particularly true in areas where compression and uplift affect older basins or ocean margins. Some of the better-known examples of basin inversion and fault reactivation have been described from commercial seismic data in areas which are just below sea-level or are of low relief (Harding 1983, Zeigler 1983, 1987). However, in collisional mountain belts (see Chapter 13), such as the Alps or Apennines, where shortening and uplift are more intense, the reactivation of pre-existing fault systems is extremely important (Gillcrist et al. 1987, Hayward & Graham 1989).

In this chapter the term *inversion* is used to describe regions which have experienced a reversal in uplift or subsidence, that is, areas which have changed from being regions of subsidence to regions of uplift, or vice versa. An area which has changed from subsidence to uplift can be considered to have been affected by *positive inversion*. An area which has changed from uplift to subsidence has undergone *negative inversion* (see also Harding 1983). Most of this chapter deals with aspects of positive inversion.

The misinterpretation of folds and thrusts as being related to thin-skinned shortening rather than the inversion of a sedimentary basin can have far reaching effects on the structural interpretation of a region, leading to: (i) use of the wrong method in section construction; (ii) incorrect calculations of the amount of orogenic shortening, and (iii) incorrect assumptions about the nature of structures at depth, both directly beneath the fold/thrust belt and further back within the hinterland of the mountain belt.

The recognition of inversion tectonics is even more important in the oil exploration industry than in academia as: (i) inversion will modify the burial history of a sedimentary basin, complicating calculations of the timing of maturation and oil generation; (ii) inversion can uplift sediments above sea-level generating a secondary porosity and/or karstification; (iii) inversion can modify the tilt of a sedimentary package, allowing different directions of fluid migration with time; (iv) inversion can reactivate older faults, changing their sealing properties and sometimes repumping fluids around the basin; and (v) inversion will form complex structures at depth and care needs to be taken to differentiate inversion structures, which reactivate moderate to steeply dipping faults, from thin-skinned thrust structures.

Figure 14.1 shows a line drawing through part of the South Celtic Sea, a Mesozoic basin which was inverted during the Early Cretaceous and Tertiary. It shows several important features of basin inversion. Note the *truncation of reflectors* at the base of the Cretaceous (Albian/Aptian) and within the Tertiary. Note also the large wavelength folds in the Cretaceous and the smaller scale folds in the Jurassic sediments in the hangingwall of one of the faults. It is also important to know how much regional, large wavelength inversion occurred and how much sediment has been stripped from the section in order to be able to produce valid basin models of the region.

ORIGINS OF INVERSION

Related to buoyancy or isostatic forces

Related to isostatic rebound. Isostatic inversion can be caused by the removal of glacial overburden or the erosion of a mountain belt. If the lithosphere has flexural strength then the regions surrounding the mountain belt, or the old ice cap, will undergo flexural uplift, decreasing away from the area of removal (Fig. 14.2). Typical examples include the uplift of molasse basins adjacent to the French and Austrian Alps, and the uplift of Palaeogene/Neogene basins on the Tibet Plateau (Coward et al. 1988).

Related to the diapiric action of salt. The obvious example of *diapiric inversion* is the salt pillow or diapir (see Chapter 8). Local positive and negative inversion can occur during burial by the movement of salt up or down the dip of a tilted fault block (Fig. 14.3). Figure 14.4 shows a seismic section from offshore Angola, where successive sedimentary packages show different onlap directions recording the migration history of the salt at depth (Fig. 14.5).

---

* This Chapter should be read after Chapters 6, 11, 12 and 13.

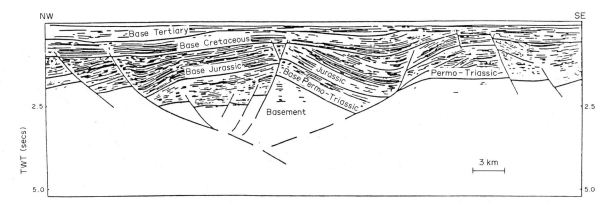

Fig. 14.1. Example of inversion tectonics from the South Celtic Sea. TWT, two-way travel time; PTR, Permo-Triassic.

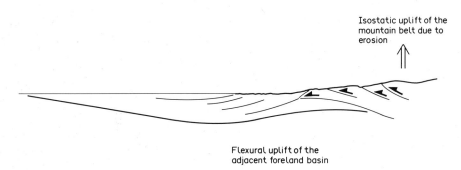

Fig. 14.2. Sketch section to illustrate inversion of a foreland basin associated with the unroofing of a mountain belt. Arrows indicate thrusts.

Inversion structures related to salt tectonics are likely to be small in scale and easy to identify from their shape, size, location and their relationship to evaporite layers. The inversion will affect only the rocks above the salt; the pre-salt succession will be unaffected.

Related to lithospheric heating. Where plates migrate across hot spots they can undergo surface uplift of 1–2 km. The resulting *thermal domes* may be up to 2000 km across, that is the wavelength of the inversion structures is very large. Where the uplift is associated with fracturing and volcanism, *magmatic underplating* can occur causing prolonged uplift. During *thermal cooling* of the lithosphere the region previously overlying the hot spot will subside. However magmatic underplating thickens the crust and hence maintains the uplift following lithospheric cooling.

An example of regional inversion related to hot spot development is that associated with Palaeocene magmatic activity in the North Atlantic (Fig. 14.6). Igneous activity along a NW–SE trend associated with this hot spot caused underplating and uplift of the western margin of the British Isles. Similarly the Mid-North Sea High, which separates the Northern and Southern Permian basins in the North Sea, overlies a region of Permo-Carboniferous intrusions and may have maintained uplift during the Permo-Triassic due to igneous underplating.

Related to extensional faulting. Where there has been heterogeneous stretching of the lithosphere and hence *extension-related inversion*, *footwall uplift* of rotated fault blocks and rift-flank uplift at the margins of large

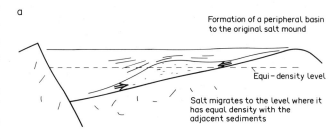

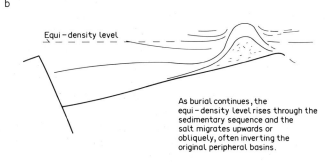

Fig. 14.3. (a) Migration of salt causing inversion of minor basins. The salt will migrate up or down slope towards the level of equal density with adjacent sediments. For a pile of average sediments (e.g. silty sands) this depth is c. 600 m (Roberts *et al*. 1989). (b) Along the flanks of a basin or half-graben, the salt will migrate up or down slope to this level and then, with continued burial, will gradually either rise diapirically to form a salt wall or pillar or migrate up slope. As the salt migrates up-slope it will locally invert small basins that formed peripheral to the earlier salt mound.

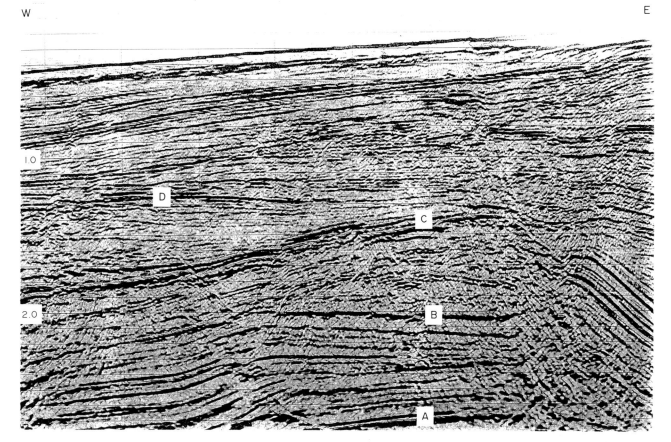

Fig. 14.4. Seismic section from offshore Angola (courtesy of Conoco Angola Ltd) showing variable uplift/subsidence related to salt migration at depth. Reflector A marks the top of the evaporite sequence. Reflector B marks the top of the post-salt sedimentary package deposited before salt migration. Note the onlaps and thickening of this sedimentary package to the west, away from a palaeo-high possibly formed above a tilted fault block in the basement. Reflector C marks the top of the sedimentary package formed by salt withdrawal from a palaeo-high. Reflector D marks the top of a package of sediments formed in a peripheral basin to a salt mound formed by salt migration back to the palaeo-high. Above reflector D, the sediments again thicken towards the palaeo-high recording subsequent migration of the salt during listric faulting. The dipping reflectors in the eastern part of the section (right) record part of the rollover fold on to a listric fault formed during the post-D evacuation of the salt. Vertical scale in 0.1 seconds (two-way travel time). Horizontal scale approximately equal to vertical scale.

grabens may occur (Fig. 14.7) (Keen 1987) (see Chapter 11). Stretching, involving offset of upper and lower lithospheric extension, causes uplift above regions of lower plate stretch (see Coward 1986) (Fig. 14.8). Where the lithosphere has flexural strength during stretching, footwall uplift will occur (Kusznir et al. 1991). Examples occur along the flanks of most grabens, for example, along the flanks of the Gulf of Suez and the present-day Red Sea, and along the flanks of the North Sea in the late Jurassic and Cretaceous. Figure 14.9 shows a seismic section with tilting and positive inversion of the western part of the Viking graben in the northern North Sea.

Related to horizontal plate movements — tectonic inversion

Related to changes in plate motion causing collisional processes. Plate collision, especially oblique plate collision, is one of the most common causes of tectonic inversion. Almost all zones of continental collision show some evidence of *collision-related inversion tectonics* as well as thin-skinned thrusting (see Chapter 6). Typical examples include the Alpine inversion of the Dauphinois basin in the Western Alps and the closure of the Variscan basin in Northwest Europe. Figure 14.10 shows a section through the Western Alps (Coward et al. 1991) to illustrate the Cenozoic uplift, tightening and local thrust reactivation of a Laiassic half-graben above the Belledonne and Pelvoux basement massifs. Most of the basement massifs in the Western Alps are uplifted Mesozoic extensional fault blocks (Gillcrist et al. 1987). Thin-skinned thrust tectonics affect the post-rift sedimentary sequence (Middle Jurassic–Upper Cretaceous) while more steeply dipping faults involving basement rocks affect the deeper part of the section.

Inversion related to changes of relative plate motion are not confined to areas of continent/continent collision. Large-scale inversion structures occur along most of the NW Pacific margin, related to closure of Palaeogene back-arc basins. They are particularly prominant in the East China Sea and can also be traced

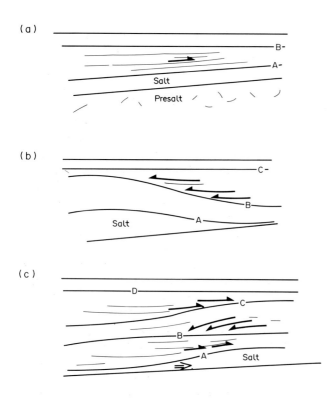

Fig. 14.5. Simplified subsidence history of the region shown in Fig. 14.4. See text for details.

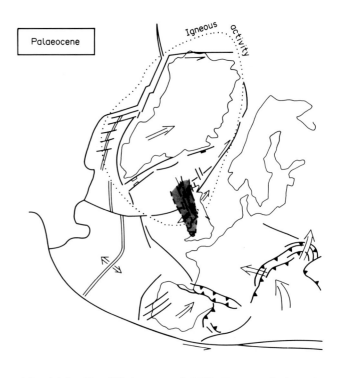

Fig. 14.6. Simplified map of NW Europe during the Palaeocene showing the region of igneous activity overlying the mantle hot spot (after White & McKenzie 1989). The NW-trending lines over Britain represent the trends of dyke intrusions associated with several igneous centres in and offshore western Scotland. These dykes and their associated faults indicate NE–SW extension at this time. The shading shows the region of inversion probably associated with underplating caused by deep intrusions.

south to Taiwan and north to the Sea of Japan.

Related to strike-slip tectonics. Strike-slip related inversion can occur at the restraining bends or offsets along major *transcurrent faults* (see Chapter 12). Examples include (1) the Permo-Carboniferous inversion of the Orcadian basin, northern Scotland, related to right-lateral movement along the Great Glen fault, and (2) inversion of the Rhone graben in the southern Baltic Sea by left-lateral movement along the Tornquist line.

Related to rotational block faulting. Shortening can occur across the block between reactivated bounding faults (Fig. 14.11). If the boundary conditions are fixed so that there is no lateral expansion or contraction, and if the original extensional faults rotate so that they eventually lie closer to the orientation of the shear couple, the blocks will need to narrow and lengthen (Fig. 14.11). If the blocks rotate so that the original extensional faults rotate away from the shear system, the blocks will narrow and shorten. The amount of

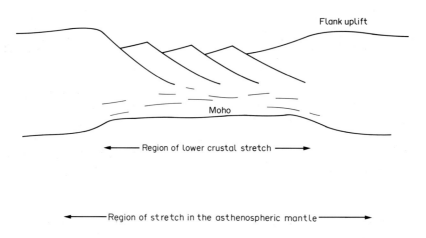

Fig. 14.7. Sketch illustrating flank uplift at the edges of an extensional graben.

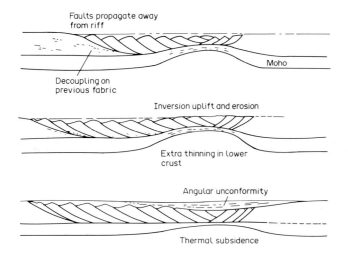

Fig. 14.8. Sketch sections illustrating heterogeneous stretching of the lithosphere, with a wide zone of extension in the upper crust and a narrow zone of extension in the lower crust and upper mantle. Less subsidence or even uplift occurs above the region of lower lithospheric stretching followed by more extreme thermal subsidence (after Coward 1986).

block narrowing or *widening* can be calculated from the relationship:

$$l_1/l_0 = \sin(\alpha - \psi)/\sin \alpha$$

where α is the original angle made by the faults, relative to the plane of the regional shear and ω is the angle of rotation (Fig. 14.11). This relationship is summarized in graphical form in Fig. 14.12.

Thus a diffuse shear system which rotates older crustal blocks can cause extension or *rotational-block-faulting related inversion*. The inversion can occur distant from the plate collisional zone and may not be directly related to the continent/continent collision. Examples include the inverted Carboniferous fault blocks in the southern North Sea (Coward 1992). Coward suggested that during the Early Carboniferous a NE-trending right-lateral shear couple affected the Caledonian structures of southern Britain, causing rotation from a WNW trend to a NW trend, widening the blocks and allowing stretching in the order of $\beta = 1.1 - 1.15$. During the Late Carboniferous, the blocks were rotated by a similar trending left-lateral shear couple, causing tightening across the blocks and inversion of the half-grabens. A similar model can be used to explain the Variscan inversion of the Palaeozoic basins in South Wales, where the originally E–W

Fig. 14.9. Seismic line illustrating Cretaceous uplift on the western flanks of the Viking graben, in the vicinity of the Tern Field (courtesy of Nopec). Vertical scale in seconds (two-way travel time). Note tilt and truncation of the pre-rift sediments (Upper-Palaeozoic-Jurassic) and onlap of the post-rift sediments (Cretaceous).

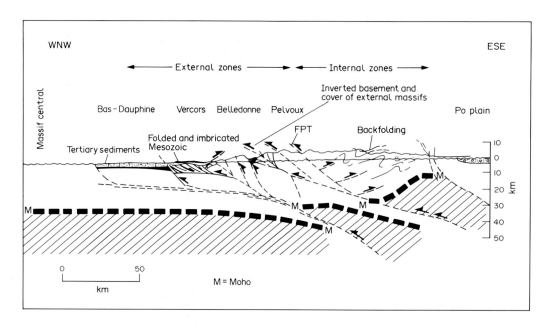

Fig. 14.10. A simplified section through the Western Alps showing the uplifted extensional fault blocks of the Belledonne and Pelvoux Massifs (after Coward et al. 1991).

trending fault blocks were rotated by 30–40° during the Variscan orogeny, associated with a NW-trending right-lateral shear couple. Devonian and Carboniferous syn-rift and post-rift sediments were expelled from the half-grabens in a series of folds and thrusts. Similarly, the Palmyrides in Syria were uplifted by inversion tectonics with c. 20° anti-clockwise rotation associated with a major N–S trending left-lateral shear system linked to the opening of the Red Sea. Block rotation is probably most dramatic in southern California, where originally N–S trending fault blocks have been rotated through 90° by a shear couple associated with the San Andreas fault zone. The basins initially opened as the fault blocks rotated towards 90° to the trend of the shear couple and then closed with subsequent additional rotation, producing uplift and thrust zones such as the Transverse Ranges.

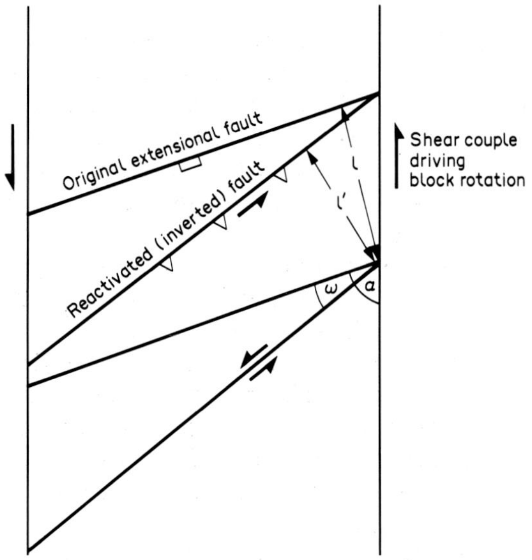

CONTROLS ON THE GEOMETRY OF TECTONIC INVERSION

Strength of the crust

A possible *strength profile* through the crust and lithospheric mantle is shown in Fig. 14.13(a), assuming simple rheologies dependent on the presence of

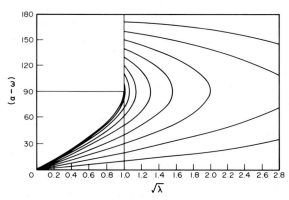

Fig. 14.11. Model showing the strains across fault blocks deformed by a shear couple. See text for details.

Fig. 14.12. Graphical solution to Fig. 14.11 showing the shortening or extension ($\sqrt{\lambda}$) across a block related to the initial orientation of the block relative to the simple shear (α) and the amount of rotation (ω). Depending on the orientation of the blocks, the blocks will widen ($\sqrt{\lambda}>1$) resulting in additional extension, or narrow ($\sqrt{\lambda}<1$) resulting in inversion. Note that to maintain constant surface area, a corresponding change in length ($=1/\sqrt{\lambda}$) should occur along the long axes of the blocks, unless the tectonic situation involves regional extension or contraction.

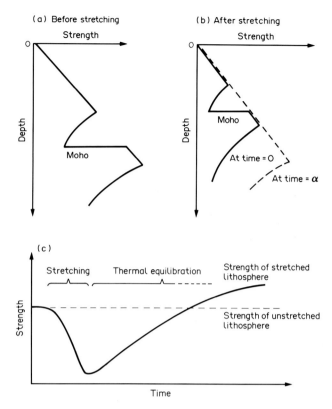

Fig. 14.13. (a) Strength/depth profile through the lithosphere before stretching. (b) Strength/depth profile following stretching at time $t = 0$ (solid line) and at time $t = \alpha$, after thermal equilibration (pecked line). (c) Strength/time profile for homogeneously stretched lithosphere.

particular minerals, such as wet quartz in the upper crust, and olivine in the mantle (also see Chapter 10). The upper part of the crust will behave in a brittle manner, the strength increasing with confining pressure and depth. In the lower crust, ductile deformation will dominate and the strength will decrease with increasing temperature with depth. The local strength may be controlled by a variety of minerals and impurities and hence the curves will be more complex. However, low strength regions will occur in the lower crust. The *lithospheric strength* is critically controlled by lithospheric thickness; thin lithosphere is strong, thick lithosphere is weak. Hence stretching initially weakens the lithosphere but after temperature equilibration, stretched lithosphere becomes stronger than unstretched lithosphere (Figs 13b & c).

If compression occurs soon after stretching, then the stretched lithosphere will invert easily. However, if there is a long time gap (e.g. >100 Ma) between stretching and compression, then the stretched basins and continental margins will be stronger than the adjacent unstretched lithosphere. This pattern is well developed in the Western Alps where the Dauphinois basin shows examples of inversion during the Eocene but acted as a more rigid mass during the Oligo-Miocene.

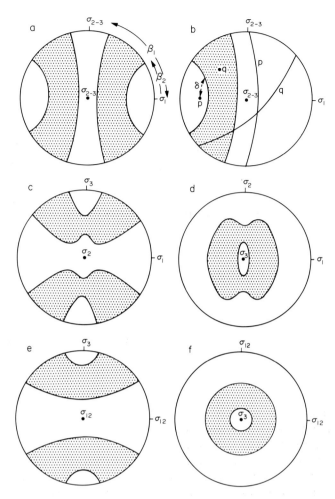

Fig. 14.14. Stereographic diagrams to show regions where poles to planes should plot (shaded) to allow failure instead of new fractures developing in the solid rock (a) and (b) for stresses $\sigma_1 > \sigma_2 = \sigma_3$. (c) and (d) for stresses $\sigma_1 > \sigma_2 > \sigma_3$. (e) and (f) for $\sigma_1 = \sigma_2 > \sigma_3$ (from Gillcrist *et al.* 1987).

Fault reactivation

The capability of a normal fault to reactivate largely depends on its orientation relative to the principal compressive stress (Fig. 14.14) (see Chapter 5). Steeply dipping normal faults are difficult to reactivate by full frontal inversion, unless they have a very low cohesion and low sliding friction. Gently dipping normal faults are easy to invert. It is possible to reactivate steeper faults by oblique or strike-slip movements especially where the minimum compressive stress is horizontal (Fig. 14.14e) (Gillcrist *et al.* 1987).

The normal faults, which rotated to gentle dips during extension, fail preferentially during inversion. In the case of a fault whose dip changes down-dip, that is, a *listric fault*, the flat part could fail during inversion, while fold or secondary fault structures form at the foot of the steeper ramp.

Assuming horizontal compression, then normal faults with dips of 40–60° would only fail if the coefficient of sliding friction was low. During the development of a thrust wedge, the maximum compressive stress would be inclined to the horizontal and hence the older normal faults would be at a greater angle to the compressive stress and thus would be less likely to fail. A pre-existing fault is likely to have a lower cohesive strength than the intact rock, but some fault gouges may be much stronger and the deposition of quartz on the fault plane may have a hardening effect.

GEOMETRY OF INVERTED NORMAL FAULTS

Where the original normal faults are *growth faults* then tectonic inversion can lead to a reversal in stratigraphic separation down the dip of the fault (Fig. 14.15). In the upper levels the net displacement is one of reverse fault geometry or compressive folding while at depth the net displacement may be one of extension. This pattern can often be observed both along the strike, and down the dip, of a fault (e.g. Williams *et al.* 1989) (Fig. 14.16), and can be analyzed using the *stratigraphic separation diagram*. The null point, that is, the position of no stratigraphic separation, can be mapped on sections and maps and the projection of the null line calculated in three dimensions.

It is possible to quantify the relative magnitudes of contractional and extensional movement measured in the plane of the section using the position of the null point in the syn-rift section (Williams *et al.* 1989). The *inversion ratio* (R_i) is defined as the ratio of contractional to extensional displacement and can be calculated from

$$R_i = d_c/d_h = 1 - d_e/d_h$$

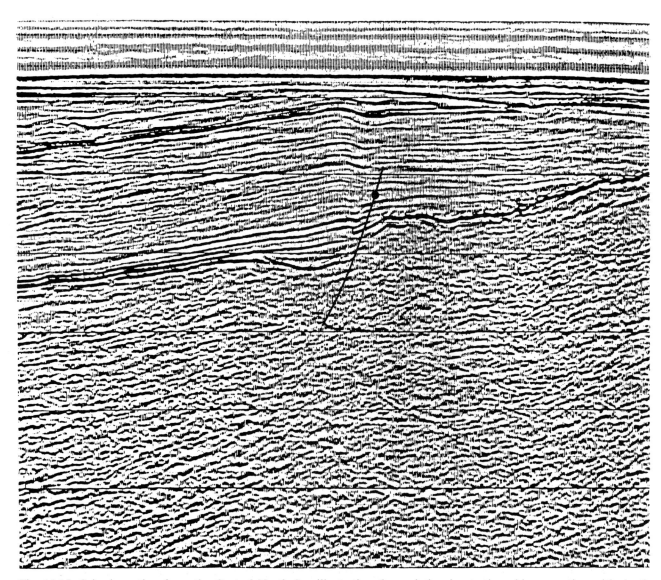

Fig. 14.15. Seismic section from the Central North Sea illustrating the variation in stratigraphic separation with depth associated with inversion of a normal fault system. Note that the high-level rocks (post-rift) show net contraction while the deeper-level rocks show net extension. Between the two zones there is a null point, a point of zero net displacement. The solid line shows the fault, the circle the null point. Following rifting and thermal subsidence, the null point lies at the top of the syn-rift sediments. During inversion the null point migrates down through the syn-rift. After total inversion all the beds, including the pre-rift sediments, show net contraction. From Biddle & Rudolph (1988).

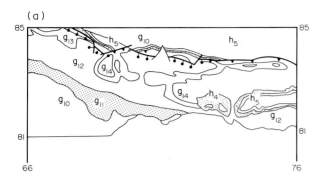

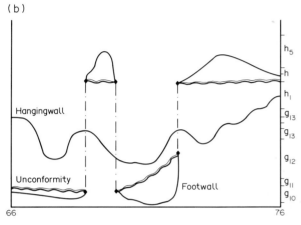

Fig. 14.16. (a) Simplified geological map of South Dorset near Weymouth, to show the variation in fault mode from normal to reverse as a result of inversion. $g_{10}-g_{14}$ refer to Upper Jurassic strata, while h_1 to h_5 refer to Lower Cretaceous strata. (b) Stratigraphic separation diagram constructed for this region. Note that some of the Upper Jurassic/Lower Cretaceous stratigraphy is missing as a result of inversion coupled with erosion. The missing stratigraphy is represented by dashed/dotted lines in the footwall curve (from Cooper et al. 1989). Numbers refer to U.K. National Grid Lines in kilometres.

where d_h is the thickness of the syn-rift sequence, d_c the thickness of the syn-rift sequence in contraction and d_e the thickness of the syn-rift sequence below the null point. Where $R_i = 1$, the null point is at the base of the syn-rift sequence and total inversion has occurred.

Where there is sedimentation during inversion, then the sediment package may thicken away from the fault wall, leading to pinch outs and onlaps towards the fault. Figure 14.17 shows an example from an inverted Devonian half-graben from the West Orkney basin, north of Scotland.

Often new folds or faults will form (Fig. 14.18). These structures commonly develop during inversion as a result of the decrease in bed length due to (i) reversal in displacement up the fault, and (ii) change in dip of the faults. Block faults may back-rotate during inversion until their dip reaches a critical value for reactivation. This back-rotation will be more pronounced if the displacements are oblique; if they are highly oblique to strike-slip motion then the faults could possibly rotate to almost vertical. During back-rotation the syn-rift fill will be squeezed out in a series of folds and thrusts. Figure 14.19 shows an array of folds generated above inverted normal faults from the southern North Sea.

During inversion, the post-rift cover will shorten into a series of folds or thrusts which may detach close to the syn-rift/post-rift boundary and may appear as thin-skinned detachment structures. Alternatively, the thrusts may steepen downwards to link with the earlier extensional faults.

In regions of inversion tectonics care needs to be taken with section construction and section balancing. For example, in regions of thin-skinned thrusting, the excess area (A in Fig. 14.20) can be used to find the *depth of detachment*, that is, the depth to detachment is given by $A/d1$ where $d1$ is the change in length. However, in a region of inversion tectonics the excess area records the amount of sediment deposited in the original half-graben. The excess area (A) increases as the half-graben inverts and the null point migrates down through the syn-rift sediments. Thus the excess area records the sediment fill of the syn-rift sediments following deposition of those sediments which have been inverted to reach the null point. After complete inversion and in some cases after partial inversion, when all the sediments show some shortening across the fault, the excess area records a combination of the amount of syn-rift sediment and the level to the regional depth to detachment.

Section restoration techniques require knowledge of the depth to the top of the syn-rift sediments and the rates of extensional fault growth in the syn-rift sequence. Techniques of section balancing using line-length restoration require knowledge of the rate of change of bed length with depth in the syn-rift sequence.

Listric faults will reactivate easily at depth, but the steeper parts of the faults will not reactivate leading to the development of folds on the hangingwalls and *'short-cut' faults* on the hangingwalls or footwalls (Fig. 14.21). These footwall faults commonly lead to the development of *'floating islands'* of pre-rift material, bounded by the original normal fault and by the short-cut. The short-cuts may splay into imbricate fans with fore- and back-thrust vergence, leading to what has often been termed a *'flower structure'* (Fig. 14.22). Typical flower structures occur where the faults splay from one, often steeply dipping, fault at depth. They can form from branching Riedel shears associated with strike-slip faults, but many inversion structures show a similar geometry. As tectonic inversion generally involves some strike-slip or oblique-slip movement, there is sometimes some confusion about the kinematics of flower structures (also see Chapter 12).

Most faults are listric to some extent, due to the growth of rotational straight faults through a sedimentary package, the compaction of straight faults or the bending of large faults due to isostatic forces (Kusznir et al. 1991, Coward 1992). Hence short-cut styles of thrust geometry are extremely common in tectonic inversion. Some of the short-cut thrusts may be mistaken for thin-skinned structures (e.g. Fig. 14.22a), while others may generate systems of *fore-thrusts* and

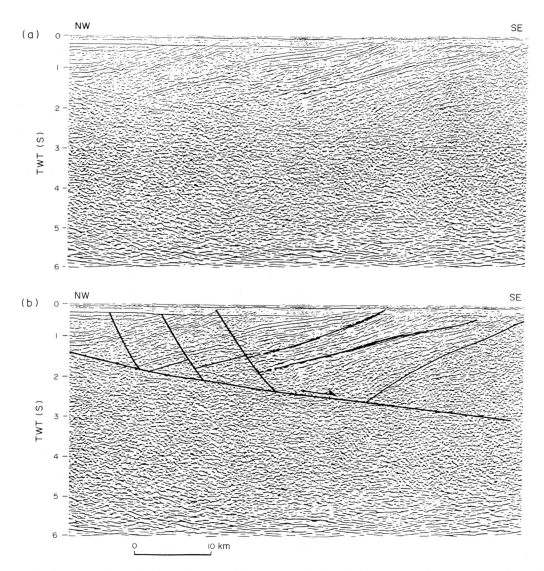

Fig. 14.17. Seismic section (a) and interpretation (b) from part of the West Orkney basin, North Scotland showing the sedimentary package thickening away from the fault in the syn-inversion sediments. TWT, two-way travel time in seconds.

back-thrusts similar to those seen at mountain fronts (Fig. 14.22b).

Schematic illustrations of the more common inversion geometries encountered in the French Alps are shown in Fig. 14.23. At the pre-rift and syn-rift stratigraphic levels some faults still show net extension. Others, however, show net contraction and can form an imbricate reverse fault zone which can only be differentiated from a thin-skinned thrust zone by the high cut-off angles between faults and bedding.

Examples of these *high-angle imbricate zones*, developed from the reworking of closely spaced extensional faults, can be seen near the summit of the Plateau d'Emparis, Pelvoux, and on the eastern side of the Belledonne Massif, where they can easily be mistaken for thin-skinned thrusts (cf. Butler 1983).

Alpine inversion tectonics also leads to the development of small-scale *lateral escape structures*, where rocks can be squeezed sideways into earlier basins (Fig. 14.24). During extension, *lateral ramps* offset

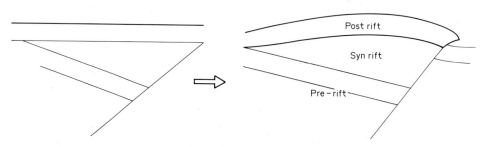

Fig. 14.18. Development of a fold in the post-rift and upper part of the syn-rift sediments associated with inversion and shortening of a half-graben.

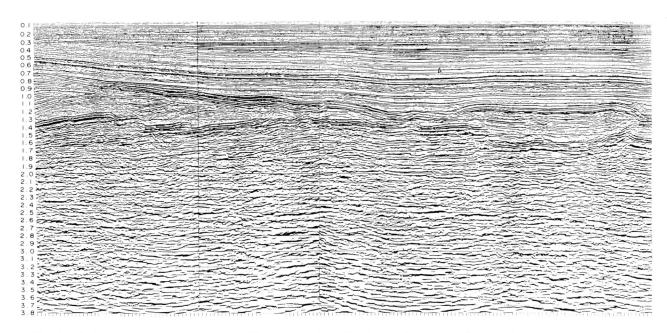

Fig. 14.19. Seismic section through Quad 53 of the southern North Sea showing a train of folds and reverse faults associated with Tertiary inversion of a Mesozoic half-graben. Note that the inversion is coeval in each half-graben (from Badley *et al.* 1989). Vertical scale in seconds (two-way travel time).

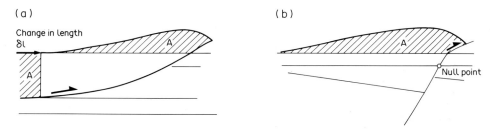

Fig. 14.20. (a) Diagram illustrating the excess area method for calculating the depth to detachment in thrust faults and (b) the origin of the excess area in inverted terrains.

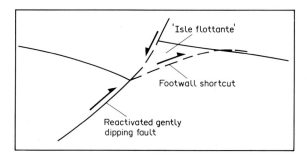

Fig. 14.21. Model for inversion showing the reactivation of a gently dipping normal fault with associated buttressing by the steeper part of the fault, leading to a new short-cut fault in the original footwall.

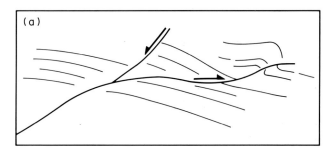

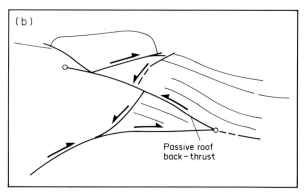

Fig. 14.22. Inversion geometries showing the development of (a) fore-thrusts and (b) back-thrust systems associated with the inversion of a normal fault at depth.

zones of thinned crust, forming a tooth-like margin to a basin. Where these lateral ramps are offset, before or during inversion, then the teeth of the stretched crust on one margin of the basin may not fit back into the sockets on the opposite margin, leading to local lateral expulsion of material. Inversion can lead to strain complexities where material does not extrude vertically by crustal thickening, but also escapes laterally, so that three-dimensional strains and incremental strain

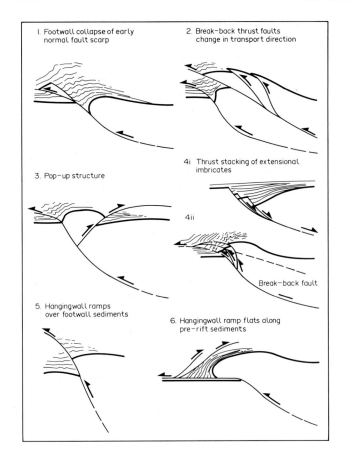
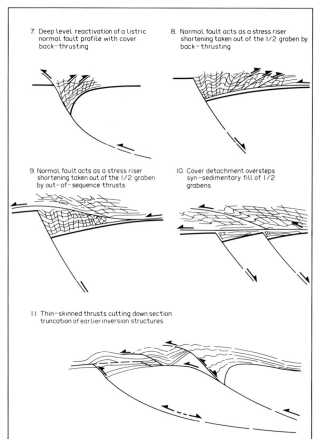

Fig. 14.23. Schematic illustrations of the more common inversion geometries encountered in the French Alps (from Coward et al. 1991).

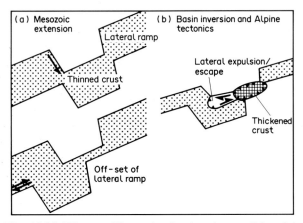

Fig. 14.24. Diagram illustrating in plan the possible origin of lateral movements during inversion tectonics (after Coward et al. 1991).

histories vary markedly over a small area. Buttressing by basin bounding faults may lead to local pure shear strains and to lateral expulsion. Thus in the Alps, the incremental strain work of Gourlay (1986) and Spencer (1992) shows a pronounced change in extension direction with time, which can be related to an increase in lateral expulsion with time. The most prominent lateral expulsion occurs close to the northwest edge of the Valais Zone (Fig. 14.25), an important basin boundary during the Cretaceous, and another example lies close to the southeast boundary of the Belledonne fault block.

Section-balancing techniques which ignore basin inversion can lead to errors in the interpretation of the deep structure of a thrust-fold belt and indeed of the entire mountain belt. Consider for example the data shown in Fig. 14.26, which represents the edge of a thrust-fold belt. This section is simplified from examples in the Sulaiman and Kohat Ranges, Pakistan. The section shows a layer of sediments gently folded, thrusted and uplifted to form a *mountain-front monocline* (also see Chapter 18) against the foreland basin. No thrusts emerge into the foreland basin and all the shortening within the thrust-fold belt has to be transferred onto a back-thrust beneath the frontal monocline of the mountain belt. The thin-skinned model (Fig. 14.27a) assumes that the uplift of the thrust-fold belt is a result of thrust imbrication at depth. As the shortening within this deep-level imbricate/duplex zone far exceeds the shortening seen in the upper layers of the thrust-fold belt, a passive-roof back-thrust has to be postulated beneath the upper layer of the thrust-fold zone. As the basement is not imbricated by the thin-skinned thrusts, it must continue x kilometres back beneath the hinterland of the thrust-fold belt, where x is the shortening on the imbricate/duplex zone. The thick-skinned model (Fig. 14.27b) assumes that much of the uplift beneath the thrust-fold belt is a result

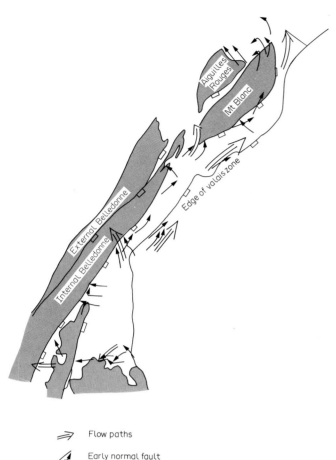

between folds produced by thin-skinned tectonics and folds produced by inversion. If one or more of the following criteria hold, then inversion tectonics should be considered viable for the region. (1) Compressive structures show simultaneous growth. In thin-skinned thrust zones the structures generally propagate towards the foreland, hence increasing the size of the thrust wedge with time, while in thick-skinned or inversion tectonic regimes, the original normal faults often reactivate simultaneously (see Fig. 14.19). (2) There is a lack of a mountain belt or surface slope which could drive the thin-skinned tectonics. (3) There is independent evidence for the presence of an older basin, for example, the presence of null points on sections or maps or the rapid change in thickness or facies of the sediments.

Examples of reverse faults and folds which have previously been interpreted as thin-skinned thrusts, but which may warrant reinterpretation as thick-skinned inversion structures, include parts of the Apennines of Italy, the Palmyrides of Syria, the Zagros Ranges in Iraq and western Iran (Ameen 1992), and the Kirthar, Sulaiman and Kohat Ranges in western Pakistan.

Fig. 14.25. Map of the external massifs of the Western Alps, showing the positions of the Mesozoic normal faults. The incremental extension directions and their variations in time and space are shown, based on Spencer (1992). Suggested flow paths for the internal thrust sheets are shown, suggesting lateral escape from the buttressing effect of the Mesozoic normal faults.

of shortening of a sedimentary basin and expulsion of syn-rift sediments. There is no necessity for basement to continue back beneath the hinterland of the thrust-fold belt. Thus it is important to be able to differentiate

QUANTIFYING AMOUNTS OF INVERSION

Though it is relatively easy to quantify the degree of compressional reactivation of extension faults and define an inversion ratio (R_i) on a particular fault, it is often difficult to quantify the amount of uplift associated with regional basin inversion. Burial data, including porosity changes with depth, vitrinite reflectance data and spore and conodont coloration data can be used to estimate how much material has been removed from a sedimentary column (Higgs & Beese 1986, Roberts 1989). Fission track studies give estimates of the maximum burial temperature, the amount of material removed and the likely time of uplift (e.g. Green 1986). Regional chronostratigraphic studies, together with burial data and regional structural studies may indicate the lateral extent of the inversion.

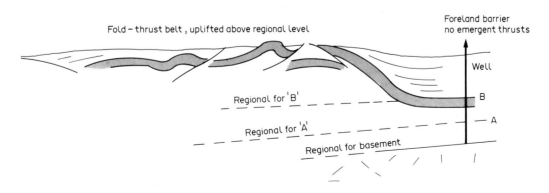

Fig. 14.26. Diagram illustrating the typical structure of a mountain front, based on the structural styles encountered in the frontal regions of the Sulaiman and Kohat Ranges in western Pakistan. The sedimentary package 'B' has been uplifted above its regional level. The problem is to determine the structure at depth beneath the thrust-fold belt.

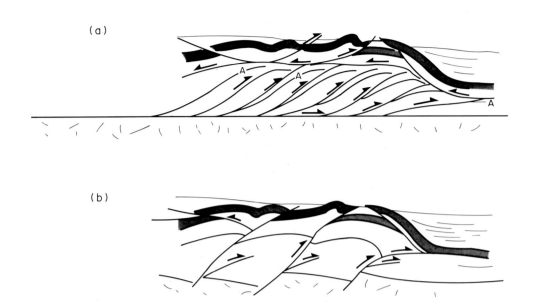

Fig. 14.27. (a) Solution to the structure shown in Fig. 14.24 assuming a thin-skinned thrust model, involving one or more deep duplex zones and a considerable amount of shortening. A passive-roof back-thrust is required beneath the upper part of the thrust-fold zone. (b) Solution to the structure shown in Fig. 14.26 assuming a thick-skinned model involving reactivation of earlier extensional faults. The uplift of the thrust-fold zone can be related to uplift of a thick package of syn-rift sediments.

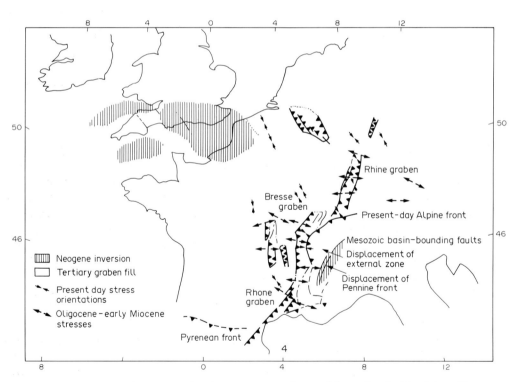

Fig. 14.28. Map showing areas of Palaeogene extension and inversion in Northwest Europe.

Thus in the Celtic Sea, vitinite reflectance data and the configuration of the top Cretaceous reflector indicate uplift of 1–2 km. Uplift of the adjacent onshore regions by as much as 2 km is shown by spore colouration data (Roberts 1989). Similarly, in the Manx basin and the Lake Distict in northwest England, fission track data document uplift of early Tertiary age, of an amount between 1–2 km and 2–4 km depending on the postulated geothermal gradient (Green 1986).

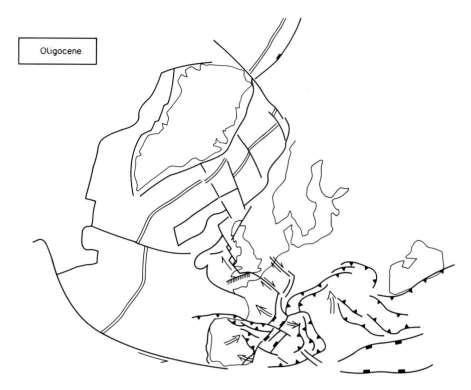

Fig. 14.29. Map of Northwest Europe showing regional tectonics in the Oligocene and the development of regions of tectonic inversion at compressional jogs on regional strike-slip faults. These faults were reactivated by intraplate strains caused by variations in spreading rate along the Atlantic Ocean and the Rhone–Rhine grabens that link the two spreading systems.

EXAMPLES AND CAUSES OF TECTONIC INVERSION IN NORTHWEST EUROPE

According to Zeigler (1982, 1987) much of the inversion in Northwest Europe was due to Alpine collision. Zeigler (op. cit.) and Beach (1988) envisaged the inversion to be linked by some sub-horizontal deep crustal shear zone, equivalent to a 'flat' in a thin-skinned thrust zone. This argument is rejected here because: (1) inversion structures in Northwest Europe occur from the Middle Cretaceous to the Neogene and cannot easily be matched with pulses of Pyrenean or Alpine compression; (2) inversion structures are locally diachronous; they formed at slightly different ages in different parts of the North Sea; (3) inversion structures are found throughout Northwest Europe and possibly the most dramatic structures are found close to the continental shelf, west of Shetland; and (4) during the Oligocene, when tectonic inversion produced important *forced folds* (Chapter 10) in southern England, the Rhone and Bresse grabens were formed in France. This was a region of continental extension separating a region of inversion in southern Britain from a region of Alpine thrusting to the south (Fig. 14.28).

The deformation of the Alpine basins in western France is obviously related to Alpine plate collision, but the causes of inversion tectonics in southern Britain and the North Sea must be related to some form of intraplate deformation.

(1) *Due to reactivations of older basement lineaments, transferring displacement across NW Europe.* The Bray-Southwest England – St Georges Channel lineament began as a Variscan microplate boundary. Right-lateral movement along this lineament, which linked opening of the North Atlantic with the tips of the Bresse – Rhone – Ligurian rift systems, would cause inversion at compressional offsets such as the Purbeck anticline. Where the offset was dilational, a new Tertiary basin formed, for example in the southern half of Cardigan Bay (Fig. 14.29). Other major right-lateral systems produced the Weald anticline, the inversion along the Sole Pit system and inversion episodes west of Shetland.

(2) *Cretaceous inversion in the southern North Sea involving the simultaneous reactivation of numerous Mesozoic normal faults.* Local small-scale structures suggest shear along the fauls. The southern North Sea was probably affected by a shear couple related to phases of North Atlantic – Biscay opening, and this shear couple rotated early fault blocks, causing inversion similar in style to Permo-Carboniferous inversion structures in the same region.

Therefore, intraplate inversion can be linked to plate boundaries by shear couples developed along pre-existing fault zones, for example, the Bray – Devon system, or wide zones of diffuse shear causing block rotation and inversion. These shear couples can be linked to extensional or contractional plate margins. Hence collision tectonics are not a pre-requisite of inversion.

REFERENCES

Ameen, M. S. 1992. The effect of basement tectonics on hydrocarbon generation, migration and accumulation in northern Iraq. *Bull. Am. Ass. Petrol. Geol.* **76**, 356–370.

Badley, M. E., Price, J. D. & Backshall, L. C. 1989. Inversion, reactivated faults and related structures: seismic examples from the southern North Sea. In: *Inversion Tectonics* (edited by Cooper, M. A. & Williams, G. D.) *Spec. Publs geol. Soc. Lond.* **44**, 201–219.

Biddle, K. T. & Rudolph, T. W. 1988. Early Tertiary structured inversion in the Stord Basin, Norwegian North Sea. *J. geol. Soc. London*, **145**, 603–611.

Butler, R. W. H. 1983. Balanced cross sections and their implications for the deep structure of the NW Alps. *J. Struct. Geol.* **5**, 125–137.

Coward, M. P. 1986. Heterogeneous stretching, simple shear and basin development. *Earth Planet. Sci. Lett.* **80**, 325–336.

Coward, M. P. 1992. Structural interpretation with emphasis on extensional tectonics. *Japec Course No. 122*.

Coward, M. P., Gillcrist, R. & Trudgill, B. 1991. Extensional structures and their tectonic inversion in the Western Alps. In: *The Geometry of Normal Faults* (edited by Roberts, A. M., Yielding, G. & Freeman, B.). *Spec. Publs geol. Soc. Lond.* **56**, 93–112.

Gillcrist, R., Coward, M. P. & Mugnier, J. L. 1987. Structural inversion, examples from the Alpine Foreland and the French Alps. *Geodinimica Acta* **1**, 5–34.

Green, P. F. 1986. On the thermal tectonic evolution of Northern England: evidence from fission track analysis. *Geol. Mag.* **123**, 493–506.

Harding, T. P. 1983. Seismic characteristics and identification of negative flower structures, positive flower structures and positive structural inversion. *Bull. Am. Ass. Petrol. Geol.* **69**, 582–600.

Hayward, A. B. & Graham, R. H. 1989. Some geometrical characteristics of inversion. In: *Inversion Tectonics* (edited by Cooper, M. A. & Williams, G. D.). *Spec. Publs geol. Soc. Lond.* **44**, 17–40.

Higgs, K. & Beese, A. P. 1986. A Jurassic microflora from the Colbond Clay of Cloyne, County Cork. *Irish J. Earth Sci.* **7**, 99–109.

Keen, C. E. 1987. Some important consequences of lithospheric extension. In: *Continental Extensional Tectonics* (edited by Coward, M. P., Dewey, J. F. & Hancock, P. L.). *Spec. Publs geol. Soc. Lond.* **28**, 67–73.

Kusznir, N. J., Marsden, G. & Egan, S. S. 1991. A flexural cantilever simple shear/pure shear model of continental extension: application to the Jeanne d'Arc basin, Grand Banks and Viking Graben, North Sea. In: *The Geometry of Normal Faults* (edited by Roberts, A. M., Yielding, G. & Freeman, B.). *Spec. Publs geol. Soc. Lond.* **56**, 41–60.

Roberts, A. M., Price, J. D. & Olsen, T. S. 1990. Late Jurassic half-graben control on the siting and the structure of hydrocarbon accumulations: UK/Norwegian Central Graben. In: *Tectonic Events Responsible for Britain's Oil and Gas Reserves* (edited by Hardman, R. F. P. & Brooks, J). *Spec. Publs geol. Soc. Lond.* **55**, 229–257.

Roberts, D. G. 1989. Basin inversion in and around the British Isles. In: *Inversion Tectonics* (edited by Cooper, M. A. & Williams, G. D). *Spec. Publs geol. Soc. Lond.* **44**, 131–150.

Williams, G. D., Powell, C. M. & Cooper, M. A. 1989. Geometry and kinematics of inversion tectonics. In: *Inversion Tectonics* (edited by Cooper, M. A. & Williams, G. D.). *Spec. Publs geol. Soc. Lond.* **44**, 3–16.

Zeigler, P. A. 1983. Inverted basins in the Alpine Foreland. In: *Seismic Expression of Structural Styles* (edited by Bally, A. W.). *Am. Ass. Petrol. Geol. Stud. Geol.* **15**, 3.3.3–3.3.12.

Zeigler, P. A. 1987. Compressional intra-plate tectonics in the Alpine Foreland. *Tectonophysics* **137**, 420 pp.

CHAPTER 15

Suspect Terranes

WES GIBBONS

INTRODUCTION AND HISTORICAL BACKGROUND

THE THEME of this chapter is the accretion and dispersal, especially by strike-slip faulting, of fault-bounded tectono-stratigraphic units in orogenic belts. Initial application of plate tectonic models to orogenic belts, such as the classic work by Dewey (1969), emphasized the use of sequential cross-sections to illustrate the evolution of the orogen through time. These cross-sections were typically drawn roughly orthogonal to the regional strike of the orogen and did not address the possibility of large-scale lateral movements in and out of the plane of section. The main focus of such interpretations was the known timing of orogenic events, combined with the positions of key tectono-stratigraphic units such as ophiolitic suture zones.

Although initially highly successful in furthering our understanding of ancient orogens such as the Appalachian–Caledonian system, the use of orthogonal plate tectonic models in the Pacific (Cordilleran) margin of North America failed to produce a consistent interpretation applicable along the strike of the orogen. The main problem in western North America is the presence of a network of strike-parallel faults, at least some (and probably many) of which have a significant strike-slip component to their movement history. A more pragmatic approach to the problem of interpreting Cordilleran geology had been championed by Irwin as early as 1960. Working in the Klamath Mountains of Northern California, Irwin realized that the first stage in interpreting the geology of this region required the recognition of geologically separate belts. Each of these belts possessed a distinctive stratigraphy and structure and, in some cases, the belts were further subdivided into 'sub-belts' which, by 1972, Irwin was calling 'terranes'. By this time a similar collage of fault-bounded units was being recognized in other parts of the orogen, such as Alaska (Berg et al. 1972).

An increasing emphasis on the ubiquity of strike-parallel faults along the west coast of North America led towards an explanation of a long-standing problem in Cordilleran geology, namely the presence of exotic fossils within fault-bounded units. In 1950 Thompson and Wheeler had shown the distribution of exotic Tethyan fusilinids in northwestern U.S.A. The obviously far-travelled nature of these fusilinid-bearing rocks was subsequently incorporated into a *Wilsonian cycle* interpretation of Pacific opening and closure (Wilson 1968), based on the successful Iapetus model for the Atlantic region. The Wilsonian model, undermined by a lack of evidence for Pacific closure, was superseded by the argument that the Permian fusilinids had formed at the equator and moved across the ocean floor on the oceanic plate en route to subsequent accretion to the American continent (Monger & Ross 1971). Continued acquisition of more palaeobiological data on exotic faunas, combined with palaeomagnetic work (Irving & Yole 1972, Packer & Stone 1972, 1974) generated increasing interest in the possibility of large-scale movements of rock units both across the oceans and along the continental margin. This latter concept came of age in 1977 when a compelling case, combining independent geophysical, lithostratigraphic and palaeontological lines of evidence, was presented to support the idea of 1000s of kilometres of movement for the fault-bounded unit named Wrangellia. The interpretation was based essentially on a combination of low-latitude palaeomagnetic results and equatorial palaeoclimatic data derived from the interpretation of Triassic carbonates and high diversity faunas. Wrangellia became recognized as a classic example of a far-travelled unit that originated in the Pacific Ocean, moved east to become accreted to the North American plate, and was subsequently moved by strike-slip faulting along the continental margin (Fig. 15.1).

Unlike Wrangellia, many other fault-bounded units occurring within the Western Cordillera were initially only suspected of being far travelled; unequivocal evidence was lacking so that such units came to be known as *suspect terranes*. It had become increasingly clear that large numbers of tectonostratigraphic units in the Western Cordillera, especially those situated west of the 0.704 contour of initial strontium isotope ratios (Kistler & Peterman 1978), were 'suspect' (Fig. 15.2). By 1982 over 50 suspect terranes had been identified in California alone (McWilliams & Howell 1982), and Jones et al. (1982) were arguing that the entire western side of North America had been "grafted on by the piecemeal addition of large, prefabricated blocks of crust most of which were carried 1000s of kilometres east and north from their sites of origins in the Pacific Basin''.

By the early 1980s, therefore, these ideas had become firmly rooted in the thinking of geologists in western North America. The terrane concept was quickly exported to other orogens, dispersed by articles such as Jones *et al.* (1982), Smith (1983), and Kerr (1983). The Appalachian orogen, on the east side of North America, long recognized as divisible into a series of 'zones' and 'blocks' came also to be treated as a mosaic of suspect terranes (e.g. Drake & Morgan 1981, Zen 1981, 1983, Keppie 1982, Williams & Hatcher 1983). A similar approach was applied to the British and Irish Caledonides (Bluck 1983, Gibbons 1983, Gibbons & Gayer 1985, Hutton 1987), and other orogenic belts such as the Tasmanides of Eastern Australia (e.g. Cawood 1983) and New Zealand (e.g. Howell 1980). This has catalyzed a new and continuing spate of research activity in the tectonic analysis of these ancient orogens.

DEFINITIONS

The word *terrane*, used on its own without reference to the suspect terrane concept merely refers to an area that possesses its own distinctive geology: the area or surface over which a particular rock or group of rocks is

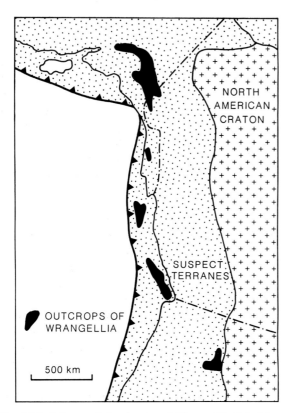

Fig. 15.1. Dispersed fragments of Wrangellia (black) — a disjunct oceanic terrane displaced along the North American craton. Collage of suspect terranes accreted to the cratonic margin shown stippled.

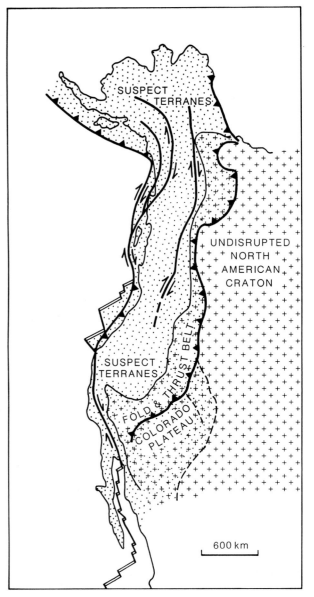

Fig. 15.2. Cordilleran suspect terranes (stipple) in northwestern America lying between the plate margin (trench, transform fault or spreading ridge) and the cratonic margin (stipple and crosses) disrupted by orogenic events over the last 100 Ma (modified after Howell 1989).

prevalent (*Longmans Dictionary of the English Language 1984*). A definition of the term terrane as used in the western Cordillera was provided by Irwin in 1972: "an association of geologic features, such as stratigraphic formations, intrusive rocks, mineral deposits and tectonic history, some or all of which lend a distinguishing character to a particular tract of rocks and which differ from those of an adjacent terrane".

Some dictionaries define terrane as a synonym of terrain, although others will distinguish between the two words by emphasizing the more normal geological usage of the spelling terrane (e.g. *Longmans 1984, Shorter Oxford English Dictionary*, 3rd edn, 1983). Some British workers have resisted the adoption of the 'terrane' spelling, viewing this as an unwelcome and unnecessary American importation. Such an attitude seems pointless as there is value in distinguishing terrane as a geological term as distinct from the more general geographical 'terrain'.

Irwin's 1972 terrane definition did not explicitly state what was to become the single most important characteristic of the Cordilleran terrane collage: that each terrane in this region is bounded by faults. This emphasis on faulted margins was primarily a result of increasing suspicion that many, and perhaps most, of the Cordilleran terranes had undergone significant strike-slip transport along the North American cratonic margin. Thus, following the work of Berg et al. (1972) and others, the tectonic isolation of Cordilleran terranes was encapsulated by the simplest and most widely used definition of such *suspect terranes* as: "fault bounded geological entities of regional extent each characterized by a geological history that is different from that of adjacent terranes" (Coney et al. 1980, McWilliam & Howell 1982). Implicit in this definition was the suspicion that such a terrane, once recognized, may well be highly displaced with respect to both adjacent terranes and to the nearest craton. In this context the use of the word *craton* refers to that part of North America that was not deformed at the time of terrane displacement and accretion (cf. Chapter 10).

The terms *allochthonous terrane* and *exotic terrane* have been employed by several authors to describe terranes with proven large displacements (e.g. McWilliams & Howell 1982, Ben Avraham & Nur 1983). The term allochthonous (Dewey et al. 1991) is synonymous with the simpler and more immediately intelligible term displaced (cf. Jones et al. 1977). A *displaced (or allochthonous) terrane* may be defined as a fault bounded unit that has been displaced with respect to adjacent terranes so that there is a striking contrast in the geology across the terrane boundaries. A *transcurrent displaced terrane* (shortened to *transcurrent terrane*) is a faulted bounded unit that has been juxtaposed against adjacent areas by predominantly strike-slip faulting so that there is a striking contrast across the terrane boundaries. Use of the term *exotic terrane* is best restricted to those displaced terranes that possess a geology which is alien to the cratonic margin against which they are now accreted, for example, a continental fragment or oceanic seamount that has moved large distances across an ocean before impinging upon a cratonic margin. Use of all these qualifying adjectives such as displaced, transcurrent and exotic, should represent an improved understanding of what was initially classed as a suspect terrane, that is, the suspect terrane has been proven to be guilty of significant fault displacement.

Various collective nouns have been employed to describe a group of terranes. Most of such terms are informal, such as '*family of terranes*' (Kaplan Morris et al. 1986), '*terrane collage*' (many authors), '*terrane complex*' (Roberts 1988), and '*superterrane*' (e.g. Butler et al. 1989, Gibbons 1990). The term *superterrane* used to describe a unit comprising more than one terrane has the advantage that it is analogous to the lithostratigraphic use of group and supergroup.

Because the vast majority of suspect terranes are defined primarily on their stratigraphy and structural history several authors have employed the term *tectonostratigraphic terrane* (e.g. Howell 1980, Schermer et al. 1984, Bishop et al. 1985, Haston et al. 1989). Such a term does not add anything new to the definition of suspect terrane and, although harmless, seems unnecessary. It is part of the process of terrane definition that a distinctive tectonostratigraphic unit with tectonic boundaries be recognized as a suspect terrane.

Attempts to define other terrane types have met with varied success. An example is provided by the four-fold classification of Jones et al. (1982) (repeated in later publications such as Schermer et al. 1984). This organizes terranes into *stratified terranes* (possessing a coherent lithostratigraphy); *disrupted terranes* (mélanges); *metamorphic terranes* (overprinted by terrane-wide metamorphism) and *composite terranes*. These four categories, however, do not cover all terrane types and, more importantly, are not mutually exclusive — a stratified terrane, for example, might include disrupted olistostromic formations, be overprinted by low-grade metamorphism, and form part of a composite terrane. In essence a suspect terrane is defined by all aspects of its geology, and a proliferation of new formal categories is unhelpful. If the main characteristic of the geology of a terrane can be encapsulated informally by a word such as plutonic or gneissic, then this is helpful and self-explanatory (although only sometimes possible).

The term composite terrane, however, continues to be used commonly in the literature to describe a particular type of superterrane. This term was introduced in the pioneering terrane map produced by Berg et al. (1978) to cover the northern part of the North American Cordillera. A *composite terrane* comprises two or more terranes that amalgamated to form a superterrane prior to its accretion to a continental margin. An example of a composite terrane is provided by the Santa Lucia–Orocopia superterrane in California (the Santa Lucia–Orocopia allochthon of Kaplan Morris et al. 1986). This composite unit amalgamated in Cretaceous times and was subsequently accreted to southern California by Early Tertiary times.

Essential to the terrane concept are the three terms amalgamation, accretion and dispersion (Fig. 15.3). *Tectonic amalgamation* refers to the arrival by faulting of one terrane against another (an event commonly referred to as '*docking*'). A modern example of an area in which terranes are actively amalgamating is provided by the Phillipines archipelago. *Terrane accretion* represents the time at which a terrane first adheres to a continental margin, such as the recent accretion of Taiwan to the continental margin of China. The term is particularly useful in describing the tectonic event when suspect terranes that have ridden passively upon an oceanic plate impinge upon a continental margin. *Transcurrent terrane dispersal* occurs when previously accreted terranes are faulted into smaller pieces and

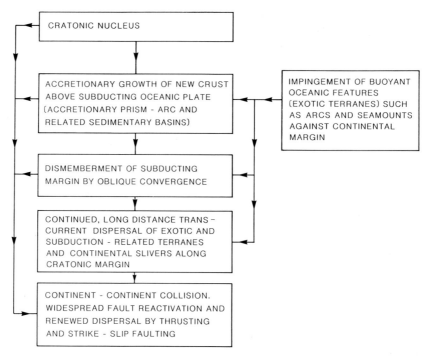

Fig. 15.3. Flow chart illustrating typical events involved in accretionary continental growth and transcurrent terrane dispersal. An active continental margin may undergo some or all of these events as subduction progresses.

scattered along the continental margin by predominantly strike-slip faulting. The term is commonly used to describe the effects of lateral movements along the interface between continents and oceans, although a similar result can be induced during continent–continent collision. Present-day continued northward movement of India under Tibet, for example, is inducing strike-slip dispersion of terranes in Northern and Eastern China (see Chapter 13). The term *disjunct terrane* has been employed to describe the stratigraphically correlative but spatially separated pieces of one originally unified terrane such as Wrangellia — a dispersed terrane now spread over greater than 30° of latitude from an original separation of less than 7° (Jones *et al.* 1977, Schermer *et al.* 1984, Plafker *et al.* 1989). Presumably each isolated part of such a dispersed terrane will undergo a second (and perhaps more) phase of accretion to the continental margin once faulting has ceased. *Terrane dispersal* need not be effected only by strike-slip faulting — previously accreted terranes may be subsequently dispersed by collision-related thrusts. Another form of terrane dispersal may be effected prior to accretion by rifting at spreading ridges or triple junctions; this can, for example, separate continental fragments that will subsequently be accreted to an active continental margin.

The acquisition of more data during the 1980s on some of the better-known displaced terranes enabled increasingly detailed interpretations of their tracking history to be prepared. An example is provided by McLaughlin *et al.* (1988) on the Coast Range ophiolite of California. This ophiolitic terrane is interpreted as having formed in a supra-subduction setting at equatorial latitudes during Jurassic times (169–163 Ma). It was subsequently superimposed by Late Jurassic arc magmatism (156–145 Ma) and accreted to the North American margin around 145 Ma. By Palaeocene times (c. 90 Ma) the Coast Range rocks had been depositionally overlain by Great Valley strata and underplated (see Chapters 9 and 13) by Franciscan Complex accretionary mélanges. The terrane was then dispersed by at least two phases of Cenozoic right-lateral transcurrent faulting, the later phase being associated with the propagation of the late Cenozoic San Andreas transform fault system. As emphasized by Şengör (1988) it is important to attempt to distinguish between the initial accretion of rock units to a host continent, and any subsequent dispersal along the continental margin of already assembled terrane collages. One obvious consequence of transcurrent terrane dispersal is that slivers of accretionary prism material may be moved long distances before docking against some originally distant arc. Such dispersal may involve several subduction complexes, and faulting will not necessarily be strictly parallel to the trench. Any ancient juxtaposition of coeval arc and accretionary prism rocks (the 'paired metamorphic belts' of Miyashiro) cannot therefore be assumed to record the original subduction polarity. The implications of conclusions such as these are particularly significant in ancient orogens where the present outcrop pattern is likely to have been produced by the final strike-slip dispersal of the orogen (e.g. Bobyarchick 1988).

TERRANE ORIGINS AND KINEMATICS

The development of the suspect terrane concept led to the recognition that many potential modern analogues exist on the floors of present-day oceans. Ridges, rises,

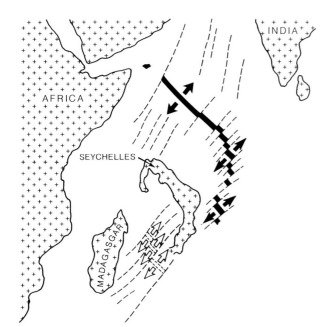

Fig. 15.4. The separation of the Seychelles continental sliver from the Indian craton took place at about 65 Ma (magnetic anomaly 28) when the oceanic spreading ridge jumped north from between Madagascar and the Seychelles into the Indian–Seychelles continental margin.

plateaux, and volcanic seamounts are eventually destined to be carried with the host oceanic plate to a destructive plate margin (Ben Avraham et al. 1981, Ben Avraham & Nur 1983). At the subduction zone these relatively buoyant units are likely to be transferred to the hangingwall of the overriding plate. It has been estimated that something like 10% of ocean floor comprises atypical, upstanding units suitable for eventual accretion to a continental margin (Ben Avraham & Nur 1983). Travel rates across oceans of potential suspect terranes, relative to the bounding continents, can be remarkably high. Plate motion studies in the Pacific, based on magnetic lineation, palaeoclimatic and hotspot trajectory data reveal relative speeds of over 120 mm/yr in some cases. (Note that all such rates are given here in mm/yr as this corresponds directly with km/Ma.) Reconstructions by Engebretson et al. (1987), for example, estimate northward movements for the Kula plate (a totally subducted oceanic plate in the northeast Pacific) at times to have been in the order of 1° latitude per million years. Zonenshain et al. (1987) estimate even higher relative rates of nearly 13,000 km travel for the Kula Plate between 130 and 50 Ma at an average speed of 164 mm/yr. These latter authors also estimate that continental fragments riding buoyantly upon oceanic crust have travelled nearly 10,000 km across the Pacific Ocean before accreting to the Eurasian continental margin.

All of the present oceans contain prominent masses of continental crust rifted from larger continents: examples include Madagascar and the Seychelles in the Indian Ocean, the Rockall Plateau in the Atlantic Ocean, and the Lord Howe Rise in the southwest Pacific Ocean. Continental crust is weaker than oceanic crust so that along continental margins rifting or transcurrent faulting will favour the development of long, thin slices of continent a short distance inboard from the ocean (Vink et al. 1984). The separation of Madagascar from the Seychelles took place about 65 Ma when the Indian oceanic spreading ridge jumped into the Seychelles–Indian continental margin (Fig. 15.4). The Lord Howe Rise and Norfolk Ridge, each measuring in excess of 1000 km long and 100 km wide, were produced by the thinning and rifting of eastern Australian continental crust. A third example is provided by the development of the Dead Sea transform system which owes its origin to the abandonment of the Gulf of Suez in favour of thicker continental crust east of the Mediterranean (Steckler 1989). Perhaps the most obvious example of this process of slicing continents into linear terranes is that of Baja California.

The area around Baja California is one of the best studied active transcurrent terrane boundaries, and is being produced by the progressive segmentation and replacement of the ocean spreading ridge by a dextral strike-slip fault system (Fig. 15.5). Analysis of the

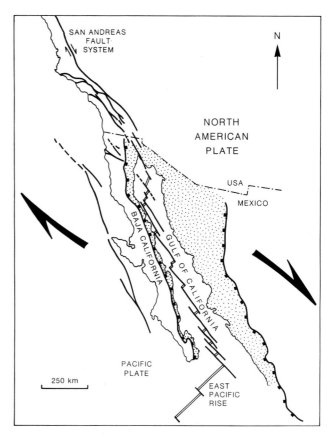

Fig. 15.5. The boundary of the southeastern edge of the Pacific plate shows the progressive dismemberment of the East Pacific rise spreading ridge through the Gulf of California to produce the dextral transform San Andreas fault system in California. Baja California is moving northwestwards relative to the North American continent by about 64 mm/yr. A wide zone of extension (stippled area bounded by normal fault scarps ▬▬▬) across the Gulf of California reflects the transtensional nature of this plate boundary (after Stock & Hodges 1989).

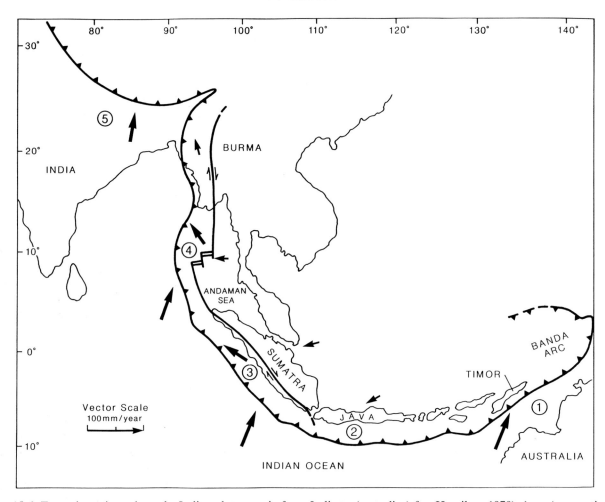

Fig. 15.6. Tectonic settings along the Indian plate margin from India to Australia (after Hamilton 1979). 1 = Arc–continent collision: obduction of Banda arc over Australian cratonic margin. 2 = Orthogonal oceanic plate subduction beneath Java. 3 = Oblique subduction involving the dismemberment of Sumatra by dextral strike-slip faulting and production of a transcurrent terrane between arc and trench. 4 = Extreme oblique subduction produces supra-subduction ophiolite by oceanic spreading in the Andaman Sea. 5 = Continent–continent collision between India and Asia.

anomaly pattern south of Baja California has shown that the peninsula has been moving northwestward at an average speed of 60 mm/yr over the last 4 Ma (Larson et al. 1968). Similar results have been obtained from magnetic measurements made at the mouth of the Gulf (66–49 mm/yr). Earthquake data collected during this century have resulted in seismic moment determinations that estimate an average movement of 65 mm/yr (Reichle et al. 1976), and this rate has been supported by satellite geodetic data that demonstrate movements of 64 mm/yr (Ness et al. 1985). The new geodetic technology using satellite laser ranging (SCR) and very long baseline interferometry (VLBI) provides extremely powerful tools with which to analyze relative global plate motions, but for the more localized scales needed for measuring present transcurrent terrane displacements then electronic distance methods (EDM) are more appropriate. Recent EDM results from a network spanning the 150 km-wide Gulf of California (including several islands in the Gulf) show the plate boundary to be a wide zone of dextral shearing with an overall relative movement of 80 ± 30 mm/yr across the Gulf (Ortlieb et al. 1989). Relative movement is being taken up within a zone of weakened lithosphere over 100 km wide within which elastic strain is spatially distributed across the Gulf. The width of this elastic strain accumulation zone decreases towards and into the San Andreas Fault System to some 60 km in Southern California and 20 km in Central California (Fig. 15.5). Continued movement of the 'Baja' sliver at its present speed would result in the dispersal of this terrane along the margin of Northwest America and accretion somewhere against the continental edge of Alaska in less than 80 Ma.

The primary driving force behind the slivering and dispersal of linear terranes along a continental margin is oblique subduction (see Chapters 9, 12 and 13). Plate convergence is a complex process with changes taking place rapidly along destructive margins in both space and time. The present-day interaction between the Indian plate and the Indonesian region varies from orthogonal oceanic plate subduction beneath Java at rates of around 68 mm/yr, through increasing degrees of obliquity northwards through Sumatra into the Andaman Sea (Fig. 15.6) (Hamilton 1979). In Sumatra the obliquity is expressed by a prominent right-lateral fault system that slices the length of the island along the active volcanic arc. In the Andaman Sea the obliquity is

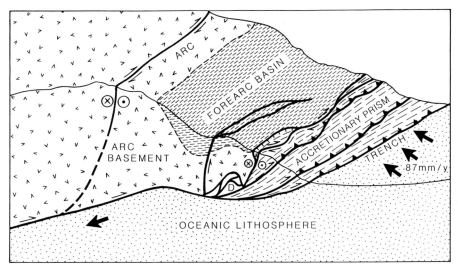

Fig. 15.7. Present-day transpressional subduction along the Central Aleutian trench (after Ryan & Scholl 1989). The accretionary prism-arc system is being disrupted into a series of slivers by right-lateral strike-slip faulting during oblique plate convergence. D = postulated strike-slip duplex involving arc basement rocks and inducing uplft of forearc sediments.

so great as to induce extension above the subducting slab and the production of supra-subduction ophiolite beneath a spreading ridge in the forearc basin (Hamilton 1979, Moores et al. 1984). Further north, subduction activity continues to decline within the essentially transcurrent belt that connects with the main Himalayan collision zone (see Chapter 13). To the west of Java, the impingement and ongoing subduction of Australian continental crust beneath the Banda arc provides the classic modern example of early continent-arc collision. All of these events are taking place along the same convergent margin simultaneously. In ten million years from today, at the present rate of convergence, 680 km of oceanic crust will have been subducted beneath Java, and western Sumatra could have been displaced over 500 km northwards towards Burma. The present complexity of the Indonesian region north and east of the Indian plate margin, riddled as it is with potential suspect terranes, has been used as a modern analogue for the more mature terrane collage of Northwestern America (Silver & Smith 1983).

The transcurrent 'slivering' of active margins during oblique subduction is not confined to the magmatic arc although this will be thick, hot, and relatively weak, so providing a focus for strike-slip faulting. Several presently active margins show transcurrent faulting both behind the arc and, more especially, in the forearc. Jarrard (1986), Karig et al. (1986) and Hansen (1988) and others have emphasized the importance of strike-slip faulting in the forearc area, with Jarrard using the term '*forearc slivers*' (e.g. the area between western Sumatra and the trench on Fig. 15.6). A modern example of this process is provided by the accretionary prism of the Central Aleutian arc (Ryan & Scholl 1989) (Fig. 15.7). The offscraped material in this latter example is being disrupted by a network of high angle faults that run parallel to the trench continuously for at least 300 km. Avé Lallement & Oldow (1989) introduced the term *transpressional terrane* to describe such slivers that are "bound by the subduction zone and by arc-parallel strike-slip faults which may occur in the forearc, volcanic arc or backarc regions". Some of these arc-parallel terranes, however, are under net extension, for example, Baja California (Stock & Hodges 1989), and the eastern sector of the Andreanof block in the Aleutian subduction zone (Ryan & Scholl 1989). The term *transcurrent terrane,* as previously defined, is therefore preferable to transpressional. The steep fault boundaries to these nascent suspect terranes are likely to become more gently inclined with depth, curving into an underlying subduction décollement level. Whilst the most important parameter governing this process is the obliquity of subduction, the strength of the overriding plate, the degree of coupling between the plates, and the dip of the subduction zone will all have an influence.

Study of present-day convergent margins, where oblique subduction seems an inevitable consequence of changing plate motions through time, requires an equally mobilistic expectation of suspect terrane accretion and dispersal in ancient orogens. While such expectations are relatively easily realized by an examination of the Western Cordillera, where terrane dispersal movements are still active and where such movements have been south to north across latitude, there has been less willingness to view older orogens such as the Appalachian/Caledonian and Variscan belts with the same degree of lateral mobility. Furthermore, existing models for some other orogenic belts such as the Western Alps (Chapter 13) require much slower aggregate movements through time than is typical of present rates at modern active margins (mm rather than cm/yr). Although it is possible that this is a correct interpretation, resulting perhaps from the much nar-

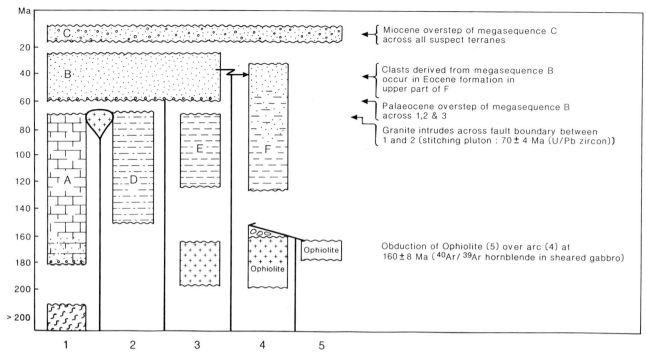

Fig. 15.8. Comparative stratigraphic column for 5 hypothetical suspect terranes. The diagram shows the docking of two suspect terranes (2 and 3) against the cratonic margin (1) in Late Cretaceous–Early Paleogene times. The presence of a Jurassic arc in terrane 3 records the initial accretion of this terrane to the cratonic margin, with both 2 and 3 subsequently being dispersed transcurrently along the margin until Early Cenozoic times. Terranes 4 and 5 record Jurassic amalgamation of two oceanic terranes to form a composite unit docked against the continent (as an exotic composite terrane) in Eocene times. 1 = Cratonic basement (Precambrian) overlain unconformably by two marine shelf megasequences (A = Late Jurassic/Early Cretaceous; B = Palaeogene) and Neogene continental clastics. 2 = Continental margin basinal sediments (late Mesozoic megasequence D) overlain by Cenozoic megasequences B and C. 3 = Lower–Middle Jurassic arc overlain unconformably by Cretaceous marine megasequence E and Cenozoic megasequences B and C. 4 = Middle Jurassic arc on Lower Jurassic ophiolitic basement and overlain by olistostrome beneath obducted Middle Jurassic ophiolite. These ophiolitic rocks are subsequently overlain by oceanic sediment megasequence F, the upper part of which shows provenancial linkage by clasts derived from megasequence B.

rower oceans involved in Alpine closure, the likelihood remains that a more mobilistic approach involving large lateral movements needs to be considered here also.

TERRANE IDENTIFICATION AND CHARACTERISATION

The first step in identifying suspect terranes within an orogenic belt involves field mapping and the development of expertise in the local stratigraphy of an area. From such expertise an appreciation will develop regarding the characteristic geology of each terrane and the position and nature of its faulted margins. Once a fault bounded suspect terrane has been identified, research should concentrate first on the determination of some connection between it and other (perhaps disjunct) areas within the orogen. A useful approach to this involves the preparation of comparative stratigraphic columns for a terrane and its neighbouring areas (Figs 15.8 and 15.9). Such *terrane assembly diagrams* graphically illustrate when there was first some geological connection between contiguous terranes.

Direct age constraints on the docking of one suspect terrane against its neighbouring areas can be provided by: (1) sedimentary or igneous extrusive overstep sequences across the terrane boundary; (2) radiometric ages of the fault rocks defining the terrane boundary (a maximum age will be given by fossils or radiometric dates from rocks affected by such faulting); (3) radiometric ages of igneous bodies that intrude across the terrane boundary — the term *stitching pluton* is commonly used to describe plutonic intrusions that straddle a terrane boundary, a good example being provided by the Ackley Granite intruding across the Gander/Avalon boundary in Newfoundland (Blenkinsop et al. 1976). Other lines of evidence that may be used to constrain terrane-docking times include radiometric age data on minerals grown during metamorphic events seen on both sides of a terrane boundary, and identification of clasts eroded from distinctive lithologies that characterise a particular terrane (e.g. provenance studies in ophiolite clasts in sediments derived from ophiolites exposed on the other side of a bounding fault).

Once defined, most suspect terranes are given a name that relates to a type locality or area, for example, the Pine Mountain terrane of Hooper & Hatcher (1988); and the Wrangellia terrane (named after the Wrangell Mountains of Southern Alaska) of Jones et al. (1977). A proliferation of such names within terrane collages,

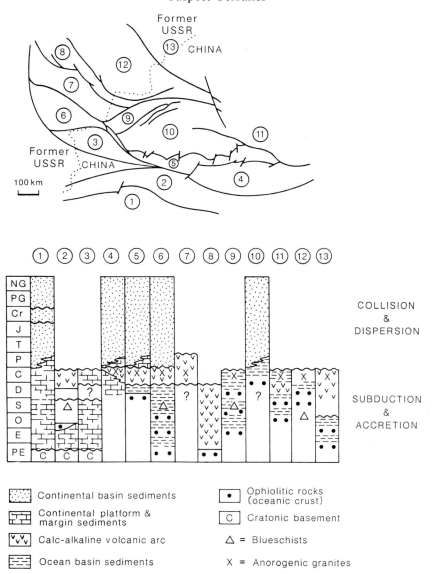

Fig. 15.9. Preliminary small-scale terrane map and stratigraphic correlation chart for 13 suspect terranes in Central Asia (after Coleman 1989). Note the distinction between those suspect terranes with continental or oceanic basement, and the emphasis on the tectonic setting for each suspect terrane (oceanic/arc/continental margin or interior). Several of these suspect terranes are likely to be upgraded to superterrane status as more detailed work is synthesised. Evidence for Palaeozoic subduction and accretionary continental growth has been disrupted by the Cenozoic collision of India with Asia. Thus it is likely that the Cenozoic collision and terrane dispersal is superimposed upon, and reactivates, an older collage of suspect terranes.

however, makes for indigestible reading for those not closely familiar with the area in question. Giving geographical names to terranes does not address the problem of relating each terrane to an appropriate or likely tectonic setting. Terrane maps should therefore not only clearly depict the outcrop, name and fault boundaries for each suspect terrane, but the colours or shading should attempt to provide some genetic interpretation for each terrane; for example, ophiolitic, accretionary prism, forearc, arc, continental cratonic, etc. The presence within a suspect terrane of a basement of oceanic or continental crust should be clearly indicated where known. Recent examples exhibiting this approach include Coleman (1989) and Keppie & Dallmeyer (1989).

Once the credentials of a suspect terrane have been fully investigated and its geological isolation proven, it is necessary to search for data that might indicate large-scale transport before it reached its present position. Although previous interpretations of an area inevitably colour ones perceptions (*"I'd never have seen it if I hadn't believed it"*), the logical approach to terrane analysis requires a lack of preconceived ideas regarding fault movement history. It is just as unacceptable to assume large travel distances for a suspect terrane as to deduce that because such large displacements have not been proven they probably did not take place. The 'guilty until proven innocent' maxim is better toned down in such a way that a suspect terrane remains suspected of having travelled large distances until it can be proven otherwise. An appreciation of the kinematic realities of present-day plate motion, combined with an unbiased and strictly logical approach, are prerequisites for the interpretation of suspect terranes. The following section outlines the types of evidence most commonly used in the prosecution of suspect terranes, that is, those lines of evidence that provide the most compelling data in support of large displacements.

Some terrane boundaries are more obvious and easy to identify than others. Slivers of ophiolitic rocks and mélanges, especially if at blueschist or eclogite grade (e.g. Erdmer 1987), are strong indications of important suture zones along which major fault displacements are likely to have occurred. The ophiolitic Scottish Highland Border Complex in the Caledonides, and the blueschists within the Himalayan suture zones both represent slices of exotic material preserved along terrane boundaries. Of almost equal potential importance is the presence of major, steep, ductile shear zones along suspect terrane boundaries (see Chapter 1). Previous work on suspect terranes has sometimes failed to concentrate on, and fully describe, fault rocks exposed along the terrane boundaries. This is especially important where the boundaries expose ductile shear zones within which kinematic indicators such as S/C fabrics, asymmetric boudins, rotated porphyroclasts etc. may be present (e.g. Simpson & Schmid 1983). The important advances made in interpreting asymmetric fault rock fabrics over the last decade link closely into the methods of terrane analysis. Mylonitic rocks have a tectonic significance far greater than their narrow outcrops might suggest. Recognition and interpretation of ductile shear zones, combined with radiometric age data, are especially important within high-grade metamorphic basement terranes where other evidence may be lacking; for example, the network of schistose suspect terrane boundaries recently recognized within the Archaean gneisses in southern West Greenland (Nutman et al. 1989).

Many of the World's classic transcurrent terrane boundaries such as the Alpine fault in New Zealand, the Median Tectonic Line in Japan, and the Dover fault in Newfoundland are wide (from a few hundred metres to a few kilometres), steep, *ductile shear zones* containing *mylonitic L/S tectonites* with gently plunging synkinematic mineral lineations. Such shear zones will show a continuum of events from early ductile shearing at high temperatures (commonly producing mylonitic schists) through to brittle, cataclastic textures produced at shallower crustal levels. *Pseudotachylite veins* may cut mylonitic fabrics and be themselves reworked within crudely foliated or unfoliated cataclastic breccias and gouge zones. Such a spectrum of *fault-rock* production may result simply from the progressive uplift of deep-seated shear zone rocks on one side of an obliquely transcurrent fault, or may represent some more complex polyphase history of strike-slip and dip-slip movements.

A suspect terrane boundary does not need to be a steep structure, and many such boundaries are gently inclined faults variously interpreted as thrusts, normal faults or low-angle strike-slip faults. Such low-angle tectonic contacts may be produced either during transcurrent dispersion, by the development of flower structures (see Chapter 12) rooting into a steep shear zone, by the presence of a listric geometry to strike-slip faults, or by later compressional or extensional events reworking older faults. The final accretion of a transcurrent terrane to a continental margin may be effected by inboard obduction of the terrane during some compressional orogenic event that will obscure the evidence for earlier strike-slip translation.

In addition to the use of fault rocks in determining the relative movement direction along a terrane boundary, many mylonitic schists of greenschist or higher grade include lithologies suitable for radiometric age dating. $^{40}Ar/^{39}Ar$ mineral ages on micas within recrystallized sediments and granitic rocks, and amphiboles in sheared basic rocks, are particularly useful in recording late-kinematic uplift ages of such shear zones and so provide a minimum age constraint on ductile shearing.

Palaeomagnetic techniques continue to provide a key tool in the elucidation of terrane movements. Such methods can reveal not only shifts across palaeolatitude through time (using magnetic inclination data), but also record wholesale *rotation of a terrane* as it progresses on its journey. Magnetic declination data, effectively providing a remanent compass orientation, consistently reveal a clockwise rotation of outboard terranes relative to cratonic North America during right-lateral dispersal along the continental margin. Although clearly successful in the Cordilleran orogen where terrane movements have been south to north, the technique has been much less successful in the Appalachian/Caledonian orogen. Not only were Appalachian/Caledonian movements closer to east to west than south to north, but the rocks were much older and show more complex thermal and structural histories. Even in the Cordilleran belt the interpretation of palaeomagnetic data may be controversial, especially when data are obtained from plutonic rocks that may have undergone wholesale tilting (e.g. Butler et al. 1989). In addition, palaeomagnetic methods may be used not only to detect large terrane displacements but as evidence for the exact opposite. Renne & Scott (1988), for example, present a palaeomagnetic case against an exotic origin for what they consider a *native terrane* in the southeastern Klamath Mountains of California. Ideally, the palaeomagnetic case should be based on data from different laboratories and combined with other lines of geological evidence.

The presence within a suspect terrane of fossils that are clearly incompatible with coeval species found inboard on the craton raises the degree of suspicion that large fault displacements may have been involved. The most quoted example is provided by the large Palaeozoic foraminifera known as fusilinids that by Permian times occurred within two major families (e.g Monger & Price 1982). One of these families (*Verbeekinidae*) is recorded in highest densities in the equatorial Palaeotethys region and only occurs at high latitudes in suspect terranes scattered around the circum-Pacific margin. The other fusilinid family

(*Schwagerinidae*) occurs at high latitude on the cratonic margin. Palaeontological data are not, however, always unequivocal. In the southeastern Klamath Mountains of California, for example, brachiopod, molluscan, coral, and fusilinid faunas have been exhaustively studied but as yet no consensus has been reached concerning the overall faunal provinciality. Inconsistent interpretations of these faunas perhaps suggest that endemism among so called Tethyan faunas was less pronounced than has generally been believed (Renne & Scott 1988). Similar problems exist in other, older orogens — recent work in the Irish Caledonides, for example, emphasizes the lack of definite correlation between Ordovician faunas found in outboard suspect terranes and those typical of the established cratonic margin provinces associated with Laurentia, Baltica and Gondwana (Harper & Parkes 1989). Many such faunas are typically oceanic or marginal in nature (Fortey & Cocks 1988), with different genera showing different degrees of mobility across oceans, latitude and communities.

Sedimentological studies commonly play a key role in compiling data on suspect terranes. Sudden changes in the lithostratigraphy of coeval sediment packages across faults, with the truncation of major sediment supply systems, are common across displaced terrane boundaries. Detailed palaeocurrent analysis and increasingly sophisticated clast provenance and sedimentary geochemistry studies may reveal missing source areas incompatible with present outcrops (e.g. Bluck 1983, Haughton 1988). Radiometric ages from clasts can be an especially interesting method with which to investigate the provenance of sedimentary detritus (e.g. Elders 1987, Dempster & Bluck 1989, Haughton *et al.* 1990). Increasingly refined radiometric techniques such as using the $^{40}Ar/^{39}Ar$ laser probe to produce age population data on detrital micas and hornblende (Kelly & Bluck 1989) open new possibilities for terrane studies, and commonly reveal completely unforeseen complexities. The work of Kelley & Bluck (1989), for example, has shown that fresh andesitic clasts with Ordovician sediments from the Southern Uplands of Scotland were not, as had previously been assumed, eroded from a coeval arc. Instead, the radiometric data require the andesitic fragments to have already been over 80 Ma old before their incorporation into the sediments. However, given the complexities of modern sedimentary supply systems at active plate margins such results should not perhaps seem so surprising. Many of the sediments currently being supplied to the Sumatra–Java trench, for example, have travelled thousands of kilometres from the Himalayan mountains via the Ganges/Brahmaputra river system, and the Bengal submarine fan. The accretionary prism (see Chapter 9) presently being constructed, primarily from these sediments, therefore has a large exotic component that could be difficult to interpret if discovered in some more ancient, highly dispersed terrane collage.

The nature of basement rocks within a suspect terrane commonly bestows a distinctive identity. Such a basement may be distinguished on the basis of general lithology, radiometric ages, geochemistry (especially isotopes), and geophysical (aeromagnetic, gravity, deep seismic) data. Marked changes in the basement commonly occur across terrane boundaries. In the British Caledonides, for example, the prominent Highland Boundary fault system contains slivers of ophiolitic material (Highland Border Complex) and abruptly terminates the regional outcrop of Late Proterozoic metamorphic basement (Dalradian Supergroup). Similarly, exposures of Monian basement in northwest Wales terminate against the steep Menai Strait fault system within which slivers of ophiolitic blueschist are preserved (Gibbons 1987). There is not, however, always concordance between field data and indirect methods such as isotope geochemistry. Isotopic data on plutonic rocks cropping out in the northern British Caledonides, for example, show no significant change across the Highland Border fault system — the main change being across the next terrane boundary to the south (the Southern Uplands fault system) (Thirlwall 1989). This discovery can be interpreted in several ways that need not disagree with the field observations; for example, the source area for plutons on either side of the Highland Boundary fault system may be beneath a gently dipping terrane-boundary shear zone at depth. Alternatively, tectonic dispersal of suspect terrane slivers subparallel to the strike of an orogen may not necessarily result in radically different deep-crustal basements being juxtaposed at every terrane boundary.

The effect of dominantly vertical, rather than strike-slip, movements along a fault system can juxtapose basement and cover sequences and so produce a sudden, dramatic change in geology across the fault. Even without a basement-cover relationship, dip-slip juxtaposition of deeper metamorphic levels against low-grade or unmetamorphosed strata can lead to equally striking changes across faults. Most transcurrent terrane boundaries will show some component of dip-slip movement; along the transpressional Alpine fault in New Zealand, the southeastern side (around Mount Cook) has been estimated to be rising at a rate in the order of 10 mm/yr. Under an average geothermal gradient such rapid movements would need less than 2 million years to bring up amphibolite facies rocks against unmetamorphosed strata. This effect is enhanced along such transpressional boundaries where the heat flow is increased due to advection carrying the heat upwards more rapidly than it can be dissipated laterally (Cooper 1980, Koons 1987). It is not enough therefore to assume large-scale transcurrent movements simply on the basis of striking contrasts in geology across major faults. The problem is heightened by the fact that transcurrent terrane boundaries are likely to provide a focus for subsequent dip-slip reactivation which will tend to obscure the evidence for early lateral movements. For example, one of the most prominent fault systems in Southern Britain (the Welsh Borderlands fault system) shows a long history of Phanerozoic reactivation so that evidence for its earliest movements, although suspected to be transcurrent, have been eradicated (Woodcock & Gibbons 1988).

ACCRETION, DISPERSAL AND OROGENESIS

Studies of the World's major orogens continue to reveal how each belt, although generated over a long period (perhaps 100 Ma) is punctuated by relatively short-lived (e.g. <10 Ma) tectono-metamorphic events that affect all or part of the orogen. Such events are commonly referred to as separate orogenies such as the Grampian and Acadian orogenies in the Appalachian/Caledonian system, and the Nevadan and Laramide orogenies in the Western Cordillera. Given that orogenic belts are produced in response to oceanic plate subduction, the production of orogenic crescendi and diminuendi within and inboard of the plate margin presumably reflects major changes in plate interaction. Prominent orogenic pulses may therefore be produced by such events as the initiation of subduction, the arrival and attempted subduction of relatively buoyant areas riding on the oceanic crust, a rapid change in the angle of the subducting slab and the rate or direction of convergence, and subduction of a spreading centre. The Laramide orogeny, for example, has been related to a rapid decrease in the subducting slab angle beneath North America (e.g. Coney & Reynolds 1977). Similarly, the subduction and orogenic history of western South America has been related to variations in the angle of the subducting slab in both space and time (e.g. gentle angle of less than 30° yields compression and no magmatism) (Megard 1987).

The accretion of exotic terranes is widely considered to be an important mechanism initiating pulses of orogenic activity that are felt far inboard within the host continent. The 145 Ma accretion of the Coast Range Ophiolite arc terrane to the cratonic margin of North America, for example, was coincident with the Nevadan orogeny, a major plate reorganization in the East Pacific basin, and the temporary cessation of magmatism (McLaughlin et al. 1988). A more recent example involving the accretion of an oceanic volcanic arc is provided by Taiwan. This arc terrane, generated on the oceanic Phillipine plate, has collided obliquely with the continental margin of China. Similarly, the collision of the Indonesian Banda arc with northern Australia, producing the Wetar back-arc thrust belt behind Timor (Breen et al. 1989), provides a classic modern example of arc obduction over a cratonic shelf margin.

It is possible to recognize major changes in plate interaction and attempt to relate these to pulses of orogenic activity by studying the *apparent polar wander (APW) path* record for each of the cratonic active margins. The APW path for the North American craton, for example, reveals a series of abrupt changes in plate velocity and direction (May et al. 1989). The craton appears suddenly to have moved rapidly northwards during late Jurassic to Early Cretaceous

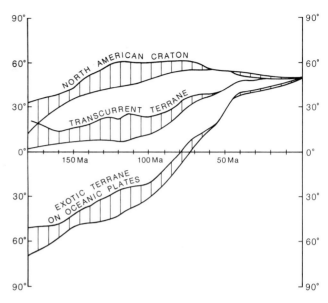

Fig. 15.10. Three scenarios for the absolute latitudinal shifts of a point presently situated at the North American continental margin at 48°N (strait of Juan de Fuca) (after Ave Lallement and Oldow 1988). The uncertainty for each curve derives from using two data sets (Engebretson et al. 1985, Debiche et al. 1987). The greatest latitudinal shift is shown if the point were part of an exotic terrane attached to the Pacific, Kula and Farallon oceanic plates. Note that if the point had been part of a transcurrent terrane this would have been moving northwards more slowly than the American craton from 180–120 Ma and so could have been dispersed up to 3500 km sinistrally southeastwards along the continental margin before a later (post-100 Ma) reversal of relative movement sense produced up to 5000 km of displacement back northwestwards in a dextral sense.

time (c. 150–135 Ma) at a speed of over 150 mm/yr (possibly as high as 230 mm/yr). This phase of northwards movement is interpreted by several authors as a time when suspect terranes were being moved relatively southwards down the cratonic margin (e.g. Avé Lallemant & Oldow 1988) (Fig. 15.10). Such an interpretation accords with the earlier models of Engebretson et al. (1985) and with an increasing body of structural and, to a lesser extent, palaeomagnetic and lithostratigraphic data. The inception of rapid northward movement and left-lateral terrane dispersal coincided with the collapse of a back-arc basin during the Nevadan orogeny in the Klamath Mountains. The question as to whether the compressive orogenic movements occurring at this time were due to arc-continent collision and/or a change in relative plate motion remains unresolved. One important corollary of this suspected early left-lateral terrane dispersal along the North American margin is that during the later, well-documented, phase of right-lateral dispersion (mainly from the late Cretaceous to Eocene) some suspect terranes would have moved back along the margin. The possibility therefore exists for some Cordilleran suspect terranes having been displaced over 2000 km sinistrally before the later dextral regime, also involving thousands of kilometres displacement, took over. The potential complexity induced by such relative reversals of lateral fault movements along previously active orogenic margins is daunting to consider.

CONCLUSIONS

In conclusion, the application of the suspect terrane concept represents no more than a natural progression in our understanding of plate tectonic processes. This does not, however, diminish the importance of recognizing suspect terranes because the degree of potential lateral transport along orogenic belts is not a trivial problem. The main value of the terrane concept is in the appreciation it has given us of the consequences of oblique subduction in the form of lateral movements along active plate margins. A consideration of the speed at which many transcurrently displaced terranes have moved (>50 km/Ma), combined with a realization of the time involved (>100 Ma) in the construction of orogenic belts, makes large-scale transcurrent faulting along the margins of large oceans inevitable. Extra complexities are added by the heterogeneity of the ocean floor and the effects of local collisional terrane accretion during subduction. Ancient orogens produced on the margins of wide oceans, however, will preserve less of the early history of subduction and accretion, and more of the final dispersion of a mosaic of displaced terranes along the orogenic strike. The results of transcurrent terrane dispersal along a continental margin are likely to be as complex as the suspect terrane concept is, in essence, simple.

REFERENCES

Ave Lallement, H. G. & Oldow, J. S. 1988. Early Mesozoic southward migration of Cordilleran transpressional terranes. *Tectonics* **7**, 1057–1075.

Ben-Avraham, Z. & Nur, A. 1983. An introductory overview to the concept of displaced terranes. *Can. J. Earth Sci.* **20**, 994–999.

Ben-Avraham, Z. & Nur, A. 1987. Effects of collisions at trenches on oceanic ridges and passive margins. *Am. geophys. U. geodynam. Ser.* **18**, 9–18.

Ben-Avraham, Z., Nur, A., Jones, D. & Cox, A. 1981. Continental accretion and orogeny: oceanic plateaus to allochthonous terranes. *Science* **213**, 47–54.

Berg, H. C., Jones, D. L. & Richter, D. H. 1972. Gravina-Nutzotin belt — tectonic significance of an upper Mesozoic sedimentary and volcanic sequence in southern and southeastern Alaska. *U.S. geol. Surv. Prof. Pap.* **800-D**, D1–24.

Berg, H. C., Jones, D. L. & Coney, P. J. 1978. Map showing pre-Cenozoic tectonostratigraphic terranes of southeastern Alaska and adjacent areas. *U.S. geol. Surv. Open File Rep.* **78–1085**, scale 1:1,000,000.

Blenkinsop, J., Cueman, P. F. & Bell, K. 1976. Age relationships along the Hermitage Bay – Dover Fault system, Newfoundland. *Nature* **262**, 377–378.

Bluck, B. J. 1983. Role of the Midland Valley of Scotland in the Caledonian orogeny. *Trans. R. Soc. Edinb.* **74**, 119–136.

Bobyarchick, A. R. 1988. Location and geometry of Alleghanian dispersal-related strike-slip faults in the southern Appalachians. *Geology* **16**, 915–919.

Breen, N. A., Silver, E. A. & Roof, S. 1989. The Wetar Back Arc Thrust Belt, Eastern Indonesia: the effect of accretion against an irregularly shaped arc. *Tectonics* **8**, 85–98.

Butler, R. F., Gehrels, G. E., McClellend, W. C., May, S. R. & Klepacki, D. 1989. Discordant palaeomagnetic poles from the Canadian Coast Plutonic Complex: regional tilt rather than large-scale displacement? *Geology* **17**, 691–694.

Cawood, P. A. 1983. Accretionary tectonics and terrane dispersal within the New England fold belt, Eastern Australia, Proceedings of the Circum-Pacific terrane conference. *Stanford Univ. Publ. geol. Ser.* **18**, 50–52.

Clarke, D. B. & Halliday, A. N. 1985. Sm/Nd isotopic investigation of the age and origin of the Meguma Zone metasedimentary rocks. *Can. J. Earth Sci.* **22**, 102–107.

Coleman, R. G. 1989. Continental growth of Northwest China. *Tectonics* **8**, 621–635.

Coney, P. J., 1987. Circum-Pacific tectogenesis in the North American Cordillera. *Am. geophys. U. geodynam. Ser.* **18**, 59–69.

Coney, P. J. & Reynolds, S. J. 1977. Cordilleran Benioff Zones. *Nature* **270**, 403–406.

Coney, P. J., Jones, D. L. & Monger, J. W. H. 1980. Cordilleran suspect terranes. *Nature* **288**, 329–333.

Cooper, A. F. 1989. Retrograde alteration of chromium kyanite in metachert and amphibolite whiteschist from the Southern Alps, New Zealand, with implications for uplift on the Alpine Faults. *Contrib. Mineral. Petrol.* **75**, 153–164.

Debiche, M. G., Cox, A. & Engebretson, D. 1987. The motion of allochthonous terranes across the North Pacific basin. *Spec. Pap. geol. Soc. Am.* **207**, 1–49.

Dempster, T. J. & Bluck, B. J. 1989. The age and origin of boulders in the Highland Border Complex: constraints on terrane movements. *J. geol. Soc. Lond.* **146**, 377–379.

Dewey, J. F. 1969. Evolution of the Caledonian–Appalachian orogen. *Nature* **222**, 124–129.

Dewey, J. F., Gass, I. G., Curry, G. B., Harris, N. B. W. & Şengör, A. M. C. 1991. *Allochthonous Terranes*. Cambridge University Press, Cambridge.

Elders, C. F. 1987. The provenance of granite boulders in conglomerates of the northern and central belts of the southern uplands of Scotland. *J. geol. Soc. Lond.* **144**, 853–863.

Engebretson, D. C., Cox, A. & Gordon, R. G. 1985. Relative motions between oceanic and continental plates in the Pacific basin. *J. geophys. Res.* **89**, 10291–10310.

Engebretson, D. C., Cox, A. & Debiche, M. 1987. Reconstructions, plate interactions, trajectories of oceanic and continental plates in the Pacific Ocean. *Am. geophys. U. geodynam. Ser.* **18**, 19–27.

Erdmer, P. 1987. Blueschist and eclogite in mylonitic allochthons, Ross River and Watson Lake areas, southeastern Yukon. *Can. J. Earth Sci.* **24**, 1439–1449.

Fortey, R. A. & Cocks, L. R. M. 1988. Arenig to Llandovery faunal distributions in the Caledonides, In: *The Caledonian-Appalachian Orogen* (edited by Harris, A. L. & Fettes, D. J.). *Spec. Publs geol. Soc. Lond.* **38**, 233–246.

Gibbons, W. 1983. Stratigraphy, subduction and strike-slip faulting in the Mona Complex of North Wales — a review. *Proc. Geol. Ass.* **94**, 147–163.

Gibbons, W. 1990. Transcurrent ductile shear zones and the dispersal of the Avalon Superterrane, In: *The Cadomian Orogeny* (edited by D'Lemos, R. S., Strachan, R. A. & Topley, C. G.). *Spec. Publs geol. Soc. Lond.* **51**, 407–423.

Gibbons, W. & Gayer, R. A. 1985. British Caledonian Terranes. *Earth Evolution Sciences, Monograph Series* **1**, 3–16.

Hamilton, W. J. 1979. Tectonics of the Indonesian region. *U. S. geol. Surv. Prof. Pap.* **1078**.

Hansen, W. L. 1988. A model for terrane accretion: Yukon-Tanana and Slide Mountain terranes, northwest North America. *Tectonics* **7**, 1167–1177.

Harper, D. A. P. & Parkes, M. A. 1989. Palaeontological constraints on the definition and development of Irish Caledonide terranes. *J. geol. Soc. Lond.* **146**, 413–416.

Haston, R. B., Luyendyk, B. P., Landis, C. A. & Coombs, D. S. 1989. Palaeomagnetism and question of original location of the Permian Brook Street Terrane. New Zealand. *Tectonics* **8**, 791–802.

Haughton, P. D. W. 1988. A cryptic Caledonian flysch terrane in Scotland: *J. geol. Soc. Lond.* **145**, 685–703.

Haughton, P. D. W., Rogers, G. & Halliday, A. N. 1990. Provenance of Lower Old Red Sandstone Conglomerate, SE Kincardineshire: evidence for the timing of Caledonian terrane accretion in central Scotland. *J. geol. Soc. Lond.* **147**, 105–120.

Hooper, R. J. & Hatcher, R. D. 1988. Pine Mountain terrane, a complex window in the Georgia and Alabama Piedmont; evidence from the eastern termination. *Geology* **16**, 307–310.

Howell, D. G. 1980. Mesozoic accretion of exotic terranes along the New Zealand segment of Gondwanaland. *Geology* **8**, 487–491.

Howell, D. G. 1989. *Tectonics of Suspect Terranes: Mountain Building and Continental Growth.* Chapman and Hall London, 1–232.

Hutton, D. H. W. 1987. Strike-slip terranes and a model for the evolution of the British and Irish Caledonides. *Geol. Mag.* **124**, 405–425.

Irving, E. & Yole, R. W. 1972. Palaeomagnetism and the kinematic history of mafic and ultramafic rocks in fold mountain belts. *Ottawa Earth Phys. Branch Publ.* **42**, 87–95.

Irwin, W. P. 1960. Geological reconnaissance of the northern Coast Ranges and Klamath Mountains, California. *Bull. Calif. Div. Mines* **179**, 1–80.

Irwin, W. P. 1972. Terranes of the western Paleozoic and Triassic belt in the southern Klamath Mountains, California. *U.S. geol. Surv. Prof. Pap.* **800-C**, 103–111.

Jarrard, R. D. 1986. Terrane motion by strike-slip faulting of forearc slivers. *Geology* **14**, 780–783.

Jones, D. L., Silberling, N. J. & Hillhouse, J. 1977. Wrangellia — a displaced terrane in northwestern North America. *Can. J. Earth Sci.* **14**, 2565–2577.

Jones, D. L., Cox, A., Coney, P. & Beck, M. 1982. The growth of western North America. *Scient. Am.* **247**, 70–84.

Kaplan Morris, L., Lund, S. P. & Bottjer, D. J. 1986. Palaeolatitude drift history of displaced terranes in southern and Baja California. *Nature* **321**, 844–847.

Karig, D. E., Sarewitz, D. R. & Haeck, G. D. 1986. Role of strike-slip faulting in the evolution of allochthonous terranes in the Phillipines. *Geology* **14**, 852–855.

Kelley, S. & Bluck, B. J. 1989. Detrital mineral ages from the Southern Uplands using $^{40}Ar/^{39}Ar$ laser probe. *J. geol. Soc. Lond.* **146**, 401–403.

Keppie, J. D. 1982. Terranes in the northern Appalachians. *Nova Scotia Dep. Mines Energy*, scale 1:5,000,000.

Keppie, J. D. & Dallmeyer, R. D. 1989. Tectonic map of pre-Mesozoic terranes in Circum-Atlantic Orogen. *Int. geol. Correl. Prog. Proj. 233: Terranes in the circum-Atlantic Palaeozoic orogens.*

Kerr, R. A. 1983. Suspect terranes and continental growth. *Science* **222**, 36–38.

Kistler, R. W. & Peterman, Z. E. 1978. Reconstruction of crustal blocks of California on the basis of initial Sr isotopic compositions of Mesozoic plutons. *U. S. geol. Surv. Prof. Pap.* **1061**, 1–27.

Koons, P. D. 1987. Some thermal and mechanical consequences of rapid uplift: an example from the Southern Alps, New Zealand. *Earth Planet Sci. Lett.* **86**, 307–319.

Larson, R. L., Menard, H. W. & Smith, S. 1968. Gulf of California: A result of ocean floor spreading and transform faulting. *Science* **161**, 781–784.

May, S. R., Beck, M. E. Jr & Butler, R. 1989. North American apparent polar wander, plate motion, and left-oblique convergence: Late Jurassic–Early Cretaceous orogenic consequences. *Tectonics* **8**, 443–451.

McLaughlin, R. J., Blake, M. C. Jr., Griscom, A., Blome, C. D. & Murchey, B. 1988. Tectonics of formation, translation, and dispersal of the Coast Range ophiolite of California. *Tectonics* **7**, 1033–1056.

McWilliams, M. O. & Howell, D. G. 1982. Exotic terranes of western California. *Nature* **297**, 215–217.

Mégard, F. 1987. Cordilleran Andes and Marginal Andes: a review of Andean geology north of the Arica elbow (18°S). *Am. geophys. U. geodynam. Ser.* **18**, 71–95.

Monger, J. W. H. & Ross, C. A. 1971. Distribution of fusilinaceans in the western Canadian Cordillera. *Can. J. Earth Sci.* **8**, 259–278.

Monger, J. W. H., Price, R. A. & Tempelman-Kluit, D. J. 1982. Tectonic accretion and the origin of two major metamorphic and plutonic welts in the Canadian Cordillera. *Geology* **10**, 70–75.

Moores, E. M., Robinson, P. T., Malpas, J. & Xenophonotos, C. 1984. Model for the origin of the Troodos Massif, Cyprus, and other mid-east ophiolites. *Geology* **12**, 500–503.

Noel, J. R., Spariosu, D. J. & Dallmeyer R. D. 1988. Palaeomagnetism and $^{40}Ar/^{39}Ar$ ages from the Carolina slate belt, Albemarle, North Carolina: Implications for terrane amalgamation with North America. *Geology* **16**, 64–68.

Nutman, A. P., Friend, C. R. L., Baadsgaard, H. & McGregor, V. R. 1989. Evolution and assembly of Archaen gneiss terranes in the Godthabsfjorn region, southern West Greenland: structural, metamorphic and isotopic evidence. *Tectonics* **8**, 573–589.

Opdyke, N. D., Jones, D. S., MacFadden, B. J., Smith, D. L., Mueller, P. A. & Shuster, R. D. 1987. Florida as an exotic terrane: Palaeomagnetic and geochronologic investigation of Lower Palaeozoic rocks from the subsurface of Florida. *Geology* **15**, 900–903.

Ortlieb, L., Ruegg, J. C., Angelier, J., Colletta, B., Kasser, M. & Lesage, P. 1989. Geodetic and tectonic analyses along an active plate boundary: the Central Gulf of California. *Tectonics* **8**, 429–441.

Packer, D. R. & Stone, D. B. 1972. An Alaskan Jurassic palaeomagnetic pole and the Alaskan orocline. *Nature.* **237**, 25–26.

Packer, D. R. & Store, D. B. 1974. Palaeomagnetism of Jurassic rocks from southern Alaska, and the tectonic implications. *Can. J. Earth Sci.* **11**, 976–997.

Plafker, G., Blome, C. D. & Silberling, N. J. 1989. Reinterpretation of lower Mesozoic rocks on the Chilkat Peninsula, Alaska, as a displaced fragment of Wrangellia. *Geology* **17**, 3–6.

Reichle, M. S., Sharman, G. F. & Brune, J. N. 1976. Sonobuoy and teleseismic studies of the Gulf of California transform fault earthquake sequence. *Bull. seism. Soc. Am.* **66**, 1623–1641.

Renne, P. R. & Scott, G. R. 1988. Structural chronology, oroclinal deformation and tectonic evolution of the southeastern Klamath Mountains, California. *Tectonics* **7**, 1223–1242.

Roberts, D. 1988. The terrane concept and the Scandanavian Caledonides: a synthesis: *Norges Geol. Undersøk. Bull.* **413**, 93–99.

Roeske, S. M., Mattinson, J. M. & Armstrong, R. L. 1989.

Isotopic ages of glaucophane schists on the Kodiak Islands, southern Alaska, and their implications for the Mesozoic tectonic history of the Border Ranges fault system. *Bull. geol. Soc. Am.* **101**, 1021–1037.

Ryan, H. F. & Scholl, D. W. 1989. The evolution of forearc structures along an oblique convergent margin, central Aleutian arc. *Tectonics* **8**, 497–516.

Scneibner, E. 1987. Paleozoic tectonic development of Eastern Australia in relation to the Pacific region. *Am. geophys. U. Geodynam. Ser.* **18**, 133–165.

Schermer, E. R., Howell, D. G. & Jones, D. L. 1984. The origin of allochthonous terranes: perspectives on the growth and shaping of continents. *A. Rev. Earth Planet. Sci.* **12**, 107–131.

Şengör, A. M. C. 1988. Evolution of thought on thrust faulting and the Alpine-Himalayan system. *Geol. För. Stockh. Förh.* **110**, 416.

Silver, E. A. & Smith, R. B. 1983. Comparison of terrane accretion in modern SE Asia and the Mesozoic North American Cordillera. *Geology* **11**, 198–202.

Simpson, C. & Schmid, S. M. 1983. An evaluation of criteria to deduce the sense of movement in sheared rocks. *Bull. geol. Soc. Am.* **94**, 1281–1288.

Smith, P. J. 1983. Suspect terranes. *Nature* **305**, 475–476.

Sporli, K. B. 1987. Development of the New Zealand Microcontinent. *Am. geophys. U. Geodynam. Ser.* **18**, 115–132.

Steckler, M. S. 1989. The role of lithospheric strength variations in the formation of allochthonous terranes. *Abstr. R. Soc. Lond. Discuss. Meet. Allochthonous Terranes.*

Stock, J. M. & Hodges, K. V. 1989. Pre-Pliocene extension around the Gulf of California and the transfer of Baja California to the Pacific plate. *Tectonics* **8**, 99–115.

Thirlwall, M. F. 1989. Movement on proposed terrane boundaries in Northern Britain: constraints from Ordovician–Devonian igneous rocks. *J. geol. Soc. Lond.* **146**, 373–376.

Vink, G. E., Morgan, W. J. & Wu-ling Zhao. 1984. Preferential rifting of continents: a source of displaced terranes. *J. geophys. Res.* **89**, 10,072–10,076.

Wheeler, R. L. & Bollinger, G. A. 1984. Seismicity and suspect terranes in the southeastern United States. *Geology* **12**, 323–326.

Williams, H. 1984. Miogeoclines and suspect terranes of the Celdonian-Appalachian Orogen: tectonic patterns in the North Atlantic region. *Can. J. Earth Sci.* **21**, 887–901.

Williams, H. & Hatcher, R. D. 1982. Suspect terranes and the accretionary history of the Appalachian orogen. *Geology* **10**, 530–536.

Woodcock, N. & Gibbons, W. 1988. Is the Welsh Borderlands Fault System a terrane boundary? *J. geol. Soc. Lond.* **145**, 915–924.

Zen, E-An. 1988. Evidence for accreted terranes and the effect of metamorphism. *Am. J. Sci.* **288-A**, 1–15.

Zonenshain, L. P., Kononov, M. V. & Savostin, L. A. 1987. Pacific and Kula/Eurasia relative motions during the last 130 Ma and their bearing on orogenesis in northeast Asia. *Am. geophys. U. Geodynam. Ser.* **18**, 29–47.

CHAPTER 16

Tectonosedimentation: with Examples from the Tertiary-Recent of Southeast Japan

KEVIN T. PICKERING and ASAHIKO TAIRA

INTRODUCTION

SEDIMENTARY successions deposited in basins at destructive plate margins provide an important record of orogenic processes, and the resulting stratigraphies. This chapter examines various aspects of the tectonosedimentary evolution of the modern Japanese arc-related deposits in the arc–arc collision zone, southeast Japan, referred to as the Izu collision zone (ICZ). The ICZ encompasses the region of Neogene-Recent arc–arc collision between the mainland Japanese, or Honshu, arc and the Izu-Bonin island arc immediately west of the trench–trench–trench triple junction off southeast Japan, and records the uplift and incremental accretion by crustal imbrication of arc crust and associated sediments. The pre-collision stratigraphy of the palaeo-Izu-Bonin arc, pre-3–2.5 Ma, appears to be principally controlled by tectonic processes, whereas the post-collision stratigraphy of the plate-boundary infill, particularly the younger parts of the Kazusa Group forearc-basin fill, during the past 0.8 Ma, can be interpreted to show a strong glacio-eustatic signature.

The fluid activity and its relationship to the regional tectonics in the Nankai, Japan and Izu-Bonin (Ogasawara) forearcs is also reviewed. The ICZ region shows abundant evidence for fluid activity in the progressively deforming and lithifying sediments, both in the onland successions and in the present collision zone.

This chapter is based particularly on papers by Taira *et al.* (1989), Pickering *et al.* (1990), Soh *et al.* (1991), Taira & Pickering (1991), and Pickering *et al.* (1993), but it also includes a new sequence stratigraphic synthesis of the Neogene of Central Honshu in order to compare local/regional tectonics with the Exxon group's proposed eustatic controls on depositional systems. The emphasis throughout is on tectonics and sedimentation, the role of fluids in dewatering, lithifying and progressively deforming sediments, and stratigraphy.

This site-specific approach to the theme of tectonosedimentation is taken for two reasons. First, relationships between sedimentation and tectonic activity and hydrogeology are currently at the forefront of much thinking about the evolution of accretionary prisms. Second, many reviews of the general topic of tectonics and sedimentation, often emphasizing terrestrial sediments, have already been published. This chapter should be read after Chapter 7 "Prelithification Deformation" and Chapter 9 "Arc-Trench Tectonics". An understanding of "Linked Fault Systems" (Chapter 5) and "Continental Collision" (Chapter 13) is also assumed.

Two types of ocean plate convergence occur along the Pacific margin of Japan (Fig. 16.1): (a) *subduction of normal-thickness oceanic crust*, as in the Nankai or Japan trenches, and (b) *arc–arc collision-accretion*, as in the Izu collision zone (ICZ). In the Nankai trough, a thick succession of trench turbidites and hemipelagites have been incorporated into a well-developed *accretionary prism*. This accretionary process is analogous to that which led to the formation of most of the basement of the Japanese island arc or Honshu arc. The prolonged history of oceanic plate subduction recorded in the Honshu arc, since about 200 Ma, has produced a crustal section 30 km thick (*c.* 10 kb), comprising mainly supracrustal rocks and some igneous arc material, in which oceanic crustal rocks do not exceed more than 30%.

Accretionary wedges created by oceanic plate subduction, such as that associated with the Nankai trough, contain virtually no crustal slices, even at deeper structural levels, and the layer III gabbroic rocks do not constitute an important component of the accreted material in the Honshu arc. Instead, the basement comprises exclusively oceanic supracrustal sequences.

Intra-plate deformation of a subducting arc may change the plate boundary position, so that basin inversion occurs in the intra-oceanic basins — something that is documented in this chapter for the Izu collision zone. The dimensions of accreted arc segments are probably governed largely by the critical spacing of zones of intra-plate deformation, including the location and orientation of transform faults intersecting the arc basement. Figure 16.2 summarizes the plate-tectonic history of the Japanese arc systems since the late Permian. The main tectonic events were as follows.

(1) During the late Permian–Triassic, there was the

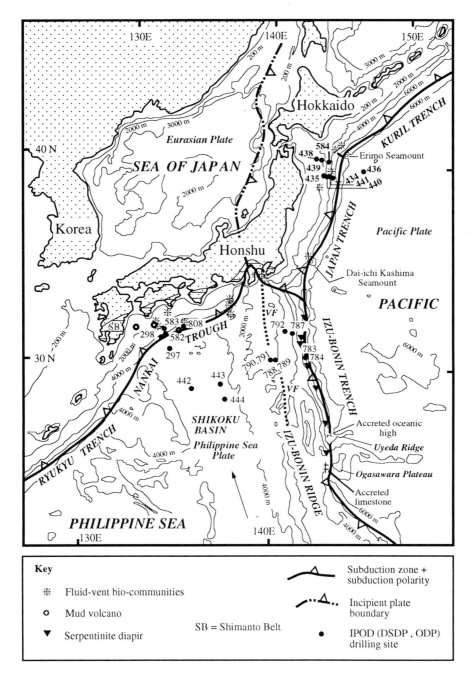

Fig. 16.1. Tectonic setting of the Japanese subduction zones and location of DSDP and ODP sites, volcanic front of Izu-Bonin arc (pecked line labelled VF), probable fluid-venting related biological communities (asterisks), serpentinite diapirs, and location of selected mud volcanoes referred to in this paper. SB, Shimanto belt. Convergence vector of the Philippine Sea plate from Seno (1989) is shown as an arrow.

development of an accretionary prism along the southeast Asian continental margin, already the site of considerable Phanerozoic terrane accretion (e.g. Yangtze, Sino-Korean, Tarim and Bureya terranes) against the Siberian continent. Notable accretionary events include the partial obduction of a seamount chain as the Akiyoshi Limestone terrane in western Honshu (Kanmera & Sano 1986, Kanmera et al. 1990). The Sangun metamorphic belt represents part of this Permian accretionary prism.

(2) During the Middle and late Jurassic, additional extensive accretionary events occurred, with the subduction-accretion of seamounts, and the development of substantial accretionary prisms. In the late Jurassic,

microcontinental blocks with Silurian–Triassic rocks (Kurosegawa and South Kitakami terranes) were accreted, probably associated with the large-scale folding of the Sambagawa metamorphic belt. The late Jurassic appears to have involved considerable sinistral strike-slip along the Median Tectonic Line (MTL) and Kurosegawa Tectonic Line (KTL); there are significant faunal and floral difference across the MTL and KTL. Oblique subduction and strike-slip faulting continued throughout the late Jurassic and early Cretaceous.

(3) The late Cretaceous (particularly the Campanian) witnessed the formation of a large accretionary prism to form the Shimanto and Hidaka belts. This intense phase of accretionary prism growth may have been associated

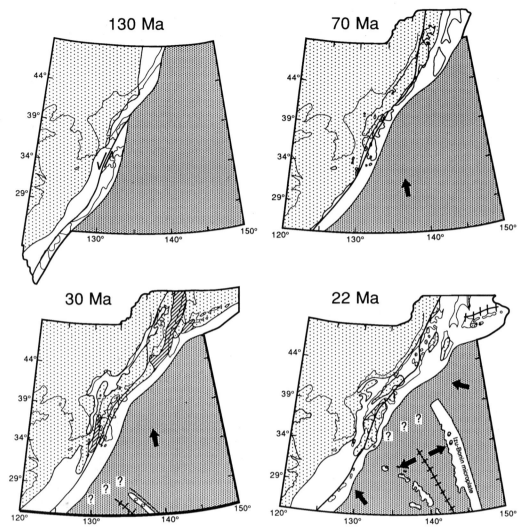

Fig. 16.2. Plate-tectonic evolution of Japan since c. 130 Ma (modified after Taira 1986, Hibbard & Karig 1990, Koyama 1991).

with underplating causing uplift of the Sambagawa metamorphic belt (Taira et al. 1989).

(4) The main Tertiary collision events included:

(a) Early Palaeogene collision-accretion of the Eastern against the Western (Okhotsk) Hokkaido terranes, and the final emplacement of the Yezo Group upon the Horokanai ophiolitic basement, probably from an original forearc basin setting (Kiminami et al. 1985);

(b) Late Palaeogene – early Miocene rifting to initiate the Sea of Japan, associated with a change from non-marine, including freshwater lacustrine, environments;

(c) Oligocene – Miocene backarc rifting behind the Izu-Bonin arc to produce the Shikoku marginal basin and western Kyushu – Palau ridge. The continental margin of southern Honshu and Shikoku was probably dominated by strike-slip tectonics;

(d) At about 15 – 14 Ma, the Sea of Japan and Okhotsk Sea marginal basins underwent a phase of rapid seafloor spreading, although opening began much earlier at c. 26 Ma: northern Honshu and Hokkaido rotated anticlockwise, whereas Kyushu, Shikoku and southern Honshu rotated clockwise. These events appear to have occurred within the space of only a few million years. A consequence of the opening of the Sea of Okhotsk was to cause strong E – W compression in Hokkaido to form a mountain belt, and uplift the Hidaka metamorphic belt. Geological evidence from the Shimanto belt has led Hibbard & Karig (1990) to postulate that: (i) during the opening of the Shikoku marginal basin, from 26 – 15 Ma, it was separated from the Japanese margin by an extension of the Pacific plate, and (ii) the Shikoku basin spreading centre collided with southwest Japan at c. 15 Ma, at which time the TTT triple junction was probably initiated. In this scenario, a transform boundary is interpreted to separate the Pacific plate from the northern edge of the Shikoku basin, on the Philippine Sea plate (Hibbard & Karig 1990). Koyama (1991) summarizes the palaeomagnetic evidence, including data from the recent ODP Leg 126, to support a clockwise rotation of the Shikoku basin and associated Izu-Bonin island arc (and Kyushu-Palau Ridge), associated with the opening of the Sea of Japan. The palaeomagnetic data from onland and marine sites shows large (30 – 100°) clockwise deflections of declinations in Eocene to middle Miocene rocks, consistent with large clockwise rotation and northward drift (greater than 10° of latitude) for the

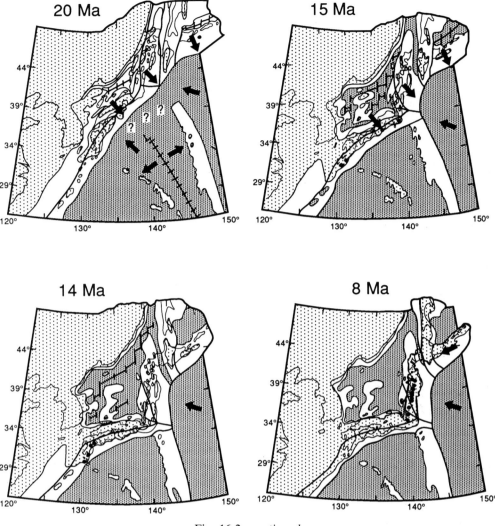

Fig. 16.2. *continued*

entire Philippine Sea plate between 27–17 Ma (Koyama 1991). In Fig. 16.2, we show plate reconstructions using data from Taira (1986), Hibbard & Karig (1990), and Koyama (1991).

(e) During the late Miocene, the Izu-Bonin arc began to impinge against the Honshu arc to generate the Izu collision zone (ICZ). Episodic collision-accretion occurred to amalgamate imbricated crustal slices from the Izu-Bonin arc onto the leading edge of the southeast part of the Honshu arc. In southeast Honshu, the arc–arc collision, which is still continuing, caused the bending of the geological belts, such as in the Fossa Magna; and

(f) Late Miocene–Recent backarc oblique, dextral, rifting in the Okinawa trough, with the propagation of the rift system as far north as Kyushu.

(5) Present-day intraplate or inter-microplate deformation is occurring along the eastern margin of the Sea of Japan (probably since the Pliocene), possibly to initiate the closure of this basin along an incipient E-dipping subduction zone (Tamaki & Honza 1985). There is abundant evidence for neotectonic activity associated with the accretionary systems of Japan, something that has been very well documented by the submersible work of the Franco-Japanese KAIKO Project (e.g. Cadet *et al.* 1987a, b, Le Pichon *et al.* 1987a, b, c). Intraplate deformation is also taking place in the ICZ, for example behind the Zenisu ridge to initiate a new plate boundary and accrete the northern segment of the Izu-Bonin arc onto the Honshu arc. Presently, the Izu-Bonin arc appears to be undergoing a major phase of extension with active normal faulting occurring, for example, the major extensional faulting in a graben running along the entire axis of the arc. It is the tectono sedimentary aspects of the ICZ that are the focus for much of this paper. The following section, however, concerns the nature of fluids in the *forearcs* of the Japanese subduction zones because of their paramount importance in controlling sediment deformation processes and indeed depositional processes following modification of the seafloor.

SEDIMENT DEFORMATION AND FLUID ACTIVITY IN THE JAPANESE FOREARCS

Several important fluid pathways (see Chapter 7) can be identified within forearc accretionary prisms (see Tarney *et al.* 1991): (a) *pervasive advection of fluids* through granular or fracture permeability; (b) *fluid*

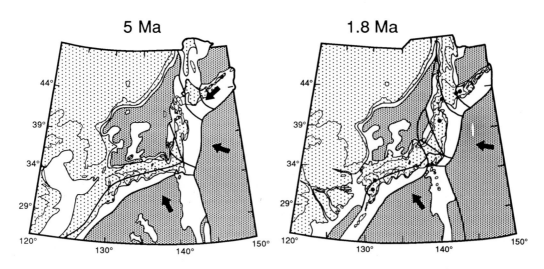

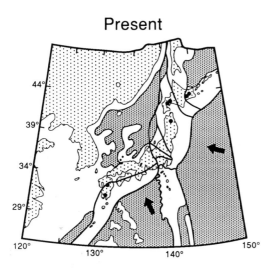

Fig. 16.2. continued

migration through mesoscopic structures such as *veins, dykes, deformation bands* and *'annealed' faults* (see Chapter 5); (c) *concentrated fluid flow* through specific *low-permeability channels*, such as fault zones and *décollement horizons*, and (d) *fluid transfer* associated with large-scale material transfer, as in *diapirism*. These four mechanisms are interactive, although their relative importance varies between forearcs. In the Nankai trough forearc, for example, channellized fluid venting can be important associated with frontal deformation and accretion, large-scale fluid advection being more significant in landward sites. The same may be true for other Japanese forearcs, but the nature of fluid advection is poorly understood in the Japan and Izu-Bonin trench forearcs.

A variety of sediment and crustal deformation, associated with fluid activity, has been observed in the present forearc slopes, trenches and ancient onland outcrops in the Japanese island arcs. The sources for these fluids include the subducting slab and sediments, accreted and subcreted sediments, cover sequences, basement magma-related fluids and meteoric water.

The Nankai forearc represents a typical clastic-dominated accretionary prism (Fig. 16.3), where the expulsion of pore fluids from sediments seems to have occurred intermittently, through both channellized and diffusive mechanisms, some of which appears to be pulsed. *Mud diapirs* occur within the trench, near the toe and upper slope of the accretionary prism (Fig. 16.4). The Japan trench forearc is characterized by *tectono-gravity slope instability* and collapse, resulting in zones of fluid escape. The Izu-Bonin trench forearc contains recently discovered serpentinite diapirs in which the mud matrix includes blocks of basalt, and even blueschist. Vein structures have been reported from both the Japan and Izu-Bonin trench forearcs. A well-developed *bottom-simulating reflector* (BSR) is only recorded from the Nankai trench forearc at about 200 mbsf, being the shallowest known BSR in an accretionary prism. In the ICZ, where there is arc–arc collision, fluid migration and expulsion (*fluid venting*) occurs in a wide zone of distributed deformation, including the oceanic plate side. Fluid and *sediment injection structures* are common on all scales. Fossil *Calyptogena* beds in the Pliocene basins of the collision zone provide good examples of the tectono-stratigraphic

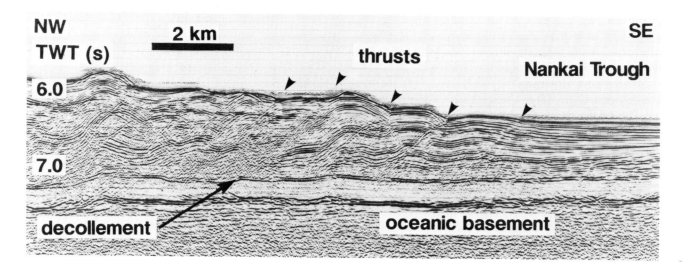

Fig. 16.3. Seismic line NT62–8 across the Nankai accretionary prism in the vicinity of ODP Leg 131 Site 808. TWT = two-way travel time in seconds.

control and geometry of fluid activity associated with venting along palaeo-fault zones. Thus, fluid activity in forearcs is an important contemporary process that has perhaps been under-estimated in the geological column.

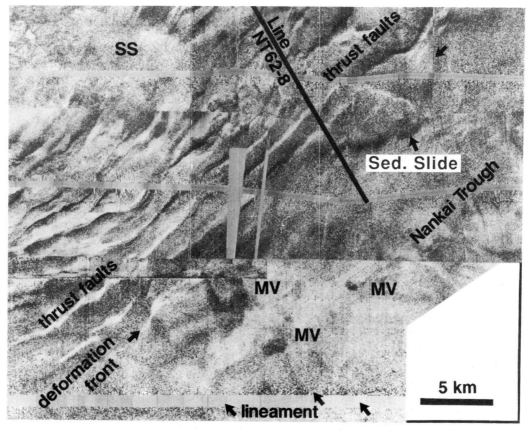

Fig. 16.4. IZANAGI Sidescan sonar image of the seafloor in the vicinity of ODP Leg 131 Site 808. Position of seismic line NT62–8 (Fig. 16.3); MV, mud volcano; SS and sed. slide, submarine slide; lower-slope thrusts intersect sea floor; lineations on trench outer slope are intersection of cross-faults with sea floor.

In the Izu-Bonin forearc, an *Alvin* submersible dive has shown the presence of *chemical chimneys* associated with fluid venting. These chimneys are made of carbonate, as calcite and aragonite, together with silicate (a new Mg-silicate). Fryer *et al.* (1990a, b) considered that a biogenic source in the Mariana forearc fluids is unlikely to have been responsible for producing these chimneys. Instead, they suggested that methane was generated at depth beneath the forearc wedge, possibly by the interaction between mantle material and water from the subducting slab, including both sediment-hosted water and slab-hosted water. A large flux of fluid from the subducting slab to the forearc wedge should play a major role in creating the forearc thermal structure, metamorphism and arc magmatism.

Thermally, the forearc region is characterized by a higher heat flux (increasing towards the *volcanic front*) compared to the trench. Honda & Uyeda (1983) suggested that the heat provided by convection within the mantle wedge will be relatively uniformly distributed when the forearc crust is thin, but that a concentrated heat flux will result when the crust is thick. It seems, therefore, that the thermal influence from the mantle wedge is greater in the Izu-Bonin forearc (as expressed by the occurrence of serpentinite diapirism) than in the Japan and Nankai trench forearcs. ODP Legs 125 and 126 encountered abundant dark vein-like structures in the sediments, typically 3–5 cm long, subvertical to bedding, in some cases S-shaped, and with bifurcating terminations (Taylor *et al.* 1990). *Sediment dykes* and other fluidization structures were also reported from the forearc sediments. Some vein structure occurs at depths of less than 50 m below the seafloor, suggesting a phase of early fluid migration. Recently, Kimura *et al.* (1989) documented vein structure farther south from within decimetres of the sea floor, in unconsolidated volcaniclastic muds, near the junction of the Mariana and Yap trenches. Such observations provide an important constraint on the nature of sediment deformation in forearcs and arc collision zones but, as yet, their variability and nature remains controversial and unresolved. The important observation from the Izu-Bonin accretionary system is that fluid activity within the forearc is currently taking place and involves fluid migration from a deep-seated, thermogenic, source.

One of the principal sources of fluids in accretionary complexes is the release of loosely, and more tightly, bound structural water from sediments, especially clay minerals, during compaction and early diagenesis. The general decrease in porosity with depth, as seen in ODP Leg 131 Site 808 (Fig. 16.5), provides an indication of the volumes of trapped seawater that are released during the early phases of compaction when primary porosity is lost. The illitization of smectites, for example, will release relatively large volumes of low-chloride fluids. In ODP Leg 131 Site 808, near the toe of the Nankai accretionary prism, the décollement horizon is at a temperature of about 90°C (Taira *et al.* 1991), and within the smectite–illite transformation field, therefore a substantial component of the low-chloride anomaly owes its origin to this early diagenetic mineralogical change, from about 530 mbsf.

The physical–chemical nature and reasons for the specific localization of the décollement in accretionary prisms is a fundamental problem. Initial results from ODP Leg 131 indicated that the location of the décollement, an overpressured, dynamically-sealed c. 20 m-thick horizon of brecciated muds, was not controlled by a lithological change. A re-evaluation of the data, together with new shore-based geochemical analyses of the hemipelagic mudstones (Pickering *et al.* 1993a), reveals a hitherto unrecognized lithological change that may have influenced the localization of the décollement. The décollement is also coincident with a marked change in sediment accumulation rate. The lower part of the décollement comprises carbonate muds. The décollement is also associated with an anomaly in the heavy rate earth elements (REE) relative to the lighter REE (Pickering *et al.* 1993a). The REE pattern (Fig. 16.6), associated with the carbonate muds, is interpreted as having a provenance in an horizon rich in smectite clays (undergoing illitization), that shows a depletion in the LREE relative to the HREE: because of the greater solubility of the LREE. The bromine and iodine spikes in the interstitial water samples at the décollement could facilitate the preferential solution of LREE as halide complexes, because of their increased solubility relative to the HREE. This re-evaluation of the data suggests that the position of the décollement may be linked to a significant lithological change that was not originally appreciated because of the brecciated nature of the zone. The location of the décollement appears to be associated with an interval of partly degraded ash-rich sediments.

Below the décollement horizon at about 1060–1110 mbsf there is an interval of umber-like metaliferous sediments, identified by Pickering *et al.* (1993a), which are particularly enriched in the REE (Fig. 16.6). The subcretion and/or subduction of such sediments may provide important geochemical tracers in the forearc and arc magmas, an area of study requiring more research.

The three Japanese forearc regions show a wide variety of tectonic style and fluid activity which is summarized in Fig. 16.7 and Table 16.1. The pervasiveness of fluid activity in the Nankai, Japan and Izu-Bonin forearcs demonstrate that the fluid flux can play an important role in shallow to deep level tectonics, and the thermal structure of forearc regions.

The main morphotectonic characteristics of the Nankai trough, Japan trench, and Izu-Bonin trench forearcs are: (1) the Nankai trough forearc is an accretionary system dominated by frontal accretion, and with a long history of development as a fold-thrust belt; (2) the Japan trench forearc is predominantly erosional, mainly by slope failure and sliding, and has a pronounced subsidence history. The basement of the trench landward slope is composed of Cretaceous

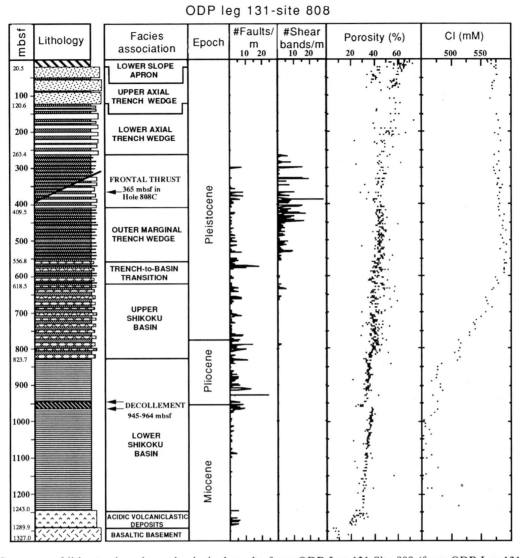

Fig. 16.5. Summary of lithostratigraphy and principal results from ODP Leg 131 Site 808 (from ODP Leg 131 Shipboard Scientific Party in Taira, Hill et al. 1991).

consolidated sediments; (3) the Izu-Bonin trench forearc appears to be more stable than the other two forearcs, with evidence of local oceanic crustal accretion. The basement is composed of an older lava – volcaniclastic complex.

The fluid pathways and permeability variations within these three forearcs appear to be different. In the Nankai forearc, concentrated fluid expulsion occurs at the toe of the prism and upper trench landward slope; the rest of the prism is dominated by slow fluid advection as sediments gradually dewater. In the Japan trench forearc, concentrated fluid flows occur in relation to normal faulting, and there appears to be no pervasive advection from depth. The décollement horizon could be over-pressured and subject to relatively lower frictional forces. The forward migration of fluids along the décollement horizon produces over-pressuring at the trench landward lower-slope, which probably triggers large-scale sediment failure. In the Izu-Bonin forearc, slow fluid advection may occur throughout the permeable forearc basement, and the décollement horizon may not be over-pressured. The rapid loss of water from the décollement, through permeable arc-basement, and the correspondingly enhanced frictional effects could explain the reason for the accretion of oceanic highs in the Izu-Bonin forearc, and also provide an explanation for the relatively steep trench landward-slope, assuming the applicability of the *Coulomb wedge model* (Davis et al. 1983) (also see Chapters 9 and 13).

It is possible to estimate the volume of sediments that are incorporated into the subduction zone in the three forearc regions discussed in this chapter. The parameters used are the thickness of sediments incorporated into the subduction zone, and the rate of subduction. In the Nankai forearc, the mean thickness of sediments is about 1000 m, including trench wedge and hemipelagites. The rate of subduction off Shikoku is about 3 cm a^{-1}. In the Japan trench, the estimate is not so simple due to the development of *horst-and-graben* structure (see Chapter 9). We estimate a mean vertical displacement of graben blocks of about 300 m, based on seismic profiles (KAIKO I Research Group 1986). The thickness of the pelagic and hemipelagic

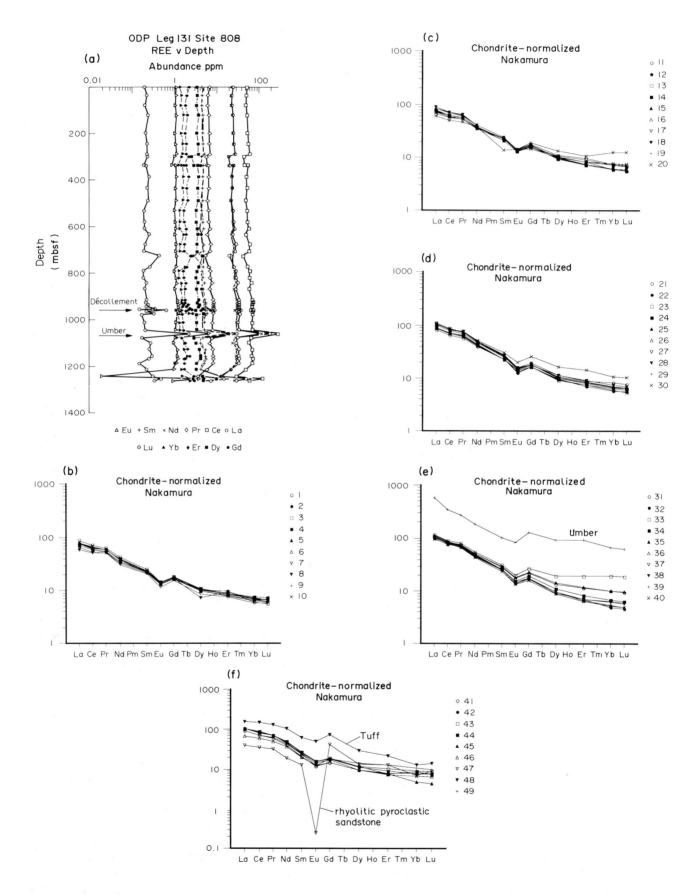

Fig. 16.6. (a) REE abundances (ppm) versus depth (metres below the sea floor, mbsf) from pilot study of representative muds at Ocean Drilling Program Leg 131 Site 808 near the toe of Nankai accretionary prism, off-shore from Shikoku, Japan. (b) – (f): REE chondrite-normalized data. $N = 49$ representative samples of muds, including some tuffs/pyroclastic sands for comparative purposes. In the 20 m thick décollement horizon, 3 out of 4 samples show a significant depletion in the light REE, relative to the heavy REE. After Pickering et al. (1993a). See text for explanation.

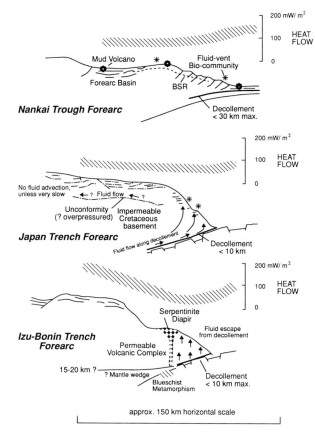

Fig. 16.7. Summary of the plumbing and heat flow patterns in the Nankai, Japan and Izu-Bonin accretionary prisms (from Taira & Pickering 1991).

cover on the Pacific seafloor is about 600 m. Because the horst-and-graben structures are about equally developed, the mean thickness of sediments subducted is c. 750 m, assuming that the grabens are filled by debris at the trench. The subduction rate is approximately 10 cm a^{-1}. In the Izu-Bonin forearc, the sediment thickness on the oceanic crust is about 400 m. The mean displacement of the normal faults are similar to those in the Japan trench, that is, about 300 m, giving a mean sediment thickness of 550 m that can potentially be subducted. The subduction rate is 6 cm a^{-1}. These estimates, although they are first approximations and ignore the porosity variations, suggest that the rate at which sediment subduction occurs in the three forearcs, expressed in $m^3 m^{-1}$ width are as follows: Nankai: 30 $m^3 a^{-1}$; Japan: 75 $m^3 a^{-1}$; Izu-Bonin: 33 $m^3 a^{-1}$ (Taira & Pickering 1991).

These values show that the volumes are the same order of magnitude in the three forearcs: the greater value for the Japan trench forearc possibly occurs because it receives more water than the other forearcs. The sediment volumes involved in the Japan trench and Izu-Bonin trench forearcs, where sediments are subducted directly underneath the trench landward slope without frontal accretion, are roughly comparable to, or probably larger than, that for the Nankai accretionary prism in which frontal accretion is a major process. These contrasting accretionary processes of predominantly *frontal accretion* (Nankai) and *subcretion* (Japan and Izu-Bonin) should provide very different mechanisms for sediment dewatering and fluid migration paths.

Amongst the possible fluid sources, geochemical data suggest that magma-related fluids are probably relatively insignificant in the case of the Japanese forearcs. An input from meteoric water may be important, as for example is the case in the Peru margin, but little is known of its significance in the Japanese forearcs.

The Japanese forearcs and trenches exhibit considerable evidence for diffusive and channellized, focused, fluid flow and venting. There are three different modes of fluid migration in forearcs: (1) *diffusive advection* through porous media; (2) *channellized flow* in some permeable conduits, and (3) *migration of fluids* related to host material transfer such as diapirism. Again, the relative importance of these mechanisms and their temporal nature (transient versus steady-state), remains unclear in the three forearcs. We suggest that the fluid pathways will, to a large extent, be controlled by the nature of the forearc basement and the volume of subducted sediments.

Probably, the best evidence for diffusive fluid flow is the presence of a BSR. The widespread occurrence of a BSR in the Nankai trough forearc suggests that fluid migration occurs throughout most of the trench landward slope region. The absence of a BSR in the Japan and Izu-Bonin trench forearcs does not preclude fluid migration throughout the landward slope. There may be different fluid circulation patterns in the lower landward slope region of the Japan and Izu-Bonin forearcs.

The Japan trench forearc is underlain by Cretaceous accretionary prism and forearc sediments. The outcrops of equivalent rocks in Hokkaido suggests that the concealed prism is composed of highly consolidated and impermeable rocks, mainly deformed flysch deposits. Such strata may act as an impermeable seal on any subducted water-rich sediments: the reason why there is no BSR in the Japan trench forearc may be a consequence of insufficient water being supplied to the overlying sediments to cause pervasive fluid advection. As pointed out by von Huene & Culotta (1989), the existence of a long and continuous seismic image of the décollement horizon in the Japan trench forearc suggests an over-pressured zone. Thus, the décollement horizon with an upper seal, should be a conduit for fluid flow. As there is insufficient material to accrete above the décollement, no internal deformation related to accretion occurs. Instead, the forward migration of fluids should produce an over-pressured toe in the lower landward slope. This may be the reason why the lower-trench slope is so weak in the Japan trench prism, which is subject to repeated sediment mass failure.

In the Izu-Bonin arc, the basement of the trench landward slope is mainly composed of Eocene–Oligocene volcanic rocks, including volcaniclastics, hyaloclastic breccias, and pillow and sheet lavas (Taylor *et al.* 1990); locally, there is accreted oceanic material.

Table 16.1. Comparison of morpho-tectonics and hydrogeologic features of the three forearcs

	Nankai	Japan	Izu-Bonin
Trench sediments and tectonics (Subduction rate)	1–2 km thick turbidites and hemipelagites. No horst-and-graben on outer swell (3 cm/yr)	500 m pelagites and hemipelagites. Horst-and-graben. (10 cm/yr^{-1})	500 m thick pelagites and ashes. Horst-and-graben. (6 cm/yr^{-1})
Shape of trench landward slope	Single slope or gentle lower slope and steep upper slope	Very steep lower slope and gentle upper slope	Steep lower slope and gentle upper slope
Tectonics of trench landward slope	Sediment accretion	Gravity failure	Partial accretion of oceanic high
Forearc rocks	Accretionary prism (mostly turbidites and hemipelagites)	Cretaceous sedimentary rocks overlain by Cenozic sequence	Eocene–Oligocene volanic complex overlain by Neogene volcaniclastics
Vent-related biological communities	Occur at the toe region and upper slope	Occur at the lower trench landward slope and trench-slope break	Not known
Diapiric features	Mud volcanoes at the toe regions, upper slope and forearc basin	Not known	Serpentinite diapirs at the mid-slope bench
Bottom simulating reflectors	Well developed in the trench landward slope	No development	No development
Heat flow	Landward decrease (200–50 mW/m^2) High at the toe	Uniform (20–50 mW/m^2)	Landward increase (20–80 mW/m^2)
Vein structure	Not known	Known	Known
Hydrogeologic interpretation	Concentrated flow at the toe and upper slope. Slow advection in entire prism. Over-pressured décollement	Over-pressured décollement with forward fluid flow. Over-pressured lower slope	Less over-pressured décollement. Slow advection through forearc basement. Migration of deep-seated fluids through serpentinite diapir

After Taira & Pickering (1991).

We believe that the Izu-Bonin forearc has essentially achieved a steady state, with neither voluminous accretion nor major gravity tectonics to modify the volume of the forearc. It is also reasonable to assume that the Izu-Bonin basement comprises very permeable materials. In such a setting, subducted sediments would be expected to dewater rapidly. We therefore speculate that in the Izu-Bonin forearc, a water-saturated décollement horizon is even less likely to develop than in the Japan trench forearc. As material in the Izu-Bonin forearc is generally permeable, massive sediment failure, as slides, is uncommon. A corollary of these arguments is that the décollement in the Izu-Bonin forearc has a greater frictional resistance compared to that in the Nankai and Japan trench forearcs. The absence of a BSR in the Izu-Bonin forearc may be due to lithologic control in organic-poor volcaniclastic rocks, which may explain why the accretion of an oceanic high is taking place at the Izu-Bonin forearc.

The foregoing discussion leads to two general problems concerning the deeper seismicity and thermal conditions in the forearcs. Firstly, a characteristic of subduction zones in general is that there is virtually no seismic activity along the subduction zone to depths of 20–30 km (Yoshii 1979, Fukao 1979). Such seismically quiet zones may exist because they are subject to smaller frictional effects due to the presence of fluids (Shimamoto 1985). The amount of subducted fluids may be a function of the rate of dewatering in the shallow part of the prism, at depths of up to about 5 km, although the linkage between shallow dewatering processes and the availability of fluids from greater depths is unclear. Very little data are currently available on this problem.

The second general problem concerns the thermal structure of accretionary prisms. Reck (1987) noted that the heat flow in the Japan trench forearc appears too high to produce blueschist metamorphism at depth. He suggested that the high heat flux in the forearc is a manifestation of extensive fluid migration to the surface. The calculated temperature at a depth of 30 km in the three forearc case studies is: 600°C in Nankai, 300°C in the Japan trench forearc, and 700°C in the Izu-Bonin forearc (Uyeda et al. unpublished data).

Although the surface heat flow measurements in the forearcs are too few to make any conclusive statement, these temperatures suggest that Reck's hypothesis may be valid. The recent recovery of blueschist from the Izu-Bonin forearc suggests that blueschist metamorphism has been occurring at depth. The temperature of subduction zones is believed to be determined primarily by the age of oceanic lithosphere, and calculations of shear heating along the slip zones, whereas energy transfer by fluid flux has not been fully evaluated.

NEOGENE BASIN EVOLUTION IN THE IZU COLLISION ZONE, SOUTHEAST CENTRAL JAPAN

The present configuration of the Izu-Bonin arc comprises, from east to west, a 7–9 km deep prominent *trench*, a *forearc region* including a *trench landward-slope, outer-arc high*, a *volcanic arc*, an active *backarc-rift*, and the Shikoku *marginal basin*. The *forearc basin* is filled by volcaniclastic and hemipelagic sediments banked behind the outer-arc high. In the southern part of the Izu-Bonin forearc, the outer-arc high is called the Ogasawara ridge and includes the Ogasawara Islands. Several mature, dendritic, submarine canyon systems have developed across the Izu-Bonin forearc basins and the outer-arc high. These canyons have incised as much as 1 km into the 1.5–4 km thick sedimentary succession.

The evolution of the arc system started with the initiation of westward subduction of the Pacific Plate in the early–middle Eocene (Uyeda & Ben-Avraham 1972). An episode of middle Oligocene to early Miocene rifting and sea floor spreading generated the Shikoku basin and divided the arc in two, the present Izu-Bonin arc and the remnant, submerged, Kyushu-Palau ridge. The present incipient rifting of the Izu-Ogasawara arc began in the late Pliocene to early Pleistocene (Taylor *et al.* 1990). The Izu-Bonin forearc has undergone only minor deformation since the Eocene, when subduction commenced (Honza & Tamaki 1985).

The Izu-Bonin forearc is characterized by the *offscraping* of an oceanic high at the toe of the *inner-trench wall*. A SeaBeam survey by the Japanese Maritime Safety Agency (Iwabuchi *et al.* 1988) of the Uyeda ridge (Smoot & Heffner 1986), shows a linear, E–W trending, topographic high about 150 km in length, 18 km wide, and with a maximum elevation of 4.2 km, to the north of the Ogasawara plateau. The ridge is being subducted at the Izu-Bonin trench where water depths are in excess of 9 km. At the toe of the landward wall, there is a prominent 8 by 4 km high at about 5 km landward from the trench axis. This topographic feature is oblique to the linear and smooth N–S trending cliff line of the trench landward slope. It also shows a continuous magnetic anomaly from the ocean side and no associated gravity anomaly, suggesting that this portion is a continuation of the oceanic crust. The linear high is probably an offscraped segment of the Uyeda ridge at the toe of the inner trench wall (Iwabuchi *et al.* 1988).

A segment of the Ogasawara plateau is subducting at the southern Izu-Bonin trench. A prominent topographic high, with an unusual shape occurs in the toe of the inner trench wall at this boundary, as the rectangular-shaped, 50 by 20 km, Hahajima Seamount, with about 2 km of relief from the surrounding oceanic floor. Dredge samples from this seamount are ultramafics. To the southeast of this seamount, there is a steep, N–S trending cliff interpreted as the toe of the trench landward slope. At a depth of 3.3 km, which is about 1 km above the trench axis, a deep-tow survey by JAMSTEC (Monma *et al.* 1990) revealed rock exposures of whitish-coloured rocks, and from dredging these were shown to include late Cretaceous, Turonian, nannofossil limestones. Micritic limestones of this age have not previously been recorded from the Izu-Bonin forearc, but they are comparable to the limestones that veneer the Ogasawara plateau. It seems likely that at least part of this trench landward slope may comprise accreted Ogasawara plateau crust. By extrapolation, it can be inferred that the nearby Hahajima Seamount, with its flat-topped shape, is probably accreted Ogasawara plateau material. Thus, active offscraping of an oceanic high is taking place at the Izu-Bonin trench.

On the mid-slope bench of the trench landward slope, extending from the Izu-Bonin to the Mariana trench, there are some serpentinite seamounts ranging from 5 km to more than 20 km in diameter, and from several hundred to more than 1400 m in height (Ishii 1985). These seamounts are spaced at intervals of 15–60 km along the bench situated less than 50 km from the trench axis.

Two sites on ODP Leg 125 were drilled on the seamounts within the Izu-Bonin forearc (Torishima Seamount), and three sites into Conical Seamount in the Mariana forearc. Primary mantle material was recovered as mainly highly depleted, tectonized, harzburgite which underwent medium-grade metamorphism. The fluids contain significant amounts of hydrocarbon, with methane concentrations of up to $400\,\mu l\,l^{-1}$ of sediments.

In the north, the Izu collision zone (ICZ) has experienced several million years of progressive tectonic deformation, probably initiated about 15 Ma in relation to the opening of the Japan Sea in the Middle Miocene (Itoh 1986). The thickness of the crust to the Moho in the ICZ reaches approximately 40 km (Ishibashi 1988), although the crustal thickness of normal Izu-Bonin arc is about 15–20 km (Murauchi *et al.* 1968, Hotta 1970); this increase in crustal thickness can be ascribed to *crustal imbrication*.

The collision-related deformation resulted in the formation of tectonic crustal segments up to several tens of kilometres in dimension and bounded by thrusts, between which there are volcanic and volcaniclastic successions derived from the Izu-Bonin arc (Figs 16.8 and 16.9). The Mineoka ophiolite (and subsurface

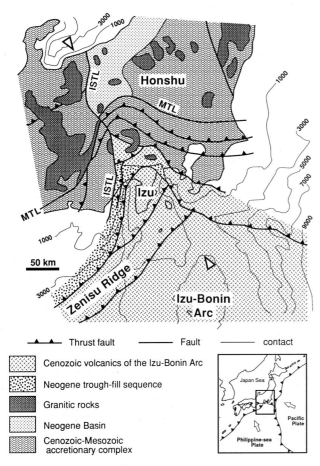

Fig. 16.8. Geotectonic map of the southeastern part of the Honshu (Izu collision zone), showing flexure of geologic belts and modern basins as a result of the collision of the Izu-Bonin arc with the Honshu arc. NNW–SSE crustal profile between A–A' shown in Fig. 16.9.

forearc ophiolite remains unclear, but obduction was probably initiated by the collision of the Honshu and Izu-Bonin arcs at about 15 Ma when the ICZ was generated.

During *collision-accretion*, relatively short-lived foreland basins formed and were infilled by overall coarsening-upward sedimentary sequences bounded by major thrusts (see also Chapter 13). From north to south, the basin-fill sediments tend to become younger in age and range from Middle Miocene to Pleistocene. The thrust movements associated with the generation of the foreland basins are interpreted to have shifted southward with time as a result of the incremental migration of the plate boundary caused by the incorporation of the frontal segment of the Izu-Bonin arc onto the Honshu arc side (Taira *et al.* 1989).

An example of a foreland basin infill in southeast Japan is the Ashigara Group. The late Pleistocene Ashigara Group is exposed on the boundary between the Izu Peninsula, the Izu-Bonin ridge, and the Tanzawa Mountains on the Honshu arc, and is more than 5 km thick. The group represents an overall shallowing-up stratigraphy, and comprises the Hinata (Neishi), Seto, Hata and Shiozawa Formations in ascending order. The basement remains unknown. The lower Hinata Formation comprises thin-bedded turbidites with tuffaceous sandstone layers, and benthic foraminifera, suggesting that the Hata Group was deposited in an upper bathyal environment, at a depth of about 1000–2000 m. The Seto Formation comprises deep-marine, resedimented conglomerates and thin-bedded turbidites, interpreted as channel fill and levée sediments in a submarine fan. The Hata Formation consists of mudstones and thin-bedded turbidites with sediment slides and debrites, and is interpreted as a slope to shelf-edge deposit. Benthic foraminifera and mollusc fossils suggest the sedimentary environment of the Hata Formation was upper bathyal, ranging from 300 to 100 m water depth. The uppermost Shiozawa Formation consists of conglomerates with minor sandstone and mudstone, interpreted as alluvial fan

segments as expressed in the Mineoka–Hayama uplift zone) appears to represent a slice of Eocene basement from the leading edge of the Izu-Bonin arc, obducted as a forearc ophiolite onto the leading collisional edge of the Honshu arc. The precise age of obduction of this

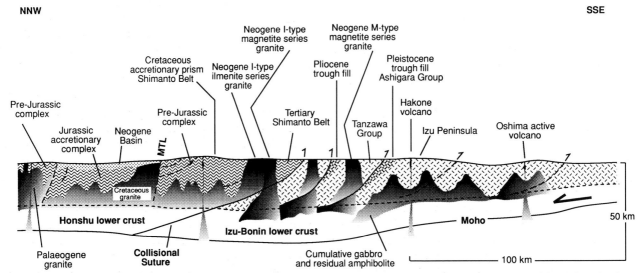

Fig. 16.9. Interpreted N–S cross-section through the crust from the Honshu to the Izu-Bonin arcs. Position of Moho from deep seismic profiles. Cross-section constructed between A and A' shown in Fig. 16.8.

deposits. The Shiozawa Formation, up to 2.3 km thick, was deposited in about 0.4 Ma (Huchon & Kitazato 1984). Most of the detritus filling the Ashigara Group basin was derived from the frontal part of the Honshu arc as the thrust stack rose, but minor volcaniclastics were provided from the Izu-Bonin ridge, as the foreland. The sediments are entirely lithified despite their late Pleistocene, age.

At present, four major slices of the Izu-Bonin crust can be recognized around the collision zone between the Izu-Bonin arc (ridge) and the Honshu arcs. The slices of Izu-Bonin arc have been successively tilted northwards as the crustal segments of arc material attempted to subduct below the Honshu arc. Each slice is bounded by a major thrust zone or an elongate fault-controlled sedimentary basin filled by volcanics and/or volcaniclastic sediments.

The collision-related deformation extends from the ICZ to the Philippine Sea plate in the northeastern Shikoku basin (Chamot-Rooke & Le Pichon 1989). A few tens of kilometres seaward of the Nankai trough, a linear topographic high, the Zenisu ridge, runs from the northeastern Shikoku basin to the Izu-Bonin arc (Figs 16.10a & b), and is interpreted as a consequence of the rupture and uplift of the Shikoku basin oceanic crust due to intraplate compressional tectonics (Le Pichon et al. 1987a, Lallemant et al. 1989, Taira et al. 1989). A probable future consequence is that the intra-oceanic South Zenisu basin will be initiated as a new trench to the south of the Zenisu ridge. Le Pichon et al. (1987a) have proposed that the marginal thrust, on the northern margin of the South Zenisu basin, is currently changing from an intraplate fault to an incipient plate-boundary thrust as collision-accretion processes continue. Volcaniclastic material, derived from the Izu-Bonin ridge to the east/northeast, has accumulated in the intra-oceanic South Zenisu basin.

Sediment slides, and other types of surficial wet-sediment deformation structures (see Chapter 7) are particularly common in the Neogene ICZ deposits onland. Seismic reflection profiles and sidescan sonar records from the present submarine slopes associated with the Honshu and Izu-Bonin arcs also reveal large-scale sediment mass failure (Fig. 16.11).

The sedimentary setting of the Miocene-Pliocene Miura-Boso region is comparable to the Zenisu ridge region today. There are striking similarities between the history of the Neogene arc–arc collision zone and the evolution of the present ICZ, even as far as identifying a palaeo-Zenisu ridge (the Hayama-Mineoka uplift).

The Tertiary stratigraphy in the Miura-Boso region shows a tripartite division as the Mineoka, Hayama-Hota, and Miura Groups in order of decreasing age; they are overlain unconformably by the Kazusa and Shimosa Groups (Figs 16.12a & b). Pre-Tertiary basement is absent in the region. The Quaternary Kazusa and Shimosa Groups comprise mainly shallow-marine siliclastics (Katsura 1984). The Kazusa Group is up to about 3500 m thick in the Kanto Plain to the north of the Boso Peninsula, and shows an overall shallowing-upward sequence.

Figures 16.13 (a & b) summarize the stratigraphy of the Northern and Southern basins which developed on a template of oceanic basaltic basement of the Izu-Bonin crust prior to accretion. The Mineoka and Hayama-Hota Groups crop out in the hilly areas and form the core of a horst (Hayama-Mineoka uplift belt), whereas the Miura Group is exposed in the lowlands associated with an unconformable Quaternary cover. The Oligocene–Eocene Mineoka Group comprises mainly siliceous shale with thin-bedded turbidites, chert and siliceous tuff (Ogawa & Taniguchi 1988). The group includes ophiolitic rocks of basaltic and related material within sheared ultramafic rocks as tectonic blocks (ophiolite and/or serpentinite melange) (the Mineoka ophiolite of Uchida & Arai 1978, Arai et al. 1983, Ogawa & Taniguchi 1988). The Mineoka Group has been severely affected by tectonic compression, with the rocks of the Mineoka Group being thrusted and folded. Ogawa & Taniguchi (1988) proposed that the Mineoka ophiolite was emplaced in the Eocene as a fragment of delaminated oceanic crust with seamouts. Taira et al. (1989) reinterpreted the Mineoka Ophiolite as the product, at least in part, of *accreted serpentinite diapirs* in the forearc region of the Izu-Bonin ridge. We believe, however, that the Mineoka ophiolite is a forearc ophiolite, of Eocene oceanic crust, probably from the leading edge of the Izu-Bonin island arc, with initial obduction occurring during the early phases of arc-continent collision about 15 Ma (see below).

The Lower to Middle Miocene Hayama-Hota Group crops out north and south of the Eocene Mineoka Group. Most of the strata are steeply dipping, and tightly folded and thrust, with much of the disruption having occurred while the sediments were still wet. Two lithologic units are recognized in the Hayama-Hota Group: the northern and southern Hayama-Hota Group, separated by the Mineoka Group. The northern Hayama-Hota Group comprises thick-bedded turbidites, including abundant terrigenous detritus, probably derived from the Honshu arc, but also with small amounts of volcaniclastic material (Nakajima et al. 1981, Suzuki et al. 1984, Ogawa & Taniguchi 1988). In contrast, the southern (or main) Hayama-Hoto Group consists of tuffaceous hemipelagic and/or pelagic sediments and thick-bedded dacitic tuff. Small amounts of coarse clastic rocks, derived from the Mineoka Group, occur along the marginal portions. The southern group contains little terrigenous detritus from the Honshu arc. From the difference in lithologic characteristics, particularly the amount of terrigenous material, Ogawa & Taniguchi (1988) suggested that the southern Hayama-Hota Group was deposited far from the sedimentary basin of the northern Hayama-Hota Group, the latter being sited close to the Honshu arc.

The basin-fill sediments constitute the Middle Miocene to late Pliocene Miura Group, and comprise sedimentary successions between the top-bounding Kurotaki unconformity and the underlying Tagoegawa

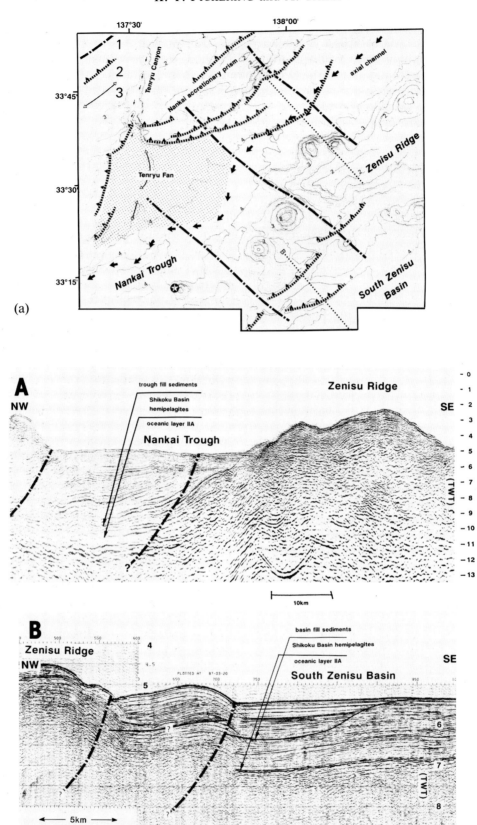

Fig. 16.10. (a) Geomorphologic and tectonic map in the Zenisu Ridge region. Note the development of thrust and transverse fault systems. Dredge site KH86–5, *Hakuho-maru* cruise, is shown by a closed circle. 100 m counter interval. Key: (1) strike-slip, transverse, faults, (2) thrusts, (3) anticlines. (b) Seismic profile across the Nankai trough, Zenisu ridge and South Zenisu basin. Profile lines are shown in Fig. 16.10a. (A) representative seismic profile across the Nankai trough and Zenisu ridge. Note the interpreted level of the boundaries between the trench fill sediments, hemipelagic sediments and oceanic layer IIA. (B) seismic profile across the South Zenisu basin, showing the three principal tectono-stratigraphic layers and deformation due to thrusting. After Soh *et al.* (1991).

Unconformity (Mitsunashi et al. 1976, Eto 1986a, b). The Kurotaki unconformity is of regional extent across the Oiso area, and the Miura and Boso Peninsulas. Above the Tagoegawa unconformity, the Miura Group onlaps the underlying Hayama-Hota Group in the western part of the study area (e.g. Eto 1986a), and the contact becomes apparently conformable in the east, suggesting that here there is an essentially conformable transition from the underlying Hayama-Hota Group in this area (Nakajima et al. 1981).

The formation of the Hayama-Mineoka uplift belt as an intraplate rupture, during arc–arc collision, may correspond to the c. 11–10 Ma synchronous transition from the hemipelagic Hayama-Hota Group to the basin-fill Miura Group in the Northern and Southern basins. During the development of the Hayama-Mineoka uplift zone, two deep-marine sedimentary basins were formed, a Northern and a Southern basin. The former basin was sited on the plate boundary, and was filled by Honshu arc-derived detritus and Izu-Bonin arc-derived material. In contrast, the Southern basin was an intra-oceanic basin on the forearc of the Izu-Bonin ridge. The lapilli and scoria of low-K tholeiite, and tuffaceous mudstone, were deposited in the Southern basin, without Honshu arc-derived material. Small *graben-like supraridge basins* in the Hayama-Mineoka uplift zone accumulated detritus derived from the underlying Mineoka and Hayama Groups, together with shallow-marine limestone. Ultramafic breccias and chert pebbles, derived from the Mineoka ophiolite, were deposited in the supraridge basins with shallow-marine faunas. Similar lithologic variation can be recognized between the Nankai trough and the South Zenisu basin around the present Zenisu ridge. Lithologic variations in neighbouring tectonic segments with terrigenous, detritus-free or detritus-dominant and ophiolitic (plus island arc) sequences, represent important features in arc–arc collision zones.

The Hayama-Mineoka uplift zone was destroyed before the Kurotaki unconformity (c. 3–2.5 Ma), because the Hasse Formation in the Southern basin received Honshu arc-derived material at that time. The rheological model of Chamot-Rooke & Le Pichon (1989) suggests continuous uplift at the *intraplate ridge* (Zenisu ridge) and subsidence at the plate boundary basin (Nankai trough) from the time of initial lithospheric rupture. As the geohistory and subsidence analysis of the Northern basin suggests that the basement of the Northern basin continuously subsided during the development of the Hayama-Mineoka uplift zone, the model proposed by Chamot-Rooke & Le Pichon (1989) can explain the development of the Northern basin under a compressive, collisional, stress field. The ancient, Mio-Pliocene Hayama-Mineoka uplift zone and the modern Zenisu ridge reveal the repetitious style of tectonic fragmentation of arc crust and basin formation during arc–arc collision.

The Miura Group (Yamaguchi et al. 1983, Eto et al. 1987) ranges from late Miocene–Pliocene (10–3 Ma) and comprises deep shelf to basinal sediments, mainly scoriaceous to pumiceous volcaniclastics derived from the proto Izu-Bonin arc. These sediments, now exposed onland represent part of the fossil Neogene accretionary complex resulting from arc–arc collision in the ICZ. The sediments accumulated in relatively short-lived foreland basins, filled by overall coarsening and shallowing upward sequences developed along the major basin-bounding thrust faults during the incremental incorporation of segments of the Izu-Bonin arc onto the Honshu arc (Taira et al. 1989).

The Miura Group is divided into two successions: a Middle Miocene to early Pliocene fine-grained, and a late Pliocene coarse-grained, volcaniclastic succession. The lithologic characteristics and basin morphology are different in the northern and southern basins, even in age-equivalent strata, demonstrating the existence of two disparate basins. The southern and northern stratigraphies are defined in terms of Southern and Northern basins, respectively (Fig. 16.14). In order to simplify stratigraphical nomenclature, Soh et al. (1991) defined the entire Middle Miocene to early Pliocene stratigraphies in the Southern basin as the Misaki Formation, and that in the Northern basin as the Zushi Formation, also the early to late Pliocene sequences in the Northern basin as the Ikego Formation, and that in the Southern basin as the Hasse (Hatsuse) Formation. The Oiso Formation (Otsuka 1929), exposed at the Oiso coast, west of the Miura Peninsula, is included within the Misaki Formation, because: (a) the Oiso Formation is lithologically similar to the Misaki Formation rather than the Zushi Formation, particularly the presence of scoria fall beds not seen in the Zushi Formation, and (b) the Komayama Group of alkaline pillow basalts and deep-marine deposits, interpreted as the continuation of the Hayama and/or Mineoka Groups, is exposed to the north of the outcrop of the Oiso Formation.

The chronology of deformation, at least in the lower part, of the Miura Group was: (a) normal faulting, some of which was syndepositional, prior to the accretion of the Miura block from the Izu-Bonin arc onto the Honshu arc, coeval with (b) phases of 'vein structure' development, perpendicular to layering and coplanar with the essentially N–S regional compression direction and inferred strike of the palaeo-forearc slope; (c) a major phase of layer-parallel shortening, primarily by bedding-parallel thrusts with considerable duplication of the stratigraphy, that was earlier than but also partially coeval with (d) folding and mainly reverse, conjugate, faulting at high angles to bedding, and (e) late-stage regional open folding, with normal and reverse oblique-slip faulting related to dextral transpression and transtension, after accretion of the Miura block at the leading edge of the Honshu arc. Palaeomagnetic studies by Yoshida et al. (1984) suggest an approximately 30° clockwise rotation of the Miura Peninsula, or Miura block, after deposition of the Miura Group, that is, since 3–2.5 Ma.

An important aspect of the faulting observed in the Misaki Formation is that an early phase of intense layer-parallel shortening and duplication by essentially

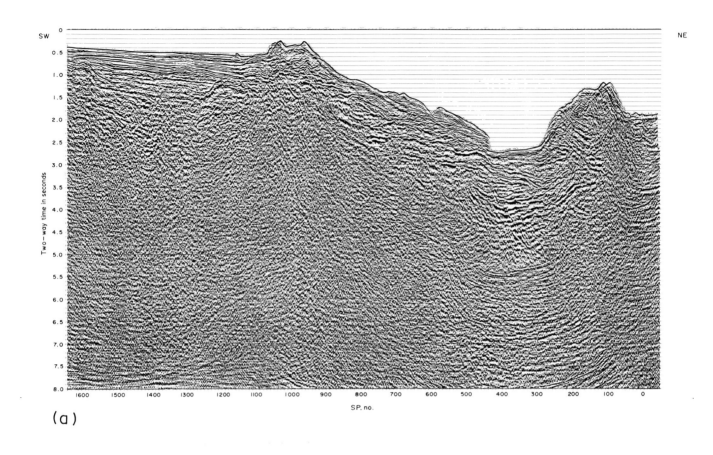

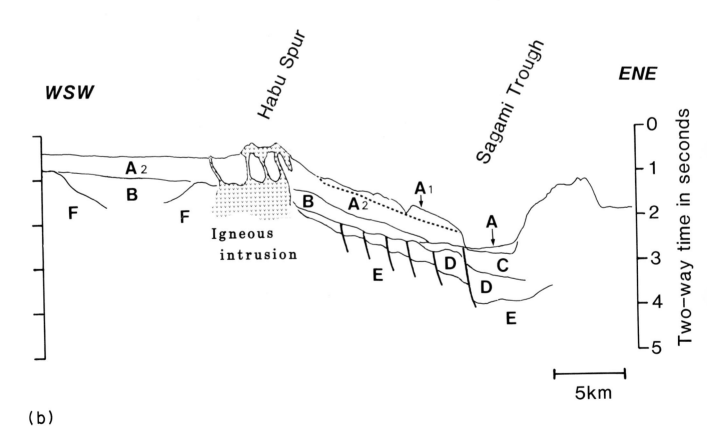

Fig. 16.11. Seismic profile (a) and line interpretation (b) of gravity-driven sediment slides on the northern part of the Izu-Bonin arc.

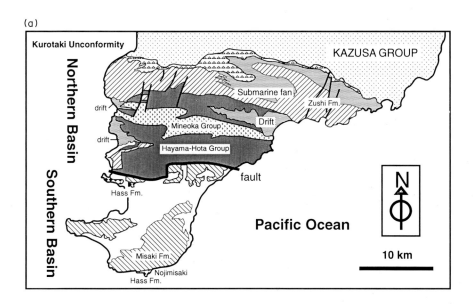

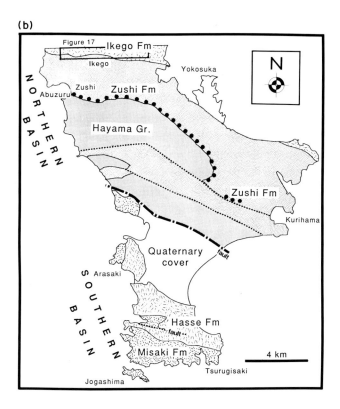

Fig. 16.12.(a) Geological map of Boso Peninsula, showing Northern and Southern basins. Graben fills and the marginal facies are exposed in the area shown stippled. (b) Geological map of the Miura Peninsula, showing the Northern and Southern basins. Marginal facies of the Izu-Bonin arc-derived coarse clastic material is exposed in the dotted area.

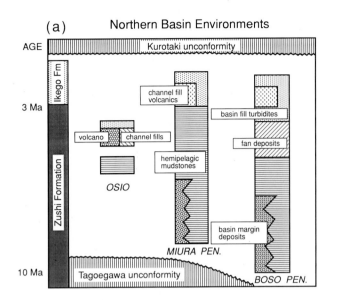
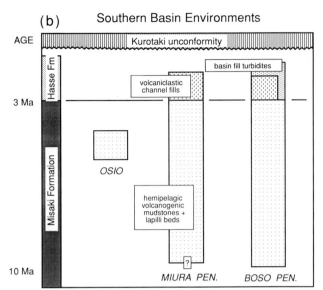

Fig. 16.13.(a) Lithofacies section of the Northern basin. Note the lateral lithofacies variation in the Zushi Formation in the Oiso area, Miura and Boso Peninsulas during the Miocene–Pliocene (c. 5–4 Ma). (b) Lithofacies section for the Southern basin. Note homogeneous lithologies of the Misaki and Hasse Formations in the Southern basin from the Oiso area in the west to the Boso Peninsula in the east.

bedding-parallel thrusts preceded deposition of the Hasse Formation, because the basal erosional unconformity at Jogashima (southern Miura Peninsula) can be observed to truncate such thrusts. We therefore propose that the compressional deformation history of the Miura Group was progressive and synchronous with deposition. The early layer-parallel thrusts tend not to be associated with fluid-related reaction rims (oxidation fronts, generally of ferric minerals), unlike many of the later high-angle faults. We attribute this difference to

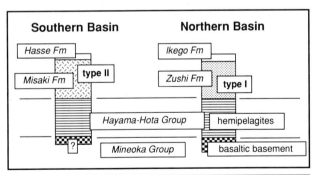

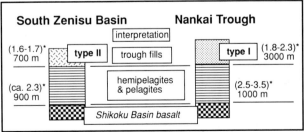

* RMS velocity and layer thickness

type I: trough fills derived from two provinces on different plates
type II: intra-plate trough fills derived from the oceanic plate

Fig. 16.14. Comparative schematic sections of the Miocene–Pliocene Boso-Miura region and modern Zenisu region. Note the similarities between the Southern basin and the South Zenisu basin, and also between the Northern basin and Nankai trough.

the fact that, in general, the early low-angle faults were reducing porosity in the dewatering and lithifying sediments but did not act as long-lived fluid pathways; the later high-angle faults, however, were probably significant fluid conduits for the dewatering sediments at a more advanced stage of lithification. Present-day fresh and marine waters preferentially concentrate along these later stage faults, further enhancing the chemical weathering along the fractures.

In general, the vein structure predates virtually all of the faulting, although there are a few rare cases where the vein structure may cut layer-parallel thusts — which may also be very early structures. Vein structure occurs within blocks of sediment caught up in early, chaotic, sediment slide deposits (Pickering *et al.* 1990). Such data suggest that the vein structure formed early in the deformational history of the dewatering sediments, and probably at rather shallow depths of burial. The early vein structures and later injection structures were formed under conditions of elevated pore-fluid pressures in sediments undergoing lithification and progressive deformation. SEM back-scatter studies show high average atomic number contrasts between the infill of the vein structure and the surrounding sediments (Pickering *et al.* 1990). The clasts and microfabric, however, appear to be similar. Some reorientation of the platy minerals is seen at the margins of the vein structure. Stable carbon and oxygen isotope studies, together with geochemical probe data, are consistent with the pore fluids being derived from trapped sea water in which certain elements, such as S, Fe and probably Ca, were preferentially concentrated in the vein structure and wet-sediment injections.

Detailed descriptions of *vein structure* can be found in Moore (1986) and Chapters 5 and 7. Vein structures are very common in the Japan and Izu-Bonin forearcs, but appear absent to rare in the Nankai forearc. This may be due to the lithological differences between the various forearcs. Vein structure is well developed in deep-marine siliceous (radiolarian-rich), tuffaceous, mudrocks that are so common in the Japan and Izu-Bonin, but not the Nankai, forearcs. There have been many interpretations of vein structure (e.g. Knipe 1986, Lundberg & Moore 1986). Cowan (1982), who was the first to describe vein structure from the Middle America trench, interpreted it as extensional fractures. Knipe (1986) interpreted the structure as forming in response to gravity-induced downslope failure of sediment. Knipe (1986) also noted the possibility that veins may develop or be modified to a sigmoidal geometry as a result of shear parallel to bedding in unconsolidated slope sediments (Kimura et al. 1989).

The systematic vein structure is interpreted as an early manifestation of fluid movement associated with forearc extension within muddy tuffaceous sediments on the relatively steep submarine slopes within the Neogene–Recent ICZ. The vein structures appear to be spatially associated with early faults, possibly suggesting a causal link between fluid flow and fault activity in sediments that still contained a significant amount of pore fluids.

Southern basin

The stratigraphy of the Southern basin is well exposed for more than 80 km along the WNW–ESE trending coastline from the Oiso area in the west to the Boso Peninsula in the east and across the southern part of Miura Peninsula. The southern part of the Southern basin is presently below sea level. The original stratigraphic relationship between the underlying Hayama-Hota Group and the overlying Misaki Formation is obscured everywhere by faults. No distinct marginal facies occur along the boundary. The oldest age of the Misaki Group is c. 10 Ma (Ando et al. 1989). The Misaki Formation is estimated as approximately 1200 m thick, but stratigraphic duplication by bedding-parallel thrusts and the presence of wet-sediment injections allow only a crude estimation of the thickness.

The Misaki Formation chiefly consists of alternations of tuffaceous mud and lapilli layers intercalated with thick-bedded pumiceous tuff layers. Abundant wet-sediment deformation, together with complex faulting and folding (Kodama 1968, Ogawa et al. 1985), have modified, but rarely obscured, primary sedimentary structures.

Lapilli beds (mafic volcaniclastics) first occur in the Misaki Formation. The lapilli beds appear to pass laterally from more proximal deposits on the Miura Peninsula to distal equivalents on the Boso Peninsula (Akamine et al. 1956). Most of the lapilli beds in the Misaki Formation are interpreted as sediment gravity-flow deposits, but some primary, water-lain, fall deposits are recognized on the Miura Peninsula and the Oiso area. There are also current-rippled siltstones and fine-grained sandstones showing evidence of reworking by either deep, vigorous, thermohaline currents, possibly contour currents, and/or deep internal waves, possibly related to tides.

Bulk-rock chemical analyses of the lapilli (interpreted as primary, water-lain, fall deposits) show differences in chemical composition between the Miura Peninsula and the Oiso area (Fig. 16.15). The lapilli from the Miura Peninsula are characterized by low-K tholeiite series basaltic andesite, which has a geochemical signature similar to that of the modern Mihara-yama volcano, located at the volcanic front (Fujii et al. 1988). The morphology of the lapilli, SEM studies of the glass shards, and the presence of accretionary lapilli, suggest that the lapilli beds were the products of mafic, phreatomagmatic, Surtseyan-type eruptions (Soh et al. 1989). In contrast, the bulk chemical composition of the lapilli in the Oiso area is characterized by low-alkaline high-alumina basalt series, similar to those of present Izu monogenetic volcanoes (IMV), as proposed by Hamuro (1985). The mixed zone or boundary in chemical composition was located between the Oiso area and the Miura Peninsula, now submerged. Also, it appears that, during deposition of the Misaki Formation, the volcanic front was sited between the Oiso area and the Miura Peninsula. The present volcanic front runs in a N–S direction to the west of the Oiso area, and Miura Peninsula (Fig. 16.16).

Benthic foraminifera suggest palaeodepths in the Southern basin varied, but tended to deepen eastward (Kitazato 1986, Nakao et al. 1986, Ando et al. 1989). *Cubicides wuellestorfi, Melonis parkerae, Osangularia culter, Parareloides bradyi, Pullenia bulloides,* and *Uvigerina proboscidea,* for example, occur on the Miura Peninsula where the estimated palaeodepth ranged from middle to lower bathyal environments (Ando et al. 1989, J. Young, British Museum, personal communication 1989). Biogenic siliceous debris, mainly diatomaceous (as seen in the SEM), and nannofossils are fairly common, The faunal assemblage comprises (J. Young, British Museum, personal communication 1989): *Reticulofenestra pseudoumbilicus, Coccolithus pelagicus, Calcidiscus leptocoporous, C. macintyrei, Helicosphaera carteri, Sphenolithus abies, Discoaster variabilis, D. surculus,* and *D. pentaradiatus* (one specimen only found): the age interpretation of late Miocene–early Pliocene (Biozone NN 14–15) is based on the co-occurrence of *R. Pseudoumbilicus, S. abies*, and *D. pentaradiatus,* together with an absence of *D. quinqueramus,* and the fact that many of the *C. pelagicus* specimens possess a well-developed bridge. At the time of deposition of the uppermost Misaki Formation, the common occurrence of *C. pelagicus*, the absence of *D. brouweri*, and the low abundance of *S. abies*, are suggestive of cooler water temperatures (J. Young, British Museum, personal communication 1989). On the Boso Peninsula, the depositional environment of the Misaki Formation has been interpreted as below the calcite compensation depth, CCD (> about 4000 m), based on the dissolution of the tests of calcareous foraminifera (Nakao et al. 1986). In

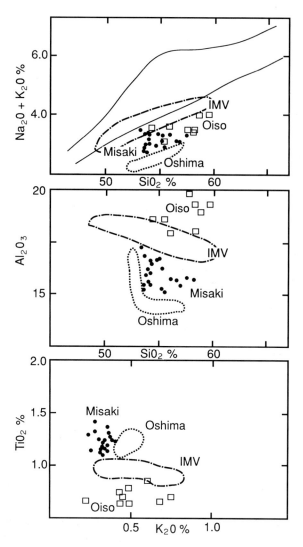

Fig. 16.15. Plots of $Na_2O + K_2O$ vs SiO_2, Al_2O_3 vs SiO_2 and TiO_2 vs K_2O for lapilli from the Misaki Peninsula and the Oiso areas. Note the similarities for the lapilli between Misaki Peninsula and the active volcano of Oshima, and between the Izu monogenetic volcanoes (IMV) and the Oiso area. The data suggest that the Misaki area was located on the forearc-basin side at the time of accumulation of the Misaki Formation, whereas the Oiso area was situated behind the volcanic front. After Soh et al. (1991).

volcaniclastic unit comprises, from bottom to the top, sediment gravity-flow deposits and cross-bedded strata. Subordinate sediment gravity-flow deposits occur towards the bottom of the volcaniclastic unit, associated with a basal erosional surface and many internal discordances, suggesting that the volcaniclastic deposits are canyon or channel fill sediments.

On the southern tip of Boso Peninsula, the Nojimasaki Conglomerate, part of the Hasse Formation, contains angular andesite clasts, and some pebbles of granitoid composition, similar in chemical composition to that of the Tanzawa Mountain, together with reddish recrystallized chert, similar in lithology to that of the Cretaceous Shimanto Group in the Honshu arc. The siliceous shale clasts contain Permian and Triassic radiolaria such as *Follicullus, Pseudostylosphaera* and *Triassocampe* (S. Saito, personal communication 1990) showing that some detritus was derived from the Honshu arc (probably the Chichibu Group), and not only from the relatively young Izu-Bonin arc side. These gravels represent the first stratigraphic occurrence of Honshu arc-derived sediment to reach the Southern basin.

Northern basin

The Northern basin is exposed from Katsura on the Boso Peninsula in the east, to the Oiso area in the west, and is filled by the Zushi and Ikego Formations. The Zushi Formation mainly comprises hemipelagic mudstones, very thin-bedded cross-laminated siltstones (possible low-concentration turbidity-current deposits and the products of various semi-permanent to episodic bottom currents), and small amounts of volcaniclastic sandstone turbidites: the lithology and sedimentary facies of the Zushi Formation, however, varies laterally. For example, 4–5 Ma coarse clastic detritus of submarine fan and/or channel-fill sediments are recognized in the Oiso area and Boso Peninsula, but similar contemporaneous coarse clastic detritus is absent on the Miura Peninsula.

In the western, Oiso area, there are two sedimentary facies-associations: hemipelagic or pelagic mudstones and thin-bedded turbidites interpreted as the deposits of the landward slope and basin plain, and submarine channel volcanic and terrigenous coarse clastic resedimented conglomerates (Ito 1988). The channel-fill conglomerates are subrounded to subangular andesite and lapilli tuff, and some black shale and sandstone clasts are also included. An *in situ* calc-alkaline andesitic submarine volcano is also recognized (Takatoriyama Formation, Ito 1988). The Zushi Formation on the Miura Peninsula is lithologically homogeneous, and consists mainly of tuffaceous hemipelagic mudstone intercalated with thin-bedded turbidites. Volcaniclastic conglomerates containing basaltic andesite and dacite clasts, occur along the southern margin (e.g. Tagoegawa Conglomerates, Eto 1986b).

There are three lithologic units on the Boso Peninsula

contrast, there is no evidence in the Oiso area for such deep-water environments, but instead wood fragments (<10 cm long) are common in the succession (Otsuka 1929), suggesting that the sedimentary environment of the Oiso area was relatively near to the mouth of a river or beach.

The lithologic boundary between the Misaki and the overlying Hasse (Hatsuse) Formations is well defined everywhere, although there is no exposure of age-equivalent strata in the Oiso area to the west. The Hasse Formation comprises three main lithologic units: coarse-grained, cross-bedded volcaniclastic sediments, well-bedded turbidites, and pebble/cobble-sandstone debris-flow deposits. The Hasse Formation attains a thickness of over 300 m. It has been suggested that both lithologic units are age-equivalent, at least in the Boso Peninsula (Kotake 1988). The coarse-grained

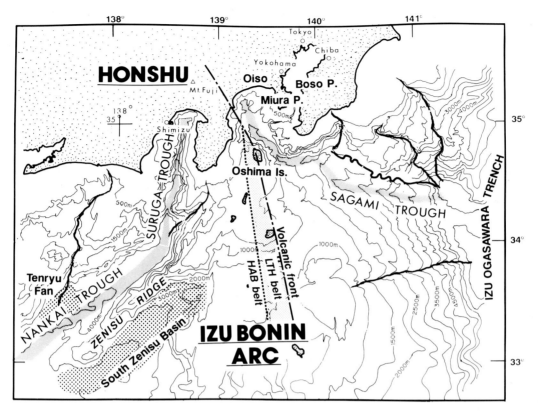

Fig. 16.16. Geomorphologic index map around the southeastern part of Honshu arc, central Japan, to show the location of the present volcanic front and geochemical zonation of volcanic rocks along the volcanic front (after Takahashi 1986). LTH, low-K tholeiite zone; HAB, low-alkaline high-alumina basalt zone. Modified after Soh et al. (1991).

which, in ascending order, are: hemipelagic mudstones (the Amatsu, Kiyosumi formations), submarine fan complex (the Kiyosuma Formation), and basin-fill turbidites (the Anno Formation) (Mitsunashi et al. 1976, Nakajima et al. 1981). The coarse clastic unit (the lowest portion of Kinone Formation; sensu Nakajima et al. 1981) occurs as a marginal facies along the Hayama-Mineoka uplift belt. The lower hemipelagic mudstone, with a small amount of thin-bedded sandstones, is interpreted as a distal facies of the Zushi Formation on the Miura Peninsula. The middle submarine fan complex comprises thick-bedded sandstone turbidites interpreted as a main channel fill and fan lobes (Tokuhashi 1979, 1989). Palaeocurrent analysis shows that the coarse clastic detritus was transported from the north, that is, the Honshu arc (Kitazato 1986, Tokuhashi 1979), and the clast composition in the coarse clastic detritus supports this interpretation (Tokuhashi 1979).

The palaeodepth of the Zushi Formation on the Miura Peninsula, estimated from benthic foraminifera, ranges from middle to upper bathyal environments (Eto et al. 1987). To the east, on the Boso Peninsula, the upper portion of the Zushi Formation (the submarine fan complex and basin-fill turbidites) was deposited in middle to upper bathyal environments, respectively (Hatta & Tokuhashi 1984). The lithology of the Zushi Formation suggest that the palaeo-bathymetry of the Northern basin was probably essentially the same from west to east, which is consistent with the present palaeobathymetric pattern of the Honshu arc deepening southward (Kitazato 1986).

The overlying Pliocene Ikego Formation is divided into two lithologic units; a volcaniclastic and a thin-bedded turbidite unit. The volcaniclastics consist of sediment gravity-flow deposits and chaotic slide deposits, interpreted as channel-fill sediments. A thin-bedded turbidite unit comfortably overlies the lower volcaniclastics and/or the underlying Zushi Formation. The Ikego Formation contains *Uvigerina* spp. and *Melonis pompilioides*, suggesting upper bathyal environments (Eto et al. 1987). Benthic foraminifera suggest that the Northern basin shallowed with time.

On the Boso Peninsula, in the Miocene–Pliocene Northern basin, the deep-marine volcanic sandstones and conglomerates of the Ikego Formation contain a giant clam colony of *Calyptogena*, (Niitsuma et al. 1989). Deep-water fluid-venting related biological communities characterized by *Calyptogena* are now known to occur along the present Japanese subduction zones (Boulegue et al. 1987, Ohta & Laubier 1987, Fujioka & Taira 1989). The sedimentary setting of these modern examples appears limited to fault zones within the landward slopes of trenches, other plate boundaries, and zones of intra-oceanic plate deformation. Calcite cements obtained from the sandstones within the Ikego Formation have extremely light carbon isotopic ratios of $\delta^{13}C = -19.7‰$ to $-49.2‰$ (Soh et al. 1991), suggesting an origin from biogenic CH_4 due to fluid seepage. Detailed mapping shows that the *Calyptogena* colony in the Ikego Formation probably formed on fault-related talus (chaotic) deposits along a canyon wall trending NNW–SSE (Fig. 16.17). This trend is normal to the general strike of the Miura Group and Hayama-Mineoka uplift belt, possibly indicating active syn-sedimentary transverse faulting. The Hayama-Mineoka uplift belt was probably a barrier to sediment dispersal between the Southern and Northern basins, except for channellized links where transverse faults cut the submarine ridge to enable Honshu arc-derived detritus to reach the Southern basin.

Palaeogeography and basin history

The Miura Group records the evolution of plate-boundary sedimentary basins, ranging from intra-oceanic to slope and/or forearc basins. In the Middle Miocene to early Pliocene, the interpretation of the Southern basin is that it was an intra-oceanic foreland basin developed on the Izu-Bonin arc, while the Northern basin is thought to have been a plate boundary basin, much like the present Suruga trough. A topographic high developed between both basins as an intra-oceanic thrust ridge due to the rupture of the Izu-Bonin arc. The highly deformed sediments in this ridge, for example the Hayama Group that contains mélanges, represent an earlier accretionary prism. As a result of the plate boundary migration, both the Northern and Southern basins were incorporated into the Honshu arc after the formation of the Kurotaki unconformity at *c.* 3–2.5 Ma. Figure 16.18 summarizes the interpreted plate settings of the Northern and Southern basins in the latest Miocene.

During the deposition of the Misaki Formation, the Honshu arc underwent rapid uplift as a result of the collision of the Izu-Bonin ridge with the Honshu arc, to release large amounts of terrigenous detritus into the Northern basin, but with little terrigenous materials reaching the Southern basin.

The Southern basin crossed the low-K tholeiite and low-alkaline high-alumina basalt arc volcanic zones. The present Izu monogenetic volcanic series around the northern portion of the Izu-Bonin arc is sited about 10 km behind the present volcanic front, and the low-alkaline tholeiite series domain disappears around Hakone Volcano, located on the northern end of the Izu-Bonin Ridge. This suggests that the basin formed obliquely to the arc volcanic front, that is, obliquely to the trench. In addition, the presence of large amounts of lapilli on the Miura Peninsula suggests that the low-K tholeiite series volcanism developed behind the Southern basin. From the present setting around the Izu Peninsula, it seems that the low-K tholeiite volcanoes only developed on the Izu-Bonin ridge side.

The palaeo-bathymetry of the Southern basin, estimated from benthic foraminifera and sedimentary facies, deepened eastward. Kitazato (1986) has shown that the late Miocene palaeo-bathymetry around the Honshu arc side deepened southward from a coastal setting (beach) to about 3000 m water depth. An eastward deepening is typical of the present forearc to the Izu-Bonin Ridge, ranging from emergent volcanic islands to an 8-km-deep trench. The Southern sedimentary basin originated as a forearc basin on the Izu-Bonin ridge, as seen in the underlying Hayama-Hota Group (around the Southern basin) composed entirely of detritus-free, volcaniclastic hemipelagic sediments. The first occurrence of Honshu arc-derived detritus is recorded in the *c.* 4–3 Ma Hasse Formation, suggesting that by that time the plate boundary had begun to migrate to the Southern basin or even farther south.

The conformable transition between the Hota Group and the overlying Amatsu Formation is important for understanding the origin and development of the Northern basin. The lithologic characteristics led Ogawa & Taniguchi (1988) to interpret the upper Hayama-Hota Group as the deposits of the mixed zone of detritus provided both from the Honshu arc and the Izu-Bonin ridge and, therefore, above the plate boundary between the Izu-Bonin and Honshu arcs, that is, the trench-fill. The amount of terrigenous clastics in the Northern basin that were derived from the Honshu arc, increased upwards. The Kiyosumi Formation on the Boso Peninsula, for example, was derived entirely from the Honshu arc (Tokuhashi 1979). Limited amounts of detritus of low-K tholeiite origin (Izu-Bonin arc-derived material, Tagoegawa Conglomerate and

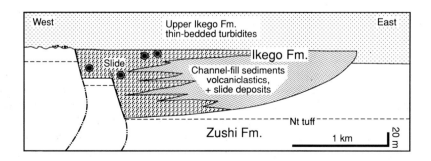

Fig. 16.17. Schematic cross section of the Ikego Formation on the Miura Peninsula. Stars show the occurrence of *Calyptogena* colonies. Channel or canyon formation was related to the development of syn-sedimentary transverse fault movements associated with fluid venting. Modified after Soh *et al.* (1991).

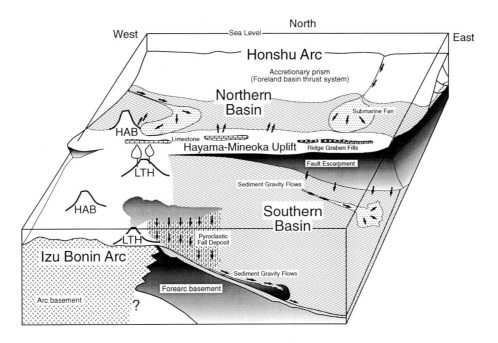

Fig. 16.18. Reconstruction of the sedimentary environments associated with the Northern and Southern basins c. 5–4 Ma.

Ikego Formation) are also recognized in the Northern basin. A similar dual supply system, derived from two different plates, is recognized today in the arc-continent collision basin of the coast range of Taiwan (e.g. Teng & Lo 1985, Teng & Chen 1988). Similarly, it is possible to interpret the Northern basin as having been located on the plate boundary between both arcs.

Gravity and geomagnetic anomaly data provide additional information on the geological structure of the Hayama-Mineoka uplift belt (Tonouchi 1981, Morijiri et al. 1987). Narrow negative anomaly belts extend E–W along the Hayama-Mineoka uplift belt, interpreted by Morijiri et al. (1987) as indicating that the Mineoka ophiolite dips steeply northwards. The Bouguer anomaly contours become closely spaced along the boundary between the Hota-Hayama Group and the Southern basin, suggesting the presence of a fault along the southern margin of the Hayama-Mineoka uplift belt (Ueda et al. 1987). The idea of such a faulted boundary is supported by the development of distinct marginal facies within the Miura Group, and the observed faulted contacts between the overlying Miura Group and the underlying Hota Group.

Koike (1957) proposed that during the late Miocene the Hayama-Mineoka uplift belt was a topographic high (ridge) extending WNW–ESE. Shallow marine calcirudites, consisting of fragments of *Fabellum* sp., *Balanus* sp., and calcareous algae, probably covered the submarine Hayama-Mineoka uplift belt (e.g. Shimoyamaguchi Member, Eto 1986b). On the Boso Peninsula, graben-like depressions formed in the axial part of the Hayama-Mineoka uplift belt. Talus deposits infilled these depressions as breccias and debrites, including a shallow marine fauna of *Chlamys akihoensis*, *C. vesiculosa* and *Conus* sp. (e.g. Okuzure Formation, Otsuka & Koike 1949). The shallowest part of the uplift belt was probably less than c. 200 m water depth. It appears that the Mineoka uplift belt was a submarine horst.

Large-scale sediment slides are recognized within the basins, particularly the Southern basin. In the c. 3–0.9 Ma Chikura Group in the southern Boso Peninsula, for example, there is a submarine slide (Mera and Hata Formations) that is at least 700 m thick, can be mapped over 10 km, and involved up to 2 km of sediment translation towards the NNW, that is, arcwards (Ito & Sugiyama 1989, fig. 16). The timing of this sizeable submarine sliding event has been constrained to between 0.9 and 0.65 Ma, and the slide sheet is interpreted as having occurred towards the NNW above a landward-dipping back-slope of a thrust (Ito & Sugiyama 1989). The sliding event occurred during a major glacio-eustatic sea-level fall (Haq et al. 1987), which may have been a contributary factor in creating unstable slopes.

Modern analogues for the Miocene–Pliocene sedimentation of southeast Japan occur offshore from the Honshu arc around the present-day Zenisu ridge. The Zenisu ridge is a linear topographic high on the northern Izu-Bonin arc with a northern plate-boundary basin parallel to the ridges as the Nankai-Suruga basin systems, and the South Zenisu basin as an intra-oceanic basin. The Zenisu ridge extends northeastward, in continuity with one of the numerous en échelon ridges of the Izu-Bonin arc. To the southwest, the Zenisu ridge gradually dies out into the Shikoku basin. Distinct NNW–SSE oriented transverse faults are recognized on the ridge (Aoki et al. 1982, Ueda et al. 1985, Le Pichon et al. 1987a).

The boundary between the Zenisu ridge and the South Zenisu basin is a very steep escarpment with a topographic relief of 900 m. Seismic profiles show that the basement of the Zenisu ridge is tilted 3–5° northwards (Ueda et al. 1985). Several unnamed in situ volcanoes, ranging from 5 to 10 km in diameter, are recognized on the northern flank of the Zenisu ridge. Dredge samples obtained from a probable inactive volcano (position: lat. 33°15'; long. 137°40'), are calc-alkaline hornblende dacite lavas of island arc origin (Hernandez & Lallemant 1988, Tamura & Ishii 1988).

In the vicinity of the Zenisu ridge, the Nankai trough comprises the Nankai axial channel and the Tenryu submarine fan system. The Nankai axial channel systems is the main supply system in the Nankai trough, and is still active, being supplied by sediments from the up-dip Suruga trough off the eastern Shikoku area. The Tenryu submarine fan is 30 km in radius, and is probably presently inactive, although it is actively deforming.

The compressive tectonics in the Zenisu ridge region were first pointed out by Aoki et al. (1982) through the interpretation of multichannel seismic lines. They observed a reverse fault on the southern flank of the Zenisu ridge, dipping SE with a vertical offset of the oceanic crust reaching 2.5 km at the base of the ridge. Closely-spaced single channel profiling and SeaBeam mapping has also revealed the compressive tectonics at the southern scarp of the Zenisu ridge, with the presence of an active *fold mound*, thrust sheet topography, and uplift and tilting of the Shikoku basin oceanic crust and overlying hemipelagic sediment (Le Pichon et al. 1987a, Lallemant et al. 1989). A submersible dive observation using *Nautile* (Le Pichon et al. 1987a), also identified a 'vent' ecosystem of *Calyptogena* bivalves, suggesting localized pore-water seepage. In addition, present intra-oceanic shortening in the South Zenisu basin is recognized by active seismicity along this belt.

Three distinctive seismic units are defined in the South Zenisu basin (Fig. 16.14), based on their RMS velocity, that is, 1.6–1.7, 2.3 and 3.5 km s^{-1}, from bottom to top. Using the acoustic characteristics and the samples obtained during a *Tansei-maru* cruise and KAIKO *Nautile* dives (Le Pichon et al. 1987a, table 1), they can be interpreted as the oceanic basement, the Shikoku basin hemipelagic cover and the volcaniclastic thin-bedded turbidites, from bottom to top. The seismic profile also suggests that the trench wedge occurs in the upper volcaniclastic turbidite layer, which is progressively deformed northwestward. The basin fill is c. 1500 m thick. It should be noted that the thickness of the Shikoku basin hemipelagic sediments does not change laterally from the Zenisu ridge to the South Zenisu basin. The upper volcaniclastic sediments are interpreted as derived from the Izu-Bonin ridge along the axial channel of the elongate basin. No Honshu arc-derived detritus is present in the South Zenisu basin. Accordingly, the South Zenisu basin is likely to have been formed after deposition of the upper volcaniclastic turbidites, as a sedimentary basin due to intra plate deformation. The sedimentary sequence in this basin is comparable to that of the Oligocene–Pliocene Southern basin, ranging from the Mineoka Group to the Misaki Formation.

Although direct determination of the lowest acoustic basement age is not possible (no basement rocks were collected following dredging and dives), there is indirect data; for example, magnetic lineations and their correlation with the geomagnetic time-scale. A lineated pattern of N140 to N160°E trending magnetic anomalies can be recognized across the ridge, locally disrupted by small transform faults (Chamot-Rooke et al. 1987). The identification of the magnetic lineations suggests an anomaly of 6B to 6C age (22.7 to 24.3 Ma) in agreement with the spreading history of the Shikoku basin inferred from its western side (Lallemant et al. 1989). These indirect age determinations are reasonably consistent and suggest an early Miocene age, that is, the very early spreading stage of the Shikoku basin. The hemipelagic mudstone samples obtained by the submersible *Nautile* and by dredging using R/V *Tansei-maru*, from the southern escarpment of the Zenisu ridge, which probably expose the middle layer, appear to range from early Miocene to Middle Miocene (Le Pichon et al. 1987a). The upper layer comprises mainly thin-bedded volcaniclastic turbidites confined to the trough basin, and suggest that deposition of the upper layer defines the time of activation of the trough basin as a result of the rupture and uplift of the Shikoku basin oceanic crust.

The lower two layers of the Shikoku basin hemipelagites and oceanic layer II are much denser than the equivalents in the South Zenisu basin. In contrast, the upper layer is very similar in RMS velocity to that of the South Zenisu basin: the upper layer in the Nankai trough, however, is different in seismic signature to that of South Zenisu basin. Taira & Niitsuma (1986) showed that the fill sediments of the Nankai trough are variable in lithology, that they were longitudinally supplied from the drainage of the Fujikawa River and Mt. Fuji, that is a mixture in composition between the Honshu-derived and Izu-Bonin-derived materials. Some detritus derived from the Honshu arc is recognized as a lateral supply system (Tenryu Fan).

The age of the boundary between the middle hemipelagic layer and the upper volcaniclastic cover is poorly constrained. Therefore, the precise timing of the Zenisu ridge remains unknown. Chamot-Rooke & Le Pichon (1989) investigated the formation of the Zenisu ridge in terms of the compressive mechanical failure of a thin perfectly-elastic plate, and from their rheological model they suggested that the formation of the Zenisu ridge is likely to be younger than 4 Ma (Chamot-Rooke & Le Pichon 1989, fig. 11).

The evolutionary processes of the Zenisu ridge may lead to a seaward jump of the consuming plate boundary further south. Therefore, part of the oceanic crust basement to the Zenisu ridge, the submarine volcanoes on the ridge, and trough-fill sediments around the Zenisu ridge, may be incorporated into the Honshu arc as ophiolite and associated trough-fill sequences, like the Hayama Mineoka uplift zone and the Mio-Pliocene sequences mentioned above.

SEQUENCE STRATIGRAPHY IN SOUTHEAST JAPAN AND THE QUATERNARY KAZUSA GROUP

The concepts and methods in *sequence stratigraphy* are now routinely applied to the integrated study of many sedimentary basins (e.g. van Wagoner et al. 1988, 1990). Sequence stratigraphy is the study of rock relationships within a chronostratigraphic framework of repetitive, genetically linked strata bounded by surfaces of erosion or non-deposition, or their correlative conformities (van Wagoner et al. 1988); alternatively, sequence stratigraphy is the study of genetically related facies within a framework of chronostratigraphically significant surfaces, and in which the sequence is the fundamental stratal unit for sequence stratigraphic analysis (van Wagoner et al. 1990).

Although philosophically similar in many respects to the Exxon approach, Galloway (1989a, 1989b) proposed using depositional sequences. Galloway (1989a) pointed out that: "The most readily identifiable surface is the transgressive surface... The second most easily recognizable surface in outcrops is the surface of maximum flooding..." (Haq et al. 1987). On this basis, Galloway (1989a) put forward the concept of a genetic stratigraphic sequence. A *genetic stratigraphic sequence* is a sedimentary product of a depositional episode. Each sequence comprises a (1) progradational facies-association; (2) aggradational facies-association, and (3) retrogradational or transgressive facies-association. Genetic sequences are bounded by a sedimentary veneer or surface that records the depositional hiatus that occurs over much of the transgressed shelf and adjacent slope during maximum marine flooding.

Galloway (1989a, b) applied his genetic stratigraphic sequence approach to the Cenozoic basin of the northwest Gulf of Mexico, in which sequences were defined by regional marine flooding surfaces. Galloway (1989b) made the following observations. (1) Continental margins are characterized by repetitive episodes of basin-margin offlap, punctuated by transgressive events associated with marine flooding of the depositional platform. (2) Continental-margin outbuilding is concentrated at one or more shelf-edge deltaic depocentres separated by interdeltaic bights. Depocentres remained fixed during a depositional episode, but commonly relocate during transgression and flooding. (3) A distinct syndepositonal structural style in prograding continental margins results in sporadic uplift of a basin-fringing peripheral bulge and accentuates preservation of shelf-margin facies along zones of extensional normal faulting and enhanced subsidence. Galloway (1989b) found that the early Cenozoic Gulf Coast sequences appear to be most closely related to tectonic events of the intraplate source terrane, which control the rate and location of sediment supply, and the basin-margin response to sediment loading. In contrast, he found that the late Cenozoic sequences appear to more closely reflect the proposed Exxon eustatic curve.

Other approaches to sequence stratigraphy include the work of Einsele (1985), who considered the response of sediments to sea-level changes in differing storm-dominated margins and epeiric seas, particularly the Mesozoic epicontinental, mud-dominated, seas in the slowly subsiding basins of Germany. Such basins contrast markedly with the rapidly subsiding shelf-margin seas subject to rapid changes in sea level that are glacio-eustatically driven, and from which much of the Exxon philosophy is based. He pointed out that the base level to which sediments aggrade is the storm wave base rather than sea level *sensu stricto*. Einsele (1985) believes that sediment accumulation patterns are particular to depositional sites and conditions. Perhaps, the principal difference between the Exxon model and Einsele's (1985) perspective is that: (1) regressions were gradual rather than abrupt; (2) purely aggradation depositional units reach their maximum thickness where basin subsidence is most rapid, for example in the centre of basins, and (3) the base level to which sediments may aggrade is the storm wave base.

Eustatic sea-level changes are generated by: (1) changes in the volume of mid-ocean ridges and oceanic spreading centres on a timescale of 10s of millions of years; (2) tectonic (orogenic) activity, typically on a timescale of 10s of millions of years; (3) transmission of intraplate and interplate stress, possibly at high-frequency timescale of about 10^5 years to millions of years; but such processes are poorly understood; and (4) glacio-eustasy, in which ice-cap construction may occur in time periods over 10^5 years and melting in 10^4 years. The total melting of a Greenland-sized ice-cap would cause global sea-level to rise by about 6 m, whereas if the entire Antarctica ice sheet melted, then the rise would be approximately 60 m. Isostatic rebound, for example from the melting of extensive ice sheets and glaciers, will cause relatively local changes in land elevation, on a timescale of $10^3 - 10^4$ years. Thermal expansion and contraction of the ocean waters, associated with global climate change, will exert a relatively small, but important, control on sea level.

There have been various published 'global/eustatic' sea-level curves mainly produced by the Exxon group (e.g. Haq et al. 1987, 1988). The evidence for these curves has never been made freely and publically available for the academic community to rigorously test. Indeed, in many areas of the world where detailed, high-resolution, sequence stratigraphy has been undertaken, little support for the construction of the *Exxon eustatic curve* has been forthcoming. Miall (1992) tested the premise of the Exxon cycle chart (Haq et al. 1988), that there exists a globally correlatable suite of third-order eustatic cycles, and concluded that it remains unproven. In detail, Miall (1992) concluded: "The implied precision of the Exxon global cycle chart is not supportable, because it is greater than that of the best available chronostratigraphic techniques, such as those used to construct the global standard time scale.

Stratigraphy of Miura & Boso Peninsulas, SE Japan

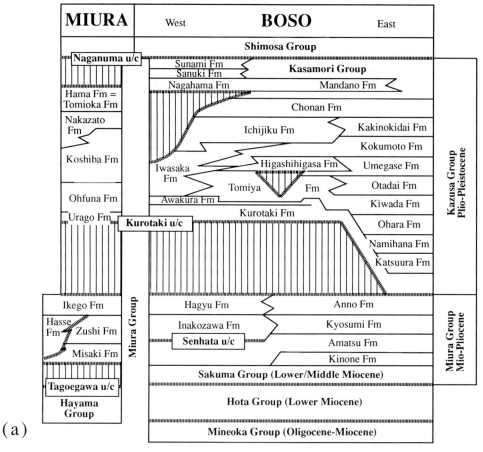

Fig. 16.19. Palaeogene–Neogene stratigraphy of southeast Japan. (a) Formations; (b) Interpreted sedimentary environments.

Correlations of new stratigraphic sections with the Exxon chart will always succeed, because there are so many sequence-boundary events from which to choose. This is demonstrated by the use of four synthetic sections constructed from tables of random numbers. A minimum of 77% successful correlations of random events with the Exxon chart was achieved. The existing cycle chart represents an amalgam of regional and local tectonic events and probably also includes unrecognizable miscorrelations. It is of questionable value as an independent standard of geologic time."

The concepts and approach provided by sequence stratigraphy have again linked sedimentology and stratigraphy in an exciting way, after a period of divorce brought about by the facies approach to sedimentary environments which was so much an integral part of the philosophy of the 1970s. Although the global synchronicity, or correlatability, of events has been undermined (see discussion by Miall 1992), the principles of sequence stratigraphy, the associated ways of dividing up stratigraphic units, and interpreting many of the causal mechanisms, remain valid. For these reasons, a thorough understanding of the basic principles of sequence stratigraphy remains important. Research work must still be focused on producing a eustatic sea-level curve, and in understanding the magnitude and causes of sea-level change. Chemostratigraphic techniques and improved chronostratigraphic dates are paving the way forward.

Accommodation space, that is, the space available to be filled by sediment, is controlled by sediment supply, eustasy and tectonic subsidence/uplift. The fundamental causal mechanisms that control the higher frequency sequence generation remain poorly understood. It is possible to envisage a ternary diagram that shows the principal factors controlling basin infilling as end-members, that is, sediment supply, subsidence history and eustatic change in sea level (Galloway 1989a). Such a ternary scheme is useful because it leads to a conceptual classification of input-dominated, tectonic-dominated, and eustatic-dominated sedimentary basins.

Perhaps the best natural laboratory in which to test many of the paradigms in sequence stratigraphy exists at young active plate margins where there is a good opportunity to dissemble tectonic from climatically induced changes in sea level and, therefore, relative base level. The Neogene forearc basins of SE Japan provide one such opportunity.

Figure 16.19 summarizes the Palaeogene–Neogene stratigraphy of southeast Japan, and emphasizes those components of the stratigraphy that appear to be

Stratigraphy of Miura & Boso Peninsulas, SE Japan
Sedimentary Environments

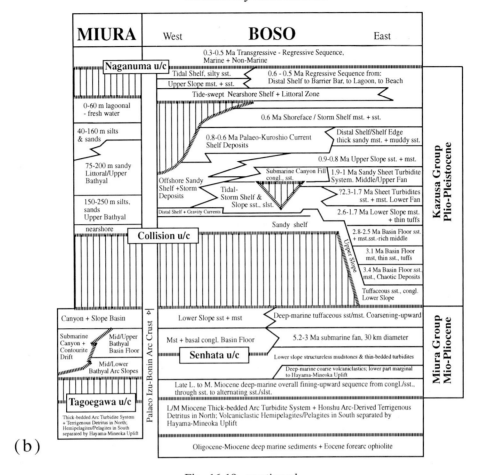

Fig. 16.19. *continued*

associated with relative low-stands of sea level in this region. The pre-accretionary stratigraphy of the Miura-Boso blocks (i.e. 3 – 2.5 Ma) does not conform to the Haq *et al.* (1987) Exxon-type curves for global sea-level fluctuations, and is best explained principally in the context of plate-tectonic processes during crustal delamination and accretion of successive segments of Izu-Bonin arc crust. In the ICZ, the post-accretion stratigraphy of the Kazusa and Shimosa Groups, above the Kurotaki unconformity, could be interpreted within the context of alternating low- and high-stand systems related to glacio-eustasy but before the last major global cooling at c. 0.8 Ma. Such an approach must be considered very speculative.

The Kazusa Group (see Katsura 1984) is an overall regressive sequence, up to 3500 m in thickness, from deep-marine turbidites to shallow-marine and littoral zone deposits (Fig. 16.20). The Kazusa Group accumulated in a tectonically active forearc basin, where maximum rates of subsidence have been estimated by Kaizuke (1987) as 2 m ka^{-1}. In the Kazusa Group, there is a good match between the youngest three major eustatic lowstands in sea level shown on the Exxon eustatic sea-level curve (Haq *et al.* 1987) and the development of deep-marine lowstand wedges and/or enhanced bottom-water current activity; the other parts of the stratigraphy being more explicable as having formed during intermediate or relative highstands of sea level (Fig. 16.21). The three depositional systems that may have formed during major falls in relative sea level, either tectonically or glacio-eustatically driven, are identified on the Boso Peninsula as follows:

2.8 Ma, accumulation of the sandstone-rich parts of deep-marine, 220 m thick, Ohara Formation, as a lowstand wedge;

1.6 – 1.4 Ma, deposition of the 540 m thick, sandstone-rich, deep-marine Otadai Formation and the immediately overlying 525 m thick, sandstone-rich, deep-marine Umegase Formation, as a lowstand wedge. Synchronous with the accumulation of the Umegase system, submarine canyons were excavated to form 'Vail type-1' unconformities, and then filled during the rising sea level, as the up to 200 m thick, gravel-rich, Higashihigasa Formation;

0.8 – 0.6 Ma, deposition of the up to 400 m thick, mainly cross-stratified, sandy Ichijiku Formation, which accumulated on the shelf in response to enhanced contour-current (palaeo-Kuroshio current) activity in water depths estimated as between 50 – 130 m (Nakayama & Masuda 1989), probably due to the more vigorous ventilation of the oceans during a phase of deteriorating global climate. The sand waves are interpreted as having been deposited in the northeastern part of the palaeo-straits called the Hayama-Mineoka Straits (Nakayama & Masuda 1989).

In the younger sediments of the Kazusa Group, a more confident interpretation of glacio-eustatically-controlled sea-level changes is possible. Recent studies by Ito (1992), have involved the recognition of six depositional systems (up to 700 m in aggregate thickness) in the upper part of the Kazusa Group (Kokumoto Formation and younger). Using the stable oxygen-isotope sea-level index and the age framework for the upper part of the Kazusa Group (c. 0.72 – 0.45 Ma), Ito (1992) identified a periodicity in the depositional cycles of between 20,000 and 90,000 years. Furthermore, Ito (1992) has demonstrated that these high-frequency depositional sequences can be interpreted to result from the interaction between fast rates of sediment accumulation and rapid rates of glacio-eustatic sea-level change during the middle Pleistocene. During the middle Pleistocene, the average rate of glacio-eustatic sea-level change, estimated from the empirical relationship between stable oxygen-isotope records and sea levels was c. 3 m ka^{-1} (0.11 – 0.1‰ equivalent to a change in sea level of 10 m; Fairbanks and Matthews 1978, Chappell & Shackleton 1986). During the middle Pleistocene, the glacio-eustatic changes in sea level were faster than the tectonically controlled vertical movements, therefore, the development of type-1 sequence boundaries (Van Wagoner et al. 1988) was probably a consequence of glacio-eustasy (Ito 1992).

The Pleistocene Shimosa Group overlies the Kazusa Group and was deposited in the forearc basin of the palaeo-Tokyo Bay which was initiated about 500,000 years ago. The Shimosa Group comprises eight stratigraphic units as transgressive – regressive cycles of non-marine and shallow-marine sediments that were deposited under a substantial glacio-eustatic control (Ito & Masuda 1989, Kondo 1989, Nakazato et al. 1989). Farther west, near Shizuoka in Central Honshu, the Plio – Pleistocene Kakegawa Group was deposited in a forearc basin (paleo-Enshu-nada basin), with a shelf less than 10 km wide, and is a major transgressive – regressive cycle with a strong glacio-eustatic signature (Ishibashi 1989). For example, the major regressive event (associated with a significant cooling in seawater temperature, as deduced from faunal data) begins about 1.6 Ma and correlates well with the 1.7 Ma fall in eustatic sea level shown on the Haq et al. (1987) third-order curve. This 'fit', however, does not rule out a tectonic control (Ishibashi 1989).

Stratigraphically below the Kazusa Group, in the Hasse Formation, and uppermost Misaki Formation (the upper part of the Miura Group), there is a dramatic change in the style of sedimentation at about 3 Ma (roughly coincident with accretion of the Miura block to Honshu), from a sequence dominated by background hemipelagites to a succession which is almost entirely reworked, coarse-grained, volcaniclastics with the fines winnowed away. On the Exxon eustatic sea-level curve, this change in depositional character coincides with a major lowstand, and the substantial growth of polar ice — the oxygen-isotope record for the Atlantic and Pacific ocean rims suggesting a stepped deterioration in global climate, with substantial cooling events in: (1) the early Eocene from the thermal maximum at c. 55 Ma; (2) in the earliest Oligocene at c. 37 – 36 Ma, attributable to an increase in global ice volume (C13R Subchron at 39.5 Ma, Wei 1991), probably marking the onset of an Antarctica ice cap, and (3) in the Middle Miocene at c. 14 Ma (Berger & Meyer 1987; Compton et al. 1990, fig. 1). The resulting increase in thermal gradient from the pole to equator, would be expected to enhance bottom-water activity, with stronger thermohaline circulation – contour currents. At about 3 – 2.5 Ma, we interpret the palaeo-Kuroshio (contour) current, flowing from west to east, to have become significantly more vigorous at depth (up to hundreds of metres), with the result that coarse-grained megaripples and small dunes developed under this current (upper Misaki and Hasse Formations).

A study of Pacific basin marine molluscan faunas by Tsuchi (1990) has shown that following the 16 Ma climatic optimum in the Pacific, there was an abrupt cooling of seawater temperatures on a pan-Pacific scale since 15 Ma. This was then followed by episodes of relatively warm seawater at 6 and 3 Ma on the Pacific side of Japan. Since 3 Ma, Tsuchi's (1990) data supports a phased decrease of seawater temperatures, which coincides with the deposition of the Hasse Formation in the Miura Group, and the accumulation of the Kazusa Group. The Neogene ocean-water circulation patterns were similar to the present, with the warm-water Kuroshio current flowing from SW to NE, and the cold-water Oyashio current flowing from N to S: both currents interact east of mainland Japan in the region of the TTT triple junction, which could explain the strong bidirectionality of some of the palaeocurrent data, from either the WSW or ENE in the Hasse Formation (Fig. 16.22) (Pickering unpublished data). Today, the Kuroshio Current flow can be relatively fast. In the Osumi Strait which is 35 km wide at its most narrow, off southern Kyushu, the Kuroshio current is presently active in water depths up to 80 – 100 m where surface current speeds reach a maximum 2 – 5 knots. Under this current, flowing from SW to NE, megaripples have developed with low relief and wavelengths of 4 – 28 m (Ikehara 1989). In this area, there are also sandwaves of medium-grained sand, including pebbles, with wavelengths from 24 to 136 m and heights of 1.2 – 7.2 m (Ikehara 1989). Nakayama & Masuda (1989) interpret the Pleistocene depositional environment of the Ichijiku Formation, in the Kazusa Group, as a palaeo-strait between the mainland of Japan (Honshu arc), and a submarine ridge, the Hayama-Mineoka uplift (= present-day Osumi Strait south of Kyushu).

In the Pliocene Hasse Formation, which was deposited under the influence of strong, deep-water, ocean bottom currents (including the formation of

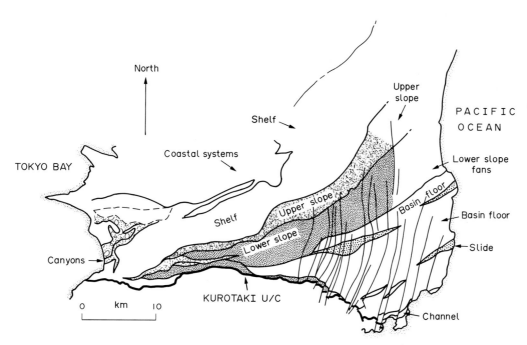

Fig. 16.20. Stratigraphy of Kazusa Group above the Kurotaki unconformity interpreted as sedimentary environments. Note that this post-collisional stratigraphy represents an overall shallowing-upward, regressive, system above a major unconformity. High-angle faults have relatively small horizontal offsets.

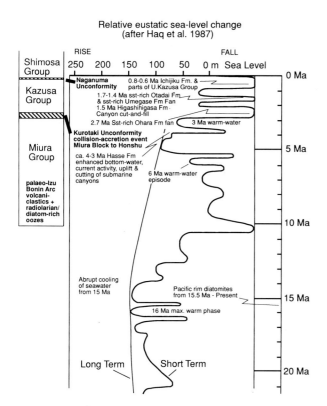

Fig. 16.21. Possible lowstand turbidite systems in the Miura Group, Miura Peninsula, and the Kazusa Group, Boso Peninsula, compared with their temporal position on the eustatic sea-level curve of Haq *et al.* (1987). Only major events are indicated. Note that the reliability of the Exxon eustatic sea-level curve is questioned in this paper, although for the past 3 Ma, where significant high-latitude ice is well documented and glacio-eustasy accepted, the comparison may be valid. See text for discussion.

bidirectional cross-stratification similar in appearance to tidal currents), the finer-grained pumiceous sediment was mainly winnowed out, leaving a contourite complex up to a few hundred metres thick, and preferentially preserved in the narrow reaches of the ICZ. Much of the Hasse Formation appears to have accumulated within submarine canyons, as at Jogashima on the southern tip of the Miura Peninsula; the canyon erosion being accelerated during the low-stand in sea level inferred for this time. The mean velocity of the palaeo-Kuroshio current would have been further enhanced by the narrow sea straits in the ICZ as the current became focused from the Nankai trough area to the ICZ, both between volcanic edifices and the Izu-Bonin arc and the mainland. Our palaeocurrent analysis of the upper Misaki Formation and Hasse Formation supports strong, deep reworking currents flowing from the west (WSW) towards the east (ENE). Thus, at the time of accelerated tectonic activity, as the Miura block was being uplifted and accreted to Honshu, so a major oceanographic event conspired to change the nature of the deep-water stratigraphy from background hemipelagic sedimentation to contourite bedforms. There is no faunal evidence to suggest that the tectonic uplift associated with accretion resulted in a change out of upper bathyal depths for the upper Misaki Formation and Hasse Formation, although any tectonically controlled vertical uplift of the Miura block could have been of the order of 1 km, by analogy with the present Zenisu ridge. At the time of accretion, 3–2.5 Ma, the tectonic and glacio-eustatic controls were operating in tandem, thereby making it difficult to dissemble the relative importance of these processes in generating the stratigraphy.

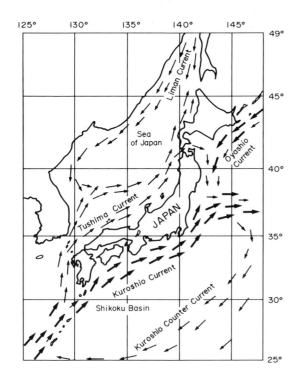

Fig. 16.22. Present-day flow of the warm-water Kuroshio current and the cold-water Oyashio current. During cold-climate periods, the strength of these currents was greater, with enhanced thermohaline ventilation and the preferential development of tractional bedforms in deep water.

As a caveat to the above arguments, even in a well-dated stratigraphy such as that considered in this paper, the error bars in dating the stratigraphic components, together with the inherent uncertainties in the significance of the Exxon sea-level curve and the underpinning rationale used to erect the coastal onlap curve, make it very easy to force-fit the data to the hypothesis. The above correlations between the Exxon eustatic curve and the stratigraphy of the Kazusa Group may be spurious. Indeed, the high-resolution oxygen-isotope curve for the western Pacific (ODP Site 677), based on benthonic and planktonic foraminifera, through the entire Pleistocene and upper Pliocene (to 2.6 Ma), appear to show clearly only the onset of the last major Pleistocene, glaciation at about 0.8 Ma (fig. 7, Shackleton et al. 1990). Thus, this oxygen-isotope signal from the Pacific does not appear to support the notion that the eustatic falls in sea level proposed by the Exxon Group for the time periods about 1.4–1.6 Ma and 2 Ma are a consequence of glacio-eustasy. Of course, this line of reasoning does not invalidate the Exxon curve for these time intervals (Haq et al. 1987), because other plate-tectonic mechanisms may prove to exert a high-frequency control on global sea level (e.g. cf. Cathles & Hallam 1991) — it merely suggests that, in many cases, glacio-eustasy probably was not the driving mechanism.

In summary, we have shown that it is possible to interpret the stratigraphy of southeast central Japan in the context of sequence stratigraphy and even within the context of the Exxon eustatic sea-level curve for the past 3 Ma, but until further research is completed (in progress), we remain sceptical about its reliability and applicability. We believe that much of the early Neogene stratigraphy can be accounted for by tectonic processes, whereas, in contrast, the later forearc-basin fill (Kazusa-Shimosa Groups) is well-explained by glacio-eustatic changes in sea level. The jury is out and has yet to return its verdict.

ENDPIECE

This chapter has examined various aspects of tectonics and sedimentation by using examples from the forearc regions of the Japanese arcs, especially the ICZ. The principal point of the chapter has been to show that tectonics and sedimentation cannot be considered in isolation from the role of pore-fluids in progressively dewatering, deforming and lithifying sediments, nor in the absence of the sequence-stratigraphic framework. In these respects, the Neogene-Recent geology of the Japanese island forearcs is second to none.

REFERENCES

Akamine, H., Iwai, S., Koike, K., Naruse, H., Omori, T., Seki, Y., Suzuki, Y. & Watanabe, K. 1956. On the Miura Group in the Miura Peninsula. *Chikyukagaku Earth Sci.* **3**, 1–8 (in Japanese).

Ando. J., Tanaka, Y. & Hasegawa, S. 1989. Sedimentary environment of the Miura Group in the southern area of the Miura Peninsula. *Abstr. Vol. 6 96th A. Meet. geol. Soc. Japan*, 216, (in Japanese).

Aoki, Y., Tamano, T. & Kato, S. 1982. Detailed structure of the Nankai Trough from migrated seismic sections. In: *Studies in Continental Margin Geology* (edited by Watkins, J. S. & Drake, C. L.). *Mem. Am. Ass. Petrol. Geol.* **34**, 309–322.

Arai, S., Ito, T. & Ozawa, K. 1983. Ultramafic-mafic clastic rocks from the Mineoka Belt, central Japan. *J. geol. Soc. Jap.* **89**, 287–297 (in Japanese with English abstract).

Berger, W. H. & Meyer, L. A. 1987. Cenozoic paleoceanography 1986: an introduction. *Paleoceanography* **2**, 613–623.

Boulegue, J., Benedetti, E. L., Dron, D., Mariotti, A. & Lettolle, R. 1987. Geochemical and biogeochemical observation on the biological communities associated with fluid venting in the Nankai Trough and Japan Trench subduction zones. *Earth Planet. Sci. Lett.* **83**, 329–342.

Cadet, J.-P., Kobayashi, K., Lallemant, S., Jolivet, L., Aubouin, J., Boulegue, J., Dubois, J., Hotta, H., Ishii, T., Konishi, K., Niitsuma, N. & Shimamura, H. 1987a. Deep scientific dives in the Japan and Kuril Trenches. *Earth Planet. Sci. Lett.* **83**, 313–328.

Cadet, J.-P., Kobayashi, K., Aubouin, J., Boulegue, J., Deplus, C., Dubois, J., Von Huene, R., Jolivet, L., Kanazawa, T., Kasahara, J., Koizumi, K., Lallemant, S., Nakamura, Y., Pautot, G., Suyehiro, K., Tani, S., Tokuyama, H. & Yamazaki, T. 1987b. The Japan Trench and its juncture with the Kuril Trench: cruise results of the Kaiko project, Leg 3. *Earth Planet. Sci. Lett.* **83**, 267–284.

Cathles, L. M. & Hallam, A. 1991. Stress-induced changes in plate density, Vail sequences, epeirogeny, and short-lived global sea level fluctuations. *Tectonics* **10**, 659–671.

Chamot-Rooke, N. & Le Pichon, X. 1989. Zenisu Ridge: mechanical model of formation. *Tectonophysics* **160**, 175–193.

Chamot-Rooke, N., Renard, V. & Le Pichon, X. 1987. Magnetic anomalies in the Shikoku Basin: a new interpretation. *Earth Planet. Sci. Lett.* **83**, 214–228.

Chappell, J. & Shackleton, N. J. 1986. Oxygen isotopes and sea level. *Nature*, **324**, 137–140.

Compton, J. S., Snyder, S. W. & Hodell, D. A. 1990. Phosphogenesis and weathering of shelf sediments from the southeastern United States: implications for Miocene $\delta^{13}C$ excursions and global cooling. *Geology* **18**, 1227–1230.

Cowan, D. S. 1982. Origin of vein structure in slope sediments on the inner slope of the Middle America Trench off Guatemala. In: *Initial Reports of the Deep Sea Drilling Project*, US Government Printing Office, Washington DC. **67**, 645–649.

Davis, D. M., Suppe, J. & Dahlen, F. A. 1983. Mechanics of fold-and-thrust belts and accretionary wedges. *J. geophys. Res.* **88**, 1153–1172.

Einsele, G. 1985. Responses of sediments to sea-level changes in differing subsiding storm-dominated marginal and epeiric basins. In: *Sedimentary and Evolutionary Cycles* (edited by Bayer, U. & Seilacher, A.). 68–112. Springer, Berlin.

Eto, T. 1986a. Stratigraphy of the Hayama Group in the Miura Peninsula, Japan. *Sci. Rep. Yokohama Natn. Univ.* Section II **33**, 67–103 (in Japanese with English abstract).

Eto, T. 1986b. Stratigraphic study of the Miura and Kazusa Groups in the Miura Peninsula. *Sci. Rep. Yokohama Natn. Univ.* Section II **33**, 107–132 (in Japanese with English abstract).

Eto, T., Oda, M., Hasegawa, S., Honda, N. & Funayama, M. 1987. Geologic age and paleoenvironment based upon microfossils of the Cenozoic sequence in the middle and northern parts of the Miura Peninsula. *Sci. Rep. Yokohama Natn. Univ.* Section II **34**, 41–57 (in Japanese with English abstract).

Fairbanks, R. G. & Matthews, R. K. 1978. The marine oxygen isotope record in Pleistocene coral, Barbados, West Indies. *Quat. Res.* **10**, 181–196.

Fryer, P., Pearce, J. A., Stokking, L. B. et al. 1990a. *Proc. Ocean Drilling Prog. Init. Rep.* College Station, Texas, **125**.

Fryer, P., Saboda, K., Johnson, L., Mackay, M. E., Moore, G. F. & Stoffer, P. 1990b. Conical Seamount: SeaMARK II, Alvin submersible and seismic-reflection studies. In: *Proceedings of the Ocean Drilling Program, Initial Reports* (edited by Fryer, P., Pearce, J. A. & Stokking, L. B. et al.). College Station, Texas, **125**, 69–80.

Fujii, T., Aramaki, S., Kaneko, T., Ozawa, K., Kawanabe, Y. & Fukuoka, T. 1988. Petrology of the lavas and ejecta of the November, 1986 eruption of Izu-Oshima. *Bull. volcan. Soc. Jap.* **33**, S235–S254 (in Japanese with English abstract).

Fujioka, K. & Taira, A. 1989. Tectono-sedimentary settings of seep biological communities — a synthesis from the Japanese subduction zones. In: *Sedimentary Facies at the Active Plate Margin* (edited by Taira, A. & Masuda, F.). TERRAPUB, Tokyo 577–602.

Fukao, Y. 1979. Tsunami earthquakes and subduction processes near deep-sea trenches. *J. geophy. Res.* **84**, 2303–2314.

Galloway, W. E. 1989a. Genetic stratigraphic sequences in basin analysis I: architecture and genesis of flooding-surface bounded depositional units. *Bull. Am. Ass. Petrol. Geol.* **73**, 125–142.

Galloway, W. E. 1989b. Genetic stratigraphic sequences in basin analysis II: application to northwest Gulf of Mexico Cenozoic basin. *Bull. Am. Ass. Petrol. Geol.* **73**, 143–154.

Hamuro, K. 1985. Petrology of the Higashi-Izu Monogenic Volcano Group. *Bull. Earthquake Inst. Univ. Tokyo* **60**, 335–400.

Haq, U. B., Hardenbol, J. & Vail, P. R. 1987. Chronology of fluctuating sea levels since the Triassic. *Science* **235**, 1156–1167.

Haq, B. U., Hardenbol, J. & Vail, P. R. 1988. Mesozoic and Cenozoic chronostratigraphy and eustatic cycles. In: *Sea-Level Research: An Integrated Approach*, (edited by Wilgus, C. K. et al.) *Spec. Publs Soc. Econ. Paleont. Mineral.* **42**, 71–108.

Hatta, A. & Tokuhashi, S. 1984. On the foraminiferal assemblage in the hemipelagic mudstone of the Kiyosumi and Anno Formations, Boso Peninsula, Japan. *NOM (J. Osaka Micropaleont.)* **12**, 17–32 (in Japanese with English abstract).

Hernandez, L. & Lallemant, S. 1988. Preliminary report on a dacitic lava dredged on the Zenisu Ridge. In: *Preliminary Report Hakuho-maru Cruise KH86–5*. Ocean Research Institute University of Tokyo, 270–284.

Hibbard, J. P. & Karig, D. E. 1990. Alternative plate model for the early Miocene evolution of the southwest Japan margin. *Geology* **18**, 170–179.

Honda, S. & Uyeda, S. 1983. Thermal process in subduction zones — a review and preliminary approach on the origin of arc volcanism. In: *Arc Volcanism — Physics and Tectonics* (edited by Shimozuru, D. & Yokoyama, I.). TERRAPUB, Tokyo 117–140.

Honza, E. & Tamaki, K. 1985. Bonin arc. In: *The Ocean Basins and Margins: The Pacific Ocean* (edited by Nairn, A. E. M., Stehli, F. G. & Uyeda, S.). Plenum Press, New York. **7**, 459–499.

Hotta, H. 1970. A crustal section across the Izu-Ogasawara arc and trench. *J. Phys. Earth* **18**, 125–142.

Huchon, P. & Kitazato, H. 1984. Collision of the Izu block with Central Japan during the Quaternary and geological evolution of the Ashigara area. *Tectonophysics* **110**, 201–210.

Ikehara, K. 1989. The Kuroshio-generated bedform system in the Osumi Strait, southern Kyushu, Japan. In: *Sedimentary Facies in the Active Plate Margin*, (edited by Taira, A. & Masuda, F.). TERRAPUB, Tokyo, 261–273.

Ishii, T. 1985. Dredged samples from the Ogasawara fore-arc seamount or 'Ogasawara Paleoland' forearc ophiolite. In: *Formation of Active Ocean Margins* (edited by Nasu, N. et al.). TERRAPUB, Tokyo, 307–342.

Ishibashi, K. 1988. 'Kanagawaken seibuokijisin' and earthquake prediction 1. *Kagaku Sci.* **58**, 537–547 (in Japanese).

Ishibashi, M. 1989. Sea-level controlled shallow-marine systems in the Plio-Pleistocene Kakegawa Group, Shizuoka, Central Honshu, Japan: comparison of transgressive and regressive phases. In: *Sedimentary Facies in the Active Plate Margin* (edited by Taira, A. & Masuda, F.). TERRAPUB, Tokyo, 345–363.

Ito, M. 1988. Neogene depositional history in Oiso hill. Development of Okinoyama bank chain on landward slope of Sagami Trough, central Honshu, Japan. *J. geol. Soc. Jap.* **92**, 47–64.

Ito, M. 1989. The Itsukaichimachi Group: a Middle Miocene strike-slip basin-fill in the southwestern margin of the Kanto Mountains, Central Honshu, Japan. In: *Sedimentary Facies in the Active Plate Margin* (edited by Taira, A. & Masuda, F.). TERRAPUB, Tokyo. 659–673.

Ito, M. 1992. High-frequency depositional sequences of the upper part of the Kazusa Group, a middle Pleistocene forearc basin fill in Boso Peninsula, Japan. *Sediment. Geol.* **76**, 155–175.

Ito, M. & Masuda, F. 1989. Petrofacies of paleo-Tokyo Bay sands, the Upper Pleistocene of Central Honshu, Japan. In: *Sedimentary Facies in the Active Plate Margin* (edited by Taira, A. & Masuda, F.). TERRAPUB, Tokyo. 179–196.

Ito, T. & Sugiyama, S. 1989. Basal structures of the Pleistocene Chikura submarine sliding sheet, Central Japan. In: *Sedimentary Facies in the Active Plate Margin* (edited by Taira, A. & Masuda, F.). TERRAPUB, Tokyo 511–528.

Itoh, Y. 1986. Differential rotation of Northeastern part of Southwest Japan — paleomagnetism of Early to late Miocene rocks from Yatsuo Area in Chichibu District. *J. Geomag. Geoelec.* **38**, 325–334.

Iwabuchi, Y., Kato, S. & Kasuga, S. 1987. A Pacific ridge accretion onto the Philippine Sea Plate across the Izu-Ogasawara Trench. *Mod. Sea Bottom Res.* Japan Hydrographic Association **7**, 121–126. (in Japanese with English abstract).

Kaiko I Research Group 1986. Topography and Structure of Trenches Around Japan. In: *Data Atlas of Franco-Japanese Kaiko Project, Phase I* (edited by Taira, A. & Tokuyama, H.). University of Tokyo Press, Tokyo, 305 pp.

Kaiko II Research Group 1987. 6000 *Meters Deep: A Trip to the Japanese Trenches*. Tokyo University Press, Tokyo. 105 pp.

Kaizuka, S. 1987. Quaternary crustal movements in Kanto, Japan. *J. Geog.* **96**, 51–68 (in Japanese).

Kanmera, K. & Sano, H. 1986. Stratigraphy and structural relationship among Pre-Jurassic accretionary and collisional system in Akiyoshi Terrane. *International Symposium on Pre-Jurassic East Asia, IGCP Project 224, Guidebook for Excursion.* Osaka City University, Japan, 50–88.

Kanmera, K., Sano, S. & Isozaki, Y. 1990. Akiyoshi Terrane. In: *Pre-Cretaceous Terranes of Japan* (edited by Ichikawa, K., Mizutani, S., Hara, I., Hada, S. & Yao, A.). *Publ. IGCP Proj. 224, Pre-Jurassic Evolution of Eastern Asia.* Osaka City University, Japan, 49–62.

Karig, D. E. & Lundberg, N. 1990. Deformation bands from the toe of the Nankai accretionary prism. *J. geophys. Res.* **95**, 9099–9109.

Katsura, Y. 1984. Depositional environments of the Plio-Pleistocene Kazusa Group, Boso Peninsula, Japan. *Sci. Rep. Inst. Geosci. Univ. Tsukuba, Series B* **4**, 69–104.

Kiminami, K., Kito, N. & Tajika, J. 1985. Mesozoic Group in Hokkaido — Stratigraphy and age and their significance. *J. Ass. geol. Collab. Jap.* (*Chikyu Kagaku*) **39**, 1–17.

Kimura, G., Koga, K. & Fujioka, K. 1989. Deformed soft sediments at the junction between the Mariana and Yap trenches. *J. Struct. Geol.* **11**, 463–472.

Kitazato, H. 1986. Evolution of the paleogeography in the Southern Fossa Magna Region. *Earth Monthly* (*Chikyu*) **88**, 605–611 (in Japanese).

Knipe, R. J. 1986. Microstructural evolution of vein arrays preserved in Deep Sea Drilling Project cores from the Japan Trench, Leg 57. In: *Structural Fabrics in Deep Sea Drilling Cores from Forearcs*, (edited by Moore, J. C.). *Mem. geol. Soc. Am.* **166**, 75–87.

Kodama, K. 1968. An analytical study on the minor faults in the Jogashima Island. *J. geol. Soc. Jap.* **74**, 265–278 (in Japanese with English abstract).

Koike, K. 1957. Geostructural history around the South Kwanto Region. *Chikyu Kagaku Earth Sci.* **24**, 1-18 (in Japanese).

Kondo, Y. 1989. Faunal condensation in early phases of glacio-eustatic sea-level rise, found in the Middle to late Pleistocene Shimosa Group, Boso Peninsula, Central Japan. In: *Sedimentary Facies in the Active Plate Margin* (edited by Taira, A. & Masuda, F.). TERRAPUB, Tokyo, 197–212.

Kotake, N. 1988. Upper Cenozoic marine sediments in southern part of the Boso Peninsula. *J. geol. Soc. Jap.* **94**, 187–206.

Koyama, M. 1991. Tectonic evolution of the Philippine Sea Plate based on paleomagnetic results. *J. Geog.* **100**, 628–641 (in Japanese).

Lallemant, S., Chamot-Rooke, N., Le Pichon, X. & Rangin, C. 1989a. Zenisu Ridge: a deep intraoceanic thrust related to subduction, off Southwest Japan. *Tectonophysics* **160**, 151–174.

Lallemant, S., Culotta, R. & Von Huene, R. 1989b. Subduction of the Daiichi Kashima Seamount in the Japan Trench. *Tectonophysics* **160**, 231–247.

Le Pichon, X., Iiyama, T., Chamley, H., Charvet, J., Faure, M., Konishi, K., Lallemant, S., Okada, H., Rangin, C., Renard, V., Taira, A., Urabe, T. & Uyeda, S. 1987a. Nankai Trough and Zenisu Ridge: a deep-submersible survey. *Earth Planet. Sci. Lett.* **83**, 285–299.

Le Pichon, X., Iiyama, T., Chamley, H., Charvet, J., Faure, M., Fujimoto, H., Furuta, T., Ida, Y., Kagami, H., Lallemant, S., Leggett, J., Murata, A., Okada, H., Rangin, C., Renard, V., Taira, A. & Tokuyama, H. 1987b. Nankai Trough and the fossil Shikoku Ridge: results of Box 6 Kaiko survey. *Earth Planet. Sci. Lett.* **83**, 186–198.

Le Pichon, X., Iiyama, T., Chamley, H., Charvet, J., Faure, M., Fujimoto, H., Furuta, T., Ida, Y., Kagami, H., Lallemant, S., Leggett, J., Murata, A., Okada, H., Rangin, C., Renard, V., Taira, A. & Tokuyama, H. 1987c. The eastern and western ends of Nankai Trough: results of Box 5 and Box 7 Kaiko survey. *Earth Planet. Sci. Lett.* **83**, 199–213.

Lundberg, N. & Moore, J. C. 1986. Macroscopic structural features in Deep Sea Drilling Project cores from forearc regions. In: *Structural Fabrics in Deep Sea Drilling Project Cores From Forearcs* (edited by Moore, J. C.). *Mem. geol. Soc. Am.* **166**, 13–44.

Miall, A. D. 1986. Eustatic sea level changes interpreted from seismic stratigraphy: a critique of the methodology with particular reference to the North Sea Jurassic record. *Bull. Am. Ass. Petrol. Geol.* **70**, 131–137.

Miall, A. D. 1992. Exxon global cycle chart: An event for every occasion? *Geology* **20**, 787–790.

Mitsunashi, S., Kikuchi, T., Suzuki, Y., Hirayama, J., Nakajima, M., Oka, S., Kodama, K., Horiuchi, M., Katsurashima, S., Miyashita, M., Yazaki, K., Kageyama, K., Nasu, N., Kagami, H., Honza, E., Kimura, M., Nirei, H., Higuchi, S., Hara, I., Kono, K., Endo, A., Kawashima, S. & Aoki, S. 1976. *Geology of the Tokyo Bay and Adjacent Areas, 1: 100,000*. Geological Survey Japan, 91 pp. (in Japanese with English abstract).

Momma, H., Naka, J. & Matsumoto, T. 1990. Preliminary report of deep tow surveys in the Hachijo depression, Kaikata Seamount and Hahajima Seamount (DK 88–3–IZU). *Tech. Rep. Jap. Mar. Sci. Technol. Center.* (*JAMSTEC*) **23**, 219–236 (in Japanese with English abstract).

Moore, J. C. (ed.) 1986. Structural fabrics in Deep Sea Drilling cores from forearcs. *Mem. geol. Soc. Am.* **66**.

Morijiri, R., Kinoshita, H. & Nago, T. 1987. On the geophysical survey of the Mineoka Ophiolite belt, southern part of the Boso Peninsula, Chiba Prefecture, Japan. *A. Rep. Inst. Mar. Ecosystem Univ. Chiba* **7**, 24–26 (in Japanese).

Murauchi, S. *et al.* (13 Authors) 1968. Crustal structure of the Philippine Sea. *J. geophys. Res.* **73**, 3143–3171.

Nakajima, T., Makimoto, H., Hirayama, J. & Tokuhashi, S. 1981. *Geology of the Kamogawa District. Quadrangle Series, scale 1: 50,000.* Geological Survey Japan. (in Japanese with English abstract).

Nakayama, N. & Masuda, F. 1989. Ocean current-controlled sedimentary facies of the Pleistocene Ichijiku Formation, Kazusa Group, Boso Peninsula, Japan. In: *Sedimentary Facies in the Active Plate Margin* (edited by Taira, A. & Masuda, F.). TERRAPUB, Tokyo, 275–293.

Nakao, S., Kotake, N. & Niitsuma, N. 1986. Geology of the Ishido area in the southern part of the Boso Peninsula, central Japan. *Geosci. Rep. Univ. Shizuoka* **12**, 209–238 (in Japanese with English abstract).

Nakazato, H., Sato, H. & Masuda, F. 1989. Coastal eolian dune deposits of the Pleistocene Shimosa Group in Chiba, Japan. In: *Sedimentary Facies in the Active Plate Margin* (edited by Taira, A. & Masuda, F.). TERRAPUB, Tokyo, 131–141.

Niitsuma, N., Matsushima, Y. & Hirata, D. 1989. Abyssal molluscan colony of *Calyptogena* in the Pliocene strata of the Miura Peninsula, central Japan. *Palaeogeog. Palaeoclimat. Palaeoecol.* **71**, 193–203.

Ogawa, Y. & Taniguchi, H. 1988. Geology and tectonics of the Miura-Boso Peninsula and the adjacent area. *Mod. Geol.* **12**, 147–168.

Ogawa, Y., Horiuchi, K., Taniguchi, H. & Naka, J. 1985. Collision of the Izu arc with Honshu and the effects of oblique subduction in the Miura-Boso Peninsulas. *Tectonophysics* **119**, 349–379.

Ohta, S. & Laubier, L. 1987. Deep biological communities in the subduction zone of Japan from bottom photographs taken during 'Nautile' dives in the Kaiko Project. *Earth Planet. Sci. Lett.* **83**, 329–342.

Otsuka, Y. 1929. On the stratigraphy around the Oiso area. *J. geol. Soc. Jap.* **36**, 435–456.

Otsuka, Y. & Koike, K. 1949. Geology of the central portion of the Boso Peninsula. *Found. Nat. Sci. Rep. Univ. Tokyo* **2**, 31–32 (in Japanese).

Pickering, K. T., Agar, S. M. & Prior, D. J. 1990. Vein structure and the role of pore fluids in early wet-sediment deformation, Late Miocene volcaniclastics, Miura Group, SE Japan. In: *Deformation Mechanisms, Rheology and Tectonics* (edited by Knipe, R. J. & Rutter, E. H.). *Spec. Publs geol. Soc. Lond.* **54**, 417–430.

Pickering, K. T., Marsh, N. & Dickie, B. 1993a. Inorganic major, trace and rare earth element analyses of the muds and mudstones from Site 808. In: *Proceedings of the Ocean Drilling Program, Volume 131B*. Ocean Drilling Program, College Station, Texas, 427–450.

Pickering, K. T., Underwood, M. B. & Taira, A. 1993b. Stratigraphic synthesis of the DSDP-ODP sites in the Shikoku Basin, Nankai Trough, and accretionary prism. In: *Proceedings of the Ocean Drilling Program, Volume 131B*. Ocean Drilling Program, College Station, Texas, 313–330.

Reck, B. H. 1987. Implications of measured thermal gradients for water movement through the northeast Japan accretionary prism. *J. geophys. Res.* **92**, 3683–3690.

Saito, S. 1992. Stratigraphy of Cenozoic strata in the southern terminus area of Boso Peninsula, Central Japan. *Contrib. Inst. Geol. Palaeont. Tohoku Univ.* **93**, 1–37.

Shackleton, N. J., Berger, A. & Peltier, W. R. 1990. An alternative astronomical calibration of the lower Pleistocene timescale based on ODP Site 677. *Trans. R. Soc. Edinb. Earth Sci.* **81**, 251–261.

Shimamoto, T. 1985. The origin of large or great thrust-type earthquakes along subducting plate boundaries. *Tectonophysics* **119**, 37–65.

Smoot, N. C. & Heflner, K. J. 1989. Bathymetry and possible tectonic interaction of the Uyeda Ridge with its environment. *Tectonophysics* **124**, 23–36.

Soh, W., Taira, A., Ogawa, Y., Taniguchi, H., Pickering, K. T. & Stow, D. 1989. Submarine depositional processes for volcaniclastic sediments in the Mio-Pliocene Misaki Formation, Miura Group, central Japan. In: *Sedimentary Facies in the Active Plate Margin* (edited by Taira, A. & Masuda, F.). TERRAPUB, Tokyo, 619–630.

Soh, W., Pickering, K. T., Taira, A. & Tokuyama, H. 1991. Basin evolution in the arc–arc Izu collision zone, Mio-Pliocene Miura Group, central Japan. *J. geol. Soc. Lond.* **148**, 317–330.

Suzuki, Y., Kondo, K. & Saito, T. 1984. Latest Eocene planktonic foraminifera from the Mineoka Group, Boso Peninsula. *J. geol. Soc. Jap.* **90**, 497–499 (in Japanese).

Taira, A. 1986. *Newton* **6**, 56–87.

Taira, A. & Niitsuma, N. 1986. Turbidite sedimentation in the Nankai trough as interpreted from magnetic fabric, grain size, and detrital modal analyses. In: *Initial Reports Deep Sea Drilling Project* (edited by Kagami, H., Karig, D. E., Coulbourn, W. C. *et al.*). US Government Printing Office, Washington, DC **87**, 611–632.

Taira, A. & Pickering, K. T. 1991. Sediment deformation and fluid activity in the Nankai, Izu-Bonin and Japan forearc slopes and trenches. *Phil. Trans. R. Soc. Lond.* **A335**, 289–313.

Taira, A., Katto, J., Tashiro, M., Okamura, M. & Kodama, K. 1988. The Shimanto Belt in Shikoku, Japan — Evolution of Cretaceous to Miocene accretionary prism. *Mod. Geol.* **12**, 5–46.

Taira, A. Tokuyama, H. & Soh, W. 1989. Accretion tectonics and evolution of Japan. In: *The Evolution of the Pacific Ocean Margins* (edited by Zvi Ben-Avraham). Oxford University Press, Oxford, 100–123.

Taira, A., Hill, I. & Firth, J. *et al.* 1991. *Proceedings of the Ocean Drilling Program, Initial Reports*. Ocean Drilling Program, College Station, Texas, **131**.

Takahashi, M. 1986. Collision tectonics of the Southern Fossa Magna Region from the magmatic activity. *Earth Monthly (Gekkan Chikyu)* **8**, 586–591.

Tamaki, K. & Honza, E. 1985. Incipient subduction and obduction along the eastern margin of the Japan Sea. *Tectonophysics* **119**, 381–404.

Tamura, Y. & Ishii. T. 1988. Description of samples from Takuyo–daiichi sea mount and Zenisu ridge, during KH86–5 cruise. In: *Preliminary Report Hakuho-maru Cruise KH86–5*. Ocean Research Institute, University of Tokyo, 167–172.

Tarney, J., Pickering, K. T., Knipe, R. J. & Dewey, J. F. (eds) 1991. *The Behaviour and Influence of Fluids in Subduction Zones*. The Royal Society, London. 418 pp.

Taylor, B., Fujioka, K. *et al.* 1990. *Proceedings of the Ocean Drilling Program, Initial Reports*. Ocean Drilling Program, College Station, Texas, **126**.

Teng, L. S. & Chen, W. S. 1988. Stratigraphy and geologic history of the coastal range, eastern Taiwan. In: *Field Guide Book: Symposium on the Arc-Continent Collision and Orogenic Sedimentation in the Eastern Taiwan and Ancient Analogs* (edited by Teng, L. S., Lundberg, N. & Lee, C. W.). 4–1–4–25.

Teng, L. S. & Lo, H. J. 1985. Sedimentary sequence in the island arc settings of the Coast Range, eastern Taiwan. *Geol. Acta* **23**, 77–98.

Tokuhashi, S. 1979. Three dimensional analysis of a large sandy-flysh body, Mio-Pliocene Kiyosumi Formation, Boso Peninsula, Japan. *Mem. Faculty Sci. Kyoto Univ. Geol. Mineral.* **46**, 1–66.

Tokuhashi, S. 1989. Two stages of submarine fan sedimentation in an ancient forearc basin, Central Japan. In: *Sedimentary Facies in the Active Plate Margin* (edited by Taira, A. & Masuda, F.). TERRAPUB, Tokyo, 439–468.

Tonouchi, S. 1981. Palaeomagnetic and geotectonic investigation of ophiolite sites and surrounding rocks in South Central Honshu, Japan. Thesis for Tokyo University.

Tsuchi, R. 1990. Neogene events in Japan and the Pacific. *Palaeogeog. Palaeoclimat. Palaeont.* **77**, 355–365.

Uchida, T. & Arai, S. 1978. Petrology of ultramafic rocks from the Boso Peninsula and Miura Peninsula. *J. geol. Soc. Jap.* **84**, 561–570.

Ueda, Y., Tozaki, T. & Kaneko, T. 1985. Geomagnetic anomalies around the Zenisu Ridge and the Suruga Trough and their tectonic implication. *Rep. Hydrographic Res. Maritime Safety Agency Jap.* **20**, 83–107 (in Japanese with English abstract).

Ueda, Y., Nakagawa, H., Hiraiwa, T., Asao, T. & Kubota, R. 1987. Gravity anomalies and derived subterranean structure on/around Tokyo Bay and southern Kanto District. *Rep. Hydrographic Res. Maritime Safety Agency Jap.* **22**, 179–206 (in Japanese with English abstract).

Uyeda, S. & Ben-Avraham, Z. 1972. Origin and development of the Philippine Sea. *Nature* **240** 176–178.

Van Wagoner, J. C., Posamentier, H. W., Mitchum, R. M. Jr., Vail, P. R. & Sarg, J. F. 1988. An overview of the fundamentals of sequence stratigraphy and key definitions. In: *Sea-Level Changes: An Integrated Approach* (edited by Wilgus, C. K., Hastings, B. S., Kendall, C. G. St. C., Posamentier, H. W., Ross, C. A. & Van Wagoner, J. C.). *Spec. Publs Soc. Econ. Palaeont. Mineral.* **42**, 39–45.

Van Wagoner, J. C., Mitchum, R. M., Campion, K. M. & Rahmanian, V. D. 1990. Siliciclastic sequence stratigraphy in well logs, cores, and outcrops. *Am. Ass. Petrol. Geol. Meth. Exploration Ser.* **7**, 55 pp.

Von Huene, R. & Culotta, R. 1989. Tectonic erosion at the front of the Japan Trench convergent margin. *Tectonophysics* **160**, 75–90.

Wei, W. 1991. Evidence for an earliest Oligocene abrupt cooling in the surface waters of the Southern Ocean. *Geology* **19**, 780–783.

Yamaguchi, T., Matsushima, Y., Hirata, D., Arai, S., Ito, T., Murata, A., Machida, Y., Arai, Y., Takayanagi, F., Oda, Y., Okada, H. & Kitazato, A. 1983. An unconformity between the Hatsuse and Miyata Formations in Shinomiyata, Miura City. *Nat. Hist. Rep. Kanagawa* **4**, 87–93 (in Japanese).

Yoshii, T. 1979. Compilation of geophysical data around the Japanese Islands (1). *Bull. Earthquake Res. Inst. Univ. Tokyo* **54**, 75–117.

Yoshida, S., Shibuya, H., Torii, M. & Sasajima, S. 1984. Post-Miocene clockwise rotation of the Miura Peninsula and its adjacent area. *J. Geomagnet. Geoelec.* **36**, 579–584.

CHAPTER 17

Archaean Tectonics

JOHN S. MYERS and ALFRED KRÖNER

INTRODUCTION

THE ARCHAEAN includes about half the geological history of the Earth, from the oldest known rocks at about 4000 Ma up to 2500 Ma. Taylor & McLennan (1985) consider that over two-thirds of the Earth's sialic crust was formed during this time. Archaean sialic crust survived intact in cratonic terrains that have not been subjected to much deformation in later times. Such terrains occur in North America, Greenland, Finland, Siberia, China, India, Africa, Australia, Antarctica and Brazil. Archaean crust also occurs as tectonic slices in Proterozoic and Phanerozoic orogens, and may form the basement beneath many Proterozoic and Phanerozoic basins and platform cover sequences. Much of the Archaean crust in younger orogens, however, is modified by tectonic, magmatic and metamorphic processes (so-called *reworking*) that occurred after 2500 Ma. This chapter is restricted to tectonic processes that occurred during the Archaean, and these are best seen in rocks of Archaean cratons where there was no subsequent ductile deformation (also see Chapter 10).

In the 1960s and early 1970s it was widely believed that Archaean tectonic processes were fundamentally different from those of the Proterozoic and Phanerozoic. The Archaean crust was thought to have been more ductile and dominated by diapiric, vertical tectonic structures (*sag-tectonics*, Goodwin 1981). Plate tectonic processes were generally not thought to have operated, and only a few mobilistic models were proposed (e.g. Talbot 1973, Burke *et al.* 1976, Langford & Murin 1976). This chapter reviews and reconsiders some of these ideas; it presents well-exposed examples of Archaean structures and their tectonic interpretations, and demonstrates that Archaean tectonics were really not as different from younger tectonics as was previously believed.

Archaean cratons have been broadly divided into two kinds of *terrain* (see Windley 1984). (1) *Granite–greenstone terrains* in which narrow strips of *greenstones* (metavolcanic and metasedimentary rocks) wrap around dome-like composite bodies of granitoid rocks often called *batholiths* (following MacGregor 1951). Regional metamorphism is generally in prograde greenschist to amphibolite facies. Deformation intensity and metamorphic grade are generally higher in the vicinity of major granite-greenstone contacts. (2) *High-grade gneiss terrains* dominated by heterogeneous quartzofeldspathic gneisses with complex fold interference structures and regional metamorphism in upper amphibolite to granulite facies.

Although this subdivision is useful as a broad generalization, it has led to a polarization of ideas on Archaean tectonics and crustal evolution. The two kinds of terrain were thought to have formed in quite different ways, and there was considerable speculation about the relation between the high- and low-grade terrains and their development (see Windley 1984).

Recent detailed mapping and precise geochronology of Archaean rocks in the Superior province of Canada, southern Africa, and Western Australia has shown that there are gradations between the two kinds of terrain. Some high-grade gneiss terrains represent contemporaneous, more highly metamorphosed equivalents of adjacent granite–greenstone terrains (Gee *et al.* 1986, Kröner 1991). In some other cases it has been shown that high-grade gneiss terrains represent older sialic crust than adjacent granite–greenstone terrains (Compston & Kröner 1988, Myers 1988), and that the two kinds of terrain were either tectonically juxtaposed or formed as a basement and cover.

The tectonic evolution of granite–greenstone terrains was widely thought to have been controlled by *granite diapirism*. The greenstones were related to vertical tectonic processes and were either considered to represent primordial crust that sagged into a substrate of rising granitoid material (Anhaeusser 1973, Glikson 1979), or formed in rifts as a volcanic and sedimentary cover to a thin unstable sialic crust (Goodwin 1977, Groves & Batt 1984, Ayres & Thurston 1985, Ludden *et al.* 1986). This partly mobile sialic basement rose as diapirs (Schwerdtner & Lumbers 1980), and there was further deposition of greenstones in arcuate basins between the rising granitoid diapirs (Gorman *et al.* 1978, Hickman 1984).

Greenstones have also been interpreted as remnants of back-arc basins equivalent to the Phanerozoic Rocas Verdes Complex of southern Chile (Tarney *et al.* 1976), and the associated granitoid rocks have been linked to

Cordilleran batholiths (Windley & Smith 1976). These and other interpretations are discussed by Windley (1984), Park (1982) and Kröner (1985, 1991).

The following sections outline some of the main kinds of large-scale structures seen in Archaean rocks and current interpretations of their tectonic evolution. They demonstrate that the main tectonic structures of both high- and low-grade terrains are broadly similar.

STRUCTURES IN HIGH-GRADE GNEISS TERRAINS

The predominant rock type in Archaean high-grade terrains is a banded grey gneiss, derived by deformation and concomitant recrystallization of plutonic rocks of the so-called *TTG-suite* (tonalite – trondhjemite – granodiorite) (Fig. 17.1). These gneisses are in many cases interlayered with rocks of volcanic origin (greenstone equivalents) (Fig. 17.2) or with metasedimentary rocks, and may be associated with several generations of basic dykes. The ubiquitous parallelism of planar structures in rocks of most high-grade terrains is not unique to the Archaean and is the result of structural transposition during progressive deformation at high metamorphic grade (Myers 1978, Passchier *et al.* 1990).

Detailed mapping of well-exposed high-grade gneiss terrains has revealed the dominance of *early flat-lying structures* marked by rock units intercalated on all scales by thrusting and recumbent folding, and sheet-like intrusions (Bridgwater *et al.* 1974, 1976, Myers 1976, Chadwick & Nutman 1979, James & Black 1981, Jackson 1984, Passchier *et al.* 1990). These structures were subsequently refolded by upright folds.

Greenland provides the most extensive outcrops of Archaean high-grade gneisses where both small- and large-scale structures are well displayed in rugged relief up to 2000 m high. Much of the gneiss complex in West Greenland has been mapped in detail by the Geological

Fig. 17.1. Grey tonalitic, *c.* 3560 Ma, Ngwane Gneiss (formerly called the Bimodal Suite) cut by an undeformed network of *c.* 3100 Ma pegmatite veins in the bed of the Usutu River, southwest of Sidvokodvo, Swaziland. The thin pegmatite layers in the gneiss reflect strong deformation that transposed an early vein network, by folding and rotation, into sub-parallel layers.

Fig. 17.2. Tightly folded black layers of amphibolite (derived from basaltic volcanics) at Tovqussaq nunâ, Greenland (see Fig. 17.3). The volcanic rocks were intruded by sheets of tonalite and then strongly deformed. Layers were attenuated and transposed into sub-parallelism, and were subsequently deformed again into tight folds and recrystallized in granulite facies at *c.* 3000 Ma.

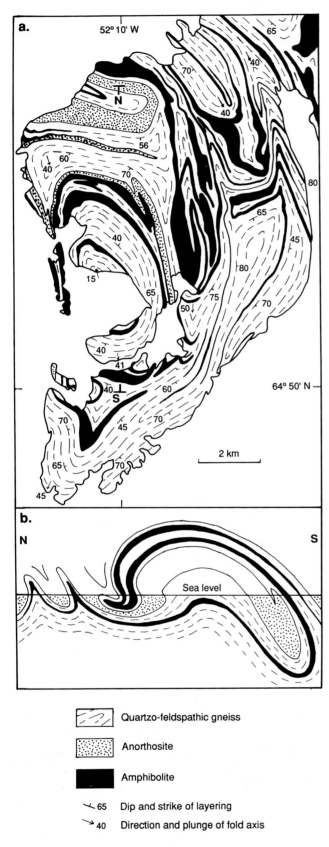

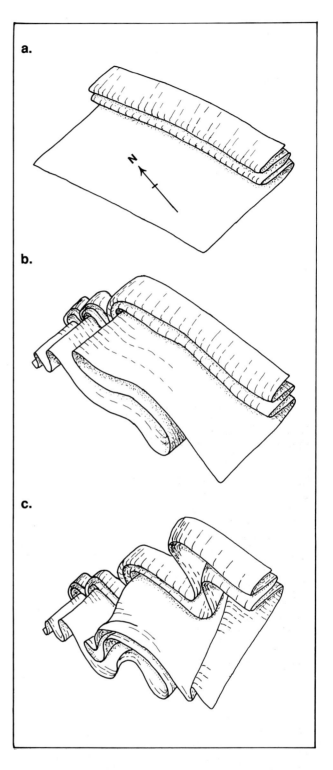

Fig. 17.3. (a) Map showing large-scale fold interference structures on the peninsula of Tovqussaq nunâ in Greenland (located on Fig. 17.5) compiled from Berthelsen (1960, plate 4), Bridgwater et al. (1976, fig. 68) and Garde (1989). (b) Simplified cross-section along the N–S line in (a) showing recumbent isoclinal folds refolded by upright folds forming the Tovqussaq dome. After Berthelsen (1960, fig. 74).

Fig. 17.4. Diagrams illustrating the evolution of the fold structures at Tovqussaq nunâ after Berthelsen (1960, fig. 78). (a) Isoclinal folds with NNW–SSE axes. (b) Isoclinal folds refolded by recumbent folds with WSW–ENE axes. (c) Recumbent folds locally refolded, forming a large-scale domal fold interference structure in the southwest.

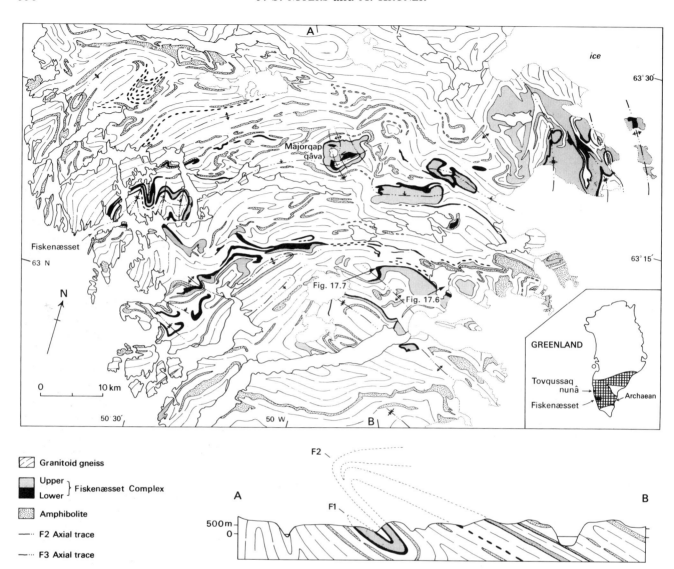

Fig. 17.5. Map and cross-section of large-scale fold interference structures in the Fiskenaesset region of Greenland after Myers (1981). Locations of Figs 17.6 and 17.7 are shown.

Fig. 17.6. Recumbent isoclinal fold of amphibolite (black) and anorthosite (white) on a 400 m high cliff on the edge of the inland ice cap 55 km east of Fiskenaesset, Greenland (see Fig. 17.5). View to the northeast.

Fig. 17.7. Section of layering formed by a combination of igneous layering, intrusive sheets and strong deformation, seen on a 300 m high cliff, 40 km east of Fiskenaesset, Greenland (see Fig. 17.5). The upper third of the cliff marked 'A' consists of leucogabbro with compositional igneous layering (visible below the letter A), sub-parallel to intrusive sheets and tectonic layering seen in the lower parts of the cliff. The middle part of the cliff comprises black layers of gabbro and ultramafic rocks disrupted by intrusive sheets of tonalite. The lower third of the cliff marked 'T' consists of strongly deformed tonalite and a few thin layers of gabbroic and ultramafic rocks. The rocks were recrystallized in amphibolite and granulite facies during the deformation and the tonalite is now gneissose with a prominent layering marked by pegmatite layers that have been deformed into sub-parallelism.

Survey of Greenland (1:20 000 scale mapping published at 1:100 000).

The first geometric analysis of large-scale Archaean structures in Greenland was made by Berthelsen (1960). He demonstrated the presence of three superimposed fold phases, with two generations of large-scale recumbent folds refolded by upright structures into broad domal antiforms and tight synforms (Figs 17.3 and 17.4)

In the Fiskenaesset region of Greenland (Fig. 17.5) a distinct igneous stratigraphy and abundant primary way-up indicators in a sill-like gabbro-anorthosite intrusion called the Fiskenaesset Complex enabled the facing directions of early recumbent folds to be determined (Myers 1981). The recumbent folding (Fig. 17.6) was accompanied by thrust-stacking of the stratigraphy and the syntectonic emplacement of sheets of tonalite (Fig. 17.7). This led to considerable crustal thickening and was followed by two sets of upright folds that formed large-scale *dome-and-basin fold interference structures* of the previous flat-lying structures (Fig. 17.8) (Myers 1976, 1981). Regional metamorphism culminated in granulite facies, and was followed by patchy retrogression to amphibolite facies and the development of large-scale steeply inclined shear zones.

It is significant that although the early recumbent structures in the Fiskenaesset region have fold amplitudes in excess of 30 km, no deep crustal rocks are exposed. The presence, throughout the whole region, of a single layered intrusion of gabbro-anorthosite, split and thickened by thrusts and concordant sheets of tonalite, suggests that the whole region represents a relatively thin slice of mid-crustal rocks that was tectonically and magmatically thickened by décollement on deeper crustal rocks (perhaps more mafic and ultramafic granulites), that are nowhere visible.

The structures described above appear to be typical of much of the Archaean gneiss complex of Greenland (Bridgwater *et al.* 1976). Amphibolites and metasedimentary rocks that may be high-grade equivalents of greenstones are widespread as intensely deformed concordant layers within the tonalitic gneisses. In a few places, lower-grade greenstone sequences occur that unconformably overlie a gneissose basement. They have been deformed together with the basement into large-scale upright fold interference structures (Fig. 17.9).

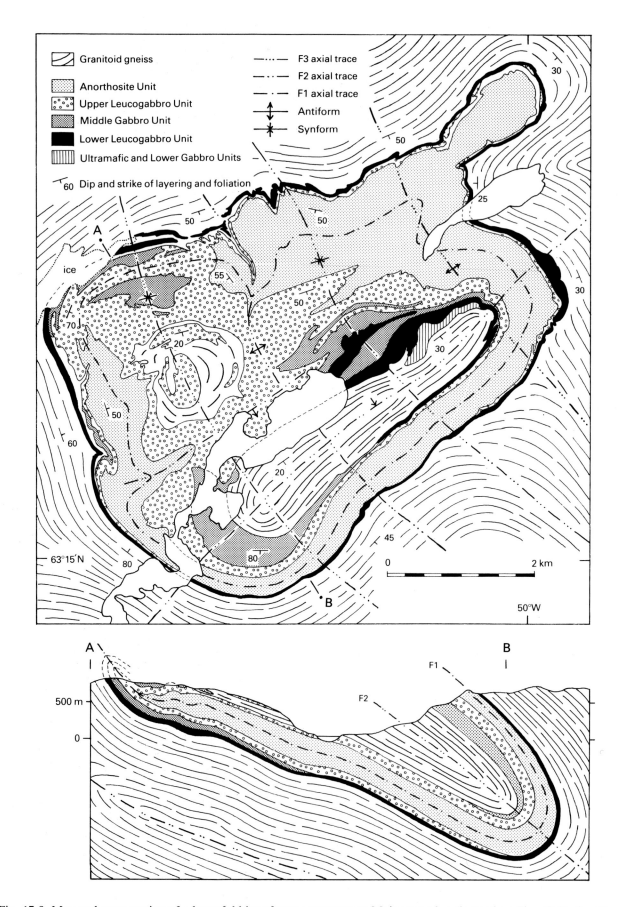

Fig. 17.8. Map and cross-section of a large fold interference structure at Majorqap qâva (located on Fig. 17.5) after Myers (1981). They show a recumbent syncline of c. 2860 Ma anorthosite and tonalitic gneiss refolded twice into a synformal fold interference structure. The rocks were subsequently recrystallized in granulite facies at c. 2800 Ma.

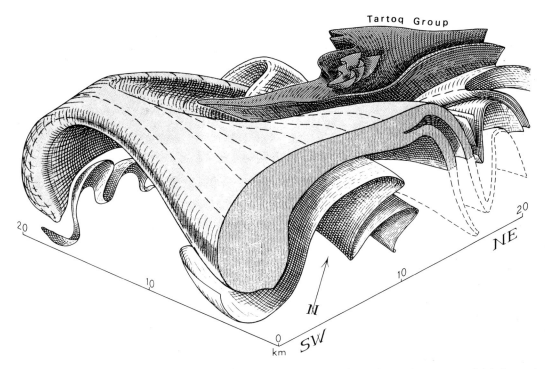

Fig. 17.9. Interpretation of Archaean structures, with nappes in a tonalitic gneissose basement refolded together with an overlying metavolcanic cover (Tartoq Group), 200 km SSE of Fiskenaesset, Greenland, after Berthelsen & Henriksen (1975) and Bridgwater *et al.* (1976). (Reprinted from Bridgwater *et al.* (1976) in *Geology of Greenland* edited by Escher, A. and Watt, W. S. 1976, fig. 66 © Geological Survey of Greenland and Meddelelser om Grønland.)

Jackson (1984) reported similar structures from the early Archaean Ancient Gneiss Complex of Swaziland, southern Africa, where recumbent isoclinal folds in banded orthogneisses of the 3640–3550 Ma Bimodal Suite were intruded at 3450 Ma by a coarse porphyritic trondhjemite which itself became involved in later deformation and was transformed into a gneiss. Kröner *et al.* (1989, 1991) have shown on the basis of precise U/Pb zircon dating that several generations of TTG were emplaced into these gneisses between 3400 and 3200 Ma. However, the heterogeneous gneisses that were sampled were intensely deformed and the relative age relations of the various components were not recognized in the field.

James & Black (1981) described several sets of tight recumbent folds from an early Archaean polymetamorphic high-grade terrain in Enderby Land, Antarctica.

STRUCTURES IN GRANITE–GREENSTONE TERRAINS

Extensive granite–greenstone terrains are best exposed in the Superior and Slave Provinces of Canada, the Zimbabwe and Kaapvaal Cratons of southern Africa, the Dharwar craton of India, the Aldan Shield of Siberia, the NE part of the Baltic Shield in Europe, and in the Pilbara and Yilgarn Cratons of Western Australia.

MacGregor (1951) interpreted all the Archaean granitoid rocks of Zimbabwe as batholiths emplaced by diapirism into a cover of greenstones (Fig. 17.10). Deformation of the greenstones was attributed to the diapiric rise of the granitoid magmas (also see Chapter 8). This interpretation was widely followed in other granite–greenstone terrains and was the main tectonic explanation of the supposedly distinct

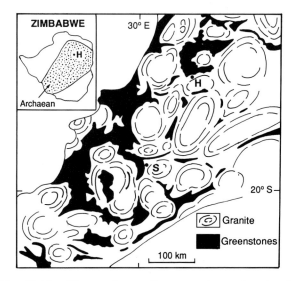

Fig. 17.10. Map of granite–greenstone terrain in Zimbabwe after MacGregor (1951) who considered the structures to be a result of the diapiric rise of granites into the greenstones. H locates Harare and S locates Shurugwi.

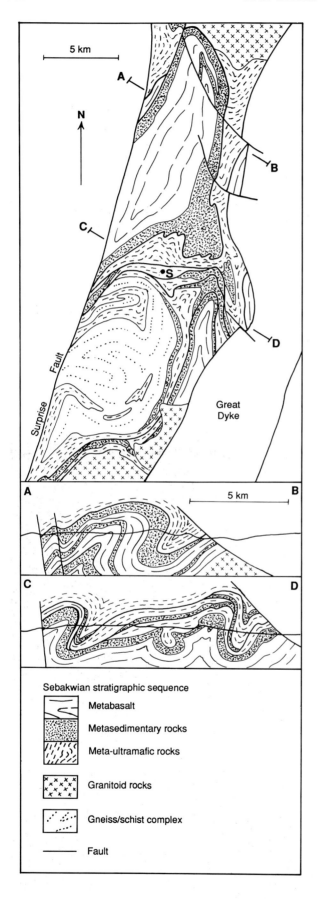

Fig. 17.11. Map and cross-sections A–B and C–D showing a folded, thrust-repeated, inverted sequence of Sebakwian rocks around Shurugwi in Zimbabwe after Stowe (1974, 1984). Interpreted by Stowe as the lower limb of an allochthonous nappe.

Archaean structure of these regions. Advocates of this model differed only in their interpretations of the nature of the diapirism as either magmatic (Anhaeusser 1975, 1984, Glikson 1979) or solid state (Drury 1977, Gee 1979, Ramberg 1980, Hickman 1984).

This model of essentially vertical tectonics is now in doubt following the recognition of major thrusts, duplex structures and nappes in many greenstone belts, and of the tectonic interleaving of greenstone units of diverse lithology and age (Stowe 1974, 1984, Bickle *et al.* 1980, Platt 1980, de Wit 1982, Lamb 1982, Myers & Watkins 1985, Swager *et al.* 1990, Corfu & Ayres 1991, Tomkinson & King 1991) (Figs 17.11 and 17.12).

Recent detailed mapping has led to the questioning of the diapiric hypothesis and the conclusion that the main structures of granite–greenstone terrains reflect repeated folding associated with regional horizontal shortening (Platt 1980, Lamb 1984, Snowden 1984, Myers & Watkins 1985, Daigneault *et al.* 1990, Schwerdtner 1990). If any individual pluton was emplaced diapirically it was of only minor significance. Most greenstone belts examined showed evidence of early flat-lying structures, thrusts and nappes, that were refolded by later upright structures (Figs 17.13 and 17.14) (Platt 1980, Snowden 1984, Myers & Watkins 1985). Rather than forming steep-sided plutons, the granitoid rocks of some so-called 'batholiths' were shown to be large sill-like bodies (Hunter 1973, Tomkinson & King 1991, p.75) emplaced concordantly within the greenstones (Myers & Watkins 1985), similar in form to the main granitoid intrusions of many high-grade gneiss terrains (Myers 1976, 1981) (Figs 17.5 and 17.7).

Krogstad *et al.* (1989) described four phases of folding from the c.2700 Ma Kolar greenstone belt in southern India and showed that the first two generations of tight and isoclinal folds were related to an episode of strong E–W sub-horizontal compression. Foliations in the gneisses on both sides of the belt are parallel to the foliation in the greenstones.

Most structural analyses published in the 1980s and early 1990s contrast with those from the 1970s. The recent work suggests that the main tectonic structures of granite–greenstone terrains appear to be broadly similar to those of high-grade gneiss terrains. This supports the view that these two types of terrain evolved in similar tectonic settings but at different crustal levels (Percival 1989, Kröner 1991).

Archaean Tectonics

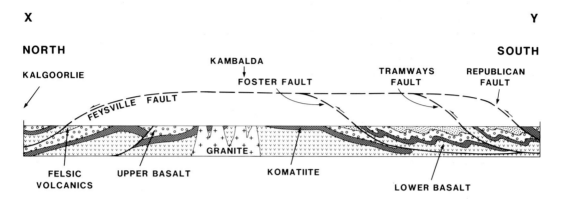

Fig. 17.12. Cross-section showing early thrusts and duplex structures in the c. 2700 Ma granite–greenstone Kalgoorlie Terrane of the Yilgarn Craton, Australia, after Swager et al. (1990). Located by the section line X–Y on Fig. 17.16.

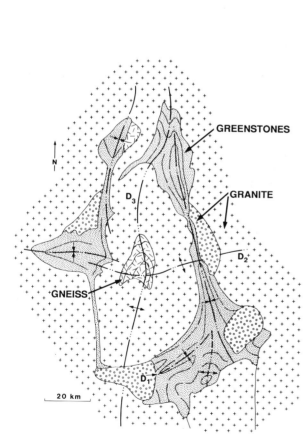

Fig. 17.13. Map of large-scale fold interference structure in granite–greenstone terrain of the Yilgarn Craton, Australia. The structure resulted from two generations of upright antiformal folds refolding previously flat-lying tectonic structures. After Myers & Watkins (1985), located by the letter Z on Fig. 17.16.

Fig. 17.14. Small-scale fold interference structures in a c. 2900 Ma pegmatite-banded gneiss in the core of the large-scale fold interference structure shown on Fig. 17.13. The pen is 14 cm long.

TECTONIC MODELS

Most recent interpretations of Archaean structures relate them to plate tectonic processes. This is not surprising because palaeomagnetic data for Archaean rocks have now established that the ancient cratons moved independently of each other with average velocities that were not much different from those of today (Kröner *et al.* 1984). A broad spectrum of plate tectonic situations have been described and examples are outlined below.

Rifted continents

Burke *et al.* (1985a) proposed that the Pongola Supergroup formed in what may be the oldest well-preserved *continental rift* on the Kaapvaal Craton in southern Africa at about 3000 Ma. This metamorphosed sequence of bimodal metavolcanics and clastic sedimentary rocks was followed by the eruption of the Ventersdorp Supergroup basalts and andesites in another rift system just prior to, and during, the collision of the Kaapvaal and Zimbabwe Cratons between *c.* 2700 and 2600 Ma (Burke *et al.* 1985b). The broadly similar Fortescue Group in Western Australia formed a flood basalt sequence on older granite–greenstone terrain of the Pilbara Craton that was being extended by rifting between 2765 and 2687 Ma (Arndt *et al.* 1991).

Most of these rocks are only mildly deformed and were subjected to a low-greenschist grade of metamorphism, and so sedimentary and volcanic structures and textures are abundantly preserved.

Amalgamated volcanic arcs

It has been proposed that the granite–greenstone terrains of the Canadian Superior Province and the Australian Yilgarn Craton comprise an amalgamation of several *tectonostratigraphic terranes* (Hoffman 1989, Card 1990, Myers 1990, 1991, Swager *et al.* 1990, Williams 1990, also see Chapter 15) (Figs 17.15 and 17.16). The terranes reflect oceanic, island arc, and continental arc volcanism, and involve *back-arc basins, island arcs, fore-arc accretionary sedimentary prisms* and *oceanic plateaux* (see Chapter 9). In the Superior Province the terranes formed during at least two episodes at 3000–2800 Ma and 2720–2710 Ma and were successively accreted from north to south at 2725, 2705, and 2695 Ma (Fig. 17.15) (Card 1990). The diverse terranes of the Yilgarn Craton of Australia also appear to have been swept together and amalgamated between 2750 and 2650 Ma (Fig. 17.16) (Myers 1992).

The Canadian Slave Province is now interpreted as comprising a continent that collided and amalgamated with island arcs and sedimentary accretionary prisms at about 2650 Ma (Fyson & Helmstaedt 1988, Kusky 1990). Gneiss terranes of as yet uncertain tectonic setting are also recognized to have been amalgamated in West Greenland between 2750 and 2650 Ma (Friend *et al.* 1989, Nutman *et al.* 1989).

A number of major fault zones have recently been recognized within the stratigraphy of the Barberton greenstone belt in the Kaapvaal Craton of southern Africa (Tomkinson & King 1991). The faults separate greenstone sequences with different stratigraphy and age, and suggest that the Barberton greenstone belt may represent a number of amalgamated terranes.

In the Dhawar Craton of southern India, Krogstad *et al.* (1989) have shown that the sialic crust on the east and west sides of the Kolar greenstone belt represents two different terranes. The western terrane comprises >3200 Ma felsic gneiss intruded by monzonitic to granitic rocks whereas the eastern terrane mainly consists of *c.* 2530 Ma granodioritic gneisses. The Kolar greenstone belt of komatiitic and basaltic rocks is thought to represent part of an intervening ocean basin.

Continent–arc and continent–continent collisions

The oldest known well-documented example of tectonics possibly related to *subduction* processes is seen in the Ancient Gneiss Complex of Swaziland and adjacent Barberton granite–greenstone terrain (Jackson *et al.* 1987, Hunter 1991, Hunter *et al.* 1992, in press, de Wit *et al.* 1992). Major orogeny involving northward thrusting of island arc and rift greenstones, and the emplacement of thick sheets of tonalite, trondhjemite, granodiorite and granite occurred during several tectono-magmatic episodes between *c.* 3450 and *c.* 3100 Ma (Fig. 17.17) (see also Kröner *et al.* 1991). The Fig Tree Group of the Barberton greenstone belt, and later coarse clastic foreland basin deposits represented by the Moodies Group, were caught up in the thrusting and folding (Jackson *et al.* 1987).

The best preserved evidence for Archaean *continent–continent collision* (see Chapter 13) is probably seen in rocks of the Limpopo mobile belt of southern Africa where collision between the Kaapvaal and Zimbabwe Cratons at *c.* 2640 Ma led to intense deformation and granulite facies metamorphism in a complex collision zone comprising an *Andean-type volcanic arc* followed by a fold-and-thrust belt on the margin of the Kaapvaal Craton (Burke *et al.* 1986).

In Western Australia, a 3730–3300 Ma continental remnant called the Narryer terrane collided and amalgamated between 2750 and 2650 Ma with the Murchison terrane comprising 3000–2700 Ma granites and greenstones, and a sialic basement of 3500–3000 Ma gneisses (Myers 1988, 1992) (Fig. 17.16).

Continent–continent or arc–continent collision was probably also responsible for the formation of extensive late Archaean granulite-facies assemblages including voluminous charnockites in a broad E–W trending

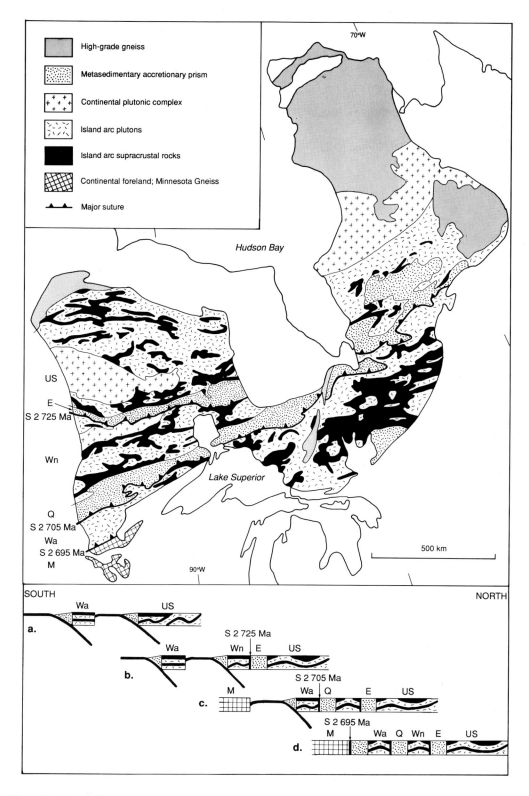

Fig. 17.15. Map showing major tectonostratigraphic terranes of the Superior Province, Canada, after Hoffman (1989) and Card (1990). Diagrammatic sections below (a – d) illustrate a plate-tectonic model with progressive north to south assembly of these terranes after Langford & Morin (1976) and Hoffman (1989). Individual terranes are indicated by letters as follows: US, Uchi-Sachigo; E, English River; Wn, Wabigoon; Q, Quetico; Wa, Wawa; M, Minnesota. The letter 'S' locates major suture zones with ages given in Ma.

mobile belt along the northern margin of the North China Craton. The structures seen in these high-grade rocks are similar to those in the Limpopo belt of southern Africa. Isotopic data indicate that most rocks were generated in a juvenile setting (perhaps island arc) and were transported into the lower crust within 100 Ma of their formation (Kröner, Cui and Wang unpublished data).

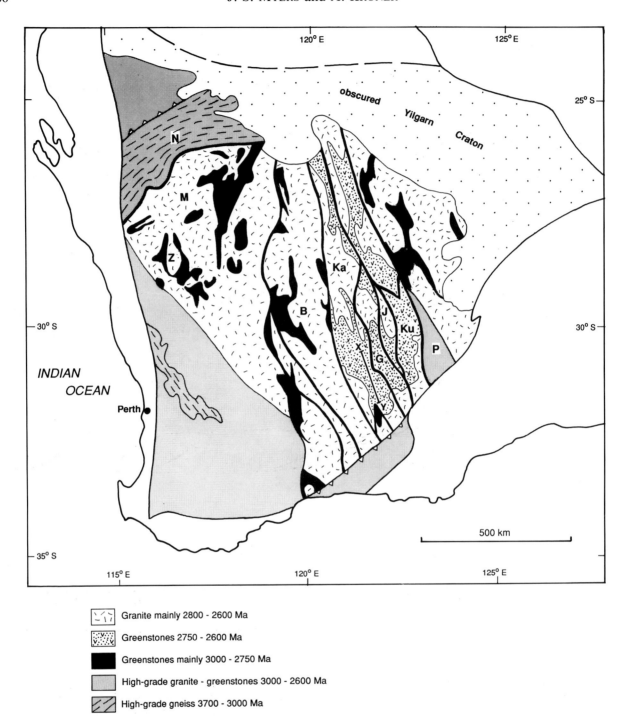

Fig. 17.16. Map showing the main components of the Yilgarn Craton, south-western Australia, after Myers (1993). The following symbols represent tectonostratigraphic terranes: N, Narryer; M, Murchison; B, Barlee; Ka, Kalgoorlie; G, Gindalbie; J, Jubilee; Ku, Kurnalpi; P, Pinjin.

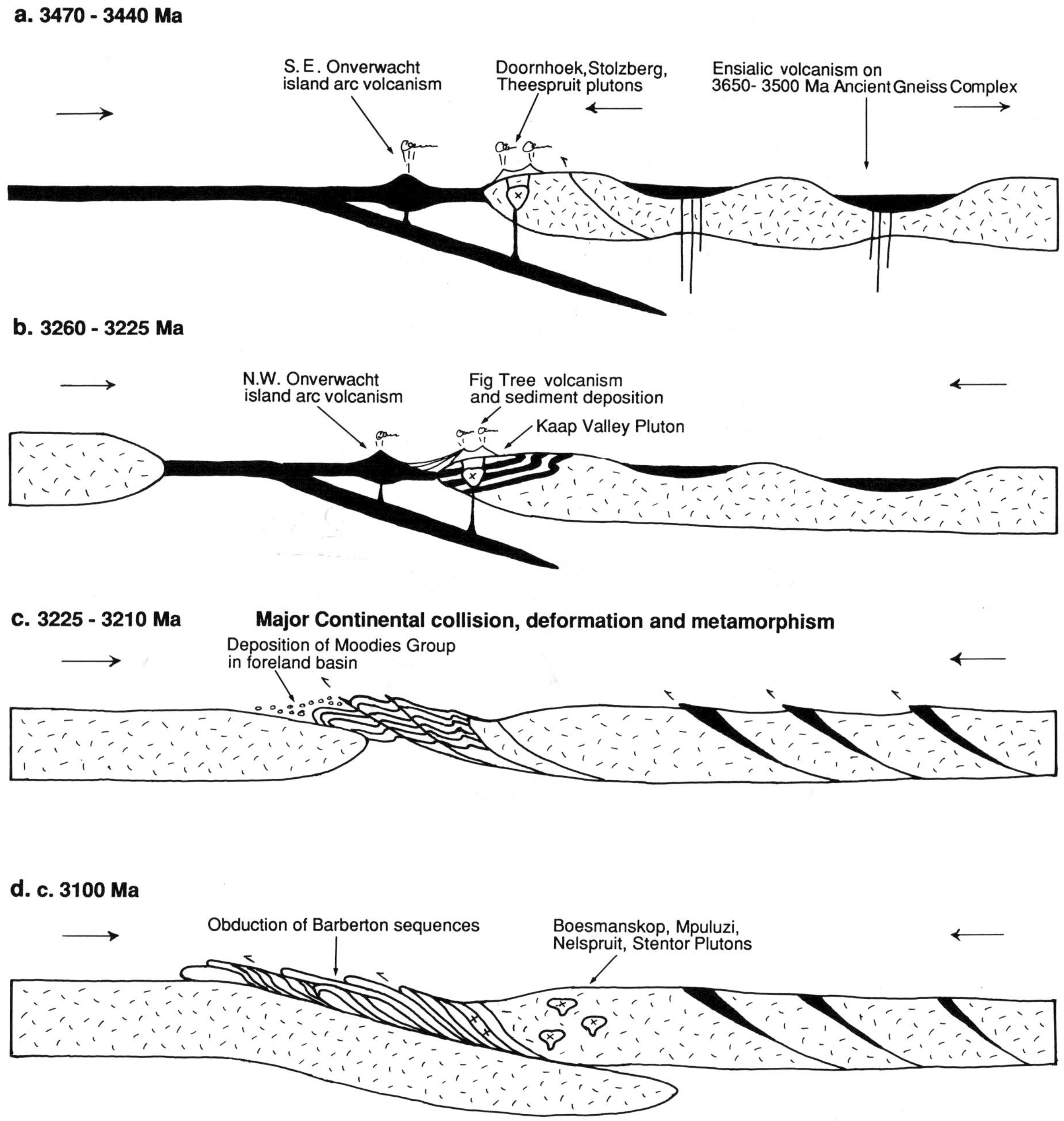

a. 3470 - 3440 Ma

S.E. Onverwacht island arc volcanism

Doornhoek, Stolzberg, Theespruit plutons

Ensialic volcanism on 3650-3500 Ma Ancient Gneiss Complex

b. 3260 - 3225 Ma

N.W. Onverwacht island arc volcanism

Fig Tree volcanism and sediment deposition

Kaap Valley Pluton

c. 3225 - 3210 Ma   Major Continental collision, deformation and metamorphism

Deposition of Moodies Group in foreland basin

d. c. 3100 Ma

Obduction of Barberton sequences

Boesmanskop, Mpuluzi, Nelspruit, Stentor Plutons

Fig. 17.17. Schematic cross-sections (a – d) illustrating the tectonic evolution of the Barberton greenstone belt and adjacent Ancient Gneiss Complex in the Kaapvaal Craton of southern Africa. Modified after Hunter *et al.* (1992) with geochronology from Ashwal (1991). The hypothesis that the northern part of the Onverwacht Group, northwest of the Inyoka fault, represents a different, younger island arc and back-arc terrane from the southern part of the Onverwacht Group is still untested by geochronology.

CONCLUSIONS

Surviving Archaean structures and tectonic styles do not differ from those of later geologic times. However, the tectonic setting responsible for the formation of the earliest continental crust (>4000 Ma) is still uncertain as no intact remnant has yet been discovered. It may not have formed by current plate tectonic processes because the very high near-surface heat flow may have been too great for large rigid plates to have been sustained. However, by at least 3600 Ma, continental segments existed that were already thick enough to permit granulite-facies assemblages to be formed, and these rocks record structures indicative of extensive horizontal shortening (Griffin *et al.* 1980, James & Black 1981, Kröner & Layer 1992).

Evidence for the independent motion of continental plates since the early Archaean suggests that shortening resulted from plate interaction, either at ancient continental margins or in plate interiors. Differences between distinctive Archaean rock associations, such as komatiites in greenstone belts and TTG gneisses, and those of later geologic times may largely reflect the thermal evolution of the Earth and may not indicate fundamentally different tectonic processes.

REFERENCES

Anhaeusser, C. R. 1973. The evolution of the early Precambrian crust of southern Africa. *Phil. Trans. R. Soc. Lond.* **A273**, 359–388.

Anhaeusser, C. R. 1975. Precambrian tectonic environments. *A. Rev. Earth Planet. Sci.* **3**, 31–53.

Anhaeusser, C. R. 1984. Structural elements of Archaean granite–greenstone terranes as exemplified by the Barberton Mountain Land, southern Africa. In: *Precambrian Tectonics Illustrated* (edited by Kröner, A. & Greiling, R.). E. Schweizerbart'sche, Stuttgart, 57–78.

Arndt, N. T., Nelson, D. R., Compston, W., Trendall, A. F. & Thorne, A. M. 1991. The age of the Fortescue Group, Hamersley Basin, Western Australia, from ion microprobe zircon U–Pb results. *Austr. J. Earth Sci.* **38**, 261–281.

Ashwal, L. D. 1991. *Two Cratons and an Orogen*, excursion guidebook and review articles for a field workshop through selected Archaean terranes of Swaziland, South Africa and Zimbabwe. IGCP Project 280. Geology Dept., University Witwatersrand, Johannesburg, 1–208.

Ayres, L. D. & Thurston P. C. 1985. Archaean supracrustal sequences in the Canadian Shield: an overview. In: *Evolution of Archaean Supracrustal Sequences* (edited by Ayres, L. D., Thurston, P. C., Card, K. D. & Weber, W.). *Geol. Soc. Can. Spec. Pap.* **28**, 343–380.

Berthelsen, A. 1960. Structural studies in the Precambrian of western Greenland: 2. Geology of Tovqussaq nunâ. *Grønlands Geol. Unders. Bull.* **25**, 1–223.

Berthelsen, A. & Henriksen, N. 1975. Geological map of Greenland 1:100 000 Ivigtut 61 V.1 Syd. *Grønlands Geol. Unders.*, Copenhagen.

Bickle, M. J., Bettenay, L. F., Boulter, C. A., Groves, D. I. & Morant, P. 1980. Horizontal tectonic interaction of an Archaean gneiss belt and greenstones, Pilbara block, Western Australia. *Geology* **8**, 525–529.

Bridgwater, D., McGregor, V. R. & Myers, J. S. 1974. A horizontal tectonic regime in the Archaean of Greenland and its implications for early crustal thickening. *Precambrian Res.* **1**, 179–197.

Bridgwater, D., Keto, L., McGregor, V. R. & Myers, J. S. 1976. Archaean gneiss complex of Greenland. In: *Geology of Greenland* (edited by Escher, A. & Watt, W. S.). Grønlands Geol. Unders., Copenhagen, 18–75.

Burke, K., Dewey, J. F. & Kidd, W. S. F. 1976. Dominance of horizontal movements, arc and microcontinental collisions during the later permobile regime. In: *The Early History of the Earth* (edited by Windley, B. F.). John Wiley, London, 113–129.

Burke, K., Kidd, W. S. F. & Kusky, T. M. 1985a. The Pongola structure of southeastern Africa: the World's oldest preserved rift? *J. Geodynam.* **2**, 35–49.

Burke, K., Kidd, W. & Kusky, T. 1985b. Is the Ventersdorp rift system of southern Africa related to a continental collision between the Kaapvaal and Zimbabwe cratons at 2.64 Ga ago? *Tectonophysics* **115**, 1–24.

Burke, K., Kidd, W. S. F. & Kusky, T. M. 1986. Archaean foreland basin tectonics in the Witwatersrand, South Africa. *Tectonics* **5**, 439–456.

Card, K. D. 1990. A review of the Superior Province of the Canadian Shield, a product of Archaean accretion. *Precambrian Res.* **48**, 99–156.

Chadwick, B. & Nutman, A. P. 1979. Archaean structural evolution in the northwest of the Buksefjorden region, southern west Greenland. *Precambrian Res.* **9**, 199–226.

Corfu, F. & Ayres, L. D. 1991. Unscrambling the stratigraphy of an Archaean greenstone belt: a U–Pb geochronological study of the Favourable Lake Belt, northwestern Ontario, Canada. *Precambrian Res.* **50**, 201–220.

Compston, W. & Kröner, A. 1988. Multiple zircon growth within early Archaean tonalitic gneiss from the Ancient Gneiss Complex, Swaziland. *Earth Planet. Sci. Lett.* **87**, 13–28.

Daigneault, R., St-Julien, P. & Allard, G. O. 1990. Tectonic evolution of the northeast portion of the Archaean Abitibi greenstone belt, Chibougamau, Quebec. *Can. J. Earth Sci.* **27**, 1714-1736.

Dewey, J. F. & Burke, K. 1973. Tibetan, Variscan and Precambrian basement reactivation: products of continental collision. *J. Geol.* **81**, 683–692.

de Wit, M. J. 1982. Gliding and overthrust nappe tectonics in the Barberton Greenstone Belt. *J. Struct. Geol.* **4**, 117-136.

de Wit, M. J., Roering, C., Hart, R. J., Armstrong, R. A., de Ronde, C. E. J., Green, W. E., Tredoux, M., Peberdy, E. & Hart, R. A. 1992. Formation of an Archaean continent. *Nature* **357**, 553–562.

Drury, S. A. 1977. Structures induced by granite diapirs in the Archaean greenstone belt at Yellowknife, Canada: implications for Archaean geotectonics. *J. Geol.* **85**, 345–358.

Friend, C. R. L., Nutman, A. P. & McGregor, V. R. 1988. Late Archaean terrane accretion in the Godthåb region, southern West Greenland. *Nature* **335**, 535–538.

Fyson, W. K. & Helmstaedt, H. 1988. Structural patterns and tectonic evolution of supracrustal domains in the Archean Slave Province, Canada. *Can. J. Earth Sci.* **25**, 301–315.

Garde, A. A. (compiler), 1989. Geological map of Greenland 1:100 000, Fiskefjord 64 V1 Nord. Grønlands Geol. Unders., Copenhagen.

Gee, R. D. 1979. Structure and tectonic style of the Western Australian Shield. *Tectonophysics* **58**, 327–369.

Gee, R. D., Myers, J. S. & Trendall, A. F. 1986. Relation between Archaean high-grade gneiss and granite–greenstone terrain in Western Australia. *Precambrian Res.* **33**, 87–102.

Glikson, A. Y. 1979. Early Precambrian tonalite–trondhjemite sialic nucleii *Earth Sci. Rev.* **15**, 1–73.

Goodwin, A. M. 1977. Archean volcanism in the Superior Province, Canadian Shield. *Geol. Ass. Can. Spec. Pap.* **16**, 205–241.

Goodwin, A. M. 1981. Precambrian perspectives. *Science* **213**, 55–61.

Gorman, B. E., Pearce, T. H. & Birkett, T. C. 1978. On the structure of Archaean greenstone belts. *Precambrian Res.* **6**, 23–41.

Griffin, W. L., McGregor, V. R., Nutman, A. P., Taylor, P. N. & Bridgwater, D. 1980. Early Archaean granulite facies metamorphism south of Ameralik, West Greenland. *Earth Planet. Sci. Lett.* **50**, 59–74.

Groves, D. I. & Batt, W. D. 1984. Spatial and temporal variations of Archaean metallogenic associations in terms of evolution of granitoid-greenstone terrains with particular emphasis on the Western Australian Shield. In: *Archaean Geochemistry* (edited by Kröner, A., Hanson, G. N. & Goodwin, A. M.). Springer, Berlin, 73–98.

Hickman, A. H. 1984. Archaean diapirism in the Pilbara Block, Western Australia. In: *Precambrian Tectonics Illustrated* (edited by Kröner, A. & Greiling, R.). E. Schweizerbart'sche, Stuttgart, 113–127.

Hoffman, P. F. 1989. Precambrian geology and tectonic history of North America. In: *The Geology of North America, Vol. A — An Overview* (edited by Bally, A. W. & Palmer, A. R.). *Geol. Soc. Am.*, 447–512.

Hunter, D. R. 1973. The granitoid rocks of the Precambrian in Swaziland. *Trans. geol. Soc. South. Afr.* **73**, 107–150.

Hunter, D. R. 1991. Crustal processes during Archaean evolution of the southeastern Kaapvaal Province. *J. Afr. Earth Sci.* **13**, 13–25.

Hunter, D. R. In press. Generation of granitoids at Archaean continental margins in southern Africa. In: *Proc. 8th Int. Conf. on Basement Tectonics*. Butte, Montana. 1988.

Hunter, D. R., Smith, R. G. & Sleigh, D. W. W. 1992. Geochemical studies of Archaean granitoid rocks in the southeastern Kaapvaal Province; implications for crustal development. *J. Afr. Earth. Sci.* **15**, 127–151.

Jackson, M. P. A. 1984. Archaean structural styles in the Ancient Gneiss Complex of Swaziland, southern Africa. In: *Precambrian Tectonics Illustrated* (edited by Kröner, A. & Greiling, R.). E. Schweizerbart'sche, Stuttgart, 1–18.

Jackson, M. P. A., Eriksson, K. A. & Harris, C. W. 1987. Early Archaean foredeep sedimentation related to crustal shortening: a reinterpretation of the Barberton sequence, southern Africa. *Tectonophysics* **136**, 197–221.

James, P. R. & Black, L. P. 1981. A review of the structural evolution and geochronology of the Archaean Napier Complex of Enderby Land, Australian Antarctic Territory. In: *Archaean Geology* (edited by Glover, J. E. & Groves, D. I.). *Spec. Publs geol. Soc. Austr.* **7**, 71–83.

Kröner, A. 1985. Evolution of the Archaean continental crust. *A. Rev. Earth Planet. Sci.* **13**, 49–74.

Kröner, A. 1991. Tectonic evolution in the Archaean and Proterozoic. *Tectonophysics* **187**, 393–410.

Kröner, A. & Layer, P. W. 1992. Crust-formation and plate motion in the early Archean. *Science*, **256**, 1405–1411.

Kröner, A., Layer, P. W. & McWilliams, M. O. 1984. Archaean palaeomagnetism: evidence for continental drift and the existence of a dipolar magnetic field since *c.* 3.5 billion years ago. *Terra Cognita* **4**, 78.

Kröner, A., Compston, W. & Williams, I. S. 1989. Growth of early Archaean crust in the Ancient Gneiss Complex of Swaziland as revealed by single zircon dating. *Tectonophysics* **161**, 271–298.

Kröner, A., Wendt, J. I., Tegtmeyer, A. R., Milisenda, C. & Compston, W. 1991. Geochronology of the Ancient Gueiss Complex, Swaziland, and implications for crustal evolution. In: *Excursion Guide for IGCP Project 280 Field Trip in southern Africa*. Rand Afrikaans University, Johannesburg, South Africa, 6–29.

Krogstad, E. J., Balakrishnan, S., Mukhopadhay, D. K., Rajamani, V. & Hanson, G. N. 1989. Plate tectonics 2.5 billion years ago: evidence at Kolar, South India. *Science*, **243**, 1337–1340.

Kusky, T. M. 1990. Evidence for Archaean ocean opening and closing in the southern Slave Province. *Tectonics* **9**, 1533-1563.

Lamb, S. H., 1984. Structures on the eastern margin of the Archaean Barberton greenstone belt, northwest Swaziland. In: *Precambrian Tectonics Illustrated* (edited by Kröner, A. & Greiling, R.). E. Schweizerbart'sche, Stuttgart, 19–39.

Langford, F. F. & Morin, J. A. 1976. The development of the Superior Province of northwestern Ontario by merging island arcs. *Am. J. Sci.* **276**, 1023–1034.

Ludden, J., Hubert, C. & Gariepy, C. 1986. The tectonic evolution of the Abitibi greenstone belt of Canada. *Geol. Mag.* **123**, 153–166.

MacGregor, A. M. 1951. Some milestones in the Precambrian of southern Africa. *Proc. geol. Soc. South Afr.* **54**, 27–71.

Myers, J. S. 1976. Granitoid sheets, thrusting, and Archaean crustal thickening in West Greenland. *Geology* **4**, 265–268.

Myers, J. S. 1978. Formation of banded gneisses by deformation of igneous rocks. *Precambrian Res.* **6**, 43–64.

Myers, J. S. 1981. The Fiskenaesset anorthosite complex — a stratigraphic key to the tectonic evolution of the West Greenland gneiss complex 3000–2800 m.y. ago. In: *Archaean Geology* (edited by Glover, J. E. & Groves, D. I.). *Spec. Publs geol. Soc. Austr.* **7**, 351–360.

Myers, J. S. 1988. Early Archaean Narryer Gneiss Complex, Yilgarn Craton, Western Australia. *Precambrian Res.* **38**, 297–307.

Myers, J. S. 1990. Precambrian tectonic evolution of part of Gondwana, southwestern Australia. *Geology* **18**, 537–540.

Myers, J. S. 1992. Tectonic evolution of the Yilgarn Craton, Western Australia. In: *The Archaean Terrains, Crustal Processes and Metallogeny* (edited by Glover, J. E. & Ho, S. E.). Geology Dept and University Extension, *University Western Australia Publ.* **22**, 265–273.

Myers, J. S. 1993. Precambrian history of the West Australian craton and adjacent orogens. *A. Rev. Earth. Planet. Sci.* **21**, 453–485.

Myers, J. S. & Watkins, K. P. 1985. Origin of granite–greenstone patterns, Yilgarn block, Western Australia. *Geology* **13**, 778–780.

Nutman, A. P., Friend, C. R. L., Baadsgaard, H. & McGregor, V. R. 1989. Evolution and assembly of Archaean gneiss terranes in the Godthåbsfjord region, southern West Greenland: structural, metamorphic and isotopic evidence. *Tectonics* **8**, 573–589.

Park, R. G. 1982. Archaean tectonics. *Geol. Rdsch.* **71**, 22–37.

Passchier, C. W., Myers, J. S. & Kröner, A. 1990. *Field Geology of High-Grade Gneiss Terrains*. Springer, Berlin.

Percival, J. A. 1989. Archaean tectonic settings in the Superior Province, Canada; a view from the bottom. In: *Granulites and Crustal Deformation*. NATO Advanced Study Institute Series. Kluver, Dordrecht.

Platt, J. P. 1980. Archaean greenstone belts: a structural test of tectonic hypotheses. *Tectonophysics* **65**, 127–150.

Ramberg, H. 1980. Diapirism and gravity collapse in the Scandinavian Caledonides. *J. geol. Soc. Lond.* **137**, 261–270.

Schwerdtner, W. M. 1990. Structural tests of diapir hypotheses in Archaean crust of Ontario. *Can. J. Earth Sci.* **27**, 387–402.

Schwerdtner, W. M. & Lubers, S. B. 1980. Major diapiric structures in the Superior and Grenville provinces of Ontario. In: *The Continental Crust and its Mineral Deposits* (edited by Strangway, D. W.). *Geol. Ass. Canada Spec. Pap.* **20**. 149–180.

Snowden, P. A. 1984. Non-diapiric batholiths in the north of the Zimbabwe Shield. In: *Precambrian Tectonics Illustrated* (edited by Kröner, A. & Greiling, R.). E. Schweizerbart'sche, Stuttgart, 135–145.

Stowe, C. W. 1974. Alpine-type structures in the Rhodesian basement complex at Selukwe. *J. geol. Soc. Lond.* **13**, 411–425.

Stowe, C. W. 1984. The early Archaean Selukwe nappe, Zimbabwe. In: *Precambrian Tectonics Illustrated* (edited by Kröner, A. & Greiling, R.). E. Schweizerbart'sche, Stuttgart, 41–56.

Swager, C., Griffin, T. J., Witt, W. K., Wyche, S., Ahmat, A. L., Hunter, W. M., & McGoldrick, P. J. 1990. Geology of the Archaean Kalgoorlie Terrane — an explanatory note. *Geol. Surv. W. Austr. Rec.* **12**, 1–55.

Talbot, C. J. 1973. A plate tectonic model for the Archaean crust. *Phil. Trans, R. Soc. Lond.* **A273**. 413–427.

Taylor, S. R. & McLennan, S. M. 1985. *The Continental Crust: Its Composition and Evolution*. Blackwell, Oxford.

Tarney, J., Dalziel, I. W. D. & de Wit, M. J. 1976. Marginal basin 'Rocas Verdes' Complex from S. Chile: a model for Archaean greenstone belt formation. In: *The Early History of the Earth* (edited by Windley, B. F.). Wiley, London, 131–146.

Tomkinson, M. J. & King, V. J. 1991. The tectonics of the Barberton Greenstone Belt — an overview. In: *Two Cratons and an Orogen,* excursion guidebook and review articles for a field workshop through selected Archaean terranes of Swaziland, South Africa and Zimbabwe (edited by Ashwal, L. D.). IGCP Project 280. Geology Dept., University Witwatersrand, Johannesburg, 69–83.

Williams, H. R. 1990. Subprovince accretion tectonics in the south-central Superior Province. *Can. J. Earth Sci.* **27**, 570–581.

Windley, B. F. 1984. *The Evolving Continents*. Wiley, London.

Windley, B. F. & Smith, J. V. 1976. Archaean high-grade complexes and modern continental margins. *Nature* **260**, 671–675.

CHAPTER 18

Neotectonics

I. S. STEWART and P. L. HANCOCK

PURPOSE AND SCOPE

THIS chapter focuses on those aspects of the study of recent structures and fault-related landforms that are special to the discipline of neotectonics, and are rarely of great significance during the investigation of palaeotectonic, that is, ancient structures. We also emphasize aspects of the contemporary stress field, particularly methods of inferring stress axis orientations. Although the contents of the chapter are intended to be self-contained, no attempt has been made to give a comprehensive account of those concepts that the discipline shares with structural geology and tectonics in general; they are fully discussed in other chapters. The reader interested in the broader aspects of neotectonics, and how the subject can be of value to geomorphologists and archaeologists is advised to consult Vita-Finzi's (1986) *Recent Earth Movements*, the only text book in the field to have appeared before 1993.

The distinctive attributes of many neotectonic studies arise from five factors.

(1) A wide range of methodologies and a variety of experts are commonly involved in a comprehensive study of the neotectonic history of a region. Palaeotectonic structures are often investigated by structural geologists working on their own.

(2) Many structures and some landforms developed in surficial deposits during earthquakes are ephemeral and require surveying by an *after-shock team* within a few days or weeks of their formation and before they are erased by natural causes or Mans' activities. The study of such relatively ephemeral structures permits fault kinematics, and earthquake fault segmentation related to a single increment of motion, to be understood.

(3) Neotectonic structures that are exposed and accessible to the field geologist were mainly formed within the uppermost 1–2 km of the crust. Most structures developed at greater depths are exposed only after a long period of uplift and denudation, and, hence, because these processes take time (say, greater than 5 million years) they are generally palaeotectonic. The preservation potential of many shallow-formed neotectonic structures is low, and, thus, they are not necessarily common in the rock record.

(4) The neotectonician has the ability to compare inferences drawn from field observations with geophysical and geodetic data about rates and mechanisms of present-day processes. For example, the orientations of stress axes inferred from populations of neotectonic faults or joints can be compared with those of the contemporary stress field, known from earthquake focal mechanisms or *in situ* stress measurements.

(5) It is possible to establish neotectonic fault displacement histories with greater precision than palaeotectonic ones because Quaternary (especially Holocene) dating techniques allow time intervals of substantially less than 1 million years to be detected.

It is, perhaps, worth reporting on the uses to which neotectonic data can be put. Firstly, the discipline contributes much to earthquake hazard reduction programmes, that is, it answers a fundamental societal need. Secondly, a knowledge of the orientation and magnitude of the contemporary stress field is of value during the appraisal of fracture permeability and, hence, subsurface fluid flow; topics of concern to petroleum geologists, hydrogeologists and those constructing underground repositories for radioactive and other hazardous waste. Thirdly, information about contemporary plate motions, rates and processes permits the palaeotectonician to erect actualistic models using uniformitarian principles. Fourthly, structures and fabrics exposed in areas of neotectonic faulting and folding, although reflecting deeper processes, record deformation mechanisms that characterize the uppermost 1–2 km of the Earth. They provide structural geologists with evidence about shallow deformation processes; evidence that is commonly lacking from inactive areas where denudation has removed the highest-level phenomena. Fifthly, an appreciation where earthquakes have occurred in the historic and prehistoric past can enlighten the interpretations of historians and archaeologists who might otherwise overlook this aspect of the physical background to cultures.

DEFINITIONS OF NEOTECTONICS AND ALLIED DISCIPLINES

In common with many disciplines that are not yet well established there is disagreement amongst its practitioners about its scope. Debate in neotectonics mainly hinges on how far back in time the prefix 'neo' indicates. The term was introduced by Obruchev (1948) to describe "the study of the young and recent movements taking place at the end of the Tertiary and the first half of the Quaternary" (quoted in Pavlides 1989). Although Mercier (also cited in Pavlides) also perceives of neotectonics as linking active and 'geological' tectonic phenomena many workers now use the word so that it embraces both contemporary movements and those that immediately preceded them. Slemmons (1991) writes: "Neotectonics can be broadly described as tectonic events and processes that have occurred in post-Miocene time . . .". Mörner (1990) takes the view that the neotectonic phase starts at different times in different places, depending on the tectonic regime. We agree with the principles of Mörner's thinking and define *neotectonics* as the branch of tectonics concerned with understanding earth movements that both occurred in the past and are continuing at the present day. The above definition takes account of much current usage and attempts to avoid arbitrarily restricting the stratigraphic scope of the subject. Thus, neotectonic structures develop in the *current tectonic regime* as defined by Muir Wood & Mallard (1992) to embrace the prevailing state of deformation within an intraplate region. The above definition of neotectonics means that there is no need for it to be regarded as synonymous with *Holocene tectonics, Quaternary tectonics* or *Neogene–Quaternary tectonics*, depending upon which lower stratigraphic bound is selected. When the word neotectonic is used as an adjective to qualify the age of a structure, its use commonly implies that the structure is interpreted as having been propagated or reactivated in a stress/strain field that has persisted without significant change of orientation to the present day. Following Wallace (1986, p. 3), whose outstandingly influential compilation of articles is entitled *Active Tectonics*, we also suggest that: "active tectonics is defined as tectonic movements that are expected to occur within a future time span of concern to society". The reader interested in the history of the usage of the term neotectonics, and the stratigraphic bounds that have been imposed to separate neo- from palaeo-, is referred to Pavlides' (1989) review of the subject.

Seismotectonics is a branch of geosciences that is concerned with the relationship between the seismological characteristics of present-day earthquakes and tectonics. *Seismogeology* is the study of the geological aspects (e.g. soft-sediment deformation, fault scarp analysis) of contemporary seismicity. Archaeological evidence for historical and pre-historical earthquakes is the concern of *archaeoseismology*. Somewhat perversely, *palaeoseismology* is not the study of earthquakes unrelated to those of the present day but concerns itself with the evidence for historic, prehistoric and other late Quaternary earthquakes. According to Hills (1956) *morphotectonics* is the study of all aspects of the relationship between geological structure and landforms; others restrict it to relationships between neotectonic structures and landforms. Here, this more specific usage is referred to as *tectonic geomorphology*, which, following the lead of Morisawa & Hack (1985) (whose book employs the phrase as its title), can be defined as the study of the relations between vertical and horizontal ground movements, and erosional and depositional processes and landforms (Bull & Wallace 1985).

CONCEPTS OF GENERAL APPLICABILITY IN NEOTECTONICS

Role of seismicity

To a large degree, the spatial extent of neotectonic deformation is mirrored by the global distribution of instrumentally recorded seismicity, developed around the turn of the century but globally coordinated only since the 1960s with the Worldwide Standardised Seismic Network (WWSSN). Thus many areas characterized by neotectonic movement are expressed as *seismically active belts*, linear or elongate zones of seismicity commonly coincident with major tectonic structures. Within seismically active belts, anomalies in the spatial distribution of seismicity might indicate *seismic gaps* — sites where significant stress is presumed to be accumulating and which, therefore, are potential sites of future rupture events. The time elapsed since the last phase of activity in these gaps possibly provides a measure of the imminency of future events, termed *seismic potential*. In addition, the distribution of seismicity within belts can identify the *partitioning of neotectonic deformation*. In subduction zones, for example, the majority of earthquakes are located within a relatively narrow zone bounded on one side by an *aseismic front* (Yoshii 1979), a sharp cut-off in seismicity some distance outboard (see Chapter 9) of the volcanic arc, and on the other by a *seismic front*, coincident with the start of the largely aseismic accretionary wedge.

The spatial partitioning of deformation along active structures is marked by *fault segmentation*, whereby fault zones can be subdivided on the basis of structural and geometric attributes into a series of inter-linked sections, termed *geometric fault segments* by Barka & Kadinskey-Cade (1988), and discussed in detail by dePolo et al. (1987). During earthquakes, one or several geometric segments may rupture, thereby defining *earthquake fault segments* (Barka & Kadinskey-Cade 1988) — discrete sections of a fault zone which fail during an individual earthquake. Such segments, commonly of the order of several tens of kilometres long within a fault zone of many hundreds of kilometres length, can be bounded by *fault barriers*; regions of high strength or low initial stress along a fault which can impede or arrest rupture propagation (Das & Aki 1977). These barriers, often expressed seismically as areas of high seismic energy release or *asperities* (Lay & Kanamori 1981, Scholz 1990), permit individual

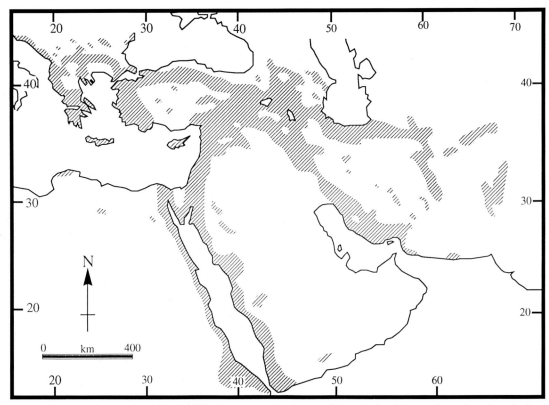

Fig. 18.1. Seismic zones (hatched) of the Middle East determined using historical evidence of earthquakes during the first 19 centuries A.D. Simplified from Ambraseys (1978).

segments of a fault zone to evolve independently of neighbouring segments.

Although instrumental seismicity provides a relatively comprehensive and accurate picture of the contemporary pattern of tectonic deformation, it is by no means complete. Areas which have exhibited little or no tectonic deformation during the duration of instrumental records, and as such have little or no seismic expression, can be shown to be actively deforming on the basis of historical seismicity, whereby written accounts or archaeological evidence of major seismic events can be used to map the incidence of earthquakes in the recent past (Ambraseys & Melville 1982, Nur 1991). This together with evidence of prehistorical seismicity, provides evidence of the changing locus of active deformation within actively deforming regions over time (Fig. 18.1). Furthermore, earthquakes are generally poor recorders of the gentle warping of the Earth's surface.

Identifying 'active' structures

As discussed earlier, assessing whether tectonic movements are likely to cause future concern to society requires the identification of faults which have moved within the duration of the current tectonic regime (CTR) (Muir Wood & Mallard 1992) and which, therefore, can be designated as *active faults*. A structure which has not moved within this period can be designated as a *non-active* or *extinct fault*, that is, it has negligible probability of experiencing displacement in the near future. This approach allows the identification of *potentially active faults* (Ambraseys 1978), pre-existing faults, expressed as lineaments, which are suitably oriented to become activated in the present CTR but which have not yet moved. In addition, although a fault cannot be considered active until it moves, many faults, even after intensive investigation, remain *unproven faults*, an intermediate category between active and extinct. The potential for an unproven fault to accommodate movement, however, can be weighted by evaluating its geometry relative to the CTR, its relationship with other structures and its associations with seismicity, making a continuum of classes from *probably active fault* to *probably extinct fault*.

Despite the need for flexible consideration of the timescale over which neotectonic processes operate, legalistic consequences require that certain neotectonic structures, particularly 'active' faults, have formal definition. Such definitions vary, however, in relation to varying levels of risk deemed acceptable to different public agencies, as well as varying degrees of activity. The U.S. Environmental Protection Agency (1981), for example, define an *active fault* as having one movement in the last 10,000 years, while the U.S. Regulatory Commission (1982) define a *capable fault* as having one displacement in 35,000 years and at least two in 500,000 years.

Such definitions of active structures focus strongly on the most recent movements along them, and necessitate an understanding of the behaviour of seismogenic faults over relatively short timescales. The frequency of earthquakes, for example, may be defined in terms of their *recurrence interval*, the time between successive rupture events on a given segment of a fault, or *repeat times*; the average time between major earthquakes within a particular region (Sykes & Quittmeyer 1981); both measures reflecting different aspects of the prevailing tectonic regime. Repeat times, for example, relate to rates of plate motion and the extent to which deformation is released seismically, while recurrence intervals relate to the duration of the *loading* or *earthquake cycle*, the progressive sequence of strain accumulation and stress release associated with fault rupture (Scholz 1990). The duration of this cycle varies markedly in different tectonic settings. In highly active interplate settings, for example, such as along subduction zones, repeat times may be the order of years and earthquake recurrence the order of a hundred years or so. Where these coincide with areas, such as Japan, with several hundred years of complete historical records of earthquakes, the temporal distribution of earthquake faulting can be well defined. In contrast, in continental areas, distributed deformation along a number of active structures result in repeat times of the orders of a few hundred years and recurrence intervals of considerably longer (Fig. 18.2). Although historical and archaelogical evidence may permit the earthquake database to be extended back several thousand years in places (Fig. 18.1), in some settings, particularly in intraplate regions where earthquakes have recurrence intervals of tens of thousands of years, past events, palaeoearthquakes, can only be identified from geological and geomorphological evidence.

Implicit in the use of recurrence intervals and repeat times is the premise that the pattern of earthquake faulting, and therefore deformation, is relatively constant over time. The pattern of both historical seismicity within many regions, however, provides evidence for temporal as well as spatial partitioning of deformation, including *migrating earthquake sequences* (Figs 18.2 and 18.3). Wallace (1984), for example, noted *fault switching* in the Basin and Range province of the western U.S.A. whereby the locus of active faulting alternated between adjacent fault zones, with different faults accommodating the bulk of the regional deformation at different times. The result is *earthquake temporal clustering* along faults, where fault movement occurs in a series of earthquakes over a geologically short period of time. Palaeoseismic studies in the northeastern Basin and Range province suggest that fault activity occurs in *seismic cycles*, involving periods of seismic activity, during which several palaeoseismic events occur within 15 ka or more, alternating with

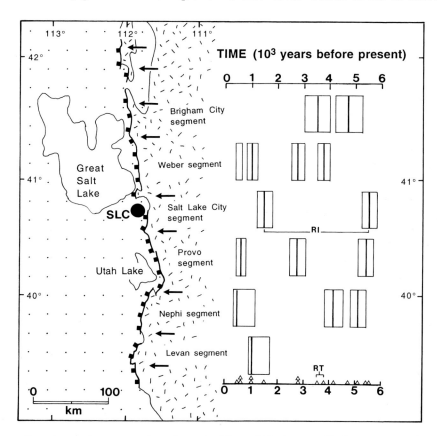

Fig. 18.2. Holocene earthquakes and segmentation along the Wasatch fault zone, Utah, eastern Basin and Range province, U.S.A. Heavy lines with solid boxes indicate fault segments (boxes on downthrown side). Arrows indicate segment boundaries. 'Graph' on the right highlights the timing of movements on the segments during the last 6000 years (after Machette *et al.* 1991). Open rectangles enclose likely time limits (radiocarbon and thermoluminescence age determinations) with bold lines indicating the best estimate for the time of faulting. Triangles on the lower axis show the temporal incidence of earthquakes across the fault zone, with the time elapsed between each event being the repeat time (RT) of the fault. In contrast, time periods separating ruptures on an individual fault segment are the recurrence interval (RI). Salt Lake City (SLC).

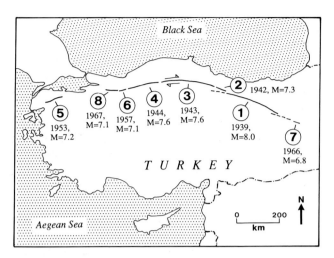

Fig. 18.3. Earthquakes generating surface faulting along the North Anatolian fault zone since 1939 until 1991. Although a general westward-directed migration of earthquakes is apparent (1–7), some individual events initiated at the western end of fault segments and propagated eastwards. The March 1992 earthquake at Erzincan (site of the 1939 event) was a surprise. After Allen (1969).

periods of seismic quiescence lasting a similar length of time (McCalpin 1993). Discriminating between inactive faults and *seismically quiescent faults*, dormant structures awaiting reactivation, therefore, is an important facet of seismic hazard assessment which can only be assessed from geological studies of palaeoearthquakes.

Assessments of the variation of earthquakes over time form an important element in understanding the long-term activity of seismogenic faults, and studies of active earthquake faulting have permitted the development of several models of earthquake recurrence and fault behaviour (Sieh 1981, Schwartz & Coppersmith 1984). The earliest model developed, the *variable slip model*, relates to a fault with a constant slip rate, irregular displacement between events and variable earthquake sizes. In this model, two contrasting loading cycles can be envisaged (Schimazaki & Nakata 1980). The first, a *time-predictable model*, ascribed most commonly to seismic faulting in subduction zones, predicts that earthquakes occur at a critical stress level and thus the time interval between earthquakes is proportional to the size of the preceding event, permitting the recurrence interval to be estimated. The second, a *slip-predictable model* proposes that earthquakes are initiated at different stress levels but that, following rupture, stress returns to a uniform base level. As a result, the magnitude (and, therefore, the displacement) of an earthquake can be predicted, but not its timing. An alternative model, the *uniform slip model*, also assumes a constant slip rate and regular displacement during each event, but slip is accommodated by earthquakes of varying size. A modification of this model is the *characteristic earthquake model* which predicts a variable slip rate along a fault, but with displacement per event and earthquake size constant. A fourth model, the *coupled model*, assumes an overall constant slip rate

along a fault and earthquakes of variable size (Berryman & Beanland 1991). Although each segment experiences constant displacements per event, more than one segment can be activated, producing a variable but non-random pattern of activity.

Seismic and aseismic activity

The use of seismicity as a diagnostic indicator of neotectonic deformation, relies on an assumption that all, or a significant proportion of, the deformation is being released seismically. While the brittle character of the upper crust means that deformation is principally accommodated by fracturing and concomitantly the release of seismic energy, deformation may also be accommodated by aseismic processes, including *stable frictional sliding* (also referred to as *fault creep*). Discriminating between *seismic* and *aseismic deformation* is particularly important, for example, in the identification of seismic gaps, where the apparent aseismicity is interpreted as a transitory precursor to large earthquakes, but which might also reflect the localized release of stress by aseismic processes. Indeed, it is in situations such as subduction zones that aseismic deformation is likely to be particularly significant. In the Hellenic trench, for example, seismicity can only account for less than 10% of the convergence between the southern Aegean Sea and Africa, with the rest being taken up in the form of aseismic folding and faulting within the largely aseismic accretionary wedge (Jackson & McKenzie 1988, Taymaz et al. 1990). While subduction zones in general may be characterized by significant proportions of aseismic faulting, areas within the adjacent seismically active belt, such as Japan, have negligible aseismic activity (Wesnousky et al. 1982).

Active deformation in continental areas can similarly exhibit both seismic and aseismic behaviour. Aseismic movement along faults occurs by the progressive frictional sliding of partially decoupled blocks and contrasts with the *stick-slip* behaviour of earthquake-generating or *seismogenic faults*, in which very sudden slip is followed by a period of no motion during which stress is recharged. In areas of crustal shortening, deformation is commonly accommodated by the aseismic dislocation of strata along décollement horizons — sub-horizontal horizons of weak ductile material, such as salt or overpressured clays, which permit decoupling of the crustal blocks on either side of the fault and, therefore, encourage stable sliding. The extent of this *seismic coupling* can vary along faults, with parts of active faults which are essentially seismogenic, locally exhibiting aseismic behaviour. The so-called 'creeping section' of the San Andreas fault system, for example, a 180 km section of the fault in central California, is characterized by high levels of seismicity, but seismic deformation only accounts for about 5% of the 11 mm/yr of steady slip. Instead, this part of the fault exhibits the relatively rare phenomenon of *episodic slip*, whereby a uniform slip rate is maintained by prolonged periods of fault creep interrupted by small increments of seismic slip.

Distinguishing between seismic and aseismic deformation is misleading, because most tectonic deformation is accompanied by seismic strain release, and it is more appropriate to regard the two as end members of a continuum of deformation styles which can operate. The loading cycle of a seismogenic fault, for example, can comprise both seismic and aseismic components. An idealized loading cycle comprises four phases of deformation — *preseismic, coseismic, postseismic and interseismic deformation*, although individual active faults rarely exhibit complete cycles (Scholz 1990). Preseismic deformation has been demonstrated in the form of aseismic coastal uplift before some large earthquakes, such as the 1983 Sea of Japan earthquake (Mogi 1985). In addition, following coseismic deformation involving the sudden release of strain upon earthquake rupture, that rupture is often followed by a transitory aseismic relaxation in strain rate before the long-term interseismic strain accumulation is renewed.

Aseismic components of earthquake-related deformation may induce effects generally attributed to progressive aseismic deformation. Folds, for example, can develop ahead of *blind faults* propagating from depth towards the surface. Such *fault-propagation folds* (*tip-line folds*) grow by increments of coseismic displacement to produce monoclinal structures which may be the only surficial evidence of blind seismogenic faults at depth (Vita-Finzi & King 1981). Detection of such active folding is important because blind faults are capable of generating very destructive earthquakes (Yeats 1986, Stein & Yeats 1989). Assessing the role of aseismic activity in crustal deformation is dependent on the scale of monitoring. Thus the creeping section of the San Andreas fault system may be considered aseismic at a regional or global scale, but at a local scale it has a marked seismic expression. In addition to varying resolution of monitoring techniques, the relative roles of aseismic and seismic deformation along an active structure can change with time. Recent detailed dating and surveying of marine strandlines generally attributed to coseismic coastal uplift, for example, instead demonstrate high current rates of aseismic slip, suggesting many of the strandlines may be aseismic in origin, and that the relative roles of seismic and aseismic deformation vary with time (Nelson & Manley 1992). Deciphering the importance of aseismic activity in the past is particularly difficult because historical and geological evidence of past tectonic activity tend to preserve only details of very large seismogenic events which had marked surface effects. Estimates of strain rates determined from historical and palaeoseismic evidence, however, can be compared to current rates of deformation within a region to identify significant disparities which might indicate a large component of aseismic deformation (Wesnousky 1986, Ambraseys & Jackson 1990).

METHODS OF NEOTECTONIC INQUIRY

The often subtle manifestations of neotectonic deformation necessitate the use of a wide-ranging and largely disparate set of investigative approaches, which vary both in the resolution of the data they provide and the scope of the timespan they describe (Fig. 18.4 and Table 18.1). High-resolution instrumental techniques provide a precise but essentially instantaneous view of present-day Earth movements at a variety of scales. Local geophysical instrumentation networks, for example, monitor the partitioning of deformation across, or within, actively deforming structures (Prescott *et al.* 1981, Sylvester 1986), while extra-terrestrial geodetic surveying techniques and global seismological networks resolve the pattern and rate of regional and global tectonic motions (Thatcher 1986, Ward 1990). Aerial photographs and satellite imagery, too, provide instantaneous coverage of the present-day

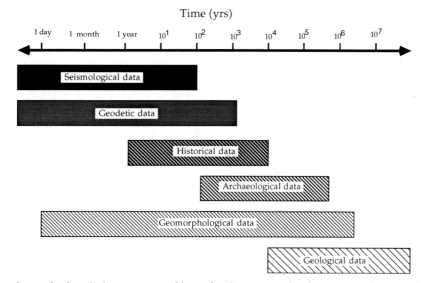

Fig. 18.4. Contrasting timescales in relation to types of investigative approaches in neotectonics. Developed from Vita-Finzi (1986).

Table 1. Techniques of acquiring local and regional neotectonic data

	Local	Regional
Geodetic	Triangulation tiltmeters, strain gauges, creepmeters	Global positioning system, very-long-baseline-interferometry, satellite laser ranging
Seismological	Microearthquake network	Worldwide seismological network
Remote sensing	Aerial photographs	Thermal imagery, radar imagery digital imagery
Geophysical	Electromagnetism	Seismic reflection, gravity anomalies
Geochemical	Electrical resistivity, radon emission	Hydrological monitoring
Historical	Eyewitness account, documentary evidence	Maps
Archaeological	Offset man-made structures	Prehistoric earthquake catalogues
Geomorphic	Fault-generated landforms	Morphometric indices, drainage patterns
Geological	Trenching	Palaeostress analysis

Earth's surface and, as a result, are useful reconnaisance tools, permitting local or regional mapping of neotectonic structures (Rothery & Drury 1984). More ambitious use of this data can be difficult, although the much improved resolution of SPOT images has allowed more sophisticated investigations (Peltzer et al. 1989). Where images of an area subject to neotectonic deformation are available for different periods of time, comparison between them can reveal specific morphological effects of that deformation (Machette 1987a).

The application of high-resolution instrumental techniques, however, is often limited by high cost or highly specific requirements, making less expensive and more flexible observational methods of neotectonic analysis more widely applicable. Such methods, which derive principally from investigation and measurement by the human observer, serve to reconstruct a less precise record of tectonic activity over a longer time span, and, therefore, are useful in evaluating the temporal distribution of neotectonic activity. They may also, however, elucidate gaps in the contemporary distribution of seismicity. In the Mediterranean region, for example, assessments of earthquake incidence derived from the historical accounts and archaeological finds have been invaluable in delimiting areas of recent tectonism which have not been identified from the pattern of instrumentally recorded seismicity (Fig. 18.1) (e.g. Ambraseys 1978, Nur 1991).

It is important to recognize that in neotectonics, a discipline which attempts to understand the spatial and temporal distribution of tectonic activity, both instrumental and observational data are important, particularly with uncertainty over the continuity of styles and rates of deformation over time.

DATING AND CHRONOLOGY

Although knowing precisely when a fault moved or a surface was tilted is critical in neotectonic studies, the evaluation of the methodologies, applicabilities, resolutions and limitations of the multitude of dating techniques applied to the study of active tectonics (Table 18.2) is outside the scope of this study. Instead, the reader is directed to excellent reviews of this subject by Pierce (1986) and Forman (1989). It is noteworthy, however, that the field of geochronology has undergone significant change in the last decade or so. Some well-established dating methods have witnessed important advances, such as the development of accelerator mass spectrometry (AMS), radiocarbon dating which requires a greatly reduced sample size and, therefore, has the potential for increased precision. In addition, other numerical dating techniques developed in the 1960s, such as thermoluminescence (TL), electron-spin resonance (ESR) and Uranium-series dating, have been refined and widely applied to neotectonic problems. Progressive improvement in the scope and limitations of established relative-dating methods, such as morphometric dating of fault scarps and soil development, has occurred along with a number of more recent developments, such as the use of cosmogenic isotopes and amino-acid geochronology, which show considerable potential as both relative and absolute dating methods, but which are only now being applied to neotectonic phenomena (Muhs 1987, Pavlich 1987).

It is also important to note that different dating methods date different aspects of neotectonic deformation. Some techniques, such as ESR, and palaeomagnetic dating of fault gouges (Grun 1992, Hailwood et al. 1992) may date actual fault movement, while other techniques, such as morphometric dating of fault scarps, and thermoluminescence of scarp-derived colluvium, record the time elapsed since a section of fault emerged at the ground surface (Forman et al. 1991). More commonly, however, dating determines the ages of geological or geomorphic units offset or unaffected by fault activity, and, therefore, provides maximum and minimum age limits, respectively, for fault activity (Allen 1986, Lagerback 1992). Where a number of such horizons have been disrupted to varying degrees by

Table 2. Summary of dating methods commonly used in neotectonic studies. The examples cited were mainly published after Pierce's (1986) review of the topic

	Dating method	Material dated	Recent studies in which method was used
Annual	Historical records	Eye witness accounts, historical documents, legends	Ambraseys & Melville (1982), Nur (1991), Ambraseys & Karcz (1992)
	Dendrochronology	Annual tree rings	Sheppard & Jacoby (1989), Van Arsdale et al. (1993)
	Varve chronology	Deformed lake sediments	Sims (1975), Adams (1982)
Radiometric	Carbon-14	Charcoal, peat and shells from offset datum horizon	Sieh & Jahns (1984), Wesnousky et al. (1991), Vita-Finzi (1987, 1992)
	Uranium-series	Fossil coral reefs, molluscs, bone, pedogenic carbonate	Edwards et al. (1988), Taylor et al. (1990), Muhs et al. (1992)
	Potassium – argon	K-bearing igneous rocks,	Martel et al. (1987)
	fission track	volcanic glass shards, zircon	Zeitler et al. (1982), Naeser (1987)
Radiologic	Uranium trend	Alluvium colluvium, loess	Muhs (1987), Muhs et al. (1989)
	Thermoluminescence	Quartz and felspar grains in fault scarp-derived colluvium	Forman et al. (1989, 1991), McCalpin & Forman (1991)
	Electron spin resonance	Quartz-bearing fault gouge	Schwarz et al. (1987), Brun (1992)
Process-oriented	Amino-acid racemization	Molluscs, skeletal material	Muhs (1987), Muhs et al. (1992)
	Lichenometry	Lichen on glacial moraines and fault scarps	Nikonov & Shebalina (1979), Hoare (1982), Wallace (1984b)
	Soil chronology	Degree of soil development on offset geomorphic surfaces	Machette (1978), Rockwell (1988), Harden & Matti (1989), Berry (1990)
	Rock weathering	Rock varnish, weathering rinds	Colman (1987), Harrington (1987)
	Slope morphometry	Fault scarps and offset erosional scarps	Wallace (1977), Nash (1980, 1986), Machette (1989)
Correlative	Stratigraphy	Scarp-derived colluvial wedges	Nelson (1992a)
	Archaeology	Pot sherds and other artifacts	King & Vita-Finzi (1984), Papanastassiou et al. (1993)
	Palynology	Offset glacial moraines	Schubert (1982)
	Palaeomagnetism	Fault gouge	Hailwood et al. (1992)

recurrent faulting, a chronology of deformation can be established that reveals the incidence of tectonic movements over time. This in turn, can be used to assess to what degree earthquake faulting is periodic, that is, whether it recurs over a regular and, therefore, predictable period of time. The most detailed geological record of earthquake activity, at Pallet Creek on the San Andreas fault, suggests that earthquake recurrence is not periodic but, instead, comprises clusters of earthquakes within a short time interval separated by much longer periods of quiescence (Sieh et al. 1989).

SEISMOTECTONICS

This section outlines the various ways in which earthquakes may be analysed to elucidate tectonic processes operating within a region. In particular, it focuses on two aspects important in constraining the characteristics of actively deforming regimes: *earthquake phenomenology* and *earthquake size*. While the former may provide an insight into the dynamic nature of tectonic deformation, the latter provides criteria with which the degree and style of deformation in different regimes can be compared.

Earthquakes rarely occur as isolated events but more commonly as part of a well-defined *earthquake sequence* (Scholz 1990). Earthquake rupture, for example, may be preceded by *foreshocks*, small earthquakes that occur in the immediate vicinity of the rupture zone and contribute to the nucleation of the main earthquake event or *mainshock* (Fig. 18.5a). Immediately following the mainshock, *aftershocks*

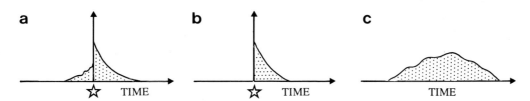

Fig. 18.5. Schematic diagram illustrating types of earthquake sequences. (a) Mainshock (star) with foreshocks and aftershocks. (b) Mainshock (star) followed by aftershocks. (c) Earthquake swarm. After Scholz (1990).

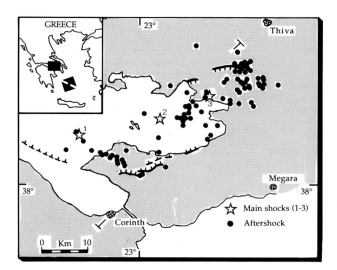

Fig. 18.6. Mainshocks and aftershocks during the 1981 Gulf of Corinth earthquake sequence. The aftershocks, which occurred mainly between ruptured fault segments, exhibit an eastwards migration over time, which mimics that of the changing locus of surface faulting, suggesting the gulf is extending towards the east. Outwards directed arrows in inset show regional stretching direction. After King *et al.* (1985).

occur throughout the rupture area (Figs 18.5a & b). These events, which are commonly an order of magnitude smaller than the mainshock, tend to nucleate at the edges of the rupture zone or at structural heterogeneities within it, and serve to relax stress concentrations produced during the main rupture (Fig. 18.6). In some settings, however, earthquakes are not preceded by foreshocks, but are characterized by an *earthquake swarm*; a protracted sequence of seismic events in which no single earthquake is dominant (Fig. 18.5c) (Sykes 1970).

In contrast to *primary earthquake ruptures*, which occur on the main fault responsible for the earthquake, rupture may occur on faults subordinate to the main fault, that is, *secondary earthquake ruptures*, or on an adjacent but unconnected fault which has been disturbed by earthquake waves radiating out from the main fault to produce *sympathetic earthquake ruptures* (Slemmons & dePolo 1986). While primary rupture can comprise a series of discrete subevents (each generating distinct seismic waves), ruptures on nearby, and often contiguous, fault surfaces generate closely spaced, but not concomitant seismic events, termed *compound earthquakes*. Depending on the number of mainshocks, such events may be called *earthquake doublets* or *multiplets*. During compound earthquakes, seismic activity can migrate along a fault zone indicating the direction of propagation, which may be *bilateral* or *unilateral*, depending on whether the rupture front propagates away from the nucleation point in one direction, or in both directions along the fault plane. During the 1981 Gulf of Corinth earthquake sequence, for example, both earthquake ruptures and their associated seismicity progressed eastward (Fig. 18.6), suggesting that the fault-bounded gulf was propagating in this direction (King *et al.* 1985).

In assessing the role of earthquakes in regional tectonic deformation, it is useful to discriminate between earthquakes in terms of their rupture dimensions, particularly *fault rupture area* (a), *fault rupture length* (L) and *fault rupture depth* (W) (Fig. 18.7). While the *schizosphere* (Scholz 1990) defines the portion of crust through which earthquakes may propagate, the *seismogenic layer* or *nucleation zone* defines a narrower range of depths at which earthquakes nucleate. The actual point of nucleation or initiation of earthquake rupture, the *hypocentre* (*focus*), can be located anywhere within the rupture surface, but the central point of that surface is defined by the *moment centroid*. The projection of the hypocentre on the Earth's surface is the *epicentre*. The contribution of *small earthquakes*, those which rupture entirely within the schizosphere, to the deformation of a region is generally regarded as minimal in comparison to that of *large earthquakes*, whose rupture dimensions equal or exceed the thickness of the seismogenic layer. It is important to note that the distinction between large and small earthquakes varies for different tectonic environments, since the greater the downdip thickness of the seismogenic layer, the greater the magnitude threshold between the large and small earthquakes (Scholz 1990).

Traditionally, earthquake size has been quantified in terms of *magnitude*, defined as the logarithm of the maximum amplitude of a specified seismic wave measured at a particular frequency and distance (100 km) from the epicentre. Different types of magnitude measurement are applicable to different conditions. *Local* or *Richter magnitude* (M_L), for example, records shallow-focus earthquakes within several hundred kilometres of the epicentre, while *surface wave magnitude* (M_S) records shallow-focus earthquakes at distances of over 600 km from the epicentre. For deep earthquakes, *body-wave magnitude* (M_b) must be used, while for earthquakes greater than $M_S = 7.5$, M_S must be described in terms of the *moment magnitude* (M_w), a measure based on the seismic energy

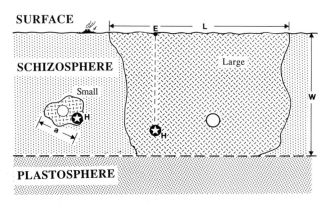

Fig. 18.7. Diagram illustrating the rupture dimensions — rupture area (a), rupture length (L) and rupture depth (W) — of small and large earthquakes, together with other earthquake parameters — hypocentre (H), epicentre (E), moment centroid (open circle). After Scholz (1990).

released by an earthquake. For most moderate, shallow-focus earthquakes, however, it is generally sufficient to take M_L, M_S and M_w to be roughly the same (Bullen & Bolt 1985). A more realistic measure of earthquake size, in the sense that it takes account of physical characteristics, is the *seismic moment* (M_o), defined as a function of fault rupture area ($L \times W$), the average displacement over the rupture surface (D) and the shear rigidity.

Estimates of earthquake size can be determined via a variety of procedures and are important in designating different levels of seismic risk. dePolo & Slemmons (1990) present an excellent review of concepts relating to determining earthquake size, but a few types of commonly used earthquake estimates are presented here. The largest earthquake determined for an area from the historical record is termed the *maximum historical earthquake*, while the largest earthquake expected to occur within a given time interval based on geological and seismological data is the *maximum earthquake*. The *maximum credible earthquake*, in contrast relates to the largest earthquake capable of occurring in an area or along a fault (California Division of Mines 1975). Often, a particular area or fault is expected to fail repeatedly with earthquakes of similar size and in a similar manner, giving rise to *characteristic earthquakes* (Schwartz & Coppersmith 1984). In addition to low-frequency seismic events, seismogenic areas are characterized by high-frequency, low-magnitude seismicity, *maximum background earthquakes*, often unrelated to specific faults.

At a given site, the first motions of P waves generated by earthquake rupture may be dilational or compressional, depending on the orientation of a fault and the sense of slip on it (also see Chapter 4). Analysis of first P-wave arrivals recorded by an array of seismometers permits the orientations of two orthogonal planes (*nodal planes*) separating first arrivals that are compressional or dilational to be estimated. The orientation of the line of intersection between these two planes, the so-called *B-axis*, is a direction normal to the direction of the slip on the fault plane. Discriminating between the *fault plane* and the *auxiliary plane*, at right angles to it, can be done on the basis of knowing which plane is geologically most likely to be the fault plane (see Fig. 4.35 in Chapter 4). When two dilational and compressional first P-wave arrival quadrants have been identified, it is known that the *P* or *compressional axis* will lie in the quadrants that have experienced first P-waves arrivals that are dilational, whereas the *T* or *tensional axis* lies in the quadrants that receive compressional P waves. Such analyses, which are commonly called *earthquake focal mechanisms*, or *fault plane solutions*, are generally made using lower-hemisphere equal-area (Schmidt/Lambert) 'stereographic' projections identical in character to those used in the analysis of observable structures. Although constructing and appraising the quality of focal mechanism from first P-wave arrivals is a job for the seismologist, appreciating their tectonic significance is relatively straightforward, permitting, for example, contrasting styles of deformation within complex tectonic regimes to be recognized (Fig. 18.8).

CHARACTERISTICS OF NEOTECTONIC STRUCTURES

Here we discuss those attributes of structures that are special to neotectonics in the sense that they are either restricted to the uppermost level of the crust and are, therefore, likely to be rapidly eroded, or mark the earliest stages of deformation that are likely to be erased after prolonged slip.

Neotectonic faults

The structural characteristics of neotectonic faults relate to a combination of *fault geometry*, that is, the along-strike and down-dip orientation of the active or principal fault plane, and *fault architecture*, the arrangement of neighbouring fault planes and fault rocks within the immediate vicinity of the main fault trace. Both these components, however, are products of the pattern of *fault propagation*, whereby an existing fault plane increases in surface area, and *fault migration*, whereby new linked fault planes develop in either the hangingwall or footwall of an existing fault plane (Stewart & Hancock 1991).

Studies of earthquake ruptures demonstrate that fault movement rarely occurs along discrete, continuous fault planes but instead it activates displacement on a broad network of interconnected and anastomosing surface fractures (Fig. 18.9). Within this relatively broad *fault zone*, the movement may be concentrated along a single narrow zone, the *principal displacement zone*, although after inspection in the field it can emerge that this too is a composite structure comprising several discontinuous *shear zones*, themselves consisting of alternating *cracks*, single narrow openings in the ground surface, and *pressure ridges*, localized topographic ridges a few metres high (Tchalenko & Ambraseys 1970). Thus what you see is a function of how closely you look. The discontinuous character of active fault zones, which is most apparent along strike-slip faults, where a map view is parallel to the direction of earthquake rupturing, can also be detected in dip-slip fault zones, and everywhere results in fault traces comprising a series of roughly continuous, linear strands or *fault segments*, separated by zones of fault intersection or termination that mark *segment boundaries*. *Segment terminations* mark points where a fault dies out along strike while *cross-fault intersections* occur where a fault segment cuts, or abuts, a pre-existing structure at a high angle to the fault zone (Bruhn et al. 1990).

Active fault traces are also characterized by marked changes in geometry or continuity. *Fault bends*, for example, mark abrupt deviations in the orientation of the fault trace, and where such a feature constitutes the junction between fault segments of contrasting

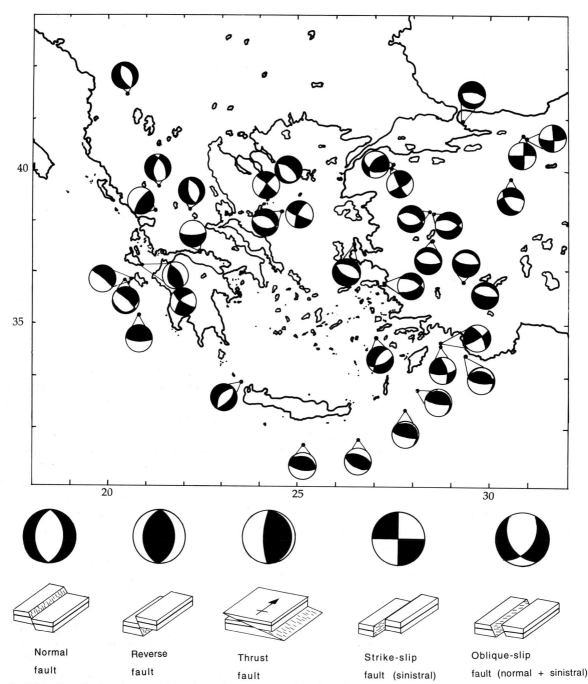

Fig. 18.8. Earthquake focal mechanisms from the Aegean region (based on McKenzie 1978) illustrating the contrasting styles of deformation within this complex tectonic regime. The lower part of the diagram shows the idealised focal mechanisms for different styles of fault movement.

orientation it is termed a *segment bend*. Similarly, the continuity of an active fault trace can be disrupted by *segment branches* where the main fault splits into two or more strands. In addition, abrupt en échelon *stepovers* (*jogs*) in the trace of the main fault may form *segment offsets* which in turn may be directly linked by a discrete *transfer fault* or an intervening zone of warping and subsidiary faulting, called a *relay ramp* (see Chapters 6, 11 and 12).

Geometric fault segmentation, that is, the subdivision of fault traces into a series of structurally and geometrically homogeneous units, is scale invariant because comparable features can be demonstrated at a variety of scales (Fig. 18.9) (Schwartz 1988, Stewart &

Hancock 1991). Recognition of geometric segments is important because earthquakes often nucleate and terminate in segment boundaries and during rupture may activate one or several geometric segments (dePolo et al. 1987, 1990). The degree to which discontinuities control rupture propagation, however, is related to their size and geometry relative to the *direction of fault rupture*. Generally, only the largest-scale discontinuities that define the longest segments along a fault zone will be important in influencing its rupture characteristics. In addition, although a similar range of segment-bounding discontinuities characterize strike-slip and dip-slip fault zones, along-strike discontinuities such as segment bends and offsets are likely to be less disruptive

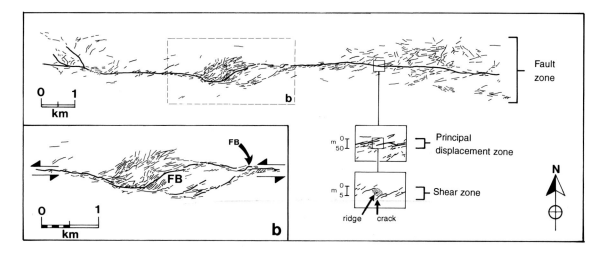

Fig. 18.9. (a) Map of fractures associated with the 1968 Dasht-e Baȳaz (Iran) earthquake highlighting the discontinuous character of surface faulting. The belt within which most of the fractures occur is called the fault zone, while the narrow zone within which most movement takes place is the principal displacement zone. This zone, however, commonly comprises several shear zones, each consisting of alternating cracks and ridges. (b) A magnified section of the M = 7.5 Dasht-e Baȳaz fault trace demonstrates the scale invariance of the pattern of surface ruptures, with features such as fault bends (FB) occurring at contrasting scales. After Tchalenko & Ambraseys (1970).

to the upward propagation of a fault rupture than to lateral rupture propagation (Wheeler 1987).

Different types of segment boundaries give rise to contrasting types of fault architecture. Where the fault planes on either side of a segment boundary are oriented in such a way that they can accommodate fault movement and not significantly disrupt the direction of fault rupture (defined by the *fault slip vector*), a *conservative barrier* (King & Yielding 1983) is formed in which deformation is concentrated along a relatively narrow zone. In contrast, where a fault plane on one side of a segment boundary cannot accommodate the sense of fault movement on an adjacent fault plane, the fault slip vector is not preserved and the intervening zone, a *non-conservative barrier* (King & Yielding 1983), transfers displacement by subsidiary faulting and volume changes. Bends or en échelon steps along strike-slip faults, for example, may be *dilational* or *antidilational* depending on whether fault movement is accommodated by area increase or reduction, respec-

tively (also see Chapter 12). At antidilational jogs or bends, slip transfer involves localised compression which is accommodated by the formation of *distributed crush breccias*, while at dilational jogs and bends, slip transfer involves localised extension, resulting in the formation of *implosion breccias*, exploded jigsaw-like breccia fragments (Fig. 18.10) (Sibson 1986). Where deformation is concentrated along a continuous fault surface, by contrast, progressive frictional wear results in the development of *attrition breccias* (Sibson 1986).

Contrasting styles of fault zone architecture are most apparent in normal faults, where uplift of the footwall preserves progressively deeper parts of the fault zone, exhuming the products of deformation processes that operated at different depths. The earliest stage of normal fault zone development occurs with the upward propagation of a *neoformed fault*, that is a fault that propagated to the surface through intact rock (Angelier 1989, Chapter 4 this volume). This propagation is accompanied by concomitant flexing of brittle strata

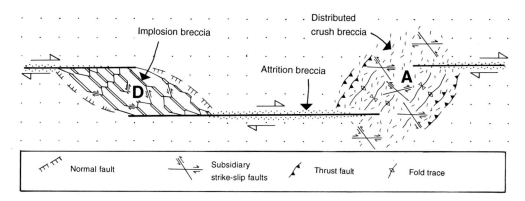

Fig. 18.10. Inferred pattern of fracturing and deformation along a segmented strike-slip fault (Sibson 1986). While dilational stepovers (D) are characterized by implosion breccias, and subsidiary normal faults and strike-slip faults, antidilational stepovers (A) are characterized by distributed crush breccias, thrusts, folds and subsidiary stike-slip faults. Along the main fault trace, frictional wear along the fault plane results in the formation of an attrition breccia.

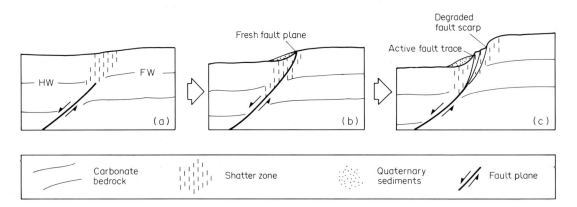

Fig. 18.11. Sequential evolution of a normal fault zone in the Aegean region. (a) Emergence of the fault at the surface is preceded by the development of a wide shatter zone in the dilational hangingwall (HW) in contrast to the footwall (FW). (b) Later increments of fault motion propagating a fault through this zone result in additional fracturing and brecciation. (c) Preferential migration of the active fault plane from the footwall towards the hangingwall of the main fault, intrafault-zone hangingwall-collapse, results in a stepped fault scarp and a layered arrangement of fault rocks. Modified from Hancock & Barka (1987).

ahead of the fault tip, resulting in *fault-precursor tipline folds* (*fault-propagation* or *drag folds*) and the formation of a *fault-precursor shatter zone*, comprising sub-vertical extension fractures that strike both parallel to and, to a lesser degree, perpendicular to the underlying fault (also see Chapter 5). At the fault tip itself, deformation ahead of a throughgoing fault plane is accommodated by intense localized fragmentation and brecciation of the adjacent shatter zone. Later propagation of a fault plane through this *fault-precursor breccia belt* results in the deformation being concentrated along the fault plane and the reconstitution of the fault-precursor breccia into a fine gouge or *attrition breccia*. On emergence at the surface, the attrition breccia forms an indurated carapace (*compact breccia sheet*) to the fault plane that protects the underlying fault-precursor breccia (*incohesive breccia belt*) from denudation (Stewart & Hancock 1990a). Comparable layered carapaces characterize the immediate footwall of normal faults with several kilometres of throw, but they comprise zones of extensive cataclasis and hydrothermal alteration (Bruhn et al. 1989). Such fluid involvement within active fault zones can give indications of the depth and extent of fracturing and brecciation. In fault zones in carbonate rocks, for example, vadose cements characterize those portions of the fault zone deformed above the water table, while phreatic cements delimit those portions deformed below the water table (Roberts et al. 1993). Overprinting of phreatic cement textures with vadose cement textures may mark the progressive uplift of the footwall portion of a fault across the water table.

An important control on the architecture of normal fault zones is the degree to which the active fault plane migrates within the zone. Studies of active normal faults demonstrate that the most recently activated or freshest fault plane within many zones commonly marks the junction between footwall bedrocks and hangingwall sediments, that is, it forms the uppermost fault plane within zones. Structural evidence suggests that this is an expression of progressive hangingwall-directed migration of fault activity within the fault zones, or *intrafault-zone hangingwall-collapse* (Stewart & Hancock 1988), which entrains slices of near-surface shatter belts and fault-precursor breccia into a footwall that becomes uplifted and denuded (Fig. 18.11).

Intrafault-zone hangingwall collapse is promoted by the tendency for normal faults to steepen near the surface in response to surficial reductions in confining pressure. As a consequence, on emergence, minor normal faults are commonly high-angle or nearly vertical fault planes, with geophysical and seismological evidence indicating that at depth they are moderately dipping (40–50°), essentially planar surfaces (see Chapters 6 and 11). At the highest crustal levels, such steepened faults commonly exploit surficial extension fractures and, as a result, are not everywhere accompanied by fault-precursor breccias. With prolonged fault movement, intrafault-zone hangingwall-collapse is generally inhibited as fault movement is increasingly concentrated on a principal fault plane along which there is intense deformation. In carbonate fault rocks, for example, this gives rise to stylobreccias, highly resistant sheets of intensely brecciated and recrystallized fault rock. According to Stewart & Hancock (1991), the degree to which intrafault-zone hangingwall-collapse is inhibited reflects the extent to which locally increased overburden pressures are maintained along a fault zone as a consequence of the occurrence of bedrock highs (salients) in its hangingwall. The locations of such *hangingwall salients* are generally coincident with sections where moderately dipping principal fault planes become emergent. Along such sections, the topographic expression of the emergent footwall block exhibits a ramp-like geometry, while along sections where salients are absent and intrafault-zone hangingwall-collapse has been encouraged, the footwall displays a step-like topography (see later).

Freshly activated or recently exhumed neotectonic fault planes are generally smooth polished surfaces or *slickensides* decorated by a suite of structural phenomena that indicate the direction and, occasionally, the sense of movement along them. The most commonly employed of these *kinematic indicators* are small-scale linear features oriented parallel to the direction of slip on the fault plane and often called *slickenside lineations* or *slickenlines* (Fig. 18.12) (also see Chapters 4 and 5). They include *striations* resulting from frictional wear along the fault and they range from fine *grooves* or *scratches* etched into the fault surface, to *debris streaks* or *trails* of comminuted material smeared along the fault plane (Means 1987). Such features do not generally define the sense of the slip direction, although deeper grooves (*tool tracks* Hancock & Barka 1987, *prod marks* Engelder 1974) can preserve a remnant of the tool (usually a small clast) in a pit at the distal end of the track. It has also been suggested that the lengths of tool tracks are equal to or less than the slip distance during a displacement event (Engelder 1974). Accretionary fibrous growth on fault surfaces may similarly record the sense of fault slip, because crystals (commonly quartz or calcite) grow parallel to the slip direction, forming asymmetric *accretion steps* whose steep risers face towards the direction of movement of the opposing block (Hancock 1985, Petit 1987, also see Chapters 4 and 5). Such contrasting features probably develop during different stages of an earthquake cycle, with frictional-wear phenomena characterizing episodes of coseismic slip, mineralization and fibre growth accompanying the continuous accumulation of interseismic slip (e.g. Power & Tullis 1989).

While different episodes of fault movement can form several generations of striations, commonly of contrasting orientations, larger-scale lineations may be inherent elements of the fault plane, and serve to record the original slip vector. Both *gutters*, narrow, flat-floored troughs, and *corrugations*, that is, large *undulations* of the fault plane, possess long axes which are several tens of metres in length and which are parallel to the general slip direction (Fig. 18.12) (Hancock & Barka 1987, Stewart & Hancock 1988, 1991).

Fault planes are also cut by a range of secondary fractures, whose trace on the fault surface are generally perpendicular to the slip direction (Petit 1987). Although the majority of these fractures are small-scale features that develop during the initiation and reactivation of the fault plane, for example *pinnate joints* (Hancock 1985, Petit 1987, and see Chapter 5), other well-defined fractures offset striated fault surfaces indicating, therefore, that their main development postdated the final increment of fault displacement. They most likely reflect stress release on fault emergence (Hancock & Barka 1987, Stewart & Hancock 1990).

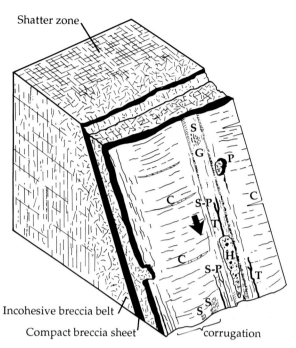

Fig. 18.12. Scaleless block diagram illustrating a range of structural phenomena associated with the uppermost levels of neotectonic normal fault zones. The uppermost layers within fault zones are underlain by alternating zone- and slope-parallel compact breccia sheets and incohesive breccia belts, within broader shatter zones. Corrugated fault planes bear a variety of slip-parallel and slip-normal structures. These structures include comb fractures (C), gutters (G), pluck holes (P), frictional-wear striae (S), slip-parallel fractures (SP), tool tracks (T), and trails of brecciated material derived from the hangingwall of the fault (Stewart & Hancock 1991).

While *comb fractures* cut fault planes at a high angle and form intersection lineations roughly normal to the slip vector, *slip-parallel fractures* give rise to intersection lineations which are roughly parallel to the slip vector (Fig. 18.13).

Neotectonic joints

Whereas it is possible to be certain that some faults are neotectonic it is more difficult to prove that a systematic joint is neotectonic. This is because such structures do not propagate at the surface and generally they do not cut surficial sediments even where they are indurated and displaced by active faults. Fissuring, rather than systematic jointing, of surficial sediments commonly accompanies earthquakes.

For a systematic joint (see Chapter 5) to be classed as neotectonic it must belong to the youngest joint set or system in a rock mass. Being the youngest joint set does not prove that it is neotectonic, but if the set being considered is not the youngest, the joints that it contains cannot be neotectonic. Joints cutting rocks deposited since the onset of the neotectonic phase, as determined from other evidence, should, of course, be neotectonic, and, thus, those cutting Neogene rocks, should be suspected of being neotectonic. In pre-Neogene rocks the time gap between the accumulation of the sequence

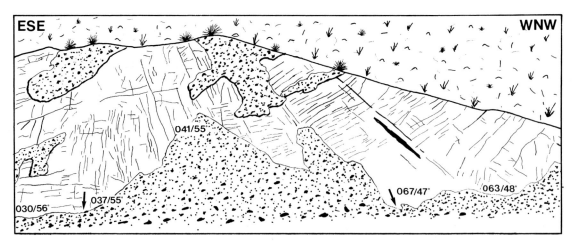

Fig. 18.13. Comb fractures and slip-parallel fractures cutting an exhumed normal fault plane near Manisa, western Turkey. The different orientations of fractures in both sets reflect a change in the slip vector (as defined by striations [arrows]), as the fault plane curves from a NNE–SSW to a ENE–WSW strike. Exhumed fault plane is approximately 6 m in height.

and the onset of the neotectonic phase is likely to be too long to argue from stratigraphic relationships alone that the joints could be neotectonic. However, this does not mean that old rocks are not cut by neotectonic joints.

The possibility that some of the joints cutting Palaeozoic sedimentary rocks in the Michigan-New York State region of northeastern U.S.A. was first raised by Holst & Foote (1981) and then by Engelder (1982). Although they demonstrated that a set of ENE-striking joints and the direction of the greatest horizontal stress (S_H) are parallel, the stratigraphic window between the youngest rocks (Palaeozoic) and the onset of the present mid-continent stress regime is large. The stratigraphic window was nearly closed by Bevan & Hancock (1986), Hancock (1987), Hancock & Engelder (1989) and Hancock (1991) who appraised the possibility that some regional joint sets cutting Cretaceous–Palaeogene rocks in southern England–northern France, Miocene rocks in the Ebro basin (Spain), and Mio-Pliocene rocks in eastern Arabia, are neotectonic. In northeastern North America, additional work on neotectonic joints by Gross & Engelder (1991) confirmed that S_H inferred from the joints correlated with the S_H direction determined from a variety of sources, including pop-ups. Engelder & Gross (1993) have also shown that although most non-systematic joints (Chapter 5) in New York State did not propagate parallel or subparallel to S_H, some mid-sections of curviplanar cross-joints linking palaeotectonic joints are locally subparallel to S_H, although where these joints abut neighbouring systematic neotectonic joints they are normal to them, and oblique to S_H.

The principal attributes of *neotectonic systematic joint networks*, and some inferences that follow from them, are as follows.

(1) Neotectonic joints belong to geometrically simple networks of single extension fracture sets or conjugate hybrid fracture sets (and spectra), everywhere dominated by a single set of systematic vertical fractures (Fig. 18.14) (see Chapter 5 for a general discussion of joints).

(2) Neotectonic joints strike in a uniform direction throughout intraplate regions of simple structure and great size (10 000 km² or more). This observation indicates that in such settings the joints are related to far-field (remote) stresses of regional tectonic significance. Even in areas of more complex structure, uniformly oriented sets of neotectonic joints are recognizable and cut older joints related to folds (Hancock & Engelder 1989).

(3) Neotectonic joints either strike parallel to the direction of S_H or, where they belong to conjugate sets or a joint spectrum, they can enclose an acute angle symmetrically about that direction as determined from independent geophysical evidence. Some 'predictions' of the direction of S_H made from neotectonic joints have been confirmed later by geophysical evidence (Hancock & Engelder 1989, Hancock 1991).

(4) Neotectonic joints are the youngest systematic joints in a rock mass although they can be cut by younger non-systematic joints. Relative age relationships between neighbouring systematic neo- and palaeo-tectonic joints can be determined where either neotectonic joints cut older joints (Fig. 18.15a), or they abut them (Fig. 18.15b).

(5) Veins parallel to exposed systematic neotectonic joints are absent or rare, suggesting that fluid pressures are unlikely to have been abnormally high during their development.

(6) Systematic neotectonic joints are multilayer fractures cutting several adjacent beds where they are competent, but rarely passing through interbedded incompetent layers (Figs 18.15a & b).

(7) Many neotectonic joints are evenly spaced fractures but some belong to narrow joint zones containing closely spaced joints separated by step-over zones (Figs 18.15a & b).

(8) Although single sets, or more rarely conjugate ones, characterize the systematic neotectonic joints described by Hancock & Engelder (1989), some

Neotectonics

networks comprise a grid-lock of coeval orthogonal extension fractures (Fig. 18.15c). Such *fracture gridlocks* are, however, common in palaeotectonic joint networks that were formed at depth, and have since experienced uplift and exhumation (see Chapter 5 for definition).

(9) Exposed neotectonic joints propagated within the uppermost 500–1000 m of the Earth's surface as a consequence of uplift permitting lateral relief, and unloading giving rise to stress release. Engelder (1982) reported borehole evidence indicating their absence at depths greater than about 500 m beneath the surface of the Appalachian Plateau. In this context, neotectonic joints are the youngest category of *unloading joints* (Engelder 1985). In some parts of the Ebro basin, neotectonic joints are absent in the highest and youngest layers that are of similar lithology to those only 50–100 m below them, and which contain well-developed systematic neotectonic sets (Hancock 1991). This observation suggests that some cover, to generate a confining pressure, is required for the propagation of shallow-formed neotectonic joints, and that they preferentially develop in near-surface channels.

A question that might be posed is: are neotectonic systematic joints being formed at depths greater than about 1000 m? Because rocks from such depths are not exposed until there has been significant uplift and exhumation, processes that take a few million years, there are, for obvious reasons, few relevant field observations. However, the report by Narr & Burress (1984) of fractures, subparallel to S_H, visible in cores from greater than 3000 m depth in North Dakota suggests that might be so. The influence of abnormally high fluid pressures could be responsible for forming hydraulically driven extension fractures at such depths. The common observation that palaeotectonic sets of small-scale fractures generally comprise both joints and veins is a pointer that jointing might occur simultaneously at two levels in the crust. 'Dry' *unloading joints* probably form in a shallow near-surface corridor, while 'wet' *tectonic joints* form at greater depths, where abnormally high fluid pressures drive fracture propagation (also see Chapter 5).

TECTONIC GEOMORPHOLOGY

Tectonic landforms

Tectonic landforms express a broad spectrum of topographic features that can be employed as indicators of the style, magnitude, and rate or timing of tectonic movements. In this discussion we discriminate between *primary tectonic landforms*, those formed as a direct result of surface displacement, and *secondary tectonic landforms*, a suite of geomorphological phenomena which, following their formation, have been offset, deformed, modified or preserved by subsequent tectonic activity. It is important to recognize, however, that the two classes of landforms can not often be readily distinguished, and both may preserve valuable records of tectonic acitivity within a region.

Primary tectonic landforms

Fault scarps are dislocations of the ground surface coincident with, or roughly coincident with, a fault plane (Stewart & Hancock 1990), and are the primary

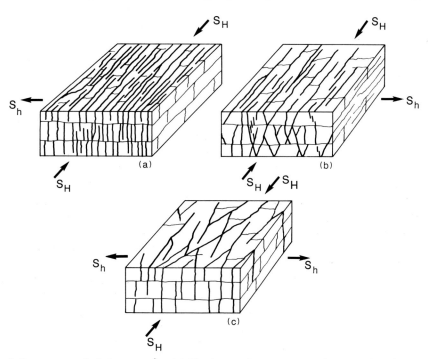

Fig. 18.14. Characteristic neotectonic joint systems. (a) Single set of systematic vertical extension joints (heavy lines) linked by non-systematic cross-joints (thin lines). (b) Spectrum of systematic vertical extension joints and steep hybrid joints linked by non-systematic cross-joints. (c) Spectrum of systematic vertical extension and hybrid joints linked by non-systematic cross-joints. S_H — greatest horizontal stress, S_h — least horizontal stress. After Hancock & Engelder (1989).

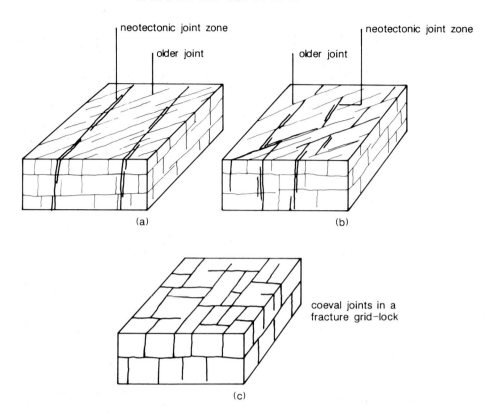

Fig. 18.15. Attributes of systematic neotectonic joints. (a) Multilayer neotectonic joint zones cutting palaeotectonic single-layer and multilayer joints. (b) Multilayer neotectonic joints and joint zones abutting palaeotectonic joints. (c) Coeval joints in a fracture grid-lock. Palaeotectonic joints commonly form such networks but they have been reported, only rarely, from neotectonic networks.

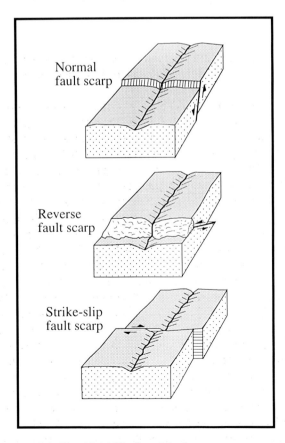

Fig. 18.16. Styles of fault scarps.

geomorphic expression of active faulting. The characteristics of fault scarps, however, vary with the amount and style of faulting and the nature of the surficial geology (Fig. 18.16). Scarps associated with strike-slip faulting or faulting in incohesive sediments, for example, are generally laterally impersistent, ephemeral centimetre-high scarplets which 'scissor' (alternate in their direction of facing) along strike, and which are best developed in areas of uneven topography (Sieh & Wallace 1987). More prominent scarps, however, generally result from dip-slip faulting (Fig. 18.16) of a level surface (piedmont) of Quaternary alluvium or colluvium at the bases of fault-generated bedrock *mountain fronts* or *range fronts*. Such *peidmont scarps* formed by normal faulting in the Basin and Range province, for example, generally possess heights of 0.5–4.0 m and extend laterally for several tens of kilometres as a result of $M_S = 6.0-7.5$ earthquakes (de Polo *et al.* 1987).

The tendency for fault reactivation to preferentially recur along the bedrock/Quaternary junction means that with repeated fault motion, single-event peidmont scarps evolve into older *multiple-event* or *composite fault scarps*, metres to a few tens of metres high. These, in turn, will develop into a few hundreds of metres high *bedrock fault escarpments*, and finally into *fault-generated range fronts*, major topographic features several hundreds of kilometres in length, and up to 1 km high. These forms are important regional expressions of active tectonism (Mayer 1986). While

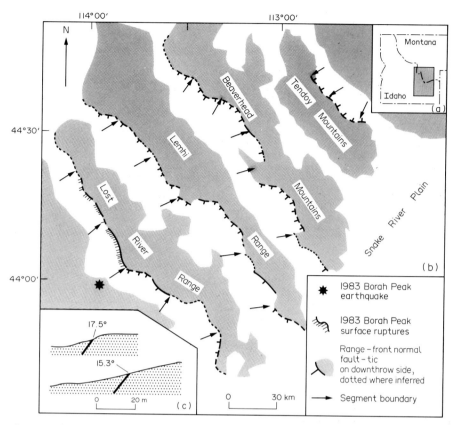

Fig. 18.17. (a) Location map. (b) Fault-generated range fronts in the northeasten Basin and Range province, western U.S.A. Where ranges are bounded by active normal faults, they comprise relatively linear sections, geometric segments separated by marked discontinuities or bends in the range front which constitute segment boundaries (arrows). Earthquake ruptures are confined to one or a number of geometric segments, as demonstrated during the 1983 Borah Peak earthquake when surface rupturing reactivated one segment, and induced secondary slip on an adjacent one. More subtle segment boundaries can be determined through the recognition of along-strike changes in the morphology of piedmont scarps along the range front base. (c) Characteristic scarp profiles. After Crone & Haller (1991).

active fault scarps along a range-front base commonly maintain sharp linearity, marked convexities and concavities in the strike of a front, termed *salients* and *embayments*, respectively, may be diagnostic of its deeper segmentation (Fig. 18.17). Similarly, segment boundaries are also commonly coincident with topographic lows along the range front, because they are areas of reduced fault slip (Fig. 18.18) (Peizhen Zhang *et al.* 1991).

Slope retreat and stream dissection of escarpments produces a series of spur ridges which abruptly terminate at the range-bounding fault in the form of *triangular facets*, moderately dipping inter-valley slopes. *Flights of faceted spurs* have been interpreted in terms of episodic uplift (Hamblin 1976) and, more recently, of distributed faulting within the range-bounding fault (Menges 1988). *Normal fault range-front* morphology is dependent on whether a *hangingwall salient* (see earlier) is present and leads to the formation of a *ramp-type scarp* or *range front*, or is absent, thus allowing a *stepped-type scarp* or *range-front* to evolve (Fig. 18.19a) (Stewart & Hancock 1990b, 1991). Range fronts formed by reverse or thrust faulting are less sensitive expressions of fault activity because they are the collapsed leading edges of overriding thrust sheets rather than the surface remnant of the range-bounding fault. *Thrust-front escarpments* tend to be more deeply embayed than their extensional equivalents (Fig. 18.19b).

Over time, fault scarps, particularly those developed in incohesive sediments, become increasingly degraded, with older scarps generally possessing a lower slope angle than younger equivalents (Fig. 18.20) (Bucknam & Anderson 1979). Fresh scarps in alluvium and colluvium, for example, generally possess a steep (>60°) free face separating undeformed *upper* and *lower original surfaces*. Over a period of 200–2000 years, however, free faces progressively retreat, shedding debris to form a moderately inclined (30°) *debris slope* and a low-angle (5–10°) *wash slope*, which accumulates at its base until the free face is buried (Wallace 1977). Subsequent degradation, between 2000 and 20,000 years, proceeds at a considerably slower rate and results in a time-dependent decline in slope angle, although after this period an equilibrium, and therefore time-dependent, form may be approached. Comparing the slope attributes of dated scarps to that of a fault scarp of unknown age is a useful relative age indicator. Such slope *morphometric dating* has been widely applied to fault scarps in the western U.S.A. where use of the methodology allows scarps to be assigned broad age categories (Nash 1980, 1986, Machette 1989).

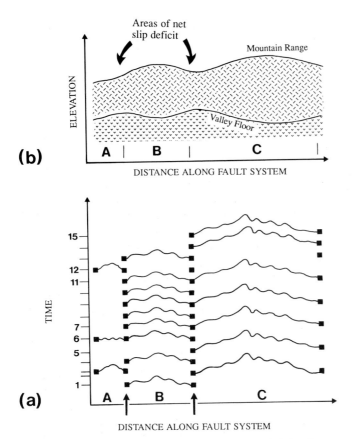

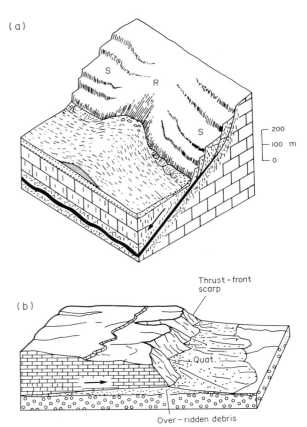

Fig. 18.18. (a) and (b) Variations in earthquake behaviour (slip) along normal faults are reflected in valley and range-front morphology. Although variable along a fault segment, fault slip generally decreases to zero at either end of earthquake ruptures. These end points often coincide with segment boundaries (solid squares), which if persistent over time accumulate net slip deficits. The resulting areas of reduced fault displacement can be manifest in the topography as areas of reduced elevation along the range front and elevated sections of the valley floor. Modified from Peizhen Zhang *et al.* (1991).

Fig. 18.19. Simplified block diagrams of range fronts developed from normal faulting in the Aegean region and thrust faulting in the Salt Ranges (Pakistan). (a) The contrasting appearances of normal fault range fronts in the Aegean region reflect the tendency of some fault zones to encourage hangingwall-directed splaying of the main fault producing a step-like front (S), while ramp-type range fronts (R) develop where such intrafault-zone hangingwall-collapse has been inhibited by bedrock highs (salients) in the hangingwall of the fault (Stewart & Hancock 1991). (b) The embayed topographic expression of thrust-front scarps reflects the fact that they are the collapsed leading edges of thrust sheets rather than the eroded remnant of the range-bounding fault itself (redrawn from Ramsay & Huber 1987).

Although *bedrock fault scarps* are less sensitive to erosion and, therefore, not amenable to this form of morphologic dating, they have a much longer preservation potential than colluvial or alluvial scarps and, hence, record a longer history of active tectonism. In less active tectonic terrains such morphologically prominent features may be the only surface record of Quaternary faulting (Mohr 1985). The degradation of such scarps is complicated by the often highly fractured and brecciated nature of the surrounding bedrock, which gives rise to highly variable scarp forms. *Aegean-type scarps*, for example, normal fault scarps underlain by carbonate bedrocks in the Aegean region possess complex slope-parallel layered architectures that comprise alternating zones of resistant and erodible fault rock. Variations in this architecture along strike, and within the fault zone, are responsible for marked changes in the rate and style of degradation giving rise to a varied assemblage of residual scarp forms. Because, in contrast to piedmont scarps, Aegean-type scarps do not evolve via an ordered and, therefore, predictable sequence of morphological change, other time-dependent criteria, such as the degree of karstification and the subaerial longevity of fault-plane phenomena, must be used to assess scarp age (Stewart 1993).

Fault scarps which are much modified by denudation and deposition so that the last remnants of the original tectonic forms have been removed, are called *residual fault scarps*. In the case of piedmont scarps, the transition to residual scarps occurs when the free face becomes fully buried, while in bedrock scarps it coincides with the complete removal of fault-plane phenomena from the fault surface (Stewart & Hancock 1990b). Although residual fault scarps can be rejuvenated by faulting, producing step-like *composite scarps*, with the complete cessation of faulting, erosional processes proceed unhindered and scarp form is controlled by the differential resistance to denudation of material on either side of a fault, and the former fault scarp becomes a *fault-line scarp*. The location of this scarp does not necessarily coincide with the fault trace, which is generally situated ahead of the scarp. Furthermore, scarp height is generally less than the neotectonic throw on the fault, and the facing direction of the scarp is not everywhere identical to the original sense of downthrow.

Geomorphic expression of buried faults

A variety of scarp-like features express *fault-propagation folds*, that is, surficially flexed strata ahead of a blind fault at depth (Fig. 18.21a). While surface flexures associated with normal faulting generally mimic the geometry of the underlying faults (Vita-Finzi & King 1985), reverse faulting warps the ground surface to produce a steep slope with an opposite sense of dip to the underlying fault plane (Fig. 18.21b). Although such *fold-limb scarps* (Stewart & Hancock 1990b) can be morphologically indistinguishable from true fault scarps, they contrast with them in that they are not coincident with the underlying fault plane on which fault motion occurred. Larger-scale fold structures can accompany recurrent reverse faulting. Discontinuous and ephemeral reverse scarplets formed during the 1980 El Asnam (Algeria) earthquake were superimposed upon a large monoclinal flexure which grew by coseismic uplift and flexural slip (King & Vita-Finzi 1981) (Fig. 18.21c), while Yeats & Lillie (1991) interpret the leading edge of the active Himalayan frontal fault as a broad anticlinal structure in the footwall of the Main Boundary thrust-front escarpment of a progressively subsiding 'dun' valley (Fig. 18.21d). Stein & Yeats (1981) demonstrated that in southern California it is possible to record the amplification of *active folds* by measuring the heights of Holcene marine terraces cut into the limbs of the Ventura anticline.

Palaeoseismic phenomena

In addition to the surface manifestation of active tectonism, the intricate architecture of the uppermost Quaternary sediments, as revealed by *trenching*; that is, excavation across a fault, or detailed geophysical imaging, can preserve a record of past tectonic activity. Indeed, such palaeoseismic phenomena constitute the primary evidence of large, prehistorical earthquakes in some areas such as inboard of subduction zone thrusts, and in intraplate settings where the responsible fault has little topographic or seismic expression.

One group of palaeoseismic phenomena are tectonic landforms which have become buried and fossilized in the geological record. Fault scarps, for example, degrade to produce a proximal wedge of poorly sorted colluvium, which becomes overlain by a distal wedge of stratified and well-sorted colluvium, following burial of the free face. *Buried fault scarps* and the stacking of scarp-derived proximal and distal colluvial wedges can produce a complex *fault-scarp stratigraphy* indicative of recurrent faulting (McCalpin 1987, Nelson 1992a). Where erosion has completely removed the fault scarp, strata offset by the causative fault can still be preserved below an erosional truncation, above which younger sediments are undeformed.

Seismic shaking, manifest at the surface in the form of *mud-volcanoes, sand-blow deposits* and possibly even *Mima mounds* (Berg 1990), can also have a subsurface expression. Thorsoon *et al.* (1986), for example, describe a complex deformation pattern associated with collapse fissures in northeastern U.S.A. which they interpret as products of the seismic ejection of water and liquefied sediment. Here, liquefaction has induced the 'rumpling' of fine sediments at depth and the injection of sand dykes, while fissure formation was accompanied by minor high-angle normal faulting and the parallel alignment of clasts in the vicinity of fissures. Comparable deformed sediments, generally called *seismites* can occur in lake sediments as a consequence of strong seismic shaking (Sims 1973, 1975, Scott & Price 1988).

Large-magnitude earthquakes can induce short-lived shoreline changes which may be preserved in the coastal stratigraphic record. Extensive coastal subsidence, transforming lowland areas into estuarine mud flats, has been documented following several large earthquakes. In these areas, alternating beds of estuarine muds and buried lowland soils indicate cycles of coseismic subsidence and postseismic shoaling (Atwater 1987, Atwater *et al.* 1992, Nelson 1992b). Similarly, extensive thin sand sheets burying lowland soils may be diagnostic of earthquake- or landslide-induced tsunami deposits (Dawson *et al.* 1988). Substantial vertical uplift of long stretches of coastline succeeds some large

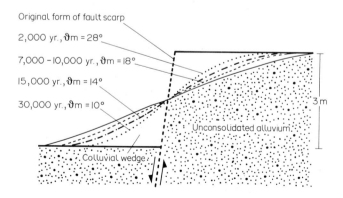

Fig. 18.20. Schematic diagram showing the original form and subsequent degradation of a fault scarp formed in unconsolidated alluvium. The maximum slope angles (in m) for the 2000, 7000–10 000 and 15 000 year old scarps are based on empirical data for scarps of known age in the eastern Basin and Range province (Bucknam & Anderson 1979), while the height in m for the 30 000 year old scarp is computed from a mathematical regression on empirical data. Vertical exaggeration is times two. Arrows denote direction of fault motion. After Crone & Haller (1991).

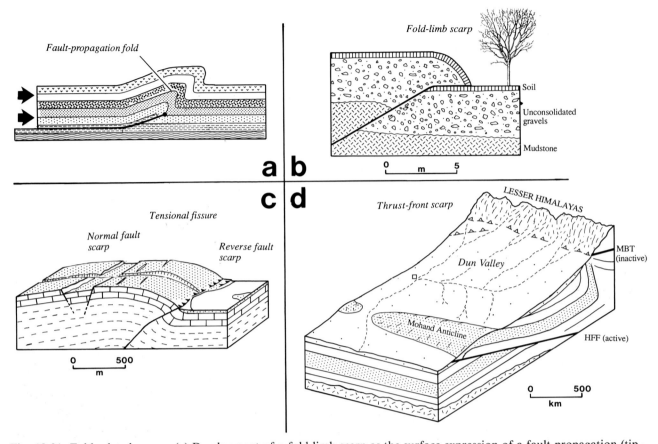

Fig. 18.21. Fold-related scarps. (a) Development of a fold-limb scarp as the surface expression of a fault-propagation (tip-line) fold related to a thrust at depth. Redrawn from Stein & Yeats (1989). (b) Fold-limb scarp formed by reverse faulting during the 1896 Riku-u earthquake (Japan). Redrawn from Matsuda *et al.* (1989). (c) Broad fold-limp scarp produced as a result of recurrent coseismic movements on a shallow thrust fault along the El Asnam (Algeria) fault zone. Redrawn from Philip & Meghraoui (1983). (d) Rising anticline developed at the leading edge of the Himalayan frontal fault (HFF), a major active fault occurring basinward of the now essentially inactive Main Boundary Thrust (MBT) which marks the highly embayed front of the Lesser Himalayas. Redrawn from Yeats & Lillie (1991).

subduction zone earthquakes that occurred on thrusts that emerge offshore.

Secondary tectonic landforms

A variety of landforms can reflect modification in response to active tectonics, with some landforms delimiting the active structures. *Fault-line valleys*, for example, occur where rivers are deflected along or restricted to the trace of a fault, while *sinkholes*, localized depressions in karstic terrains, and *fissure-ridge travertines* are aligned along faults (Altunel & Hancock 1993, Faccenna *et al.* 1993).

Geomorphic features which can be shown to have originally possessed a regular form and gradient over a wide area, or whose gradient or morphology clearly varies markedly from their modern-day equivalents, serve as important reference points from which to study rates of tilting, folding and faulting (Rockwell & Keller 1987). Evidence of horizontal ground movements can be derived from the lateral dislocation of a variety of geomorphic features: fluvial and marine terrace scarps, river channels, erosional ridge crests, the crests of glacial and alluvial fans, and lateral moraines (Fig. 18.22). *Offset stream channels* are particularly useful for recording fault displacements along strike-slip faults (e.g. Barka 1993), but generally provide only a minimum estimate of movement (Gaudemar *et al.* 1989) and can frequently be ambiguous or misleading unless care is taken to reconstruct the precise form of the palaeochannel (e.g. Wesnousky *et al.* 1990). The differential offset of linear landforms may delimit the timing and magnitudes of discrete faulting events. Figure 18.23, for example, shows a sequence of fluvial terrace scarps differentially offset by the Wellington fault, New Zealand, with older terraces recording progressively greater dextral displacements (Berryman & Beanland 1991). The pattern of offsets indicates that fault movement was characterized by dextral offsets of 3.4–4.7 m during the last five events, with a relatively constant slip rate, consistent with a characteristic earthquake model.

While vertical fault movements may be recorded in the dislocation of geomorphic surfaces such as erosion surfaces and alluvial plains, regional uplift and subsidence can be identified by a variety of geomorphic feartures of which two types, coastal and fluvial landforms are the most sensitive.

Coastal landforms are particularly valuable indicators of vertical tectonic movements because sea

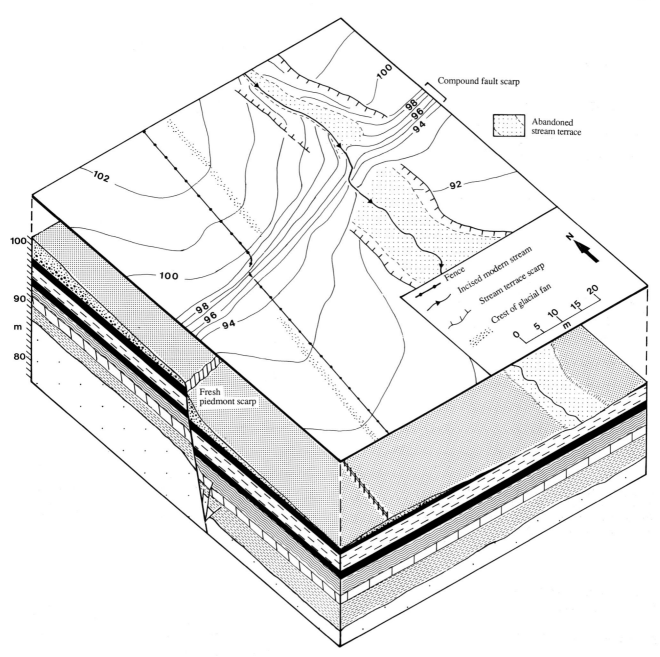

Fig. 18.22. Schematic diagram illustrating a range of geomorphological and geological features differentially offset by both lateral and vertical movement along an active fault. Older features are offset to a greater degree than younger features, while historical movement along the fault is demonstrated by a laterally dislocated fence, and a fresh piedmont fault scarp. Based loosely on Beanland & Clark (in press). Upper surface of diagram gives map view of underlying block.

level acts as a useful horizontal datum. While both depositional landforms, such as raised beaches, coral reefs, and marshes, and erosional forms, such as solutional notches and strandlines, may serve to indicate relative sea-level changes, it is often difficult to determine unequivocally whether their present-day position reflects eustatic or tectonic changes. Like sea level, the water table can also provide a useful, if ambiguous marker of vertical crustal motions. Some subsurface forms, such as *cave stalagmites* and *speleothem deposits*, for example, only accumulate where the water table has been lowered sufficiently for subterranean channels to become caves (Vita-Finzi 1986). In addition, the overprinting of vadose cements,

formed above the water table, onto phreatic cements, formed below the water table, in carbonate rocks can record the progressive uplift (or if the reverse, subsidence) of the area across the water table (Roberts et al. 1993). Occasionally, such features can provide sensitive indications of neotectonic deformation. Forti & Postpischl (1987), for example, correlate deviations in the growth direction of stalagmites with local historical earthquake activity.

River, as well as being passive markers of fault movements, can mimic the regional tectonic framework in terms of their drainage networks, as river systems adjust to active tectonics (Doornkamp 1986). The *footwall uplands* of active normal faults, for example,

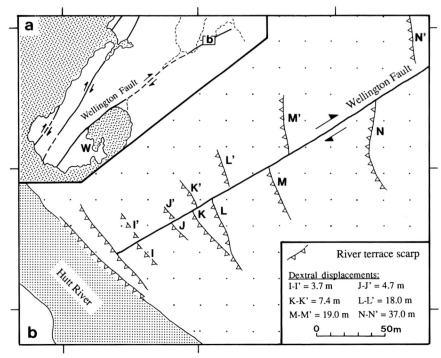

Fig. 18.23. (a) Location map. (b) Sequence of fluvial scarps (increasing in age from I to N) along the Hutt River, North Island, New Zealand, differentially offset by active strike slip movements along the Wellington Fault. From Berryman & Beanland (1991).

are commonly characterized by elongate drainage basins and a parallel arrangement of streams that flow down the face of active fault escarpments. Such *consequent streams* may extend directly to the sea (3 in Fig. 18.24) or feed *subsequent streams* or *axial channels* that flow along the base of the main fault scarps to produce a trellised drainage pattern (2 in Fig. 18.24). Uplift of the footwall uplands, however, is accompanied by backtilting of the fault block (Fig. 18.24), thereby deflecting the drainage divide towards the fault scarp and forcing streams to flow down the back slope as secondary consequents in a dendritic drainage pattern (4 in Fig. 18.24). Such backtilting is often responsible for drainage reversals, or the development of asymmetric meander belts (Leeder & Alexander 1987). *Unordered* or *insequent drainage* patterns largely develop where a chaotic pattern of faulting exists (1 in Fig. 18.24). Drainage basin characteristics can also reflect tectonic parameters. Drainage networks along active bedrock fault scarps, for example, are generally characterized by smaller basin areas and shorter principal stream lengths than networks along less active scarps, permitting the discrimination of different drainage domains, in which common lithological and tectonic attributes give rise to comparable drainage basin characteristics (Fig. 18.24) (Leeder *et al.* 1991).

Active folding and tilting, which in contrast to faulting can generate little surface expression, may be best detected in river morphology which is highly sensitive to changes in gradient (Adams 1980, Reid 1992). Rivers, may respond to changes in slope by adjusting their channel form, evolving from straight channels to meandering and then braided forms with increasing gradient. Figure 18.25, for example, shows the change in sinuousity of the Mississippi River as it traverses a number of centres of crustal uplift (Schumm 1986). More localized changes in the gradient along rivers, best expressed as discontinuites in their longitudinal profiles, can mark lithological boundaries or tectonic structures, or they might be *knick points*, breaks in slope which migrate upstream in response to headward erosion. In addition, streams crossing an area of active upwarping can be contained in *incised channels*, inducing stream deposition (aggradation) upstream and entrenchment downstream, while a stream crossing an axis of downwarping may flood its banks and induce stream incision upstream, and aggradation downstream (Schumm 1986).

Landform assemblages and morphometric indices

The tendency for active deformation to involve both vertical and horizontal ground movement, together with the complexity of geomorphic phenomena, generally make individual landforms poor indicators of tectonism. Well-developed but superficial normal fault scarps formed during the 1980 El Asnam (Algeria) earthquake, for example, served to initially obscure the overall compressional nature of the event (Fig. 18.21c). Similarly, some streams crossing the San Andreas fault are offset in a left-lateral fashion, implying the opposite sense of offset to that generally recorded. They are the product of locally anomalous stream capture (Wallace 1968). As a consequence, the overall pattern of tectonic activity along a structure should ideally be defined with reference to an assemblage of tectonic landforms. The right-lateral character of the movement along the San

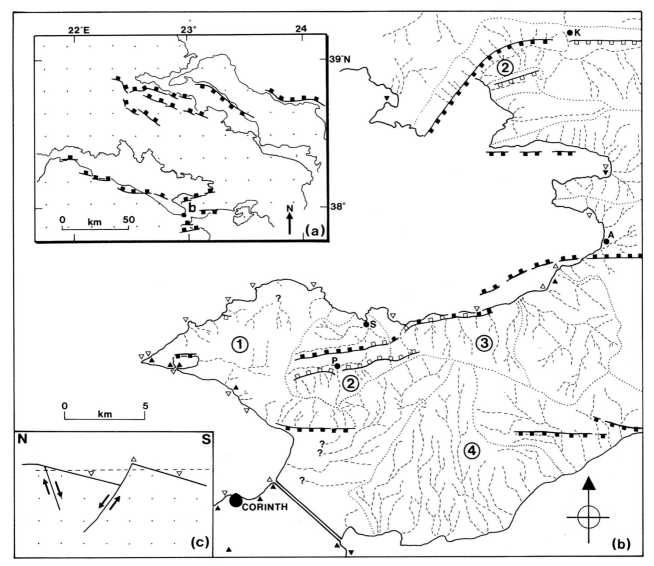

Fig. 18.24. (a) and (b) Simplified maps of the drainage network and its location in the eastern Gulf of Corinth, one of a series of NW – SE trending half-grabens in central Greece that are bounded by active normal faults (lines with solid boxes on downthrown side). Individual drainage domains, based on Leeder *et al.* (1991), are delimited by dotted lines while stream channels are shown as pecked lines. Sections of the faults reactivated during the 1981 Corinth earthquake are shown by lines with open boxes on the downthrown side, while open triangles denote the relative uplift or subsidence of the coast during these events. Solid triangles show the longer term uplift or subsidence of the coastline as determined by dating of raised shoreline fauna and historical and archaeological evidence. (c) Explanation of back tilting. See text for details.

Andreas fault at Wallace Creek (Fig. 18.26), for example, is expressed by the consistent sense of *offset of stream channels*, some of which have become completely removed from their upstream sections to form *beheaded channels*. Dating of material in these abandoned channels gives estimates of the slip rate along the fault, while the presence of young gullies unaffected by the fault, indicate that the no slip increments have occurred since the fault was last reactivated in 1857. Additional evidence for strike-slip motion is provided by a localized topographic depression or *sag pond* or *pull-apart basin* (see Chapter 12) within the overstep of adjacent fault strands, indicating localized extension consistent with right-lateral slip. Other geomorphic evidence in support of lateral displacements along the fault includes a *shutter ridge*; a topographic high displaced laterally to block drainage courses, while a prominent fault scarp marks the fault trace itself, being replaced towards the east by a broad fold-limb scarp formed by lateral bulging along the fault (Sieh & Wallace 1987).

This consideration of landform assemblages is also important where the pattern of tectonic activity revealed by young geomorphic features is at odds with that inferred from longer-lived forms. Coastal changes associated with the 1981 Gulf of Corinth (Greece) earthquake sequence, for example, indicate that areas in the footwall of an active normal fault were co-seismically uplifted while areas in its hangingwall subsided (Fig. 18.24b) (Jackson *et al.* 1982, Vita-Finzi & King 1985). Archaeological evidence and that of emergent marine strandlines and raised beaches on the onshore portions of the hangingwall block, however, attest to the longer-term uplift of these areas on the shoulders of offshore normal faults (Vita-Finzi & King 1985).

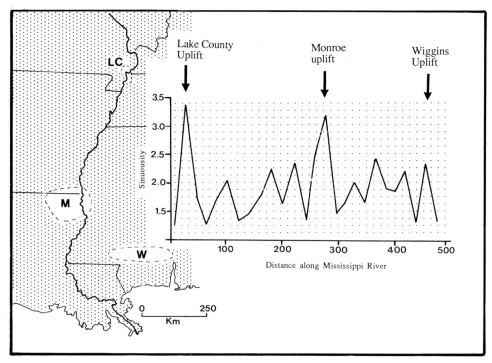

Fig. 18.25. Variations in sinuousity of the Mississippi River, southeastern U.S.A., with markedly higher values of sinuousity where the river crosses areas of active uplift. Redrawn from Schumm (1986).

The concept that the overall pattern of tectonism is best described with reference to a variety of geomorphic indicators is well illustrated by studies of fault-generated range fronts in the western U.S.A. (Rockwell et al. 1985, Keller 1986). According to these studies, the geomorphic attributes active mountain fronts are *linear mountain fronts, narrow V-shaped valleys* with steep longitudinal profiles, and *truncated ridge spurs* with well-defined *triangular facets*. Inactive fronts exhibit *embayed sinuous mountain fronts, broad U-shaped valleys* with gentle longitudinal profiles and highly *degraded spur ridges* (Bull 1987). Recognition of such phenomena permits a variety of simple morphometric indices to be erected that qualify the degree to which a front deviates from its ideal active form. Indices such as *mountain-front sinuosity* and *valley-floor width-valley height ratio* (Fig. 18.27), are relatively crude but useful measures for assessing spatial variations in the degree of tectonism within regions. Comparable approaches are now being applied to other tectonic settings (Wells et al. 1988).

POST-GLACIAL FAULTING

Post-glacial faulting relates to a particular tectonic setting whereby crustal movements occur during deglaciation and in response to crustal unloading. In

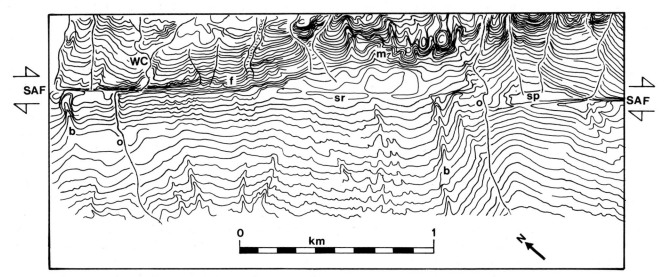

Fig. 18.26. Topographic map of the Wallace Creek (WC) area of the San Andreas fault (SAF). The map illustrates offset channels (o), beheaded channels (b), sag ponds (sp), shutter ridges (sr), a prominent fault scarp (f) and a fold-limb scarp (m) produced by recurrent right-lateral strike-slip movement along the fault. After Sieh & Wallace (1987).

Index	Definition	Derivation[1]	Measurement Procedure	Relation to tectonism	Source
Smf	Mountain-front sinuosity	Lmf / Ls	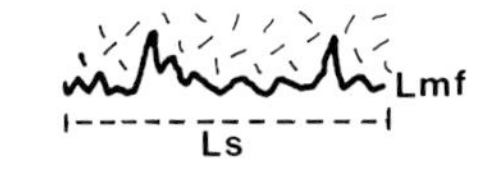	Linearity of mountain front indicates degree of active tectonism	Bull & McFadden (1977), Bull (1978)
Facet %	Mountain-front faceting	Lf / Ls	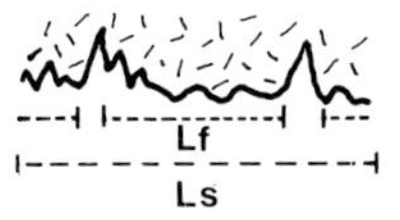	Active mountain fronts display prominent, large facets	Wells et al. (1988)
Vf	Valley floor-valley height ratio	$\dfrac{Vfw}{([Eld - Esc] + [Erd - Esc])/2}$	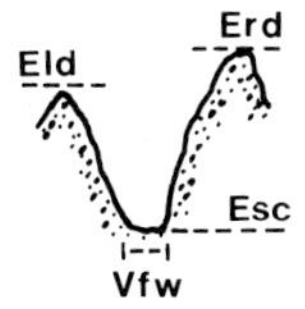	Active mountain fronts have V-shaped valleys and low Vf values	Bull & McFadden (1977), Bull (1978)
V ratio	Valley cross-section	Av / Ac		Low V ratios indicate V-shaped valleys and possible active uplift	Mayer (1986)
SL	Stream-gradient index	(ΔH / ΔL) x L		High SL values possible indicator of active mountain front	Hack (1973), Keller (1986)
K	Stream-profile concavity	area under longitudinal profile curve (shaded)	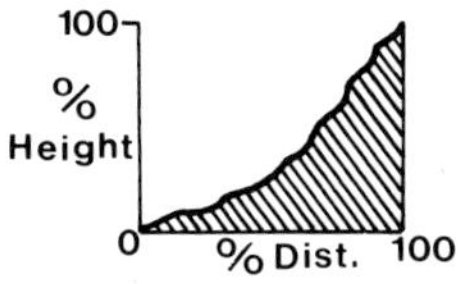	High K values indicate active base-level lowering	Shepard (1979), Wells et al. (1988)

[1] Lmf - length of mountain front along mountain-piedmont junction; Ls - straight-line length of mountain front; Lf - cumulative length of mountain front facets; Vfw - width of valley floor; Eld - height of left valley divide; Esc - elevation of valley floor; Erd - height of right valley divide; Av - area of valley in cross-section; Ac - area of semicircle with radius h; ΔH / ΔL - local stream gradient (height difference [ΔH] of stream along length of reach [ΔL]); L - total channel length from drainage divide to the centre of the reach.

Fig. 18.27. Geomorphic indices employed in the assessment of the degree of tectonic activity within a region, or along individual structures. Developed from Wells et al. (1988).

this respect, the term 'post-glacial' refers to those movements occurring after ice retreat and not to those post-dating a particular glacial stage. Thus, although the post-glacial faulting described here is largely attributed to Holocene crustal movements, post-glacial faults in general can be inherited from earlier, more pronounced, glacial episodes. Furthermore, post-glacial faulting is one extreme of a spectrum of crustal movements that accompany ice loading and unloading, and readers are directed to Mörner (1980) and Dawson (1992) for discussions on the broader effects of glacio-isostatic deformation.

Post-glacial fault movements are controlled by complex stress regimes encompassing both tectonically and glacially imposed stresses (Fig. 18.28) (Muir Wood 1989). In compressional settings, the presence of ice cover encourages local tectonic stability (Johnston 1989), but deglaciation, in contrast, induces a rapid decrease in the vertical stress relative to the horizontal stress, increasing the deviatoric stress and facilitating crustal failure. This, together with increased fluid pressures, permits movement along faults of varying orientation, although as fluid overpressuring dissipates, movement is increasingly accommodated along more favourably oriented faults. Movement commonly involves the reactivation of palaeotectonic faults rather than the initiation of neoformed faults, thus yielding a complex pattern of surface deformation (Fig. 18.29) within which the attitude of some structures can appear inconsistent with that of the contemporary regional stress field (Fenton 1991). The fact that in areas such as Sweden and Scotland, the majority of Holocene shoreline dislocations are located at or near the inferred stadial ice margin limits (Firth et al. 1993), further supports the contention that post-glacial structures are reactivated in response to deglaciation.

By their very definition, post-glacial faults may be morphologically distinct features since they commonly cut or offset terrain smoothed or stripped by glacial action (Lagerback 1992). Although it is generally to be expected that the delicate structure of morphological features are post-glacial because they are unlikely to survive erosion by large ice-sheets, many ice masses may have only a minor morphological impact (Lagerback 1992). Furthermore, similarity between post-glacial fault scarps and glacially plucked erosional scarps, requires that suspected fault scarps exhibit some characteristics diagnostic of their tectonic origin (Mohr 1985, Fenton 1991, Lagerback 1992). Firstly, *suspected fault scarps* or *ruptures* should be persistent over distances of at least 1 km, possess a marked topographic expression and disrupt pre-existing geomorphic features in a consistent fashion. Scarps should not merely be coincident with lithological banding or

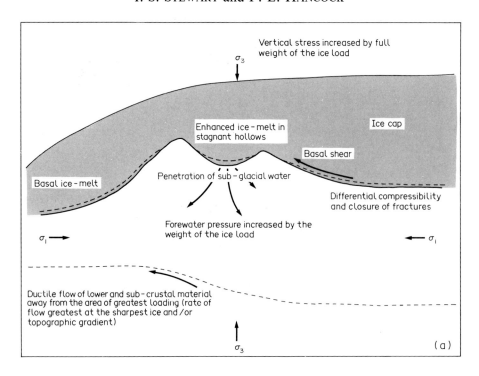

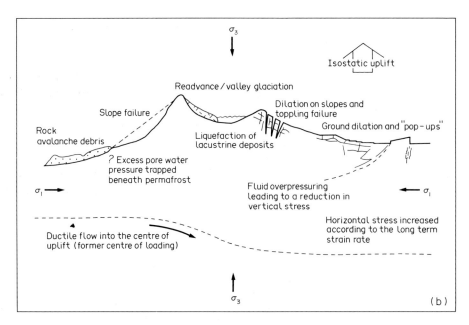

Fig. 18.28. (a) The effects of ice-cap loading on the crust in a compressive stress regime. (b) The effects of deglacial rebound on the crust in a compressive stress regime. Modified from Muir Wood (1989) and Fenton (1991).

layering within the underlying rock, and they should not display evidence of glacial modification. Instead, faults should disturb or displace late Quaternary landforms and deposits, and, ideally, occur in proximity to landforms such as eskers, meltwater channels and shorelines characteristic of a receding ice sheet. Trenching across suspected scarps is recommended to ensure that they are not the result of differential compaction or drape over existing scarps, and to determine the degree of modification by post-glacial erosion processes.

Although prominent, laterally continuous scarps, such as the 150-km-long Pärvie fault in Sweden, are clearly associated with post-glacial faulting (Lagerback 1992), post-glacial tectonic features may also be considerably more subtle. The Ungava fault rupture, for example, the first reported seismogeneic surface displacement in eastern North America, which formed during a magnitude (M_S) 6.3 earthquake in 1989 in northern Quebec, exhibits a highly variable morphology, being expressed by a discontinuous scarplet, surface cracks, deformed lake shorelines, sand volcanoes, freshly cracked boulders and a partly drained lake (Adams et al. 1991). It is likely that post-glacial ground ruptures commonly occur as chaotic fracture belts in which tensional and compressional features co-exist, comparable in style to the complex block movements that disrupt the shorelines (the so-called 'Parallel

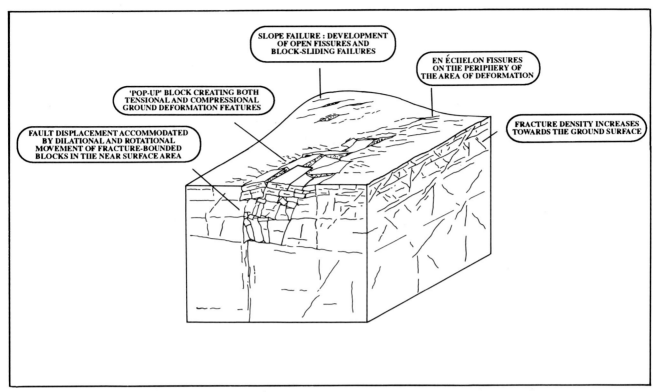

Fig. 18.29. Schematic block diagram showing the development of post-glacial surface deformation features due to movement on a non-visible fault strand. From Fenton (1991).

Roads') at Glen Roy (Sissons & Cornish 1992, Ringrose et al. 1991, Fenton 1991). Such a pattern of ground displacement has little distinct geomorphic expression, and, as a result, identification of post-glacial faulting relies on support from related deformation phenomena such as slope failures and the liquefaction of glacial and post-glacial sediments (Davenport & Ringrose 1987, Ringrose 1989).

NEOTECTONIC MOVEMENTS RELATED TO VOLCANIC ACTIVITY

Geodetic surveys of active volcanoes have revealed a range of different types of ground deformation associated with volcano-tectonic activity (Fig. 18.30) (McGuire & Saunders 1993). These deformations primarily reflect the pattern of magma storage and

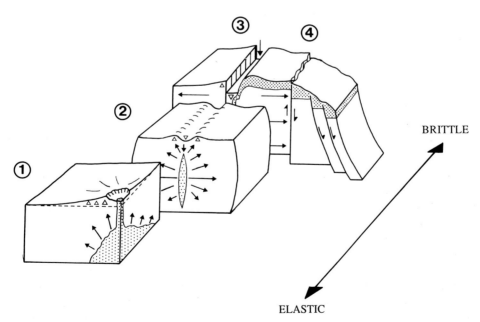

Fig. 18.30. Mechanisms responsible for ground deformation in active volcanic terrains. Movements may involve elastic deformation, such as the surface inflation/deflation above magma chambers (1) and localized ground effects above dykes (2), or brittle failure in response to dyke emplacement (3), or large-scale slope instability due to lava loading, oversteepening and faulting (4). Open triangles show sense of ground movement while arrows denote sense of internal deformation. Modified from McGuire et al. (1991).

movement within the volcano, but can be modified by the contemporary stress field within a volcano. This stress field can comprise a gravitational stress field in the upper part of the volcanic pile, superimposed on a regional tectonic stress regime within the subvolcanic basement (McGuire & Pullen 1989). In addition, such deformation is commonly accompanied by widespread earthquake activity related to stress release within the volcanic plumbing system.

Four broad categories of ground deformation can be recognized. Firstly, ground movements can occur by the inflation and deflation of the ground surface in response to the addition and later eruption of magma from a shallow reservoir. This mode of activity is well documented from the Kilauea volcano, Hawaii, where horizontal displacements of around 10 cm were detected during individual inflation/deflation cycles, although no net long-term deformation was recorded (Duffield et al. 1982b, Hoffman et al. 1990). A longer record is provided by marine strandlines at Pozzuoli in the Phlegran Fields caldera, Italy, which show alternating subsidence and uplift over the last 2000 years. Although occasionally characterized by periods of rapid uplift (350–500 mm/yr), the longer term rate of vertical motions is 12 mm/yr (Berrino et al. 1986 cited in Lajoie 1986), a similar order of magnitude to that determined from tilted Holocene shoreline terraces adjacent to the Yellowstone caldera, U.S.A. (Meyer & Locke 1986). Secondly, the subsurface horizontal propagation of dykes may be accompanied by the elastic warping of the ground surface to produce subsident troughs, with adjacent uplifted flanks, along the axis of the dykes (Murray & Pullen 1984). Thirdly, forceful emplacement and dilation of dykes can produce permanent vertical and horizontal displacements of fissure- and fault-bounded blocks (Murray & Pullen 1984, McGuire & Pullen 1989). Several metres of horizontal ground displacements have been recorded associated with episodes of dyke emplacement on Kilauea volcano, Hawaii, and Mount Etna, Italy (Swanson et al. 1976, McGuire et al. 1990). Fourthly, lava-loading of oversteepened and unbuttressed slopes, can result in the downslope sliding of tens to hundreds of metres' wide blocks of volcanic rock (Duffield et al. 1982a, Murray 1988, McGuire et al. 1991). Such large-scale, and often catastrophic, rock failure commonly occurs along faults and may be initiated by seismic activity, as demonstrated by the eruption of Mount St Helens. A number of these mechanisms are often concomitant.

The interaction between volcanic and tectonic structures in actively deforming terrains may produce a highly complex pattern of ground movements. At Mount Etna, for example, forceful emplacement of dykes in the summit conduit area is interpreted as leading to the seaward sliding of the unbuttressed eastern flank, which in turn induces episodic creep and microseismicity on a regional fault system (Stewart et al. 1993).

CONTEMPORARY TECTONIC STRESS FIELD IN THE BRITTLE UPPER CRUST

Knowledge of the orientation and relative magnitude of contemporary *in situ* stress in the *brittle upper crust* (*schizosphere*) is both of scientific interest and practical value, especially from the perspectives of earthquake hazard reduction and predicting the sub-surface flow of fluids. The *in situ* stress field is not entirely tectonic, some stresses are related to thermal effects, and others are related to local topography and superficial ground movements (e.g. Becker et al. 1990). Although some techniques permit the orientation of the three-dimensional stress field to be determined ($\sigma_1 > \sigma_2 > \sigma_3$) in many instances it is only possible to infer the orientations of the greatest horizontal stress (S_H) and the least horizontal stress (S_h). Where thrusting is occurring, S_H is σ_1 and S_h is σ_2, where there is strike-slip faulting S_H is again σ_1 but S_h is σ_3, and where there is normal faulting S_H is σ_2 and S_h is σ_3 (also see Chapter 4).

Methods of determining the orientations and/or magnitudes of stresses

Methods of inferring stress axis orientations and relative magnitudes from tectonic structures are discussed in Chapters 4 and 5. Table 18.3, which lists both geophysical and geological methods of determining stresses, also emphasizes how many more are available for the appraisal of the contemporary field, in contrast to ancient fields.

Although it is tempting to equate *P* and *T* axes determined from *earthquake focal mechanisms* (see p. 379) with σ_1 and σ_3 axes such a correlation is, as was argued by McKenzie (1969) and by Angelier in Chapter 4, not necessarily exact. Nevertheless, earthquake focal mechanisms remain a powerful way of roughly appraising the orientations of contemporary stress axes. For example, they were the most widely employed, but not always highest quality, data source used by Zoback (1992) during the preparation of the *World Stress Map* (WSM). Focal mechanisms used for compiling the WSM have their sources down to depths of about 20 km, that is, they reflect stresses in the brittle upper crust, as do those obtained from other sources used by Zoback.

A *borehole breakout elongation direction* is the horizontal direction of the ellipse which results when a formerly circular borehole becomes elliptical as a consequence of shear fracturing that leads to spalling of rock from the low pressure zones on opposite sides of the hole. Elongation directions can be determined using four-arm caliper tools, or an acoustic borehole televiewer, in holes up to 4–5 km deep; that is, at depths shallower than most earthquakes, but deeper than data derived from *in situ* or geological methods. The direction of S_H is taken to be at right angles to the elongation direction. Although the technique is fairly new (seee Cox 1970, Babcock 1978, Bell & Gough 1979, for key papers relating to the history of investigation

Table 3. Methods of determining neotectonic and palaeotectonic stresses from geophysical and geological data

Determination of contemporary stress axis orientations and/or magnitudes

Geophysical/geodetic
- earthquake focal mechanisms (fault plane solutions)
- borehole breakout elongation directions
- hydraulic fracturing (hydrofrac) tests
- *in situ* stress measurements
    - (1) strain relief (overcoring, e.g. the 'doorstopper' and triaxial methods)
    - (2) loading methods (e.g. the flatjack technique)
- ground- or satellite-based geodetic surveying

Geological/geomorphological
- attitudes of Quaternary faults on which the orientation and sense of the slip vector is known from one or more of the following: (1) striae or other lineations, (2) the displacement of geological markers, or (3) the offset of landforms
- axial trends of active folds ahead of blind thrusts
- alignments of active volcanic vents, fissures and dykes
- alignments of active fissures and ridges in Quaternary travertine deposits
- attitudes of neotectonic joints formed at shallow crustal depths
- long axes of quarry-floor buckles and other pop-ups
- offset coreholes

Determination of palaeostress axis orientations and/or magnitudes

Geological
- faults and brittle shear zones on which the orientation of a slip vector is known from one or more of the following: (1) a lineation, (2) the offset of geological markers, or (3) an array of en échelon cracks or veins within a shear zone
- aligned dykes in swarms
- roughly coeval sets of stylolitic seams and veins within the same volume of rock
- systems of single or orthogonal extension joints or systems of conjugate hybrid joints

and interpretation) it accounts for 28% of the data base used by Zoback (1992) during the preparation of the WSM.

The *hydraulic fracturing* or *hydrofrac testing* method of stress determination (e.g. Haimson & Fairhurst 1970) depends on inducing, by fluid pressurization, a hydraulic fracture to propagate from a portion of a borehole, and then measuring its azimuth by means of an impression packer. The strike of such an artificially induced hydraulic fracture, which propagates normal to σ_3, will be parallel to S_H. The method is difficult and expensive but yields high-quality results, including reliable estimates of stress magnitudes.

In situ methods, in contrast to the above techniques generally reveal stress patterns at much shallower depths, mainly in the range from a few metres to $1-2$ km. *Strain relief* methods depend upon measuring the elastic strains that small rock cores experience when they are removed from the ground. These techniques, including the triaxial and 'doorstoper' methods, are generally carried out in shallow boreholes ($1-100$ m), or mines, and, hence, the regional validity of stress orientations determined from them can be influenced by proximity to the ground, and by the small volume of rock that has been sampled. The flatjack method is a commonly used *loading method* of determining *in situ* stress. It relies on the principle of compensating an artificially induced local disturbance of the stress field by imposing an additional load, via a flatjack, a thin metal membrane filled with oil, that acts as a hydraulic fluid.

The determination of regional shortening and elongation directions from ground- or satellite-based geodetic surveying allows principal stress axis orientations to be estimated from these strains (e.g. Walcott 1978, Dragert 1986). Although geodetic techniques offer great potential they have been employed in only a few areas.

The most widely used geological method of determining contemporary stress directions and relative magnitudes is the analysis of *fault slip data* collected from active and other neotectonic faults. Wallace (1951) and Bott (1959) introduced the fundamental concepts linking slip direction and resolved shear stress, while Angelier (1984, 1990 and Chapter 4) has been a key worker from the perspectives of methodology and application. This powerful technique is available wherever slip vectors on a population of exposed faults within a relatively small volume of rock can be determined from lineations, displacement of marker lines, or offset landforms. In addition to recording geometric and kinematic field data it is also necessary to assess the evidence for the recency of faulting, using the range of observational criteria discussed earlier in this chapter and in Chapter 4. In addition, to employing active faults it is also possible to use the axial trends of *active folds* (Yeats 1986) that are amplifying ahead of blind thrusts, and are orientated roughly at right angles to the direction of S_H.

Aligned active volcanic vents and *linear trains of vents*, including *elliptical zones of eruptive vents* (Nakamura et al. 1978) are presumed to overlie

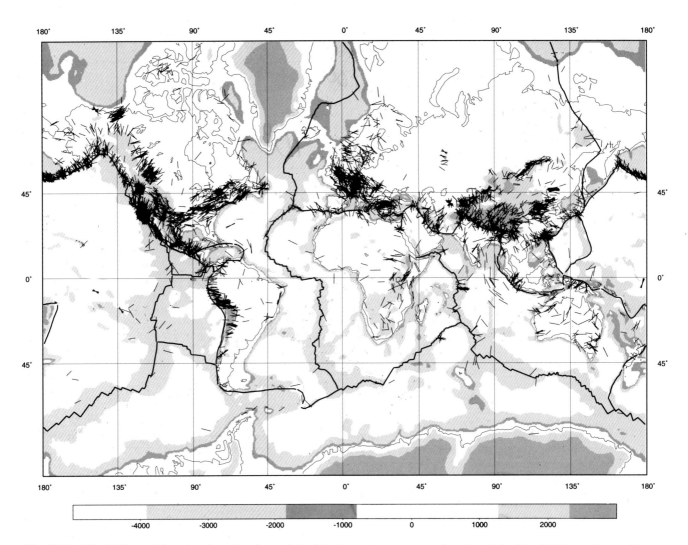

Fig. 18.31. World Stress Map showing directions of S_H. Line lengths are proportional to Zoback's (1992) quality rankings A–C. From Zoback (1992).

approximately vertical fissures that are extension fractures which propagated in the $\sigma_1 \sigma_2$ principal stress plane as a result of 'hydraulic' fracturing by magma. Thus, from such alignments the direction of S_H can be inferred to have been parallel to the trend of the alignment. *Active fissure* and *active dyke* segments in areas of present-day spreading strike roughly perpendicular to the direction of S_h (tensional), although an array of short fissure segments can be oblique to the direction of S_h as a consequence of the influence of older underlying fractures (Gudmundsson 1987). Steep normal faults, or *faulted fissures*, are surface expressions of some zones that have been extended by dyking.

Altunel & Hancock (1993) show that *travertine-filled fissures* within late Quaternary travertine fissure-ridges at Pamukkale (Turkey) strike at right angles to contemporary directions of horizontal extension. The geometry and morphology of travertine-filled fissures is similar to that of the volcanic fissures described by Gudmundsson (1987). The long axes of the ridges containing the fissures are also oriented at right angles to regional extension directions and thus can also be used for estimating stress directions where fissures are poorly exposed.

Although young faults are the most commonly used geological phenomena for determining contemporary stress axis orientations, *neotectonic joints* are of potential value because they are much more abundant and widespread than faults (Engelder & Hancock 1989, Hancock 1991). As explained in a previous section, the principal problem of using joints for this purpose is discriminating between neo- and palaeotectonic sets, and distinguishing between extension and hybrid joints (see Chapter 5).

Superficial structures known as *quarry-floor buckles* develop in some excavations after the removal of overburden (Adams 1982). Such quarry-floor buckles form most commonly where the excavation is sited in horizontal, well-layered rocks that are capable of spalling on horizontal partings, and where the magnitude of S_H is abnormally high. The long axes of quarry-floor buckles are generally oriented roughly at right angles to the direction of S_H, although their attitude may be partially controlled by pre-existing fractures. *Open-field pop-ups* are comparable geomorphic phenomena, as are *stream-valley* and *lake-floor pop-ups* (Wallach et al. 1993). *Post-glacial pop-ups* are also similar in appearance to quarry-floor buckles and are also products of stress release, but

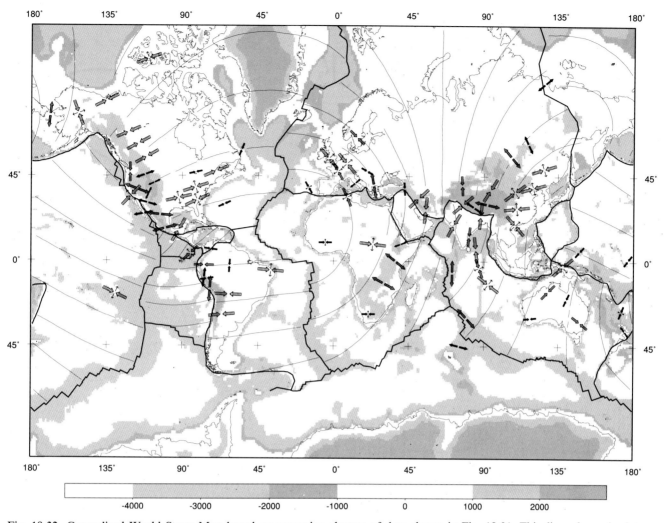

Fig. 18.32. Generalized World Stress Map based on averaging clusters of data shown in Fig. 18.31. Thin lines show absolute velocity trajectories for individual plates. A single set of inward-directed arrows indicates a thrust faulting regime, a single set of outward-directed arrows indicates a normal faulting regime, and sets of inward- and outward-directed arrows indicate a strike-slip faulting regime. The sizes of symbols are proportional to the number and consistency of the data. From Zoback (1992).

instead of the overburden that has been removed being rock it is ice which has retreated. The long axes of post-glacial pop-ups can reflect the orientation of either the contemporary stress field (e.g. Engelder & Sbar 1977, Gross & Engelder 1991), or that of older Quaternary stress fields related to isostatic rebound (Adams 1989, Adams & Bell 1991). In some areas of recent deglaciation, such rebound has reactivated pre-existing fractures, such as joints, on which a few millimetres of offset can be detected. A shortening direction calculated by Norris & Cooper (1986) from reactivated small fractures offsetting a recently deglaciated rock surface close to the Alpine fault in New Zealand is roughly parallel to that determined from earthquake focal mechanisms and geodetic measurements, made in the same region. Slip senses can also be determined from *offset coreholes* in, for example, road cuttings (Bell 1985), provided that the holes have not been moved down slope by purely superficial movements.

Variations in the contemporary stress field

The orientation of the contemporary stress field varies across the World's surface and with depth. Recording and appraising these variations within both the continents and oceans has been the focus of the *World Stress Map Project* (WSM) led by Mary Lou Zoback, and started in 1986 as part of the International Lithosphere Program. WSM project participants incorporated earthquake focal mechanisms, wellbore breakouts, fault slip data, volcanic alignments, hydraulic fracturing and overcoring information in the final compilation, assigning different quality levels to different classes of data source. Because stress patterns in *intraplate* (*midplate*) settings are simpler than those on, or adjacent to, plate boundaries there was a concentration on such settings in the project. The WSM project gave rise to a 'final' report (*J. geophys. Res.* 1992, **97** (B8), 11,703–12,013, 1987) which includes a coloured version of the World Stress Map at the scale of 1:4 million, and citations to the many relevant publications before and during the project's life.

Zoback (1992) states that most data sources suggest that principal stress planes are mainly vertical or horizontal and thus by plotting the direction of S_H the orientation of the field can be approximated on a map. Figures 18.31 and 18.32, which show a specific and a generalized WSM, respectively, illustrate how in some

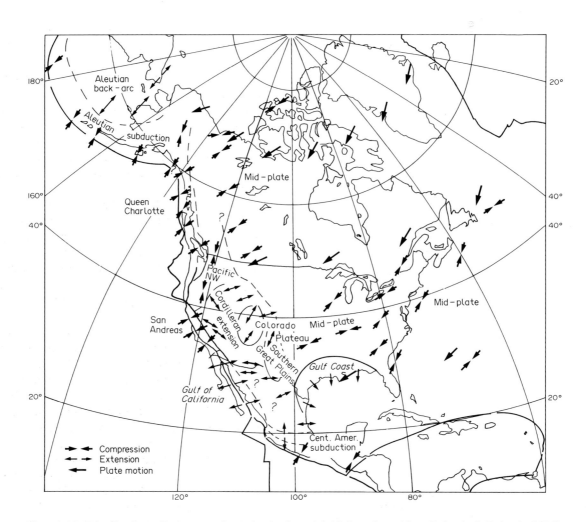

Fig. 18.33. Distribution of stress provinces in north and mid-America. After Zoback & Zoback (1991).

parts of the world S_H is aligned subparallel to the velocity trajectories of individual plates. Zoback (1992) identified from the 1:4 million stress map first- and second-order stress patterns on the basis of the size of the area within which the stress field in the brittle upper crust is uniformly oriented. *First-order stress provinces*, such as those of the midplate region of North America (Adams & Bell 1991, Zoback & Zoback 1991, Chapter 10) and western Europe (Brereton & Müller 1991) possess dimensions up to 5000 km across, that is, 20 to 200 times the thickness of the brittle upper lithosphere. Zoback (1992) notes that first-order stress provinces commonly coincide with physiographic provinces, especially in tectonically active regions. Furthermore, many of them are also characterized by S_H also being σ_1. Zoback (1992), in a critical review of the literature, considers, as have many others, that broad-scale stress fields are related to lithospheric plate driving forces, especially the negative buoyancy of subducting slabs, and the resistance to subduction. She proposes that uniformly oriented midplate fields are largely a product of ridge push and continental collision.

Second-order stress provinces are smaller, typically being less than 1000 km across, that is, roughly 5 to 10 times the thickness of the brittle upper lithosphere. Extensional regimes, in both the continents and oceans are mainly long and narrow and correspond to topographically high areas. According to Zoback (1992), they are attributed to the influence of buoyancy forces. The principal exceptions to the generalization about regions of active extension being long and narrow are the Basin and Range province in North America, and the Aegean province of the eastern Mediterranean region. Zoback (1992) thinks that other second-order stress patterns are related to flexure induced by thick sequences of sediments and post-glacial rebound, lateral density contrasts/buoyancy forces, and lateral strength contrasts. Dewey *et al.* (1986) have also postulated that loading by thrust sheets can flexurally depress adjacent forelands (also see Chapter 13). *Stress province boundaries* separating areas characterized by different directions of S_H are well established in some regions, notably in North America (Fig. 18.33).

Harper & Szymanski (1991) vividly illustrate how S_H directions inferred from wellbore breakout azimuths can, at any one site, vary markedly with depth (Fig. 18.34), and thus they emphasize that the averaging of such directional data could be misleading. By contrast, Zoback (1992) demonstrates that there is a good regional correlation between stress orientations

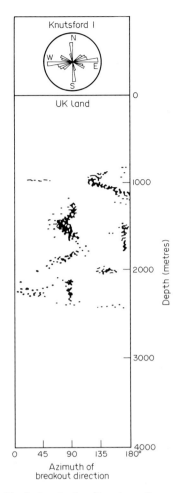

Fig. 18.34. Variation in the direction of greatest horizontal stress (S_H) with depth in the Knutsford Number 1 borehole (England) above. Azimuths of the breakout directions are shown in degrees from 0 to 180° below. After Harper & Szymanski (1991).

inferred from the near-surface and those calculated from earthquake focal mechanisms. An important, and tectonically significant, exception to this latter generalization occurs in the Jura Mountains region, where in the basement rocks beneath the basal décollement horizon, the direction of S_H is oriented uniformly NW−SE, in contrast to S_H directions in the folded cover rocks that are oriented normal to the trends of fold hinge lines, that is approximately E−W in the west and N−S in the north (Fig. 18.35) (Becker 1989). Brereton & Müller (1991) concluded that decoupling across an evaporite horizon at the base of the Mesozoic cover sequence, and above the Palaeozoic basement, is responsible for the variation of stress orientations with depth in this region.

Estimating absolute stress magnitudes is much more difficult than determining stress axis directions; for example, only about 4% of the data base for the WSM permitted such information to be inferred, mainly from overcoring data and hydrofracs. Estimating relative stress magnitudes, or the stress regime (extensional, strike-slip and thrust), is more straightforward and was achieved for a high proportion of the data points plotted on the WSM.

CONCLUSIONS

Our conclusions resemble the aspirations for the discussion of neotectonics that were set out at the start of the chapter.

(1) Investigating the distribution and history of neotectonic structures can benefit society from the perspective of increasing the data base that underpins seismic hazard assessment. Knowing the distribution of structures is clearly of significance in seismic zoning, and appraising the likely style and magnitude of faulting permits some aspects of ground rupture and shaking to be estimated.

(2) Trenching that allows a precise stratigraphy to be established via isotopic and other precise dating techniques, together with the morphometric analysis of fault scarps, are geological and geomorphic aspects of neotectonics that permit the amplitude and timing of faulting events to be estimated. These conclusions can then be used in modelling geological processes that are rate dependent.

(3) The study of exposed neotectonic structures reveals much about deformation processes that operate within the uppermost 1−2 km of the Earth's crust. Deformation products that characterize significantly deeper crustal levels can only be inspected by the field geologist after uplift and exhumation, and when the structures are extinct. Hence, direct correlation between geological structure and, for example, seismicity and absolute motions, are no longer possible.

(4) Geophysical, geodetic and geological techniques can be used to determine the orientations and relative magnitudes of contemporary stress axes. More rarely, absolute stress magnitudes can be estimated. Understanding the contemporary stress field allows a greater appreciation of lithosphere and sub-lithosphere processes.

(5) In regions of active volcanicity or post-glacial rebound, the investigation of coeval neotectonic structures enhances our understanding of processes that are usually studied by other geoscience specialists, with different concerns and methodologies.

REFERENCES

Adams, J. 1980. Active tilting of the United States mid-continent: geodetic and geomorphic evidence. *Geology* **8**, 442−446.

Adams, J. 1982. Stress-relief buckles in McFarland Quarry, Ottawa. *Can. J. Earth. Sci.* **19**, 1883−1887.

Adams, J. 1989. Postglacial faulting in eastern Canada; nature, origin and seismic hazard implications. *Tectonophysics* **163**, 323−331.

Adams, J. & Bell, J. S. 1991. Crustal stresses in Canada. In: *The Geology of North America, Decade Map Vol. 1, Neotectonics of North America* (edited by Slemmons, D. B., Engdahl, E. R., Zoback, M. D. & Blackwell, D. D.). Geological Society of America, Boulder, Colorado, 367−386.

Adams, J., Wetmiller, R. J., Hasegawa, H. & Drysdale, J. 1991. The first surface faulting from a historical earthquake in North America. *Nature* **297**, 213−214.

Aki, K. & Richards, P. 1980. *Quantitative Seismology:*

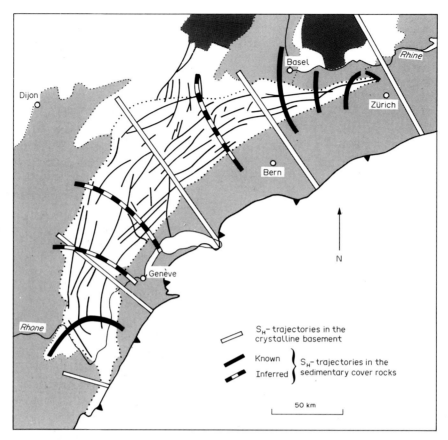

Fig. 18.35. Greatest horizontal stress trajectories in the Jura mountains region of France, Switzerland and Germany, inferred from a variety of geophysical and geological sources. Decoupling of trajectories is thought to occur at the level of an evaporite décollement horizon separating the deformed Mesozoic cover sequence from the 'rigid' Palaeozoic basement. Cenozoic sediments shown stippled, crystalline basement in black. After Becker (1989).

Theory and Methods. W. H. Freeman, San Francisco.
Allen, C. R. 1969. Active faulting in western Turkey. *Div. Geol. Sci. Calif. Inst. Tech.* **1577**, 1–32.
Altunel, E. & Hancock, P. L. 1993. Active fissuring and faulting in Quaternary travertines at Pamukkale, western Turkey. In: *Neotectonics and Active Faulting* (edited by Stewart, I. S., Vita-Finzi, C. & Owen, L. A.). *Z. Geomorph. Suppl. Vol.* **94**, 285–302.
Ambraseys, N. N. 1978. Middle East — A reappraisal of seismicity. *Q. J. Eng. Geol.* **11**, 19–32.
Ambraseys, N. N. & Jackson, J. A. 1990. Seismicity and associated strain of central Greece between 1890 and 1988. *Geophys. J. Int.* **101**, 663–708.
Ambraseys, N. N. & Karcz, I. 1992. The earthquake of 1546 in the Holy Land. *Terra Nova* **4**, 254–263.
Ambraseys, N. N. & Melville, C. P. 1982. *A History of Persian Earthquakes*. Cambridge University Press, Cambridge.
Angelier, J. 1984. Tectonic analysis of fault slip data sets. *J. geophys. Res.* **89**, 5835–5848.
Angelier, J. 1989. From orientation to magnitudes in palaeostress determinations using slip data. *J. Struct. Geol.* **11**, 37–50.
Atwater, B. F. 1987. Evidence for great Holocene earthquakes along the Outer coast of Washington State. *Science* **236**, 942–944.
Atwater, B. F., Jiménez-Núñez, H. & Vita-Finzi, C. 1992. Net Late Holocene emergence despite earthquake-induced submergence, south-central Chile. *Quat. Int.* **15/16**, 77–85.
Babcock, E. A. 1978. Measurement of subsurface fractures from dipmeter logs. *Bull. Am. Ass. Petrol. Geol.* **62**, 1111–1126.
Barka, A. A. & Kadinskey-Cade, K. 1988. Strike-slip geometry in Turkey and its influence on earthquake activity. *Tectonics* **7**, 663–684.
Barka, A. A. 1993. The North Anatolian fault zone. In: *Major Active Faults of the World: Results of IGCP Project 206* (edited by Bucknam, R. C. & Hancock, P. L.) *Ann. Tecton. Suppl. Vol.* **4**, 164–195.
Beanland, S. & Clark, M. C. In press. The Owens Valley fault zone, eastern California, and surface ruptures associated with the 1872 earthquake. *U.S. Geol. Surv. Bull.*
Becker, A. 1989. Detached neotectonic stress field in the northern Jura Mountains, Switzerland. *Geol. Rdsch.* **78**, 459–475.
Becker, A., Blenkinsop, T. G. & Hancock, P. L. 1990. Comparison and tectonic interpretation of in situ stress measurements by flatjack and doorstopper techniques in Gloucestershire, England. *Ann. Tecton.* **4**, 3–18.
Bell, J. S. 1985. Offset boreholes in the Rocky Mountains of Alberta, Canada. *Geology* **13**, 734–737.
Bell, J. S. & Gough, D. I. 1979. Northeast-southwest compressive stress in Alberta: evidence from oil wells. *Earth Planet. Sci. Lett.* **45**, 475–482.
Berg, A. W. 1990. Formation of Mima mounds: a seismic hypothesis. *Geology* **18**, 281–284.
Berry, M. E. 1990. Soil catena development on fault scarps of different ages, eastern escarpment of the Sierra Nevada. *Geomorphology* **3**, 333–350.
Berryman, K. & Beanland, S. 1991. Variation in fault behaviour in different tectonic provinces of New Zealand. In: *Characteristics of Active Faults* (edited by Hancock, P. L., Yeats, R. S. & Sanderson, D. J.). *J. Struct. Geol.* **13**, 177–190.
Bevan, T. G. & Hancock, P. L. 1986. A late Cenozoic regional mesofracture system in southern England and northern France. *J. geol. Soc. Lond.* **143**, 355–362.
Billiris, H., Paradissis, D., Veis, G. England, P., Featherstone, W., Parsons, B., Cross. P., Rands, P.,

Rayson, M., Sellers, P., Ashkenazi, V., Davison, M. Jackson, J. & Ambraseys, N. 1991. Geodetic determination of tectonic deformation in central Greece from 1900 to 1988. *Nature*, **350**, 124–129.

Bott, M. H. P. 1959. The mechanics of oblique slip faulting. *Geol. Mag.* **96**, 109–117.

Brereton, R. & Müller, B. 1991. European stress: contributions from borehole breakouts. *Phil. Trans. R. Soc.* **A337**, 165–179.

Bruhn, R. L., Yonkee, W. A. & Parry, W. T. 1990. Structural and fluid-chemical properties of seismogenic normal faults. *Tectonophysics* **175**, 139–157.

Bucknam, R. C. & Anderson, R. E. 1979. Estimation of fault-scarp ages from a scarp-height–slope-angle relationship. *Geology* **7**, 11–14.

Bull, W. B. 1978. Geomorphic tectonic activity classes of the south front of the San Gabriel Mountains, CA. *Unpubl. Final Rep., U.S. Geol. Surv. Contr.* No. 14-08-0001-G-394, 1–59.

Bull, W. B. 1987. Relative rates of long term uplift of mountain fronts. In: *Directions in Paleoseismology* (edited by Crone, A. G. & Omdahl, E. M.). *U.S. geol. Surv. Open-File Rep.* **87–693**, 192–202.

Bull, W. B. & McFadden, L. D. 1977. Tectonic geomorphology north and south of the Garlock Fault, California. In: *Geomorphology in Arid Regions* (edited by Doehring, D. O.). *Proc. 8th Ann. Geomorph. Symp.* State University of New York, Binghampton, 115–138.

Bull, W. B. & Wallace, R. E. 1986. Tectonic geomorphology. *Geology* **13**, 216.

Bullen, K. E. & Bolt, B. A. 1985. *An Introduction to the Theory of Seismology* (4th edn.) Cambridge University Press, Cambridge.

California Division of Mines and Geology 1975. Recommended guidelines for determining the maximum credible earthquake and the maximum probable earthquakes. *Calif. Div. Mines Geol. Note* **43**, 1.

Colman, S. M. 1987. Dating Quaternary deposits using weathering rinds and related phenomenon. In: *Directions in Palaeoseismology* (edited by Crone, A. G. & Omdahl, E. M.). *U.S. geol. Surv. Open-File Rep.* **87–673**, 65–69.

Cox, J. W. 1970. The high resolution dipmeter reveals dip-related borehole and formation characteristics. *11th Trans. Sp WLA, Ann. Logging Symp.* 1–25.

Crone, A. J. & Haller, K. M. 1991. Segmentation and the coseismic behaviour of Basin and Range normal faults: examples from east-central Idaho and southwestern Montana, U.S.A. In: *Characteristics of Active Faults* (edited by Hancock, P. L., Yeats, R. S. & Sanderson, D. J.). *J. Struct. Geol.* **13**, 151–164.

Das, S. & Aki, T. 1977. Fault planes with barriers: A versatile earthquake model. *J. geophys. Res.* **82**, 5658–5670.

Davenport, C. A. & Ringrose, P. S. 1987. Deformation of a Scottish Quatenary sediment sequence by strong earthquake motions. In: *Deformation of Sediments and Sedimentary Rocks* (edited by Jones, M. E. & Preston, R. M. F.). *Spec. Publs geol. Soc. Lond.* **29**, 299–314.

Dawson, A. G. 1992. *Ice Age Earth*. Routledge, London.

Dawson, A. G., Long, D. & Smith, D. E. 1988. The Storegga Slides: evidence from eastern Scotland for a possible tsunami. *Mar. Geol.* **82**, 271–276.

dePolo C. M. & Slemmons, D. B. 1990. Estimation of earthquake size for seismic hazards. In: *Neotectonics in Earthquake Evaluation* (edited by Krinitzsky, E. L. & Slemmons, D. B.). *Geol. Soc. Am. Rev. Engng. Geol.* **8**, 1–28.

dePolo, C. M., Clark, D. G., Slemmons, D. B. & Aymard, W. H. 1989. Historical Basin and Range Province surface faulting and fault segmentation. In: *Workshop on Fault Segmentation and Controls on Rupture Initiation and Termination* (edited by Schwartz, D. P. & Sibson, R. H.). *U.S. geol. Surv. Open-File Rep.* **89–315**, 131–162.

Dewey, J. F., Hempton, M. R., Kidd, W. S. F., Saroglu, F. & Sengör, A. M. C. 1986. Shortening of continental lithosphere: the neotectonics of Eastern Anatolia — a young collision zone. In: *Collision Tectonics* (edited by Coward, M. P. & Ries, A. C.). *Spec. Publs geol. Soc. Lond.* **19**, 3–36.

Doornkamp, J. C. 1986. Geomorphological approaches to the study of neotectonics. *J. geol. Soc. Lond.* **143**, 335–342.

Dragert, H. 1986. A summary of recent geodetic measurements of surface deformation on Vancouver Island. *Bull. R. Soc. N.Z.* **24**, 29–37.

Duffield, W. A., Stieltjes, L. & Varet, J. 1982a. Huge landslides blocks in the growth of Piton de la Fournaise, La Réunion, and Kilauea volcano, Hawaii. *J. Volcan. Geotherm. Res.* **12**, 147–160.

Duffield, W. A., Christiansen, R. L., Koyanagi, R. Y. & Petersen, D. W. 1982b. Storage, migration and eruption of magma at Kilauea volcano, Hawaii, 1971–1972. *J. Volcan. Geotherm. Res.* **13**, 273–307.

Edwards, R. L., Taylor, F. W., Chen, J. W. & Wasserburg, G. J. 1988. Dating earthquakes with high precision ^{230}Th ages of very young corals. *Earth Planet. Sci. Lett.* **90**, 371–381.

Engelder, J. T. 1974. Miscroscopic wear grooves on slickensides, indicators of palaeoseismicity. *J. geophys. Res.* **79**, 4387–4392.

Engelder, T. 1982. Is there a genetic relationship between selected regional joints and contemporary stress within the lithosphere of North America? *Tectonics* **1**, 161–177.

Engelder, T. 1985. Loading paths to joint propagation during a tectonic cycle: an example from the Appalachian Plateau, U.S.A. *J. Struct. Geol.* **7**, 459–476.

Engelder, T. & Gross, M. R. 1993. Curving cross joints and the lithospheric stress field in eastern North America. *Geology*, **21**, 817–820.

Engelder, T. & Sbar, M. L. 1977. The relationship between *in situ* strain relaxation and outcrop fractures in the Potsdam Sandstone, Alexandria Bay, New York. *Pageoph.* **115**, 41–55.

Faccenna, C., Florindo, F., Funicillo, R. & Lombardi, S. 1993. Tectonic setting and sinkhole features: case histories from the western central Italy. In: *Neotectonics — Recent Advances* (edited by Owen, L. A., Stewart, I. S. & Vita-Finzi, C.). *Quat. Proc.* **3**, 47–56.

Fenton, C. 1991. *Neotectonics and Palaeoseismicity in North West Scotland*. Unpubl. Ph.D. Thesis, University of Glasgow, 403 pp.

Firth, C. R., Smith, D. E. & Cullingford, R. A. 1993. Late Quaternary gacio-isostatic patterns for Scotland. In: *Neotectonics — Recent Advances* (edited by Owen, L. A., Stewart, I. S. & Vita-Finzi, C.). *Quat. Proc.* **3**, 1–14.

Forman, S. L. (ed.) 1989. Dating methods applicable to Quaternary geologic studies in the western United States. *Utah Geol. Miner. Surv.* **89–7**.

Forman, S. L., Machette, M. N., Jackson, M. E. & Maat, P. 1989. An evalution of thermoluminescence dating of palaeoearthquakes on the American Fork segment, Wasatch fault zone, Utah. *J. geophys. Res.* **94**, 1622–1630.

Forman, S. L., Nelson, A. R. & McCalpin, J. P. 1991. Thermoluminescence dating of fault-scarp-derived colluvium: deciphering the timing of palaeoearthquakes on the Weber Segment of the Wasatch Fault Zone, north-central Utah. *J. geophys, Res.* **96**, 595–605.

Forti, P. & Postpishl, D. 1986. Seismotectonics and radiometric dating of karst sediments. In: *Workshop on Historical Seismicity of the Eastern Mediterranean Region* (edited by Margottini, C. & Serva, L.). Italian Commission for Nuclear and Alternative Energy (ENEA) and International Atomic Energy Agency, Rome, 321–332.

Gaudemer, Y., Turcotte, D. L. & Tapponier, P. 1989. River offset across strike-slip faults. *Annls. Tecton.* **3**, 55–76.

Gross, M. R. & Engelder, T. 1991. A case for neotectonic joints along the Niagara Escarpment. *Tectonics*, **10**, 631–641.

Grün, R. 1992. Remarks on ESR dating of fault movements. *J. geol. Soc. Lond.* **149**, 261–264.

Gudmundsson, A. 1987. Tectonics of the Thingvellir fissure swarm, SW Iceland. *J. Struct. Geol.* **9**, 61–69.

Hack, J. T. 1973. Stream-profile analysis and stream-gradient index. *U.S. geol. Surv. J. Res.* **1**, 421–429.

Hailwood, E. A., Maddock, R. H., Ting Fung & Rutter, E. H. 1992. Palaeomagetic analysis of fault gouge and dating fault movement, Anglesey, North Wales. *J. geol. Soc. Lond.* **149**, 273–284.

Haimson, B. C. & Fairhurst, C. 1970. In situ stress determination at great depth by means of hydraulic fracturing. *11th Proc. U.S. Symp. Rock Mech.*, 559–584.

Hamblin, W. K. 1976. Patterns of displacement along the Wasatch Fault. *Geology* **4**, 619–622.

Hancock, P. L. 1985. Brittle microtectonics: principles and practice. *J. Struct. Geol.* **7**, 437–457.

Hancock, P. L. 1987. Neotectonic fractures formed during extension at shallow crustal depths. *Mem. geol. Soc. China (Taiwan)* **9**, 201–226.

Hancock, P. L. 1991. Determining contemporary stress directions from neotectonic joint systems. *Phil. Trans. R. Soc. Lond.* **A337**, 29–40.

Hancock, P. L. & Barka, A. A. 1987. Kinematic indicators on active normal faults in western Turkey. *J. Struct. Geol.* **9**, 573–584.

Hancock, P. L. & Engelder, T. 1989. Neotectonic joints. *Bull. geol. Soc. Am.* **101**, 1197–1208.

Harden, J. W. & Matti, J. C. 1989. Holocene and late Pleistocene slip rates on the San Andreas fault in Yucaipa, California, using displaced alluvial-fan deposits and soil chronology. *Bull. geol. Soc. Am.* **101**, 1107–1117.

Harper, T. R. & Szymaski, J. S. 1991. The nature and determination of stress in the accessible lithosphere. *Phil. Trans. R. Soc. Lond.* **A337**, 5–24.

Harrington, C. D. 1987. Application of rock-varnish dating of Quaternary superficial deposits in determining times of fault movements. In: *Directions in Paleoseismology* (edited by Crone, A. G. & Omdahl, E. M.). *U.S. geol. Surv. Open-File Rep.* **87–673**, 70–75.

Hills, E. S. 1956. A contribution to the morphotectonics of Australia. *J. geol. Soc. Aust.* **3**, 1–15.

Hoare, J. K. 1982. An evaluation of lichenometric methods in dating prehistoric earthquakes in the Tobin Range, Nevada. Unpubl. M.Sc. Thesis. San Francisco State University, 45 pp.

Hoffman, J. P., Ulrich, G. E. & Garcia, M. O. 1990. Horizontal ground deformation patterns and magma storage during the Puu Oo eruption of Kilauea volcano, Hawaii: episodes 22–42. *Bull. Vulcan.* **52**, 522–531.

Holst, T. B. & Foote, G. R. 1981. Joint orientation in Devonian rocks in the northern portion of the lower peninsula of Michigan. *Bull. geol. Soc. Am.* **92**, 85–93.

Jackson, J. A., Gagnepain, J., Houseman, G., King, G. C. P., Papadimitriou, P., Soufleris, C. & Viriex, J. 1982. Seismicity, normal faulting and geomorphological development of the Gulf of Corinth (Greece): the Corinth earthquakes of February and March 1981. *Earth Planet. Sci. Lett.* **57**, 37–397.

Johnston, A. C. 1989. Suppression of earthquakes by large continental ice sheets. *Nature* **303**, 467.

Keller, E. A. 1986. Investigation of active tectonics: use of surficial earth processes. In: *Active Tectonics* (compiled by Wallace, R. E.). National Academy Press, Washington, 136–147.

King, G. C. P. & Nabelek, J. 1985. Role of fault bends in the initiation and termination of earthquake rupture. *Science* **228**, 984–987.

King, G. C. P. & Vita-Finzi, C. 1981. Active folding in the Algerian earthquake of 10 October 1980. *Nature* **292**, 22–28.

King, G. C. P. & Yielding, G. 1983. The evolution of a thrust system; processes of rupture initiation, propagation and termination in the 1980 El Asnam (Algeria) earthquake. *Geophys. J. R. astr. Soc.* **77**, 915–933.

King, G. C. P., Ouyang, Z. X., Papadimitriou, P., Deschamps, A., Gagnepain, J., Houseman, G., Jackson, J.A., Soufleris, C. & Virieux, J. 1985. The evolution of the Gulf of Corinth (Greece): aftershock study of the 1981 earthquakes. *Geophys. J. R. astr. Soc.* **80**, 677–693.

Lagerback, R. 1992. Dating of Quaternary faulting in northern Sweden. *J. geol. Soc. Lond.* **149**, 285–292.

Lajoie, K. R. 1986. Coastal tectonics. In: *Active Tectonics* (compiled by Wallace, R. E.). National Academy Press, Washington, 95–124.

Lay, T. & Kanamori, H. 1981. An asperity model of great earthquake sequences. In: *Earthquake Prediction: An International Review* (edited by Simpson, D. & Richards, P.). *M. Ewing Ser.* American Geophysical Union, Washington D.C., **4**, 579–592.

Leeder, M. R. & Alexander, J. 1987. The origin and tectonic significance of asymmetric meander belts. *Sedimentology* **34**, 217–226.

Leeder, M. R., Seger, M. J. & Stark, C. P. 1991. Sedimentation and tectonic geomorphology adjacent to major active and inactive normal faults, southern Greece. *J. geol. Soc. Lond.* **148**, 331–344.

McCalpin, J. P. 1987. Geologic criteria for recognition of individual paleoseismic events in extensional environments. In: *Directions in Paleoseismology* (edited by Crone, A. G. & Omdahl, E. M.). *U.S. geol. Surv. Open-File Rep.* **87–673**, 102–114.

McCalpin, J. P. 1993. Neotectonics of the eastern Basin and Range province margin, western U.S.A. In: *Neotectonics and Active Faulting* (edited by Stewart, I. S., Vita-Finzi, C. & Owen, L. A.). *Z. Geomorph. Suppl. Vol.* **4**, 137–156.

McCalpin, J. & Forman, S. L. 1991. Late Quaternary faulting and thermoluminescence dating of the East Cache fault zone, north-central Utah. *Bull. seism. Soc. Am.* **81**, 139–161.

McGuire, W. J. & Pullen, A. D. 1989. Location and orientation of eruptive feeder-dykes at Mount Etna; influence of gravitational and regional tectonic stress regimes. *J. Volcan. Geotherm. Res.* **38**, 325–344.

McGuire, W. J. & Saunders, S. 1993. Recent earth movements at active volcanoes — a review. In: *Neotectonics – Recent Advances* (edited by Owen, L. A., Stewart, I. S. & Vita-Finzi, C.). *Quat. Proc.* **3**, 33–46.

McGuire, W. J., Pullen, A. D. & Saunders, S. J. 1990. Recent dyke-induced large-scale block movement at Mount Etna and potential slope failure. *Nature* **343**, 357–359.

McGuire, W. J., Murray, J. B., Pullen, A. D. & Saunders, S. J. 1991. Ground deformation monitoring at Mt Etna; evidence for dyke emplacement and slope stability. *J. geol. Soc. Lond.* **148**, 577–583.

McKenzie, D. P. 1969. The relation between fault plane solutions for earthquakes and the directions of the principal stresses. *Bull. seism. Soc. Am.* **59**, 591–601.

McKenzie, D. 1978. Active tectonics of the Alpine-Himalayan belt; the Aegean Sea and surrounding regions. *Geophys. J. R. astr. Soc.* **55**, 217–254.

Machette, M. N. 1978. Dating Quaternary faults in the southeastern United States by using calcic paleosols. *J. Res. U.S. geol. Surv.* **6**, 369–381.

Machette, M. N. 1987. Documentation of benchmark photographs that show the effects of the 1983 Borah Peak earthquake with some considerations for studies of scarp degradation. *Bull. seismol. Soc. Am.* **77**, 771–783.

Machette, M. 1989. Slope-morphometric dating. In: *Dating Methods Applicable to Quaternary Geologic Studies in the*

Western United States (edited by Forman, S. L.). Utah Geological and Mineral Survey **89–7**, 30–42.

Machette, M. N., Personius, S. F., Nelson, A. R., Schwartz, D. P. & Lund, W. R. 1991. The Wasatch Fault Zone — segmentation history of Holocene earthquakes. In: *Characteristics of Active Faults* (edited by Hancock, P. L., Yeats, R. S. & Sanderson, D. J.). *J. Struct. Geol.* **13**, 137–149.

Martel, S. J., Harrison, T. M. & Gillespie, A. R. 1987. Late Quaternary vertical displacement rate across the Fish Springs fault, Owens Valley Fault Zone, California. *Quat. Res.* **27**, 113–129.

Matsuda, T., Maizumi, T. & Nakata, T. 1989. Near-surface features of a thrust fault moved at the time of the 1896 Riku-u earthquake in Japan, revealed by excavation. In: *Worldwide Comparison of Characteristics of Major Active Faults*. Abstracts of the International Geological Correlation Programme Project 206 Meeting, Mammoth Lakes, California, U.S.A.

Mayer, L. 1986. Tectonic geomorphology of escarpments and mountain fronts. In: *Active Tectonics* (compiled by Wallace, R. E.). National Academic Press, Washington, 125–135.

Means, W. D. 1987. A newly recognized type of slickenside striation. *J. Struct. Geol.* **9**, 585–590.

Menges, C. M. 1988. Tectonic origin of facet benches on a normal fault-bounded mountain front: an alternative hypothesis. *Geol. Soc. Am. Abstr. Prog.* **20**, 215.

Menges, C. M. 1990. Late Quaternary fault scarps, mouintain-front landforms and Pliocene–Quaternary segmentation of the range-bounding fault-zone, Sangre de Cristo Mountains, New Mexico. *Geol. Soc. Am. Rev. Eng. Geol.* **8**, 131–156.

Meyer, G. A. & Locke, W. W. 1986. Origin and deformation of Holocene shoreline terraces, Yellowstone Lake, Wyoming. *Geology* **14**, 699–702.

Mogi, K. 1985. *Earthquake Prediction*, Academic Press, Tokyo.

Mohr, P. 1986. Possible Late Pleistocene faulting in Iar (West) Connacht, Ireland. *Geol. Mag.* **123**, 544–552.

Morisawa, M. & Hack, J. T. (eds) 1985. *Tectonic Geomorphology*. Allen & Unwin, Boston.

Mörner, N. (ed.) 1980. *Earth Rheology, Isostasy and Eustasy*. Wiley & Sons, Chichester.

Mörner, N. 1990. Neotectonics and structural geology; general introduction. *Bull. Int. Quat. Ass. Neotect. Comm.* **13**, 87.

Muhs, D. R. 1987. Applications of aminostratigraphy, strontium-isotope stratigraphy and uranium-trend dating to paleoseismology and neotectonics. In: *Directions in Paleoseismology* (edited by Crone, A. G. & Omdahl, E. M). *U.S. geol. Surv. Open-File Rep.* **87–673**, 76–83.

Muhs, D. R., Rockwell, T. & Kennedy, G. L. 1992. Late Quaternary uplift rates of marine terraces on the Pacific coast of North America, southern Oregon to Baja California Sur. *Quat. Int.* **15/16**, 121–133.

Muhs, D. R., Rosholt, J. N. & Bush, C. A. 1989. The uranium-trend dating method. Principles and applications for southern California marine terrace deposits. *Quat. Int.* **1**, 19–34.

Muir Wood, R. 1989. Fifty million years of 'passive margin' deformation in northwest Europe. In: *Earthquakes at North Atlantic Passive Margins: Neotectonics and Postglacial Rebound* (edited by Gregersen, S. & Basham, P. W.). Kluwer, Dordrecht, 7–36.

Muir Wood, R. & Mallard, D. J. 1992. When is a fault extinct? *J. geol. Soc. Lond.* **149**, 251–255.

Murray, J. B. 1988. The influence of loading by lavas on the siting of volanic eruption vents on Mt Etna. *J. Volcan. Geotherm. Res.* **35**, 121–139.

Murray, J. B. & Pullen, A. D. 1984. Three-dimensional model of the feeder conduit of the 1983 eruption of Mount Etna volcano, from ground deformation measurements. *Bull. Volcan.* **47**, 1145–1163.

Nakamura, K., Jacob, J. H. & Davies, J. N. 1978. Volcanoes as possible indicators of tectonic stress orientation — Aleutians and Alaska. *Pure Appl. Geophys.* **115**, 87–112.

Narr, W. & Burress, R. C. 1984. Origin of reservoir fractures in Little Knife Field, North Dakota. *Bull. Am. Ass. Petrol. Geol.* **68**, 1087–1100.

Naeser, C. W. 1987. The application of fission track dating to paleoseismology. In: *Directions in Paleoseismology* (edited by Crone, A. G. & Omdahl, E. M.). *U.S. geol. Surv. Open-File Rep.* **87–673**, 84–88.

Nash, D. B. 1980. Morphologic dating of degraded normal fault scarps. *J. Geol.* **88**, 353–360.

Nash, D. B. 1986. Morphologic dating and modeling degradation of fault scarps. In: *Active Tectonics* (compiled by Wallace, R. E.). National Academic Press, Washington, 181–194.

Nelson, A. R. 1992a. Lithofacies analysis of colluvial sediments — an aid in interpreting the recent history of Quaternary normal faults in the Basin and Range Province, western U.S.A. *J. sedim. Pet.* **62**, 607–621.

Nelson, A. R. 1992b. Discordant ^{14}C ages from buried tidal-marsh soils in the Cascadia subduction zone, southern Oregon coast. *Quat. Res.* **38**, 74–90.

Nelson, A. R. & Manley, W. F. 1992. Holocene coseismic and aseismic uplift fo Isla Mocha, south-central Chile. *Quat. Int.* **15/16**, 61–76.

Nikonov, A. A. & Shebalina, T.Yu. 1979. Lichenometry and earthquake-age determination in central Asia. *Nature* **280**, 675–677.

Norris, R. J. & Cooper, A. F. 1986. Small-scale fractures, glaciated surfaces, and recent strain adjacent to the Alpine fault, New Zealand. *Geology* **14**, 687–690.

Nur, A. 1991. And the walls come tumbling down. *New Sci.* **131**, 45–48.

Obruchev, V. A. 1948. Osnovnyje certy kinetiki i plastiki neotektoniki. *Izv. Akad. Nauk SSSR Ser. Geol.* **5**.

Papanastassiou, D., Maroukian, H. & Gaki-Papanastassiou, K. 1993. Morphotectonic and archaeological observations in the eastern Argive Plain (eastern Peloponnese, Greece) and their palaeoseismological implications. In: *Neotectonics and Active Faulting* (edited by Stewart, I. S., Owen, L. A. & Vita-Finzi, C.) *Z. Geomorph. Suppl. Vol.* **94**, 95–106.

Pavlich, M. J. 1987. Application of mass spectrometric measurement of ^{10}Be, ^{26}Al, ^{3}He to surficial geology. In: *Directions in Paleoseismology* (edited by Crone, A. G. & Omdahl, E. M.). *U.S. geol. Surv. Open-File Rep.* **87–673**, 39–41.

Pavlides, S. B. 1989. Looking for a definition of neotectonics. *Terra Nova* **1**, 233–235.

Peizhen Zhang, Slemmons, D. B. & Fengying Mao. 1991. Geometric pattern, rupture termination and fault segmentation of the Dixie Valley-Pleasant Valley active normal fault system, Nevada, U.S.A. In: *Characteristics of Active Faults* (edited by Hancock, P. L., Yeats, R. S. & Sanderson, D. J.). *J. Struct. Geol.* **13**, 165–176.

Peltzer, G., Tapponnier, P. & Armijo, R. 1989. Magnitude of Quaternary left-lateral displacements along the north edge of Tibet. *Science* **246**, 1285–1289.

Petit, J. P. 1987. Criteria for the sense of movement on fault surfaces in brittle rocks. *J. Struct. Geol.* **9**, 597–608.

Philip, H. & Meghraoui, M. 1983. Structural analysis and interpretation of the surface deformation of the El Asnam earthquake of October 10, 1980. *Tectonics* **2**, 17–49.

Pierce, K. L. 1986. Dating methods. In: *Active Tectonics* (compiled by Wallace, R. E.). National Academic Press, Washington, 195–214.

Power, W. L. & Tullis, T. E. 1989. The relationship between slickenside surfaces in fine-grained quartz and the seismic cycle. *J. Struct. Geol.* **11**, 879–893.

Prescott, W. H., Limowski, M. & Savage, J. C. 1981.

Geodetic measurement of crustal deformation on the San Andreas, Hayward and Calaveras Faults near San Francisco, California. *J. geophys. Res.* **86**, 10853–10869.

Ramsay, J. G. & Huber, M. J. 1987. *Folds and Fractures, Volume 2, Modern Structural Geology.* London, Academic Press.

Reid, J. B. JR. 1992. The Owens River as a tiltmeter for the Long Valley Caldera, California. *J. Geol.* **100**, 353–363.

Ringrose, P. S. 1989. Palaeoseismic (?) liquefaction event in late Quaternary lake sediment at Glen Roy, Scotland, *Terra Nova* **1**, 57–62.

Ringrose, P. S., Hancock, P. L., Fenton, C. & Davenport, C. A. 1991. Quaternary tectonic activity in Scotland. In: *Quaternary Engineering Geology* (edited by Foster, A., Culshaw, M. G., Cripps, J. C., Little, J. A. & Moon, C. F.). *Spec. Publs. Eng. Geol.* The Geological Society, London **7**, 679–686.

Roberts, G., Gawthorpe, R. & Stewart, I. 1993. Surface faulting within neotectonic normal fault zones: examples from the Gulf of Corinth fault system, central Greece. In: *Neotectonics and Active Faulting* (edited by Stewart, I. S., Vita-Finzi, C. & Owen, L. A.). *Z. Geomorph. Suppl. Vol.* **94**, 303–328.

Rockwell, T. 1988. Neotectonics of the San Cayetano Fault, Transverse Ranges, California. *Bull. geol. Soc. Am.* **100**, 500–513.

Rockwell, T. K. & Keller, E. A. 1984. Tectonic geomorphology, Quaternary chronology and paleoseismicity. In: *Developments and Applications of Geomorphology* (edited by Costa, J. E. & Fleischer, P. J.). Springer, Berlin, 203–239.

Rockwell, T. K., Keller, E. A. & Johnson, D. L. 1985. Tectonic geomorphology of alluvial fans and mountain fronts near Ventura, California. In: *Tectonic Geomorphology* (edited by Morisawa, M. & Hack, J. T.). Allen & Unwin, Boston, 183–208.

Rothery, D. A. & Drury, S. 1984. The neotectonics of the Tibetan Plateau. *Tectonics* **3**, 19–26.

Scholz, C. H. 1990. *The Mechanics of Earthquakes and Faulting.* Cambridge University Press, Cambridge.

Schubert, C. 1982. Neotectonics of the Bocono Fault, western Venezuela. *Tectonophysics* **85**, 205–220.

Schumm, S. A. 1986. Alluvial river response to active tectonics. In: *Active Tectonics* (compiled by Wallace, R. E.). National Academy Press, Washington, 80–94.

Schwartz, D. P. 1988. Geological characterization of seismic sources: moving into the 1990's. In: *Earthquake Engineering & Soil Dynamics II Proceedings; 2, Recent Advances in Ground Motion Evaluation* (edited by von Thun, J. L.). *Spec. Publs Am. Soc. Civil Engineers Geotech.* **20**, 1–42.

Schwartz, D. P. & Coppersmith, K. 1984. Fault behavior and characteristic earthquakes: examples from the Wasatch and San Andreas fault zones. *J. geophys. Res.* **89**, 5681–5689.

Schwartz, D. P. & Coppersmith, K. J. 1986. Seismic hazards: new trends in analysis using geologic data. In: *Active Tectonics* (compiled by Wallace, R. E.). National Academic Press, Washington, 215–229.

Schwarz, H. P., Buhay, W. M. & Grün, R. 1987. Electron spin resonance (ESR) dating of fault gouge. In: *Directions in Palaeoseismology* (edited by Crone, A. G. & Omdahl, E. M.). *U.S. geol. Surv. Open-File Rep.* **87–673**, 50–64.

Scott, B. & Price, S. 1988. Earthquake-induced structures in young sediments. *Tectonophysics* **147**, 165–170.

Shepherd, R. G. 1979. River channel and sediment responses to bedrock lithology and stream capture, Sandy creek drainage, central Texas. In: *Adjustments to the Fluvial System* (edited by Rhodes, D. D. & Williams, G. P.). *Proc. 10th Ann. Geomorph. Symp.* State University of New York, Binghampton, 255–276.

Sheppard, P. R. & Jacoby, G. C. 1989. Application of tree-ring analysis to paleoseismology: two case studies. *Geology* **17**, 226–229.

Shimazaki, K. & Nakata, T. 1980. Time-predictable recurrence model for large earthquakes. *Geophys. Res. Lett.* **7**, 279–282.

Sibson, R. H. 1985. Stopping earthquake ruptures at dilational jogs. *Nature* **316**, 248–251.

Sibson, R. H. 1986. Rupture interaction with fault jogs. In: *Earthquake Source Mechanics* (edited by Das, S., Boatwright, J. & Scholz, C.). *A.G.U. Geophys. Monogr.* **37**, 157–168.

Sieh, K. E. 1978. Prehistoric large earthquakes produced by slip on the San Andreas fault at Pallett Creek, California. *J. geophys. Res.* **83**, 3907–3938.

Sieh, K. 1981. A review of geological evidence for recurrence times of large earthquakes. In: *Earthquake Prediction: an International Review* (edited by Simpson, D. & Richards, P.). *M. Ewing Ser. American Geophysical Union, Washington D.C.* **4**, 209–216.

Sieh, K. E. & Jahns, R. H. 1984. Holocene activity of the San Andreas fault at Wallace Creek, California. *Bull. geol. Soc. Am.* **95**, 883–89.

Sieh, K. & Wallace, R. E. 1987. The San Andreas fault at Wallace Creek, San Luis Obispo County, California. *Geol. Soc. Am. Centennial Field Guide, Cordilleran Section*, 233–238.

Sieh, K., Stuiver, M. & Brillinger, D. 1989. A more precise chronology of earthquakes produced by the San Andreas fault in southern California. *J. geophys. Res.* **94**, 603–621.

Sims, J. D. 1973. Earthquake-induced structures in sediments of the Van Norman lake, San Fernando, California. *Science* **1/2**, 161–163.

Sims, J. D. 1975. Determining earthquake recurrence intervals from deformational structures in young lacustrine sediments. *Tectonophysics* **29**, 141–152.

Sissons, J. B. & Cornish, R. 1982. Rapid localised glacio-isostatic uplift at Glen Roy, Scotland. *Nature* **297**, 213–214.

Slemmons, D. B. 1991. Introduction. In: *The Geology of North America, Decade Map Vol. 1, Neotectonics of North America* (edited by Slemmons, D. B., Engdahl, E. R., Zoback, M. D. & Blackwell, D. D.). Geological Society of America, Boulder, Colorado, 1–20.

Slemmons, D. B. & dePolo, C. M. 1986. Evaluation of active faulting and associated hazards. In: *Active Tectonics* (compiled by Wallace, R. E.). National Academic Press, Washington, 45–62.

Stein, R. & Yeats, R. 1989. Hidden earthquakes. *Sci. Am.* **260**, 48–57.

Stewart, I. S. 1993. Sensitivity of fault-generated scarps as indicators of active tectonism: some constraints from the Aegean region. In: *Landscape Sensitivity* (edited by Thomas, D. S. G. & Allison, R. J.). Wiley, London, 129–147.

Stewart, I. S. & Hancock, P. L. 1990a. Brecciation and fracturing within neotectonic normal fault zones in the Aegean region. In: *Deformation Mechanisms, Rheology and Tectonics* (edited by Knipe, R. J. & Rutter, E. H.). *Spec. Publs geol. Soc. Lond.* **54**, 105–110.

Stewart, I. S. & Hancock, P. L. 1990b. What is a fault scarp? *Episodes* **13**, 256–263.

Stewart, I. S. & Hancock, P. L. 1991. Scales of structural heterogeneity within neotectonic normal fault zones in the Aegean region. In: *Characteristics of Active Faults* (edited by Hancock, P. L., Yeats, R. S. & Sanderson, D. J.). *J. Struct. Geol.* **13**, 191–204.

Stewart, I. S., McGuire, W., Vita-Finzi, C., Firth, C., Holmes, R. & Saunders, S. 1993. Active faulting and neotectonic deformation on the eastern flank of Mount Etna, Sicily. In: *Neotectonics and Active Faulting* (edited by Stewart, I. S., Vita-Finzi, C. & Owen, L. A.). *Z. Geomorph. Suppl. Vol.* **94**, 73–94.

Swanson, D. A., Duffield, W. A. & Fiske, R. S. 1976. Displacement of the south flank of the Kilauea Volcano: the result of forceful intrusion of magma into the rift zones. *U.S. geol. Surv. Prof. Pap.* **963**, 1–39.

Sykes, L. R. 1970. Earthquake swarms and sea-floor spreading. *J. geophys. Res.* **75**, 6598–6611.

Sykes, L. R. & Quittmeyer, R. C. 1981. Repeat times of great earthquakes along single plate boundaries. In: *Earthquake Prediction: an International Review* (edited by Simpson, D. & Richards, P.) *M. Ewing Ser. American Geophysical Union, Washington D.C.* **4**, 217–247.

Sylvester, A. G. 1986. Near-field geodesy. In: *Active Tectonics* (compiled by Wallace, R. E.). National Academy Press, Washington, 164–180.

Taylor, F. W., Edwards, R. L., Wasserburg, G. J. & Frohlich, C. 1990. Seismic recurrence intervals and timing of aseismic subduction inferred from emerged corals and reefs of central Vanuatu (New Hebrides) frontal arc. *J. geophys. Res.* **95**, 393–408.

Taymaz, T., Jackson, J. A. & Westaway, R. 1990, Earthquake mechanisms in the Hellenic Trench near Crete. *Geophys. J. Int.* **102**, 695–731.

Tchalenko, J. S. & Ambraseys, N. N. 1970. Structural analysis of the Dasht-e Bayāz (Iran) earthquake fractures. *Bull. geol. Soc. Am.* **81**, 41–60.

Thatcher, W. 1986. Geodetic measurement of active-tectonic processes. In: *Active Tectonics* (compiled by Wallace, R. E.). National Academy Press, Washington, 155–163.

Thorson, R. M., Clayton, W. S. & Seeber, L. 1986. Geological evidence for a large prehistoric earthquake in eastern Connecticut. *Geology* **14**, 463–467.

U.S. Environmental Protection Agency 1981. Standards applicable to owners and operators of hazardous waste treatment storage and disposal facilities. *Code Fed. Regul.* **40**, Parts 122.25 (1) and 264.18 (a).

U.S. Nuclear Regulatory Commission 1982. Appendix A: Seismic and geologic siting criteria for nuclear power plants. *Code of Fed. Regul.* **10**, Chap. 1, Part 100. (App A, 10, CFR 100) 1 Sept. 1982.

Van Arsdale, R., Stahle, D. & Cleaveland, M. 1993. Applications of dendrochronology to earthquake studies in the New madrid Seismic Zone of the central United States. In: *Neotectonics — Recent Advances* (edited by Owen, L. A., Stewart, I. S. & Vita-Finzi, C.). *Quat. Proc.* **3**, 85–92.

Verosub, K. L. 1987. The application of paleomagnetism to the dating of Quaternary materials. In: *Directions in Paleoseismology* (edited by Crone, A. G. & Omdahl, E. M.). *U.S. geol. Surv. Open-File Rep.* **87–673**, 89–99.

Vita-Finzi, C. 1986. *Recent Earth Movements*. Academic Press, London.

Vita-Finzi, C. 1987. ^{14}C deformation chronologies in coastal Iran, Greece and Jordan. *J. geol. Soc. Lond.* **144**, 553–560.

Vita-Finzi, C. 1992. Radiocarbon dating of late Quaternary fault segments and systems. *J. geol. Soc. Lond.* **149**, 257–260.

Vita-Finzi, C. & King, G. C. P. 1985. The seismicity, geomorphology and structural evolution of the Corinth area of Greece. *Phil. Trans. R. Soc. Lond.* **A314**, 379–407.

Walcott, R. I. 1978. Present tectonics and late Cenozoic evolution of New Zealand. *Geophys. J. R. astr. Soc.* **52**, 137–164.

Wallace, R. E. 1951. Geometry of shearing stress and relation to faulting. *J. Geol.* **59**, 118–130.

Wallace, R. E. 1977. Profiles and ages of young fault scarps, north-central Nevada. *Bull. geol. Soc. Am.* **88**, 1267–1281.

Wallace, R. E. 1984a. Patterns and timing of Late Quaternary faulting in the Great Basin province in relation to some regional tectonic features. *J. geophys. Res.* **89**, 5763–5769.

Wallace, R. E. 1984b. Faulting related to the 1915 earthquakes in Pleasant Valley, Nevada. *U.S. geol. Surv. Prof. Pap.* **1274–A**, 1–33.

Wallace, R. E. (compiler) 1986. *Active Tectonics*. National Academic Press, Washington.

Wallach, J. L., McFall, G. H., Bowlby, J. R., Mohajer, A. A., Pearce, M. & McKay, D. A. 1993. Pop-ups as geological indicators of earthquake-prone areas in intraplate eastern North America. In: *Neotectonics — Recent Advances* (edited by Owen, L. A., Stewart, I. S. & Vita-Finzi, C.). *Quat. Proc.* **3**, 67–84.

Ward, S. N. 1990. Pacific-North American plate motions: new results from Very Long Baseline Interferometry. *J. geophys. Res.* **95**, 21965–81.

Wells, S. G., Bullard, T. F., Menges, C. M., Drake, P. G., Karas, P. A., Nelson, K. I., Ritter, J. B. & Wesling, J. R. 1988. Regional variations in tectonic geomorphology along a segmented convergent plate boundary, Pacific coast of Costa Rica. *Geomorphology* **1**, 239–265.

Wesnousky, S. G. 1986. Earthquakes, Quaternary faults and seismic hazard in California. *J. geophys. Res.* **91**, 12587–12631.

Wesnousky, S. G., Scholz, C. H. & Shimazaki, K. 1982. Deformation of an island arc: rates of moment release and crustal shortening in intraplate Japan determined from seismicity and Quaternary fault data. *J. geophys. Res.* **87**, 6829–6852.

Wesnousky, S. G., Prentice, C. S. & Sieh, K. E. 1991. An offset stream channel and the rate of slip of the northern reach of San Jacinto fault zone, San Bernardino Valley, California. *Bull. geol. Soc. Am.* **103**, 700–709.

Wheeler, R. L. 1987. Boundaries between segments of normal faults — criteria for recognition and interpretation. In: *Directions in Paleoseismology* (edited by Crone, A. G. & Omdahl E. M.). *U. S. geol. Surv. Open-File Rep.* **87–673**, 385–398.

White, O. L. & Russell, D. J. 1982. High horizontal stresses in southern Ontario; their orientation and their origin. *Proc. IV Congr. Int. Ass. Eng. Geol. New Delhi* **V**, 39–54.

Yeats, R. S. 1986. Active faults related to folding. In: *Active Tectonics* (compiled by Wallace, R. E.). National Academic Press, Washington, 63–79.

Yeats, R. S. & Lillie, R. J. 1991. Contemporary tectonics of the Himalayan Frontal fault system: Folds, blind thrusts and the 1905 Kangra earthquake. In: *Characteristics of Active Faults* (edited by Hancock, P. L., Yeats, R. S. & Sanderson, D. J.). *J. Struct. Geol.* **13**, 215–225.

Yoshii, T. 1979. A detailed cross-section of the deep seismic zone beneath northeastern Honshu, Japan. *Tectonophysics* **55**, 349–360.

Zeitler, P. K., Johnson, N. M., Naeser, C. W. & Tahirkheli, R. A. K. 1982. Fission track evidence for Quaternary uplift of the Nanga Parbut region, Pakistan. *Nature* **298**, 255–257.

Zoback, M. D. & Zoback, M. L. 1991. Tectonic stress field of North America and relative plate motions. In: *The Geology of North America, Decade Map Vol. 1, Neotectonics of North America* (edited by Slemmons, D. B., Engdahl, E. R., Zoback, M. D. & Blackwell, D. D.). Geological Society of America, Boulder, Colorado, 339–366.

Zoback, M. L. 1992. First- and second-order patterns of stress in the lithosphere: the World Stress Map project. *J. geophys. Res.* **97B**, 11,703–11,728.

Thematic Index

This index lists words and phrases that refer to either deformation phenomena or deformation processes. Site- and/or time-specific examples of these phenomena or processes are not listed with the exceptions of Archaean (Chapter 17) and Neotectonic (Chapter 18) topics. The reason for excluding site- and time-specific examples is that no attempt has been made to give a coherent account of, for example, the structure of a region, or the evolution of a particular deformation zone. Entries in **bold type** indicate where a word or phrase is defined or introduced.

Accretionary prism (wedge) 145, 153, 180, **264**, 265, 320, 329
Accretion tectonics (see Suspect terrane)
Active fault (also see Neotectonics, Seismicity) 54, 372-374, **372**
Active fold 389, 399
Aegean-type scarp **388**
Aftershock **377**
Allochthon (tectonic) **162**
Allochthonous terrane **307**
Andean-type volcanic arc 364
Anticrack (also see Stylolites) **105**
Antitaxial vein fibres (also see Veins) **101**
Apparent polar wander path 316
Arc-arc collision accretion 320
Arc metamorphism 194-195
Arc-trench tectonics 180-199, 320-323, 331-344
    accreted sequence 183
    accretionary prism 180, 182, 184, 186
    accretionary prism toe 185
    anchored slab 182
    back-arc thrusting 193
    backstop 182
    coupling and decoupling **180**
    driving forces 180-181
    erosion 187-190
    extensional margins 181
    fluid movements 191, 323-332
    geothermal gradient 194-195
    metamorphism 194-195
    normal (mature) oceanic lithosphere 181
    offscraped sequence 183
    olistostrome 193
    overpressured zone 192
    palaeo-accretionary prisms 193-194
    palaeosubduction complexes 193-194
    subducting slab geometry 192-193
    subduction **180**
    tectonic accretion **182**
    tectonic mixing 193
    tilted basins 186
    trench migration 181-182
    underthrust sequence 183
    underthrusting plate 180
    volcanic front 180
Arch (swell, uplift, anteclise) 201, **204**, 205, 219
Archaean arc collisions 364
Archaean tectonics 355-369
    amalgamated volcanic arcs 364
    Archaean cratons 355
    back-arc basins 364
    batholiths 355
    continent-arc collision 364
    continent-continent collision 364
    continental rift 364
    fore-arc accretionary prisms 364
    granite diapirism 355
    granite-greenstone terrains **355**, 361-363
    island arcs 362
    oceanic plateaux 362
    reworking **355**
    sag tectonics **355**
    tectonic models 364-367
Archaeoseismology **371**

Arcs (also see Arc-trench tectonics)
    accreted serpentinite diapirs 336
    crustal imbrication 331
    inner-trench wall 331
    intraplate ridge 337
    offscraping 331
    outer-arc high 331
    trench landward-slope 331
Arrest line 51, **115**
Aseismic deformation 374
Asthenosphere **213**
Aulacogen **202**

Back-arc **181**, 193
Back-thrust 138, 277, 298
Balanced and restored cross-sections 244-245, 271
Basin (syneclise) (also see Foreland basin, Piggy-back basin, Pull-apart basin, Sag basin) **202**, 203, 204, 218-219, 224, 355
Blind fault 138, 184, **239**, 375
Block rotations 259-260, 271, 314
Boudinage 152
Brittle deformation and brittle structures 43-52, 53-100, 101-120, 121-142, 223-263, 370-409
Buck flexural model 237

C- and C'-surfaces and C/S-fabric **23**
Capable fault **372**
Cataclasis 8, **9**, 181, 381-383
Characteristic earthquake 125, **379**
Chilean type subduction **181**
Chlorite-mica stacks 153
Clastic (intrusive) sedimentary dykes and sills **108**, 187, 324, 328
Cleavage
    axial-plane fabrics 153
    disjunctive spaced cleavage 104
    en échelon cleavage surfaces **106**
    fracture cleavage 38
    pencil cleavage 38
    scaly cleavage 38
    slaty cleavage 38
    striated foliation 38
Collisional tectonics 264-288, 364-367
    Archaean collisions 364-367
    Argand number 273
    collision-related inversion 291-292
    crustal-scale thrust sheets 266, 267
    foreland 273
    hinterland 274
    P-T-t paths (thrusts) 269, 274
    post-collisional tectonics **264**
    pre-collisional tectonics **264**
    shortening calculations 271
    underplating 186, 267
Comb fracture **117**, 383
Competence and incompetence 7

Conjugate faults 61
Contemporary stress field (also see Stress) 217-218, 398-403
Continental ('A') subduction 192-193, **264**
Cordilleran suspect terrane 306
Corrugation (undulation) **383**
Coulomb wedge model 275-277, 327
Coulomb-Mohr stress relationships 45, 64, 66, 95-99, 113, 175
Crack-seal vein (also see Veins) **101**
Craton (stable zone) **200**, 200-222
Crustal lineament (see Lineament)
Crustal layers 214, 284
Crustal strength 213-215, 294-295
Current tectonic regime **371**

Dating/chronology of faulting 82-86, 376-377, 387
Deformation band 106, **152**, 324
Deformation mechanisms and processes 7-13
    Burgers vector 10
    cohesion 145
    compaction **144**
    compression **144**, 154
    consolidation **144**, 154
    constitutive equations 14
    creep **14**, 174
    critical state 147, 154
    cross slip 10
    crystal-plastic flow 8, 9-11
    crystallographic preferred orientations 16-18
    deformation path 2-6
    deformation regime 14
    deformation twinning 9
    diffusive mass transfer 8, 12-13, 150
    dislocation creep **174**
    dislocations 9, 11, 13
    ductile deformation processes 1-27
    ductile flow **6**
    dynamic recrystallization 12
    effective stress path 146
    flow laws 14, 144
    fluid pressure **144**
    frictional grain-boundary sliding 8
    grain boundary migration 12
    grain size sensitivity 15
    grain-boundary sliding 143, 154
    intercrystalline diffusive mass transport 12, 13
    overconsolidation **146**
    particulate deformation 143, 154
    peak strength 147
    plastic flow 14
    plastic limit 151
    pressure solution processes 13, 55, 103-105, 150
    reaction softening 13
    recovery **11**
    recrystallization 150
    residual strength 147
    rotation recrystallization 11-12, **12**
    salt rheology 173-175

shear strength 145
slip planes 9
solid-state diffusion 11
solution transfer creep **174**
strain hardening **147**
strain softening 7
stress corrosion 8
subgrains 11-13
superplastic flow 15
Taylor-Bishop-Hill model 16, 17
transient flow 14
transient frictional behaviour 9
underconsolidated 146
Depth-dependent stretching 225
Diagenesis 150
Diapirs and diapirism (also see Salt structures) 159-179, 324
Disjunctive spaced cleavage 104
Displaced (allochthonous) terrane **307**
Disrupted terrane 307
Docking **307**, 312
Dome-and-basin fold interference structure **359**
Domino faulting 88, 129, 132, 234-235, 241
Drape fold **210**
Ductile deformation processes 1-27
Duplex 138, 139, 259, 303
Dykes (igneous) 324, 398, 400

Earthquakes (see Seismicity)
Earthquakes focal mechanism (fault plane solution) 71-73, 79-80, **379**, 398
Effective elastic thickness 279
En échelon veins 23, 54, 60, 105, **106**
Exotic terrane 307
Extensional crenulation cleavage 23
Extensional (normal) faulting and tectonics (also see Grabens, Rifts) 121-142, 223-250
    analogue models 244
    blind normal fault 238, **239**
    Buck flexural model 237
    constant-heave Chevron construction 244
    depth-dependent stretching 225
    detachment fault 129, **236**
    domino model 88, 129, 132, 234-235, 241
    fault segment 229
    flat-ramp geometry 129, 132
    flexural model 231, 230-234
    geometric section-balancing 244-245
    gravity-driven extensional fault (slide) 129, 133, 243-246
    hangingwall subsidence 229
    horse 129
    instantaneous stretching (event) 223, 224
    large extensional faults 227-237, 246
    lateral heat-flow models 225
    marginal uplift 225
    normal drag 55, 61, 129
    normal fault **61**
    passive rotation 234
    peripheral sink 231
    planar faults 228, 232
    reverse drag 239
    scissor fault 131
    small extensional faults 237-241, 246
    soft-domino model 241
    strain partitioning 241-242
    stretching factor **223**
    tip-line loop **239**
    transfer fault 242
    Wernicke model 226
Extension fracturing (see Fractures and fracturing)
Extinct fault **372**
Exxon eustatic curve 345

Failed arm graben **208**
Fault bridge 242
Fault-line scarp **388**
Fault-line valley **390**
Fault linkage 241-242
Fault plane solution (see Earthquake focal mechanism)
Fault reactivation 63-69, 295-296
Fault rocks 9, 127, 381-383
    attrition breccia 381, 382
    cataclastic rocks **89**, 181, 381-383
    compact breccia sheet 382
    distributed crush breccia **381**
    fault-precursor breccia 382
    fault-precursor shatter zone 382
    gouge 127
    implosion breccia **381**
    incohesive breccia belt 382
    mylonite 20
Fault-propagation (precursor) (tip-line) fold 375, **382**, 389
Fault scarp 385-389
    Aegean-type scarp **388**
    bedrock fault scarp **386**
    buried fault scarp 389
    composite fault scarp **386**
    debris slope **387**
    fault-line scarp **388**
    fault scarp **385**
    fault scarp embayment/salient **387**
    fault-scarp stratigraphy 389
    range front **386**
    residual fault scarp **388**
    thrust-front scarp **387**
    triangular facet **387**
    truncated ridge spur 394
    wash slope **387**
Fault-related sinkhole 390
Fault-related speleothem deposits and stalagmites 391
Fault-related stream patterns 390-393
    beheaded stream **393**
    consequent stream **392**
    incised channel **392**
    insequent (unordered) stream **392**
    knick point **392**

offset stream 390, 393
subsequent stream **392**
Fault slip analysis 53-100, 295-296, 399
Faults and faulting (also see Extensional faulting, Neotectonic faults, Shear zones, Strike-slip faulting, Thrusting) 53-100, 121-142
    annealed fault 324
    asperity **371**
    barrier 124-125, **371**
    bend **379**
    chronology/dating of faulting 82-86, 376-377, 387
    dip direction and angle **55**
    facing direction **53**
    fault **53**
    fault bridge 127, **257**
    fault family (set) **60**
    fault pattern **60**
    fault system **60**
    fault-slip data inversion **75**, 75-82
    fault-zone architecture **379**
    gouge 127
    hard-linkage **121**
    jog (stepover) 125, **380**
    linked systems 121-142
    polyphase faulting 88-89
    propeller geometry **128**
    reactivation 63-69, 295-296, 301
    segmentation **127**, **371**, **380**
    slickenside lineation **55**, 103, 382-383
    soft-linkage **121**
    stepover (see jog)
    tip line 124
    total displacement (net separation) **56**
    transfer fault **131**, **380**
Fault wedge basin (wedge graben) 261
Fissures and related structures 108-109, 187, 324, 338, 390
    fissure **108**
    fissure-ridge travertine 390
    neptunian dyke **108**
    sediment-filled fractures **108**, 149, 153, 324, 326
    vein structure **108**, 153, 187, 324, 328
Fissure-ridge travertine 390
Flame structure 153
Flexural basins 200, 219, 277-281, 332
Flexural cantilever model 232-234
Flexural isostatic models 230
Flexural rigidity 278
Flexural upwarp 219
Floating island 297
Flower structure 136, **259**, 297
Fluids and deformation 13-14, 148-150, 190-192, 323-331
Fluid pressure and flow concepts 148-150, 323-324
    abnormal fluid pressure 148
    anisotropic permeability 149
    aquathermal pressuring 149
    biopressuring 149
    drainage **148**
    effective stress 148
    fluidization 149
    hydraulic conductivity 148
    hydrostatic pressure 148
    liquefaction 149
    overpressure 148
    permeability 148
    porosity **144**
    tectonic dewatering 148
Folds
    back fold 277
    drag fold 18-19
    fold structures 23-25, 358
    forced fold 257, 303
    parasitic fold 19
    rootless fold 211
    sheath fold 24, 187
Fold interference structure 357-361
Fold-limb scarp **389**
Foliations 22-23, 38, 104, 106, 153
Footwall collapse **273**
Footwall uplift 229, 290
Footwall upland 391
Footwall uplift 229
Fore-arc 181
Foreland 273
Foreland basin **200**, 219, 277-281, 332
Foreshock **377**
Fractal 121, 240
Fractography 115-116
Fractures and fracturing (also see Faults, Fissures, Joints, Stress, Veins)
    coefficient of internal friction 145
    cohesion 145
    conjugate shear angle 108, 117
    Coulomb-Mohr relationships 45, 64, 66, 95-99, 113, 175
    crack propagation **43**
    extension (tension) fracture 43-52, **112**, 254
    fracture **8**
    fracture grid-lock **114**
    fracture toughness **49**
    friction laws 96-97
    Griffith's contribution to fracture mechanics **44**
    hybrid fracture **113**
    Inglis's contribution to fracture mechanics **44**
    Irwin's contribution to fracture mechanics **46**
    linear elastic fracture mechanics **43**
    Mode 1 crack 51, 107, **109**, 112, 123
    Mode 2 crack **123**
    Mode 3 crack **123**
    poroelasticity **47**, 48
    shear fracture 43, **51**, **113**, 256
    stress corrosion 8, **46-47**
    subcritical crack growth **46**
    tensile strength **46**
    Wallace-Bott hypothesis **64**

Graben (also see Extensional faulting)
    graben **206**

half-graben 129, 232, 235
Granite-greenstone terrain **355**, 361-363
Gravity deformation 143-144
Greenstone belt **353**, 355-363
Growth fault 83, 171, 296
Guttenberg-Richter relationship 121, 240
Gutter **383**

Hackle marks 107, **115-116**
Half-graben 129, 232, 235
Halokinesis (also see Salt tectonics) **167**
Hangingwall ramp anticline 268
Hangingwall subsidence 229
Heat flow 214-216
High-grade gneiss terrain **355**, 356-358
    dome-and-basin fold interference structures 359
    early flat-lying structures 356
    fold interference structures 357-361, 364
    TTG-suite 356
Hinterland **200**, 275
Hinterland basin 277, **281**
Holocene tectonics **371**
Horse 129, 138, **259**
Horse-tail splay 259
Horst-and-graben structure 188, 327
Hotspot (mantle plume) 218, 290
Hybrid fracture (joint) 113
Hydraulic fracturing 47-50, 110, 399

Imbricate fan 259
Indentation (escape) tectonics 189-191, 253, 281-284, **282**
Inherited fault 63
Intracratonic fold **210**
Intraplate structures and processes 200-222, 401-403
Inversion tectonics 289-304, **289**
    buoyancy and isostatic forces 289-290
    collision-related inversion 291-292
    diapiric inversion 289
    extension-related inversion 290-291
    fault reactivation 63-69, 295-296, 301
    inversion of foreland basins 290
    inversion ratio **294**
    inverted normal fault geometries 294-298, 297
    isostatic inversion 289
    magmatic underplating 290
    negative inversion **289**
    positive inversion **289**
    rotational block-fault related inversion 292-294
    strike-slip related inversion 291
Isostatic balance 271, 281

Joints and jointing 50-51, 109-117, 383-385
    comb fracture **117**, 383
    conjugate joint system **111**

cross-joint **111**
environments of jointing 110
extension (tension) joint **112**, 114-117
faulted joint **110**
genetic classes of joint 112-114
hybrid joint **113**
hydraulic joint **110**
joint **109**
joint (fracture) grid-lock **114**, 385
joint arrest 51
joint fractography 115-116
joint periodicity index **111**
joint propagation path 50
joint sequences and ages **110**
joint set **111**
joint spacing, frequency, density and intensity **111**
joint spectrum **114**
joint style **110**
joint zone **111**
joint-elimination direction **112**
joint-interaction geometries 116
joint-system architecture **111**
joint-system geometry **111**
jointed fault **110**
joints and palaeostresses 112-114, 385
neotectonic joint **110**, 383-385, **384**
non-systematic joint **111**, 384
pinnate joint **117**, 383
release joint **110**
sealed joint **109**
systematic joint **111**
unloading joint **110**, 385

Kinematic indicators 18-25
    asymmetrical folds 18
    C'-surfaces 23
    debris streak **383**
    drag folds 18
    facets 60
    folds 23-25
    foliations 22-23
    grooves **382**
    mineral steps and fibres 60
    P shears 127, 134, 256
    parasitic folds 19
    porphyroblasts and porphyroclasts 19-21
    Riedel 23
    S-surfaces 22
    shape fabric 22
    sheath fold 24
    slickenside lineations **55**, 103, 382-383
    snowball garnets 19
    stretching faults 18
    stretching lineations 18
    stylolite peaks 60
    tension gash veins 23, 53, 60, 105, 106
    tool marks 60
    veins 25
Kink band 187

Land-flows 151
Lateral heat-flow models 225
Lateral ramps **252**, 297
Lineament 205-210, **205**, 220
Lister and Davis detachment-fault model 237
Listric faults and faulting 132, 171, 228, 232, 243, 295
Lithification **143**
Lithospheric characteristics
    lithosphere **212**
    strength 213-215, 271, 291-292, 295
    whole lithosphere failure **214**
Lithospheric plate driving forces 180-181, 215
    coupling 180
    decoupling 180
    ridge push 180, 215
    slab pull 180, 215
    trench suction 180, 215
Lithospheric simple shear 226

Magmatic underplating 290
Mainshock **377**
Marginal basin 331
Mariana-type subduction **181**
McKenzie (stretching) model 218, 223-224
Mélange 152, 193
Metamorphic core complex 236-237, **236**
Metamorphic terrane **307**
Mode 1 crack 51, 107, **109**, 112, 123
Mode 2 crack **123**
Mode 3 crack **123**
Morphotectonics 330, **371**
Mountain-front monocline 300
Mountain-front sinuosity 394
Mud and sand diapirs and volcanoes 152-153, 191-195, 324, 389
Mylonite 20

Neoformed (fault) fracture **62**, 381
Neotectonic faults (also see Faults) 214-217, 379-383
    crack 379
    fault zone 379
    hangingwall salient 382
    intrafault-zone hangingwall collapse 382
    migration 379
    principal displacement zone 379
    propagation 379
    segment 371
    shear zone 379
Neotectonic joint 383-385, **384**, 400
Neotectonics (also see Earthquakes, Seismicity, Neotectonic fault, Neotectonic joint) 370-409
    contemporary stress 398-403
    dating 376, 392, 394
    definition 371
    faults 214-217, 379-383
    joints 383-385, 400
    methodologies 375
    palaeoseismicity 389-390
    post-glacial 394-397
    seismotectonics 377-379
    tectonic geomorphology 385-397
    volcanic neotectonics 397-398, 399-400
Neptunian dyke **108**
Normal fault (also see Extensional faulting) **61**

Obduction 266
Oceanic ('B') subduction 192, 264
Offset shorelines ('Parallel Roads') 396
Offset stream channel 390, 393
Olistostrome 193

P shear 127, 134, 256
Pacific-type orogeny 194
Paired metamorphic belts 194
Palaeoseismic phenomena 389-390
Palaeostrain analysis 28-42
Palaeostress analysis (also see Stress analysis) 53-120
Palm-tree structure **136**, 259
Partitioning 282
Passive continental margin **212**, 279
Passive-roof back thrust 303
Peripheral uplift (bulge) 231
Petal 135
Piedmont scarp **386**
Piggy-back basin **281**
Pinch-and-swell structure 29
Pinnate joint **117**
Pitted pebble **105**
Plate driving forces (see Lithospheric plate driving forces)
Plateau uplift 267
Platform **200**
Plumose marking (plume structure) **115**
Porphyroblasts and porphyroclasts 19-21
Post-collisional structures 253, **264**, 281-285
Post-glacial neotectonics (faulting) 394-397
Pressure ridge 379
Pre-collisional tectonics **264**, 264-266
Prelithification deformation 143-158, 323-331
Pressure solution 13, 55, 103-105, 150
Pull-apart basin 134-135, 259-262, 283, **393**
Push-up block (mountain) 134, 260

Quaternary tectonics **371**

Ramp fault 138, 185, 268
Ramp-type scarp **387**
Range (mountain) front **386**
Reactivation (faults) 63-69, 295-296, 301
Relay fault/ramp 131-135, 242, **257**

Reverse fault **61**
Rhegmatic faulting **220**
Rhomb graben 259, 261
Rhomb horst 260
Ridge push **180**, 215
Ridge transform **251**
Riedel shear fractures 23, 60, 106, 127, 134, **254**
Riedel-within Riedel structure **256**
Rift 201, **206**, 206-208, 219
Rift pillow **208**
Roll-back (subduction roll-back) 182, 265
Rollover (anticline) 129, 152, 239, 244
Roof thrust **139**
Rootless fold **211**

Sag basin **224**
Sag pond **393**
Salt anticline **161**
Salt pillow **161**
Salt structures (see also Salt tectonics) 161-167
    allochthonous salt **162**
    bulb **162**
    crescentic and curtain folds **164**
    depopod and depotrough **167**
    enterolithic fold **164**
    half-turtle structure **167**
    intrasalt basin **167**
    marginal syncline **166**
    mushroom diapir **164**
    namakier **163**
    peripheral sink **166**
    postkinematic layer **164**
    prekinematic layer **164**
    rim syncline **166**
    salt anticline **161**
    salt canopy **162**
    salt diapir **161**
    salt dome **161**
    salt glacier **163**
    salt laccolith **163**
    salt pillow **161**
    salt plug **162**
    salt roller **162**
    salt sill **163**
    salt stock **162**
    salt suture **162**
    salt tongue **163**
    salt wall **162**
    salt welt **161**
    salt-stock canopy **162**
    salt-tongue canopy **162**
    salt-wall canopies **162**
    skirt **164**
    source layer **161**
    spine **164**
    stem **162**
    synkinematic layer **164**
    turtle structure anticline **167**
    vortex structure **164**
    welds **167**
Salt tectonics 159-179, **167**
    asymmetric intrusion **168**
    basal cutoff **173**
    buoyancy **169**
    characteristic wavelength **176**
    constrained growth **168**
    density inversion **169**
    diapirism 159-176
    dominant wavelength **176**
    downbuilding **167**
    downbuilding salt tongue **174**
    dynamic bulge **170**
    global distribution of salt structures 160
    gravity overturning **169**
    gravity spreading **169**
    halokinesis **167**
    level of neutral buoyancy **169**
    mother salt **161**
    movement cell **176**
    mushroom structure **164**
    raft tectonics **171**
    rheology of salt 173-175
    salt flats **173**
    salt fluid mechanics 176-177
    salt migration **291**
    salt ramps **173**
    salt reduction **170**
    salt rollers **172**
    salt sheet **175**
    salt withdrawal **170**
    sheet injection, spreading, thickening **173**
    spoke circulation **176**
    streamlines **176**
    structural inversion **165**
    substratum **160**
    thermal convection **169**
    unconstrained growth **168**
    unsteady flow **176**
    upbuilding **168**
Scaly clay **152**
Schizosphere **378**
Scissor fault **255**
Sediment deformation 143-158, 323-331
Sediment flows **151**
Sedimentary (clastic, intrusive) dykes and sills **108**, 149, 153, 324, 326
Sediment loading (subsidence) 219
Seismic moment **379**
Seismicity (also see Neotectonics) 121, 216-218, 240, 371-375, 377-379, 398
    aftershock **377**
    arc-trench seismicity **180**
    aseismic deformation **374**
    aseismic front **371**
    characteristic earthquake 125, **374**
    compound earthquake **378**
    coseismic deformation 230, 375
    coupled earthquake **374**
    earthquake clustering **373**

earthquake focal mechanism (fault plane solution) 71-73, 79-80, **379**, 398
earthquake loading cycle 373, 375
earthquake models 374
earthquake swarm **378**
episodic slip **374**
fault creep **374**
fault switching **373**
foreshock **377**
Gutenberg-Richter relationship 1221, 240
interseismic deformation 375
intraplate seismicity 216-218
magnitude 377
mainshock **377**
migrating earthquake sequence 373, 374
postseismic deformation 230, 375
preseismic deformation 375
quiescent fault 374
recurrence interval **373**
repeat time **373**
rupture attributes 378
schizosphere **378**
seismic belt (zone) **371**
seismic cycle 229-230, 373, 375
seismic deformation 374
seismic front **371**
seismic gap **371**
seismic slip models **374**
seismic moment **379**
seismogenic (fault) behaviour **374**
seismogenic layer 227, 378
types of maximum earthquakes **379**
Seismogenic layer 227, 378
Seismogeology **371**
Seismotectonics **371**, 377-379
Sequence stratigraphy 345-351
Serpentine diapir 188
Shear-band cleavage 23
Shear fracturing (see Fractures and fracturing)
Shear lens **259**
Shear zones 18-25, 105-108, 121-122
    brittle shear zone (fault) **106**
    brittle-ductile (semi-brittle) shear zone **105**
    deformation band (shear band) 106, **152**, 324
    ductile shear zone 18-25, **106**, 121
    en échelon cleavage surfaces **106**
    en échelon veins 23, 54, 60, 105, **106**
    Riedel shear fractures 23, 60, 106, 127, 134, **254**
    sigmoidal en échelon veins **106**
    simple shear zone 107
    transpressional and transtensional shear zones 107
Shield **200**
Short-cut fault 134, 294
Shutter ridge 393
Slab pull 180, 215
Slickolite (oblique stylolite) **55**, 105
Slickenside lineations **55**, 103, 382-383
Slide 151
Slip line 281-282
Slump and slump sheet 151, 153

Snowball garnets 19
Stem (fault) 135, 259
Stepped-type fault scarp **387**
Stitching pluton 312
Strain concepts 1-7, 28-38
    angular shear strain 1
    coaxial strain path 2
    crystallographic preferred orientation 7
    dimensional preferred orientation 7
    finite strain 5
    general three-dimensional strain 1
    heterogeneous strain 1
    homogeneous strain 1-4, **1**
    incremental strain 5
    infinitesimal strain 5
    instantaneous strain 5
    irrotational strain 4
    non-coaxial strain path 3
    plane strain 3
    principal axes of strain 1
    progressive strain 5
    pure shear 3
    rotation 2
    rotational strain 4
    shear induced vorticity 3
    simple shear 3
    spin 3, 5
    strain axes 1
    strain ellipse **1**, 28
    strain ellipsoid 1
    strain increment 5
    strain partitioning 6
    strain path 2
    total strain 5
    translation 2
    transpression 4-6, **255**
    transtension 4, **255**
    vorticity 3
Strain analysis 28-38
    bulk strain in cratons 211
    deformed markers 28, 31
    finite strain ellipsoid **28**
    Flinn diagram 36
    Fry method 34
    isochoric deformation 29
    longitudinal strain 29
    marker density method 37
    methods based on strain induced anisotropies 38
    Mohr circle construction 30
    one-dimensional analyses 28-29
    percentage extension **29**
    percentage shortening **28**
    principal stretches **28**
    reciprocal strain ellipsoid 37
    restorations 38
    Rf/ø methods 31-35, 32
    shape factor grid 33
    shear strain 29
    strain rosettes 30
    strain from flattened buckle folds 35-36

strain ratios 28
strains from twinned calcite grains 37-38
surface of no finite elongation 37
Talbot's method 37
theta curves 32
three-dimensional strain analyses 36-38
two-dimensional strain analyses 29-36
Stratigraphic pinch out 278, 281
Stratigraphic separation diagram 296
Stress analysis (also see Contemporary stress, Faults, Fractures, Joints)
   active folds 399
   borehole breakout elongation direction **398**
   burial stress **144**
   confining pressure 113, 145
   dykes 400
   dynamic clustering method **86**
   earthquake focal mechanism (fault plane solution) 71-73, 79-80, **379**, 398
   effective stress **144**
   fault slip analysis 53-100, 399
   fissures 400
   geostatic pressure 145
   hydraulic fracturing (hydrofracture) 399
   hydrostatic pressure 96, **144**
   in situ methods 399
   inverse problem 75
   least square criterion 76
   lithostatic pressure 145
   methods of determining stress 398-401
   Mohr stress analysis 95-99, 113
   non-linear least-square inversion method 80
   offset coreholes 401
   overburden pressure 145
   palaeostress **61**
   palaeostress analysis 53-100, 101-120
   palaeostress maps 90
   pore pressure 98
   pop-ups 400
   reduced stress tensor **69**, 69-82
   right dihedra (P and T dihedra) **70**
   strain relief methods 399
   stress **61**
   stress ellipsoid **65**
   stress magnitude 68, 98
   stress path 146
   stress trajectories 90-94
   uniaxial compression and extension 65
   volcanic vents 399
   World stress map 401-403
Stress province 216-218, 401-403
Stretching models 218, 223-235, 246
   Buck flexural model 237
   domino model 234-235
   instantaneous stretching (event) 223, 224
   lateral heat-flow models 225
   Lister and Davis detachment-fault model 237
   McKenzie model 218, 223-224
   pure shear models 226
   simple shear models 226

   stretching factor **223**, 224
   Wernicke model 226
Strike-slip faulting and tectonics 56, 61, 133-136, 251-263
   block rotations 259-260
   boundary transform fault **251**
   branch point 135
   bridge **257**
   cleavage-transected fold 257-258
   cognate horse **259**
   contractional duplex **259**
   domainal structure 260
   duplex **259**
   exotic horse **259**
   experimental studies 255
   extensional duplex **259**
   fan 259
   fault bend 258-259
   fault flank depression 260
   fault margin sag 260
   fault stepover (jog) **258**
   fault straight **258**
   fault-angle depression 260
   flake **259**
   flower structure **259**
   horse-tail splay 135, 259
   imbricate fan 135, 259
   in-line structure **257**
   intraplate strike-slip faulting 121-122, 200, 209-210, 281-284
   inversion 262
   lateral ramp **252**
   leaf **259**
   linkage 133
   nested domain 260
   oblique-slip zone **251**
   overlapping **257**
   P shear 127, 134, 256
   partitioning **257**
   pressure ridge (fault-slice ridge) 259, 379, 381
   principal displacement zone 135, **256**, 379, 381
   pull-apart basin 134, 135, 260-262, 283, 393
   Riedel (R) shear 23, 60, 106, 127, 134, **254**
   relay fault (zone) 135, **257**
   ridge transform fault **251**
   Riedel-within-Riedel structure **256**
   scissor fault 255
   shear lens **259**
   sidewall rip out 259
   splay fault 255
   strike-slip fault **251**
   strike-slip fault classification 253
   tear fault **252**
   tectonic setting of strike-slip zones 251-253
   tip 135
   transcurrent fault **251**, 291
   transform fault **251**
   transpression 255
   transtension 255
   trench-linked transform **252**

tulip (cactus) structure **136**, 256, **259**
underlapping **257**
wrench fault **252**
X shear **256**
Styloboudinage **105**
Stylolites (pressure solution seams) 54-59,
  104-105
  anticrack **105**
  disjunctive spaced cleavage **104**
  pitted pebble **105**
  slickolite (oblique stylolite) **55, 105**
  solution slip lineation 105
  styloboudinage **105**
  stylolite morphologies 104-105
  stylolite seam **59, 104**
  stylolite teeth (columns) and sockets **59, 104**
  stylolitic lineation **104**
Subcretion 186, 329
Subduction 180-182, 192-193, 264-265, 320
  A-subduction 192-193, **264**
  Archaean subduction 364
  B-subduction **264**
  Chilean-type subduction **181**
  coupling and decoupling **180**
  driving forces 180-182
  Mariana-type subduction **181**
  minor subduction zone **267**
  subducting (downgoing) slab **180**
  subduction 180
  subduction erosion **187**
  subduction of normal-thickness oceanic crust 320
  subduction rate 265
  subduction roll-back 182, 265
  trench suction 180
Submarine-flow 151
Suspect terranes 305-319, **305, 307**, 312, 313, 362
  accretionary continental growth 308
  apparent polar wander path 316
  composite terrane **307**
  Cordilleran suspect terrane 306
  disjunct terrane **308**
  disrupted terrane **307**
  exotic composite terrane 312
  family of terranes 307
  fore-arc-slivers 311
  native terrane 314
  stratifies terrane **307**
  superterrane **307**
  tectonic amalgamation **307**
  terrain **307**
  terrane **306**
  terrane accretion 307, 316-317
  terrane assembly diagrams **312**
  terrane characterisation 312-316
  terrane collage 307
  terrane complex **307**
  terrane dispersal 308, 316-317
  terrane identification 312-316
  terrane kinematics 308-312
  terrane origins 308-312

terrane orogenesis 316-317
terrane rotation 314
transcurrent terrane **307**, 311
transcurrent terrane dispersal 308
transpressional terrane **311**
Syndepositional (synsedimentary) fault **82**
Syn-sedimentary tectonics 144
Syneclise (see Basin)
Syntaxial vein fibres (also see Veins) **101**
Syntaxis **136**

Tear fault **252**
Tectonic erosion in arc-trench settings 187-190
Tectonic escape (see Indentation tectonics)
Tectonic geomorphology **371**, 385-397
Tectonic landforms 385-397
Tectonosedimentation 320-354
Tectonostratigraphic terrane 307
Tensile fracturing (also see Fractures and fracturing) 8
Terrane (see Suspect terrane)
Thick-skinned deformation 138, 271
Thin-skinned deformation 138, 271
Thrusting and thrust tectonics 53-100, 136-139, 266-
  271, 274-277
  back fold 277
  back thrust 298
  backward-propagating thrust **137**
  blind thrust 138, 184
  break-back forethrust systems 277
  Coulomb wedge model 275-277, 327
  critically tapered wedge 276
  crustal-scale thrust sheet 266
  décollement 138
  depth of detachment 294
  displacement path 285
  emergent thrust 138
  erosional window 138
  flat-ramp geometry 138
  footwall collapse **273**
  foreland propagation 272
  forelimb thrust 138
  forethrust 297, 301
  forward-propagating thrust **137**
  frontal ramp 138
  frontal thrust package 275
  hangingwall ramp anticline 269
  hinterland-dipping thrust 139
  imbricate structure 184
  inter-thrusting wedge **285**
  lateral ramp 136, **138**, 298
  oblique ramp 138
  out-of-sequence thrust 185
  P-T-t paths of thrusts 269
  piggy-back thrusting 124, 186, 273
  pop-up 138
  thick- and thin-skinned thrusting 136, 138, 271
  thrust kinematics 266
  thrust linkage 136
  thrust wedge 274, 275

triangle zone 138
Tool mark 60
Transcurrent fault **251**, 291
Transfer fault (zone) **129**, 131-132, 241-242, **252**
Transform fault **251**
Transpression 4, 6, 107, 127, 129, **255**
Transtension 4, 127, 129, **255**
Trench 331
Trench migration 181-182
Trench suction 180, 215
Trenching of earthquake faults 376, 396
Triangle zone 138
Tulip structure **136, 259**

Underplating 186, 264, 267
Uplift and thickening (also see Arch) 219-220, 271-274
    asymmetric thickening 273
    flexural upwarp 219
    footwall collapse **272**
    isostatic balance 271
    piggy-back thrusting 272
    plateau uplift 267
    thermal phase of uplift 272
    two-layer thickening model 274
    uplift and subsidence rates 272

Valley-floor width-height ratio 394
Veins 25, 101-103, 324, 383
    accretionary mineral step 60, 383
    antiaxial vein-fibres **101**
    composite vein **102**
    crinoid-type fibres **103**
    dilatant vein **101**, 105
    en échelon veins 23, 54, 60, 105, **106**
    fibre-sheet vein (shear vein) 103
    ghost fabrics in veins **101**
    granular texture in veins **101**
    inclusion bands and trails **102**
    massive- or granular-fill veins **101**
    parallel fibrous mineral grains 101
    pressure shadows and veins 102
    pyrite-type fibres **103**
    replacement vein **101**
    stretched vein fibres **102**
    vein structure 108, 153, 187, 324, 338
    wall-rock inclusions in veins **101**
Vein structure 108, 153, 187, 324, 338
Volcanic arc 180-199, 331
Volcanic front 180, 326
Volcanic neotectonics 397-398, 399-400

Watterson's fault growth model 239
Web structure **152**
Wernicke extensional model 226
World Wide Standardised Seismic Network 371
Wrench fault **252**